Pearson New International Edition

Automation, Production Systems, and
Computer-Integrated Manufacturing
Mikell P. Groover
Third Edition

PEARSON

Pearson Education Limited
Edinburgh Gate
Harlow
Essex CM20 2JE
England and Associated Companies throughout the world

Visit us on the World Wide Web at: www.pearsoned.co.uk

© Pearson Education Limited 2014

ISBN 10: 1-292-02592-1
ISBN 13: 978-1-292-02592-6

British Library Cataloguing-in-Publication Data
A catalogue record for this book is available from the British Library

Printed in the United States of America

Table of Contents

Chapter 1

Introduction

CHAPTER CONTENTS

The systems aspects of manufacturing are more important than ever today. The word manufacturing was originally derived from two Latin words, *manus* (hand) and *factus* (make), so that the combination means *made by hand*. This was the way manufacturing was accomplished when the word first appeared in the English language around 1567. Commercial goods of those times were made by hand. The methods were handicraft, accomplished in small shops, and the goods were relatively simple, at least by today's standards. As many years passed, factories were developed, with many workers at a single site, and the work had to be organized using machines rather than handicraft techniques. The products became more complex, and so did the processes. Workers had to specialize in their tasks. Rather than overseeing the fabrication of the entire product, they were responsible for only a small part of the total work. More up-front planning was required, and more coordination of the operations was needed to keep track of progress in the factories. The systems of production, which rely on many separate but interacting functions, were evolving.

Today, the systems of production are indispensable in manufacturing. Modern manufacturing enterprises that manage these production systems must cope with the economic realities of the modern world. These realities include the following:

- *Globalization.* Once underdeveloped countries like China, India, and Mexico are becoming major players in manufacturing, due largely to their high populations and low labor costs. Other regions of the world with low labor costs include Latin America, Eastern Europe, and Southeast Asia, and the countries in these regions have also become important suppliers of manufactured goods.
- *International outsourcing.* Parts and products once made in the United States by American companies are now being made offshore (overseas, so that cargo ships are required to deliver the items) or near-shore (in Mexico or Central America, so that rail and truck deliveries are possible). In general, international outsourcing means loss of jobs in the United States.
- *Local outsourcing.* Companies can also outsource by using suppliers within the United States. Reasons why companies elect local outsourcing include: (1) benefits from using suppliers that specialize in certain production technologies, (2) lower labor rates in smaller companies, and (3) limitations of available in-house manufacturing capabilities.
- *Contract manufacturing.* This refers to companies that specialize in manufacturing entire products, not just parts, under contract to other companies. Contract manufacturers specialize in efficient production techniques, freeing their customers to specialize in the design and marketing of the products.
- *Trend toward the service sector in the U.S. economy.* There has been a gradual erosion of direct labor jobs in manufacturing while jobs in service industries (e.g., healthcare, food services, retail) have increased in numbers.
- *Quality expectations.* Customers, both consumer and corporate, demand that the products they purchase are of the highest quality. Perfect quality is the expectation.
- *The need for operational efficiency.* To be successful, U.S. manufacturers must be efficient in their operations to overcome the labor cost advantage enjoyed by their international competitors. In some cases, the labor cost advantage is a factor of ten.

This book is all about the production systems that are used to manufacture products and the parts assembled into those products. The emphasis is on the systems and technologies used in the United States, but the United States certainly does not possess a monopoly on them. Just as the U.S. economy has grown during the last 120 years to be an industrial powerhouse through its manufacturing might, other countries have copied our methods, in some cases improving on them, to become formidable competitors in the world of production. Japan, Germany, and South Korea stand as examples of this competition. China is fast rising to become perhaps the largest economy in the world during the latter half of the 21st

century, thanks to its strengths in manufacturing. India is following a similar path. Manufacturing is important in all of these countries, including the United States.

In this introductory chapter, we provide an overview of production systems and how they are sometimes automated and computerized. In subsequent chapters, we examine how manufacturers can successfully compete by employing modern manufacturing approaches and technologies. The approaches and technologies include the following:

- *Automation.* The use of automated equipment compensates for the labor cost disadvantage relative to international competitors. Automation reduces labor costs, decreases production cycle times, and increases product quality and consistency.
- *Material handling technologies.* Manufacturing usually involves a sequence of activities performed at different locations in the plant. The work must be transported, stored, and tracked as it moves through the plant.
- *Manufacturing systems.* These involve the integration and coordination of multiple automated and/or manual workstations through the use of material handling technologies to achieve a synergistic effect compared to the independent operation of individual workstations. Examples include production lines, manufacturing cells, and automated assembly systems.
- *Flexible manufacturing.* Much of the outsourcing to international competitors, such as China and Mexico, has involved high-volume consumer goods production. Flexibility in manufacturing allows U.S. manufacturers to compete effectively in the low-volume/high-mix product categories.
- *Quality programs.* Manufacturers must employ techniques such as statistical quality control and Six Sigma to achieve the quality levels expected by their customers.
- *Computer integrated manufacturing* (CIM). The technologies include computer-aided design (CAD), computer-aided manufacturing (CAM), and computer networks to integrate manufacturing and logistics operations.
- *Lean production.* Accomplishing more work with fewer resources is the general goal of lean production, which involves techniques to increase labor productivity and operational efficiency.

Let us begin our overview of production systems by defining the term. We then examine automation and computer integrated manufacturing and the roles they play in production systems. In Chapter 2, we examine the manufacturing operations that the production systems are intended to accomplish.

1.1 PRODUCTION SYSTEMS

A production system is a collection of people, equipment, and procedures organized to perform the manufacturing operations of a company (or other organization). Production systems can be divided into two categories or levels as indicated in Figure 1.1:

1. *Facilities.* The facilities of the production system consist of the factory, the equipment in the factory, and the way the equipment is organized.
2. *Manufacturing support systems.* This is the set of procedures used by the company to manage production and to solve the technical and logistics problems encountered in

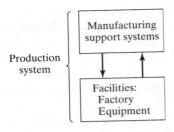

Figure 1.1 The production system consists of facilities and manufacturing support systems.

ordering materials, moving the work through the factory, and ensuring that products meet quality standards. Product design and certain business functions are included among the manufacturing support systems.

In modern manufacturing operations, portions of the production system are automated and/or computerized. However, production systems include people. People make these systems work. In general, direct labor people *(blue collar workers)* are responsible for operating the facilities, and professional staff people *(white collar workers)* are responsible for the manufacturing support systems.

1.1.1 The Facilities

The facilities in the production system are the factory, production machines and tooling, material handling equipment, inspection equipment, and computer systems that control the manufacturing operations. Facilities also include the *plant layout*, which refers to the way the equipment is physically arranged in the factory. The equipment is usually organized into logical groupings, and we refer to these equipment arrangements and the workers who operate them as the *manufacturing systems* in the factory. Manufacturing systems can be individual work cells, consisting of a single production machine and worker assigned to that machine. We more commonly think of manufacturing systems as groups of machines and workers, for example, a production line. The manufacturing systems come in direct physical contact with the parts and/or assemblies being made. They "touch" the product. In terms of the human participation in the processes performed by the manufacturing systems, three basic categories can be distinguished, as depicted in Figure 1.2: (a) manual work systems, (b) worker-machine systems, and (c) automated systems.

Manual Work Systems. A manual work system consists of one or more workers performing one or more tasks without the aid of powered tools. Manual material handling tasks are common activities in manual work systems. Production tasks commonly require the use of hand tools. A *hand tool* is a small tool that is manually operated by the strength and skill of the human user. When using hand tools, a *workholder* is often employed to grasp the work part and position it securely during processing. Examples of production-related manual tasks involving the use of hand tools include

- A machinist using a file to round the edges of a rectangular part that has just been milled
- A quality control inspector using a micrometer to measure the diameter of a shaft
- A material handling worker using a dolly to move cartons in a warehouse
- A team of assembly workers putting together a piece of machinery using hand tools.

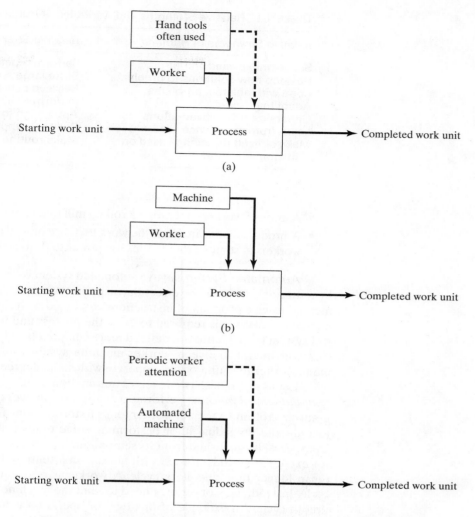

Figure 1.2 Three categories of manufacturing systems: (a) manual work system, (b) worker-machine system, and (c) automated system.

Worker-Machine Systems. In a worker-machine system, a human worker operates powered equipment, such as a machine tool or other production machine. This is one of the most widely used manufacturing systems. Worker-machine systems include combinations of one or more workers and one or more pieces of equipment. The workers and machines are combined to take advantage of their relative strengths and attributes, which are listed in Table 1.1. Examples of worker-machine systems include the following:

- A machinist operating an engine lathe in a tool room to fabricate a part for a custom-designed product
- A fitter and an industrial robot working together in an arc–welding work cell

TABLE 1.1 Relative Strengths and Attributes of Humans and Machines

Relative Strengths of Humans	*Relative Strengths of Machines*
Sense unexpected stimuli	Perform repetitive tasks consistently
Develop new solutions to problems	Store large amounts of data
Cope with abstract problems	Retrieve data from memory reliably
Adapt to change	Perform multiple tasks simultaneously
Generalize from observations	Apply high forces and power
Learn from experience	Perform simple computations quickly
Make difficult decisions based on incomplete data	Make routine decisions quickly

- A crew of workers operating a rolling mill that converts hot steel slabs into flat plates
- A production line in which the work units are moved by mechanized conveyor and the workers at some of the stations use power tools to accomplish their assembly tasks.

Automated Systems.   An automated system is one in which a process is performed by a machine without the direct participation of a human worker. Automation is implemented using a program of instructions combined with a control system that executes the instructions. Power is required to drive the process and to operate the program and control system (these terms are defined more completely in Chapter 4). There is not always a clear distinction between worker-machine systems and automated systems, because many worker-machine systems operate with some degree of automation.

Let us distinguish two levels of automation: semi-automated and fully automated. A *semi-automated machine* performs a portion of the work cycle under some form of program control, and a human worker tends to the machine for the remainder of the cycle, by loading and unloading it, or performing some other task each cycle. A *fully automated machine* is distinguished from its semi-automated counterpart by the capacity to operate for extended periods of time with no human attention. By extended periods of time, we mean longer than one work cycle. A worker is not required to be present during each cycle. Instead, the worker may need to tend the machine every tenth cycle, or every hundredth cycle. An example of this type of operation is found in many injection molding plants, where the molding machines run on automatic cycles, but periodically the molded parts at the machine must be collected by a worker.

In certain fully automated processes, one or more workers are required to be present to continuously monitor the operation, and make sure that it performs according to the intended specifications. Examples of these kinds of automated processes include complex chemical processes, oil refineries, and nuclear power plants. The workers do not actively participate in the process except to make occasional adjustments in the equipment settings, to perform periodic maintenance, and to spring into action if something goes wrong.

1.1.2 Manufacturing Support Systems

To operate the production facilities efficiently, a company must organize itself to design the processes and equipment, plan and control the production orders, and satisfy product quality requirements. These functions are accomplished by manufacturing support systems—people and procedures by which a company manages its production operations.

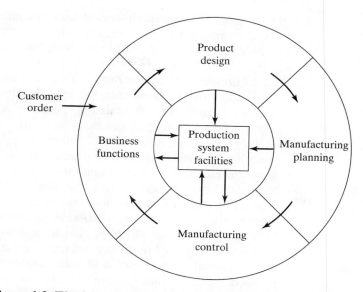

Figure 1.3 The information-processing cycle in a typical manufacturing firm.

Most of these support systems do not directly contact the product, but they plan and control its progress through the factory.

Manufacturing support involves a cycle of information-processing activities, as illustrated in Figure 1.3. The production system facilities described in Section 1.1.1 are pictured in the center of the figure. The information-processing cycle can be described as consisting of four functions: (1) business functions, (2) product design, (3) manufacturing planning, and (4) manufacturing control.

Business Functions. The business functions are the principal means of communicating with the customer. They are, therefore, the beginning and the end of the information-processing cycle. Included in this category are sales and marketing, sales forecasting, order entry, cost accounting, and customer billing.

The order to produce a product typically originates from the customer and proceeds into the company through the sales and marketing department of the firm. The production order will be in one of the following forms: (1) an order to manufacture an item to the customer's specifications, (2) a customer order to buy one or more of the manufacturer's proprietary products, or (3) an internal company order based on a forecast of future demand for a proprietary product.

Product Design. If the product is to be manufactured to customer design, the design will have been provided by the customer. The manufacturer's product design department will not be involved. If the product is to be produced to customer specifications, the manufacturer's product design department may be contracted to do the design work for the product as well as to manufacture it.

If the product is proprietary, the manufacturing firm is responsible for its development and design. The cycle of events that initiates a new product design often originates in the sales and marketing department; the information flow is indicated in Figure 1.3. The departments of the firm that are organized to accomplish product design might include research and development, design engineering, and perhaps a prototype shop.

Manufacturing Planning. The information and documentation that constitute the product design flows into the manufacturing planning function. The information-processing activities in manufacturing planning include process planning, master scheduling, requirements planning, and capacity planning.

Process planning consists of determining the sequence of individual processing and assembly operations needed to produce the part. The manufacturing engineering and industrial engineering departments are responsible for planning the processes and related technical details. Manufacturing planning includes logistics issues, commonly known as production planning. The authorization to produce the product must be translated into the master production schedule. The *master production schedule* is a listing of the products to be made, the dates on which they are to be delivered, and the quantities of each. Months are traditionally used to specify deliveries in the master schedule. Based on this schedule, the individual components and subassemblies that make up each product must be planned. Raw materials must be purchased or requisitioned from storage, purchased parts must be ordered from suppliers, and all of these items must be planned so that they are available when needed. This entire task is called *material requirements planning*. In addition, the master schedule must not list more quantities of products than the factory is capable of producing each month with its given number of machines and manpower. A function called *capacity planning* is concerned with planning the manpower and machine resources of the firm.

Manufacturing Control. Manufacturing control is concerned with managing and controlling the physical operations in the factory to implement the manufacturing plans. The flow of information is from planning to control as indicated in Figure 1.3. Information also flows back and forth between manufacturing control and the factory operations. Included in the manufacturing control function are shop floor control, inventory control, and quality control.

Shop floor control deals with the problem of monitoring the progress of the product as it is being processed, assembled, moved, and inspected in the factory. Shop floor control is concerned with inventory in the sense that the materials being processed in the factory are work-in-process inventory. Thus, shop floor control and inventory control overlap to some extent.

Inventory control attempts to strike a proper balance between the risk of too little inventory (with possible stock-outs of materials) and the carrying cost of too much inventory. It deals with such issues as deciding the right quantities of materials to order and when to reorder a given item when stock is low.

The function of *quality control* is to ensure that the quality of the product and its components meet the standards specified by the product designer. To accomplish its mission, quality control depends on inspection activities performed in the factory at various times during the manufacture of the product. Also, raw materials and component parts from outside sources are sometimes inspected when they are received, and final inspection and testing of the finished product is performed to ensure functional quality and appearance. Quality control also includes data collection and problem-solving approaches to address process problems related to quality. Examples of these approaches are statistical process control (SPC) and Six Sigma.

1.2 AUTOMATION IN PRODUCTION SYSTEMS

Some components of the firm's production system are likely to be automated, whereas others will be operated manually or clerically. The automated elements of the production system can be separated into two categories: (1) automation of the manufacturing systems in the factory, and (2) computerization of the manufacturing support systems. In modern production systems, the two categories overlap to some extent, because the automated manufacturing systems operating on the factory floor are themselves usually implemented by computer systems and connected to the computerized manufacturing support systems and management information system operating at the plant and enterprise levels. The term computer-integrated manufacturing is used to indicate this extensive use of computers in production systems. The two categories of automation are shown in Figure 1.4 as an overlay on Figure 1.1.

1.2.1 Automated Manufacturing Systems

Automated manufacturing systems operate in the factory on the physical product. They perform operations such as processing, assembly, inspection, and material handling, in some cases accomplishing more than one of these operations in the same system. They are called automated because they perform their operations with a reduced level of human participation compared with the corresponding manual process. In some highly automated systems, there is virtually no human participation. Examples of automated manufacturing systems include:

- Automated machine tools that process parts
- Transfer lines that perform a series of machining operations
- Automated assembly systems
- Manufacturing systems that use industrial robots to perform processing or assembly operations
- Automatic material handling and storage systems to integrate manufacturing operations
- Automatic inspection systems for quality control.

Automated manufacturing systems can be classified into three basic types: (1) fixed automation, (2) programmable automation, and (3) flexible automation. They generally operate as fully automated systems although semi-automated systems are common in

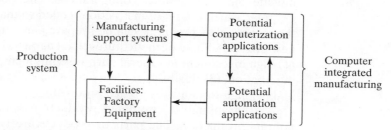

Figure 1.4 Opportunities for automation and computerization in a production system.

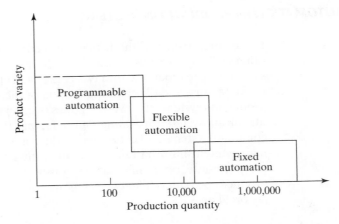

Figure 1.5 Three types of automation relative to production quantity and product variety.

programmable automation. The relative positions of the three types of automation for different production volumes and product varieties are depicted in Figure 1.5.

Fixed Automation. Fixed automation is a system in which the sequence of processing (or assembly) operations is fixed by the equipment configuration. Each operation in the sequence is usually simple, involving perhaps a plain linear or rotational motion or an uncomplicated combination of the two, such as the feeding of a rotating spindle. It is the integration and coordination of many such operations into one piece of equipment that makes the system complex. Typical features of fixed automation are (1) high initial investment for custom-engineered equipment, (2) high production rates, and (3) relative inflexibility of the equipment to accommodate product variety.

The economic justification for fixed automation is found in products that are produced in very large quantities and at high production rates. The high initial cost of the equipment can be spread over a very large number of units, thus making the unit cost attractive compared with alternative methods of production. Examples of fixed automation include machining transfer lines and automated assembly machines.

Programmable Automation. In programmable automation, the production equipment is designed with the capability to change the sequence of operations to accommodate different product configurations. The operation sequence is controlled by a *program*, which is a set of instructions coded so that they can be read and interpreted by the system. New programs can be prepared and entered into the equipment to produce new products. Some of the features that characterize programmable automation include (1) high investment in general purpose equipment, (2) lower production rates than fixed automation, (3) flexibility to deal with variations and changes in product configuration, and (4) high suitability for batch production.

Programmable automated production systems are used in low- and medium-volume production. The parts or products are typically made in batches. To produce each new batch of a different product, the system must be reprogrammed with the set of machine instructions that correspond to the new product. The physical setup of the machine must also be changed: Tools must be loaded, fixtures must be attached to the machine table, and the required machine settings must be entered. This changeover procedure takes time. Consequently, the typical cycle for a given product includes a period during which the setup and reprogramming takes place, followed by a period in which the parts in the batch are produced. Examples of programmable automation include numerically controlled (NC) machine tools, industrial robots, and programmable logic controllers.

Flexible Automation. Flexible automation is an extension of programmable automation. A flexible automated system is capable of producing a variety of parts (or products) with virtually no time lost for changeovers from one part style to the next. There is no lost production time while reprogramming the system and altering the physical setup (tooling, fixtures, machine settings). Accordingly, the system can produce various mixes and schedules of parts or products instead of requiring that they be made in batches. What makes flexible automation possible is that the differences between parts processed by the system are not significant, so the amount of changeover required between styles is minimal. The features of flexible automation include (1) high investment for a custom-engineered system, (2) continuous production of variable mixtures of products, (3) medium production rates, and (4) flexibility to deal with product design variations. Examples of flexible automation are the flexible manufacturing systems for performing machining operations. The first of these systems was installed in the late 1960s.

1.2.2 Computerized Manufacturing Support Systems

Automation of the manufacturing support systems is aimed at reducing the amount of manual and clerical effort in product design, manufacturing planning and control, and the business functions of the firm. Nearly all modern manufacturing support systems are implemented using computers. Indeed, computer technology is used to implement automation of the manufacturing systems in the factory as well. The term *computer-integrated manufacturing* (CIM) denotes the pervasive use of computer systems to design the products, plan the production, control the operations, and perform the various information-processing functions needed in a manufacturing firm. True CIM involves integrating all of these functions in one system that operates throughout the enterprise. Other terms are used to identify specific elements of the CIM system. For example, *computer-aided design* (CAD) denotes the use of computer systems to support the product design function. *Computer-aided manufacturing* (CAM) denotes the use of computer systems to perform functions related to manufacturing engineering, such as process planning and numerical control part programming. Some computer systems perform both CAD and CAM, and so the term *CAD/CAM* is used to indicate the integration of the two into one system.

Computer-integrated manufacturing involves the information-processing activities that provide the data and knowledge required to successfully produce the product. They are accomplished to implement the four basic manufacturing support functions identified earlier: (1) business functions, (2) product design, (3) manufacturing planning, and (4) manufacturing control. These four functions form a cycle of events that must accompany the physical production activities but do not directly touch the product.

1.2.3 Reasons for Automating

Companies undertake projects in manufacturing automation and computer-integrated manufacturing for a variety of good reasons. Some of the reasons used to justify automation are the following:

1. *To increase labor productivity.* Automating a manufacturing operation usually increases production rate and labor productivity. This means greater output per hour of labor input.

2. *To reduce labor cost.* Ever-increasing labor cost has been and continues to be the trend in the world's industrialized societies. Consequently, higher investment in automation has become economically justifiable to replace manual operations. Machines are increasingly being substituted for human labor to reduce unit product cost.

3. *To mitigate the effects of labor shortages.* There is a general shortage of labor in many advanced nations, and this has stimulated the development of automated operations as a substitute for labor.

4. *To reduce or eliminate routine manual and clerical tasks.* An argument can be put forth that there is social value in automating operations that are routine, boring, fatiguing, and possibly irksome. Automating such tasks improves the general level of working conditions.

5. *To improve worker safety.* Automating a given operation and transferring the worker from active participation in the process to a monitoring role, or removing the worker from the operation altogether, makes the work safer. The safety and physical well-being of the worker has become a national objective with the enactment of the Occupational Safety and Health Act (OSHA) in 1970. This has provided an impetus for automation.

6. *To improve product quality.* Automation not only results in higher production rates than manual operation, it also performs the manufacturing process with greater uniformity and conformity to quality specifications.

7. *To reduce manufacturing lead time.* Automation helps reduce the elapsed time between customer order and product delivery, providing a competitive advantage to the manufacturer for future orders. By reducing manufacturing lead time, the manufacturer also reduces work-in-process inventory.

8. *To accomplish processes that cannot be done manually.* Certain operations cannot be accomplished without the aid of a machine. These processes require precision, miniaturization, or complexity of geometry that cannot be achieved manually. Examples include certain integrated circuit fabrication operations, rapid prototyping processes based on computer graphics (CAD) models, and the machining of complex, mathematically defined surfaces using computer numerical control. These processes can only be realized by computer controlled systems.

9. *To avoid the high cost of not automating.* There is a significant competitive advantage gained in automating a manufacturing plant. The advantage cannot easily be demonstrated on a company's project authorization form. The benefits of automation often show up in unexpected and intangible ways, such as in improved quality, higher sales, better labor relations, and better company image. Companies that do not automate are likely to find themselves at a competitive disadvantage with their customers, their employees, and the general public.

1.3 MANUAL LABOR IN PRODUCTION SYSTEMS

Is there a place for manual labor in the modern production system? The answer is yes. Even in a highly automated production system, humans are still a necessary component of the manufacturing enterprise. For the foreseeable future, people will be required to manage and maintain the plant, even in those cases where they do not participate directly in its manufacturing operations. Let us separate our discussion of the labor issue into two parts, corresponding to our previous distinction between facilities and manufacturing support: (1) manual labor in factory operations and (2) labor in manufacturing support systems.

1.3.1 Manual Labor in Factory Operations

There is no denying that the long-term trend in manufacturing is toward greater use of automated machines to substitute for manual labor. This has been true throughout human history, and there is every reason to believe the trend will continue. It has been made possible by applying advances in technology to factory operations. In parallel and sometimes in conflict with this technologically driven trend are issues of economics that continue to find reasons for employing manual labor in manufacturing operations.

Certainly one of the current economic realities in the world is that there are countries whose average hourly wage rates are so low that most automation projects are impossible to justify strictly on the basis of cost reduction. These countries include China, India, Russia, Mexico, and many other countries in Eastern Europe, Southeast Asia, and Central America. With the passage of the North American Free Trade Agreement (NAFTA), the North American continent has become one large labor pool. Within this pool, Mexico's labor rate is an order of magnitude less than that in the United States. U.S. corporate executives who make decisions on factory locations and the outsourcing of work must reckon with this economic reality.

In addition to the labor cost issue, there are other reasons, ultimately based on economics, that make the use of manual labor a feasible alternative to automation. Humans possess certain attributes that give them an advantage over machines in certain situations and certain kinds of tasks (Table 1.1). A number of situations can be listed in which manual labor is preferred over automation:

- *Task is technologically too difficult to automate.* Certain tasks are very difficult (either technologically or economically) to automate. Reasons for the difficulty include (1) problems with physical access to the work location, (2) adjustments required in the task, (3) manual dexterity requirements, and (4) demands on hand-eye coordination. Manual labor is used to perform the tasks in these cases. Examples include automobile final assembly lines where many final trim operations are accomplished by human workers, inspection tasks that require judgment to assess quality, or material handling tasks that involve flexible or fragile materials.

- *Short product life cycle.* If the product must be designed and introduced in a short period of time to meet a near-term window of opportunity in the marketplace, or if the product is anticipated to be on the market for a relatively short period, then a manufacturing method designed around manual labor allows for a much faster product launch than does an automated method. Tooling for manual production can be fabricated in much less time and at much lower cost than comparable automation tooling.

- *Customized product.* If the customer requires a one-of-a-kind item with unique features, manual labor may have the advantage as the appropriate production resource because of its versatility and adaptability. Humans are more flexible than any automated machine.
- *Ups and downs in demand.* Changes in demand for a product necessitate changes in production output levels. Such changes are more easily made when manual labor is used as the means of production. An automated manufacturing system has a fixed cost associated with its investment. If output is reduced, that fixed cost must be spread over fewer units, driving up the unit cost of the product. On the other hand, an automated system has an ultimate upper limit on its output capacity. It cannot produce more than its rated capacity. By contrast, manual labor can be added or reduced as needed to meet demand, and the associated cost of the resource is in direct proportion to its employment. Manual labor can be used to augment the output of an existing automated system during those periods when demand exceeds the capacity of the automated system.
- *Need to reduce risk of product failure.* A company introducing a new product to the market never knows for sure what the ultimate success of that product will be. Some products will have long life cycles, while others will be on the market for relatively short lives. The use of manual labor as the productive resource at the beginning of the product's life reduces the company's risk of losing a significant investment in automation if the product fails to achieve a long market life. In Section 1.4.3, we discuss an automation migration strategy that is suitable for introducing a new product.
- *Lack of capital.* Companies are sometimes forced to use manual labor in their production operations when they lack the capital to invest in automated equipment.

1.3.2 Labor in Manufacturing Support Systems

In manufacturing support functions, many of the routine manual and clerical tasks can be automated using computer systems. Certain production planning activities are better accomplished by computers than by clerks. Material requirements planning (MRP, Section 25.2) is an example. In material requirements planning, order releases are generated for component parts and raw materials based on the master production schedule for final products. This requires a massive amount of data processing that is best suited to computer automation. Many commercial software packages are available to perform MRP. With few exceptions, companies that need to accomplish MRP rely on computers. Humans are still required to interpret and implement the output of these MRP computations and to otherwise manage the production planning function.

In modern production systems, the computer is used as an aid in performing virtually all manufacturing support activities. Computer-aided design systems are used in product design. The human designer is still required to do the creative work. The CAD system is a tool that augments the designer's creative talents. Computer-aided process planning systems are used by manufacturing engineers to plan the production methods and routings. In these examples, humans are integral components in the operation of the manufacturing support functions, and the computer-aided systems are tools to increase productivity and improve quality. CAD and CAM systems rarely operate completely in automatic mode.

Humans will continue to be needed in manufacturing support systems, even as the level of automation in these systems increases. People will be needed to do the decision making, learning, engineering, evaluating, managing, and other functions for which humans are much better suited than machines, according to Table 1.1. Even if all of the manufacturing systems in the factory are automated, there will still be a need for the following kinds of work to be performed:

- *Equipment maintenance.* Skilled technicians will be required to maintain and repair the automated systems in the factory when these systems break down. To improve the reliability of the automated systems, preventive maintenance programs are implemented.
- *Programming and computer operation.* There will be a continual demand to upgrade software, install new versions of software packages, and execute the programs. It is anticipated that much of the routine process planning, numerical control part programming, and robot programming may be highly automated using artificial intelligence (AI) in the future. But the AI programs must be developed and operated by people.
- *Engineering project work.* The computer-automated and integrated factory is likely never to be finished. There will be a continual need to upgrade production machines, design tooling, solve technical problems, and undertake continuous improvement projects. These activities require the skills of engineers working in the factory.
- *Plant management.* Someone must be responsible for running the factory. There will be a limited staff of professional managers and engineers who are responsible for plant operations. There is likely to be an increased emphasis on managers' technical skills compared with traditional factory management positions, where the emphasis is on personnel skills.

1.4 AUTOMATION PRINCIPLES AND STRATEGIES

The preceding discussion leads us to conclude that automation is not always the right answer for a given production situation. A certain caution and respect must be observed in applying automation technologies. In this section, we offer three approaches for dealing with automation projects:[1] (1) the USA Principle, (2) Ten Strategies for Automation and Process Improvement, and (3) an Automation Migration Strategy.

1.4.1 The USA Principle

The USA Principle is a common sense approach to automation and process improvement projects. Similar procedures have been suggested in the manufacturing and automation trade literature, but none has a more captivating title than this one. USA stands for (1) understand the existing process, (2) simplify the process, and (3) automate the process. A statement of the USA principle appears in an article published by the American Production and Inventory Control Society [6]. The article is concerned with implementing

---

[1]There are additional approaches not discussed here, but in which the reader may be interested; for example, the ten steps to integrated manufacturing production systems discussed in J. Black's book: *The Design of the Factory with a Future* [1].

enterprise resource planning (ERP, Section 25.6.2), but the USA approach is so general that it is applicable to nearly any automation project. Going through each step of the procedure for an automation project may in fact reveal that simplifying the process is sufficient and automation is not necessary.

Understand the Existing Process.   The first step in the USA approach is to comprehend the current process in all of its details. What are the inputs? What are the outputs? What exactly happens to the work unit between input and output? What is the function of the process? How does it add value to the product? What are the upstream and downstream operations in the production sequence, and can they be combined with the process under consideration?

Some of the traditional industrial engineering charting tools used in methods analysis are useful in this regard, such as the operation chart and the flow process chart [4]. Application of these tools to the existing process provides a model of the process that can be analyzed and searched for weaknesses (and strengths). The number of steps in the process, the number and placement of inspections, the number of moves and delays experienced by the work unit, and the time spent in storage can be ascertained by these charting techniques.

Mathematical models of the process may also be useful to indicate relationships between input parameters and output variables. What are the important output variables? How are these output variables affected by inputs to the process, such as raw material properties, process settings, operating parameters, and environmental conditions? This information may be valuable in identifying what output variables need to be measured for feedback purposes and in formulating algorithms for automatic process control.

Simplify the Process.   Once the existing process is understood, then the search begins for ways to simplify. This often involves a checklist of questions about the existing process. What is the purpose of this step or this transport? Is this step necessary? Can this step be eliminated? Does this step use the most appropriate technology? How can this step be simplified? Are there unnecessary steps in the process that might be eliminated without detracting from function?

Some of the ten strategies for automation and production systems (Section 1.4.2) can help simplify the process. Can steps be combined? Can steps be performed simultaneously? Can steps be integrated into a manually operated production line?

Automate the Process.   Once the process has been reduced to its simplest form, then automation can be considered. The possible forms of automation include those listed in the ten strategies discussed in the following section. An automation migration strategy (Section 1.4.3) might be implemented for a new product that has not yet proven itself.

1.4.2   Ten Strategies for Automation and Process Improvement

Following the USA Principle is a good first step in any automation project. As suggested previously, it may turn out that automation of the process is unnecessary or cannot be cost justified after the process has been simplified.

If automation seems a feasible solution to improving productivity, quality, or other measure of performance, then the following ten strategies provide a road map to search for these improvements. These ten strategies were originally published in my first book.[2] They seem as relevant and appropriate today as they did in 1980. We refer to them as strategies for automation and process improvement because some of them are applicable whether the process is a candidate for automation or just for simplification.

1. *Specialization of operations.* The first strategy involves the use of special-purpose equipment designed to perform one operation with the greatest possible efficiency. This is analogous to the specialization of labor, which is employed to improve labor productivity.

2. *Combined operations.* Production occurs as a sequence of operations. Complex parts may require dozens or even hundreds of processing steps. The strategy of combined operations involves reducing the number of distinct production machines or workstations through which the part must be routed. This is accomplished by performing more than one operation at a given machine, thereby reducing the number of separate machines needed. Since each machine typically involves a setup, setup time can usually be saved by this strategy. Material handling effort, nonoperation time, waiting time, and manufacturing lead time are all reduced.

3. *Simultaneous operations.* A logical extension of the combined operations strategy is to simultaneously perform the operations that are combined at one workstation. In effect, two or more processing (or assembly) operations are being performed simultaneously on the same workpart, thus reducing total processing time.

4. *Integration of operations.* This strategy involves linking several workstations together into a single integrated mechanism, using automated work handling devices to transfer parts between stations. In effect, this reduces the number of separate work centers through which the product must be scheduled. With more than one workstation, several parts can be processed simultaneously, thereby increasing the overall output of the system.

5. *Increased flexibility.* This strategy attempts to achieve maximum utilization of equipment for job shop and medium-volume situations by using the same equipment for a variety of parts or products. It involves the use of flexible automation concepts (Section 1.2.1). Prime objectives are to reduce setup time and programming time for the production machine. This normally translates into lower manufacturing lead time and less work-in-process.

6. *Improved material handling and storage.* A great opportunity for reducing nonproductive time exists in the use of automated material handling and storage systems. Typical benefits include reduced work-in-process and shorter manufacturing lead times.

7. *On-line inspection.* Inspection for quality of work is traditionally performed after the process is completed. This means that any poor-quality product has already been produced by the time it is inspected. Incorporating inspection into the manufacturing process permits corrections to the process as the product is being made. This reduces scrap and brings the overall quality of the product closer to the nominal specifications intended by the designer.

[2]M. P. Groover, *Automation, Production Systems, and Computer-Aided Manufacturing,* Prentice Hall, Englewood Cliffs, NJ, 1980.

8. *Process control and optimization*. This includes a wide range of control schemes intended to operate the individual processes and associated equipment more efficiently. By this strategy, the individual process times can be reduced and product quality can be improved.

9. *Plant operations control*. Whereas the previous strategy was concerned with the control of the individual manufacturing process, this strategy is concerned with control at the plant level. It attempts to manage and coordinate the aggregate operations in the plant more efficiently. Its implementation usually involves a high level of computer networking within the factory.

10. *Computer-integrated manufacturing* (CIM). Taking the previous strategy one level higher, we have the integration of factory operations with engineering design and the business functions of the firm. CIM involves extensive use of computer applications, computer data bases, and computer networking throughout the enterprise.

The ten strategies constitute a checklist of possibilities for improving the production system through automation or simplification. They should not be considered mutually exclusive. For most situations, multiple strategies can be implemented in one improvement project. The reader will see these strategies implemented in the many systems discussed throughout the book.

1.4.3 Automation Migration Strategy

Owing to competitive pressures in the marketplace, a company often needs to introduce a new product in the shortest possible time. As mentioned previously, the easiest and least expensive way to accomplish this objective is to design a manual production method, using a sequence of workstations operating independently. The tooling for a manual method can be fabricated quickly and at low cost. If more than a single set of workstations is required to make the product in sufficient quantities, as is often the case, then the manual cell is replicated as many times as needed to meet demand. If the product turns out to be successful, and high future demand is anticipated, then it makes sense for the company to automate production. The improvements are often carried out in phases. Many companies have an *automation migration strategy*, that is, a formalized plan for evolving the manufacturing systems used to produce new products as demand grows. A typical automation migration strategy is the following:

Phase 1: *Manual production* using single-station manned cells operating independently. This is used for introduction of the new product for reasons already mentioned: quick and low-cost tooling to get started.

Phase 2: *Automated production* using single-station automated cells operating independently. As demand for the product grows, and it becomes clear that automation can be justified, then the single stations are automated to reduce labor and increase production rate. Work units are still moved between workstations manually.

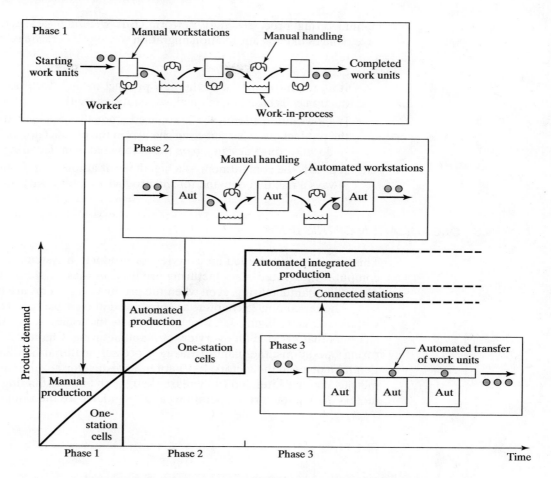

Figure 1.6 A typical automation migration strategy. Phase 1: manual production with single independent workstations. Phase 2: automated production stations with manual handling between stations. Phase 3: automated integrated production with automated handling between stations. Key: Aut = automated workstation.

Phase 3: *Automated integrated production* using a multistation automated system with serial operations and automated transfer of work units between stations. When the company is certain that the product will be produced in mass quantities and for several years, then integration of the single-station automated cells is warranted to further reduce labor and increase production rate.

This strategy is illustrated in Figure 1.6. Details of the automation migration strategy vary from company to company, depending on the types of products they make and the manufacturing processes they perform. But well-managed manufacturing companies have policies like the automation migration strategy. There are several advantages of such a strategy:

- It allows introduction of the new product in the shortest possible time, since production cells based on manual workstations are the easiest to design and implement
- It allows automation to be introduced gradually (in planned phases), as demand for the product grows, engineering changes in the product are made, and time is provided to do a thorough design job on the automated manufacturing system
- It avoids the commitment to a high level of automation from the start, since there is always a risk that demand for the product will not justify it.

1.5 ORGANIZATION OF THIS BOOK

This chapter has provided an overview of production systems and how automation and computer integrated manufacturing are used in these systems. We see that people are needed in manufacturing, even when the production systems are highly automated.

The remaining 25 chapters are organized into six parts. Let us describe the six parts with reference to Figure 1.7, which shows how the topics fit together. Part I includes two chapters that provide an overview of manufacturing. Chapter 2 is a survey of manufacturing operations: the manufacturing processes, material handling, and other activities that take place in the factory. In Chapter 3, we develop several mathematical models and metrics that are intended to increase the reader's understanding of the issues and parameters in manufacturing operations and to underscore their quantitative nature.

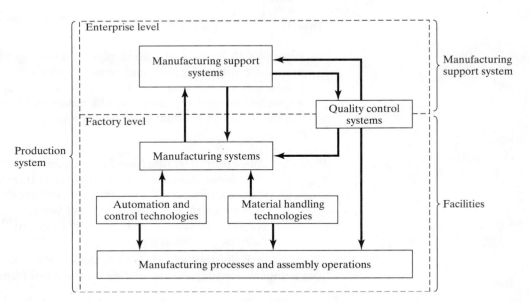

Figure 1.7 Overview and relationships among the six parts of the book.

Part II consists of six chapters that are concerned with automation technologies. Whereas Chapter 1 discusses automation in general terms, Part II describes the technical details. Automation relies heavily on control systems, so Part II is called "Automation and Control Technologies." These technologies include numerical control, industrial robotics, and programmable logic controllers.

Part III includes three chapters on material handling technologies that are used primarily in factories and warehouses. The technologies involve equipment for transporting materials, storing them, and automatically identifying them for material control purposes.

Part IV is concerned with the integration of automation technologies and material handling technologies into manufacturing systems—those that operate in the factory and touch the product. Some of these manufacturing systems are highly automated, while others rely largely on manual labor. Part IV contains seven chapters, covering single-station work cells, production lines, assembly systems, cellular manufacturing, and flexible manufacturing systems.

The importance of quality control must not be overlooked in modern production systems. Part V covers this topic, dealing with statistical process control and inspection issues. We describe some of the significant inspection technologies here, such as machine vision and coordinate measuring machines. As suggested in Figure 1.7, quality control (QC) systems include elements of both facilities and manufacturing support systems. QC is an enterprise-level function, but it has equipment and procedures that operate in the factory.

Finally, Part VI addresses the remaining manufacturing support functions in the production system. We include a chapter on product design and how it is supported by computer-aided design systems. The second chapter in Part VI is concerned with process planning and how it is automated by computer-aided process planning. Here we also discuss concurrent engineering and design for manufacturing. Chapter 25 covers production planning and control, including topics such as material requirements planning (MRP, mentioned earlier in Chapter 1), manufacturing resource planning (MRP II), and enterprise resource planning (ERP). Our book concludes with a chapter on just-in-time and lean production—approaches that modern manufacturing companies are using to run their businesses.

REFERENCES

[1] BLACK, J. T., *The Design of the Factory with a Future,* McGraw-Hill, Inc., NY, 1991.

[2] ENGARDIO, P., "A New World Economy," *BusinessWeek*, August 22/29, 2005, pp 52–61.

[3] GROOVER, M. P., *Fundamentals of Modern Manufacturing: Materials, Processes, and Systems,* 3d ed., John Wiley & Sons, Inc., Hoboken, NJ, 2007.

[4] GROOVER, M. P., *Work Systems and the Methods, Measurement, and Management of Work,* Pearson/Prentice Hall, Upper Saddle River, NJ, 2007.

[5] HARRINGTON, J., *Computer Integrated Manufacturing,* Industrial Press, Inc., NY, 1973.

[6] KAPP, K. M., "The USA Principle," *APICS — The Performance Advantage*, June 1997, pp 62–66.

[7] SPANGLER, T., R. MAHAJAN, S. PUCKETT, and D. STAKEM, "Manual Labor — Advantages, When and Where?" *MSE 427 Term Paper*, Lehigh University, 1998.

[8] WEBER, A., "Managing the Reality of Offshore Assembly, *Assembly*, March 2005, pp 56–67.

[9] WEBER, A., "Automation vs. Outsourcing," *Assembly*, December 2006, pp 46–55.

[10] ZAKARIA, F., "Does the Future Belong to China?" *Newsweek*, May 9, 2005, pp 26–40.

REVIEW QUESTIONS

1.1 What are some of the realities mentioned at the beginning of the chapter with which modern manufacturing enterprises must cope? Name four.

1.2 What is a production system?

1.3 Production systems can be divided into two categories or levels. Name and briefly define the two levels.

1.4 What are manufacturing systems, and how are they distinguished from production systems?

1.5 Manufacturing systems are divided into three categories, according to worker participation. Name the three categories.

1.6 What are the four functions included within the scope of manufacturing support systems?

1.7 Three basic types of automation are defined in the text. What is fixed automation and what are some of its features?

1.8 What is programmable automation and what are some of its features?

1.9 What is flexible automation and what are some of its features?

1.10 What is computer integrated manufacturing?

1.11 What are some of the reasons why companies automate their operations? Nine reasons are given in the text. Name five.

1.12 Identify three situations in which manual labor is preferred over automation.

1.13 Human workers will be needed in factory operations, even in the most highly automated operations. The text identifies at least four types of work for which humans will be needed. Name three.

1.14 What is the USA Principle? What does each of the letters stand for?

1.15 The text lists ten strategies for automation and process improvement. Identify five of these strategies.

1.16 What is an automation migration strategy?

1.17 What are the three phases of a typical automation migration strategy?

Manufacturing Operations[1]

CHAPTER CONTENTS

Manufacturing can be defined as the application of physical and chemical processes to alter the geometry, properties, and/or appearance of a given starting material to make parts or products; manufacturing also includes the joining of multiple parts to make assembled products. The processes that accomplish manufacturing involve a combination of machinery, tools, power, and manual labor, as depicted in Figure 2.1(a). Manufacturing is

---

[1]The chapter introduction and Sections 2.1 and 2.2 are based on M. P. Groover, *Fundamentals of Modern Manufacturing: Materials, Processes, and Systems*, Chapter 1.

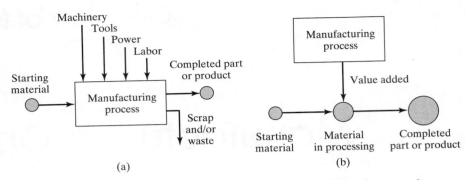

Figure 2.1 Alternative definitions of manufacturing (a) as a technological process, and (b) as an economic process.

almost always carried out as a sequence of operations. Each successive operation brings the material closer to the desired final state.

From an economic viewpoint, manufacturing is concerned with the transformation of materials into items of greater value by means of one or more processing and/or assembly operations, as depicted in Figure 2.1(b). The key point is that manufacturing *adds value* to the material by changing its shape or properties or by combining it with other materials that also have been altered. When iron ore is converted into steel, value is added. When sand is transformed into glass, value is added. When petroleum is refined into plastic, value is added. And when plastic is molded into the complex geometry of a patio chair, it is made even more valuable.

In this chapter, we provide a survey of manufacturing operations. We begin by examining the industries that are engaged in manufacturing and the types of products they produce. We then discuss fabrication and assembly processes used in manufacturing as well as the activities that support these processes, such as material handling and inspection. Next, several product parameters such as production quantity, product variety, and product complexity are introduced, and we explore the influence they have on production operations and the facilities required to accomplish those operations. The chapter concludes with a brief introduction to lean production, an approach that aims to minimize waste and maximize productivity.

The history of manufacturing includes both the development of manufacturing processes, some of which date back thousands of years, and the evolution of the production systems required to apply these processes (Historical Note 2.1). Our emphasis in this book is on the systems.

Historical Note 2.1   History of manufacturing

The history of manufacturing includes two related topics: (1) the discovery and invention of materials and processes to make things and (2) the development of systems of production. The materials and processes predate the systems by several millennia. Systems of production refer to the ways of organizing people and equipment so that production can be performed more efficiently. Some of the basic processes date as far back as the Neolithic period (circa 8000–3000 B.C.), when operations such as the following were developed: *woodworking*, *forming* and *firing* of clay pottery, *grinding* and *polishing* of stone, *spinning* of fiber and *weaving* of textiles, and *dyeing* of cloth. Metallurgy and metalworking also began during the

Neolithic, in Mesopotamia and other areas around the Mediterranean. It either spread to, or developed independently in, regions of Europe and Asia. Gold was found by early humans in relatively pure form in nature; it could be *hammered* into shape. Copper was probably the first metal to be extracted from ores, thus requiring *smelting* as a processing technique. Copper could not be readily hammered because it strain-hardened; instead, it was shaped by *casting*. Other metals used during this period were silver and tin. It was discovered that copper alloyed with tin produced a more workable metal than copper alone (casting and hammering could both be used). This heralded the important period known as the *Bronze Age* (circa 3500–1500 B.C.).

Iron was also first smelted during the Bronze Age. Meteorites may have been one source of the metal, but iron ore was also mined. The temperatures required to reduce iron ore to metal are significantly higher than for copper, which made furnace operations more difficult. Early blacksmiths learned that when certain irons (those containing small amounts of carbon) were sufficiently heated and then *quenched* (thrust into water to cool), they became very hard. This permitted the grinding of very sharp cutting edges on knives and weapons, but it also made the metal brittle. Toughness could be increased by reheating at a lower temperature, a process known as *tempering*. What we have described is, of course, the *heat treatment* of steel. The superior properties of steel caused it to succeed bronze in many applications (weaponry, agriculture, and mechanical devices). The period of its use has subsequently been named the *Iron Age* (starting around 1000 B.C.). It was not until much later, well into the nineteenth century, that the demand for steel grew significantly and more modern steelmaking techniques were developed.

The early fabrication of implements and weapons was accomplished more as crafts and trades than by manufacturing as we know it today. The ancient Romans had what might be called factories to produce weapons, scrolls, pottery, glassware, and other products of the time, but the procedures were largely based on handicraft. It was not until the *Industrial Revolution* (circa 1760–1830) that major changes began to affect the systems for making things. This period marked the beginning of the change from an economy based on agriculture and handicraft to one based on industry and manufacturing. The change began in England, where a series of important machines was invented, and steam power began to replace water, wind, and animal power. Initially, these advances gave British industry significant advantages over other nations, but eventually the revolution spread to other European countries and to the United States. The Industrial Revolution contributed to the development of manufacturing in the following ways: (1) *Watt's steam engine*, a new power-generating technology; (2) development of *machine tools*, starting with John Wilkinson's boring machine around 1775, which was used to bore the cylinder on Watt's steam engine; (3) invention of the *spinning jenny*, *power loom*, and other machinery for the textile industry, which permitted significant increases in productivity; and (4) the *factory system*, a new way of organizing large numbers of production workers based on the division of labor.

Wilkinson's boring machine is generally recognized as the beginning of machine tool technology. It was powered by waterwheel. During the period 1775–1850, other machine tools were developed for most of the conventional *machining processes*, such as *boring*, *turning*, *drilling*, *milling*, *shaping*, and *planing*. As steam power became more prevalent, it gradually became the preferred power source for most of these machine tools. It is of interest to note that many of the individual processes predate the machine tools by centuries; for example, drilling, sawing, and turning (of wood) date from ancient times.

Assembly methods were used in ancient cultures to make ships, weapons, tools, farm implements, machinery, chariots and carts, furniture, and garments. The processes included *binding* with twine and rope, *riveting* and *nailing*, and *soldering*. By around the time of Christ, *forge welding* and *adhesive bonding* had been developed. Widespread use of screws, bolts, and nuts—so common in today's assembly—required the development of machine tools, in particular, Maudsley's screw cutting lathe (1800), which could accurately form the helical threads. It was not until around 1900 that *fusion welding* processes started to be developed as assembly techniques.

While England was leading the Industrial Revolution, an important concept related to assembly technology was being introduced in the United States: *interchangeable parts* manufacture. Much credit for this concept is given to Eli Whitney (1765–1825), although its importance had been recognized by others [3]. In 1797, Whitney negotiated a contract to produce 10,000 muskets for the U.S. government. The traditional way of making guns at the time was to custom-fabricate each part for a particular gun and then hand-fit the parts together by filing. Each musket was therefore unique, and the time to make it was considerable. Whitney believed that the components could be made accurately enough to permit parts assembly without fitting. After several years of development in his Connecticut factory, he traveled to Washington in 1801 to demonstrate the principle. Before government officials, including Thomas Jefferson, he laid out components for 10 muskets and proceeded to select parts randomly to assemble the guns. No special filing or fitting was required, and all of the guns worked perfectly. The secret behind his achievement was the collection of special machines, fixtures, and gages that he had developed in his factory. Interchangeable parts manufacture required many years of development and refinement before becoming a practical reality, but it revolutionized methods of manufacturing. It is a prerequisite for mass production of assembled products. Because its origins were in the United States, interchangeable parts production came to be known as the *American System* of manufacture.

The mid and late 1800s witnessed the expansion of railroads, steam-powered ships, and other machines that created a growing need for iron and steel. New methods for producing steel were developed to meet this demand. Also during this period, several consumer products were developed, including the sewing machine, bicycle, and automobile. To meet the mass demand for these products, more efficient production methods were required. Some historians identify developments during this period as the *Second Industrial Revolution*, characterized in terms of its effects on production systems by the following: (1) mass production, (2) assembly lines, (3) the scientific management movement, and (4) electrification of factories.

Mass production was primarily an American phenomenon. Its motivation was the mass market that existed in the United States. Population in the United States in 1900 was 76 million and growing. By 1920 it exceeded 106 million. Such a large population, larger than any western European country, created a demand for large numbers of products. Mass production provided those products. Certainly one of the important technologies of mass production was the *assembly line*, introduced by Henry Ford (1863–1947) in 1913 at his Highland Park plant (Historical Note 15.1).. The assembly line made mass production of complex consumer products possible. Use of assembly line methods permitted Ford to sell a Model T automobile for less than $500 in 1916, thus making ownership of cars feasible for a large segment of the American population.

The *scientific management* movement started in the late 1800s in the United States in response to the need to plan and control the activities of growing numbers of production workers. The movement was led by Frederick W. Taylor (1856–1915), Frank Gilbreath (1868–1924) and his wife Lilian (1878–1972), and others. Scientific management included: (1) *motion study*, aimed at finding the best method to perform a given task; (2) *time study*, to establish work standards for a job; (3) extensive use of *standards* in industry; (4) the *piece rate system* and similar labor incentive plans; and (5) use of data collection, record keeping, and cost accounting in factory operations.

In 1881, *electrification* began with the first electric power generating station being built in New York City, and soon electric motors were being used as the power source to operate factory machinery. This was a far more convenient power delivery system than the steam engine, which required overhead belts to mechanically distribute power to the machines. By 1920, electricity had overtaken steam as the principal power source in U.S. factories. Electrification also motivated many new inventions that have affected manufacturing operations and production systems. The twentieth century was a time of more technological advances than all previous centuries combined. Many of these developments have resulted in the *automation* of manufacturing. Historical notes on some of these advances in automation are included in this book.

2.1 MANUFACTURING INDUSTRIES AND PRODUCTS

Manufacturing is an important commercial activity, carried out by companies that sell products to customers. The type of manufacturing performed by a company depends on the kinds of products it makes. Let us first take a look at the scope of the manufacturing industries and then consider their products.

Manufacturing Industries. Industry consists of enterprises and organizations that produce and/or supply goods and/or services. Industries can be classified as primary, secondary, and tertiary. *Primary industries* are those that cultivate and exploit natural resources, such as agriculture and mining. *Secondary industries* convert the outputs of the primary industries into products. Manufacturing is the principal activity in this category, but the secondary industries also include construction and power utilities. *Tertiary industries* constitute the service sector of the economy. A list of specific industries in these categories is presented in Table 2.1.

In this book, we are concerned with the secondary industries (middle column in Table 2.1), which are composed of the companies engaged in manufacturing. It is useful to distinguish the process industries from the industries that make discrete parts and products. The process industries include chemicals, pharmaceuticals, petroleum, basic metals, food, beverages, and electric power generation. The discrete product industries include automobiles, aircraft, appliances, computers, machinery, and the component parts from

TABLE 2.1 Specific Industries in the Primary, Secondary, and Tertiary Categories, Based Roughly on the International Standard Industrial Classification (ISIC) Used by the United Nations

Primary	Secondary	Tertiary (Service)
Agriculture	Aerospace	Banking
Forestry	Apparel	Communications
Fishing	Automotive	Education
Livestock	Basic metals	Entertainment
Quarrying	Beverages	Financial services
Mining	Building materials	Government
Petroleum	Chemicals	Health and medical
	Computers	Hotels
	Construction	Information
	Consumer appliances	Insurance
	Electronics	Legal services
	Equipment	Real estate
	Fabricated metals	Repair and maintenance
	Food processing	Restaurants
	Glass, ceramics	Retail trade
	Heavy machinery	Tourism
	Paper	Transportation
	Petroleum refining	Wholesale trade
	Pharmaceuticals	
	Plastics (shaping)	
	Power utilities	
	Publishing	
	Textiles	
	Tire and rubber	
	Wood and furniture	

TABLE 2.2 International Standard Industrial Classification (ISIC) Codes for Various Industries in the Manufacturing Sector

Basic Code	Products Manufactured
31	Food, beverages (alcoholic and nonalcoholic), tobacco
32	Textiles, clothing, leather goods, fur products
33	Wood and wood products (e.g., furniture), cork products
34	Paper, paper products, printing, publishing, bookbinding
35	Chemicals, coal, petroleum, plastic, rubber, products made from these materials, pharmaceuticals
36	Ceramics (including glass), nonmetallic mineral products (e.g., cement)
37	Basic metals (e.g., steel, aluminum, etc.)
38	Fabricated metal products, machinery, equipment (e.g., aircraft, cameras, computers and other office equipment, machinery, motor vehicles, tools, televisions)
39	Other manufactured goods (e.g., jewelry, musical instruments, sporting goods, toys)

which these products are assembled. The International Standard Industrial Classification (ISIC) of industries according to types of products manufactured is listed in Table 2.2. In general, the process industries are included within ISIC codes 31–37, and the discrete product manufacturing industries are included in ISIC codes 38 and 39. However, it must be acknowledged that many of the products made by the process industries are finally sold to the consumer in discrete units. For example, beverages are sold in bottles and cans. Pharmaceuticals are often purchased as pills and capsules.

Production operations in the process industries and the discrete product industries can be divided into continuous production and batch production. The differences are shown in Figure 2.2.

Continuous production occurs when the production equipment is used exclusively for the given product, and the output of the product is uninterrupted. In the process industries, continuous production means that the process is carried out on a continuous stream of material, with no interruptions in the output flow, as suggested by Figure 2.2(a). The material being processed is likely to be in the form of a liquid, gas, powder, or similar physical state. In the discrete manufacturing industries, continuous production means 100% dedication of the production equipment to the part or product, with no breaks for product changeovers. The individual units of production are identifiable, as in Figure 2.2(b).

Batch production occurs when the materials are processed in finite amounts or quantities. The finite amount or quantity of material is called a *batch* in both the process and discrete manufacturing industries. Batch production is discontinuous because there are interruptions in production between batches. The reason for using batch production is because the nature of the process requires that only a finite amount of material can be accommodated at one time (e.g., the amount of material might be limited by the size of the container used in processing) and/or because there are differences between the parts or products made in different batches (e.g., a batch of 20 units of part A followed by a batch of 50 units of part B in a machining operation, and a setup changeover is required between batches because of differences in the tooling and fixturing required). The differences in batch production between the process and discrete manufacturing industries are portrayed in Figure 2.2(c) and (d). Batch production in the process industries generally means that the starting materials are in liquid or bulk form, and they are processed altogether as a unit. By contrast, in the discrete manufacturing industries, a batch is a certain quantity of work units,

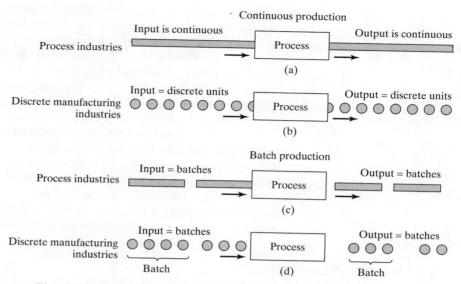

Figure 2.2 Continuous and batch production in the process and discrete manufacturing industries, including (a) continuous production in the process industries, (b) continuous production in the discrete manufacturing industries, (c) batch production in the process industries, and (d) batch production in the discrete manufacturing industries.

and the work units are usually processed one at a time rather than all together at once. The number of parts in a batch can range from as few as one to as many as thousands of units.

Manufactured Products. As indicated in Table 2.2, the secondary industries include food, beverages, textiles, wood, paper, publishing, chemicals, and basic metals (ISIC codes 31–37). The scope of our book is primarily directed at the industries that produce discrete products (ISIC codes 38 and 39). The two groups interact with each other, and many of the concepts and systems discussed in the book are applicable to the process industries, but our

TABLE 2.3 Manufacturing Industries Whose Products Are Likely to Be Produced by the Production Systems Discussed in This Book

Industry	Typical Products
Aerospace	Commercial and military aircraft
Automotive	Cars, trucks, buses, motorcycles
Computers	Mainframe and personal computers
Consumer appliances	Large and small household appliances
Electronics	TVs, DVD players, audio equipment, video game consoles
Equipment	Industrial machinery, railroad equipment
Fabricated metals	Machined parts, metal stampings, tools
Glass, ceramics	Glass products, ceramic tools, pottery
Heavy machinery	Machine tools, construction equipment
Plastics (shaping)	Plastic moldings, extrusions
Tire and rubber	Tires, shoe soles, tennis balls

attention is mainly on the production of discrete hardware, which ranges from nuts and bolts to cars, airplanes, and digital computers. Table 2.3 lists the manufacturing industries and corresponding products for which the production systems in this book are most applicable.

Final products made by the industries listed in Table 2.3 can be divided into two major classes: consumer goods and capital goods. *Consumer goods* are products purchased directly by consumers, such as cars, personal computers, TVs, tires, toys, and tennis rackets. *Capital goods* are products purchased by other companies to produce goods and supply services. Examples of capital goods include commercial aircraft, mainframe computers, machine tools, railroad equipment, and construction machinery.

In addition to final products, which are usually assembled, there are companies in industry whose business is primarily to produce materials, components, and supplies for the companies that make the final products. Examples of these items include sheet steel, bar stock, metal stampings, machined parts, plastic moldings, cutting tools, dies, molds, and lubricants. Thus, the manufacturing industries consist of a complex infrastructure with various categories and layers of intermediate suppliers with whom the final consumer never deals.

2.2 MANUFACTURING OPERATIONS

There are certain basic activities that must be carried out in a factory to convert raw materials into finished products. Limiting our scope to a plant engaged in making discrete products, the factory activities are (1) processing and assembly operations, (2) material handling, (3) inspection and test, and (4) coordination and control.

The first three activities are the physical activities that "touch" the product as it is being made. Processing and assembly operations alter the geometry, properties, and/or appearance of the work unit. They add value to the product. The product must be moved from one operation to the next in the manufacturing sequence, and it must be inspected and/or tested to ensure high quality. It is sometimes argued that material handling and inspection activities do not add value to the product. However, material handling and inspection may be required to accomplish the necessary processing and assembly operations, for example, loading parts into a production machine and assuring that a starting work unit is of acceptable quality before processing begins. Our viewpoint is that value is added through the totality of manufacturing operations performed on the product. All unnecessary operations, whether they are processing, assembly, material handling, or inspection, must be eliminated from the sequence of steps needed to complete a given product. We return to these ideas in our discussion of lean production in Section 2.5.

2.2.1 Processing and Assembly Operations

Manufacturing processes can be divided into two basic types: (1) processing operations and (2) assembly operations. A *processing operation* transforms a work material from one state of completion to a more advanced state that is closer to the final desired part or product. It adds value by changing the geometry, properties, or appearance of the starting material. In general, processing operations are performed on discrete workparts, but some processing operations are also applicable to assembled items, for example, painting a welded sheet metal car body. An *assembly operation* joins two or more components to create a new entity, which is called an assembly, subassembly, or some other term that refers to the specific joining process. Figure 2.3 shows a classification of manufacturing processes and how they divide into various categories.

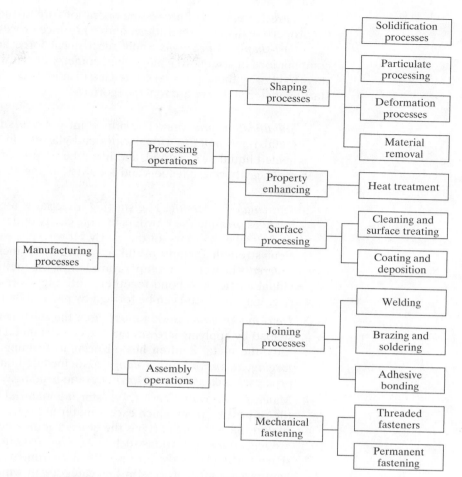

Figure 2.3 Classification of manufacturing processes.

Processing Operations. A processing operation uses energy to alter a work-part's shape, physical properties, or appearance to add value to the material. The forms of energy include mechanical, thermal, electrical, and chemical. The energy is applied in a controlled way by means of machinery and tooling. Human energy may also be required, but human workers are generally employed to control the machines, to oversee the operations, and to load and unload parts before and after each cycle of operation. A general model of a processing operation is illustrated in Figure 2.1(a). Material is fed into the process, energy is applied by the machinery and tooling to transform the material, and the completed workpart exits the process. As shown in our model, most production operations produce waste or scrap, either as a natural byproduct of the process (e.g., removing material as in machining) or in the form of occasional defective pieces. A desirable objective in manufacturing is to reduce waste in either of these forms.

More than one processing operation is usually required to transform the starting material into final form. The operations are performed in the particular sequence to achieve the geometry and/or condition defined by the design specification.

Three categories of processing operations are distinguished: (1) shaping operations, (2) property-enhancing operations, and (3) surface processing operations.

Part-shaping operations apply mechanical force and/or heat or other forms and combinations of energy to change the geometry of the work material. There are various ways to classify these processes. The classification used here is based on the state of the starting material. There are four categories:

1. *Solidification processes.* The important processes in this category are *casting* (for metals) and *molding* (for plastics and glasses), in which the starting material is a heated liquid or semifluid, and it can be poured or otherwise forced to flow into a mold cavity where it cools and solidifies, taking a solid shape that is the same as the cavity.

2. *Particulate processing.* The starting material is a powder. The common technique involves *pressing* the powders in a die cavity under high pressure to cause the powders to take the shape of the cavity. However, the compacted workpart lacks sufficient strength for any useful application. To increase strength, the part is then *sintered*—heated to a temperature below the melting point, which causes the individual particles to bond together. Both metals (powder metallurgy) and ceramics (e.g., clay products) can be formed by particulate processing.

3. *Deformation processes.* In most cases, the starting material is a ductile metal that is shaped by applying stresses that exceed the metal's yield strength. To increase ductility, the metal is often heated prior to forming. Deformation processes include *forging*, *extrusion*, and *rolling*. Also included in this category are sheet metal processes such as *drawing*, *forming*, and *bending*.

4. *Material removal processes.* The starting material is solid (commonly a metal, ductile or brittle), from which excess material is removed from the starting workpiece so that the resulting part has the desired geometry. Most important in this category are *machining* operations such as *turning*, *drilling*, and *milling*, accomplished using sharp cutting tools that are harder and stronger than the work metal. *Grinding* is another common process in this category, in which an abrasive grinding wheel is used to remove material. Other material removal processes are known as *nontraditional processes* because they do not use traditional cutting and grinding tools. Instead, they are based on lasers, electron beams, chemical erosion, electric discharge, or electrochemical energy.

Property-enhancing operations are designed to improve mechanical or physical properties of the work material. The most important property-enhancing operations involve heat treatments, which include various temperature-induced strengthening and/or toughening processes for metals and glasses. Sintering of powdered metals and ceramics, mentioned previously, is also a heat treatment, which strengthens a pressed powder workpart. Property-enhancing operations do not alter part shape, except unintentionally in some cases, for example, warping of a metal part during heat treatment or shrinkage of a ceramic part during sintering.

Surface processing operations include (1) cleaning, (2) surface treatments, and (3) coating and thin film deposition processes. *Cleaning* includes both chemical and mechanical processes to remove dirt, oil, and other contaminants from the surface. *Surface treatments* include mechanical working, such as shot peening and sand blasting, and physical processes like diffusion and ion implantation. *Coating* and *thin film deposition* processes apply a coating of material to the exterior surface of the workpart. Common coating processes include *electroplating, anodizing* of aluminum, and organic coating (call it *painting*). Thin film deposition processes include *physical vapor deposition* and *chemical vapor deposition* to form extremely thin coatings of various substances. Several surface processing operations have been adapted to fabricate semiconductor materials (most commonly silicon) into integrated circuits for microelectronics. These processes include chemical vapor deposition, physical vapor deposition, and oxidation. They are applied to very localized areas on the surface of a thin wafer of silicon (or other semiconductor material) to create the microscopic circuit.

Assembly Operations. The second basic type of manufacturing operation is assembly, in which two or more separate parts are joined to form a new entity. Components of the new entity are connected together either permanently or semipermanently. Permanent joining processes include *welding, brazing, soldering,* and *adhesive bonding.* They combine parts by forming a joint that cannot be easily disconnected. Mechanical assembly methods are available to fasten two or more parts together in a joint that can be conveniently disassembled. The use of *threaded fasteners* (e.g., screws, bolts, nuts) are important traditional methods in this category. Other mechanical assembly techniques that form a permanent connection include *rivets, press fitting,* and *expansion fits.* Special assembly methods are used in electronics. Some of the methods are identical to or adaptations of the above techniques. For example, soldering is widely used in electronics assembly. Electronics assembly is concerned primarily with the assembly of components (e.g., integrated circuit packages) to printed circuit boards to produce the complex circuits used in so many of today's products.

2.2.2 Other Factory Operations

Other activities that must be performed in the factory include material handling and storage, inspection and testing, and coordination and control.

Material Handling and Storage. A means of moving and storing materials between processing and/or assembly operations is usually required. In most manufacturing plants, materials spend more time being moved and stored than being processed. In some cases, the majority of the labor cost in the factory is consumed in handling, moving, and storing materials. It is important that this function be carried out as efficiently as possible. In Part III of our book, we consider the material handling and storage technologies that are used in factory operations.

Eugene Merchant, an advocate and spokesman for the machine tool industry for many years, observed that materials in a typical metal machining batch factory or job shop spend more time waiting or being moved than being processed [4]. His observation is illustrated in Figure 2.4. About 95% of a part's time is spent either moving or waiting (temporary storage). Only 5% of its time is spent on the machine tool. Of this 5%, less than 30% of the time on the machine (1.5% of the total time of the part) is time during which actual cutting is taking place. The remaining 70% (3.5% of the total) is required for loading and unloading, part handling and positioning, tool positioning, gaging, and other elements of non-processing time. These time proportions indicate the significance of material handling and storage in a typical factory.

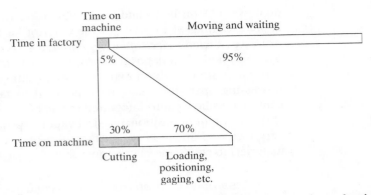

Figure 2.4 How time is spent by a typical part in a batch production machine shop [4].

Inspection and Testing. Inspection and testing are quality control activities. The purpose of inspection is to determine whether the manufactured product meets the established design standards and specifications. For example, inspection examines whether the actual dimensions of a mechanical part are within the tolerances indicated on the engineering drawing for the part. Testing is generally concerned with the functional specifications of the final product rather than with the individual parts that go into the product. For example, final testing of the product ensures that it functions and operates in the manner specified by the product designer. In Part V of this text, we examine inspection and testing.

Coordination and Control. Coordination and control in manufacturing include both the regulation of individual processing and assembly operations and the management of plant level activities. Control at the process level involves the achievement of certain performance objectives by properly manipulating the inputs and other parameters of the process. Control at the process level is discussed in Part II of the book.

Control at the plant level includes effective use of labor, maintenance of the equipment, moving materials in the factory, controlling inventory, shipping products of good quality on schedule, and keeping plant operating costs to a minimum. The manufacturing control function at the plant level represents the major point of intersection between the physical operations in the factory and the information processing activities that occur in production. We discuss many of these plant and enterprise level control functions in Parts V and VI.

2.3 PRODUCTION FACILITIES

A manufacturing company attempts to organize its facilities in the most efficient way to serve the particular mission of each plant. Over the years, certain types of production facilities have come to be recognized as the most appropriate way to organize for a given type of manufacturing. Of course, one of the most important factors that determine the type of manufacturing is the type of products that are made. As mentioned previously, this book is concerned primarily with the production of discrete parts and products. The quantity of parts and/or products made by a factory has a very significant influence on its facilities and the way manufacturing is organized. *Production quantity* refers to the number of units of a given part or product produced annually by the plant. The annual part or product quantities produced in a given factory can be classified into three ranges:

1. *Low production*: Quantities in the range of 1 to 100 units.
2. *Medium production*: Quantities in the range of 100 to 10,000 units.
3. *High production*: Production quantities are 10,000 to millions of units.

The boundaries between the three ranges are somewhat arbitrary (author's judgment). Depending on the types of products, these boundaries may shift by an order of magnitude or so.

Some plants produce a variety of different product types, each type being made in low or medium quantities. Other plants specialize in high production of only one product type. It is instructive to identify product variety as a parameter distinct from production quantity. *Product variety* refers to the different product designs or types that are produced in a plant. Different products have different shapes and sizes and styles, they perform different functions, they are sometimes intended for different markets, some have more components than others, and so forth. The number of different product types made each year can be counted. When the number of product types made in a factory is high, this indicates high product variety.

There is an inverse correlation between product variety and production quantity in terms of factory operations. When product variety is high, production quantity tends to be low; and vice versa. This relationship is depicted in Figure 2.5. Manufacturing plants tend to specialize in a combination of production quantity and product variety that lies somewhere inside the diagonal band in Figure 2.5. In general, a given factory tends to be limited to the product variety value that is correlated with that production quantity.

Although we have identified product variety as a quantitative parameter (the number of different product types made by the plant or company), this parameter is much less exact than production quantity, because details on how much the designs differ are not captured

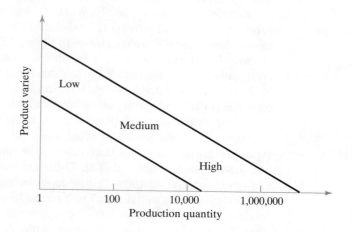

Figure 2.5 Relationship between product variety and production quantity in discrete product manufacturing.

simply by the number of different designs. The differences between an automobile and an air conditioner are far greater than between an air conditioner and a heat pump. Products can be different, but the extent of the differences may be small or great. The automotive industry provides some examples to illustrate this point. Each of the U.S. automotive companies produces cars with two or three different nameplates in the same assembly plant, although the body styles and other design features are nearly the same. In different plants, the same company builds heavy trucks. Let us use the terms "hard" and "soft" to describe these differences in product variety. *Hard product variety* is when the products differ substantially. In an assembled product, hard variety is characterized by a low proportion of common parts among the products; in many cases, there are no common parts. The difference between a car and a truck is hard. *Soft product variety* is when there are only small differences between products, such as the differences between car models made on the same production line. There is a high proportion of common parts among assembled products whose variety is soft. The variety between different product categories tends to be hard; the variety between different models within the same product category tends to be soft.

We can use the three production quantity ranges to identify three basic categories of production plants. Although there are variations in the work organization within each category, usually depending on the amount of product variety, this is nevertheless a reasonable way to classify factories for the purpose of our discussion.

2.3.1 Low Production

The type of production facility usually associated with the quantity range of 1 to 100 units/year is the *job shop*, which makes low quantities of specialized and customized products. The products are typically complex, such as space capsules, aircraft, and special machinery. Job shop production can also include fabricating the component parts for the products. Customer orders for these kinds of items are often special, and repeat orders may never occur. Equipment in a job shop is general purpose and the labor force is highly skilled.

A job shop must be designed for maximum flexibility to deal with the wide part and product variations encountered (hard product variety). If the product is large and heavy, and therefore difficult to move in the factory, it typically remains in a single location, at least during its final assembly. Workers and processing equipment are brought to the product, rather than moving the product to the equipment. This type of layout is referred to as a *fixed-position layout*, shown in Figure 2.6(a), in which the product remains in a single location during its entire fabrication. Examples of such products include ships, aircraft, railway locomotives, and heavy machinery. In actual practice, these items are usually built in large modules at single locations, and then the completed modules are brought together for final assembly using large-capacity cranes.

The individual parts that comprise these large products are often made in factories that have a *process layout*, in which the equipment is arranged according to function or type. The lathes are in one department, the milling machines are in another department, and so on, as in Figure 2.6(b). Different parts, each requiring a different operation sequence, are routed through the departments in the particular order needed for their processing, usually in batches. The process layout is noted for its flexibility; it can accommodate a great variety of alternative operation sequences for different part configurations. Its disadvantage is that the machinery and methods to produce a part are not designed for high efficiency. Much material handling is required to move parts between departments, so in-process inventory tends to be high.

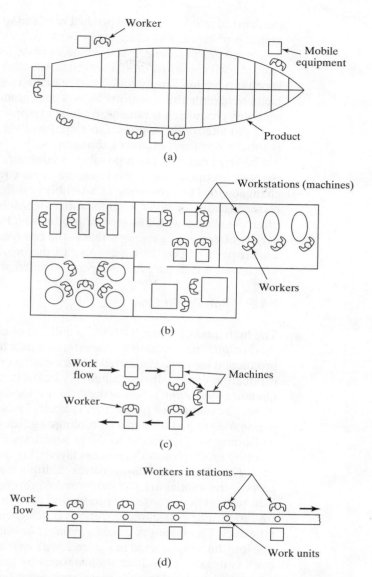

Figure 2.6 Various types of plant layout: (a) fixed-position layout, (b) process layout, (c) cellular layout, and (d) product layout.

2.3.2 Medium Production

In the medium quantity range (100–10,000 units annually), we distinguish between two different types of facility, depending on product variety. When product variety is hard, the traditional approach is *batch production*, in which a batch of one product is made, after which the facility is changed over to produce a batch of the next product, and so on. Orders for each product are frequently repeated. The production rate of the equipment is greater than the demand rate for any single product type, and so the same equipment can be shared among multiple products. The changeover between production runs takes time. Called the *setup time* or *changeover time*, it is the time to change tooling and to set up and reprogram the machinery. This is lost production time, which is a disadvantage of batch manufacturing. Batch production is commonly used in make-to-stock situations, in which items are manufactured to replenish inventory that has been gradually depleted by demand. The equipment for batch production is usually arranged in a process layout, Figure 2.6(b).

An alternative approach to medium range production is possible if product variety is soft. In this case, extensive changeovers between one product style and the next may not be required. It is often possible to configure the equipment so that groups of similar parts or products can be made on the same equipment without significant lost time for changeovers. The processing or assembly of different parts or products is accomplished in cells consisting of several workstations or machines. The term *cellular manufacturing* is often associated with this type of production. Each cell is designed to produce a limited variety of part configurations; that is, the cell specializes in the production of a given set of similar parts or products, according to the principles of *group technology* (Chapter 18). The layout is called a *cellular layout*, depicted in Figure 2.6(c).

2.3.3 High Production

The high quantity range (10,000 to millions of units per year) is often referred to as *mass production*. The situation is characterized by a high demand rate for the product, and the production facility is dedicated to the manufacture of that product. Two categories of mass production can be distinguished: (1) quantity production and (2) flow line production. *Quantity production* involves the mass production of single parts on single pieces of equipment. The method of production typically involves standard machines (such as stamping presses) equipped with special tooling (e.g., dies and material handling devices), in effect dedicating the equipment to the production of one part type. The typical layout used in quantity production is the process layout, Figure 2.6(b).

Flow line production involves multiple workstations arranged in sequence, and the parts or assemblies are physically moved through the sequence to complete the product. The workstations consist of production machines and/or workers equipped with specialized tools. The collection of stations is designed specifically for the product to maximize efficiency. The layout is called a *product layout*, and the workstations are arranged into one long line, as depicted in Figure 2.6(d), or into a series of connected line segments. The work is usually moved between stations by powered conveyor. At each station, a small amount of the total work is completed on each unit of product.

The most familiar example of flow line production is the assembly line, associated with products such as cars and household appliances. The pure case of flow line production is where there is no variation in the products made on the line. Every product is identical, and the line is referred to as a *single-model production line*. However, to successfully market a given product, it is often necessary to introduce model variations so that individual customers can choose the exact style and options that appeal to them. From a production viewpoint, the model differences represent a case of soft product variety. The term *mixed-model production line* applies to those situations where there is soft variety in the products made on the line. Modern automobile assembly is an example. Cars coming off the assembly line have variations in options and trim representing different models (and, in many cases, different nameplates) of the same basic car design. Other examples

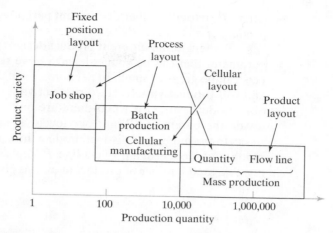

Figure 2.7 Types of facilities and layouts used for different levels of production quantity and product variety.

include small and major appliances. The Boeing Commercial Airplane Company uses production line techniques to assemble its 737 model.

Much of our discussion of the types of production facilities is summarized in Figure 2.7, which adds detail to Figure 2.5 by identifying the types of production facilities and plant layouts used. As the figure shows, some overlap exists among the different facility types.

2.4 PRODUCT/PRODUCTION RELATIONSHIPS

As noted in the preceding section, companies organize their production facilities and manufacturing systems in the most efficient manner for the particular products they make. It is instructive to recognize that there are certain product parameters that are influential in determining how the products are manufactured. Let us consider the following parameters: (1) production quantity, (2) product variety, (3) product complexity (of assembled products), and (4) part complexity.

2.4.1 Production Quantity and Product Variety

We previously discussed production quantity and product variety in Section 2.3. Let us now develop a set of symbols to represent these important parameters. Let Q = production quantity and P = product variety. Thus we can discuss product variety and production quantity relationships as PQ relationships.

Q refers to the number of units of a given part or product that are produced annually by a plant. Our interest includes both the quantities of each individual part or product style and the total quantity of all styles. Let us identify each part or product style by using the subscript j, so that Q_j = annual quantity of style j. Then let Q_f = total quantity of all parts or products made in the factory (the subscript f refers to factory). Q_j and Q_f are related as

$$Q_f = \sum_{j=1}^{P} Q_j \qquad (2.1)$$

where P = total number of different part or product styles, and j is a subscript to identify products, $j = 1, 2, \ldots, P$.

P refers to the different product designs or types that are produced in a plant. It is a parameter that can be counted, and yet we recognize that the difference between products can be great or small. In Section 2.3, we distinguished between hard product variety and soft product variety. Hard product variety is when the products differ substantially. Soft product variety is when there are only small differences between products. Let us divide the parameter P into two levels, as in a tree structure. Call them P_1 and P_2. P_1 refers to the number of distinct product lines produced by the factory, and P_2 refers to the number of models in a product line. P_1 represents hard product variety, and P_2 soft variety. The total number of product models is given by

$$P = \sum_{j=1}^{P_1} P_{2j} \tag{2.2}$$

where the subscript j identifies the product line: $j = 1, 2, \ldots, P_1$.

EXAMPLE 2.1 Product Lines and Product Models

A company specializes in home entertainment products. It produces only TVs and audio systems. Thus $P_1 = 2$. In its TV line it offers 15 different models, and in its audio line it offers 5 models. Thus for TVs, $P_2 = 15$, and for audio systems, $P_2 = 5$. The totality of product models offered is given by Eq. (2.2):

$$P = \sum_{j=1}^{2} P_{2j} = 15 + 5 = 20$$

2.4.2 Product and Part Complexity

How complex is each product made in the plant? Product complexity is a complicated issue. It has both qualitative and quantitative aspects. Let us try to deal with it using quantitative measures. For an assembled product, one possible indicator of *product complexity* is its number of components—the more parts, the more complex the product is. This is easily demonstrated by comparing the numbers of components in various assembled products, as in Table 2.4. Our list demonstrates that the more components a product has, the more complex it tends to be.

For a manufactured component, a possible measure of *part complexity* is the number of processing steps required to produce it. An integrated circuit, which is technically a monolithic silicon chip with localized alterations in its surface chemistry, requires hundreds of processing steps in its fabrication. Although it may measure only 12 mm (0.5 in) on a side and 0.5-mm (0.020 in) thick, its complexity is orders of magnitude greater than a round washer of 12-mm (1/2-in) outside diameter, stamped out of 0.8-mm (1/32-in) thick stainless steel in one step. In Table 2.5, we have compiled a list of manufactured parts with the typical number of processing operations required for each.

So, we have complexity of an assembled product defined as the number of distinct components; let n_p = the number of parts per product. And we have processing complexity of each part as the number of operations required to make it; let n_o = the number of operations or processing steps to make a part. We can draw some distinctions among production

TABLE 2.4 Typical Number of Separate Components in Various Assembled Products (Compiled from [1], [3], and Other Sources)

Product (Approx. Date or Circa)	Approx. Number of Components
Mechanical pencil (modern)	10
Ball bearing (modern)	20
Rifle (1800)	50
Sewing machine (1875)	150
Bicycle chain	300
Bicycle (modern)	750
Early automobile (1910)	2000
Automobile (modern)	20,000
Commercial airplane (1930)	100,000
Commercial airplane (modern)	1,000,000
Space shuttle (modern)	10,000,000

TABLE 2.5 Typical Number of Processing Operations Required To Fabricate Various Parts

Part	Approx. Number of Processing Operations	Typical Processing Operations Used
Plastic molded part	1	Injection molding
Washer (stainless steel)	1	Stamping
Washer (plated steel)	2	Stamping, electroplating
Forged part	3	Heating, forging, trimming
Pump shaft	10	Machining (from bar stock)
Coated carbide cutting tool	15	Pressing, sintering, coating, grinding
Pump housing, machined	20	Casting, machining
V-6 engine block	50	Casting, machining
Integrated circuit chip	75	Photolithography, various thermal and chemical processes

plants on the basis of n_p and n_o. As defined in Table 2.6, three different types of plant can be identified: parts producers, pure assembly plants, and vertically integrated plants.

Let us develop some simple relationships among the parameters P, Q, n_p, and n_o that indicate the level of activity in a manufacturing plant. We will ignore the differences between P_1 and P_2 here, although Eq. (2.2) could be used to convert these parameters into the corresponding P value. The total number of products made annually in a plant is the sum of the quantities of the individual product designs, as expressed in previous Eq. (2.1). Assuming that the products are all assembled and that all component parts used in these products are made in the plant (no purchased components), then the total number of parts manufactured by the plant per year is given by

$$n_{pf} = \sum_{j=1}^{P} Q_j n_{pj} \qquad (2.3)$$

where n_{pf} = total number of parts made in the factory (pc/yr), Q_j = annual quantity of product style j (products/yr), and n_{pj} = number of parts in product j (pc/product).

TABLE 2.6 Production Plants Distinguished by n_p and n_o Values

	$n_p = 1$	$n_p > 1$
$n_o > 1$	*Parts producer.* This plant makes individual components; each component requires multiple processing steps. No assembly, thus $n_p = 1$.	*Vertically integrated plant.* The plant makes all its parts and assembles them into its final products; thus, $n_o > 1$, $n_p > 1$.
$n_o = 1$	*Handicraft shop.* Not really a production plant. Makes one part per year; thus, $n_o = 1$, $n_p = 1$.	*Assembly plant.* A pure assembly plant produces no parts. It purchases all parts from suppliers. One operation is required to assemble each part to the product, thus $n_o = 1$.

Finally, if all parts are manufactured in the plant, then the total number of processing operations performed by the plant is given by

$$n_{of} = \sum_{j=1}^{P} Q_j n_{pj} \sum_{k=1}^{n_{pj}} n_{ojk} \qquad (2.4)$$

where n_{of} = total number of operation cycles performed in the factory (ops/yr), and n_{ojk} = number of processing operations for each part k, summed over the number of parts in product j, n_{pj}. Parameter n_{of} provides a numerical value for the total activity level in the factory.

We might try to simplify this to better conceptualize the situation by using average values for the four parameters P, Q, n_p, and n_o. In effect, we are assuming that the number of product designs P are produced in equal quantities Q, that all products have the same number of components n_p, and that all components require an equal number of processing steps n_o. In this case, the total number of product units produced by the factory is given by

$$Q_f = PQ \qquad (2.5)$$

The total number of parts produced by the factory is given by

$$n_{pf} = PQn_p \qquad (2.6)$$

The total number of manufacturing operations performed by the factory is given by

$$n_{of} = PQn_p n_o \qquad (2.7)$$

Using these simplified equations, consider the following example.

EXAMPLE 2.2   A Production System Problem

Suppose a company has designed a new product line and is planning to build a new plant to manufacture this product line. The new line consists of 100 different product types, and for each product type the company wants to produce 10,000 units annually. The products average 1000 components each, and the average number of processing steps required for each component is 10. All parts will be made in the factory. Each processing step takes an average of 1 min. Determine (a) how many products, (b) how many parts, and (c) how many production operations will be required each year, and (d) how many workers will be needed for the plant, if it operates one eight-hour shift for 250 day/yr?

Solution: (a) The total number of units to be produced by the factory is given by

$$Q = PQ = 100 \times 10,000 = 1,000,000 \text{ products annually.}$$

(b) The total number of parts produced is

$$n_{pf} = PQn_p = 1,000,000 \times 1000 = 1,000,000,000 \text{ parts annually.}$$

(c) The number of distinct production operations is

$$n_{of} = PQn_pn_o = 1,000,000,000 \times 10 = 10,000,000,000 \text{ operations.}$$

(d) First consider the total time to perform these operations. If each operation takes 1 min (1/60 hr),

$$\text{Total time} = 10,000,000,000 \times 1/60 = 166,666,667 \text{ hr}$$

If each worker works 2000 hr/yr (250 days/yr $\times$ 8 hr/day), then the total number of workers required is

$$w = \frac{166,666,667}{2000} = 83,333 \text{ workers.}$$

The factory in our example is a fully integrated factory. It would be a big factory. The number of workers we have calculated only includes direct labor. Add indirect labor, staff, and management, and the number increases to well over 100,000 employees. Imagine the parking lot. And inside the factory, the logistics problems of dealing with all of the products, parts, and operations would be overwhelming. No organization in its right mind would consider building or operating such a plant today—not even the federal government.

2.4.3 Limitations and Capabilities of a Manufacturing Plant

Companies do not attempt the kind of factory in our example. Instead, today's factories are designed with much more specific missions. Referred to as *focused factories*, they are plants that concentrate "on a limited, concise, manageable set of products, technologies, volumes, and markets" [5]. It is a recognition that a manufacturing plant cannot do everything. It must limit its mission to a certain scope of products and activities in which it can best compete. Its size is typically about 500 workers or fewer, although the number may vary for different types of products and manufacturing operations.

Let us consider how a plant, or its parent company, limits the scope of its manufacturing operations and production systems. In limiting its scope, the plant in effect makes a set of deliberate decisions about what it will not try to do. Certainly one way to limit a plant's scope is to avoid being a fully integrated factory, at least to the extent of our Example 2.2. Instead, the plant specializes in being either a parts producer or an assembly plant. Just as it decides what it will not do, the plant must also decide on the specific technologies, products, and volumes in which it will specialize. These decisions determine the plant's intended manufacturing capability. *Manufacturing capability* refers to the technical and physical limitations of a manufacturing firm and each of its plants. We can identify several dimensions of this capability: (1) technological processing capability, (2) physical size and weight of product, and (3) production capacity.

Technological Processing Capability.   The technological processing capability of a plant (or company) is its available set of manufacturing processes. Certain plants perform machining operations, others roll steel billets into sheet stock, and others build automobiles. A machine shop cannot roll steel, and a rolling mill cannot build cars. The underlying feature that distinguishes these plants is the set of processes they can perform. Technological processing capability is closely related to the material being processed. Certain manufacturing processes are suited to certain materials, while other processes are suited to other materials. By specializing in a certain process or group of processes, the plant is simultaneously specializing in a certain material type or range of materials.

Technological processing capability includes not only the physical processes, but also the expertise possessed by plant personnel in these processing technologies. Companies are limited by their available processes. They must focus on designing and manufacturing products for which their technological processing capability provides a competitive advantage.

Physical Product Limitations.   A second aspect of manufacturing capability is imposed by the physical product. Given a plant with a certain set of processes, there are size and weight limitations on the products that can be accommodated in the plant. Big, heavy products are difficult to move. To move products, the plant must be equipped with cranes of large load capacity. Smaller parts and products made in large quantities can be moved by conveyor or fork lift truck. The limitation on product size and weight extends to the physical capacity of the manufacturing equipment as well. Production machines come in different sizes. Larger machines can be used to process larger parts. Smaller machines limit the size of the work that can be processed. The set of production equipment, material handling, storage capability, and plant size must be planned for products that lie within a certain size and weight range.

Production Capacity.   A third limitation on a plant's manufacturing capability is the production quantity that can be produced in a given time period (e.g., month or year). Production capacity is defined as the maximum rate of production per period that a plant can achieve under assumed operating conditions. The operating conditions refer to the number of shifts per week, hours per shift, direct labor manning levels in the plant, and similar conditions under which the plant has been designed to operate. These factors represent inputs to the manufacturing plant. Given these inputs, how much output can the factory produce?

Plant capacity is often measured in terms of output units, such as annual tons of steel produced by a steel mill, or number of cars produced by a final assembly plant. In these cases, the outputs are homogeneous, more or less. In cases where the output units are not homogeneous, other factors may be more appropriate measures, such as available labor hours of productive capacity in a machine shop that produces a variety of parts.

2.5   LEAN PRODUCTION

Today, manufacturing companies and their plants must operate efficiently and effectively to remain competitive in the global economy. One of the general approaches that has been successful in these efforts is lean production. In this section, we provide a brief description of this approach because it is so widely used in manufacturing operations.

Lean production means operating the factory with the minimum possible resources and yet maximizing the amount of work that is accomplished with these resources. Resources include workers, equipment, time, space, and materials. Lean production also implies completing the products in the minimum possible time and achieving a very high level of quality, so that the customer is completely satisfied. In short, lean production means doing more with less, and doing it better.

Manufacturing operations often include many wasteful activities, that is, activities that do not really add value to the product, such as producing defective parts, producing more parts than are needed, handling materials unnecessarily, and having workers waiting. Manufacturing activities can be divided into three categories according to the value they contribute to the part or product being made:

1. *Value-adding activities*. These are work activities that contribute real value to the work unit. They include processing and assembly operations that alter the part or product in a way that the customer can recognize and appreciate.

2. *Auxiliary activities*. These are activities that support the value-adding activities but do not themselves contribute value to the part or product. They are necessary because without them, the value-adding activities could not be accomplished. Auxiliary activities include loading and unloading a production machine that performs a value-adding process.

3. *Wasteful activities*. These are activities that do not add value nor do they support the value-adding activities. If these activities were not performed, there would be no adverse effect on the product.

Lean production works by eliminating the wasteful activities so that only the value-adding and auxiliary activities are performed. When this is done, it means that fewer resources are required, the work is completed in less time, and higher quality is achieved in the final product. Some of the programs associated with lean production are the following:

- *Just-in-time delivery of parts*. This refers to the manner in which parts are moved through the production system when a sequence of manufacturing operations is required to make them. In the ideal just-in-time system, each part is delivered to the downstream workstation immediately before that part is needed at the station. This discipline minimizes the amount of work-in-process inventory between stations, and it motivates a high level of quality in the parts that are made.

- *Worker involvement*. Workers in a lean environment are assigned greater responsibilities and are provided with training that allows them to be flexible in the work they can do. Also, workers participate in worker teams to solve problems faced by the company. The underlying philosophy is the opposite of dividing the total workload into many different job classifications and having work rules that forbid workers from performing tasks outside their job classifications.

- *Continuous improvement*. This involves an unending search for ways to make improvements in products and manufacturing operations. It is usually accomplished by worker teams who cooperate to develop solutions to production and quality problems.

- *Reduced setup times*. Methods engineering is used to minimize the time needed to change over from one setup to the next in batch production. This allows batch sizes to be smaller, thereby reducing the amount of work-in-process in the factory.

- *Stopping the process when something is wrong.* Production machines are designed to stop automatically when a defective part is made, or the required quantity has been completed, or some other abnormal operation is detected. This increases part quality and avoids overproduction.
- *Error prevention.* This refers to the use of low-cost devices and design features at each workstation that prevent errors from occurring. Common errors in production include omitting processing or assembly steps, locating a part in a fixture incorrectly, and using an improper tool.
- *Total productive maintenance.* This is a program that includes preventive maintenance and other procedures to avoid machine breakdowns that disrupt production operations. A central feature of the program is to have the worker at the machine perform minor repairs and maintenance procedures.

A more complete enumeration and discussion of the programs in a lean production system are provided in Chapter 26.

REFERENCES

[1] BLACK, J. T., *The Design of the Factory with a Future*, McGraw-Hill, Inc., NY, 1991.

[2] GROOVER, M. P., *Fundamentals of Modern Manufacturing: Materials, Processes, and Systems*, 3d ed., John Wiley & Sons, Inc., Hoboken, NJ, 2007.

[3] HOUNSHELL, D. A., *From the American System to Mass Production, 1800–1932*, The Johns Hopkins University Press, Baltimore, MD, 1984.

[4] MERCHANT, M. E., The Inexorable Push for Automated Production, *Production Engineering*, January 1977, pp 45–46.

[5] SKINNER, W., "The Focused Factory," *Harvard Business Review*, May–June 1974, pp 113–121.

REVIEW QUESTIONS

2.1 What is manufacturing?

2.2 What are the three basic industry categories?

2.3 What is the difference between consumer goods and capital goods?

2.4 What is the difference between a processing operation and an assembly operation?

2.5 Name the four categories of part-shaping operations, based on the state of the starting work material.

2.6 Assembly operations can be classified as permanent joining methods and mechanical assembly. What are the four types of permanent joining methods?

2.7 What is the difference between hard product variety and soft product variety?

2.8 What type of production does a job shop perform?

2.9 Flow line production is associated with which one of the following layout types: (a) cellular layout, (b) fixed-position layout, (c) process layout, or (d) product layout?

2.10 What is the difference between a single-model production line and a mixed-model production line?

2.11 What is meant by the term *technological processing capability*?

2.12 What is lean production?

2.13 In lean production, what is just-in-time delivery of parts?

2.14 In lean production, what does worker involvement mean?

2.15 In lean production, what does continuous improvement mean, and how is it usually accomplished?

PROBLEMS

2.1 A plant produces three product lines: A, B, and C. There are six models within product line A, four models within B, and eight within C. Average annual production quantities of each A model is 500 units, 700 units for each B model, and 1100 units for each C model. Determine the values of (a) P and (b) Q_f for this plant.

2.2 The ABC Company is planning a new product line and will build a new plant to manufacture the parts for this product line. The product line will include 50 different models. Annual production of each model is expected to be 1000 units. Each product will be assembled of 400 components. All processing of parts will be accomplished in one factory. There are an average of six processing steps required to produce each component, and each processing step takes 1.0 minute (includes an allowance for setup time and part handling). All processing operations are performed at workstations, each of which includes a production machine and a human worker. If each workstation requires a floor space of 250 ft^2, and the factory operates one shift (2000 hr/yr), determine (a) how many production operations, (b) how much floor space, and (c) how many workers will be required in the plant.

2.3 The XYZ Company is planning to introduce a new product line and will build a new factory to produce the parts and assembly the final products for the product line. The new product line will include 100 different models. Annual production of each model is expected to be 1000 units. Each product will be assembled of 600 components. All processing of parts and assembly of products will be accomplished in one factory. There are an average of 10 processing steps required to produce each component, and each processing step takes 30 sec. (includes an allowance for setup time and part handling). Each final unit of product takes 3.0 hours to assemble. All processing operations are performed at work cells that include a production machine and a human worker. Products are assembled on single workstations consisting of two workers each. If each work cell and each workstation require 200 ft^2, and the factory operates one shift (2000 hr/yr), determine: (a) how many production operations, (b) how much floor space, and (c) how many workers will be required in the plant.

2.4 If the company in Problem 2.3 were to operate three shifts (6000 hr/yr) instead of one shift, determine the answers to (a), (b), and (c).

Chapter 3

Manufacturing Models and Metrics

CHAPTER CONTENTS

Successful manufacturing companies use a variety of metrics to help manage their operations. Quantitative metrics provide a company with the means to track performance in successive periods (e.g., months and years), try out new technologies and new systems to determine their merits, identify problems with performance, compare alternative methods, and make good decisions. Manufacturing metrics can be divided into two basic categories: (1) production performance measures and (2) manufacturing costs. Metrics that indicate production performance include production rate, plant capacity, proportion uptime on equipment (a reliability measure), and manufacturing lead time. Manufacturing costs that are important to a company include labor and material costs, the costs of producing its products, and the cost of operating a given piece of equipment. In this chapter, we define these metrics and show how they are calculated.

From Chapter 3 of *Automation, Production Systems, and Computer-Integrated Manufacturing*, Third Edition.
Mikell P. Groover. Copyright © 2008 by Pearson Education, Inc. Publishing as Prentice Hall. All rights reserved.

3.1 *MATHEMATICAL MODELS OF PRODUCTION PERFORMANCE*

Many aspects of manufacturing are quantitative. We encountered some of these quantitative aspects in the previous chapter in the four product parameters: production quantity Q, product variety P, number of parts per product n_p, and number of operations to produce a part n_o. In this section, we define several additional parameters and variables that are measured quantitatively. We also develop mathematical models that can be used to define and calculate these parameters. In subsequent chapters, we refer back to these definitions in our discussion of specific topics in automation and production systems.

3.1.1 Production Rate

Production rate for an individual processing or assembly operation is usually expressed as an hourly rate, that is, work units completed per hour (pc/hr). Let us consider how the production rate is determined for the three types of production: batch production, job shop production, and mass production. Our starting point is the cycle time.

Cycle Time. For any production operation, the *cycle time* T_c is defined as the time that one work unit spends being processed or assembled. It is the time between when one work unit begins processing (or assembly) and when the next unit begins. T_c is the time an individual part spends at the machine, but not all of this time is productive (recall the Merchant study, Section 2.2.2). In a typical processing operation, such as machining, T_c consists of (1) actual machining operation time, (2) workpart handling time, and (3) tool handling time per workpiece. As an equation, this can be expressed as:

$$T_c = T_o + T_h + T_{th} \tag{3.1}$$

where T_c = cycle time (min/pc), T_o = time of the actual processing or assembly operation (min/pc), T_h = handling time (min/pc), and T_{th} = tool handling time (min/pc). The tool handling time consists of time spent changing tools when they wear out, time changing from one tool to the next, tool indexing time for indexable inserts or for tools on a turret lathe or turret drill, tool repositioning for a next pass, and so on. Some of these tool handling activities do not occur every cycle; therefore, they must be spread over the number of parts between their occurrences to obtain an average time per workpiece.

Each of the terms, T_o, T_h, and T_{th}, has its counterpart in other types of discrete-item production. There is a portion of the cycle when the part is actually being processed (T_o); there is a portion of the cycle when the part is being handled (T_h); and there is, on average, a portion when the tooling is being adjusted or changed (T_{th}). Accordingly, we can generalize Eq. (3.1) to cover most processing operations in manufacturing.

Batch and Job Shop Production. In batch production, the time to process one batch consisting of Q work units is the sum of the setup time and processing time; that is,

$$T_b = T_{su} + QT_c \tag{3.2}$$

where T_b = batch processing time (min), T_{su} = setup time to prepare for the batch (min), Q = batch quantity (pc), and T_c = cycle time per work unit (min/cycle). We assume that one work unit is completed each cycle and so T_c also has units of min/pc. If more than one part is produced each cycle, then Eq. (3.2) must be adjusted accordingly. Dividing batch time by batch quantity, we have the average production time per work unit T_p for the given machine:

$$T_p = \frac{T_b}{Q} \tag{3.3}$$

The average production rate for the machine is simply the reciprocal of production time. It is usually expressed as an hourly rate:

$$R_p = \frac{60}{T_p} \tag{3.4}$$

where R_p = hourly production rate (pc/hr), T_p = average production time per minute (min/pc), and the constant 60 converts minutes to hours.

For job shop production when quantity $Q = 1$, the production time per work unit is the sum of setup and cycle times:

$$T_p = T_{su} + T_c \tag{3.5}$$

For job shop production when the quantity is greater than one, the production rate is determined as in the batch production case discussed above.

Mass Production. For quantity type mass production, we can say that the production rate equals the cycle rate of the machine (reciprocal of operation cycle time) after production is underway and the effects of setup time become insignificant. That is, as Q becomes very large, $(T_{su}/Q) \to 0$ and

$$R_p \to R_c = \frac{60}{T_c} \tag{3.6}$$

where R_c = operation cycle rate of the machine (pc/hr), and T_c = operation cycle time (min/pc).

For flow line mass production, the production rate approximates the cycle rate of the production line, again neglecting setup time. However, the operation of production lines is complicated by the interdependence of the workstations on the line. One complication is that it is usually impossible to divide the total work equally among all of the workstations on the line; therefore, one station ends up with the longest operation time, and this station sets the pace for the entire line. The term *bottleneck station* is sometimes used to refer to this station. Also included in the cycle time is the time to move parts from one station to the next at the end of each operation. In many production lines, all work units on the line are moved simultaneously, each to its respective next station. Taking these factors into account, the cycle time of a production line is the longest processing (or assembly) time plus the time to transfer work units between stations. This can be expressed as

$$T_c = T_r + \text{Max } T_o \tag{3.7}$$

where T_c = cycle time of the production line (min/cycle), T_r = time to transfer work units between stations each cycle (min/cycle), and Max T_o = operation time at the bottleneck station (the maximum of the operation times for all stations on the line, min/cycle). Theoretically, the production rate can be determined by taking the reciprocal of T_c as

$$R_c = \frac{60}{T_c} \tag{3.8}$$

where R_c = theoretical or ideal production rate, but let us call it the cycle rate to be more precise (cycles/hr), and T_c = ideal cycle time from Eq. (3.7) (min/cycle).

Production lines are of two basic types: (1) manual and (2) automated. In the operation of automated production lines, another complicating factor is reliability. Poor reliability reduces the available production time on the line. This results from the interdependence of workstations in an automated line, in which the entire line is forced to stop when one station breaks down. The actual average production rate R_p is reduced to a value that is often substantially below the ideal R_c given by Eq. (3.8). We discuss reliability and some of its terminology in Section 3.1.3. The effect of reliability on automated production lines is examined in Chapters 16 and 17.

3.1.2 Production Capacity

We mentioned production capacity in our discussion of manufacturing capabilities (Section 2.4.3). Production capacity is defined as the maximum rate of output that a production facility (or production line, work center, or group of work centers) is able to produce under a given set of assumed operating conditions. The production facility usually refers to a plant or factory, and so the term *plant capacity* is often used for this measure. As mentioned before, the assumed operating conditions refer to the number of shifts per day (one, two, or three), number of days in the week (or month) that the plant operates, employment levels, and so forth.

The number of hours of plant operation per week is a critical issue in defining plant capacity. For continuous chemical production in which the reactions occur at elevated temperatures, the plant is usually operated 24 hr/day, 7 day/wk. For an automobile assembly plant, capacity is typically defined as one or two shifts. In the manufacture of discrete parts and products, a growing trend is to define plant capacity for the full 7-day week, 24 hr/day. This is the maximum time available (168 hr/wk), and if the plant operates fewer hours than the maximum, then its maximum possible capacity is not being fully utilized.

Quantitative measures of plant capacity can be developed based on the production rate models derived earlier. Let PC = the production capacity of a given facility under consideration. Let the measure of capacity = the number of units produced per week. Let n = the number of machines or work centers in the facility. A *work center* is a manufacturing system in the plant typically consisting of one worker and one machine. It might also be one automated machine with no worker, or multiple workers working together on a production line. It is capable of producing at a rate R_p unit/hr, as defined in Section 3.1.1. Each work center operates for a certain number of hours per shift H_{sh} (an 8 hr/shift is typical in production). Provision for setup time is included in R_p, according to Eq. (3.4). Let S_w denote the number of shifts per week. These parameters can be combined to calculate the production capacity of the facility,

$$PC = nS_w H_{sh} R_p \qquad (3.9)$$

where PC = weekly production capacity of the facility (output units/wk), n = number of work centers working in parallel producing in the facility, S_w = number of shifts per period (shift/wk), H_{sh} = hr/shift (hr), and R_p = hourly production rate of each work center (output units/hr). Although we have used a week as the time period of interest, Eq. (3.9) can easily be adapted to other periods (months, years, etc.). As in previous equations, our assumption is that the units processed through the group of work centers are homogeneous, and therefore the value of R_p is the same for all units produced.

EXAMPLE 3.1   Production Capacity

The turret lathe section has six machines, all devoted to the production of the same part. The section operates 10 shift/wk. The number of hours per shift averages 8.0. Average production rate of each machine is 17 unit/hr. Determine the weekly production capacity of the turret lathe section.

Solution: From Eq. (3.9),

$$PC = 6(10)(8.0)(17) = 8160 \text{ output unit/wk}$$

If we include the possibility that each work unit requires n_o operations in its processing sequence, with each operation requiring a new setup on either the same or a different machine, then the plant capacity equation must be amended as

$$PC = \frac{n S_w H_{sh} R_p}{n_o} \tag{3.10}$$

where n_o = number of distinct operations through which work units are routed, and the other terms have the same meaning as before.

Equation (3.10) indicates the operating parameters that affect plant capacity. Changes that can be made to increase or decrease plant capacity over the short term are:

1. Change the number of shifts per week S_w. For example, Saturday shifts might be authorized to temporarily increase capacity.
2. Change the number of hours worked per shift H_{sh}. For example, overtime on each regular shift might be authorized to increase capacity.

Over the intermediate or longer term, the following changes can be made to increase plant capacity:

3. Increase the number of work centers, n, in the shop. This might be done by using equipment that was formerly not in use, acquiring new machines, and hiring new workers. Decreasing capacity is easier, except for the social and economic impact: Workers must be laid off and machines decommissioned.
4. Increase the production rate, R_p, by making improvements in methods or process technology.
5. Reduce the number of operations, n_o, required per work unit by using combined operations, simultaneous operations, or integration of operations (Section 1.4.2, strategies 2, 3, and 4).

This capacity model assumes that all n machines are producing 100% of the time, and there are no bottleneck operations due to variations in process routings to inhibit smooth flow of work through the plant. In real batch production machine shops where each product has a different operation sequence, it is unlikely that the work distribution among the productive resources (machines) can be perfectly balanced. Consequently, there are some operations that are fully utilized while other operations occasionally stand idle waiting for work. Let us examine the effect of utilization.

3.1.3 Utilization and Availability

Utilization refers to the amount of output of a production facility relative to its capacity. Expressing this as an equation,

$$U = \frac{Q}{PC} \qquad (3.11)$$

where U = utilization of the facility, Q = actual quantity produced by the facility during a given time period (i.e., pc/wk), and PC = production capacity for the same period (pc/wk).

Utilization can be assessed for an entire plant, a single machine in the plant, or any other productive resource (i.e., labor). For convenience, it is often defined as the proportion of time that the facility is operating relative to the time available under the definition of capacity. Utilization is usually expressed as a percentage.

EXAMPLE 3.2 Utilization

A production machine operates 80 hr/wk (2 shifts, 5 days) at full capacity. Its production rate is 20 unit/hr. During a certain week, the machine produced 1000 parts and was idle the remaining time. (a) Determine the production capacity of the machine. (b) What was the utilization of the machine during the week under consideration?

Solution: (a) The capacity of the machine can be determined using the assumed 80-hr week as follows:

$$PC = 80(20) = 1600 \text{ unit/wk}$$

(b) Utilization can be determined as the ratio of the number of parts made by the machine relative to its capacity.

$$U = 1000/1600 = 0.625 \quad (62.5\%)$$

The alternative way of assessing utilization is by the time during the week that the machine was actually used. To produce 1000 units, the machine was operated

$$H = \frac{1000 \text{ pc}}{20 \text{ pc/hr}} = 50 \text{ hr}$$

Utilization is defined relative to the 80 hr available.

$$U = 50/80 = 0.625 \quad (62.5\%)$$

Availability is a common measure of reliability for equipment. It is especially appropriate for automated production equipment. Availability is defined using two other reliability terms, *mean time between failures* (MTBF) and *mean time to repair* (MTTR). The MTBF is the average length of time the piece of equipment runs between breakdowns, and MTTR is the average time required to service the equipment and put it back into operation when a breakdown occurs, as illustrated in Figure 3.1. Availability is defined as

$$A = \frac{MTBF - MTTR}{MTBF} \qquad (3.12)$$

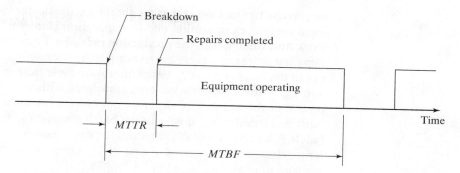

Figure 3.1 Time scale showing *MTBF* and *MTTR* used to define availability *A*.

where A = availability, $MTBF$ = mean time between failures (hr), and $MTTR$ = mean time to repair (hr). Availability is typically expressed as a percentage. When a piece of equipment is brand new (and being debugged), and later when it begins to age, its availability tends to be lower.

EXAMPLE 3.3   Effect of Utilization and Availability on Plant Capacity

Consider previous Example 3.1. Suppose the same data from that example were applicable, but that the availability of the machines A = 90%, and the utilization of the machines U = 80%. Given this additional data, compute the expected plant output.

Solution: Previous Eq. (3.9) can be altered to include availability and utilization as

$$Q = AU(nS_wH_{sh}R_p) \tag{3.13}$$

where A = availability and U = utilization. Combining the previous and new data, we have

$$Q = 0.90(0.80)(6)(10)(8.0)(17) = 5875 \text{ output unit/wk}$$

3.1.4   Manufacturing Lead Time

In the competitive environment of modern business, the ability of a manufacturing firm to deliver a product to the customer in the shortest possible time often wins the order. This time is referred to as the manufacturing lead time. Specifically, we define *manufacturing lead time* (MLT) as the total time required to process a given part or product through the plant, including any lost time due to delays, time spent in storage, reliability problems, and so on. Let us examine the components of MLT.

Production usually consists of a sequence of individual processing and assembly operations. Between the operations are material handling, storage, inspections, and other nonproductive activities. Let us therefore divide the activities of production into two main categories, operation and nonoperation elements. An operation is performed on a work unit when it is in the production machine. The nonoperation elements include handling, temporary storage, inspections, and other sources of delay when the work unit is not in the machine. Let T_c = the operation cycle time at a given machine or workstation, and T_{no} = the nonoperation time associated with the same machine. Further, let us suppose that the number of separate operations (machines) through which the work unit must be routed = n_o. If we assume batch production, then there are Q work units in the batch. A setup is generally required to prepare each production machine for the particular product, which requires a time = T_{su}. Given these terms, we can define manufacturing lead time as

$$MLT_j = \sum_{i=1}^{n_{oj}} (T_{suji} + Q_j T_{cji} + T_{noji})$$ (3.14)

where MLT_j = manufacturing lead time for part or product j (min), T_{suji} = setup time for operation i (min), Q_j = quantity of part or product j in the batch being processed (pc), T_{cji} = operation cycle time for operation i (min/pc), T_{noji} = nonoperation time associated with operation i (min), and i indicates the operation sequence in the processing; $i = 1, 2, \ldots, n_{oj}$. The MLT equation does not include the time the raw workpart spends in storage before its turn in the production schedule begins.

To simplify and generalize our model, let us assume that all setup times, operation cycle times, and nonoperation times are equal for the n_{oj} machines. Further, let us suppose that the batch quantities of all parts or products processed through the plant are equal and that they are all processed through the same number of machines, so that $n_{oj} = n_o$. With these simplifications, Eq. (3.14) becomes:

$$MLT = n_o(T_{su} + QT_c + T_{no})$$ (3.15)

where MLT = average manufacturing lead time for a part or product (min). In an actual batch production factory, which this equation is intended to represent, the terms n_o, Q, T_{su}, T_c, and T_{no} would vary by product and by operation. These variations can be accounted for by using properly weighted average values of the various terms. The averaging procedure is explained in the Appendix at the end of this chapter.

EXAMPLE 3.4   Manufacturing Lead Time

A certain part is produced in a batch size of 100 units. The batch must be routed through five operations to complete the processing of the parts. Average setup time is 3 hr/operation, and average operation time is 6 min (0.1 hr). Average nonoperation time due to handling, delays, inspections, etc., is 7 hours for each operation. Determine how many days it will take to complete the batch, assuming the plant runs one 8-hr shift/day.

Solution: The manufacturing lead time is computed from Eq. (3.15)

$$MLT = 5(3 + 100 \times 0.1 + 7) = 100 \text{ hours}$$

At 8 hr/day, this amounts to $100/8 = 12.5$ days.

Equation (3.15) can be adapted for job shop production and mass production by making adjustments in the parameter values. For a job shop in which the batch size is one ($Q = 1$), Eq. (3.15) becomes

$$MLT = n_o(T_{su} + T_c + T_{no}) \qquad (3.16)$$

For mass production, the Q term in Eq. (3.15) is very large and dominates the other terms. In the case of quantity type mass production in which a large number of units are made on a single machine ($n_o = 1$), the MLT simply becomes the operation cycle time for the machine after the setup has been completed and production begins.

For flow line mass production, the entire production line is set up in advance. Also, the nonoperation time between processing steps is simply the transfer time T_r to move the part or product from one workstation to the next. If the workstations are integrated so that all stations are processing their own respective work units, then the time to accomplish all of the operations is the time it takes each work unit to progress through all of the stations on the line. The station with the longest operation time sets the pace for all stations:

$$MLT = n_o(T_r + \text{Max } T_o) = n_o T_c \qquad (3.17)$$

where MLT = time between start and completion of a given work unit on the line (min), n_o = number of operations on the line; T_r = transfer time (min), Max T_o = operation time at the bottleneck station (min), and T_c = cycle time of the production line (min/pc). $T_c = T_r + \text{Max } T_o$ from Eq. (3.7). Since the number of stations is equal to the number of operations ($n = n_o$), Eq. (3.17) can also be stated as

$$MLT = n(T_r + \text{Max } T_o) = n T_c \qquad (3.18)$$

where the symbols have the same meaning as above, and we have substituted n (number of workstations or machines) for number of operations n_o.

3.1.5 Work-in-Process

Work-in-process (WIP) is the quantity of parts or products currently located in the factory that either are being processed or are between processing operations. WIP is inventory that is in the state of being transformed from raw material to finished product. An approximate measure of work-in-process can be obtained from the following, using terms previously defined:

$$WIP = \frac{AU(PC)(MLT)}{S_w H_{sh}} \qquad (3.19)$$

where WIP = work-in-process in the facility (pc), A = availability, U = utilization, PC = production capacity of the facility (pc/wk), MLT = manufacturing lead time, (wk), S_w = number of shifts per week (shift/wk), and H_{sh} = hours per shift (hr/shift). Equation (3.19) states that the level of WIP equals the rate at which parts flow through the factory multiplied by the length of time the parts spend in the factory. The time units for $(PC)/S_w H_{sh}$ must be consistent with the units for MLT. We examine the costs of this in-process inventory in Section 25.5.2.

Work-in-process represents an investment by the firm, but one that cannot be turned into revenue until all processing has been completed. Many manufacturing companies sustain major costs because work remains in-process in the factory too long.

3.2 MANUFACTURING COSTS

Decisions on automation and production systems are usually based on the relative costs of alternatives. In this section we examine how these costs and cost factors are determined.

3.2.1 Fixed and Variable Costs

Manufacturing costs can be classified into two major categories: (1) fixed costs and (2) variable costs. A *fixed cost* is one that remains constant for any level of production output. Examples include the cost of the factory building and production equipment, insurance, and property taxes. All of the fixed costs can be expressed as annual amounts. Expenses such as insurance and property taxes occur naturally as annual costs. Capital investments such as building and equipment can be converted to their equivalent uniform annual costs using interest rate factors.

A *variable cost* is one that varies in proportion to the level of production output. As output increases, variable cost increases. Examples include direct labor, raw materials, and electric power to operate the production equipment. The ideal concept of variable cost is that it is directly proportional to output level. When fixed cost and variable cost are added, we have the following total cost equation:

$$TC = FC + VC(Q) \tag{3.20}$$

where TC = total annual cost ($/yr), FC = fixed annual cost ($/yr), VC = variable cost ($/pc), and Q = annual quantity produced (pc/yr).

When comparing automated and manual production methods (Section 1.4), it is typical that the fixed cost of the automated method is high relative to the manual method, and the variable cost of automation is low relative to the manual method, as pictured in Figure 3.2. Consequently, the manual method has a cost advantage in the low quantity range, while automation has an advantage for high quantities. This reinforces the arguments presented in Section 1.4.1 on the appropriateness of manual labor for certain production situations.

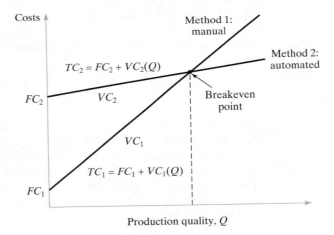

Figure 3.2 Fixed and variable costs as a function of production output for manual and automated production methods.

3.2.2 Direct Labor, Material, and Overhead

Fixed versus variable are not the only possible classifications of costs in manufacturing. An alternative classification separates costs into (1) direct labor, (2) material, and (3) overhead. This is often a more convenient way to analyze costs in production. The *direct labor cost* is the sum of the wages and benefits paid to the workers who operate the production equipment and perform the processing and assembly tasks. The *material cost* is the cost of all raw materials used to make the product. In the case of a stamping plant, the raw material consists of the steel sheet stock used to make stampings. For the rolling mill that made the sheet stock, the raw material is the iron ore or scrap iron out of which the sheet is rolled. In the case of an assembled product, materials include component parts manufactured by supplier firms. Thus, the definition of "raw material" depends on the company. The final product of one company can be the raw material for another company. In terms of fixed and variable costs, direct labor and material must be considered as variable costs.

Overhead costs are all of the other expenses associated with running the manufacturing firm. Overhead divides into two categories: (1) factory overhead and (2) corporate overhead. *Factory overhead* consists of the costs of operating the factory other than direct labor and materials, such as the factory expenses listed in Table 3.1. Factory overhead is treated as fixed cost, although some of the items in our list could be correlated with the output level of the plant. *Corporate overhead* is the cost not related to the company's manufacturing activities, such as the corporate expenses in Table 3.2. Many companies operate more than one factory, and this is one of the reasons for dividing overhead into factory and corporate categories. Different factories may have significantly different factory overhead expenses.

J Black [1] provides some typical percentages for the different types of manufacturing and corporate expenses. These are presented in Figure 3.3. We might make several observations about these data. First, total manufacturing cost represents only about 40% of the product's selling price. Corporate overhead expenses and total manufacturing cost are about equal. Second, materials (including purchased parts) make up the largest percentage of total manufacturing cost, at around 50%. And third, direct labor is a relatively small proportion of total manufacturing cost: 12% of manufacturing cost and only about 5% of final selling price.

Overhead costs can be allocated according to a number of different bases, including direct labor cost, material cost, direct labor hours, and space. Most common in industry is

TABLE 3.1 Typical Factory Overhead Expenses

Plant supervision	Applicable taxes	Factory depreciation
Line foreman	Insurance	Equipment depreciation
Maintenance crew	Heat and air conditioning	Fringe benefits
Custodial services	Light	Material handling
Security personnel	Power for machinery	Shipping and receiving
Tool crib attendant	Payroll services	Clerical support

TABLE 3.2 Typical Corporate Overhead Expenses

Corporate executives	Engineering	Applicable taxes
Sales and marketing	Research and development	Office space
Accounting department	Other support personnel	Security personnel
Finance department	Insurance	Heat and air conditioning
Legal counsel	Fringe benefits	Lighting

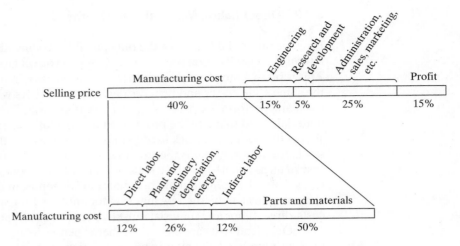

Figure 3.3 Breakdown of costs for a manufactured product [1].

direct labor cost, which we will use here to illustrate how overheads are allocated and subsequently used to compute factors such as selling price of the product.

The allocation procedure (simplified) is as follows. For the most recent year (or several years), all costs are compiled and classified into four categories: (1) direct labor, (2) material, (3) factory overhead, and (4) corporate overhead. The objective is to determine an *overhead rate* (also called *burden rate*) that could be used in the following year to allocate overhead costs to a process or product as a function of the direct labor costs associated with that process or product. In our treatment, separate overhead rates will be developed for factory and corporate overheads. The *factory overhead rate* is calculated as the ratio of factory overhead expenses (category 3) to direct labor expenses (category 1); that is,

$$FOHR = \frac{FOHC}{DLC} \tag{3.21}$$

where $FOHR$ = factory overhead rate, $FOHC$ = annual factory overhead costs ($/yr), and DLC = annual direct labor costs ($/yr).

The *corporate overhead rate* is the ratio of corporate overhead expenses (category 4) to direct labor expenses:

$$COHR = \frac{COHC}{DLC} \tag{3.22}$$

where $COHR$ = corporate overhead rate, $COHC$ = annual corporate overhead costs ($/yr), and DLC = annual direct labor costs ($/yr). Both rates are often expressed as percentages. If material cost were used as the allocation basis, then material cost would be used as the denominator in both ratios. Let us present two examples to illustrate (1) how overhead rates are determined and (2) how they are used to estimate manufacturing cost and establish selling price.

EXAMPLE 3.5 Determining Overhead Rates

Suppose that all costs have been compiled for a certain manufacturing firm for last year. The summary is shown in the table below. The company operates two different manufacturing plants plus a corporate headquarters. Determine

(a) the factory overhead rate for each plant, and (b) the corporate overhead rate. These rates will be used by the firm to predict the following year's expenses.

Expense Category	Plant 1 ($)	Plant 2 ($)	Headquarters ($)	Totals ($)
Direct labor	800,000	400,000		1,200,000
Materials	2,500,000	1,500,000		4,000,000
Factory expense	2,000,000	1,100,000		3,100,000
Corporate expense			7,200,000	7,200,000
Totals	5,300,000	3,000,000	7,200,000	15,500,000

Solution: (a) A separate factory overhead rate must be determined for each plant. For plant 1, we have:

$$FOHR_1 = \frac{\$2,000,000}{\$800,000} = 2.5 = 250\%$$

For plant 2,

$$FOHR_2 = \frac{\$1,100,000}{\$400,000} = 2.75 = 275\%$$

(b) The corporate overhead rate is based on the total labor cost at both plants.

$$COHR = \frac{\$7,200,000}{\$1,200,000} = 6.0 = 600\%$$

EXAMPLE 3.6 Estimating Manufacturing Costs and Establishing Selling Price

A customer order of 50 parts is to be processed through plant 1 of the previous example. Raw materials and tooling are supplied by the customer. The total time for processing the parts (including setup and other direct labor) is 100 hr. Direct labor cost is $10.00/hr. The factory overhead rate is 250% and the corporate overhead rate is 600%. (a) Compute the cost of the job. (b) What price should be quoted to a potential customer if the company uses a 10% markup?

Solution: (a) The direct labor cost for the job is $(100 \text{ hr})(\$10.00/\text{hr}) = \$1000$.
The allocated factory overhead charge, at 250% of direct labor, is $(\$1000)(2.50) = \$2500$. The total factory cost of the job, including allocated factory overhead $= \$1000 + \$2500 = \$3500$.
The allocated corporate overhead charge, at 600% of direct labor, is $(\$1000)(6.00) = \$6000$. The total cost of the job including corporate overhead $= \$3500 + \$6000 = \$9500$.

(b) If the company uses a 10% markup, the price quoted to the customer would be $(1.10)(\$9500) = \$10,450$.

3.2.3 Cost of Equipment Usage

The trouble with overhead rates as we have developed them here is that they are based on labor cost alone. A machine operator who runs an old, small engine lathe whose book value is zero will be costed at the same overhead rate as an operator running a new CNC turning center just purchased for $500,000. Obviously, the time on the machining center is more productive and should be valued at a higher rate. If differences in rates of different production machines are not recognized, manufacturing costs will not be accurately measured by the overhead rate structure.

To deal with this difficulty, it is appropriate to divide the cost of a worker running a machine into two components: (1) direct labor and (2) machine. Associated with each is an applicable overhead rate. These costs apply not to the entire factory operations, but to individual work centers. A work center can be any of the following: (1) one worker and one machine, (2) one worker and several machines, (3) several workers operating one machine, or (4) several workers and machines. In any of these cases, it is advantageous to separate the labor expense from the machine expense in estimating total production costs.

The direct labor cost consists of the wages and benefits paid to operate the work center. Applicable factory overhead expenses allocated to direct labor cost might include state taxes, certain fringe benefits, and line supervision. The machine annual cost is the initial cost of the machine apportioned over the life of the asset at the appropriate rate of return used by the firm. This is done using the capital recovery factor, as

$$UAC = IC(A/P, i, n) \tag{3.23}$$

where UAC = equivalent uniform annual cost ($/yr); IC = initial cost of the machine ($); and $(A/P, i, n)$ = capital recovery factor that converts initial cost at year 0 into a series of equivalent uniform annual year-end values, where i = annual interest rate and n = number of years in the service life of the equipment. For given values of i and n, $(A/P,i,n)$ can be computed as follows:

$$(A/P, i, n) = \frac{i(1 + i)^n}{(1 + i)^n - 1} \tag{3.24}$$

Values of $(A/P, i, n)$ can also be found in interest tables that are widely available.

The uniform annual cost can be expressed as an hourly rate by dividing the annual cost by the number of annual hours of equipment use. The machine overhead rate is based on those factory expenses that are directly assignable to the machine. These include power to drive the machine, floor space, maintenance and repair expenses, and so on. In separating the factory overhead items in Table 3.1 between labor and machine, judgment must be used; admittedly, the judgment is sometimes arbitrary. The total cost rate for the work center is the sum of labor and machine costs. This can be summarized as follows for a work center consisting of one worker and one machine:

$$C_o = C_L(1 + FOHR_L) + C_m(1 + FOHR_m) \tag{3.25}$$

where C_o = hourly rate to operate the work center ($/hr), C_L = direct labor wage rate ($/hr), $FOHR_L$ = factory overhead rate for labor, C_m = machine hourly rate ($/hr), and $FOHR_m$ = factory overhead rate applicable to the machine.

It is the author's opinion that corporate overhead expenses should not be included in the analysis when comparing production methods. Including them serves no purpose other than to dramatically increase the costs of the alternatives. The fact is that these corporate overhead expenses are present whether or not any of the alternatives is selected. On the other hand, when estimating costs for pricing decisions, corporate overhead should be included because over the long run, these costs must be recovered through revenues generated from selling products.

EXAMPLE 3.7   Hourly Cost of a Work Center

The following data are given for a work center consisting of one worker and one machine: direct labor rate = $10.00/hr, applicable factory overhead rate on labor = 60%, capital investment in machine = $100,000, service life of the machine = 8 yr, rate of return = 20%, salvage value in 8 yr = 0, and applicable factory overhead rate on machine = 50%. The work center will be operated one 8-hr shift, 250 day/yr. Determine the appropriate hourly rate for the work center.

Solution: Labor cost per hour = $C_L(1 + FOHR_L)$ = $10.00(1 + 0.60) = $16.00/hr. The investment cost of the machine must be annualized, using an 8-yr service life and a rate of return = 20%. First we compute the capital recovery factor:

$$(A/P, 20\%, 8) = \frac{0.20(1 + 0.20)^8}{(1 + 0.20)^8 - 1} = \frac{0.20(4.2998)}{4.2998 - 1} = 0.2606$$

Now the uniform annual cost for the $100,000 initial cost can be determined:

$$UAC = \$100,000(A/P, 20\%, 8) = 100,000(0.2606) = \$26,060.00/yr$$

The number of hours per year = (8 hr/day)(250 day/yr) = 2000 hr/yr. Dividing this into UAC gives 26,060/2000 = $13.03/hr. Applying the factory overhead rate, we have

$$C_m(1 + FOHR_m) = \$13.03(1 + 0.50) = \$19.55/hr$$

Total cost rate for the work center is

$$C_o = 16.00 + 19.55 = \$35.55/hr.$$

REFERENCES

[1]   BLACK, J T., *The Design of the Factory with a Future*, McGraw-Hill, Inc., NY, 1991.

[2]   GROOVER, M. P., *Fundamentals of Modern Manufacturing: Materials, Processes, and Systems*, 3d ed., John Wiley & Sons, Inc., Hoboken, NJ, 2007.

REVIEW QUESTIONS

3.1   What is the cycle time in a manufacturing operation?

3.2   What is a bottleneck station?

3.3   What is production capacity?

3.4 How can plant capacity be increased or decreased in the short term?

3.5 What is utilization in a manufacturing plant?

3.6 What is availability?

3.7 What is manufacturing lead time?

3.8 What is work-in-process?

3.9 How are fixed costs distinguished from variable costs in manufacturing?

3.10 Name five typical factory overhead expenses

3.11 Name five typical corporate overhead expenses

PROBLEMS

Production Concepts and Mathematical Models

3.1 A certain part is routed through six machines in a batch production plant. The setup and operation times for each machine are given in the table below. The batch size is 100 and the average nonoperation time per machine is 12 hours. Determine (a) manufacturing lead time and (b) production rate for operation 3.

Machine	Setup Time (hr.)	Operation Time (min.)
1	4	5.0
2	2	3.5
3	8	10.0
4	3	1.9
5	3	4.1
6	4	2.5

3.2 Suppose the part in the previous problem is made in very large quantities on a production line in which an automated work handling system is used to transfer parts between machines. Transfer time between stations = 15 s. The total time required to set up the entire line is 150 hours. Assume that the operation times at the individual machines remain the same. Determine (a) manufacturing lead time for a part coming off the line, (b) production rate for operation 3, and (c) theoretical production rate for the entire production line.

3.3 The average part produced in a certain batch manufacturing plant must be processed sequentially through six machines on average. Twenty (20) new batches of parts are launched each week. Average operation time = 6 minutes, average setup time = 5 hours, average batch size = 25 parts, and average nonoperation time per batch = 10 hr/machine. There are 18 machines in the plant working in parallel. Each of the machines can be set up for any type of job processed in the plant. The plant operates an average of 70 production hours per week. Scrap rate is negligible. Determine (a) manufacturing lead time for an average part, (b) plant capacity, and (c) plant utilization. (d) How would you expect the nonoperation time to be affected by the plant utilization?

3.4 Based on the data in the previous problem and your answers to that problem, determine the average level of work-in-process (number of parts-in-process) in the plant.

3.5 An average of 20 new orders are started through a certain factory each month. On average, an order consists of 50 parts that are processed sequentially through 10 machines in the factory. The operation time per machine for each part = 15 min. The nonoperation time per order at each machine averages 8 hours, and the required setup time per order = 4 hours. There are a total of 25 machines in the factory working in parallel. Each of the machines can be set up for any type of job processed in the plant. Only 80% of the machines are operational at any time (the other 20% are in repair or maintenance). The plant operates 160 hours per month. However, the plant manager complains that a total of 100 overtime machine-hours must be authorized each month in order to keep up with the production schedule. (a) What is the manufacturing lead time for an average order? (b) What is the plant capacity (on a monthly basis) and why must the overtime be authorized? (c) What is the utilization of the plant according to the definition given in the text? (d) Determine the average level of work-in-process (number of parts-in-process) in the plant.

3.6 The mean time between failures for a certain production machine is 250 hours, and the mean time to repair is 6 hours. Determine the availability of the machine.

3.7 One million units of a certain product are to be manufactured annually on dedicated production machines that run 24 hours per day, 5 days per week, 50 weeks per year. (a) If the cycle time of a machine to produce one part is 1.0 minute, how many of the dedicated machines will be required to keep up with demand? Assume that availability, utilization, and worker efficiency = 100%, and that no setup time will be lost. (b) Solve part (a) except that availability = 0.90.

3.8 The mean time between failures and mean time to repair in a certain department of the factory are 400 hours and 8 hours, respectively. The department operates 25 machines during one 8-hour shift per day, 5 days per week, 52 weeks per year. Each time a machine breaks down, it costs the company $200 per hour (per machine) in lost revenue. A proposal has been submitted to install a preventive maintenance program in this department. In this program, preventive maintenance would be performed on the machines during the evening so that there will be no interruptions to production during the regular shift. The effect of this program is expected to be that the average MTBF will double, and half of the emergency repair time normally accomplished during the day shift will be performed during the evening shift. The cost of the maintenance crew will be $1500 per week. However, a reduction of maintenance personnel on the day shift will result in a savings during the regular shift of $700 per week. (a) Compute the availability of machines in the department both before and after the preventive maintenance program is installed. (b) Determine how many total hours per year the 25 machines in the department are under repair both before and after the preventive maintenance program is installed. In this part and in part (c), ignore effects of queueing of the machines that might have to wait for a maintenance crew. (c) Will the preventive maintenance program pay for itself in terms of savings in the cost of lost revenues?

3.9 There are nine machines in the automatic lathe section of a certain machine shop. The setup time on an automatic lathe averages 6 hours. The average batch size for parts processed through the section is 90. The average operation time = 8.0 minutes. Under shop rules, an operator can be assigned to run up to three machines. Accordingly, there are three operators in the section for the nine lathes. In addition to the lathe operators, there are two setup workers who perform machine setups exclusively. These setup workers are kept busy the full shift. The section runs one 8-hour shift per day, 6 days per week. However, an average of 15% of the production time is lost is lost due to machine breakdowns. Scrap losses are negligible. The production control manager claims that the capacity of the section should be 1836 pieces per week. However, the actual output averages only 1440 units per week. What is the problem? Recommend a solution.

3.10 A certain job shop specializes in one-of-a-kind orders dealing with parts of medium-to-high complexity. A typical part is processed sequentially through ten machines in batch sizes of one. The shop contains a total of eight conventional machine tools and operates 35 hours per week of production time. The machine tools are interchangeable in the sense that they can be set up for any operation required on any of the parts. Average time values on the part are:

machining time per machine = 0.5 hour, work handling time per machine = 0.3 hour, tool change time per machine = 0.2 hour, setup time per machine = 6 hours, and nonoperation time per machine = 12 hours. A new programmable machine has been purchased by the shop that is capable of performing all ten operations in a single setup. The programming of the machine for this part will require 20 hours; however, the programming can be done off-line, without tying up the machine. The setup time will be 10 hours. The total machining time will be reduced to 80% of its previous value due to advanced tool control algorithms; the work handling time will be the same as for one machine; and the total tool change time will be reduced by 50% because it will be accomplished automatically under program control. For the one machine, nonoperation time is expected to be 12 hours. (a) Determine the manufacturing lead time for the traditional method and for the new method. (b) Compute the plant capacity for the following alternatives: (i) a job shop containing the eight traditional machines, and (ii) a job shop containing two of the new programmable machines. Assume the typical jobs are represented by the data given above. (c) Determine the average level of work-in-process for the two alternatives in part (b), if the alternative shops operate at full capacity. (d) Identify which of the ten automation strategies (Section 1.5.2) are represented (or probably represented) by the new machine.

3.11 A factory produces cardboard boxes. The production sequence consists of three operations: (1) cutting, (2) indenting, and (3) printing. There are three machines in the factory, one for each operation. The machines are 100% reliable and operate as follows when operating at 100% utilization: (1) In *cutting*, large rolls of cardboard are fed into the cutting machine and cut into blanks. Each large roll contains enough material for 4,000 blanks. Production cycle time = 0.03 minute/blank during a production run, but it takes 35 minutes to change rolls between runs. (2) In *indenting*, indentation lines are pressed into the blanks to allow the blanks to be bent into boxes later. The blanks from the previous cutting operation are divided and consolidated into batches whose starting quantity = 2,000 blanks. Indenting is performed at 4.5 minutes per 100 blanks. Time to change dies on the indentation machine = 30 min. (3) In *printing*, the indented blanks are printed with labels for a particular customer. The blanks from the previous indenting operation are divided and consolidated into batches whose starting quantity = 1,000 blanks. Printing cycle rate = 30 blanks per minute. Between batches, changeover of the printing plates is required, which takes 20 minutes. In-process inventory is allowed to build up between machines 1 and 2, and between machines 2 and 3, so that the machines can operate independently as much as possible. Based on this data and information, determine the maximum possible output of this factory during a 40-hour week, in completed blanks/week (completed blanks have been cut, indented, and printed). Assume steady state operation, not startup.

Costs of Manufacturing Operations

3.12 Theoretically, any given production plant has an optimum output level. Suppose a certain production plant has annual fixed costs FC = $2,000,000. Variable cost VC is functionally related to annual output Q in a manner that can be described by the function $VC = \$12 + \$0.005Q$. Total annual cost is given by $TC = FC + VC \times Q$. The unit sales price for one production unit P = $250. (a) Determine the value of Q that minimizes unit cost UC, where $UC = TC/Q$; and compute the annual profit earned by the plant at this quantity. (b) Determine the value of Q that maximizes the annual profit earned by the plant; and compute the annual profit earned by the plant at this quantity.

3.13 Costs have been compiled for a certain manufacturing company for the most recent year. The summary is shown in the table below. The company operates two different manufacturing plants, plus a corporate headquarters. Determine (a) the factory overhead rate for each plant, and (b) the corporate overhead rate. The firm will use these rates to predict costs for the following year.

Expense Category	Plant 1	Plant 2	Corporate Headquarters
Direct labor	$1,000,000	$1,750,000	
Materials	$3,500,000	$4,000,000	
Factory expense	$1,300,000	$2,300,000	
Corporate expense			$5,000,000

3.14 The hourly rate for a certain work center is to be determined based on the following data: direct labor rate = $15.00/hr; applicable factory overhead rate on labor = 35%; capital investment in machine = $200,000; service life of the machine = 5 years; rate of return = 15%; salvage value in 5 years = zero; and applicable factory overhead rate on machine = 40%. The work center will be operated two 8-hour shifts, 250 days per year. Determine the appropriate hourly rate for the work center.

3.15 In the previous problem if the work load for the cell can only justify a one shift operation, determine the appropriate hourly rate for the work center.

3.16 In the operation of a certain production machine, one worker is required at a direct labor rate of $10/hr. Applicable labor factory overhead rate = 50%. Capital investment in the system = $250,000, expected service life = 10 years, no salvage value at the end of that period, and the applicable machine factory overhead rate = 30%. The work cell will operate 2000 hr/yr. Use a rate of return of 25% to determine the appropriate hourly rate for this work cell.

3.17 In the previous problem, suppose that the machine will be operated three shifts, or 6000 hr/yr, instead of 2000 hr/yr. Note the effect of increased machine utilization on the hourly rate compared to the rate determined in the previous problem.

3.18 The break-even point is to be determined for two production methods, one a manual method and the other automated. The manual method requires two workers at $9.00/hr each. Together, they produce at a rate of 36 units/hr. The automated method has an initial cost of $125,000, a 4-year service life, no salvage value, and annual maintenance costs = $3000. No labor (except for maintenance) is required to operate the machine, but the power required to run the machine is 50 kW (when running). Cost of electric power is $0.05/kWh. If the production rate for the automated machine is 100 units/hr, determine the break-even point for the two methods, using a rate of return = 25%.

APPENDIX A3: AVERAGING PROCEDURES FOR PRODUCTION MODELS

As indicated in our presentation of the production models in Section 1.1, special averaging procedures are required to reduce the inherent variations in actual factory data to single parameter values used in our equations. This appendix explains the averaging procedures.

A straight arithmetic average is used to compute the value of batch quantity Q and the number of operations (machines) in the process routing n_o. Let n_Q = number of batches of the various part or product styles to be considered. This might be the number of batches processed through the plant during a certain time period (i.e., week, month, year), or it might be a sample of size n_Q taken from this time period for analysis purposes. The average batch quantity is given by

$$Q = \frac{\sum_{j=1}^{n_Q} Q_j}{n_Q} \qquad (A3.1)$$

where Q = average batch quantity, pc; Q_j = batch quantity for part or product style j of the total n_Q batches or styles being considered, pc, where $j = 1, 2, \ldots, n_Q$. The average number of operations in the process routing is a similar computation:

$$n_o = \frac{\sum\limits_{j=1}^{n_Q} n_{oj}}{n_Q} \tag{A3.2}$$

where n_o = average number of operations in all process routings under consideration, n_{oj} = number of operations in the process routing of part or product style j, and n_Q = number of batches under consideration.

When factory data are used to assess the terms T_{su}, T_c, and T_{no}, weighted averages must be used. To calculate the grand average setup time for n_Q different part or product styles, we first compute the average setup time for each style; that is,

$$T_{suj} = \frac{\sum\limits_{k=1}^{n_{oj}} T_{sujk}}{n_{oj}} \tag{A3.3}$$

where T_{suj} = average setup time for part or product style j, min; T_{sujk} = setup time for operation k in the processing sequence for part or product style j, min; where $k = 1, 2, \ldots, n_{oj}$; and n_{oj} = number of operations in the processing sequence for part or product style j. Using the n_Q values of T_{suj} calculated from the above equation, we can now compute the grand average setup time for all styles, given by

$$T_{su} = \frac{\sum\limits_{j=1}^{n_Q} n_{oj} T_{suj}}{\sum\limits_{j=1}^{n_Q} n_{oj}} \tag{A3.4}$$

where T_{su} = setup time grand average for all n_Q part or product styles included in the group of interest, min; and the other terms are defined above.

A similar procedure is used to obtain grand averages for operation cycle time T_c and nonoperation time T_{no}. Considering cycle time first,

$$T_{cj} = \frac{\sum\limits_{k=1}^{n_{oj}} T_{cjk}}{n_{oj}} \tag{A3.5}$$

where T_{cj} = average operation cycle time for part or product style j, min; T_{cjk} = cycle time for operation k in the processing sequence for part or product style j, where $k = 1, 2, \ldots, n_{oj}$, min; and n_{oj} = number of operations in the processing sequence for style j. The grand average cycle time for all n_Q styles is given by

$$T_c = \frac{\sum\limits_{j=1}^{n_Q} n_{oj} T_{cj}}{\sum\limits_{j=1}^{n_Q} n_{oj}} \tag{A3.6}$$

where T_c = operation cycle time grand average for all n_Q part or product styles being considered, min; and the other terms are defined above. The same forms of equation apply for nonoperation time T_{no},

$$T_{noj} = \frac{\sum_{k=1}^{n_{oj}} T_{nojk}}{n_{oj}} \tag{A3.7}$$

where T_{noj} = average nonoperation time for part or product style j, min; T_{nojk} = nonoperation time for operation k in the processing sequence for part or product style j, min. The grand average for all styles (batches) is

$$T_{no} = \frac{\sum_{j=1}^{n_Q} n_{oj} T_{noj}}{\sum_{j=1}^{n_Q} n_{oj}} \tag{A3.8}$$

where T_{no} = nonoperation time grand average for all parts or products considered, min; and other terms are defined above.

Introduction to Automation

CHAPTER CONTENTS

Automation can be defined as the technology by which a process or procedure is accomplished without human assistance. It is implemented using a *program of instructions* combined with a *control system* that executes the instructions. To automate a process, *power* is required, both to drive the process itself and to operate the program and control system. Although automation can be applied in a wide variety of areas, it is most closely associated with the manufacturing industries. It was in the context of manufacturing that the term was originally coined by an engineering manager at Ford Motor Company in 1946 to describe the variety of automatic transfer devices and feed mechanisms that had been installed in Ford's production plants (Historical Note 4.1). It is ironic that nearly all modern applications of automation are controlled by computer technologies that were not available in 1946.

In this part of the book, we examine technologies that have been developed to automate manufacturing operations. The position of automation and control technologies in

From Chapter 4 of *Automation, Production Systems, and Computer-Integrated Manufacturing*, Third Edition.
Mikell P. Groover. Copyright © 2008 by Pearson Education, Inc. Publishing as Prentice Hall. All rights reserved.

Historical Note 4.1   History of automation

The history of automation can be traced to the development of basic mechanical devices, such as the wheel (circa 3200 B.C.), lever, winch (circa 600 B.C.), cam (circa 1000), screw (1405), and gear in ancient and medieval times. These basic devices were refined and used to construct the mechanisms in waterwheels, windmills (circa 650), and steam engines (1765). These machines generated the power to operate other machinery of various kinds, such as flour mills (circa 85 B.C.), weaving machines (flying shuttle, 1733), machine tools (boring mill, 1775), steamboats (1787), and railroad locomotives (1803). Power, and the capacity to generate it and transmit it to operate a process, is one of the three basic elements of an automated system.

After his first steam engine in 1765, James Watt and his partner, Matthew Boulton, made several improvements in the design. One of the improvements was the flying-ball governor (around 1785), which provided feedback to control the throttle of the engine. The governor consisted of a ball on the end of a hinged lever attached to the rotating shaft. The lever was connected to the throttle valve. As the speed of the rotating shaft increased, the ball was forced to move outward by centrifugal force; this in turn caused the lever to reduce the valve opening and slow the motor speed. As rotational speed decreased, the ball and lever relaxed, thus allowing the valve to open. The flying-ball governor was one of the first examples in engineering of feedback control, an important type of control system—the second basic element of an automated system.

The third basic element of an automated system is the program of instructions that directs the actions of the system or machine. One of the first examples of machine programming was the Jacquard loom, invented around 1800. This loom was a machine for weaving cloth from yarn. The program of instructions that determined the weaving pattern of the cloth consisted of a metal plate containing holes. The hole pattern in the plate directed the shuttle motions of the loom, which in turn determined the weaving pattern. Different hole patterns yielded different cloth patterns. Thus, the Jacquard loom was a programmable machine, one of the first of its kind.

By the early 1800s, the three basic elements of automated systems—power source, controls, and programmable machines—had been developed, although these elements were primitive by today's standards. It took many years of refinement and many new inventions and developments, both in these basic elements as well as in the enabling infrastructure of the manufacturing industries, before fully automated production systems became a common reality. Important examples of these inventions and developments include *interchangeable parts* (circa 1800, Historical Note 1.1); *electrification* (starting in 1881); the *moving assembly line* (1913, Historical Note 15.1); mechanized *transfer lines* for mass production, whose programs were fixed by their hardware configuration (1924, Historical Note 16.1); a mathematical theory of *control systems* (1930s and 1940s); and the MARK I electromechanical *computer* at Harvard University (1944). These inventions and developments had all been realized by the end of World War II.

Since 1945, many new inventions and developments have contributed significantly to automation technology. Del Harder coined the word *automation* around 1946 in reference to the many automatic devices that the Ford Motor Company had developed for its production lines. The first electronic digital computer was developed at University of Pennsylvania in 1946. The first *numerical control* machine tool was developed and demonstrated in 1952 at Massachusetts Institute of Technology based on a concept proposed by John Parsons and Frank Stulen (Historical Note 7.1). By the late 1960s and early 1970s, digital computers were being connected to machine tools. In 1954, the first *industrial robot* was designed and in 1961 it was patented by George Devol (Historical Note 8.1). The first commercial robot was installed to unload parts in a die casting operation in 1961. In the late 1960s, the first *flexible manufacturing system* in the United States was installed at Ingersoll Rand Company to perform machining operations on a variety of parts (Historical Note 19.1). Around 1969, the first *programmable logic controller* was introduced (Historical Note 9.1). In 1978, the first commercial *personal computer* (PC) was introduced by Apple Computer, although a similar product had been introduced in kit form as early as 1975.

Developments in computer technology were made possible by advances in electronics, including the *transistor* (1948), *hard disk* for computer memory (1956), *integrated circuits* (1960), the *microprocessor* (1971), *random access memory* (1984), megabyte capacity memory chips (circa 1990), and the *Pentium* microprocessors (1993). Software developments related to automation have been equally important, including the *FORTRAN* computer programming language (1955), the *APT* programming language for numerical control (NC) machine tools (1961), the *UNIX* operating system (1969), the *VAL* language for robot programming (1979), Microsoft *Windows* (1985), and the *JAVA* programming language (1995). Advances and enhancements in these technologies continue.

the larger production system is shown in Figure 4.1. In the present chapter, we provide an overview of automation: What are the elements of an automated system? What are some of the advanced features beyond the basic elements? And what are the levels in an enterprise where automation can be applied? In the following two chapters, we discuss industrial control systems and the hardware components of these systems. These two chapters serve as a foundation for the remaining chapters in our coverage of automation and control technologies. These technologies are (1) numerical control (Chapter 7), (2) industrial robotics (Chapter 8), and (3) programmable logic controllers (Chapter 9).

4.1 BASIC ELEMENTS OF AN AUTOMATED SYSTEM

An automated system consists of three basic elements: (1) power to accomplish the process and operate the system, (2) a program of instructions to direct the process, and (3) a control system to actuate the instructions. The relationship amongst these elements is illustrated in Figure 4.2. All systems that qualify as being automated include these three basic elements in one form or another.

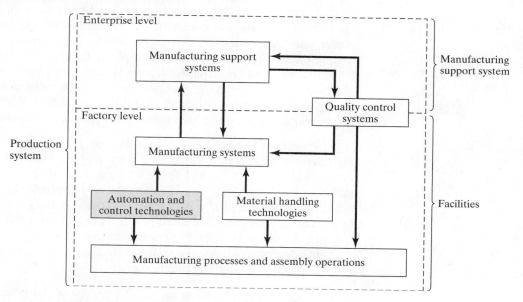

Figure 4.1 Automation and control technologies in the production system.

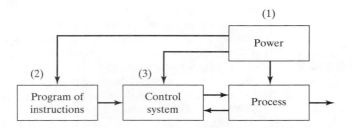

Figure 4.2 Elements of an automated system: (1) power, (2) program of instructions, and (3) control systems.

4.1.1 Power to Accomplish the Automated Process

An automated system is used to operate some process, and power is required to drive the process as well as the controls. The principal source of power in automated systems is electricity. Electrical power has many advantages in automated as well as nonautomated processes:

- Electrical power is widely available at moderate cost. It is an important part of our industrial infrastructure.
- Electrical power can be readily converted to alternative energy forms: mechanical, thermal, light, acoustic, hydraulic, and pneumatic.
- Electrical power at low levels can be used to accomplish functions such as signal transmission, information processing, and data storage and communication.
- Electrical energy can be stored in long-life batteries for use in locations where an external source of electrical power is not conveniently available.

Alternative power sources include fossil fuels, solar energy, water, and wind. However, their exclusive use is rare in automated systems. In many cases when alternative power sources are used to drive the process itself, electrical power is used for the controls that automate the operation. For example, in casting or heat treatment, the furnace may be heated by fossil fuels, but the control system to regulate temperature and time cycle is electrical. In other cases, the energy from these alternative sources is converted to electric power to operate both the process and its automation. When solar energy is used as a power source for an automated system, it is generally converted in this way.

Power for the Process.   In production, the term *process* refers to the manufacturing operation that is performed on a work unit. In Table 4.1, a list of common manufacturing processes is compiled along with the form of power required and the resulting action on the work unit. Most of the power in manufacturing plants is consumed by these kinds of operations. The "power form" indicated in the middle column of the table refers to the energy that is applied directly to the process. As indicated above, the power source for each operation is often converted from electricity.

In addition to driving the manufacturing process itself, power is also required for the following material handling functions:

- *Loading and unloading the work unit.* All of the processes listed in Table 4.1 are accomplished on discrete parts. These parts must be moved into the proper position and orientation for the process to be performed, and power is required for

TABLE 4.1 Common Manufacturing Processes and Their Power Requirements

Process	Power Form	Action Accomplished
Casting	Thermal	Melting the metal before pouring into a mold cavity where solidification occurs.
Electric discharge machining (EDM)	Electrical	Metal removal is accomplished by a series of discrete electrical discharges between electrode (tool) and workpiece. The electric discharges cause very high localized temperatures that melt the metal.
Forging	Mechanical	Metal workpart is deformed by opposing dies. Workparts are often heated in advance of deformation, thus thermal power is also required.
Heat treating	Thermal	Metallic work unit is heated to temperature below melting point to effect microstructural changes.
Injection molding	Thermal and mechanical	Heat is used to raise temperature of polymer to highly plastic consistency, and mechanical force is used to inject the polymer melt into a mold cavity.
Laser beam cutting	Light and thermal	A highly coherent light beam is used to cut material by vaporization and melting.
Machining	Mechanical	Cutting of metal is accomplished by relative motion between tool and workpiece.
Sheet metal punching and blanking	Mechanical	Mechanical power is used to shear metal sheets and plates.
Welding	Thermal (maybe mechanical)	Most welding processes use heat to cause fusion and coalescence of two (or more) metal parts at their contacting surfaces. Some welding processes also apply mechanical pressure to the surfaces.

this transport and placement function. At the conclusion of the process, the work unit must similarly be removed. If the process is completely automated, then some form of mechanized power is used. If the process is manually operated or semiautomated, then human power may be used to position and locate the work unit.

- *Material transport between operations.* In addition to loading and unloading at a given operation, the work units must be moved between operations. We consider the material handling technologies associated with this transport function in Chapter 10.

Power for Automation. Above and beyond the basic power requirements for the manufacturing operation, additional power is required for automation. The additional power is used for the following functions:

- *Controller unit.* Modern industrial controllers are based on digital computers, which require electrical power to read the program of instructions, make the control calculations, and execute the instructions by transmitting the proper commands to the actuating devices.

- *Power to actuate the control signals.* The commands sent by the controller unit are carried out by means of electromechanical devices, such as switches and motors, called *actuators* (Section 6.2). The commands are generally transmitted by means of low-voltage control signals. To accomplish the commands, the actuators require more power, and so the control signals must be amplified to provide the proper power level for the actuating device.

- *Data acquisition and information processing.* In most control systems, data must be collected from the process and used as input to the control algorithms. In addition, a requirement of the process may include keeping records of process performance or product quality. These data acquisition and record keeping functions require power, although in modest amounts.

4.1.2 Program of Instructions

The actions performed by an automated process are defined by a program of instructions. Whether the manufacturing operation involves low, medium, or high production, each part or product made in the operation requires one or more processing steps that are unique to that part or product. These processing steps are performed during a work cycle. A new part is completed during each work cycle (in some manufacturing operations, more than one part is produced during the work cycle; e.g., a plastic injection molding operation may produce multiple parts each cycle using a multiple cavity mold). The particular processing steps for the work cycle are specified in a *work cycle program*. Work cycle programs are called *part programs* in numerical control (Chapter 7). Other process control applications use different names for this type of program.

Work Cycle Programs.   In the simplest automated processes, the work cycle consists of essentially one step, which is to maintain a single process parameter at a defined level, for example, maintain the temperature of a furnace at a designated value for the duration of a heat treatment cycle. (We assume that loading and unloading of the work units into and from the furnace is performed manually and is therefore not part of the automatic cycle.) In this case, programming simply involves setting the temperature dial on the furnace. To change the program, the operator simply changes the temperature setting. In an extension of this simple case, the single-step process is defined by more than one process parameter, for example, a furnace in which both temperature and atmosphere are controlled.

In more complicated systems, the process involves a work cycle consisting of multiple steps that are repeated with no deviation from one cycle to the next. Most discrete part manufacturing operations are in this category. A typical sequence of steps (simplified) is the following: (1) load the part into the production machine, (2) perform the process, and (3) unload the part. During each step, there are one or more activities that involve changes in one or more process parameters. *Process parameters* are inputs to the process, such as temperature setting of a furnace, coordinate axis value in a positioning system, valve opened or closed in a fluid flow system, and motor on or off. Process parameters are distinguished from *process variables*, which are outputs from the process; for example, the actual temperature of the furnace, the actual position of the axis, the actual flow rate of the fluid in the pipe, and the rotational speed of the motor. As our list of examples suggests, the changes in process parameter values may be continuous (gradual changes during the processing step; for example, gradually increasing temperature during a heat treatment cycle) or discrete (stepwise changes, for example, on/off). Different process parameters may be involved in each step.

EXAMPLE 4.1   An Automated Turning Operation

Consider an automated turning operation that generates a cone-shaped product. Assume the system is automated and that a robot is used to load and unload the work unit. The work cycle consists of the following steps: (1) load starting workpiece, (2) position cutting tool prior to turning, (3) turn, (4) reposition tool to a safe location at end of turning, and (5) unload finished workpiece. Identify the activities and process parameters for each step of the operation.

Solution: In step (1), the activities consist of the robot manipulator reaching for the raw workpart, lifting and positioning the part into the chuck jaws of the lathe, then retreating to a safe position to await unloading. The process parameters for these activities are the axis values of the robot manipulator (which change continuously), the gripper value (open or closed), and the chuck jaw value (open or closed).

In step (2), the activity is the movement of the cutting tool to a "ready" position. The process parameters associated with this activity are the x- and z-axis position of the tool.

Step (3) is the turning operation. It requires the simultaneous control of three process parameters: rotational speed of the workpiece (rev/min), feed (mm/rev), and radial distance of the cutting tool from the axis of rotation. To cut the conical shape, radial distance must be changed continuously at a constant rate for each revolution of the workpiece. For a consistent finish on the surface, the rotational speed must be continuously adjusted to maintain a constant surface speed (m/min); and for equal feed marks on the surface, the feed must be set at a constant value. Depending on the angle of the cone, multiple turning passes may be required to gradually generate the desired contour. Each pass represents an additional step in the sequence.

Steps (4) and (5) are the reverse of steps (2) and (1), respectively, and the process parameters are the same.

Many production operations consist of multiple steps, sometimes more complicated than our turning example. Examples of these operations include automatic screw machine cycles, sheet metal stamping operations, plastic injection molding, and die casting. Each of these manufacturing processes has been used for many decades. In earlier versions of these operations, the work cycles were controlled by hardware components, such as limit switches, timers, cams, and electromechanical relays. In effect, the hardware components and their arrangements served as the program of instructions that directed the sequence of steps in the processing cycle. Although these devices were quite adequate in performing their sequencing function, they suffered from the following disadvantages: (1) They often required considerable time to design and fabricate, forcing the production equipment to be used for batch production only; (2) making even minor changes in the program was difficult and time consuming; and (3) the program was in a physical form that was not readily compatible with computer data processing and communication.

Modern controllers used in automated systems are based on digital computers. Instead of cams, timers, relays, and other hardware devices, the programs for computer-controlled equipment are contained in magnetic tape, diskettes, compact disks (CD-ROMs), computer memory, and other modern storage technologies. Virtually all new equipment that perform the above mass production operations are designed with some type of computer controller to execute their respective processing cycles. The use of digital computers as the process controller allows improvements and upgrades to be made in the control programs, such as the addition of control functions not foreseen during initial equipment design. These kinds of control changes are often difficult to make with the previous hardware devices.

A work cycle may include manual steps, in which the operator performs certain activities during the work cycle, and the automated system performs the rest. A common example is the loading and unloading of parts by the operator into and from a numerical control machine between machining cycles, while the machine performs the cutting operation under part program control. Initiation of the cutting operation of each cycle is triggered by the operator activating a "start" button after the part has been loaded.

Decision-Making in the Programmed Work Cycle. In our previous discussion of automated work cycles, the only two features of the work cycle are (1) the number and sequence of processing steps, and (2) the process parameter changes in each step. Each work cycle consists of the same steps and associated process parameter changes with no variation from one cycle to the next. The program of instructions is repeated each work cycle without deviation. In fact, many automated manufacturing operations require decisions to be made during the programmed work cycle to cope with variations in the cycle. In many cases, the variations are routine elements of the cycle, and the corresponding instructions for dealing with them are incorporated into the regular part program. These cases include

- *Operator interaction.* Although the program of instructions is intended to be carried out without human interaction, the controller unit may require input data from a human operator in order to function. For example, in an automated engraving operation, the operator may have to enter the alphanumeric characters that are to be engraved on the work unit (e.g., plaque, trophy, belt buckle). After the characters are entered, the system accomplishes the engraving automatically. (An everyday example of operator interaction with an automated system is a bank customer using an automated teller machine. The customer must enter the codes indicating what transaction the teller machine must accomplish.)
- *Different part or product styles processed by the system.* In this instance, the automated system is programmed to perform different work cycles on different part or product styles. An example is an industrial robot that performs a series of spot welding operations on car bodies in a final assembly plant. These plants are often designed to build different body styles on the same automated assembly line, such as two-door and four-door sedans. As each car body enters a given welding station on the line, sensors identify which style it is, and the robot performs the correct series of welds for that style.
- *Variations in the starting work units.* In some manufacturing operations the starting work units are not consistent. A good example is a sand casting as the starting work unit in a machining operation. The dimensional variations in the raw castings sometimes necessitate an extra machining pass to bring the machined dimension to the specified value. The part program must be coded to allow for the additional pass when necessary.

In all of these examples, the routine variations can be accommodated in the regular work cycle program. The program can be designed to respond to sensor or operator inputs by executing the appropriate subroutine corresponding to the input. In other cases, the variations in the work cycle are not routine at all. They are infrequent and unexpected, such as the failure of an equipment component. In these instances, the program must include contingency procedures or modifications in the sequence to cope with conditions that lie outside the normal routine. We discuss these measures later in the chapter in the context of advanced automation functions (Section 4.2).

Various production situations and work cycle programs have been discussed here. Let us attempt to summarize the features of work cycle programs (part programs) used to direct the operations of an automated system:

- *Number of steps in work cycle.* How many distinct steps or work elements are included in the work cycle? A typical sequence in discrete production operations is (1) load, (2), process, (3) unload.
- *Manual participation in the work cycle.* Is a human worker required to perform certain steps in the work cycle, such as loading and unloading the production machine, or is the work cycle fully automated?
- *Process parameters.* How many process parameters must be controlled during each step? Are the process parameters continuous or discrete? How are the process parameters actuated? Do the parameters need to be changed during the step, for example, a positioning system whose axis values change during the processing step?
- *Operator interaction.* For example, is the operator required to enter processing data for each work cycle?
- *Variations in part or product styles.* Are the work units identical each cycle, as in mass production (fixed automation) or batch production (programmable automation), or are different part or product styles processed each cycle (flexible automation)?
- *Variations in starting work units.* Variations can occur in starting dimensions or materials. If the variations are significant, some adjustments may be required in the work cycle.

4.1.3 Control System

The control element of the automated system executes the program of instructions. The control system causes the process to accomplish its defined function, to carry out some manufacturing operation. Let us provide a brief introduction to control systems here. The following chapter describes this important industrial technology in more detail.

The controls in an automated system can be either closed loop or open loop. A *closed loop control system*, also known as a *feedback control system*, is one in which the output variable is compared with an input parameter, and any difference between the two is used to drive the output into agreement with the input. As shown in Figure 4.3, a closed loop

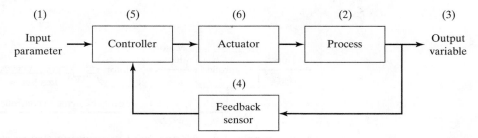

Figure 4.3 A feedback control system.

Figure 4.4   An open loop control system.

control system consists of six basic elements: (1) input parameter, (2) process, (3) output variable, (4) feedback sensor, (5) controller, and (6) actuator. The *input parameter*, often referred to as the *set point*, represents the desired value of the output. In a home temperature control system, the set point is the desired thermostat setting. The *process* is the operation or function being controlled. In particular, it is the *output variable* that is being controlled in the loop. In the present discussion, the process of interest is usually a manufacturing operation, and the output variable is some process variable, perhaps a critical performance measure in the process, such as temperature or force or flow rate. A *sensor* is used to measure the output variable and close the loop between input and output. Sensors perform the feedback function in a closed loop control system. The *controller* compares the output with the input and makes the required adjustment in the process to reduce the difference between them. The adjustment is accomplished using one or more *actuators*, which are the hardware devices that physically carry out the control actions, such as electric motors or flow valves. It should be mentioned that our model in Figure 4.3 shows only one loop. Most industrial processes require multiple loops, one for each process variable that must be controlled.

In contrast to the closed loop control system, an *open loop control system* operates without the feedback loop, as in Figure 4.4. In this case, the controls operate without measuring the output variable, so no comparison is made between the actual value of the output and the desired input parameter. The controller relies on an accurate model of the effect of its actuator on the process variable. With an open loop system, there is always the risk that the actuator will not have the intended effect on the process, and that is the disadvantage of an open loop system. Its advantage is that it is generally simpler and less expensive than a closed loop system. Open loop systems are usually appropriate when the following conditions apply: (1) the actions performed by the control system are simple, (2) the actuating function is very reliable, and (3) any reaction forces opposing the actuator are small enough to have no effect on the actuation. If these characteristics are not applicable, then a closed loop control system may be more appropriate.

Consider the difference between a closed loop and open loop system for the case of a positioning system. Positioning systems are common in manufacturing to locate a workpart relative to a tool or workhead. Figure 4.5 illustrates the case of a closed loop

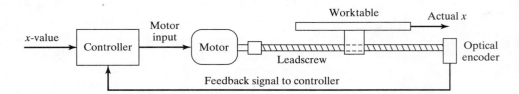

Figure 4.5   A (one-axis) positioning system consisting of a leadscrew driven by a dc servomotor.

positioning system. In operation, the system is directed to move the worktable to a specified location as defined by a coordinate value in a Cartesian (or other) coordinate system. Most positioning systems have at least two axes (e.g., an x–y positioning table) with a control system for each axis, but our diagram only illustrates one of these axes. A dc servomotor connected to a leadscrew is a common actuator for each axis. A signal indicating the coordinate value (e.g., x-value) is sent from the controller to the motor that drives the leadscrew, whose rotation is converted into linear motion of the positioning table. As the table moves closer to the desired x-coordinate value, the difference between the actual x-position and the input x-value is reduced. The actual x-position is measured by a feedback sensor (e.g., an optical encoder). The controller continues to drive the motor until the actual table position corresponds to the input position value.

For the open loop case, the diagram for the positioning system would be similar to the preceding, except that no feedback loop is present and a stepper motor is used in place of the dc servomotor. A stepper motor is designed to rotate a precise fraction of a turn for each pulse received from the controller. Since the motor shaft is connected to the leadscrew, and the leadscrew drives the worktable, each pulse converts into a small constant linear movement of the table. To move the table a desired distance, the number of pulses corresponding to that distance is sent to the motor. Given the proper application, whose characteristics match the preceding list of operating conditions, an open loop positioning system works with high reliability.

We consider the engineering analysis of closed loop and open loop positioning systems in the context of numerical control in a subsequent chapter (Section 7.5).

4.2 ADVANCED AUTOMATION FUNCTIONS

In addition to executing work cycle programs, an automated system may be capable of executing advanced functions that are not specific to a particular work unit. In general, the functions are concerned with enhancing the safety and performance of the equipment. Advanced automation functions include the following: (1) safety monitoring, (2) maintenance and repair diagnostics, and (3) error detection and recovery.

Advanced automation functions are made possible by special subroutines included in the program of instructions. In some cases, the functions provide information only and do not involve any physical actions by the control system, for example, reporting a list of preventive maintenance tasks that should be accomplished. Any actions taken on the basis of this report are decided by the human operators and managers of the system and not by the system itself. In other cases, the program of instructions must be physically executed by the control system using available actuators. A simple example of this case is a safety monitoring system that sounds an alarm when a human worker gets dangerously close to the automated equipment.

4.2.1 Safety Monitoring

One of the significant reasons for automating a manufacturing operation is to remove workers from a hazardous working environment. An automated system is often installed to perform a potentially dangerous operation that would otherwise be accomplished manually by human workers. However, even in automated systems, workers are still needed to service the system, at periodic intervals if not full-time. Accordingly, it is important that the automated system be designed to operate safely when workers are in attendance. In addition, it is essential that the automated system carry out its process in a way that is not self-destructive. Thus, there are two reasons for providing an automated system with a safety monitoring capability: (1) to protect human workers in the vicinity of the system, and (2) to protect the equipment associated with the system.

Safety monitoring means more than the conventional safety measures taken in a manufacturing operation, such as protective shields around the operation or the kinds of manual devices that might be utilized by human workers, such as emergency stop buttons. *Safety monitoring* in an automated system involves the use of sensors to track the system's operation and identify conditions and events that are unsafe or potentially unsafe. The safety monitoring system is programmed to respond to unsafe conditions in some appropriate way. Possible responses to various hazards might include one or more of the following:

- completely stopping the automated system,
- sounding an alarm,
- reducing the operating speed of the process,
- taking corrective actions to recover from the safety violation.

This last response is the most sophisticated and is suggestive of an intelligent machine performing some advanced strategy. This kind of response is applicable to a variety of possible mishaps, not necessarily confined to safety issues, and is called error detection and recovery (Section 4.2.3).

Sensors for safety monitoring range from very simple devices to highly sophisticated systems. The topic of sensor technology is discussed in Chapter 6 (Section 6.1). The following list suggests some of the possible sensors and their applications for safety monitoring:

- Limit switches to detect proper positioning of a part in a workholding device so that the processing cycle can begin.
- Photoelectric sensors triggered by the interruption of a light beam; this could be used to indicate that a part is in the proper position or to detect the presence of a human intruder in the work cell.
- Temperature sensors to indicate that a metal workpart is hot enough to proceed with a hot forging operation. If the workpart is not sufficiently heated, then the metal's ductility might be too low, and the forging dies might be damaged during the operation.
- Heat or smoke detectors to sense fire hazards.
- Pressure-sensitive floor pads to detect human intruders in the work cell.
- Machine vision systems to perform surveillance of the automated system and its surroundings.

It should be mentioned that a given safety monitoring system is limited in its ability to respond to hazardous conditions by the possible irregularities that have been foreseen by the system designer. If the designer has not anticipated a particular hazard, and consequently has not provided the system with the sensing capability to detect that hazard, then the safety monitoring system cannot recognize the event if and when it occurs.

4.2.2 Maintenance and Repair Diagnostics

Modern automated production systems are becoming increasingly complex and sophisticated, complicating the problem of maintaining and repairing them. Maintenance and repair diagnostics refers to the capabilities of an automated system to assist in identifying the source of potential or actual malfunctions and failures of the system. Three modes of operation are typical of a modern maintenance and repair diagnostics subsystem:

1. *Status monitoring.* In the status monitoring mode, the diagnostic subsystem monitors and records the status of key sensors and parameters of the system during normal operation. On request, the diagnostics subsystem can display any of these values and provide an interpretation of current system status, perhaps warning of an imminent failure.

2. *Failure diagnostics.* The failure diagnostics mode is invoked when a malfunction or failure occurs. Its purpose is to interpret the current values of the monitored variables and to analyze the recorded values preceding the failure so that the cause of the failure can be identified.

3. *Recommendation of repair procedure.* In the third mode of operation, the subsystem recommends to the repair crew the steps that should be taken to effect repairs. Methods for developing the recommendations are sometimes based on the use of expert systems in which the collective judgments of many repair experts are pooled and incorporated into a computer program that uses artificial intelligence techniques.

Status monitoring serves two important functions in machine diagnostics: (1) providing information for diagnosing a current failure and (2) providing data to predict a future malfunction or failure. First, when a failure of the equipment has occurred, it is usually difficult for the repair crew to determine the reason for the failure and what steps should be taken to make repairs. It is often helpful to reconstruct the events leading up to the failure. The computer is programmed to monitor and record the variables and to draw logical inferences from their values about the reason for the malfunction. This diagnosis helps the repair personnel make the necessary repairs and replace the appropriate components. This is especially helpful in electronic repairs where it is often difficult to determine on the basis of visual inspection which components have failed.

The second function of status monitoring is to identify signs of an impending failure, so that the affected components can be replaced before failure actually causes the system to go down. These part replacements can be made during the night shift or another time when the process is not operating, so the system experiences no loss of regular operation.

4.2.3 Error Detection and Recovery

In the operation of any automated system, there are hardware malfunctions and unexpected events during operation. These events can result in costly delays and loss of production until the problem has been corrected and regular operation is restored. Traditionally, equipment malfunctions are corrected by human workers, perhaps with the aid of a maintenance and repair diagnostics subroutine. With the increased use of computer control for manufacturing processes, there is a trend toward using the control computer not only to diagnose the malfunctions but also to automatically take the necessary corrective action to restore the system to normal operation. The term *error detection and recovery* is used when the computer performs these functions.

Error Detection. The error detection step uses the automated system's available sensors to determine when a deviation or malfunction has occurred, interpret the sensor signal(s), and classify the error. Design of the error detection subsystem must begin with a systematic enumeration of the possible errors that can occur during system operation. The errors in a manufacturing process tend to be very application specific. They must be anticipated in advance in order to select sensors that will enable their detection.

In analyzing a given production operation, the possible errors can be classified into one of three general categories: (1) random errors, (2) systematic errors, and (3) aberrations. *Random errors* occur as a result of the normal stochastic nature of the process. These errors occur when the process is in statistical control (Section 20.3). Large variations in part dimensions, even when the production process is in statistical control, can cause problems in downstream operations. By detecting these deviations on a part-by-part basis, corrective action can be taken in subsequent operations. *Systematic errors* are those that result from some assignable cause such as a change in raw material properties or drift in an equipment setting. These errors usually cause the product to deviate from specifications so as to be unacceptable in quality terms. Finally, the third type of error, *aberrations*, results from either an equipment failure or a human mistake. Examples of equipment failures include fracture of a mechanical shear pin, bursts in a hydraulic line, rupture of a pressure vessel, and sudden failure of a cutting tool. Examples of human mistakes include errors in the control program, improper fixture setups, and substitution of the wrong raw materials.

The two main design problems in error detection are (1) anticipating all of the possible errors that can occur in a given process, and (2) specifying the appropriate sensor systems and associated interpretive software so that the system is capable of recognizing each error. Solving the first problem requires a systematic evaluation of the possibilities under each of the three error classifications. If the error has not been anticipated, then the error detection subsystem cannot correctly detect and identify it.

EXAMPLE 4.2   Error Detection in an Automated Machining Cell

Consider an automated cell consisting of a CNC machine tool, a parts storage unit, and a robot for loading and unloading the parts between the machine and the storage unit. Possible errors that might affect this system can be divided into the following categories: (1) machine and process, (2) cutting tools, (3) workholding fixture, (4) part storage unit, and (5) load/unload robot. Develop a list of possible errors (deviations and malfunctions) that might be included in each of these five categories.

Solution: The following is a list of the possible errors in the machining cell for each of the five categories:

- *Machine and process.* Possible errors include loss of power, power overload, thermal deflection, cutting temperature too high, vibration, no coolant, chip fouling, wrong part program, and defective part.

- *Cutting tools.* Possible errors include tool breakage, tool wear-out, vibration, tool not present, and wrong tool.

- *Workholding fixture.* Possible errors include part not in fixture, clamps not actuated, part dislodged during machining, part deflection during machining, part breakage, and chips causing location problems.

- *Part storage unit.* Possible errors include workpart not present, wrong workpart, and oversized or undersized workpart.
- *Load/unload robot.* Possible errors include improper grasping of workpart, dropping of workpart, and no part present at pickup.

Error Recovery. Error recovery is concerned with applying the necessary corrective action to overcome the error and bring the system back to normal operation. The problem of designing an error recovery system focuses on devising appropriate strategies and procedures that will either correct or compensate for the variety of errors that can occur in the process. Generally, a specific recovery strategy and procedure must be designed for each different error. The types of strategies can be classified as follows:

1. *Make adjustments at the end of the current work cycle.* When the current work cycle is completed, the part program branches to a corrective action subroutine specifically designed for the error detected, executes the subroutine, and then returns to the work cycle program. This action reflects a low level of urgency and is most commonly associated with random errors in the process.

2. *Make adjustments during the current cycle.* This generally indicates a higher level of urgency than the preceding type. In this case, the action to correct or compensate for the detected error is initiated as soon as the error is detected. However, it must be possible to accomplish the designated corrective action while the work cycle is still being executed.

3. *Stop the process to invoke corrective action.* In this case, the deviation or malfunction requires that the work cycle be suspended during corrective action. It is assumed that the system is capable of automatically recovering from the error without human assistance. At the end of the corrective action, the regular work cycle is continued.

4. *Stop the process and call for help.* In this case, the error cannot be resolved through automated recovery procedures. This situation arises because (1) the automated cell is not enabled to correct the problem or (2) the error cannot be classified into the predefined list of errors. In either case, human assistance is required to correct the problem and restore the system to fully automated operation.

Error detection and recovery requires an interrupt system (Section 5.3.2). When an error in the process is sensed and identified, an interrupt in the current program execution is invoked to branch to the appropriate recovery subroutine. This is done either at the end of the current cycle (type 1 above) or immediately (types 2, 3, and 4). At the completion of the recovery procedure, program execution reverts back to normal operation.

EXAMPLE 4.3   Error Recovery in an Automated Machining Cell

For the automated cell of Example 4.2, develop a list of possible corrective actions that might be taken by the system to address some of the errors.

TABLE 4.2   Error Recovery in an Automated Machining Cell: Possible Corrective Actions That Might Be Taken in Response to Errors Detected During the Operation

Error Detected	Possible Corrective Action to Recover
Part dimensions deviating due to thermal deflection of machine tool	Adjust coordinates in part program to compensate (category 1 corrective action)
Part dropped by robot during pickup	Reach for another part (category 2 corrective action)
Starting workpart is oversized	Adjust part program to take a preliminary machining pass across the work surface (category 2 corrective action)
Chatter (tool vibration)	Increase or decrease cutting speed to change harmonic frequency (category 2 corrective action)
Cutting temperature too high	Reduce cutting speed (category 2 corrective action)
Cutting tool failed	Replace cutting tool with another sharp tool (category 3 corrective action).
No more parts in parts storage unit	Call operator to resupply starting workparts (category 4 corrective action)
Chips fouling machining operation	Call operator to clear chips from work area (category 4 corrective action)

Solution: A list of possible corrective actions is presented in Table 4.2.

4.3   LEVELS OF AUTOMATION

The concept of automated systems can be applied to various levels of factory operations. One normally associates automation with the individual production machines. However, the production machine itself is made up of subsystems that may themselves be automated. For example, one of the important automation technologies we discuss in this part of the book is numerical control (Chapter 7). A modern numerical control (NC) machine tool is an automated system. However, the NC machine itself is composed of multiple control systems. Any NC machine has at least two axes of motion, and some machines have up to five axes. Each of these axes operates as a positioning system, as described in Section 4.1.3, and is, in effect, itself an automated system. Similarly, a NC machine is often part of a larger manufacturing system, and the larger system may itself be automated. For example, two or three machine tools may be connected by an automated part handling system operating under computer control. The machine tools also receive instructions (e.g., part programs) from the computer. Thus we have three levels of automation and control included here (the positioning system level, the machine tool level, and the manufacturing system level). For our purposes in this text, we can identify five possible levels of automation in a production plant. They are defined below, and their hierarchy is depicted in Figure 4.6.

1. *Device level.* This is the lowest level in our automation hierarchy. It includes the actuators, sensors, and other hardware components that comprise the machine level. The devices are combined into the individual control loops of the machine, for example, the feedback control loop for one axis of a CNC machine or one joint of an industrial robot.

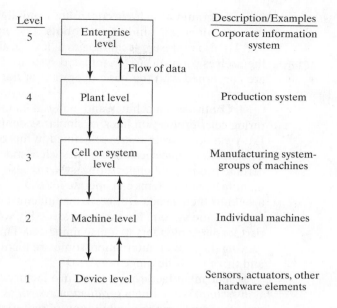

Level		Description/Examples
5	Enterprise level	Corporate information system
4	Plant level	Production system
3	Cell or system level	Manufacturing system-groups of machines
2	Machine level	Individual machines
1	Device level	Sensors, actuators, other hardware elements

Figure 4.6   Five levels of automation and control in manufacturing.

2. *Machine level.* Hardware at the device level is assembled into individual machines. Examples include CNC machine tools and similar production equipment, industrial robots, powered conveyors, and automated guided vehicles. Control functions at this level include performing the sequence of steps in the program of instructions in the correct order and making sure that each step is properly executed.

3. *Cell or system level.* This is the manufacturing cell or system level, which operates under instructions from the plant level. A manufacturing cell or system is a group of machines or workstations connected and supported by a material handling system, computer, and other equipment appropriate to the manufacturing process. Production lines are included in this level. Functions include part dispatching and machine loading, coordination among machines and material handling system, and collecting and evaluating inspection data.

4. *Plant level.* This is the factory or production systems level. It receives instructions from the corporate information system and translates them into operational plans for production. Likely functions include order processing, process planning, inventory control, purchasing, material requirements planning, shop floor control, and quality control.

5. *Enterprise level.* This is the highest level, consisting of the corporate information system. It is concerned with all of the functions necessary to manage the company: marketing and sales, accounting, design, research, aggregate planning, and master production scheduling.

Most of the technologies discussed in this part of the book are at level 2 (the machine level), although we discuss level 1 automation technologies (the devices that make up a control system) in Chapter 6. The level 2 technologies include the individual controllers (e.g., programmable logic controllers and digital computer controllers), numerical control machines, and industrial robots. The material handling equipment discussed in Part III also represent technologies at level 2, although some of the handling equipment are themselves sophisticated automated systems. The automation and control issues at level 2 are concerned with the basic operation of the equipment and the physical processes they perform.

Controllers, machines, and material handling equipment are combined into manufacturing cells, production lines, or similar systems, which make up level 3, considered in Part IV. A *manufacturing system* is defined in this book as a collection of integrated equipment designed for some special mission, such as machining a defined part family or assembling a certain product. Manufacturing systems also include people. Certain highly automated manufacturing systems can operate for extended periods of time without humans present to attend to their needs. But most manufacturing systems include workers as important elements of the system, for example, assembly workers on a conveyorized production line or part loaders/unloaders in a machining cell. Thus, manufacturing systems are designed with varying degrees of automation; some are highly automated, others are completely manual, and there is a wide range between.

The manufacturing systems in a factory are components of a larger system, a production system. We define a *production system* as the people, equipment, and procedures that are organized for the combination of materials and processes that comprise a company's manufacturing operations. Production systems are at level 4, the plant level, while manufacturing systems are at level 3 in our automation hierarchy. Production systems include not only the groups of machines and workstations in the factory but also the support procedures that make them work. These procedures include production control, inventory control, material requirements planning, shop floor control, and quality control. These systems are discussed in Parts IV and V. They are often implemented not only at the plant level but also at the corporate level (level 5).

REFERENCES

[1]   BOUCHER, T. O., *Computer Automation in Manufacturing*, Chapman & Hall, London, UK, 1996.

[2]   GROOVER, M. P., "Automation," *Encyclopaedia Britannica, Macropaedia*, 15th ed, Chicago, IL, 1992. Vol. 14, pp. 548–557.

[3]   GROOVER, M. P., "Automation," *Handbook of Design, Manufacturing, and Automation*, R. C. Dorf and A. Kusiak (eds.), John Wiley & Sons, Inc., NY, 1994, pp. 3–21.

[4]   GROOVER, M. P., "Industrial Control Systems," *Maynard's Industrial Engineering Handbook*, 5th ed., K. Zandin (ed.), McGraw-Hill Book Company, NY, (2001).

[5]   PLATT, R., *Smithsonian Visual Timeline of Inventions* (London: Dorling Kindersley Ltd., 1994).

[6]   "The Power of Invention," *Newsweek Special Issue*, Winter 1997–98 (pp. 6–79)

REVIEW QUESTIONS

4.1   What is automation?

4.2   An automated system consists of what three basic elements?

4.3   What is the difference between a process parameter and a process variable?

4.4 What are two reasons decision-making is required in a programmed work cycle?

4.5 What is the difference between a closed loop control system and an open loop control system?

4.6 What is safety monitoring in an automated system?

4.7 What is error detection and recovery in an automated system?

4.8 Name three of the four possible strategies in error recovery.

4.9 Identify the five levels of automation in a production plant.

Chapter 5

Industrial Control Systems

CHAPTER CONTENTS

The control system is one of the three basic components of an automated system (Section 4.1). In this chapter, we examine industrial control systems, in particular how digital computers are used to implement the control function in production. *Industrial control* is defined here as the automatic regulation of unit operations and their associated equipment as well as the integration and coordination of the unit operations into the larger production system. In the context of our book, the term *unit operations* usually refers to manufacturing operations; however, the term also applies to the operation of material handling and other industrial equipment. Let us begin our chapter by comparing the application of industrial control in the processing industries with its application in the discrete manufacturing industries.

From Chapter 5 of *Automation, Production Systems, and Computer-Integrated Manufacturing*, Third Edition.
Mikell P. Groover. Copyright © 2008 by Pearson Education, Inc. Publishing as Prentice Hall. All rights reserved.

5.1 PROCESS INDUSTRIES VERSUS DISCRETE MANUFACTURING INDUSTRIES

In our previous discussion of industry types in Chapter 2, we divided industries and their production operations into two basic categories: (1) process industries and (2) discrete manufacturing industries. Process industries perform their production operations on *amounts* of materials, because the materials tend to be liquids, gases, powders, and similar materials, whereas discrete manufacturing industries perform their operations on *quantities* of materials, because the materials tend to be discrete parts and products. The kinds of unit operations performed on the materials are different in the two industry categories. Some of the typical unit operations in each category are listed in Table 5.1.

5.1.1 Levels of Automation in the Two Industries

The levels of automation (Section 4.3) in the two industries are compared in Table 5.2. The significant differences are seen in the low and intermediate levels. At the device level, there are differences in the types of actuators and sensors used in the two industry categories, simply because the processes and equipment are different. In the process industries, the devices are used mostly for the control loops in chemical, thermal, or similar processing operations, whereas in discrete manufacturing, the devices control the mechanical actions of machines. At the next level up, the difference is that unit operations are controlled in the process industries, and machines are controlled in the discrete manufacturing operations. At the third level, the difference is between control of interconnected unit processing operations and interconnected machines. At the upper levels (plant and enterprise), the control issues are similar, allowing for the fact that the products and processes are different.

5.1.2 Variables and Parameters in the Two Industries

The distinction between process industries and discrete manufacturing industries extends to the variables and parameters that characterize the respective production operations. The reader will recall from the previous chapter (Section 4.1.2) that we defined variables as outputs of the process and parameters as inputs to the process. In the process industries, the variables and parameters of interest tend to be continuous, whereas in discrete manufacturing, they tend to be discrete. Let us explain the differences with reference to Figure 5.1.

TABLE 5.1 Typical Unit Operations in the Process Industries and Discrete Manufacturing Industries

Typical Unit Operations in the Process Industries	Typical Unit Operations in the Discrete Manufacturing Industries
Chemical reactions	Casting
Comminution	Forging
Deposition (e.g., chemical vapor deposition)	Extrusion
	Machining
Distillation	Mechanical assembly
Mixing and blending of ingredients	Plastic molding
Separation of ingredients	Sheet metal stamping

TABLE 5.2 Levels of Automation in the Process Industries and Discrete Manufacturing Industries

Level	Level of Automation in the Process Industries	Level of Automation in the Discrete Manufacturing Industries
5	*Corporate level*—management information system, strategic planning, high-level management of enterprise	*Corporate level*—management information system, strategic planning, high-level management of enterprise
4	*Plant level*—scheduling, tracking materials, equipment monitoring	*Plant or factory level*—scheduling, tracking work-in-process, routing parts through machines, machine utilization
3	*Supervisory control level*—control and coordination of several interconnected unit operations that make up the total process	*Manufacturing cell or system level*—control and coordination of groups of machines and supporting equipment working in coordination, including material handling equipment
2	*Regulatory control level*—control of unit operations	*Machine level*—production machines and workstations for discrete part and product manufacture
1	*Device level*—sensors and actuators comprising the basic control loops for unit operations	*Device level*—sensors and actuators to accomplish control of machine actions

A *continuous variable* (or parameter) is one that is uninterrupted as time proceeds, at least during the manufacturing operation. A continuous variable is generally considered to be *analog*, which means it can take on any value within a certain range. The variable is not restricted to a discrete set of values. Production operations in both the process industries and discrete parts manufacturing are characterized by continuous variables. Examples include force, temperature, flow rate, pressure, and velocity. All of these variables

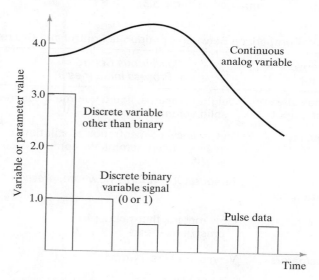

Figure 5.1 Continuous and discrete variables and parameters in manufacturing operations.

(whichever ones apply to a given production process) are continuous over time during the process, and they can take on any of an infinite number of possible values within a certain practical range.

A *discrete variable* (or parameter) is one that can take on only certain values within a given range. The most common type of discrete variable is *binary*, meaning it can take on either of two possible values, ON or OFF, open or closed, and so on. Examples of discrete binary variables and parameters in manufacturing include limit switch open or closed, motor on or off, and workpart present or not present in a fixture. Not all discrete variables (and parameters) are binary. Other possibilities are variables that can take on more than two possible values but less than an infinite number, that is, *discrete other than binary*. Examples include daily piece counts in a production operation and the display of a digital tachometer. A special form of discrete variable (and parameter) is *pulse data*, which consist of a train of pulses as shown in Figure 5.1. As a discrete variable, a pulse train might be used to indicate piece counts, for example, parts passing on a conveyor activate a photocell to produce a pulse for each part detected. As a process parameter, a pulse train might be used to drive a stepper motor.

5.2 CONTINUOUS VERSUS DISCRETE CONTROL

Industrial control systems used in the process industries have tended to emphasize the control of continuous variables and parameters. By contrast, the manufacturing industries produce discrete parts and products, and their controllers have tended to emphasize discrete variables and parameters. Just as we have two basic types of variables and parameters that characterize production operations, we also have two basic types of control: (1) *continuous control*, in which the variables and parameters are continuous and analog; and (2) *discrete control*, in which the variables and parameters are discrete, mostly binary discrete. Some of the differences between continuous control and discrete control are summarized in Table 5.3.

TABLE 5.3 Comparison Between Continuous Control and Discrete Control

Comparison Factor	Continuous Control in Process Industries	Discrete Control in Discrete Manufacturing Industries
Typical measures of product output	Weight measures, liquid volume measures, solid volume measures	Number of parts, number of products
Typical quality measures	Consistency, concentration of solution, absence of contaminants, conformance to specification	Dimensions, surface finish, appearance, absence of defects, product reliability
Typical variables and parameters	Temperature, volume flow rate, pressure	Position, velocity, acceleration, force
Typical sensors	Flow meters, thermocouples, pressure sensors	Limit switches, photoelectric sensors, strain gages, piezoelectric sensors
Typical actuators	Valves, heaters, pumps	Switches, motors, pistons
Typical process time constants	Seconds, minutes, hours	Less than a second

In reality, most operations in the process and discrete manufacturing industries include both continuous and discrete variables and parameters. Consequently, many industrial controllers are designed with the capability to receive, operate on, and transmit both types of signals and data. In Chapter 6, we discuss the various types of signals and data in industrial control systems and how the data are converted for use by digital computer controllers.

To complicate matters, since digital computers began replacing analog controllers in continuous process control applications around 1960, continuous process variables are no longer measured continuously. Instead, they are sampled periodically, in effect creating a discrete sampled-data system that approximates the actual continuous system. Similarly, the control signals sent to the process are typically stepwise functions that approximate the previous continuous control signals transmitted by analog controllers. Hence, in digital computer process control, even continuous variables and parameters possess characteristics of discrete data, and these characteristics must be considered in the design of the computer-process interface and the control algorithms used by the controller.

5.2.1 Continuous Control Systems

In continuous control, the usual objective is to maintain the value of an output variable at a desired level, similar to the operation of a feedback control system as defined in the previous chapter (Section 4.1.3). However, most continuous processes in the practical world consist of many separate feedback loops, all of which have to be controlled and coordinated to maintain the output variable at the desired value. Examples of continuous processes are the following:

- Control of the output of a chemical reaction that depends on temperature, pressure, and input flow rates of several reactants. All of these variables and/or parameters are continuous.
- Control of the position of a workpart relative to a cutting tool in a contour milling operation in which complex curved surfaces are generated. The position of the part is defined by x-, y-, and z-coordinate values. As the part moves, the x, y, and z values can be considered as continuous variables and/or parameters that change over time to machine the part.

There are several ways to achieve the control objective in a continuous process control system. In the following paragraphs, we survey the most prominent categories.

Regulatory Control. In regulatory control, the objective is to maintain process performance at a certain level or within a given tolerance band of that level. This is appropriate, for example, when the performance attribute is some measure of product quality, and it is important to keep the quality at the specified level or within a specified range. In many applications, the performance measure of the process, sometimes called the *index of performance*, must be calculated based on several output variables of the process. Except for this feature, regulatory control is to the overall process what feedback control is to an individual control loop in the process, as suggested by Figure 5.2.

The trouble with regulatory control (and also with a simple feedback control loop) is that compensating action is taken only after a disturbance has affected the process output. An error must be present for any control action to be taken. The presence of an error

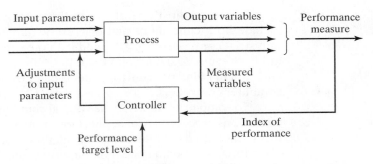

Figure 5.2   Regulatory control.

means that the output of the process is different from the desired value. The following control mode, feedforward control, addresses this issue.

Feedforward Control.   The strategy in feedforward control is to anticipate the effect of disturbances that will upset the process by sensing them and compensating for them before they can affect the process. As shown in Figure 5.3, the feedforward control elements sense the presence of a disturbance and take corrective action by adjusting a process parameter that compensates for any effect the disturbance will have on the process. In the ideal case, the compensation is completely effective. However, complete compensation is unlikely because of imperfections in the feedback measurements, actuator operations, and control algorithms, so feedforward control is usually combined with feedback control, as shown in our figure. Regulatory and feedforward control are more closely associated with the process industries than with discrete product manufacturing.

Steady-State Optimization.   This term refers to a class of optimization techniques in which the process exhibits the following characteristics: (1) there is a well-defined index of performance, such as product cost, production rate, or process yield; (2) the

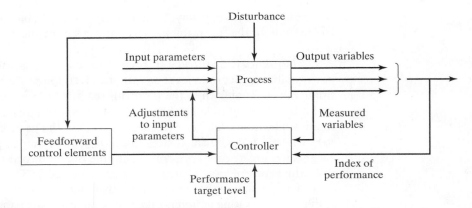

Figure 5.3   Feedforward control, combined with feedback control.

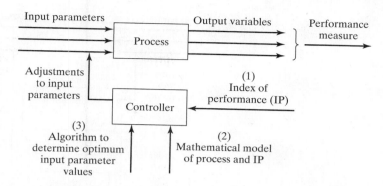

Figure 5.4   Steady-state (open loop) optimal control.

relationship between the process variables and the index of performance is known; and (3) the values of the system parameters that optimize the index of performance can be determined mathematically. When these characteristics apply, the control algorithm is designed to make adjustments in the process parameters to drive the process toward the optimal state. The control system is open loop, as seen in Figure 5.4. Several mathematical techniques are available for solving steady-state optimal control problems, including differential calculus, calculus of variations, and various mathematical programming methods.

Adaptive Control.   Steady-state optimal control operates as an open loop system. It works successfully when there are no disturbances that invalidate the known relationship between process parameters and process performance. When such disturbances are present in the application, a self-correcting form of optimal control can be used, called adaptive control. Adaptive control combines feedback control and optimal control by measuring the relevant process variables during operation (as in feedback control) and using a control algorithm that attempts to optimize some index of performance (as in optimal control).

Adaptive control is distinguished from feedback control and steady-state optimal control by its unique capability to cope with a time-varying environment. It is not unusual for a system to operate in an environment that changes over time and for the changes to have a potential effect on system performance. If the internal parameters or mechanisms of the system are fixed, as in feedback control or optimal control, the system may perform quite differently in one type of environment than in another. An adaptive control system is designed to compensate for its changing environment by monitoring its own performance and altering some aspect of its control mechanism to achieve optimal or near-optimal performance. In a production process, the "time-varying environment" consists of the variations in processing variables, raw materials, tooling, atmospheric conditions, and the like, any of which may affect performance.

The general configuration of an adaptive control system is illustrated in Figure 5.5. To evaluate its performance and respond accordingly, an adaptive control system performs three functions, as shown in the figure:

1. *Identification function.* In this function, the current value of the index of performance of the system is determined, based on measurements collected from the process. Since the environment changes over time, system performance also

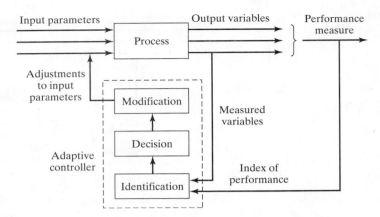

Figure 5.5 Configuration of an adaptive control system.

changes. Accordingly, the identification function must be accomplished more or less continuously over time during system operation.

2. *Decision function.* Once system performance has been determined, the next function is to decide what changes should be made to improve performance. The decision function is implemented by means of the adaptive system's programmed algorithm. Depending on this algorithm, the decision may be to change one or more input parameters to the process, to alter some of the internal parameters of the controller, or to make other changes.

3. *Modification function.* The third function of adaptive control is to implement the decision. Whereas decision is a logic function, modification is concerned with physical changes in the system. It involves hardware rather than software. In modification, the system parameters or process inputs are altered using available actuators to drive the system toward a more optimal state.

Adaptive control is most applicable at levels 2 and 3 in our automation hierarchy (Table 5.2). Adaptive control has been the subject of research and development for several decades; it was originally motivated by problems of high-speed flight control in the age of jet aircraft. The principles have been applied in other areas as well, including manufacturing. One notable example is *adaptive control machining*, in which changes in process variables such as cutting force, power, and vibration are used to effect control over process parameters such as cutting speed and feed rate.

On-Line Search Strategies. On-line search strategies can be used to address a special class of adaptive control problem in which the decision function cannot be sufficiently defined; that is, the relationship between the input parameters and the index of performance is not known, or not known well enough to use adaptive control as previously described. Therefore, it is not possible to decide on the changes in the internal parameters of the system to produce the desired performance improvement. Instead, experiments must be performed on the process. Small systematic changes are made in the input parameters of the process to observe the effect of these changes on the output variables. Based on the results of these experiments, larger changes are made in the input parameters to drive the process toward improved performance.

On-line search strategies include a variety of schemes to explore the effects of changes in process parameters, ranging from trial-and-error techniques to gradient methods. All of the schemes attempt to determine which input parameters cause the greatest positive effect on the index of performance and then move the process in that direction. There is little evidence that on-line search techniques are used much in discrete parts manufacturing. Their applications are more common in the continuous process industries.

Other Specialized Techniques. Other specialized techniques include strategies that are currently evolving in control theory and computer science. Examples include learning systems, expert systems, neural networks, and other artificial intelligence methods for process control.

5.2.2 Discrete Control Systems

In discrete control, the parameters and variables of the system are changed at discrete moments in time. The changes involve variables and parameters that are also discrete, typically binary (ON/OFF). The changes are defined in advance by means of a program of instructions, for example, a work cycle program (Section 4.1.2). The changes are executed either because the state of the system has changed or because a certain amount of time has elapsed. These two cases can be distinguished as (1) event-driven changes or (2) time-driven changes [2].

An *event-driven change* is executed by the controller in response to some event that has caused the state of the system to be altered. The change can be to initiate an operation or terminate an operation, start a motor or stop it, open a valve or close it, and so forth. Examples of event-driven changes are

- A robot loads a workpart into the fixture, and the part is sensed by a limit switch. Sensing the part's presence is the event that alters the system state. The event-driven change is that the automatic machining cycle can now commence.
- The diminishing level of plastic molding compound in the hopper of an injection molding machine triggers a low-level switch, which in turn opens a valve to start the flow of new plastic into the hopper. When the level of plastic reaches the high-level switch, this triggers the valve to close, thus stopping the flow of pellets into the hopper.
- Counting parts moving along a conveyor past an optical sensor is an event-driven system. Each part moving past the sensor is an event that drives the counter.

A *time-driven change* is executed by the control system either at a specific point in time or after a certain time lapse has occurred. As before, the change usually consists of starting something or stopping something, and the time when the change occurs is important. Examples of time-driven changes are

- In factories with specific starting times and ending times for the shift and uniform break periods for all workers, the "shop clock" is set to sound a bell at specific moments during the day to indicate these start and stop times.
- Heat treating operations must be carried out for a certain length of time. An automated heat treating cycle consists of automatic loading of parts into the furnace (perhaps by a robot) and then unloading after the parts have been heated for the specified length of time.

- In the operation of a washing machine, once the laundry tub has been filled to the preset level, the agitation cycle continues for a length of time set on the controls. When this time is up, the timer stops the agitation and initiates draining of the tub. (By comparison with the agitation cycle, filling the laundry tub with water is event-driven. Filling continues until the proper level is sensed, which causes the inlet valve to close.)

The two types of change correspond to two different types of discrete control, called combinational logic control and sequential control. *Combinational logic control* is used to control the execution of event-driven changes, and *sequential control* is used to manage time-driven changes. These types of control are discussed in our expanded coverage of discrete control in Chapter 9.

Discrete control is widely used in discrete manufacturing as well as the process industries. In discrete manufacturing, it is used to control the operation of conveyors and other material transport systems (Chapter 10), automated storage systems (Chapter 11), stand-alone production machines (Chapter 14), automated transfer lines (Chapter 16), automated assembly systems (Chapter 17), and flexible manufacturing systems (Chapter 19). All of these systems operate by following a well-defined sequence of start-and-stop actions, such as powered feed motions, parts transfers between workstations, and on-line automated inspections.

In the process industries, discrete control is associated more with batch processing than with continuous processes. In a typical batch processing operation, each batch of starting ingredients is subjected to a cycle of processing steps that involves changes in process parameters (e.g., temperature and pressure changes), possible flow from one container to another during the cycle, and finally packaging. The packaging step differs depending on the product. For foods, packaging may involve canning or boxing. For chemicals, it means filling containers with the liquid product. And for pharmaceuticals, it may involve filling bottles with medicine tablets. In batch process control, the objective is to manage the sequence and timing of processing steps as well as to regulate the process parameters in each step. Accordingly, batch process control typically includes both continuous control as well as discrete control.

5.3 COMPUTER PROCESS CONTROL

The use of digital computers to control industrial processes had its origins in the continuous process industries in the late 1950s (Historical Note 5.1). Prior to that point, analog controllers were used to implement continuous control, and relay systems were used to implement discrete control. At that time, computer technology was in its infancy, and the only computers available for process control were large, expensive mainframes. Compared with today's technology, the digital computers of the 1950s were slow, unreliable, and not well suited to process control applications. The computers that were installed sometimes cost more than the processes they controlled. Around 1960, digital computers started replacing analog controllers in continuous process control applications, and around 1970, programmable logic controllers started replacing relay banks in discrete control applications. Advances in computer technology since the 1960s and 1970s have resulted in the development of the microprocessor. Today, virtually all industrial processes, certainly new installations, are controlled by digital computers based on microprocessor technology. Microprocessor-based controllers are discussed in Section 5.3.3.

Historical Note 5.1   Computer process control [1, 7].

Control of industrial processes by digital computers can be traced to the process industries in the late 1950s and early 1960s. These industries, such as oil refineries and chemical plants, use high-volume continuous production processes characterized by many variables and associated control loops. The processes had traditionally been controlled by analog devices, each loop having its own set point value and in most instances operating independently of other loops. Any coordination of the process was accomplished in a central control room, where workers adjusted the individual settings, attempting to achieve stability and economy in the process. The cost of the analog devices for all of the control loops was considerable, and the human coordination of the process was less than optimal. The commercial development of the digital computer in the 1950s offered the opportunity to replace some of the analog control devices with the computer.

The first known attempt to use a digital computer for process control was at a Texaco refinery in Port Arthur, Texas in the late 1950s. Texaco had been contacted in 1956 by computer manufacturer Thomson Ramo Woodridge (TRW), and a feasibility study was conducted on a polymerization unit at the refinery. The computer control system went on line in March 1959. The control application involved 26 flows, 72 temperatures, three pressures, and three compositions. This pioneering work did not escape the notice of other companies in the process industries as well as other computer companies. The process industries saw computer process control as a means of automation, and the computer companies saw a potential market for their products.

The available computers in the late 1950s were not reliable, and most of the subsequent process control installations operated by either printing out instructions for the operator or by making adjustments in the set points of analog controllers, thereby reducing the risk of process downtime due to computer problems. The latter mode of operation was called *set point control*. By March 1961, a total of 37 computer process control systems had been installed. Much experience was gained from these early installations. The *interrupt feature* (Section 5.3.2), by which the computer suspends current program execution to quickly respond to a process need, was developed during this period.

The first *direct digital control* (DDC) system (Section 5.3.3), in which certain analog devices are replaced by the computer, was installed by Imperial Chemical Industries in England in 1962. In this implementation, 224 process variables were measured, and 129 actuators (valves) were controlled. Improvements in DDC technology were made, and additional systems were installed during the 1960s. Advantages of DDC noted during this time included (1) cost savings by eliminating analog instrumentation, (2) simplified operator display panels, and (3) flexibility due to reprogramming capability.

Computer technology was advancing, leading to the development of the *minicomputer* in the late 1960s. Process control applications were easier to justify using these smaller, less expensive computers. Development of the *microcomputer* in the early 1970s continued this trend. Lower cost process control hardware and interface equipment (such as analog-to-digital converters) were becoming available due to the larger markets made possible by low-cost computer controllers.

Most of the developments in computer process control up to this time were biased toward the process industries rather than discrete part and product manufacturing. Just as analog devices had been used to automate process industry operations, relay banks were widely used to satisfy the discrete process control (ON/OFF) requirements in manufacturing automation. The *programmable logic controller* (PLC), a control computer designed for discrete process control, was developed in the early 1970s (Historical Note 9.1). Also, *numerical control* (NC) machine tools (Historical Note 7.1) and industrial *robots* (Historical Note 8.1), technologies that preceded computer control, started to be designed with digital computers as their controllers.

The availability of low-cost microcomputers and programmable logic controllers resulted in a growing number of installations in which a process was controlled by multiple computers networked together. The term *distributed control* was used for this kind of system, the first of which was a product offered by Honeywell in 1975. In the early 1990s, *personal computers* (PCs) began to be utilized on the factory floor, sometimes to provide scheduling and engineering data to shop floor personnel, in other cases as the operator interface to processes controlled by PLCs. Today, a growing number of PCs are being used to directly control manufacturing operations.

In this section on computer process control, we identify the requirements placed on the computer in industrial control applications. We then examine the capabilities that have been incorporated into the control computer to address these requirements, and finally we survey the various forms of computer control used in industry.

5.3.1 Control Requirements

Whether the application involves continuous control, discrete control, or both, there are certain basic requirements that tend to be common to nearly all process control applications. By and large, they are concerned with the need to communicate and interact with the process on a real-time basis. A *real-time controller* is a controller that is able to respond to the process within a short enough time period that process performance is not degraded. Real-time control usually requires the controller to be capable of *multitasking*, which means coping with multiple tasks concurrently without the tasks interfering with one another.

There are two basic requirements that must be managed by the controller to achieve real-time control:

1. *Process-initiated interrupts.* The controller must be able to respond to incoming signals from the process. Depending on the relative importance of the signals, the computer may need to interrupt execution of a current program to service a higher priority need of the process. A process-initiated interrupt is often triggered by abnormal operating conditions, indicating that some corrective action must be taken promptly.

2. *Timer-initiated actions.* The controller must be capable of executing certain actions at specified points in time. Timer-initiated actions can be generated at regular time intervals, ranging from very low values (e.g., $100 \ \mu s$) to several minutes, or they can be generated at distinct points in time. Typical timer-initiated actions in process control include (1) scanning sensor values from the process at regular sampling intervals, (2) turning on and off switches, motors, and other binary devices associated with the process at discrete points in time during the work cycle, (3) displaying performance data on the operator's console at regular times during a production run, and (4) recomputing optimal process parameter values at specified times.

These two requirements correspond to the two types of changes mentioned previously in the context of discrete control systems: (1) event-driven changes and (2) time-driven changes.

In addition to these basic requirements, the control computer must also deal with other types of interruptions and events. These include the following:

3. *Computer commands to process.* In addition to receiving incoming signals from the process, the control computer must send control signals to the process to accomplish a corrective action. These output signals may actuate a certain hardware device or readjust a set point in a control loop.

4. *System- and program-initiated events.* These are events related to the computer system itself. They are similar to the kinds of computer operations associated with business and engineering applications of computers. A *system-initiated event* involves communications among computers and peripheral devices linked together in a network. In these multiple computer networks, feedback signals, control commands, and other data must be transferred back and forth among the computers in the overall control of the process. A *program-initiated event* occurs when the program calls for some non-process-related action, such as the printing or display of reports on a printer or monitor. In process control, system- and program-initiated events generally occupy a low level of priority compared with process interrupts, commands to the process, and timer-initiated events.

5. *Operator-initiated events.* Finally, the control computer must be able to accept input from operating personnel. Operator-initiated events include (1) entering new programs; (2) editing existing programs; (3) entering customer data, order number, or startup instructions for the next production run; (4) requesting process data; and (5) calling for emergency stops.

5.3.2 Capabilities of Computer Control

The above requirements can be satisfied by providing the controller with certain capabilities that allow it to interact on a real-time basis with the process and the operator. The capabilities are (1) polling, (2) interlocks, (3) interrupt system, and (4) exception handling.

Polling (Data Sampling).   In computer process control, polling refers to the periodic sampling of data that indicates the status of the process. When the data consist of a continuous analog signal, sampling means that the continuous signal is substituted with a series of numerical values that represent the continuous signal at discrete moments in time. The same kind of substitution holds for discrete data, except that the number of possible numerical values the data can take on is more limited—certainly the case with binary data. We discuss the techniques by which continuous and discrete data are entered into and transmitted from the computer in Chapter 6. Other names for polling include *sampling* and *scanning*.

In some systems, the polling procedure simply requests whether any changes have occurred in the data since the last polling cycle and then collects only the new data from the process. This tends to shorten the cycle time required for polling. Issues related to polling include

1. *Polling frequency.* This is the reciprocal of the time interval between data collections.
2. *Polling order.* The polling order is the sequence in which the different data collection points of the process are sampled.
3. *Polling format.* This refers to the manner in which the sampling procedure is designed. The alternatives include (a) entering all new data from all sensors and other devices every polling cycle; (b) updating the control system only with data that have changed since the last polling cycle; or (c) using *high-level and low-level scanning,* or

conditional scanning, in which only certain key data are collected each polling cycle (high-level scanning), but if the data indicates some irregularity in the process, a low-level scan is undertaken to collect more complete data to ascertain the source of the irregularity.

These issues become increasingly critical with very dynamic processes in which changes in process status occur rapidly.

Interlocks.　An interlock is a safeguard mechanism for coordinating the activities of two or more devices and preventing one device from interfering with the other(s). In process control, interlocks provide a means by which the controller is able to sequence the activities in a work cell, ensuring that the actions of one piece of equipment are completed before the next piece of equipment begins its activity. Interlocks work by regulating the flow of control signals back and forth between the controller and the external devices.

There are two types of interlocks, input interlocks and output interlocks, where input and output are defined relative to the controller. An *input interlock* is a signal that originates from an external device (e.g., a limit switch, sensor, or production machine) and that is sent to the controller. Input interlocks can be used for either of the following functions:

1. To proceed with the execution of the work cycle program. For example, the production machine communicates a signal to the controller that it has completed its processing of the part. This signal constitutes an input interlock indicating that the controller can now proceed to the next step in the work cycle, which is to unload the part.
2. To interrupt the execution of the work cycle program. For example, while unloading the part from the machine, the robot accidentally drops the part. The sensor in its gripper transmits an interlock signal to the controller indicating that the regular work cycle sequence should be interrupted until corrective action is taken.

An *output interlock* is a signal sent from the controller to some external device. It is used to control the activities of each external device and to coordinate their operation with that of the other equipment in the cell. For example, an output interlock can be used to send a control signal to a production machine to begin its automatic cycle after the workpart has been loaded into it.

Interrupt System.　Closely related to interlocks is the interrupt system. As suggested by our discussion of input interlocks, there are occasions when it becomes necessary for the process or operator to interrupt the regular controller operation to deal with more pressing matters. All computer systems are capable of being interrupted, if nothing else, by turning off the power. A more sophisticated interrupt system is required for process control applications. An *interrupt system* is a computer control feature that permits the execution of the current program to be suspended to execute another program or subroutine in response to an incoming signal indicating a higher priority event. Upon receipt of an interrupt signal, the computer system transfers to a predetermined subroutine designed to deal with the specific interrupt. The status of the current program is remembered so that its execution can be resumed when servicing of the interrupt has been completed.

TABLE 5.4 Possible Priority Levels in an Interrupt System

Priority Level	Computer Function
1 (lowest priority)	Most operator inputs
2	System and program interrupts
3	Timer interrupts
4	Commands to process
5	Process interrupts
6 (highest priority)	Emergency stop (operator input)

Interrupt conditions can be classified as internal or external. *Internal interrupts* are generated by the computer system itself. These include timer-initiated events, such as polling of data from sensors connected to the process, or sending commands to the process at specific points in clock time. System- and program-initiated interrupts are also classified as internal because they are generated within the system. *External interrupts* are external to the computer system; they include process-initiated interrupts and operator inputs.

An interrupt system is required in process control because it is essential that more important programs (ones with higher priority) be executed before less important programs (ones with lower priorities). The system designer must decide what level of priority should be attached to each control function. A higher priority function can interrupt a lower priority function. A function at a given priority level cannot interrupt a function at the same priority level. The number of priority levels and the relative importance of the functions depend on the requirements of the individual process control situation. For example, emergency shutdown of a process because of safety hazards would occupy a very high priority level, even if it is an operator-initiated interrupt. Most operator inputs would have low priorities.

One possible organization of priority rankings for process control functions is shown in Table 5.4. Of course, the priority system may have more or fewer than the number of levels shown here, depending on the control situation. For example, some process interrupts may be more important than others, and some system interrupts may take precedence over certain process interrupts, thus requiring more than the six levels indicated in our table.

To respond to the various levels of priority defined for a given control application, an interrupt system can have one or more interrupt levels. A *single-level interrupt system* has only two modes of operation: normal mode and interrupt mode. The normal mode can be interrupted, but the interrupt mode cannot. This means that overlapping interrupts are serviced on a first-come, first-served basis, which could have potentially hazardous consequences if an important process interrupt was forced to wait its turn while a series of less important operator and system interrupts were serviced. A *multilevel interrupt system* has a normal operating mode plus more than one interrupt level. The normal mode can be interrupted by any interrupt level, but the interrupt levels have relative priorities that determine which functions can interrupt others. Example 5.1 illustrates the difference between the single-level and multilevel interrupt systems.

EXAMPLE 5.1 Single-Level Versus Multilevel Interrupt Systems

Three interrupts representing tasks of three different priority levels arrive for service in the reverse order of their respective priorities. Task 1 with the lowest priority, arrives first. Soon after, higher priority Task 2 arrives. And soon after

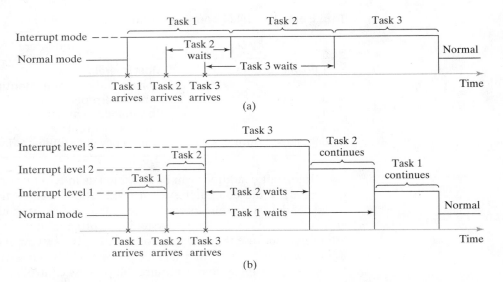

Figure 5.6 Response of the computer control system in Example 5.1 to three priority interrupts for (a) a single-level interrupt system and (b) a multilevel interrupt system. Task 3 is the highest level priority. Task 1 is the lowest level. Tasks arrive for servicing in the order 1, then 2, then 3. In (a), Task 3 must wait until Tasks 1 and 2 have been completed. In (b), Task 3 interrupts execution of Task 2, whose priority level allowed it to interrupt Task 1.

that, highest priority Task 3 arrives. How would the computer control system respond under (a) a single-level interrupt system and (b) a multilevel interrupt system?

Solution: The response of the system for the two interrupt systems is shown in Figure 5.6.

Exception Handling. In process control, an *exception* is an event that is outside the normal or desired operation of the process or control system. Dealing with the exception is an essential function in industrial process control and generally occupies a major portion of the control algorithm. The need for exception handling may be indicated through the normal polling procedure or by the interrupt system. Examples of events that may invoke exception handling routines include

- product quality problem,
- process variables operating outside their normal ranges,
- shortage of raw materials or supplies necessary to sustain the process,
- hazardous conditions such as a fire,
- controller malfunction.

In effect, exception handling is a form of error detection and recovery, discussed in the context of advanced automation capabilities (Section 4.2.3).

5.3.3 Forms of Computer Process Control

There are various ways in which computers can be used to control a process. First, we can distinguish between process monitoring and process control as illustrated in Figure 5.7. In process monitoring, the computer is used to simply collect data from the process, while in process control, the computer regulates the process. In some process control implementations, certain actions are implemented by the control computer that do not require feedback data to be collected from the process. This is open loop control. However, in most cases, some form of feedback or interlocking is required to ensure that the control instructions have been properly carried out. This more common situation is closed loop control.

In this section, we survey the various forms of computer process monitoring and control, all but one of which are commonly used in industry today. Direct digital control (DDC) represents a transitory phase in the evolution of computer process control technology. In its pure form, it is no longer used today. However, we briefly describe DDC to reveal the opportunities it contributed. Distributed control systems, often implemented using personal computers, are the most recent means of implementing computer process control.

Computer Process Monitoring.   Computer process monitoring is one of the ways in which the computer can be interfaced with a process. It involves the use of the computer to observe the process and associated equipment and to collect and record data from the operation. The computer is not used to directly control the process. Control remains in the hands of humans who use the data to guide them in managing and operating the process. The data collected by the computer in computer process monitoring can generally be classified into three categories:

1. *Process data.* These are measured values of input parameters and output variables that indicate process performance. When the values are found to indicate a problem, the human operator takes corrective action.

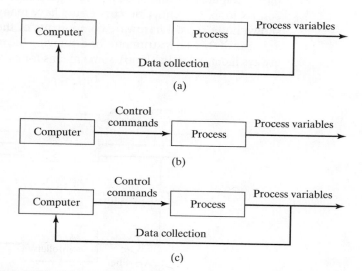

Figure 5.7   (a) process monitoring, (b) open loop process control, and (c) closed loop process control.

2. *Equipment data.* These data indicate the status of the equipment in the work cell. The data are used to monitor machine utilization, schedule tool changes, avoid machine breakdowns, diagnose equipment malfunctions, and plan preventive maintenance.

3. *Product data.* Government regulations require certain manufacturing industries to collect and preserve production data on their products. The pharmaceutical and medical supply industries are prime examples. Computer monitoring is the most convenient means of satisfying these regulations. A firm may also want to collect product data for its own use.

Collecting data from factory operations can be accomplished by any of several means. Shop data can be entered by workers through manual terminals located throughout the plant or can be collected automatically by means of limit switches, sensor systems, bar code readers, or other devices. Sensors are described in Chapter 6. Automatic identification and data collection technologies are discussed in Chapter 12. The collection and use of production data in factory operations for scheduling and tracking purposes is called *shop floor control*, covered in Chapter 25.

Direct Digital Control. DDC was certainly one of the important steps in the development of computer process control. Let us briefly examine this computer control mode and its limitations, which motivated improvements leading to modern computer control technology. DDC is a computer process control system in which certain components in a conventional analog control system are replaced by the digital computer. The regulation of the process is accomplished by the digital computer on a time-shared, sampled-data basis rather than by the many individual analog components working in a dedicated continuous manner. With DDC, the computer calculates the desired values of the input parameters and set points, and these values are applied through a direct link to the process, hence the name "direct digital" control.

The difference between direct digital control and analog control can be seen by comparing Figures 5.8 and 5.9. The first figure shows the instrumentation for a typical analog control loop. The entire process would have many individual control loops, but only one is shown here. Typical hardware components of the analog control loop include the sensor and transducer, an instrument for displaying the output variable (such an instrument is not always included in the loop), some means for establishing the set point of the loop (shown

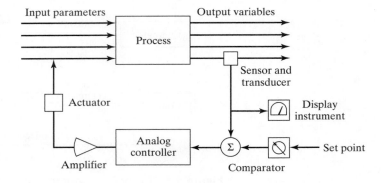

Figure 5.8   A typical analog control loop.

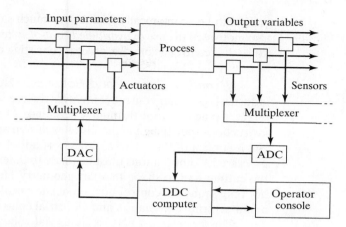

Figure 5.9 Components of a DDC system.

as a dial in the figure, suggesting that the setting is determined by a human operator), a comparator (to compare set point with measured output variable), the analog controller, an amplifier, and the actuator that determines the input parameter to the process.

In the DDC system (Figure 5.9), some of the control loop components remain unchanged, including (probably) the sensor and transducer as well as the amplifier and actuator. Components likely to be replaced in DDC include the analog controller, recording and display instruments, set point dials, and comparator. New components in the loop include the digital computer, analog-to-digital and digital-to-analog converters (ADCs and DACs), and multiplexers to share data from different control loops with the same computer.

DDC was originally conceived as a more efficient means of performing the same kinds of control actions as the analog components it replaced. However, the practice of simply using the digital computer to imitate the operation of analog controllers seems to have been a transitional phase in computer process control. Additional opportunities for the control computer were soon recognized, including:

- *More control options than traditional analog.* With digital computer control, it is possible to perform more complex control algorithms than with the conventional proportional-integral-derivative control modes used by analog controllers; for example, on/off control or nonlinearities in the control functions can be implemented.

- *Integration and optimization of multiple loops.* This is the ability to integrate feedback measurements from multiple loops and to implement optimizing strategies to improve overall process performance.

- *Ability to edit the control programs.* Using a digital computer makes it relatively easy to change the control algorithm when necessary by simply reprogramming the computer. Reprogramming the analog control loop is likely to require hardware changes that are more costly and less convenient.

These enhancements have rendered the original concept of direct digital control more or less obsolete. In addition, computer technology itself has progressed dramatically so that much smaller and less expensive yet more powerful computers are available for process control than the large mainframes available in the early 1960s. This has allowed computer process control to be economically justified for much smaller scale processes and equipment. It has also motivated the use of *distributed control systems,* in which a network of microcomputers is utilized to control a complex process consisting of multiple unit operations and/or machines.

Numerical Control and Robotics. Numerical control (NC) is another form of industrial computer control. It involves the use of the computer (again, a microcomputer) to direct a machine tool through a sequence of processing steps defined by a program of instructions specifying the details of each step and their sequence. The distinctive feature of NC is control of the relative position of a tool with respect to the object (workpart) being processed. Computations must be made to determine the trajectory that will be followed by the cutting tool to shape the part geometry. Hence, NC requires the controller to execute not only sequence control but geometric calculations as well. Because of its importance in manufacturing automation and industrial control, NC is covered in detail in Chapter 7.

Closely related to NC is industrial robotics, in which the joints of the manipulator (robot arm) are controlled to move the end of the arm through a sequence of positions during the work cycle. As in NC, the controller must perform calculations during the work cycle to implement motion interpolation, feedback control, and other functions. In addition, a robotic work cell usually includes other equipment besides the robot, and the activities of the other equipment in the work cell must be coordinated with those of the robot. This coordination is achieved using interlocks. We discuss industrial robotics in Chapter 8.

Programmable Logic Controllers. Programmable logic controllers (PLCs) were introduced around 1970 as an improvement on the electromechanical relay controllers used at the time to implement discrete control in the discrete manufacturing industries. The evolution of PLCs has been facilitated by advances in computer technology, and present-day PLCs are capable of much more than the 1970s controllers. We can define a modern *programmable logic controller* as a microprocessor-based controller that uses stored instructions in programmable memory to implement logic, sequencing, timing, counting, and arithmetic control functions for controlling machines and processes. Today's PLCs are used for both continuous control and discrete control applications in both the process industries and discrete manufacturing. We cover PLCs and the kinds of control they are used to implement in Chapter 9.

Supervisory Control. The term *supervisory control* is usually associated with the process industries, but the concept applies equally well to discrete manufacturing automation, where it corresponds to cell or system level control. Supervisory control represents a higher level of control than DDC, NC, and PLCs. In general, these other types of control systems are interfaced directly to the process. By contrast, supervisory control is often superimposed on these process-level control systems and directs their operations. The relationship between supervisory control and the process-level control techniques is illustrated in Figure 5.10.

In the context of the process industries, *supervisory control* denotes a control system that manages the activities of a number of integrated unit operations to achieve certain economic objectives for the process. In some applications, supervisory control is not much more than regulatory control or feedforward control. In other applications, the supervisory control system is designed to implement optimal or adaptive control. It seeks to optimize some well-defined objective function, which is usually based on economic criteria such as yield, production rate, cost, quality, or other objectives that pertain to process performance.

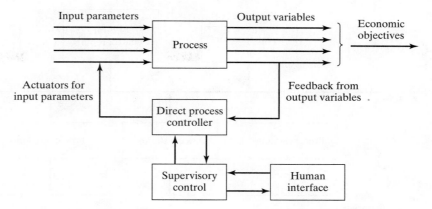

Figure 5.10 Supervisory control superimposed on other process-level control systems.

In the context of discrete manufacturing, *supervisory control* can be defined as the control system that directs and coordinates the activities of several interacting pieces of equipment in a manufacturing cell or system, such as a group of machines interconnected by a material handling system. Again, the objectives of supervisory control are motivated by economic considerations. The control objectives might include minimizing part or product costs by determining optimum operating conditions, maximizing machine utilization through efficient scheduling, or minimizing tooling costs by tracking tool lives and scheduling tool changes.

It is tempting to conceptualize a supervisory control system as being completely automated, so that the system operates with no human interference or assistance. But in virtually all cases, supervisory control systems are designed to allow interaction with human operators, and the responsibility for control is shared between the controller and the human. The relative proportions of responsibility differ, depending on the application.

Distributed Control Systems.   With the development of the microprocessor, it became feasible to connect multiple microcomputers together to share and distribute the process control workload. The term *distributed control system* (DCS) is used to describe such a configuration, which consists of the following components and features [8]:

- Multiple process control stations located throughout the plant to control the individual loops and devices of the process.
- A central control room equipped with operator stations, where supervisory control of the plant occurs.
- Local operator stations distributed throughout the plant. This provides the DCS with redundancy. If a control failure occurs in the central control room, the local operator stations take over the central control functions. If a local operator station fails, the other local operator stations assume the functions of the failed station.
- All process and operator stations interact with each other by means of a communications network, or data highway, as it is often called.

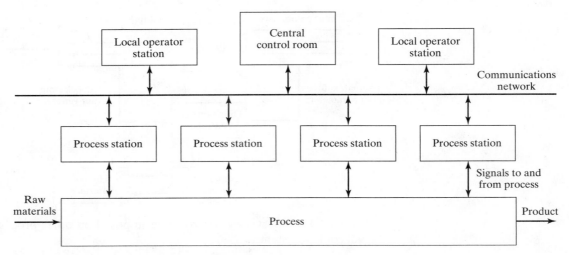

Figure 5.11   Distributed control system.

These components are illustrated in a typical configuration of a distributed process control system presented in Figure 5.11. There are a number of benefits and advantages of the DCS: (1) A DCS can be installed for a given application in a very basic configuration, then enhanced and expanded as needed in the future; (2) since the system consists of multiple computers, this facilitates parallel multitasking; (3) because of its multiple computers, a DCS has built-in redundancy; (4) control cabling is reduced compared with a central computer control configuration; and (5) networking provides process information throughout the enterprise for more efficient plant and process management.

Development of DCSs started around 1970. One of the first commercial systems was Honeywell's TDC 2000, introduced in 1975 [1]. The first DCS applications were in the process industries. In the discrete manufacturing industries, programmable logic controllers were introduced about the same time. The concept of distributed control applies equally well to PLCs; that is, multiple PLCs located throughout a factory to control individual pieces of equipment but integrated by means of a common communications network. Introduction of the PC shortly after the DCS and PLC, and its subsequent increase in computing power and reduction in cost over the years, have stimulated a significant growth in the adoption of PC-based DCSs for process control applications.

PCs in Process Control.   Today, PCs dominate the computer world. They have become the standard tool by which business is conducted, whether in manufacturing or in the service sector. Thus, it is no surprise that PCs are being used in growing numbers in process control applications. Two basic categories of PC implementations in process control can be distinguished: (1) operator interface and (2) direct control. Whether used as the operator interface or for direct control, PCs are likely to be networked with other computers to create distributed control systems.

When used as the operator interface, the PC is interfaced to one or more PLCs or other devices (possibly other microcomputers) that directly control the process. Personal computers have been used to perform the operator interface function since the early 1980s. In this function, the computer performs certain monitoring and supervisory control functions, but it does not directly control the process. Some advantages of using a PC only as the operator interface are that (1) the PC provides a user-friendly interface for the operator; (2) the PC can be used for all of the conventional computing and data processing functions that PCs traditionally perform; (3) the PLC or other device that is directly controlling the process is isolated from the PC, so a PC failure will not disrupt control of the process; and (4) the computer can be easily upgraded as PC technology advances and capabilities improve, while the PLC control software and connections with the process can remain in place.

The second way of implementing PCs in process control is *direct control*, which means that the PC is interfaced directly to the process and controls its operations in real time. The traditional thinking has been that it is too risky to permit the PC to directly control the production operation. If the computer were to fail, the uncontrolled operation might stop working, produce a defective product, or become unsafe. Another factor is that conventional PCs, equipped with the usual business-oriented operating system and applications software, are designed for computing and data processing functions, not for process control. They are not intended to be interfaced with an external process in the manner necessary for real-time process control. Finally, most PCs are designed to be used in an office environment, not in the harsh factory atmosphere.

Recent advances in both PC technology and available software have challenged this traditional thinking. Starting in the early 1990s, PCs have been installed at an accelerating pace for direct control of industrial processes. Several factors have enabled this trend:

- Widespread familiarity with PCs. User-friendly software for the home and business has certainly contributed to the popularity of PCs. There is a growing expectation among workers that they will be provided with a computer in their workplace, even if that workplace is in the factory.
- Availability of high performance PCs, capable of satisfying the demanding requirements of process control (Section 5.3.1).
- Trend toward *open architecture philosophy* in control systems design, in which vendors of control hardware and software agree to comply with published standards that allow their products to be interoperable. This means that components from different vendors can be interconnected in the same system. The traditional philosophy had been for each vendor to design proprietary systems, requiring the user to purchase the complete hardware and software package from one supplier. Open architecture allows the user a wider choice of products in the design of a given process control application.
- Availability of PC operating systems that facilitate real-time control, multitasking, and networking. At the same time, these systems provide the user friendliness of the desktop PC and most of the power of an engineering workstation. Installed in the factory, a PC equipped with the appropriate software can perform multiple functions simultaneously, such as data logging, trend analysis, and displaying an animated view of the process as it proceeds, all while reserving a portion of its CPU capacity for direct control of the process.

Regarding the factory environment issue, this can be addressed by using industrial-grade PCs, which are equipped with enclosures designed for the rugged plant environment. Compared with the previously discussed PC/PLC configuration, in which the PC is used only as the operator interface, there is a cost savings from installing one PC for direct control rather than a PC plus a PLC. A related issue is data integration: Setting up a data link between a PC and a PLC is more complex than when the data are all in one PC.

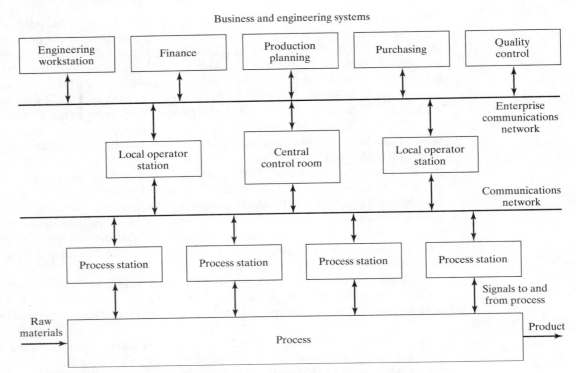

Business and engineering systems

Figure 5.12   Enterprise-wide PC-based DCS.

Enterprise-Wide Integration of Factory Data.   The most recent progression in PC-based distributed control is enterprise-wide integration of factory operations data, as depicted in Figure 5.12. This trend is consistent with modern information management and worker empowerment philosophies. These philosophies assume fewer levels of company management and greater responsibilities for front-line workers in sales, order scheduling, and production. The networking technologies that allow such integration are available. The latest PC operating systems provide a number of built-in and optional features for connecting the industrial control system in the factory to enterprise-wide business systems and supporting data exchange between various applications (e.g., allowing data collected in the plant to be used in analysis packages, such as spreadsheets). The term *enterprise resource planning* (ERP) refers to a computer software system that achieves company-wide integration of not only factory data but of all the other data required to execute the business functions of the organization. A key feature of ERP is the use of a single central database that can be accessed from anywhere in the company. Some of the details of ERP are discussed in Chapter 25 (Section 25.6.2).

Following are some of the capabilities that are enabled by making process data available throughout the enterprise:

1. Managers can have more direct access to factory floor operations.

2. Production planners can use the most current data on times and production rates in scheduling future orders.

3. Sales personnel can provide realistic estimates on delivery dates to customers, based on current shop loading.

4. Order trackers are able to provide inquiring customers with current status information on their orders.

5. Quality control personnel are made aware of real or potential quality problems on current orders, based on access to quality performance histories from previous orders.

6. Cost accounting has access to the most recent production cost data.

7. Production personnel can access part and product design details to clarify ambiguities and do their job more effectively.

REFERENCES

[1] ASTROM, K. J., and WITTENMARK, B., *Computer-Controlled Systems—Theory and Design*, 3d ed., Prentice Hall, Upper Saddle River, NJ, 1997.

[2] BATESON, R. N., *Introduction to Control System Technology*, 7th ed., Prentice Hall, Upper Saddle River, NJ, 2002.

[3] BOUCHER, T. O., *Computer Automation in Manufacturing*, Chapman & Hall, London, UK, 1996.

[4] CAWLFIELD, D., "PC-Based Direct Control Flattens Control Hierarchy, Opens Information Flow," *Instrumentation & Control Systems*, September 1997, pp 61–67.

[5] GROOVER, M. P., "Industrial Control Systems," *Maynard's Industrial Engineering Handbook*, 5th ed., K. Zandin (ed.), McGraw-Hill Book Company, NY, 2001.

[6] HIRSH, D., "Acquiring and Sharing Data Seamlessly," *Instrumentation and Control Systems*, October 1997, pp. 25–35.

[7] OLSSON, G., and G. PIANI, *Computer Systems for Automation and Control*, Prentice Hall, London, UK, 1992.

[8] PLATT, G., *Process Control: A Primer for the Nonspecialist and the Newcomer*, 2d ed., Instrument Society of America, Research Triangle Park, NC, 1998.

[9] RULLAN, A., "Programmable Logic Controllers versus Personal Computers for Process Control," *Computers and Industrial Engineering*, Vol. 33, Nos. 1–2, 1997, pp. 421–424.

[10] STENERSON, J., *Fundamentals of Programmable Logic Controllers, Sensors, and Communications*, 3d ed., Pearson/Prentice Hall, Upper Saddle River, NJ, 2004.

REVIEW QUESTIONS

5.1 What is industrial control?

5.2 What is the difference between a continuous variable and a discrete variable?

5.3 Name and briefly define each of the three different types of discrete variables.

5.4 What is the difference between a continuous control system and a discrete control system?

5.5 What is feedforward control?

5.6 What is adaptive control?

5.7 What are the three functions of adaptive control?

5.8 What is the difference between an event-driven change and a time-driven change in discrete control?

5.9 What are the two basic requirements that must be managed by the controller to achieve real-time control?

5.10 What is polling in computer process control?

5.11 What is an interlock? What are the two types of interlocks in industrial control?

5.12 What is an interrupt system in computer process control?

5.13 What is computer process monitoring?

5.14 What is direct digital control (DDC), and why is it no longer used in industrial process control applications?

5.15 Are programmable logic controllers (PLCs) more closely associated with the process industries or the discrete manufacturing industries?

5.16 What is a distributed control system?

5.17 What is the open architecture philosophy in control systems design?

Chapter 6

Hardware Components for Automation and Process Control

CHAPTER CONTENTS

To implement automation and process control, the control computer must collect data from and transmit signals to the production process. In Section 5.1.2, process variables and parameters were classified as continuous or discrete, with several subcategories in the discrete class. The digital computer operates on digital (binary) data, whereas at least some of the data from the process are continuous and analog. Accommodations for this difference must be made in the computer-process interface. The components required to implement this interface are the following:

1. Sensors for measuring continuous and discrete process variables.
2. Actuators that drive continuous and discrete process parameters.
3. Devices that convert continuous analog signals to digital data.

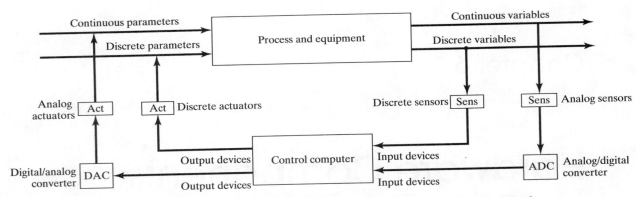

Figure 6.1 The computer process control system, showing the various types of components required to interface the process with the computer.

4. devices that convert digital data into analog signals
5. input/output devices for discrete data.

Figure 6.1 shows the overall configuration of the process control system and how these five component categories are used to interface the process with the computer. This model represents the general arrangement of most of the material handling systems and manufacturing systems described in Chapters 10 through 19. Our present chapter is organized around the five categories of components.

6.1 SENSORS

A wide variety of sensors is available for collecting data from the manufacturing process for use in feedback control. A sensor is a *transducer*, which is a device that converts a physical variable of one form into another form that is more useful for the given application. In particular, a *sensor* is a device that converts a physical stimulus or variable of interest (such as temperature, force, pressure, or displacement) into a more convenient form (usually an electrical quantity such as voltage) for the purpose of measuring the stimulus. The conversion process quantifies the variable, so that it can be interpreted as a numerical value.

Sensors can be classified in various ways, the most relevant of which for our treatment is by category of stimulus or physical variable that is to be measured, as presented in Table 6.1. For each category, there may be multiple variables that can be measured, as indicated in the right-hand column. These variables are typical of those found in industrial processes.

In addition to the type of stimulus, sensors are also classified as analog or discrete, consistent with our classification of process variables in Chapter 5. An *analog* measuring device produces a continuous analog signal such as electrical voltage, whose value varies in an analogous manner with the variable being measured. Examples are thermocouples, strain gages, and potentiometers. The output signal from an analog measuring device must be converted to digital data by an analog-to-digital converter (Section 6.3) in order to be used by a digital computer.

TABLE 6.1 Stimulus Categories and Associated Physical Variables

Stimulus Category	Examples of Physical Variables
Mechanical	Position (displacement, linear and angular), velocity, acceleration, force, torque, pressure, stress, strain, mass, density
Electrical	Voltage, current, charge, resistance, conductivity, capacitance
Thermal	Temperature, heat, heat flow, thermal conductivity, specific heat
Radiation	Type of radiation (e.g., gamma rays, X-rays, visible light), intensity, wavelength
Magnetic	Magnetic field, flux, conductivity, permeability
Chemical	Component identities, concentration, pH levels, presence of toxic ingredients, pollutants

Sources: Based on similar tables in [6] and [7].

A *discrete* measuring device produces an output that can have only certain values. Discrete sensor devices are often divided into two categories: binary and digital. A *binary* measuring device produces an on/off signal. The most common devices operate by closing an electrical contact from a normally open position. Limit switches operate in this manner. Other binary sensors include photoelectric sensors and proximity switches. A *digital* measuring device produces a digital output signal, either in the form of a set of parallel status bits (e.g., a photoelectric sensor array) or a series of pulses that can be counted (e.g., an optical encoder). In either case, the digital signal represents the quantity to be measured. Digital transducers are becoming increasingly common because they are easy to read when used as stand-alone measuring instruments and because they are compatible with digital computer systems. Many of the common sensors and measuring devices used in industrial control systems are listed alphabetically in Table 6.2. A significant trend in sensor technology has been the development of very small sensors. The term *microsensor* refers to measuring devices whose physical features have dimensions in the micron range, where 1 micron ($1\ \mu$m) $= 10^{-6}$ m. Microsensors are usually fabricated out of silicon using processing techniques associated with integrated circuit manufacture.

Sensors are distinguished as active or passive. An *active sensor* responds to the stimulus without the need for any external power. An example is a thermocouple, which responds to an increase in temperature by generating a small voltage (millivolt range) that is functionally related to temperature (in the ideal, its voltage is directly proportional to temperature). A *passive sensor* requires an external source of power in order to operate. A thermistor illustrates this case. It also measures temperature, but its operation requires an electric current to be passed through it. As the temperature increases, the thermistor's electrical resistance is altered. The resistance can be measured, and related back to temperature.

For each sensor, there is a *transfer function*, which is the relationship between the value of the physical stimulus and the value of the signal produced by the sensor in response to the stimulus. The transfer function is the input/output relationship. The stimulus is the input, and the signal generated by the device is the output. The transfer function can be simply expressed as follows:

$$S = f(s) \tag{6.1}$$

where $S =$ the output signal (usually voltage), $s =$ the stimulus, and $f(s)$ is the functional relationship between them.

TABLE 6.2 Common Measuring Devices Used in Automation

Measuring Device	Description
Accelerometer	Analog device used to measure vibration and shock. Can be based on various physical phenomena (e.g., capacitive, piezoresistive, piezoelectric).
Ammeter	Analog device that measures the strength of an electrical current.
Bimetallic switch	Binary switch that uses a bimetallic coil to open and close electrical contact as a result of temperature change. A *bimetallic coil* consists of two metal strips of different thermal expansion coefficients bonded together.
Bimetallic thermometer	Analog temperature measuring device consisting of bimetallic coil (see definition above) that changes shape in response to temperature change. Shape change of coil can be calibrated to indicate temperature.
Dynamometer	Analog device used to measure force, power, or torque. Can be based on various physical phenomena (e.g., strain gage, piezoelectric effect).
Float transducer	Float attached to lever arm. Pivoting movement of lever arm can be used to measure liquid level in vessel (analog device) or to activate contact switch (binary device).
Fluid flow sensor	Analog measurement of liquid flow rate, usually based on pressure difference between flow in two pipes of different diameter.
Fluid flow switch	Binary switch similar to limit switch but activated by increase in fluid pressure rather than by contacting object.
Linear variable differential transformer	Analog position sensor consisting of primary coil opposite two secondary coils separated by a magnetic core. When primary coil is energized, induced voltage in secondary coil is function of core position. Can also be adapted to measure force or pressure.
Limit switch (mechanical)	Binary contact sensor in which lever arm or pushbutton closes (or opens) an electrical contact.
Manometer	Analog device used to measure pressure of gas or liquid. Based on comparison of known and unknown pressure forces. A *barometer* is a specific type of manometer used to measure atmospheric pressure.
Ohmmeter	Analog device that measures electrical resistance.
Optical encoder	Digital device used to measure position and/or speed, consisting of a slotted disk separating a light source from a photocell. As disk rotates, photocell senses light through slots as a series of pulses. Number and frequency of pulses are proportional (respectively) to position and speed of shaft connected to disk. Can be adapted for linear as well as rotational measurements. (The optical encoder is described in more detail in Section 7.5.2 on numerical control positioning systems.)
Photoelectric sensor array	Digital sensor consisting of linear series of photoelectric switches. Array is designed to indicate height or size of object interrupting some but not all of the light beams.
Photoelectric switch	Binary noncontact sensor (switch) consisting of emitter (light source) and receiver (photocell) triggered by interruption of light beam. Two common types: (1) *transmitted type,* in which object blocks light beam between emitter and receiver; and (2) *retroreflective type,* in which emitter and receiver are located in one device and beam is reflected off remote reflector except when object breaks the reflected light beam.
Photometer	Analog sensor that measures illumination and light intensity. Can be based on various photodetector devices, including photodiodes, phototransistors, and photoresistors.
Piezoelectric transducer	Analog device based on piezoelectric effect of certain materials (e.g., quartz) in which an electrical charge is produced when the material is deformed. Charge can be measured and is proportional to deformation. Can be used to measure force, pressure, and acceleration.
Potentiometer	Analog position sensor consisting of resistor and contact slider. Position of slider on resistor determines measured resistance. Available for both linear and rotational (angular) measurements.

(continued)

TABLE 6.2 Common Measuring Devices Used in Automation *(continued)*

Measuring Device	Description
Proximity switch	Binary noncontact sensor is triggered when nearby object induces changes in electromagnetic field. Can be based on any of several physical principles, including inductance, capacitance, ultrasonics, and optics.
Radiation pyrometer	Analog temperature-measuring device that senses electromagnetic radiation in the visible and infrared range of spectrum.
Resistance-temperature detector	Analog temperature-measuring device based on increase in electrical resistance of a metallic material as temperature is increased.
Strain gage	Widely used analog sensor to measure force, torque, or pressure. Based on change in electrical resistance resulting from strain of a conducting material.
Tachometer	Analog device consisting of DC generator that produces an electrical voltage proportional to rotational speed.
Tactile sensor	Measuring device that indicates physical contact between two objects. Can be based on any of several physical devices such as electrical contact (for conducting materials) and piezoelectric effect.
Thermistor	Contraction of *thermal* and *resistor*. Analog temperature-measuring device based on change in electrical resistance of a semiconductor material as temperature is increased.
Thermocouple	Analog temperature-measuring device based on thermoelectric effect, in which the junction of two dissimilar metal wires emits a small voltage that is a function of the temperature of the junction. Common standard thermocouples include: chromel-alumel, iron-constantan, and chromel-constantan.
Ultrasonic range sensor	Time lapse between emission and reflection (from object) of high-frequency sound pulses is measured. Can be used to measure distance or simply to indicate presence of object.

Limit switches and other binary sensors have functional relationships that are binary, defined by the following expressions:

$$S = 1 \text{ if } s > 0 \text{ and } S = 0 \text{ if } s \leq 0 \tag{6.2}$$

The ideal functional form for an analog measuring device is a simple proportional relationship, such as

$$S = C + ms \tag{6.3}$$

where C is the output value at a stimulus value of zero, and m is the constant of proportionality between s and S. The constant m can be thought of as the *sensitivity* of the sensor. It is a measure of how much the output or response of the sensor is affected by the stimulus. For example, the sensitivity of a standard Chromel/Alumel thermocouple generates 40.6 microvolts (μV) per degree Celsius (°C). Other transfer functions have more complex mathematical forms, including differential equations that include time dynamics, which means that there is a time delay between when the stimulus occurs and when the output signal accurately indicates the value of the stimulus.

Before using any measuring device, the operator must *calibrate* it to determine the transfer function, or the inverse of the transfer function, which converts the output S into the value of the stimulus or measured variable s. The ease with which the calibration procedure can be accomplished is one criterion by which a measuring device can be evaluated. A list of desirable features of measuring devices for process control is presented in Table 6.3. Few measuring devices achieve perfect scores in all of these

TABLE 6.3 Desirable Features for Selecting Measuring Devices Used in Automated Systems

Desirable Feature	Definition and Comments
High accuracy	The measurement contains small systematic errors about the true value.
High precision	The random variability or noise in the measured value is low.
Wide operating range	The measuring device possesses high accuracy and precision over a wide range of values of the physical variable being measured.
High speed of response	The device responds-quickly to changes in the physical variable being measured. Ideally, the time lag would be zero.
Ease of calibration	Calibration of the measuring device is quick and easy.
Minimum drift	Drift refers to the gradual loss in accuracy over time. High drift requires frequent recalibration of the measuring device.
High reliability	The device is not subject to frequent malfunctions or failures during service. It is capable of operating in the potentially harsh environment of the manufacturing process where it will be applied.
Low cost	The cost to purchase (or fabricate) and install the measuring device is low relative to the value of the data provided by the sensor.

criteria, and the control system engineer must decide which features are the most important in selecting among the variety of available sensors and transducers for a given application.

6.2 ACTUATORS

In industrial control systems, an actuator is a hardware device that converts a controller command signal into a change in a physical parameter. The change in the physical parameter is usually mechanical, such as a position or velocity change. An actuator is a transducer, because it changes one type of physical quantity, such as electric current, into another type of physical quantity, such as rotational speed of an electric motor. The controller command signal is usually low level, and so an actuator may also require an *amplifier* to strengthen the signal sufficiently to drive the actuator.

Most actuators can be classified into one of three categories, according to the type of amplifier: (1) electrical, (2) hydraulic, and (3) pneumatic. *Electrical actuators* are most common; they include electric motors of various kinds, stepper motors, and solenoids. Electrical actuators can be either linear (output is linear displacement) or rotational (output is angular displacement). *Hydraulic actuators* use hydraulic fluid to amplify the controller command signal. The available devices provide either linear or rotational motion. Hydraulic actuators are often specified when large forces are required. *Pneumatic actuators* use compressed air (typically "shop air" in the factory) as the driving power. Again, both linear and rotational pneumatic actuators are available. Because of the relatively low air pressures involved, these actuators are usually limited to relatively low force applications compared with hydraulic actuators.

This section is organized into two topics: (1) electric motors, and (2) other types of actuators, including some that are powered electrically. Our coverage is not comprehensive. The purpose is to provide an introductory survey of the different types of actuators available to implement automation and process control. More complete coverage can be found in several of our references, including [2], [3], and [11].

6.2.1 Electric Motors

An electric motor converts electrical power into mechanical power. Most electric motors are rotational, and their operation can be explained with reference to Figure 6.2. The motor consists of two basic components, a stator and a rotor. The *stator* is the ring-shaped stationary component, and the *rotor* is the cylindrical part that rotates inside the stator. The rotor is assembled around a shaft that is supported by bearings, and the shaft can be coupled to machinery components such as gears, pulleys, lead screws, or spindles. Electric current supplied to the motor generates a continuously switching magnetic field that causes the rotor to rotate in its attempt to always align its poles with the opposite poles of the stator. The details relating to type of current (alternating or direct), how the continuously switching magnetic field is created, and other aspects of the motor's construction give rise to a great variety of electrical motors. The simplest and most commonly used classification is between direct current (DC) motors and alternating current (AC) motors. Within each category, there are several subcategories. Here we discuss three types that are widely used in automation and industrial control: (1) DC motors, (2) AC motors, and (3) stepper motors.

   DC Motors.   DC motors are powered by a constant current and voltage. The continuously switching magnetic field is achieved by means of a rotary switching device, called a *commutator*, which rotates with the rotor and picks up current from a set of carbon brushes that are components of the stator assembly. Its function is to continually change the relative polarity between the rotor and the stator, so that the magnetic field produces a torque to continuously turn the rotor. The use of a commutator represents the traditional construction of a DC motor. This construction is a disadvantage because it results in arcing, worn brushes, and maintenance problems. A special type of DC motor avoids the use of the commutator and brushes. Called a *brushless DC motor*, it uses solid-state circuitry to replace the brushes and commutator components. Elimination of these parts has the added benefit of reducing the inertia of the rotor assembly, allowing higher speed operation.

   DC motors are widely used for two reasons. The first is the convenience of using direct current as the power source. For example, the small electric motors in automobiles are DC because the car's battery supplies direct current. The second reason for the traditional popularity of DC motors is that their torque-speed relationships are attractive in many applications compared to AC motors.

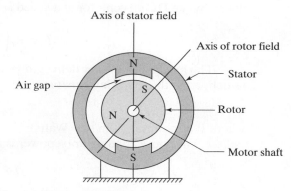

Figure 6.2   A rotating electric motor.

DC servomotors are a common type of DC motor used in mechanized and automated systems, and we will use it to represent this class of electric motors. The term *servomotor* simply means that a feedback loop is used to achieve speed control. In a DC servomotor, the stator typically consists of two permanent magnets on opposite sides of the rotor. The rotor, called the *armature* in a DC motor, consists of three sets of copper wire windings around a ferrous metal core. Input current is provided to the windings through the commutator and interacts with the magnetic field of the stator to produce the torque that drives the rotor. The magnitude of the rotor torque is a function of the current passing through the windings, and the relationship can be modeled by the following equation:

$$T = K_t I_a \tag{6.4}$$

where T = motor torque, N-m; I_a = net current flowing through the armature, A; and K_t = the motor's torque constant, N-m/A. The reason for defining I_a as the net current will be explained next.

Rotating the armature in the magnetic field of the stator produces a voltage across the armature terminals, called the back-emf. In effect, the motor acts like a generator, and the back-emf increases with rotational speed as follows:

$$E_b = K_v \omega \tag{6.5}$$

where E_b = back-emf, V; ω = angular velocity, rad/sec; and K_v = the voltage constant of the motor, V/(rad/sec). The effect of the back-emf is to reduce the current flowing through the armature windings. The angular velocity in rad/sec can be converted to the more familiar rotational speed as follows:

$$N = \frac{60\omega}{2\pi} \tag{6.6}$$

where N = rotational speed, rev/min (rpm).

Given the resistance of the armature R_a and an input voltage V_{in} supplied to the motor terminals, the resulting armature current $I_a = V_{in}/R_a$. This is the starting current and it produces a starting torque as given by Eq. (6.4). But as the armature begins to rotate, it generates the back-emf E_b, which reduces the available voltage. Thus, the actual armature current depends on the rotational speed of the rotor,

$$I_a = \frac{V_{in} - E_b}{R_a} = \frac{V_{in} - K_v \omega}{R_a} \tag{6.7}$$

where all of the terms are defined above. Combining Eqs. (6.4) and (6.7), the torque produced by the DC servomotor at a speed ω is

$$T = K_t \left(\frac{V_{in} - K_v \omega}{R_a} \right) \tag{6.8}$$

The mechanical power delivered by the motor is the product of torque and velocity, as defined in the following equation:

$$P = T\omega \tag{6.9}$$

where P = power in N-m/sec (Watts); T = motor torque, N-m; and ω = angular velocity, rad/sec. The corresponding horsepower is given by

$$HP = \frac{T\omega}{745.7} \tag{6.10}$$

where the constant 745.7 is the conversion factor 745.7 W = 1 hp.

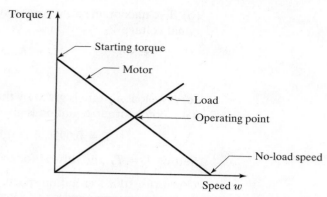

Figure 6.3 Torque-speed curve of a DC servomotor (idealized), and typical load torque relationship. The intersection of the two plots is the operating point.

The servomotor is connected either directly or through a gear reduction to a piece of machinery. The machinery may be a fan, pump, spindle, table drive, or similar mechanical apparatus. The apparatus represents the load that is driven by the motor. The load requires a certain torque to operate, and the torque is usually related to rotational speed in some way. In general, the torque increases with speed. In the simplest case, the relationship is proportional:

$$T_L = K_L\omega \tag{6.11}$$

where T_L = load torque, N-m; and K_L = the constant of proportionality between torque and angular velocity, N-m/(rad/sec). The functionality between K_L and T_L may be other than proportional, such that K_L itself depends on the angular velocity. For example, the torque required to drive a fan increases approximately as the square of the rotational speed, that is, $T_L \propto \omega^2$.

The torque developed by the motor and the torque required by the load must be balanced. That is, $T = T_L$ in steady state operation and this amount of torque is called the *operating point*. The motor torque relationship with angular velocity can be plotted as shown in Figure 6.3, called the *torque-speed curve*. Also shown in the figure is the load torque relationship. The intersection of the two plots is the operating point, which is defined by the values of torque and angular velocity.

EXAMPLE 6.1   DC servomotor operation

A DC servomotor has a torque constant K_t = 0.095 N-m/A. Its voltage constant K_v = 0.11 V/(rad/sec). The armature resistance is R_a = 1.6 ohms. A terminal voltage of 24 V is used to operate the motor. Determine (a) the starting torque generated by the motor just as the voltage is applied (this is referred to as the *stall torque*), (b) the maximum speed at a torque of zero, and (c) the operating point of the motor when it is connected to a load whose torque characteristic is given by $T_L = K_L\omega$ and K_L = 0.007 N-m/(rad/sec). Give the rotational speed in rev/min.

Solution: (a) At $\omega = 0$, the armature current is $I_a = V_{in}/R_a$ = 24/1.6 = 15 A. The corresponding torque is therefore $T = K_t I_a$ = 0.095(15) = 1.425 N-m

(b) The maximum speed is achieved when the back-emf E_b is equal to the terminal voltage V_{in}.

$$E_b = K_v \omega = 0.11 \omega = 24 \text{ V}$$

$$\omega = 24/0.11 = 218.2 \text{ rad/sec}$$

(c) The load torque is given by the equation $T_L = 0.007 \omega$
The motor torque equation is given by Eq. (6.8). Using the given data,

$$T = 0.095(24 - 0.11\omega)/1.6 = 1.425 - 0.00653\omega$$

Setting $T = T_L$ and solving for ω, we find $\omega = 105.3 \text{ rad/sec}$

Converting this to rotation speed, $N = 60(105.3)/2\pi = 1006 \text{ rev/min}$

EXAMPLE 6.2   DC servomotor power

In the previous example, what is the power delivered by the motor at the operating point? Give the answer in (a) Watts and (b) horsepower.

Solution: (a) At $\omega = 105.3$ rad/sec, and using the load torque equation,
$$T_L = 0.007 (105.3) = 0.737 \text{ N-m}$$
Power $P = T\omega = 0.737(105.3) = 77.6 \text{ W}$

(b) Horsepower $HP = 77.6/745.7 = 0.104 \text{ hp}$

Our model of DC servomotor operation neglects certain losses and inefficiencies that occur in these motors (similar losses occur in all electric motors). These losses include brush contact losses at the commutator, armature losses, windage (air drag losses at high rotational speeds of the rotor), and mechanical friction losses at the bearings. Our model also neglects the dynamics of motor operation. In fact, the inertial characteristics of the motor itself and the load that is driven by it, as well as any transmission mechanisms (e.g., gear box), would play an important role in determining how the motor operates as a function of time. Despite their limitations, the equations do illustrate one of the significant advantages of a DC servomotor, its ability to deliver a very high torque at a starting velocity of zero. In addition, it is a variable-speed motor, and its direction of rotation can be readily reversed. These are important considerations in many automation applications where the motor is called upon to frequently start and stop its rotation or to reverse direction.

AC Motors.   Although DC motors have a number of attractive features, they also have two important disadvantages: (1) the commutator and brushes used to conduct current from the stator assembly to the rotor result in maintenance problems with these motors, and (2) the most common electrical power source in industry is alternating current, not direct current. In order to use AC power to drive a DC motor, a rectifier must be added to convert the alternating current to direct current. For these reasons, AC motors are widely used in many industrial applications. They do not use brushes, and they are compatible with the predominant type of electrical power.

Alternating current motors operate by generating a rotating magnetic field in the stator, where the speed of rotation depends on the frequency of the input electrical power.

The rotor is forced to turn at the same speed as the rotating magnetic field. AC motors can be classified into two broad categories: induction motors and synchronous motors. AC *induction motors* are probably the most widely used motors in the world, due to their relatively simple construction and low manufacturing cost. In the operation of this motor type, a magnetic field is induced by the rotation of the rotor through the magnetic field of the stator. Because of this feature, the rotor in most induction motors does not need an external source of electrical power. AC *synchronous motors* operate by energizing the rotor with alternating current, which generates a magnetic field in the gap separating the rotor and the stator. This magnetic field creates a torque that turns the rotor at the same rotational speed as the magnetic forces in the stator. Synchronous motors are somewhat more complex than induction motors because they require a device called an *exciter* to initiate rotation of the rotor when power is first supplied to the motor. The exciter accelerates the rotational speed of the rotor to synchronize with that of the stator's rotating magnetic field, which is a required condition for an AC synchronous motor to function.

Both induction motors and synchronous motors operate at a constant speed that depends on the frequency of the incoming electrical power. Their applications are usually those in which running at a fixed speed is a requirement. This is a disadvantage in many automation applications because frequent speed changes are often necessary with much starting and stopping. The speed issue is sometimes addressed by using adjustable-frequency drives (called *inverters*) that control the cycle rate of the AC power to the motor. Motor speed is proportional to frequency, so changing frequency changes motor speed. Advances in solid-state electronics have also improved speed control for AC motors, and they are now competitive in some applications traditionally reserved for DC motors.

Stepper Motors. Also called *step motors* and *stepping motors*, this motor class provides rotation in the form of discrete angular displacements, called step angles. Each angular step is actuated by a discrete electrical pulse. The total angular rotation is controlled by the number of pulses received by the motor, and rotational speed is controlled by the frequency of the pulses. The step angle is related to the number of steps for the motor according to the relationship

$$\alpha = \frac{360}{n_s} \qquad (6.12)$$

where α = the step angle, degrees (°); and n_s = the number of steps for the stepper motor, which must be an integer value. Typical values for the step angle in commercially available stepper motors are 7.5°, 3.6°, and 1.8°, corresponding to 48, 100, and 200 steps (pulses) per revolution of the motor. The total angle through which the motor rotates A_m is given by

$$A_m = n_p\alpha \qquad (6.13)$$

where A_m is measured in degrees (°), n_p = the number of pulses received by the motor, and α = the step angle. The angular velocity ω and speed of rotation N are given by the expressions

$$\omega = \frac{2\pi f_p}{n_s} \qquad (6.14)$$

$$N = \frac{60 f_p}{n_s} \qquad (6.15)$$

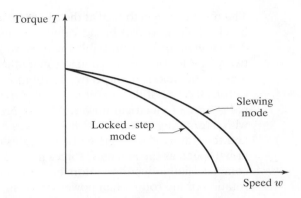

Figure 6.4 Typical torque-speed curve of a stepper motor.

where ω = angular velocity, rad/sec; N = rotational speed, rev/min; f_p = pulse frequency, pulses/sec or Hz; and n_p = the number of steps in the motor, steps/rev or pulses/rev.

The typical torque-speed relationships for a stepper motor are shown in Figure 6.4. As in the DC servomotor, torque decreases with increased rotational speeds. And because rotational speed is related to pulse frequency in the stepper motor, torque is lower at higher pulse rates. As indicated in the figure, there are two operating modes, locked-step and slewing. In the *locked-step mode*, each pulse received by the motor causes a discrete angular step to be taken; the motor starts and stops (at least approximately) with each pulse. In this mode the motor can be started and stopped, and its direction of rotation can be reversed. In the *slewing mode*, usually associated with higher speeds, the motor's rotation is more or less continuous and does not allow for stopping or reversing with each subsequent step. Nevertheless, the rotor does respond to each individual pulse; that is, the relationship between rotating speed and pulse frequency is retained in the slewing mode.

Stepper motors are used in open loop control systems for applications in which torque and power requirements are low to modest. They are widely used in machine tools and other production machines, industrial robots, x-y plotters, medical and scientific instruments, and computer peripherals. Probably the most common application is to drive the hands of analog quartz watches.

6.2.2 Other Types of Actuators

There are other types of electrical actuators in addition to motors. These include solenoids and relays, which are electromagnetic devices like electric motors, but they operate differently. There are also actuators that operate using hydraulic and pneumatic power.

Electrical Actuators Other Than Motors. The two actuators described here are solenoids and electromechanical relays. A *solenoid* consists of a movable plunger inside a stationary wire coil, as pictured in Figure 6.5. When a current is applied to the coil, it acts as a magnet, drawing the plunger into the coil. When current is switched off, a spring returns the plunger to its previous position. Linear solenoids of the type described here are often used to open and close valves in fluid flow systems, such as chemical processing equipment. In these applications, the solenoid provides a push or pull (linear) action. Rotary solenoids

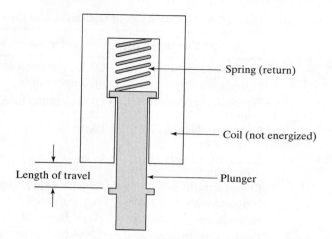

Figure 6.5 Solenoid.

are also available to provide rotary motion, usually over a limited angular range (e.g., neutral position to between 25° and 90°).

An *electromechanical relay* is an on-off electrical switch consisting of two main components, a stationary coil and a movable arm that can be made to open or close an electrical contact by means of a magnetic field that is generated when current is passed through the coil. The reason for using a relay is that it can be operated with relatively low current levels, but it opens and closes circuits that carry high currents and/or voltages. Thus, relays are a safe way to remotely switch on and off equipment that requires high amounts of electrical power.

Hydraulic and Pneumatic Actuators. These two categories of actuators are powered by pressurized fluids. Oil is used in hydraulic systems, and compressed air is used in pneumatic systems. The devices in both categories are similar in operation but different in construction due to the differences in fluid properties between oil and air. Some of the differences in properties, and their effects on the characteristics and applications of the two types of actuators, are listed in Table 6.4.

Hydraulic and pneumatic actuators are available that provide either linear or rotary motion. The cylinder, illustrated in Figure 6.6, is a common linear-motion device. The cylinder is basically a tube, and a piston is forced to slide inside the cylinder due to fluid pressure. Two types are shown in our figure: (a) single acting with spring return and (b) double acting. Although these cylinders operate in a similar way for both types of fluid power, it is more difficult to predict the speed and force characteristics of pneumatic cylinders because of the compressibility of air in these devices. For hydraulic cylinders, the fluid is incompressible, and the speed and force of the piston depend on the fluid flow rate and pressure inside the cylinder, respectively, as given by the relationships

$$v = \frac{Q}{A} \tag{6.16}$$

$$F = pA \tag{6.17}$$

where v = velocity of the piston, m/sec (in/sec); Q = volumetric flow rate, m^3/sec (in^2/sec); A = area of the cylinder cross section, m^2 (in^2); F = applied force, N (lbf); and p = fluid

TABLE 6.4 Comparison of Hydraulic and Pneumatic Systems

System Characteristic	Hydraulic System	Pneumatic System
Pressurized fluid	Oil (or water-oil emulsion)	Compressed air
Compressibility	Incompressible	Compressible
Typical fluid pressure level	20 MPa (3000 lb/in^2)	0.7 MPa (100 lb/in^2)
Forces applied by devices	High	Low
Actuation speeds of devices	Low	High
Speed control	Accurate speed control	Difficult to control accurately
Problem with fluid leaks	Yes, potential safety hazard	No problem when air leaks
Relative cost of devices	High (factor of 5 to 10 times)	Low
Device construction and manufacture	Close tolerances and good surface finishes required on components	O-rings used to prevent leaks instead of highly accurate components
Automation applications	Preferred when high forces and accurate control are required	Preferred when low cost and high speed actuation are required

pressure, N/m^2 or Pa (lb/in^2). It should be noted that in a double-acting cylinder, the area is different in the two directions due to the presence of the piston rod. When the piston is retracted into the cylinder, the cross-sectional area of the piston rod must be subtracted from the cylinder area. This means that the piston speed will be slightly greater and the applied force will be slightly less when the piston is retracting (reverse stroke) than when it is extending (forward stroke).

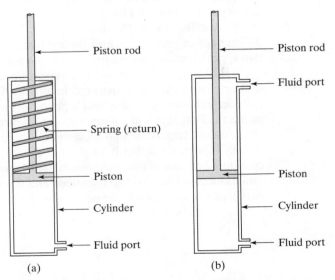

Figure 6.6 Cylinder and piston: (a) single acting with spring return and (b) double acting.

Fluid-powered rotary motors are also available to provide a continuous rotational motion. Hydraulic motors are noted for developing high torques, and pneumatic motors can be used for high-speed applications. There are several different mechanisms by which these motors operate, including the use of pistons, vanes, and turbine blades. The performance characteristics of the air-driven rotary motors are more difficult to analyze, just as we observed for the operation of the pneumatic cylinder. On the other hand, hydraulic motors have well-behaved characteristics. In general, the rotation speed of a hydraulic motor is directly proportional to the fluid flow rate, as defined in the equation

$$\omega = KQ \tag{6.18}$$

where ω = angular velocity, rad/sec; Q = volumetric fluid flow rate, m³/sec (in³/sec); and K is a constant of proportionality with units of rad/m³ (rad/in³). Angular velocity (rad/sec) can be converted to revolutions per minute (rev/min) by multiplying by $60/2\pi$.

6.3 ANALOG-TO-DIGITAL CONVERTERS

Continuous analog signals from the process must be converted into digital values to be used by the computer, and digital data generated by the computer must be converted to analog signals to be used by analog actuators. We discuss analog-to-digital conversion in this section and digital-to-analog conversion in the following section.

The procedure for converting an analog signal from the process into digital form typically consists of the following steps and hardware devices, as illustrated in Figure 6.7:

1. *Sensor and transducer.* This is the measuring device that generates the analog signal (Section 6.1).

2. *Signal conditioning.* The continuous analog signal from the transducer may require conditioning to render it into more suitable form. Common signal conditioning steps include (1) filtering to remove random noise and (2) conversion from one signal form to another, for example, converting a current into a voltage.

3. *Multiplexer.* The multiplexer is a switching device connected in series with each input channel from the process; it is used to time-share the analog-to-digital converter

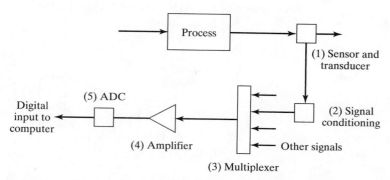

Figure 6.7 Steps in analog-to-digital conversion of continuous analog signals from process.

(ADC) among the input channels. The alternative is to have a separate ADC for each input channel, which would be costly for a large application with many input channels. Since the process variables need only be sampled periodically, using a multiplexer provides a cost-effective alternative to dedicated ADCs for each channel.

4. *Amplifier.* Amplifiers are used to scale the incoming signal up or down to be compatible with the range of the analog-to-digital converter.

5. *Analog-to-digital converter.* As its name indicates, the function of the ADC is to convert the incoming analog signal into its digital counterpart.

Let us consider the operation of the ADC, which is the heart of the conversion process. Analog-to-digital conversion occurs in three phases: (1) sampling, (2) quantization, and (3) encoding. Sampling consists of converting the continuous signal into a series of discrete analog signals at periodic intervals, as shown in Figure 6.8. In quantization, each discrete analog signal is assigned to one of a finite number of previously defined amplitude levels. The amplitude levels are discrete values of voltage ranging over the full scale of the ADC. In the encoding phase, the discrete amplitude levels obtained during quantization are converted into digital code, representing the amplitude level as a sequence of binary digits.

In selecting an analog-to-digital converter for a given application, the following factors are relevant: (1) sampling rate, (2) conversion time, (3) resolution, and (4) conversion method.

The *sampling rate* is the rate at which the continuous analog signals are sampled or polled. A higher sampling rate means that the continuous waveform of the analog signal can be more closely approximated. When the incoming signals are multiplexed, the maximum possible sampling rate for each signal is the maximum sampling rate of the ADC divided by the number of channels that are processed through the multiplexer. For example, if the maximum sampling rate of the ADC is 1000 samples/sec, and there are 10 input channels through the multiplexer, then the maximum sampling rate for each input line is 1000/10 = 100 sample/sec. (This ignores time losses due to multiplexer switching.)

The maximum possible sampling rate of an ADC is limited by the ADC conversion time. *Conversion time* of an ADC is the time interval between the application of an incoming signal and the determination of the digital value by the quantization and encoding phases of the conversion procedure. Conversion time depends on (1) the type of conversion procedure used by the ADC and (2) the number of bits n used to define the converted digital value. As n is increased, conversion time increases (bad news), but resolution of the ADC improves (good news).

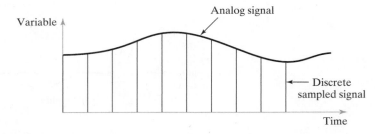

Figure 6.8 Analog signal converted into series of discrete sampled data by analog-to-digital converter.

The *resolution* of an ADC is the precision with which the analog signal is evaluated. Since the signal is represented in binary form, precision is determined by the number of quantization levels, which in turn is determined by the bit capacity of the ADC and the computer. The number of quantization levels is defined as

$$N_q = 2^n \qquad (6.19)$$

where N_q = number of quantization levels; and n = number of bits. Resolution can be defined in equation form as

$$R_{ADC} = \frac{L}{N_q - 1} = \frac{L}{2^n - 1} \qquad (6.20)$$

where R_{ADC} = resolution of the ADC, also called the *quantization-level spacing*, which is the length of each quantization level; L = full-scale range of the ADC, usually 0–10 V (the incoming signal must typically be amplified, either up or down, to this range); and N_q = the number of quantization levels, defined in Eq. (6.19).

Quantization generates an error, because the quantized digital value is likely to be different from the true value of the analog signal. The maximum possible error occurs when the true value of the analog signal is on the borderline between two adjacent quantization levels; in this case, the error is one-half the quantization-level spacing. By this reasoning, the quantization error is defined

$$\text{Quantization error} = \pm \frac{1}{2} R_{ADC} \qquad (6.21)$$

Various conversion methods are available by which to encode an analog signal into its digital equivalent. Let us discuss the most commonly used technique, called the *successive approximation method*. In this method, a series of known trial voltages are successively compared to the input signal whose value is unknown. The number of trial voltages corresponds to the number of bits used to encode the signal. The first trial voltage is half the full-scale range of the ADC, and each successive trial voltage is half the preceding value. Comparing the remainder of the input voltage with each trial voltage yields a bit value of "1" if the input exceeds the trial value and "0" if the input is less than the trial voltage. The successive bit values, multiplied by their corresponding trial voltage values, provide the encoded value of the input signal. Let us illustrate the procedure with an example.

EXAMPLE 6.3   Successive Approximation Method in Analog-to-Digital Conversion

Suppose the input signal is 6.8 V. Use the successive approximation method to encode the signal for a 6-bit register for an ADC with a full-scale range of 10 V.

Solution: The encoding procedure for the input of 6.8 V is illustrated in Figure 6.9. In the first trial, 6.8 V is compared with 5.0 V. Since 6.8 > 5.0, the first bit value is 1. Comparing the remainder (6.8 − 5.0) = 1.8 V with the second trial voltage of 2.5 V yields a 0, since 1.8 < 2.5. The third trial voltage = 1.25 V. Since 1.8 > 1.25, the third bit value is 1. The rest of the 6 bits are evaluated in the figure to yield an encoded value = 6.718 V.

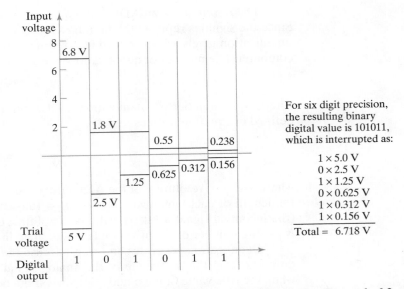

Figure 6.9 Successive approximation method applied to Example 6.3.

6.4 DIGITAL-TO-ANALOG CONVERTERS

The process performed by a digital-to-analog converter (DAC) is the reverse of the ADC process. The DAC transforms the digital output of the computer into a continuous signal to drive an analog actuator or other analog device. Digital-to-analog conversion consists of two steps: (1) *decoding*, in which the digital output of the computer is converted into a series of analog values at discrete moments in time, and (2) *data holding*, in which each successive value is changed into a continuous signal (usually electrical voltage) used to drive the analog actuator during the sampling interval.

Decoding is accomplished by transferring the digital value from the computer to a binary register that controls a reference voltage source. Each successive bit in the register controls half the voltage of the preceding bit, so that the level of the output voltage is determined by the status of the bits in the register. Thus, the output voltage is given by

$$E_o = E_{\text{ref}}\{0.5B_1 + 0.25B_2 + 0.125B_3 + \cdots + (2^n)^{-1}B_n\} \qquad (6.22)$$

where E_o = output voltage of the decoding step (V); E_{ref} = reference voltage (V); and $B_1, B_2, \ldots, B_n$ = status of successive bits in the register, 0 or 1; and n = the number of bits in the binary register.

The objective in the data holding step is to approximate the envelope formed by the data series, as illustrated in Figure 6.10. Data holding devices are classified according to the order of the extrapolation calculation used to determine the voltage output during sampling intervals. The most common extrapolator is a *zero-order hold*, in which the output voltage between sampling instants is a sequence of step signals, as in Figure 6.10(a). The voltage function during the sampling interval is constant and can be expressed very simply as

$$E(t) = E_o \qquad (6.23)$$

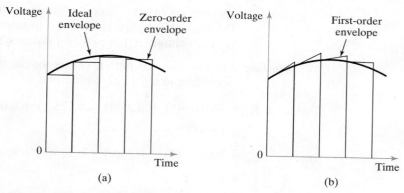

Figure 6.10 Data holding step using (a) zero-order hold and (b) first-order hold.

where $E(t)$ = voltage as a function of time t during the sampling interval (V), and E_o = voltage output from the decoding step, Eq. (6.22).

The first-order data hold is less common than the zero-order hold, but it usually approximates the envelope of the sampled data values more closely. With the *first-order hold*, the voltage function $E(t)$ during the sampling interval changes with a constant slope determined by the two preceding E_o values. Expressing this mathematically, we have

$$E(t) = E_o + \alpha t \tag{6.24}$$

where α = rate of change of $E(t)$, E_o = output voltage from Eq. (6.22) at the start of the sampling interval (V), and t = time (sec). The value of α is computed each sampling interval as

$$\alpha = \frac{E_o - E_o(-\tau)}{\tau} \tag{6.25}$$

where E_o = output voltage from Eq. (6.22) at the start of the sampling interval (V), τ = time interval between sampling instants (sec), and $E_o(-\tau)$ = value of E_o from Eq. (6.22) from the preceding sampling instant (removed backward in time by τ, V). The result of the first-order hold is illustrated in Figure 6.10(b).

EXAMPLE 6.4 Zero-Order and First-Order Data Hold for Digital-to-Analog Converter

A digital-to-analog converter uses a reference voltage of 100 V and has 6-bit precision. In three successive sampling instants, 0.5 sec apart, the data contained in the binary register are the following:

Instant	Binary Data
1	101000
2	101010
3	101101

Determine (a) the decoder output values for the three sampling instants, (b) the voltage signals between instants 2 and 3 for a zero-order hold, and (c) the voltage signals between instants 2 and 3 for a first-order hold.

Solution: (a) The decoder output values for the three sampling instants are computed according to Eq. (6.22) as follows:

Instant 1, $E_o = 100\{0.5(1) + 0.25(0) + 0.125(1) + 0.0625(0) + 0.03125(0) + 0.015625(0)\}$

$\qquad = 62.50 \text{ V}$

Instant 2, $E_o = 100\{0.5(1) + 0.25(0) + 0.125(1) + 0.0625(0) + 0.03125(1) + 0.015625(0)\}$

$\qquad = 65.63 \text{ V}$

Instant 3, $E_o = 100\{0.5(1) + 0.25(0) + 0.125(1) + 0.0625(1) + 0.03125(0) + 0.015625(1)\}$

$\qquad = 70.31 \text{ V}$

(b) The zero-order hold between sampling instants 2 and 3 yields a constant voltage $E(t) = 65.63$ V according to Eq. (6.23).

(c) The first-order hold yields a steadily increasing voltage. The slope α is given by Eq. (6.25):

$$\alpha = \frac{65.63 - 62.5}{0.5} = 6.25$$

and from Eq. (6.24), the voltage function between instants 2 and 3 is

$$E(t) = 65.63 + 6.25t$$

These values and functions are plotted in Figure 6.11. Note that the first-order hold more accurately anticipates the value of E_o at sampling instant 3 than does the zero-order hold.

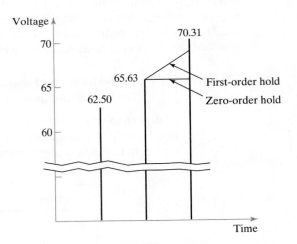

Figure 6.11 Solution to Example 6.4.

6.5 INPUT/OUTPUT DEVICES FOR DISCRETE DATA

Discrete data can be processed by a digital computer without the kinds of conversion procedures required for continuous analog signals. As indicated earlier, discrete data divide into three categories: (a) binary data, (b) discrete data other than binary, and (c) pulse data. Table 6.5 summarizes the input/output interface for the three categories of discrete data.

6.5.1 Contact Input/Output Interfaces

Contact interfaces are of two types, input and output. These interfaces read binary data from the process into the computer and send binary signals from the computer to the process, respectively. The terms input and output refer to the computer.

A *contact input interface* is a device by which binary data are read into the computer from some external source (e.g., the process). It consists of a series of simple contacts that can be either closed or open (on or off) to indicate the status of binary devices connected to the process such as limit switches (contact or no contact), valves (open or closed), or motor pushbuttons (on or off). The computer periodically scans the actual status of the contacts to update the values stored in memory.

The contact input interface can also be used to enter discrete data other than binary. This type of data is generated by devices such as a photoelectric sensor array and can be stored in a binary register consisting of multiple bits. The individual bit values (0 or 1) can be entered through the contact input interface. In effect, a certain number of contacts in the input interface are assigned to the binary register, the number of contacts being equal to the number of bits in the register. The binary number can be converted to a conventional base 10 number as needed in the application.

The *contact output interface* is the device that communicates on/off signals from the computer to the process. The contact positions are set either on or off. These positions are maintained until changed by the computer, perhaps in response to events in the process. In computer process control applications, hardware controlled by the contact output interface include alarms, indicator lights (on control panels), solenoids, and constant speed motors. The computer controls the sequence of on/off activities in a work cycle through this contact output interface.

The contact output interface can be used to transmit a discrete data value other than binary by assigning an array of contacts in the interface for that purpose. The 0 and 1 values of the contacts in the array are evaluated as a group to determine the corresponding discrete number. In effect, this procedure is the reverse of that used by the contact input interface for discrete data other than binary.

TABLE 6.5 Types of Computer Input/Output Interface for Different Types of Discrete Parameters and Variables

Type of Digital Data	Input Interface to Computer	Output Interface from Computer
Discrete data—binary (on/off)	Contact input	Contact output
Discrete data other than binary	Contact input array	Contact output array
Discrete pulse data	Pulse counters	Pulse generators

6.5.2 Pulse Counters and Generators

Discrete data can also exist in the form of a series of pulses. Such data is generated by digital transducers such as optical encoders. Pulse data are also used to control certain devices such as stepper motors.

A *pulse counter* is a device used to convert a series of pulses (call it a *pulse train*, as shown in Figure 5.1) into a digital value. The value is then entered into the computer through its input channel. The most common type of pulse counter is one that counts electrical pulses. It is constructed using sequential logic gates, called *flip-flops*, which are electronic devices that possess memory capability and that can be used to store the results of the counting procedure.

Pulse counters can be used for both counting and measurement applications. A typical counting application might add up the number of packages moving past a photoelectric sensor along a conveyor in a distribution center. A typical measurement application might indicate the rotational speed of a shaft. One possible method to accomplish the measurement is to connect the shaft to an optical encoder, which generates a certain number of electrical pulses for each rotation. To determine rotational speed, the pulse counter measures the number of pulses received during a certain time period and divides this by the time period and by the number of pulses in each revolution of the encoder. We discuss counters in the context of digital control in Section 9.1.2.

A *pulse generator* is a device that produces a series of electrical pulses whose total number and frequency are specified by the control computer. The total number of pulses might be used to drive the axis of a positioning system. The frequency of the pulse train, or pulse rate, could be used to control the rotational speed of a stepper motor. A pulse generator operates by repeatedly closing and opening an electrical contact, thus producing a sequence of discrete electrical pulses. The amplitude (voltage level) and frequency are designed to be compatible with the device being controlled.

REFERENCES

[1] ASTROM, K. J., and WITTENMARK, B., *Computer-Controlled Systems - Theory and Design*, 3d ed., Prentice Hall, Upper Saddle Rover, NJ, 1997.

[2] BATESON, R. N., *Introduction to Control System Technology*, 7th ed., Prentice Hall, Upper Saddle River, NJ, 2002.

[3] BEATY, H. W., and J. L. KIRTLEY, Jr., *Electric Motor Handbook*, McGraw-Hill Book Company, NY, 1998.

[4] BOUCHER, T. O., *Computer Automation in Manufacturing*, Chapman & Hall, London, UK, 1996.

[5] DOEBLIN, E. O., *Measurement Systems: Applications and Design*, 4th ed., McGraw-Hill, Inc., NY, 1990.

[6] FRADEN, J., *Handbook of Modern Sensors*, 3d ed., Springer-Verlag, NY, 2003.

[7] GARDNER, J. W., *Microsensors: Principles and Applications*, John Wiley & Sons, NY, 1994.

[8] GROOVER, M. P., M. WEISS, R. N. NAGEL, N. G. ODREY, and S. B. MORRIS, *Industrial Automation and Robotics*, McGraw-Hill (Primus Custom Publishing), NY, 1998.

[9] OLSSON, G., and G. PIANI, *Computer Systems for Automation and Control*, Prentice Hall, London, UK, 1992.

[10] Pessen, D. W., *Industrial Automation: Circuit Design and Components*, John Wiley & Sons, NY, 1989.

[11] Rizzoni, G., *Principles and Applications of Electrical Engineering*, 5th ed., McGraw-Hill, NY, 2007.

[12] Stenerson, J., *Fundamentals of Programmable Logic Controllers, Sensors, and Communications*, 3d ed., Pearson/Prentice Hall, Upper Saddle River, NJ, 2004.

REVIEW QUESTIONS

6.1 What is a sensor?

6.2 What is the difference between an analog sensor and a discrete sensor?

6.3 What is the difference between an active sensor and a passive sensor?

6.4 What is the transfer function of a sensor?

6.5 What is an actuator?

6.6 Nearly all actuators can be classified into one of three categories, according to type of drive power. Name the three categories.

6.7 Name the two main components of an electric motor.

6.8 In a DC motor, what is a commutator?

6.9 What are the two important disadvantages of DC electric motors that make the AC motor relatively attractive?

6.10 How is the operation of a stepper motor different from the operation of conventional DC or AC motors?

6.11 What is a solenoid?

6.12 What is the difference between a hydraulic actuator and a pneumatic actuator?

6.13 Briefly describe the three phases of the analog-to-digital conversion process.

6.14 What is the resolution of an analog-to-digital converter?

6.15 Briefly describe the two steps in the digital-to-analog conversion process.

6.16 What is the difference between a contact input interface and a contact output interface?

6.17 What is a pulse counter?

PROBLEMS

Sensors

6.1 During calibration, an Iron/Constantan thermocouple is zeroed (set to emit a zero voltage) at 0°C. At 750°C, it emits a voltage of 38.8 mV. A linear output/input relationship exists between 0°C and 750°C. Determine (a) the transfer function of the thermocouple and (b) the temperature corresponding to a voltage output of 29.6 mV.

6.2 A digital tachometer is used to determine the surface speed of a rotating workpiece in surface ft/min. Tachometers are designed to read rotational speed in rev/min, but in this case the shaft of the tachometer is directly coupled to a wheel whose outside rim is made of rubber. When the wheel rim is pressed against the surface of the rotating workpiece, the tachometer provides a reading of surface speed. The desired units for surface speed are ft/min. What is the diameter of the wheel rim that will provide a direct reading of surface speed in ft/min?

6.3 A digital flow meter operates by emitting a pulse for each unit volume of fluid flowing through it. The particular flow meter of interest here has a unit volume of 57.9 cm^3 per pulse. In a certain process control application, the flow meter emitted 6489 pulses during a period of 3.6 min. Determine (a) the total volume of fluid that flowed through the meter, (b) the flow rate of fluid flow, and (c) the pulse frequency (Hz) corresponding to a flow rate of 75,000 cm^3/min?

6.4 A tool-chip thermocouple is used to measure the cutting temperature in a turning operation. The two dissimilar metals in a tool-chip thermocouple are the tool material and the workpiece metal. During the turning operation, the chip from the work metal forms a junction with the rake face of the tool to create the thermocouple at exactly the location where temperature must be measured: at the interface between the tool and the chip. A separate calibration procedure must be performed for each combination of tool material and work metal. In the combination of interest here, the calibration curve (inverse transfer function) for a particular grade of cemented carbide tool when used to turn C1040 steel is the following: $T = 88.1E_{tc} - 127$, where T = temperature in °F, and E_{tc} = the emf output of the thermocouple in mV. (a) Revise the temperature equation so that it is in the form of a transfer function similar to that given in Eq. (6.3). What is the sensitivity of this tool-chip thermocouple? (b) During a straight turning operation, the emf output of the thermocouple was measured as 9.25 mV. What was the corresponding cutting temperature?

Actuators

6.5 A DC servomotor is used to actuate one of the axes of an x-y positioner. The motor has a torque constant of 8.75 in-lb/A and a voltage constant of 10 V/(1000 rev/min). The armature resistance is 2.0 ohms. At a given moment, the positioner table is not moving and a voltage of 20 V is applied to the motor terminals. Determine the torque (a) immediately after the voltage is applied and (b) at a rotational speed of 400 rev/min. (c) What is the maximum theoretical speed of the motor?

6.6 A DC servomotor has a torque constant = 0.088 N-m/A and a voltage constant = 0.12 V/(rad/sec). The armature resistance is 2.3 ohms. A terminal voltage of 30 V is used to operate the motor. Determine (a) the starting torque generated by the motor just as the voltage is applied, (b) the maximum speed at a torque of zero, and (c) the operating point of the motor when it is connected to a load whose torque characteristic is proportional to speed with a constant of proportionality = 0.011 N-m/(rad/sec).

6.7 In the previous problem, what is the power delivered by the motor at the operating point in units of (a) Watts and (b) horsepower?

6.8 A voltage of 24 V is applied to a DC servomotor whose torque constant = 0.115 N-m/A and voltage constant = 0.097 V/(rad/sec). Armature resistance = 1.9 ohms. The motor is directly coupled to a blower shaft for an industrial process. (a) What is the stall torque of the motor? (b) Determine the operating point of the motor if the torque-speed characteristic of the blower is given by the following equation: $T_L = K_{L1}\omega + K_{L2}\omega^2$, where T_L = load torque, N-m; ω = angular velocity, rad/sec; K_{L1} = 0.005 N-m/(rad/sec), and K_{L2} = 0.00033 N-m/(rad/sec)^2. (c) What horsepower is being generated by the motor at the operating point?

6.9 The step angle of a certain stepper motor = 1.8°. The application of interest is to rotate the motor shaft through 10 complete revolutions at an angular velocity of 20 rad/sec. Determine (a) the required number of pulses and (b) the pulse frequency to achieve the specified rotation.

6.10 A stepper motor has a step angle = 7.5°. (a) How many pulses are required for the motor to rotate through five complete revolutions? (b) What pulse frequency is required for the motor to rotate at a speed of 200 rev/min?

6.11   The shaft of a stepper motor is directly connected to a leadscrew that drives a worktable in an *x-y* positioning system. The motor has a step angle = 5°. The pitch of the lead screw is 6 mm, which means that the worktable moves in the direction of the leadscrew axis by a distance of 6 mm for each complete revolution of the screw. It is desired to move the worktable a distance of 300 mm at a top speed of 40 mm/sec. Determine (a) the number of pulses and (b) the pulse frequency required to achieve this movement.

6.12   A single-acting hydraulic cylinder with spring return has an inside diameter of 88 mm. Its application is to push pallets off of a conveyor into a storage area. The hydraulic power source can generate up to 3.2 MPa of pressure at a flow rate of 175,000 mm^3/sec to drive the piston. Determine (a) the maximum possible velocity of the piston and (b) the maximum force that can be applied by the apparatus.

6.13   A double-acting hydraulic cylinder has an inside diameter of 75 mm. The piston rod has a diameter of 14 mm. The hydraulic power source can generate up to 5.0 MPa of pressure at a flow rate of 200,000 mm^3/sec to drive the piston. (a) What are the maximum possible velocity of the piston and the maximum force that can be applied in the forward stroke? (b) What are the maximum possible velocity of the piston and the maximum force that can be applied in the reverse stroke?

6.14   A double-acting hydraulic cylinder is used to actuate a linear joint of an industrial robot. The inside diameter of the cylinder is 3.5 in. The piston rod has a diameter of 0.5 in. The hydraulic power source can generate up to 500 lb/in^2 of pressure at a flow rate of 1200 in^3/min to drive the piston. (a) Determine the maximum velocity of the piston and the maximum force that can be applied in the forward stroke. (b) Determine the maximum velocity of the piston and the maximum force that can be applied in the reverse stroke.

ADC and DAC

6.15   A continuous voltage signal is to be converted into its digital counterpart using an analog-to-digital converter. The maximum voltage range is ±30 V. The ADC has a 12-bit capacity. Determine (a) number of quantization levels, (b) resolution, (c) the spacing of each quantization level, and (d) the quantization error for this ADC.

6.16   A voltage signal with a range of zero to 115 V is to be converted by means of an ADC. Determine the minimum number of bits required to obtain a quantization error of (a) ±5 V maximum, (b) ±1 V maximum, (c) ±0.1 V maximum.

6.17   A digital-to-analog converter uses a reference voltage of 120 V dc and has eight binary digit precision. In one of the sampling instants, the data contained in the binary register = 01010101. If a zero-order hold is used to generate the output signal, determine the voltage level of that signal.

6.18   A DAC uses a reference voltage of 80 V and has six-bit precision. In four successive sampling periods, each one second long, the binary data contained in the output register were 100000, 011111, 011101, and 011010. Determine the equation for the voltage as a function of time between sampling instants 3 and 4 using (a) a zero-order hold, and (b) a first-order hold.

6.19   In the previous problem, suppose that a second-order hold were to be used to generate the output signal. The equation for the second-order hold is the following: $E(t) = E_0 + \alpha t + \beta t^2$, where E_0 = starting voltage at the beginning of the time interval. (a) For the binary data given in the previous problem, determine the values of α and β that would be used in the equation for the time interval between sampling instants 3 and 4. (b) Compare the first-order and second-order holds in anticipating the voltage at the fourth instant.

Chapter 7

Numerical Control

CHAPTER CONTENTS

From Chapter 7 of *Automation, Production Systems, and Computer-Integrated Manufacturing*, Third Edition.
Mikell P. Groover. Copyright © 2008 by Pearson Education, Inc. Publishing as Prentice Hall. All rights reserved.

Numerical control (NC) is a form of programmable automation in which the mechanical actions of a machine tool or other equipment are controlled by a program containing coded alphanumeric data. The alphanumeric data represent relative positions between a workhead and a workpart as well as other instructions needed to operate the machine. The workhead is a cutting tool or other processing apparatus, and the workpart is the object being processed. When the current job is completed, the program of instructions can be changed to process a new job. The capability to change the program makes NC suitable for low and medium production. It is much easier to write new programs than to make major alterations in the processing equipment.

Numerical control can be applied to a wide variety of processes. The applications divide into two categories: (1) machine tool applications, such as drilling, milling, turning, and other metal working; and (2) nonmachine tool applications, such as assembly, drafting, and inspection. The common operating feature of NC in all of these applications is control of the workhead movement relative to the workpart. The concept for NC dates from the late 1940s. The first NC machine was developed in 1952 (see Historical Note 7.1).

Historical Note 7.1   The First NC Machines [1], [4], [8], [10]

The development of NC owes much to the U.S. Air Force and the early aerospace industry. The first work in the area of NC is attributed to John Parsons and his associate Frank Stulen at Parsons Corporation in Traverse City, Michigan. Parsons was a contractor for the Air Force during the 1940s and had experimented with the concept of using coordinate position data contained on punched cards to define and machine the surface contours of airfoil shapes. He had named his system the *Cardamatic* milling machine, since the numerical data was stored on punched cards. Parsons and his colleagues presented the idea to the Wright-Patterson Air Force Base in 1948. The initial Air Force contract was awarded to Parsons in June 1949. A subcontract was awarded by Parsons in July 1949 to the Servomechanism Laboratories at the Massachusetts Institute of Technology to (1) perform a systems engineering study on machine tool controls and (2) develop a prototype machine tool based on the Cardamatic principle. Research commenced on the basis of this subcontract, which continued until April 1951, when a contract was signed by MIT and the Air Force to complete the development work.

Early in the project, it became clear that the required data transfer rates between the controller and the machine tool could not be achieved using punched cards, so it was proposed to use either punched paper tape or magnetic tape to store the numerical data. These and other technical details of the control system for machine tool control had been defined by June 1950. The name *numerical control* was adopted in March 1951 based on a contest sponsored by John Parsons among "MIT personnel working on the project." The first NC machine was developed by retrofitting a Cincinnati Milling Machine Company vertical Hydro-Tel milling machine (a 24-in × 60-in conventional tracer mill) that had been donated by the Air Force from surplus equipment. The controller combined analog and digital components, consisted of 292 vacuum tubes, and occupied a floor area greater than the machine tool itself. The prototype successfully performed simultaneous control of three-axis motion based on coordinate-axis data on punched binary tape. This experimental machine was in operation by March 1952.

A patent for the machine tool system entitled *Numerical Control Servo System* was filed in August 1952, and awarded in December 1962. Inventors were listed as Jay Forrester, William Pease, James McDonough, and Alfred Susskind, all Servomechanisms Lab staff during the project. It is of interest to note that a patent was also filed by John Parsons and Frank

Stulen in May 1952 for a *Motor Controlled Apparatus for Positioning Machine Tool* based on the idea of using punched cards and a mechanical rather than electronic controller. This patent was issued in January 1958. In hindsight, it is clear that the MIT research provided the prototype for subsequent developments in NC technology. As far as we know, no commercial machines were ever introduced using the Parsons-Stulen configuration.

Once the NC machine was operational in March 1952, trial parts were solicited from aircraft companies across the country to learn about the operating features and economics of NC. Several potential advantages of NC were apparent from these trials. These included good accuracy and repeatability, reduction of noncutting time in the machining cycle, and the capability to machine complex geometries. Part programming was recognized as a difficulty with the new technology. A public demonstration of the machine was held in September 1952 for machine tool builders (anticipated to be the companies that would subsequently develop products in the new technology), aircraft component producers (expected to be the principal users of NC), and other interested parties.

Reactions of the machine tool companies following the demonstrations "ranged from guarded optimism to outright negativism" [10, p. 61]. Most of the companies were concerned about a system that relied on vacuum tubes, not realizing that tubes would soon be displaced by transistors and integrated circuits. They were also worried about their staff's qualifications to maintain such equipment and were generally skeptical of the NC concept. Anticipating this reaction, the Air Force sponsored two additional tasks: (1) information dissemination to industry and (2) an economic study. The information dissemination task included many visits by Servo Lab personnel to companies in the machine tool industry as well as visits to the Lab by industry personnel to observe demonstrations of the prototype machine. The economic study showed clearly that the applications of general purpose NC machine tools were in low and medium quantity production, as opposed to Detroit-type transfer lines, which could be justified only for very large quantities.

In 1956, the Air Force decided to sponsor the development of NC machine tools at several different companies. These machines were placed in operation at various aircraft companies between 1958 and 1960. The advantages of NC soon became apparent, and the aerospace companies began placing orders for new NC machines. In some cases, they even built their own units. This served as a stimulus to the remaining machine tool companies that had not yet embraced NC. Advances in computer technology also stimulated further development. The first application of the digital computer for NC was part programming. In 1956, MIT demonstrated the feasibility of a computer-aided part programming system using an early digital computer prototype that had been developed at MIT. Based on this demonstration, the Air Force sponsored development of a part programming language. This research resulted in the development of the APT language in 1958.

The automatically programmed tool system (APT) was the brainchild of mathematician Douglas Ross, who worked in the MIT Servomechanisms Lab at the time. Recall that this project was started in the 1950s, a time when digital computer technology was in its infancy, as were the associated computer programming languages and methods. The APT project was a pioneering effort, not only in the development of NC technology, but also in computer programming concepts, computer graphics, and computer-aided design (CAD). Ross envisioned a part programming system in which (1) the user would prepare instructions for operating the machine tool using English-like words, (2) the digital computer would translate these instructions into a language that the computer could understand and process, (3) the computer would carry out the arithmetic and geometric calculations needed to execute the instructions, and (4) the computer would further process (postprocess) the instructions so that they could be interpreted by the machine tool controller. He further recognized that the programming system should be expandable for applications beyond those considered in the immediate research at MIT (milling applications).

Ross's work at MIT became a focal point for NC programming, and a project was initiated to develop a two-dimensional version of APT, with nine aircraft companies plus IBM Corporation participating in the joint effort and MIT as project coordinator. The 2D-APT system was ready for field evaluation at plants of participating companies in April 1958. Testing, debugging, and refining the programming system took approximately three years. In 1961, the Illinois Institute of Technology Research Institute (IITRI) was selected to become responsible for long-range maintenance and upgrading of APT. In 1962, IITRI announced completion of APT-III, a commercial version of APT for three-dimensional part programming. In 1974, APT was accepted as the U.S. standard for programming NC metal cutting machine tools. In 1978, it was accepted by the ISO as the international standard.

Numerical control technology was in its second decade before computers were employed to actually control machine tool motions. In the mid-1960s, the concept of *direct numerical control* (DNC) was developed, in which individual machine tools were controlled by a mainframe computer located remotely from the machines. The computer bypassed the punched tape reader, instead transmitting instructions to the machine in real time, one block at a time. The first prototype system was demonstrated in 1966 [4]. Two companies that pioneered the development of DNC were General Electric Company and Cincinnati Milling Machine Company (which changed its name to Cincinnati Milacron in 1970). Several DNC systems were demonstrated at the National Machine Tool Show in 1970.

Mainframe computers represented the state of the technology in the mid-1960s. There were no personal computers or microcomputers at that time. But the trend in computer technology was toward the use of integrated circuits of increasing levels of integration, which resulted in dramatic increases in computational performance at the same time that the size and cost of the computer were reduced. At the beginning of the 1970s, the economics were right for using a dedicated computer as the MCU. This application came to be known as *computer numerical control* (CNC). At first, minicomputers were used as the controllers; subsequently, microcomputers were used as the performance/size trend continued.

7.1 FUNDAMENTALS OF NC TECHNOLOGY

To introduce NC technology, we first define the basic components of an NC system, then describe NC coordinate systems in common use and types of motion controls used in NC.

7.1.1 Basic Components of an NC System

An NC system consists of three basic components: (1) a part program of instructions, (2) a machine control unit, and (3) processing equipment. The general relationship among the three components is illustrated in Figure 7.1.

The *part program* is the set of detailed step-by-step commands that direct the actions of the processing equipment. In machine tool applications, the person who prepares the program is called a *part programmer*. In these applications, the individual commands refer to positions of a cutting tool relative to the worktable on which the workpart is fixtured. Additional instructions are usually included, such as spindle speed, feed rate, cutting tool selection, and other functions. The program is coded on a suitable medium for submission to the machine control unit. For many years, the common medium was 1-inch wide punched tape, using a standard format that could be interpreted by the machine control unit. Today, punched tape has largely been replaced by newer storage technologies in modern machine shops. These technologies include magnetic tape, diskettes, and electronic transfer of part programs from a computer.

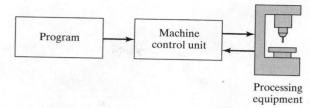

Figure 7.1 Basic components of an NC system.

In modern NC technology, the *machine control unit* (MCU) is a microcomputer and related control hardware that stores the program of instructions and executes it by converting each command into mechanical actions of the processing equipment, one command at a time. The related hardware of the MCU includes components to interface with the processing equipment and feedback control elements. The MCU also includes one or more reading devices for entering part programs into memory. Software residing in the MCU includes control system software, calculation algorithms, and translation software to convert the NC part program into a usable format for the MCU. Because the MCU is a computer, the term *computer numerical control* (CNC) is used to distinguish this type of NC from its technological predecessors that were based entirely on hard-wired electronics. Today, virtually all new MCUs are based on computer technology; hence, when we refer to NC in this chapter and elsewhere, we mean CNC.

The third basic component of an NC system is the *processing equipment* that performs the actual productive work (e.g., machining). It accomplishes the processing steps to transform the starting workpiece into a completed part. Its operation is directed by the MCU, which in turn is driven by instructions contained in the part program. In the most common example of NC, machining, the processing equipment consists of the worktable and spindle as well as the motors and controls to drive them.

7.1.2 NC Coordinate Systems

To program the NC processing equipment, a part programmer must define a standard axis system by which the position of the workhead relative to the workpart can be specified. There are two axis systems used in NC, one for flat and prismatic workparts and the other for rotational parts. Both axis systems are based on the Cartesian coordinate system.

The axis system for flat and prismatic parts consists of the three linear axes (x, y, z) in the Cartesian coordinate system, plus three rotational axes (a, b, c), as shown in Figure 7.2(a). In most machine tool applications, the x- and y-axes are used to move and position the worktable to which the part is attached, and the z-axis is used to control the vertical position of the cutting tool. Such a positioning scheme is adequate for simple NC applications such as drilling and punching of flat sheet metal. Programming these machine tools consists of little more than specifying a sequence of x-y coordinates.

The a-, b-, and c-rotational axes specify angular positions about the x-, y-, and z-axes, respectively. To distinguish positive from negative angles, the *right-hand rule* is used: Using the right hand with the thumb pointing in the positive linear axis direction $(+x, +y, \text{ or } +z)$, the fingers of the hand are curled in the positive rotational direction. The rotational axes can be used for one or both of the following: (1) orientation of the workpart to present different surfaces for machining or (2) orientation of the tool or workhead

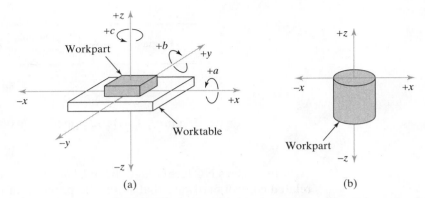

Figure 7.2 Coordinate systems used in NC (a) for flat and prismatic work and (b) for rotational work. (On most turning machines, the z-axis is horizontal rather than vertical as we have shown it.)

at some angle relative to the part. These additional axes permit machining of complex workpart geometries. Machine tools with rotational axis capability generally have either four or five axes: three linear axes plus one or two rotational axes. Most NC machine tool systems do not require all six axes.

The coordinate axes for a rotational NC system are illustrated in Figure 7.2(b). These systems are associated with NC lathes and turning centers. Although the work rotates, this is not one of the controlled axes on most turning machines. Consequently, the y-axis is not used. The path of the cutting tool relative to the rotating workpiece is defined in the x-z plane, where the x-axis is the radial location of the tool, and the z-axis is parallel to the axis of rotation of the part.

The part programmer must decide where the origin of the coordinate axis system should be located. This decision is usually based on programming convenience. For example, the origin might be located at one of the corners of the part. If the workpart is symmetrical, the zero point might be most conveniently defined at the center of symmetry. Wherever the location, this zero point is communicated to the machine tool operator. At the beginning of the job, the operator must move the cutting tool under manual control to some *target point* on the worktable, where the tool can be easily and accurately positioned. The target point has been previously referenced to the origin of the coordinate axis system by the part programmer. When the tool has been accurately positioned at the target point, the operator indicates to the MCU where the origin is located for subsequent tool movements.

7.1.3 Motion Control Systems

Some NC processes are performed at discrete locations on the workpart (e.g., drilling and spot welding). Others are carried out while the workhead is moving (e.g., turning, milling, and continuous arc welding). If the workhead is moving, it may be necessary to follow a straight line path or a circular or other curvilinear path. These different types of movement are accomplished by the motion control system, whose features are explained below.

Point-to-Point Versus Continuous Path Control. Motion control systems for NC (and robotics, Chapter 8) can be divided into two types: (1) point-to-point and

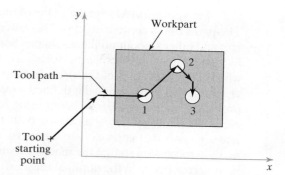

Figure 7.3 Point-to-point (positioning) control in NC. At each *x-y* position, table movement stops to perform the hole-drilling operation.

(2) continuous path. *Point-to-point systems*, also called *positioning systems*, move the worktable to a programmed location without regard for the path taken to get to that location. Once the move has been completed, some processing action is accomplished by the workhead at the location, such as drilling or punching a hole. Thus, the program consists of a series of point locations at which operations are performed, as depicted in Figure 7.3.

Continuous path systems are motion control systems capable of continuous simultaneous control of two or more axes. This provides control of the tool trajectory relative to the workpart. In this case, the tool performs the process while the worktable is moving, thus enabling the system to generate angular surfaces, two-dimensional curves, or three-dimensional contours in the workpart. This control mode is required in many milling and turning operations. A simple two-dimensional profile milling operation is shown in Figure 7.4 to illustrate continuous path control. When continuous path control is utilized to move the tool parallel to only one of the major axes of the machine tool worktable, this is called *straight-cut NC*. When continuous path control is used for simultaneous control of two or more axes in machining operations, the term *contouring* is used.

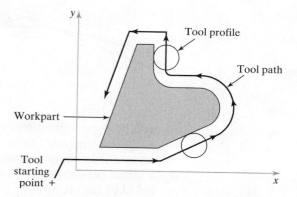

Figure 7.4 Continuous path (contouring) control in NC (*x-y* plane only). Note that cutting tool path must be offset from the part outline by a distance equal to its radius.

Interpolation Methods. One of the important aspects of contouring is interpolation. The paths that a contouring-type NC system is required to generate often consist of circular arcs and other smooth nonlinear shapes. Some of these shapes can be defined mathematically by relatively simple geometric formulas (e.g., the equation for a circle is $x^2 + y^2 = R^2$, where $R =$ the radius of the circle and the center of the circle is at the origin), whereas others cannot be mathematically defined except by approximation. In any case, a fundamental problem in generating these shapes using NC equipment is that they are continuous, whereas NC is digital. To cut along a circular path, the circle must be divided into a series of straight line segments that approximate the curve. The tool is commanded to machine each line segment in succession so that the machined surface closely matches the desired shape. The maximum error between the nominal (desired) surface and the actual (machined) surface can be controlled by the lengths of the individual line segments, as explained in Figure 7.5.

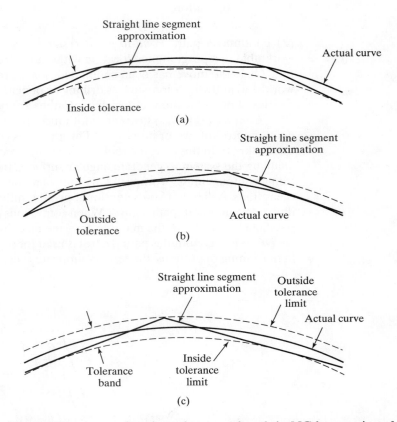

Figure 7.5 Approximation of a curved path in NC by a series of straight line segments. The accuracy of the approximation is controlled by the maximum deviation (called the tolerance) between the nominal (desired) curve and the straight line segments that are machined by the NC system. In (a) the tolerance is defined on only the inside of the nominal curve. In (b) the tolerance is defined on only the outside of the desired curve. In (c) the tolerance is defined on both the inside and outside of the desired curve.

TABLE 7.1 Numerical Control Interpolation Methods for Continuous Path Control

Linear interpolation. This is the most basic method and is used when a straight line path is to be generated in continuous path NC. Two-axis and three-axis linear interpolation routines are sometimes distinguished in practice, but conceptually they are the same. The programmer specifies the beginning point and end point of the straight line and the feed rate to be used along the straight line. The interpolator computes the feed rates for each of the two (or three) axes to achieve the specified feed rate.

Circular interpolation. This method permits programming of a circular arc by specifying the following parameters: (1) the coordinates of the starting point, (2) the coordinates of the endpoint, (3) either the center or radius of the arc, and (4) the direction of the cutter along the arc. The generated tool path consists of a series of small straight line segments (see Figure 7.5) calculated by the interpolation module. The cutter is directed to move along each line segment one by one to generate the smooth circular path. A limitation of circular interpolation is that the plane in which the circular arc exists must be a plane defined by two axes of the NC system ($x - y$, $x - z$, or $y - z$).

Helical interpolation. This method combines the circular interpolation scheme for two axes described above with linear movement of a third axis. This permits the definition of a helical path in three-dimensional space. Applications include the machining of large internal threads, either straight or tapered.

Parabolic and *cubic interpolations.* These routines provide approximations of free form curves using higher order equations. They generally require considerable computational power and are not as common as linear and circular interpolation. Most applications are in the aerospace and automotive industries for free form designs that cannot accurately and conveniently be approximated by combining linear and circular interpolations.

If the programmer were required to specify the endpoints for each of the line segments, the programming task would be extremely arduous and fraught with errors. Also, the part program would be extremely long because of the large number of points. To ease the burden, interpolation routines have been developed that calculate the intermediate points to be followed by the cutter to generate a particular mathematically defined or approximated path.

A number of interpolation methods are available to deal with the various problems encountered in generating a smooth continuous path in contouring. They include (1) linear interpolation, (2) circular interpolation, (3) helical interpolation, (4) parabolic interpolation, and (5) cubic interpolation. Each of these procedures, briefly described in Table 7.1, permits the programmer to generate machine instructions for linear or curvilinear paths using relatively few input parameters. The interpolation module in the MCU performs the calculations and directs the tool along the path. In CNC systems, the interpolator is generally accomplished by software. Linear and circular interpolators are almost always included in modern CNC systems, whereas helical interpolation is a common option. Parabolic and cubic interpolations are less common; they are only needed by machine shops that must produce complex surface contours.

Absolute Versus Incremental Positioning. Another aspect of motion control is concerned with whether positions are defined relative to the origin of the coordinate system or relative to the previous location of the tool. The two cases are called absolute positioning and incremental positioning. In absolute positioning, the workhead locations are always defined with respect to the origin of the axis system. In incremental positioning, the next workhead position is defined relative to the present location. The difference is illustrated in Figure 7.6.

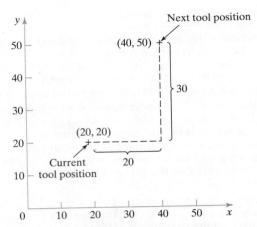

Figure 7.6 Absolute versus incremental positioning. The workhead is presently at point (20, 20) and is to be moved to point (40, 50). In absolute positioning, the move is specified by $x = 40$, $y = 50$; whereas in incremental positioning, the move is specified by $x = 20$, $y = 30$.

7.2 COMPUTER NUMERICAL CONTROL

Since the introduction of NC in 1952, there have been dramatic advances in digital computer technology. The physical size and cost of a digital computer have been significantly reduced at the same time that its computational capabilities have been substantially increased. It was logical for the makers of NC equipment to incorporate these advances in computer technology into their products, starting with large mainframe computers in the 1960s and followed by minicomputers in the 1970s and microcomputers in the 1980s. Today, NC means computer numerical control. *Computer numerical control* (CNC) is defined as an NC system whose MCU is based on a dedicated microcomputer rather than on a hard-wired controller. The latest computer controllers for CNC feature high-speed processors, large memories, solid-state flash memory, improved servos, and bus architectures [13]. Some controllers have the capability to control multiple machines in the mode of DNC (Section 7.3).

7.2.1 Features of CNC

Computer NC systems include additional features beyond what is feasible with conventional hard-wired NC. These features, many of which are standard on most CNC MCUs, include the following:

- *Storage of more than one part program.* With improvements in computer storage technology, newer CNC controllers have sufficient capacity to store multiple programs. Controller manufacturers generally offer one or more memory expansions as options to the MCU.
- *Various forms of program input.* Whereas conventional (hard-wired) MCUs are limited to punched tape as the input medium for entering part programs, CNC controllers generally possess multiple data entry capabilities, such as punched tape (if the machine shop still uses punched tape), magnetic tape, floppy diskettes, RS-232 communications with external computers, and manual data input (operator entry of program).

- *Program editing at the machine tool.* CNC permits a part program to be edited while it resides in the MCU computer memory. Hence, a program can be tested and corrected entirely at the machine site, rather than being returned to the programming office for corrections. In addition to part program corrections, editing also permits cutting conditions in the machining cycle to be optimized. After the program has been corrected and optimized, the revised version can be stored on punched tape or other media for future use.

- *Fixed cycles and programming subroutines.* The increased memory capacity and the ability to program the control computer provide the opportunity to store frequently used machining cycles as *macros* that can be called by the part program. Instead of writing the full instructions for the particular cycle into every program, a programmer includes a call statement in the part program to indicate that the macro cycle should be executed. These cycles often require that certain parameters be defined, for example, a bolt hole circle, in which the diameter of the bolt circle, the spacing of the bolt holes, and other parameters must be specified.

- *Interpolation.* Some of the interpolation schemes described in Table 7.1 are normally executed only on a CNC system because of the computational requirements. Linear and circular interpolation are sometimes hard-wired into the control unit, but helical, parabolic, and cubic interpolations are usually executed by a stored program algorithm.

- *Positioning features for setup.* Setting up the machine tool for a given workpart involves installing and aligning a fixture on the machine tool table. This must be accomplished so that the machine axes are established with respect to the workpart. The alignment task can be facilitated using certain features made possible by software options in a CNC system. *Position set* is one of these features. With position set, the operator is not required to locate the fixture on the machine table with extreme accuracy. Instead, the machine tool axes are referenced to the location of the fixture using a target point or set of target points on the work or fixture.

- *Cutter length and size compensation.* In older style controls, cutter dimensions had to be set very precisely to agree with the tool path defined in the part program. Alternative methods for ensuring accurate tool path definition have been incorporated into CNC controls. One method involves manually entering the actual tool dimensions into the MCU. These actual dimensions may differ from those originally programmed. Compensations are then automatically made in the computed tool path. Another method involves use of a tool length sensor built into the machine. In this technique, the cutter is mounted in the spindle and the sensor measures its length. This measured value is then used to correct the programmed tool path.

- *Acceleration and deceleration calculations.* This feature is applicable when the cutter moves at high feed rates. It is designed to avoid tool marks on the work surface that would be generated due to machine tool dynamics when the cutter path changes abruptly. Instead, the feed rate is smoothly decelerated in anticipation of a tool path change and then accelerated back up to the programmed feed rate after the direction change.

- *Communications interface.* With the trend toward interfacing and networking in plants today, most modern CNC controllers are equipped with a standard RS-232 or other communications interface to link the machine to other computers and computer-driven devices. This is useful for various applications, such as (1) downloading part programs from a central data file; (2) collecting operational data such as work-piece counts, cycle times, and machine utilization; and (3) interfacing with peripheral equipment, such as robots that load and unload parts.

- *Diagnostics.* Many modern CNC systems possess a diagnostics capability that monitors certain aspects of the machine tool to detect malfunctions or signs of impending malfunctions or to diagnose system breakdowns.

7.2.2 The Machine Control Unit for CNC

The MCU is the hardware that distinguishes CNC from conventional NC. The general configuration of the MCU in a CNC system is illustrated in Figure 7.7. The MCU consists of the following components and subsystems: (1) central processing unit, (2) memory, (3) I/O interface, (4) controls for machine tool axes and spindle speed, and (5) sequence controls for other machine tool functions. These subsystems are interconnected by means of a system bus, which communicates data and signals among the components of the network.

Central Processing Unit. The central processing unit (CPU) is the brain of the MCU. It manages the other components in the MCU based on software contained in main memory. The CPU can be divided into three sections: (1) control section, (2) arithmetic-logic unit, and (3) immediate access memory. The *control section* retrieves commands and data from memory and generates signals to activate other components in the MCU. In short, it sequences, coordinates, and regulates all of the activities of the MCU computer. The *arithmetic-logic unit* (ALU) consists of the circuitry to perform various calculations (addition, subtraction, multiplication), counting, and logical functions required by software residing in memory. The *immediate access memory* provides a temporary storage for data being processed by the CPU. It is connected to main memory by means of the system data bus.

Memory. The immediate access memory in the CPU is not intended for storing CNC software. A much greater storage capacity is required for the various programs and data needed to operate the CNC system. As with most other computer systems, CNC memory can be divided into two categories: (1) main memory, and (2) secondary memory. *Main memory* (also known as *primary storage*) consists of ROM (read-only memory) and RAM (random access memory) devices. Operating system software and machine

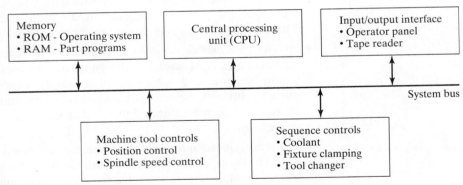

Figure 7.7 Configuration of CNC machine control unit.

interface programs (Section 7.2.3) are generally stored in ROM. These programs are usually installed by the manufacturer of the MCU. Numerical control part programs are stored in RAM devices. Current programs in RAM can be erased and replaced by new programs as jobs are changed.

High-capacity *secondary memory* (also called *auxiliary storage* or *secondary storage*) devices are used to store large programs and data files, which are transferred to main memory as needed. Common among the secondary memory devices are hard disks and portable devices that have replaced most of the punched paper tape traditionally used to store part programs. Hard disks are high-capacity storage devices that are permanently installed in the CNC machine control unit. CNC secondary memory is used to store part programs, macros, and other software.

Input/Output Interface.   The I/O interface provides communication between the various components of the CNC system, other computer systems, and the machine operator. As its name suggests, the I/O interface transmits and receives data and signals to and from external devices, several of which are indicated in Figure 7.7. The *operator control panel* is the basic interface by which the machine operator communicates to the CNC system. This is used to enter commands related to part program editing, MCU operating mode (e.g., program control vs. manual control), speeds and feeds, cutting fluid pump on/off, and similar functions. Either an alphanumeric keypad or keyboard is usually included in the operator control panel. The I/O interface also includes a display (CRT or LED) for communication of data and information from the MCU to the machine operator. The display is used to indicate current status of the program as it is being executed and to warn the operator of any malfunctions in the CNC system.

Also included in the I/O interface are one or more means of entering the part program into storage. As indicated previously, NC part programs are stored in a variety of ways. Programs can also be entered manually by the machine operator or stored at a central computer site and transmitted via local area network (LAN) to the CNC system. Whichever means is employed by the plant, a suitable device must be included in the I/O interface to allow input of the program into MCU memory.

Controls for Machine Tool Axes and Spindle Speed.   These are hardware components that control the position and velocity (feed rate) of each machine axis as well as the rotational speed of the machine tool spindle. The control signals generated by MCU must be converted to a form and power level suited to the particular position control systems used to drive the machine axes. Positioning systems can be classified as open loop or closed loop, and different hardware components are required in each case. A more detailed discussion of these hardware elements is presented in Section 7.6, together with an analysis of how they operate to achieve position and feed rate control. For our purposes here, it is sufficient to indicate that some of the hardware components are resident in the MCU.

Depending on the type of machine tool, the spindle is used to drive either (1) the workpiece or (2) a rotating cutter. Turning exemplifies the first case, whereas milling and drilling exemplify the second. Spindle speed is a programmed parameter for most CNC machine tools. Spindle speed control components in the MCU usually consist of a drive control circuit and a feedback sensor interface. The particular hardware components depend on the type of spindle drive.

Sequence Controls for Other Machine Tool Functions. In addition to control of table position, feed rate, and spindle speed, several additional functions are accomplished under part program control. These auxiliary functions are generally on/off (binary) actuations, interlocks, and discrete numerical data. A sampling of these functions is presented in Table 7.2. To avoid overloading the CPU, a programmable logic controller (Chapter 9) is sometimes used to manage the I/O interface for these auxiliary functions.

Personal Computers and the MCU. In growing numbers, personal computers (PCs) are being used in the factory to implement process control (Section 5.3.3), and CNC is no exception. Two basic configurations are being applied [7]: (1) the PC is used as a separate front-end interface for the MCU, and (2) the PC contains the motion control board and other hardware required to operate the machine tool. In the second case, the CNC control board fits into a standard slot of the PC. In either configuration, the advantage of using a PC for CNC is its flexibility to execute a variety of user software in addition to and concurrently with controlling the machine tool operation. The user software might include programs for shop-floor control, statistical process control, solid modeling, cutting tool management, and other computer-aided manufacturing software. Other benefits include improved ease of use compared with conventional CNC and ease of networking the PCs. A possible disadvantage is the lost time to retrofit the PC for CNC, particularly when installing the CNC motion controls inside the PC. Also, some machine shops prefer not to have PCs on the shop floor due to concerns about viruses and security [15].

7.2.3 CNC Software

The computer in CNC operates by means of software. There are three types of software programs used in CNC systems: (1) operating system software, (2) machine interface software, and (3) application software.

The principal function of the operating system software is to interpret the NC part programs and generate the corresponding control signals to drive the machine tool axes. It is installed by the controller manufacturer and is stored in ROM in the MCU. The operating system software consists of the following: (1) an *editor,* which permits the machine operator to input and edit NC part programs and perform other file management functions; (2) a *control program,* which decodes the part program instructions, performs interpolation and acceleration/deceleration calculations, and accomplishes other related functions to produce the coordinate control signals for each axis; and (3) an *executive program,* which manages the execution of the CNC software as well as the I/O operations of the MCU. The operating system software also includes the diagnostic routines that are available in the CNC system.

Machine interface software is used to operate the communication link between the CPU and the machine tool to accomplish the CNC auxiliary functions (Table 7.2). As previously indicated, the I/O signals associated with the auxiliary functions are sometimes implemented by means of a programmable logic controller interfaced to the MCU, so the machine interface software is often written in the form of ladder logic diagrams (Section 9.2).

Finally, the application software consists of the NC part programs that are written for machining (or other) applications in the user's plant. We postpone the topic of part programming to Section 7.6.

TABLE 7.2 Examples of CNC Auxiliary Functions Often Implemented by a Programmable Logic Controller in the MCU

CNC Auxiliary Function	Type or Classification
Coolant control	On/off output from MCU to pump
Tool changer and tool storage unit	Discrete numerical data (possible values limited to capacity of tool storage unit)
Fixture clamping device	On/off output from MCU to clamp actuator
Emergency warning or stop	On/off input to MCU from sensor; on/off output to display and alarm
Robot for part loading/unloading	Interlock to sequence loading and unloading operation; I/O signals between MCU and robot
Counters (e.g., piece counts)	Discrete numerical data (possible values limited to number of parts that can be produced in a given time period, such as a shift)

7.3 DISTRIBUTED NUMERICAL CONTROL

Historical Note 7.1 describes several ways in which digital computers have been used to implement NC. The first attempt to use a digital computer to drive the NC machine tool was called *direct numerical control* (DNC). This was in the late 1960s, before the advent of CNC. As initially implemented, DNC involved the control of a number of machine tools by a single (mainframe) computer through direct connection and in real time. Instead of using a punched tape reader to enter the part program into the MCU, the program was transmitted to the MCU directly from the computer, one block of instructions at a time. This mode of operation was referred to by the name *behind the tape reader* (BTR). The DNC computer provided instruction blocks to the machine tool on demand; when a machine needed control commands, they were communicated to it immediately. As each block was executed by the machine, the next block was transmitted. As far as the machine tool was concerned, the operation was no different from that of a conventional NC controller. In theory, DNC relieved the NC system of its least reliable components: the punched tape and tape reader.

The general configuration of a DNC system is depicted in Figure 7.8. The system consisted of four components: (1) central computer, (2) bulk memory at the central

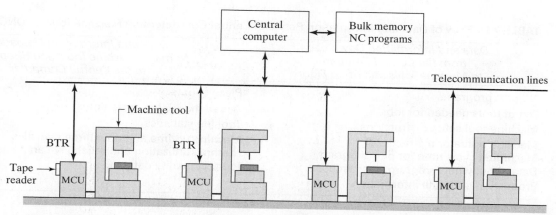

Figure 7.8 General configuration of a DNC system. Connection to MCU is behind the tape reader. Key: BTR = behind the tape reader, MCU = machine control unit.

computer site, (3) set of controlled machines, and (4) telecommunications lines to connect the machines to the central computer. In operation, the computer called the required part program from bulk memory and sent it (one block at a time) to the designated machine tool. This procedure was replicated for all machine tools under direct control of the computer. One commercially available DNC system during the 1970s claimed to be capable of controlling up to 256 machines.

In addition to transmitting data to the machines, the central computer also received data back from the machines to indicate operating performance in the shop (e.g., number of machining cycles completed, machine utilization, and breakdowns). A central objective of DNC was to achieve two-way communication between the machines and the central computer.

As the number of CNC machine installations grew during the 1970s and 1980s, a new form of DNC emerged, called *distributed numerical control* (DNC). The configuration of the new DNC is very similar to that shown in Figure 7.8 except that the central computer is connected to MCUs, which are themselves computers. This permits complete part programs, not one block at a time, to be sent to the machine tools. It also permits easier and less costly installation of the overall system, because the individual CNC machines can be put into service and distributed NC can be added later. Redundant computers improve system reliability compared with the original DNC. The new DNC permits two-way communication of data between the shop floor and the central computer, which was one of the important features included in the old DNC. However, improvements in data collection devices as well as advances in computer and communications technologies have expanded the range and flexibility of the information that can be gathered and disseminated. Some of the data and information sets included in the two-way communication flow are itemized in Table 7.3. This flow of information in DNC is similar to the information flow in shop floor control, discussed in Chapter 25.

Distributed NC systems can take on a variety of physical configurations, depending on the number of machine tools included, job complexity, security requirements, and equipment availability and preferences. There are several ways to configure a DNC system. We illustrate two types in Figure 7.9: (a) switching network and (b) LAN. Each type has several possible variations.

TABLE 7.3 Flow of Data and Information Between Central Computer and Machine Tools in DNC

Data and Information Downloaded from the Central Computer to Machine Tools and Shop Floor	Data and Information Uploaded from the Machine Tools and Shop Floor to the Central Computer
NC part programs	Piece counts
List of tools needed for job	Actual machining cycle times
Machine tool setup instructions	Tool life statistics
Machine operator instructions	Machine uptime and downtime statistics, from which
Machining cycle time for part program	machine utilization and reliability can be assessed
Data about when program was last used	Product quality data
Production schedule information	

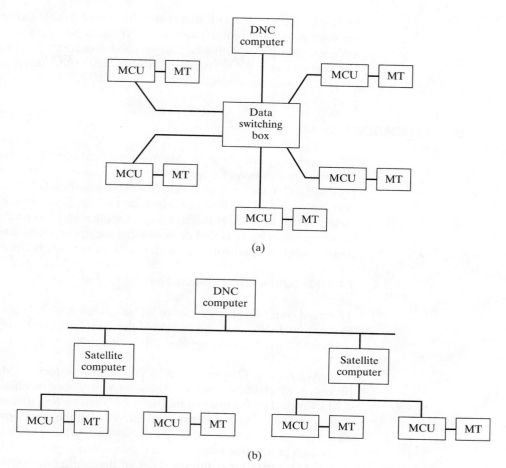

Figure 7.9 Two configurations of DNC: (a) switching network and (b) LAN. Key: MCU = machine control unit, MT = machine tool.

The switching network is the simplest DNC system to configure. It uses a data switching box to make a connection from the central computer to a given CNC machine for downloading part programs or uploading data. Transmission of programs to the MCU is accomplished through an RS-232-C connection. Virtually all commercial MCUs include the RS-232-C or compatible device as standard equipment. Use of a switching box limits the number of machines that can be included in the DNC system. The limit depends on factors such as part program complexity, frequency of service required to each machine, and capabilities of the central computer. The number of machines in the DNC system can be increased by employing a serial link RS-232-C multiplexer.

Local area networks (LANs) have been used for DNC since the early 1980s. Various network structures are used in DNC systems, among which is the centralized structure illustrated in Figure 7.9(b). In this arrangement, the computer system is organized as a hierarchy, with the central (host) computer coordinating several satellite computers that are each responsible for a number of CNC machines. Alternative LAN structures are possible, each with its relative advantages and disadvantages. Local area networks in different sections and departments of a plant are often interconnected in plant-wide and corporate-wide networks.

7.4 APPLICATIONS OF NC

The operating principle of NC has many applications. There are many industrial operations in which the position of a workhead must be controlled relative to a part or product being processed. The applications divide into two categories: (1) machine tool applications and (2) non-machine tool applications. Machine tool applications are usually associated with the metalworking industry. Non-machine tool applications comprise a diverse group of operations in other industries. It should be noted that the applications are not always identified by the name "numerical control"; this term is used principally in the machine tool industry.

7.4.1 Machine Tool Applications

The most common applications of NC are in machine tool control. Machining was the first application of NC, and it is still one of the most important commercially. In this section, we discuss NC machine tool applications with emphasis on metal machining processes.

Machining Operations and NC Machine Tools. Machining is a manufacturing process in which the geometry of the work is produced by removing excess material (Section 2.2.1). Control of the relative motion between a cutting tool and the workpiece creates the desired geometry. Machining is considered one of the most versatile processes because it can be used to create a wide variety of shapes and surface finishes. It can be performed at relatively high production rates to yield highly accurate parts at relatively low cost.

There are four common types of machining operations: (a) turning, (b) drilling, (c) milling, and (d) grinding. The four operations are shown in Figure 7.10. Each of the machining operations is carried out at a certain combination of speed, feed, and depth of cut, collectively called the *cutting conditions* for the operation. The terminology varies somewhat for grinding. These cutting conditions are illustrated in Figure 7.10 for (a) turning, (b) drilling, and (c) milling. Consider milling. The *cutting speed* is the velocity of the tool (milling cutter) relative to the work, measured in meters per minute (feet per minute). This is usually programmed into the machine as a spindle rotation speed (revolutions per minute). Cutting speed can be converted into spindle rotation speed by means of the equation

$$N = \frac{v}{\pi D} \tag{7.1}$$

where N = spindle rotation speed (rev/min), v = cutting speed (m/min, ft/min), and D = milling cutter diameter (m, ft). In milling, the *feed* usually means the size of the chip formed by each tooth in the milling cutter, often referred to as the *chip load* per tooth. This must normally be programmed into the NC machine as the feed rate (the travel rate of the machine tool table). Therefore, feed must be converted to feed rate as

$$f_r = N n_t f \tag{7.2}$$

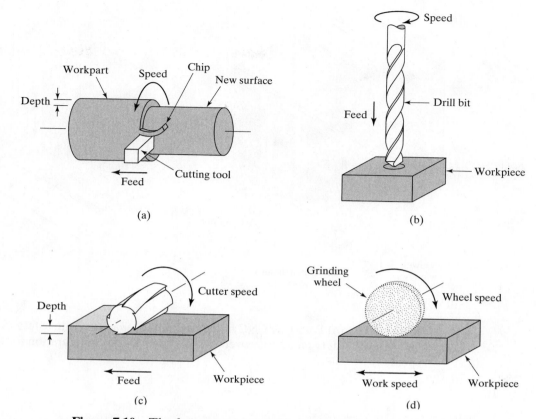

Figure 7.10 The four common machining operations are (a) turning, (b) drilling, (c) peripheral milling, and (d) surface grinding.

where f_r = feed rate (mm/min, in/min), N = rotational speed (rev/min), n_t = number of teeth on the milling cutter, and f = feed (mm/tooth, in/tooth). For a turning operation, feed is defined as the lateral movement of the cutting tool per revolution of the workpiece, so the units are millimeters per revolution (inches per revolution). *Depth of cut* is the distance the tool penetrates below the original surface of the work (mm, in). These are the parameters that must be controlled during the operation of an NC machine through motion or position commands in the part program.

Each of the four machining processes is traditionally carried out on a machine tool designed to perform that process. Turning is performed on a lathe, drilling is done on a drill press, milling on a milling machine, and so on. The common NC machine tools are listed in the following along with their typical features:

- *NC lathe*, either horizontal or vertical axis. Turning requires two-axis, continuous path control, either to produce a straight cylindrical geometry (called straight turning) or to create a profile (contour turning).
- *NC boring mill*, horizontal and vertical spindle. Boring is similar to turning, except that an internal cylinder is created instead of an external cylinder. The operation requires continuous path, two-axis control.

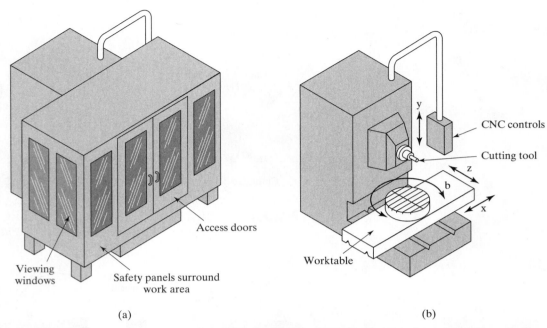

Figure 7.11 (a) Four-axis CNC horizontal milling machine with safety panels installed and (b) with safety panels removed to show typical axis configuration for the horizontal spindle.

- *NC drill press.* These machines use point-to-point control of the workhead (spindle containing the drill bit) and two axis (*x-y*) control of the worktable. Some NC drill presses have turrets containing six or eight drill bits. The turret position is programmed under NC control, allowing different drill bits to be applied to the same workpart during the machine cycle without requiring the machine operator to manually change the tool.
- *NC milling machine.* Milling machines require continuous path control to perform straight cut or contouring operations. Figure 7.11 illustrates the features of a four-axis milling machine.
- *NC cylindrical grinder.* This machine operates like a turning machine, except that the tool is a grinding wheel. It has continuous path two-axis control, similar to an NC lathe.

Numerical control has had a profound influence on the design and operation of machine tools. One of the effects is that the proportion of time spent by the machine cutting metal is significantly greater than with manually operated machines. This causes certain components such as the spindle, drive gears, and feed screws to wear more rapidly. These components must be designed to last longer on NC machines. Secondly, the addition of the electronic control unit has increased the cost of the machine, requiring higher equipment utilization. Instead of running the machine during only one shift, which is usually the schedule with manually operated machines, NC machines are often operated during two or even three shifts to obtain the required economic payback. Third, the increasing cost of labor has altered the relative roles of the human operator and the machine tool. Instead of being the highly skilled worker who controlled every aspect of part production, the NC machine operator performs only part loading and unloading, tool-changing, chip clearing, and the like. With these reduced responsibilities, one operator can often run two or three automatic machines.

The functions of the machine tool have also changed. NC machines are designed to be highly automatic and capable of combining several operations in one setup that formerly required several different machines. They are also designed to reduce the time consumed by the noncutting elements in the operation cycle, such as changing tools and loading and unloading the workpart. These changes are best exemplified by a new type of machine that did not exist prior to the development of NC: machining centers. A *machining center* is a machine tool capable of performing multiple machining operations on a single workpiece in one setup. The operations involve rotating cutters, such as milling and drilling, and the feature that enables more than one operation to be performed in one setup is automatic tool-changing. We discuss machining centers and related machine tools in our coverage of single station manufacturing cells (Section 14.3.3).

NC Application Characteristics. In general, NC technology is appropriate for low-to-medium production of medium-to-high variety product. Using the terminology of Section 2.4.1, the product is low-to-medium Q, medium-to-high P. Over many years of machine shop practice, certain part characteristics have been identified as most suited to the application of NC. These characteristics are the following:

1. *Batch production.* NC is most appropriate for parts produced in small or medium lot sizes (batch sizes ranging from one unit up to several hundred units). Dedicated automation would not be economical for these quantities because of the high fixed cost. Manual production would require many separate machine setups and would result in higher labor cost, longer lead time, and higher scrap rate.

2. *Repeat orders.* Batches of the same parts are produced at random or periodic intervals. Once the NC part program has been prepared, parts can be economically produced in subsequent batches using the same part program.

3. *Complex part geometry.* The part geometry includes complex curved surfaces such as those found on airfoils and turbine blades. Mathematically defined surfaces such as circles and helixes can also be accomplished with NC. Some of these geometries would be difficult if not impossible to achieve accurately using conventional machine tools.

4. *Much metal needs to be removed from the workpart.* This condition is often associated with a complex part geometry. The volume and weight of the final machined part is a relatively small fraction of the starting block. Such parts are common in the aircraft industry to fabricate large structural sections with low weights.

5. *Many separate machining operations on the part.* This applies to parts consisting of many machined features requiring different cutting tools, such as drilled and/or tapped holes, slots, flats, and so on. If these operations were machined by a series of manual operations, many setups would be needed. The number of setups can usually be reduced significantly using NC.

6. *The part is expensive.* This factor is often a consequence of one or more of preceding factors 3, 4, and 5. It can also result from using a high-cost starting work material. When the part is expensive, and mistakes in processing would be costly, the use of NC helps to reduce rework and scrap losses.

Although these characteristics pertain mainly to machining, they are adaptable to other production applications as well.

NC for Other Metalworking Processes. NC machine tools have been developed for other metal working processes besides machining operations. These machines include the following:

- *Punch presses* for sheet metal hole punching. The two-axis NC operation is similar to that of a drill press except that holes are produced by punching rather than drilling.
- *Presses* for sheet metal bending. Instead of cutting sheet metal, these systems bend sheet metal according to programmed commands.
- *Welding machines.* Both spot welding and continuous arc welding machines are available with automatic controls based on NC.
- *Thermal cutting machines*, such as oxyfuel cutting, laser cutting, and plasma arc cutting. The stock is usually flat; thus, two-axis control is adequate. Some laser cutting machines can cut holes in preformed sheet metal stock, requiring four- or five-axis control.
- *Tube bending machines.* Automatic tube bending machines are programmed to control the location (along the length of the tube stock) and the angle of the bend. Important applications include frames for bicycles and motorcycles.

7.4.2 Other NC Applications

The operating principle of NC has a host of other applications besides machine tool control. Some of the machines with NC-type controls that position a workhead relative to an object being processed are the following:

- *Electrical wire wrap machines.* These machines, pioneered by Gardner Denver Corporation, have been used to wrap and string wires on the back pins of electrical wiring boards to establish connections between components on the front of the board. The program of coordinate positions that define the back panel connections is determined from design data and fed to the wire wrap machine. This type of equipment has been used by computer firms and other companies in the electronics industry.
- *Component insertion machines.* This equipment is used to position and insert components on an x-y plane, usually a flat board or panel. The program specifies the x- and y-axis positions in the plane where the components are to be located. Component insertion machines find extensive applications for inserting electronic components into printed circuit boards. Machines are available for either through-hole or surface-mount applications as well as similar insertion-type mechanical assembly operations.
- *Drafting machines.* Automated drafting machines serve as one of the output devices for a CAD/CAM (computer-aided design/computer-aided manufacturing) system. The design of a product and its components is developed on the CAD/CAM system. Design iterations are developed on the graphics monitor rather than on a mechanical drafting board. When the design is finalized, the output is plotted on the drafting machine, basically a high-speed x-y plotter.

- *Coordinate measuring machines.* A coordinate measuring machine (CMM) is an inspection machine used for measuring or checking dimensions of a part. The CMM has a probe that can be manipulated in three axes and that identifies when contact is made against a part surface. The location of the probe tip is determined by the CMM control unit, thereby indicating some dimension on the part. Many coordinate measuring machines are programmed to perform automated inspections under NC. We discuss coordinate measuring machines in Section 22.4.

- *Tape laying machines for polymer composites.* The workhead of this machine is a dispenser of uncured polymer matrix composite tape. The machine is programmed to lay the tape onto the surface of a contoured mold, following a back-and-forth and crisscross pattern to build up a required thickness. The result is a multilayered panel of the same shape as the mold.

- *Filament winding machines for polymer composites.* These are similar to the preceding machine except that a filament is dipped in uncured polymer and wrapped around a rotating pattern of roughly cylindrical shape.

Additional applications of NC include cloth cutting, knitting, and riveting.

7.4.3 Advantages and Disadvantages of NC

When the production application satisfies the characteristics identified in Section 7.4.1, NC yields many advantages over manual production methods. These advantages translate into economic savings for the user company. However, NC involves more sophisticated technology than conventional production methods, and there are costs that must be considered to apply the technology effectively. In this section, we examine the advantages and disadvantages of NC.

Advantages of NC. The advantages generally attributed to NC, with emphasis on machine tool applications, are the following:

- *Nonproductive time is reduced.* NC cannot optimize the metal cutting process itself, but it can reduce the proportion of time the machine is not cutting metal. Reduction in noncutting time is achieved through fewer setups, less setup time, reduced workpiece handling time, and automatic tool changes on some NC machines. This advantage translates into labor cost savings and lower elapsed times to produce parts.

- *Greater accuracy and repeatability.* Compared with manual production methods, NC reduces or eliminates variations that are due to operator skill differences, fatigue, and other factors attributed to inherent human variabilities. Parts are made closer to nominal dimensions, and there is less dimensional variation among parts in the batch.

- *Lower scrap rates.* Because greater accuracy and repeatability are achieved, and because human errors are reduced during production, more parts are produced within tolerance. As a consequence, a lower scrap allowance can be planned into the production schedule, so fewer parts are made in each batch with the result that production time is saved.

- *Inspection requirements are reduced.* Less inspection is needed when NC is used because parts produced from the same NC part program are virtually identical. Once the program has been verified, there is no need for the high level of sampling inspection that is required when parts are produced by conventional manual methods.

Except for tool wear and equipment malfunctions, NC produces exact replicates of the part each cycle.

- *More complex part geometries are possible.* NC technology has extended the range of possible part geometries beyond what is practical with manual machining methods. This is an advantage in product design in several ways: (1) More functional features can be designed into a single part, thus reducing the total number of parts in the product and the associated cost of assembly, (2) mathematically defined surfaces can be fabricated with high precision, and (3) the limits within which the designer's imagination can wander to create new part and product geometries are expanded.
- *Engineering changes can be accommodated more gracefully.* Instead of making alterations in a complex fixture so that the part can be machined to the engineering change, revisions are made in the NC part program to accomplish the change.
- *Simpler fixtures are needed.* NC requires simpler fixtures because accurate positioning of the tool is accomplished by the NC machine tool. Tool positioning does not have to be designed into the jig.
- *Shorter manufacturing lead times.* Jobs can be set up more quickly and fewer setups are required per part when NC is used. This results in shorter elapsed time between order release and completion.
- *Reduced parts inventory.* Because fewer setups are required and job changeovers are easier and faster, NC permits production of parts in smaller lot sizes. The economic lot size is lower in NC than in conventional batch production. Average parts inventory is therefore reduced.
- *Less floor space required.* This results from the fact that fewer NC machines are required to perform the same amount of work compared to the number of conventional machine tools needed. Reduced parts inventory also contributes to lower floor space requirements.
- *Operator skill-level requirements are reduced.* Workers need fewer skills to operate an NC machine than to operate a conventional machine tool. Tending an NC machine tool usually consists only of loading and unloading parts and periodically changing tools. The machining cycle is carried out under program control. Performing a comparable machining cycle on a conventional machine requires much more participation by the operator, and a higher level of training and skill.

Disadvantages of NC. There are certain commitments to NC technology that must be made by the machine shop that installs NC equipment, and these commitments, most of which involve additional cost to the company, might be seen as disadvantages. The disadvantages of NC include the following:

- *Higher investment cost.* An NC machine tool has a higher first cost than a comparable conventional machine tool. There are several reasons why: (1) NC machines include CNC controls and electronics hardware; (2) software development costs of the CNC controls manufacturer must be included in the cost of the machine; (3) more reliable mechanical components are generally used in NC machines; and (4) NC machine tools often possess additional features not included on conventional machines, such as automatic tool changers and part changers (Section 14.3.3).
- *Higher maintenance effort.* In general, NC equipment requires more maintenance than conventional equipment, which translates to higher maintenance and repair

costs. This is due largely to the computer and other electronics that are included in a modern NC system. The maintenance staff must include personnel who are trained in maintaining and repairing this type of equipment.

- *Part programming.* NC equipment must be programmed. To be fair, it should be mentioned that process planning must be accomplished for any part, whether or not it is produced on NC equipment. However, NC part programming is a special preparation step in batch production that is absent in conventional machine shop operations.

- *Higher utilization of NC equipment.* To maximize the economic benefits of an NC machine tool, it usually must be operated for multiple shifts. This might mean adding one or two extra shifts to the plant's normal operations, with the requirement for supervision and other staff support.

7.5 ENGINEERING ANALYSIS OF NC POSITIONING SYSTEMS

An NC positioning system converts the coordinate axis values in the NC part program into relative positions of the tool and workpart during processing. Let us consider the simple positioning system shown in Figure 7.12. The system consists of a cutting tool and a worktable on which a workpart is fixtured. The table is designed to move the part relative to the tool. The worktable moves linearly by means of a rotating leadscrew, which is driven by a stepper motor or servomotor. For simplicity, we show only one axis in our sketch. To provide *x-y* capability, the system shown would be piggybacked on top of a second axis perpendicular to the first. The leadscrew has a certain pitch *p* (in/thread, mm/thread). Thus, the table moves a distance equal to the pitch for each revolution. The velocity of the worktable, which corresponds to the feed rate in a machining operation, is determined by the rotational speed of the leadscrew.

Two types of positioning control systems are used in NC systems: (a) open loop and (b) closed loop, as shown in Figure 7.13. An *open-loop system* operates without verifying that the actual position achieved in the move is the same as the desired position. A *closed-loop system* uses feedback measurements to confirm that the final position of the worktable is the location specified in the program. Open-loop systems cost less than closed-loop systems and are appropriate when the force resisting the actuating motion is minimal. Closed-loop systems are normally specified for machines that perform continuous

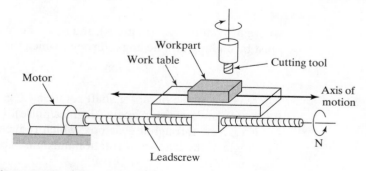

Figure 7.12   Motor and leadscrew arrangement in an NC positioning system.

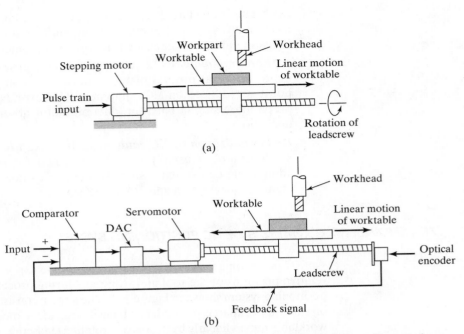

Figure 7.13 Two types of motion control in NC: (a) open loop and (b) closed loop.

path operations such as milling or turning, in which there are significant forces resisting the forward motion of the cutting tool.

7.5.1 Open-Loop Positioning Systems

An open-loop positioning system typically uses a stepper motor to rotate the leadscrew. A stepper motor is driven by a series of electrical pulses, which are generated by the MCU in an NC system. Each pulse causes the motor to rotate a fraction of one revolution, called the step angle. The possible step angles must be consistent with the relationship

$$\alpha = \frac{360}{n_s} \tag{7.3}$$

where α = step angle (degrees), and n_s = the number of step angles for the motor, which must be an integer. The angle through which the motor shaft rotates is given by

$$A_m = n_p\alpha \tag{7.4}$$

where A_m = angle of motor shaft rotation (degrees), n_p = number of pulses received by the motor, and α = step angle (degrees/pulse). The motor shaft is generally connected to the leadscrew through a gear box, which reduces the angular rotation of the leadscrew. The angle of the leadscrew rotation must take the gear ratio into account as

$$A = \frac{n_p\alpha}{r_g} \tag{7.5}$$

where A = angle of leadscrew rotation (degrees), and r_g = gear ratio, defined as the number of turns of the motor for each single turn of the leadscrew. That is,

$$r_g = \frac{A_m}{A} = \frac{N_m}{N} \tag{7.6}$$

where N_m = rotational speed of the motor (rev/min), and N = rotational speed of the leadscrew (rev/min).

The linear movement of the worktable is given by the number of full and partial rotations of the leadscrew multiplied by its pitch,

$$x = \frac{pA}{360} \tag{7.7}$$

where x = x-axis position relative to the starting position (mm, inch), p = pitch of the leadscrew (mm/rev, in/rev), and $A/360$ = number of leadscrew revolutions. The number of pulses required to achieve a specified x-position increment in a point-to-point system can be found by combining the two preceding equations as

$$n_p = \frac{360 x r_g}{p\alpha} \text{ or } \frac{n_s x r_g}{p} \tag{7.8}$$

where the second expression on the right-hand side is obtained by substituting n_s for $360/\alpha$, which is obtained by rearranging Eq. (7.3).

Control pulses are transmitted from the pulse generator at a certain frequency, which drives the worktable at a corresponding velocity or feed rate in the direction of the leadscrew axis. The rotational speed of the leadscrew depends on the frequency of the pulse train as

$$N = \frac{60 f_p}{n_s r_g} \tag{7.9}$$

where N = leadscrew rotational speed (rev/min), f_p = pulse train frequency (Hz, pulses/sec), and n_s = steps per revolution or pulses per revolution. For a two-axis table with continuous path control, the relative velocities of the axes are coordinated to achieve the desired travel direction.

The table travel speed in the direction of leadscrew axis is determined by the rotational speed as

$$v_t = f_r = Np \tag{7.10}$$

where v_t = table travel speed (mm/min, in/min), f_r = table feed rate (mm/min, in/min), N = leadscrew rotational speed (rev/min), and p = leadscrew pitch (mm/rev, in/rev).

The required pulse train frequency to drive the table at a specified linear travel rate can be obtained by combining Eqs. (7.9) and (7.10) and rearranging to solve for f_p:

$$f_p = \frac{v_t n_s r_g}{60p} \text{ or } \frac{f_r n_s r_g}{60p} \tag{7.11}$$

EXAMPLE 7.1   NC Open-Loop Positioning

The worktable of a positioning system is driven by a leadscrew whose pitch = 6.0 mm. The leadscrew is connected to the output shaft of a stepper motor through a gearbox whose ratio is 5:1 (five turns of the motor to one turn of the leadscrew). The stepper motor has 48 step angles. The table must move a distance of

250 mm from its present position at a linear velocity = 500 mm/min. Determine (a) how many pulses are required to move the table the specified distance and (b) the required motor speed and pulse rate to achieve the desired table velocity.

Solution: (a) Rearranging Eq. (7.7) to find the leadscrew rotation angle A corresponding to a distance $x = 250$ mm,

$$A = \frac{360x}{p} = \frac{360(250)}{6.0} = 15,000°$$

With 48 step angles, each step angle is

$$\alpha = \frac{360}{48} = 7.5°$$

Thus, the number of pulses to move the table 250 mm is

$$n_p = \frac{360xr_g}{p\alpha} = \frac{Ar_g}{\alpha} = \frac{15,000(5)}{7.5} = 10,000 \text{ pulses.}$$

(b) The rotational speed of the leadscrew corresponding to a table speed of 500 mm/min can be determined from Eq. (7.10):

$$N = \frac{v_t}{p} = \frac{500}{6} = 83.333 \text{ rev/min.}$$

Equation (7.6) can be used to find the motor speed:

$$N_m = r_g N = 5(83.333) = 416.667 \text{ rev/min}$$

The applied pulse rate to drive the table is given by Eq. (7.11):

$$f_p = \frac{v_t n_s r_g}{60p} = \frac{500(48)(5)}{60(6)} = 333.333 \text{ Hz}$$

7.5.2 Closed-Loop Positioning Systems

A closed-loop NC system, illustrated in Figure 7.13(b), uses servomotors and feedback measurements to ensure that the worktable is moved to the desired position. A common feedback sensor used for NC (and also for industrial robots) is the optical encoder, shown in Figure 7.14. An *optical encoder* is a device for measuring rotational speed that consists of a light source and a photodetector on either side of a disk. The disk contains slots uniformly spaced around the outside of its face. These slots allow the light source to shine through and energize the photodetector. The disk is connected to a rotating shaft whose angular position and velocity are to be measured. As the shaft rotates, the slots cause the light source to be seen by the photocell as a series of flashes. The flashes are converted into an equal number of electrical pulses. By counting the pulses and computing the frequency of the pulse train, one can determine the worktable position and velocity.

The equations that define the operation of a closed-loop NC positioning system are similar to those for an open-loop system. In the basic optical encoder, the angle between slots in the disk must satisfy the following requirement:

$$\alpha = \frac{360}{n_s} \tag{7.12}$$

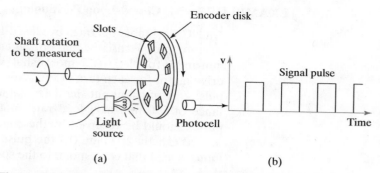

Figure 7.14 Optical encoder: (a) apparatus and (b) series of pulses emitted to measure rotation of disk.

where between slots (degrees/slot), and n_s = number of slots in the disk (slots/rev). For a certain angular rotation of the encoder shaft, the number of pulses sensed by the encoder is given by

$$n_p = \frac{A}{\alpha} \tag{7.13}$$

where n_p = pulse count emitted by the encoder, A = angle of rotation of the encoder shaft (degrees), and α = angle between slots, which converts to degrees per pulse. The pulse count can be used to determine the linear x-axis position of the worktable by factoring in the leadscrew pitch. Thus,

$$x = \frac{pn_p}{n_s} \tag{7.14}$$

where n_p and n_s are defined above, and p = leadscrew pitch (mm/rev, in/rev).

The velocity of the worktable, which is normally the feed rate in a machining operation, is obtained from the frequency of the pulse train as

$$v_t = f_r = \frac{60 p f_p}{n_s} \tag{7.15}$$

where v_t = worktable velocity (mm/min, in/min), f_r = feed rate (mm/min, in/min), f_p = frequency of the pulse train emitted by the optical encoder (Hz, pulses/sec), and the constant 60 converts worktable velocity and feed rate from millimeters per second (inches per second) to millimeters per minute (inches per minute).

The pulse train generated by the encoder is compared with the coordinate position and feed rate specified in the part program, and the difference is used by the MCU to drive a servomotor, which in turn drives the worktable. A digital-to-analog converter (Section 6.4) is used to convert the digital signals used by the MCU into a continuous analog current that powers the drive motor. Closed-loop NC systems of the type described here are appropriate when a reactionary force resists the movement of the table. Metal cutting machine tools that perform continuous path cutting operations, such as milling and turning, fall into this category.

EXAMPLE 7.2   NC Closed-Loop Positioning

An NC worktable operates by closed-loop positioning. The system consists of a servomotor, leadscrew, and optical encoder. The leadscrew has a pitch of 6.0 mm and is coupled to the motor shaft with a gear ratio of 5:1 (five turns of the drive motor for each turn of the leadscrew). The optical encoder generates 48 pulses/rev of its output shaft. The table has been programmed to move a distance of 250 mm at a feed rate = 500 mm/min. Determine (a) how many pulses should be received by the control system to verify that the table has moved exactly 250 mm, (b) the pulse rate of the encoder, and (c) the drive motor speed that correspond to the specified feed rate.

Solution: (a) Rearranging Eq. (7.14) to find n_p,

$$n_p = \frac{x n_s}{p} = \frac{250(48)}{6.0} = 2000 \text{ pulses}$$

(b) The pulse rate corresponding to 500 mm/min can be obtained by rearranging Eq. (7.15):

$$f_p = \frac{f_r n_s}{60p} = \frac{500(48)}{60(6.0)} = 66.667 \text{ Hz}$$

(c) Motor speed = table velocity (feed rate) divided by leadscrew pitch, corrected for gear ratio:

$$N_m = \frac{r_g f_r}{p} = \frac{5(500)}{6.0} = 416.667 \text{ rev/min}$$

Note that motor speed has the same numerical value as in Example 7.1 because the table velocity and motor gear ratio are the same.

7.5.3   Precision in NC Positioning

To accurately machine or otherwise process a workpart, the NC positioning system must possess a high degree of precision. Three measures of precision can be defined for an NC positioning system: (1) control resolution, (2) accuracy, and (3) repeatability. These terms are most readily explained by considering a single axis of the positioning system, as depicted in Figure 7.15. Control resolution refers to the control system's ability to divide the total range of the axis movement into closely spaced points that can be distinguished by the MCU. *Control resolution* is defined as the distance separating two adjacent addressable points in the axis movement. *Addressable points* are locations along the axis to which the worktable can be specifically directed to go. It is desirable for control resolution to be as small as possible. This depends on limitations imposed by (1) the electromechanical components of the positioning system and/or (2) the number of bits used by the controller to define the axis coordinate location.

A number of electromechanical factors affect control resolution, including leadscrew pitch, gear ratio in the drive system, and the step angle in a stepper motor for an open-loop system or the angle between slots in an encoder disk for a closed-loop system. For an open-loop positioning system driven by a stepper motor, these factors can be combined into an expression that defines control resolution as

$$CR_1 = \frac{p}{n_s r_g} \tag{7.16}$$

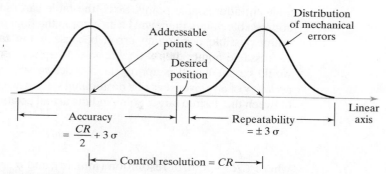

Figure 7.15 A portion of a linear positioning system axis, with definition of control resolution, accuracy, and repeatability.

where CR_1 = control resolution of the electromechanical components (mm, in), p = leadscrew pitch (mm/rev, in/rev), n_s = number of steps per revolution, and r_g = gear ratio between the motor shaft and the leadscrew as defined in Eq. (7.6). The same expression can be used for a closed-loop positioning system.

The second factor that limits control resolution is the number of bits used by the MCU to specify the axis coordinate value. For example, this limitation may be imposed by the bit storage capacity of the controller. If B = the number of bits in the storage register for the axis, then the number of control points into which the axis range can be divided = 2^B. Assuming that the control points are separated equally within the range, then

$$CR_2 = \frac{L}{2^B - 1} \qquad (7.17)$$

where CR_2 = control resolution of the computer control system (mm, in), and L = axis range (mm, in). The control resolution of the positioning system is the maximum of the two values; that is,

$$CR = \text{Max}\ \{CR_1, CR_2\} \qquad (7.18)$$

A desirable criterion is $CR_2 \leq CR_1$, meaning that the electromechanical system is the limiting factor that determines control resolution. The bit storage capacity of a modern computer controller is sufficient to satisfy this criterion except in unusual situations. Resolutions of 0.0025 mm (0.0001 in) are within the current state of NC technology.

The ability of a positioning system to move the worktable to the exact location defined by a given addressable point is limited by mechanical errors that are due to various imperfections in the mechanical system. These imperfections include play between the leadscrew and the worktable, backlash in the gears, and deflection of machine components. We assume that the mechanical errors form an unbiased normal statistical distribution about the control point whose mean $\mu = 0$. We further assume that the standard deviation σ of the distribution is constant over the range of the axis under consideration. Given these assumptions, then nearly all of the mechanical errors (99.73%) are contained within $\pm 3\sigma$ of the control point. This is pictured in Figure 7.15 for a portion of the axis range that includes two control points.

Let us now use these definitions of control resolution and mechanical error distribution to define accuracy and repeatability of a positioning system. Accuracy is defined under worst case conditions in which the desired target point lies in the middle between two adjacent addressable points. Since the table can only be moved to one or the other of the addressable points, there will be an error in the final position of the worktable. This is the maximum possible positioning error, because if the target were closer to either one of the addressable points, then the table would be moved to the closer control point and the error would be smaller. It is appropriate to define accuracy under this worst case scenario. The *accuracy* of any given axis of a positioning system is the maximum possible error that can occur between the desired target point and the actual position taken by the system. In equation form,

$$\text{Accuracy} = \frac{CR}{2} + 3\sigma \tag{7.19}$$

where CR = control resolution (mm, in), and σ = standard deviation of the error distribution. Accuracies in machine tools are generally expressed for a certain range of table travel, for example, ±0.01 mm for 250 mm (±0.0004 in. for 10 in) of table travel.

Repeatability refers to the ability of the positioning system to return to a given addressable point that has been previously programmed. This capability can be measured in terms of the location errors encountered when the system attempts to position itself at the addressable point. Location errors are a manifestation of the mechanical errors of the positioning system, which follow a normal distribution, as assumed previously. Thus, the *repeatability* of any given axis of a positioning system is ±3 standard deviations of the mechanical error distribution associated with the axis. This can be written as

$$\text{Repeatability} = \pm 3\sigma \tag{7.20}$$

The repeatability of a modern NC machine tool is around ±0.0025 mm (±0.0001 in).

EXAMPLE 7.3   Control Resolution, Accuracy, and Repeatability in NC

Suppose the mechanical inaccuracies in the open-loop positioning system of Example 7.1 are described by a normal distribution with standard deviation $\sigma = 0.005$ mm. The range of the worktable axis is 1000 mm, and there are 16 bits in the binary register used by the digital controller to store the programmed position. Other relevant parameters from Example 7.1 are the following: pitch $p = 6.0$ mm, gear ratio between motor shaft and leadscrew $r_g = 5.0$, and number of step angles in the stepper motor $n_s = 48$. Determine (a) the control resolution, (b) the accuracy, and (c) the repeatability for the positioning system.

Solution: (a) Control resolution is the greater of CR_1 and CR_2 as defined by Eqs. (7.16) and (7.17).

$$CR_1 = \frac{p}{n_s r_g} = \frac{6.0}{48(5.0)} = 0.025 \text{ mm}$$

$$CR_2 = \frac{1000}{2^{16} - 1} = \frac{1000}{65,535} = 0.01526 \text{ mm}$$

$$CR = \text{Max}\{0.025, 0.01526\} = 0.025 \text{ mm}$$

(b) Accuracy is given by Eq. (7.19):

$$\text{Accuracy} = 0.5(0.025) + 3(0.005) = 0.0275 \text{ mm}$$

(c) Repeatability = ±3(0.005) = ±0.015 mm

7.6 NC PART PROGRAMMING

NC part programming consists of planning and documenting the sequence of processing steps to be performed on an NC machine. The part programmer must have a knowledge of machining (or other processing technology for which the NC machine is designed), as well as geometry and trigonometry. The documentation portion of part programming involves the input medium used to transmit the program of instructions to the NC machine control unit (MCU). The traditional input medium dating back to the first NC machines in the 1950s is 1-inch wide punched tape. More recently, magnetic tape and floppy disks have been used for NC due to their much higher data density.

Part programming can be accomplished using a variety of procedures ranging from highly manual to highly automated methods. The methods are (1) manual part programming, (2) computer-assisted part programming, (3) part programming using CAD/CAM, and (4) manual data input.

7.6.1 Manual Part Programming

In manual part programming, the programmer prepares the NC code using a low-level machine language that is described briefly in this section and more thoroughly in Appendix A7. The coding system is based on binary numbers. This coding is the low-level machine language that can be understood by the MCU. When higher level languages are used, such as APT (Section 7.5.2 and Appendix B7), the statements in the program are converted to this basic code. NC uses a combination of the binary and decimal number systems, called the *binary-coded decimal* (BCD) system. In this coding scheme, each of the ten digits (0-9) in the decimal system is coded as a four-digit binary number, and these binary numbers are added in sequence as in the decimal number system. Conversion of the ten digits in the decimal number system into binary numbers is shown in Table 7.4. For example, the decimal value 1250 would be coded in BCD as follows:

Number Sequence	Binary Number	Decimal Value
First	0001	1000
Second	0010	200
Third	0101	50
Fourth	0000	0
Sum		1250

TABLE 7.4 Comparison of Binary and Decimal Numbers

Binary	Decimal	Binary	Decimal
0000	0	0101	5
0001	1	0110	6
0010	2	0111	7
0011	3	1000	8
0100	4	1001	9

In addition to numerical values, the NC coding system must also provide for alphabetical characters and other symbols. Eight binary digits are used to represent all of the characters required for NC part programming. Out of a sequence of characters, a word is formed. A *word* specifies a detail about the operation, such as x-position, y-position, feed rate, or spindle speed. Out of a collection of words, a block is formed. A *block* is one complete NC instruction. It specifies the destination for the move, the speed and feed of the cutting operation, and other commands that determine explicitly what the machine tool will do. For example, an instruction block for a two-axis NC milling machine would likely include the x- and y-coordinates to which the machine table should be moved, the type of motion to be performed (linear or circular interpolation), the rotational speed of the milling cutter, and the feed rate at which the milling operation should be performed.

The organization of words within a block is known as a *block format* (also called a *tape format*, because the formats were originally developed for punched tapes). Although a number of different block formats have been developed over the years, all modern controllers use the word address format, which uses a letter prefix to identify each type of word, and spaces to separate words within the block. This format also allows for variations in the order of words within the block, and omission of words from the block if their values do not change from the previous block. For example, the two commands in word address format to perform the two drilling operations illustrated in Figure 7.16 are

N001 G00 X07000 Y03000 M03

N002 Y06000

where N is the sequence number prefix, and X and Y are the prefixes for the x- and y-axes, respectively. G-words and M-words require some elaboration. G-words are called preparatory words. They consist of two numerical digits (following the "G" prefix) that prepare the MCU for the instructions and data contained in the block. For example, G00 prepares the controller for a point-to-point rapid traverse move between the previous point and the endpoint defined in the current command. M-words are used to specify miscellaneous or auxiliary functions that are available on the machine tool. The M03 in our example is used to start the spindle rotation. Other examples include stopping the spindle for a tool change, and turning the cutting fluid on or off. Of course, the particular machine tool must possess the function that is being called.

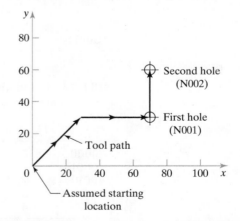

Figure 7.16 Drilling sequence for word address format example. Dimensions are in millimeters.

Words in an instruction block are intended to convey all of the commands and data needed for the machine tool to execute the move defined in the block. The words required for one machine tool type may differ from those required for a different type; for example, turning requires a different set of commands than milling. The words in a block are usually given in the following order (although the word address format allows variations in the order):

- sequence number (N-word)
- preparatory word (G-word)
- coordinates (X-, Y-, Z-words for linear axes, A-, B-, C-words for rotational axes)
- feed rate (F-word)
- spindle speed (S-word)
- tool selection (T-word)
- miscellaneous command (M-word)

For the interested reader, we have prepared Appendix A7, which describes the details of the coding system used in manual part program. Examples of programming commands are provided and the various G-words and M-words are defined.

Manual part programming can be used for both point-to-point and contouring jobs. It is most suited for point-to-point machining operations such as drilling. It can also be used for simple contouring jobs, such as milling and turning when only two axes are involved. However, for complex three-dimensional machining operations, there is an advantage in using computer-assisted part programming.

7.6.2 Computer-Assisted Part Programming

Manual part programming can be time consuming, tedious, and subject to errors for parts possessing complex geometries or requiring many machining operations. In these cases, and even for simpler jobs, it is advantageous to use computer-assisted part programming. A number of NC part programming language systems have been developed to accomplish many of the calculations that the programmer would otherwise have to do. The program is written in English-like statements that are subsequently converted to the low-level machine language. Computer-assisted part programming saves time and results in a more accurate and efficient part program. Using this programming arrangement, the various tasks are divided between the human part programmer and the computer.

The Part Programmer's Job. In computer-assisted part programming, the machining instructions are written in English-like statements that are subsequently translated by the computer into the low-level machine code that can be interpreted and executed by the machine tool controller. The two main tasks of the programmer are (1) defining the geometry of the part and (2) specifying the tool path and operation sequence.

No matter how complicated the workpart may appear, it is composed of basic geometric elements and mathematically defined surfaces. Consider our sample part in Figure 7.17. Although its appearance is somewhat irregular, the outline of the part consists of intersecting straight lines and a partial circle. The hole locations in the part can be defined in terms of the x- and y-coordinates of their centers. Nearly any component that can be conceived by a designer can be described by points, straight lines, planes, circles, cylinders, and other mathematically defined surfaces. It is the part programmer's task to identify and enumerate the geometric elements of which the part is constructed. Each element must be defined in terms of its dimensions and location relative to other elements. A few

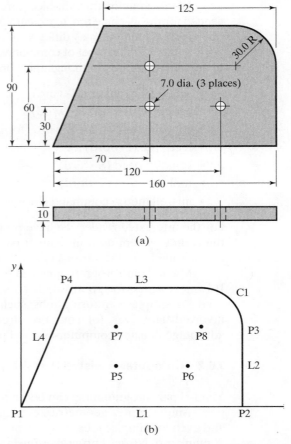

Figure 7.17 Sample part with geometry elements (points, lines, and circle) labeled for computer-assisted part programming.

examples will be instructive here to show how geometric elements are defined. We will use our sample part to illustrate, with labels of geometry elements added as shown in Figure 7.17(b). Our statements are taken from APT, which stands for automatically programmed tooling. In Appendix B7, the APT language is described in more detail.

Let us begin with the simplest geometric element, a point. The simplest way to define a point is by means of its coordinates; for example,

$$P4 = POINT/35, 90, 0$$

where the point is identified by a symbol (P4), and its coordinates are given in the order x, y, z in millimeters ($x = 35$ mm, $y = 90$ mm, and $z = 0$). A line can be defined by two points, as in the following:

$$L1 = LINE/P1, P2$$

where L1 is the line defined in the statement, and P1 and P2 are two previously defined points. And finally, a circle can be defined by its center location and radius,

$$C1 = CIRCLE/CENTER, P8, RADIUS, 30$$

where C1 is the newly defined circle, with center at previously defined point P8 and radius = 30 mm. The APT language offers many alternative ways to define points, lines, circles, and other geometric elements. A sampling of these definitions is provided in Appendix B7.

After the part geometry has been defined, the part programmer must next specify the tool path that the cutter will follow to machine the part. The tool path consists of a sequence of connected line and arc segments, using the previously defined geometry elements to guide the cutter. For example, suppose we are machining the outline of our sample part in Figure 7.17 in a profile milling operation (contouring). We have just finished cutting along surface L1 in a counterclockwise direction around the part, and the tool is presently located at the intersection of surfaces L1 and L2. The following APT statement could be used to command the tool to make a left turn from L1 onto L2 and to cut along L2:

$$GOLFT/L2, TANTO, C1$$

The tool proceeds along surface L2 until it is tangent to (TANTO) circle C1. This is a continuous path motion command. Point-to-point commands tend to be simpler; for example, the following statement directs the tool to go to a previously defined point P5:

$$GOTO/P5$$

In addition to defining part geometry and specifying tool path, the programmer must accomplish various other programming functions, such as naming the program, identifying the machine tool on which the job will be performed, specifying cutting speeds and feed rates, designating the cutter size (cutter radius, tool length, etc.), and specifying tolerances in circular interpolation.

Computer Tasks in Computer-Assisted Part Programming. The computer's role in computer-assisted part programming consists of the following tasks, performed more or less in the sequence noted: (1) input translation, (2) arithmetic and cutter offset computations, (3) editing, and (4) postprocessing. The first three tasks are carried out under the supervision of the language processing program. For example, the APT language uses a processor designed to interpret and process the words, symbols, and numbers written in APT. Other languages require their own processors. The fourth task, postprocessing, requires a separate computer program. The sequence and relationship of the tasks of the part programmer and the computer are portrayed in Figure 7.18.

The part programmer enters the program using APT or some other high-level part programming language. The *input translation* module converts the coded instructions contained in the program into computer-usable form, preparatory to further processing. In APT, input translation accomplishes the following tasks: (1) syntax check of the input code to identify errors in format, punctuation, spelling, and statement sequence; (2) assigning a

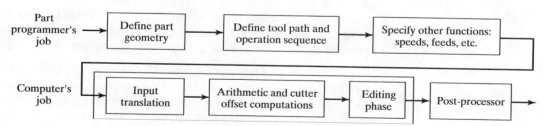

Figure 7.18 Tasks in computer-assisted part programming.

sequence number to each APT statement in the program; (3) converting geometry elements into a suitable form for computer processing; and (4) generating an intermediate file called PROFIL that is utilized in subsequent arithmetic calculations.

The *arithmetic module* consists of a set of subroutines to perform the mathematical computations required to define the part surface and generate the tool path, including compensation for cutter offset. The individual subroutines are called by the various statements used in the part programming language. The arithmetic computations are performed on the PROFIL file. The arithmetic module frees the programmer from the time-consuming and error-prone geometry and trigonometry calculations to concentrate on issues related to workpart processing. The output of this module is a file called CLFILE, which stands for "cutter location file." As its name suggests, this file consists mainly of tool path data.

During the editing stage, the computer edits the CLFILE and generates a new file called CLDATA. When printed, CLDATA provides readable data on cutter locations and machine tool operating commands. The machine tool commands can be converted to specific instructions during postprocessing. The output of the editing phase is a part program in a format that can be postprocessed for the given machine tool on which the job will be accomplished.

NC machine tool systems are different. They have different features and capabilities. High-level part programming languages, such as APT, are generally not intended for only one machine tool type. They are designed to be general purpose. Accordingly, the final task of the computer in computer-assisted part programming is *postprocessing*, in which the cutter location data and machining commands in the CLDATA file are converted into low-level code that can be interpreted by the NC controller for a specific machine tool. The output of postprocessing is a part program consisting of G-codes, x-, y-, and z-coordinates, S, F, M, and other functions in word address format. The postprocessor is separate from the high-level part programming language. A unique postprocessor must be written for each machine tool system.

7.6.3 NC Part Programming Using CAD/CAM

A *CAD/CAM system* is a computer interactive graphics system equipped with software to accomplish certain tasks in design and manufacturing and to integrate the design and manufacturing functions. We discuss CAD/CAM in Chapter 23. One of the important tasks performed on a CAD/CAM system is NC part programming. In this method of part programming, portions of the procedure usually done by the part programmer are instead done by the computer. Advantages of NC part programming using CAD/CAM include the following [12]: (1) the part program can be simulated off-line on the CAD/CAM system to verify its accuracy; (2) the time and cost of the machining operation can be determined by the CAD/CAM system; (3) the most appropriate tooling can be automatically selected for the operation; and (4) the CAD/CAM system can automatically insert the optimum values for speeds and feeds for the work material and operations.

Other advantages are described below. Recall that the two main tasks of the part programmer in computer-assisted programming are (1) defining the part geometry and (2) specifying the tool path. Advanced CAD/CAM systems automate portions of both of these tasks.

Geometry Definition Using CAD/CAM. A fundamental objective of CAD/CAM is to integrate the design engineering and manufacturing engineering functions. Certainly one of the important design functions is to design the individual components of the product. If a CAD/CAM system is used, a computer graphics model of each part is developed by the designer and stored in the CAD/CAM data base. That model contains all the geometric, dimensional, and material specifications for the part.

When the same CAD/CAM system, or a CAM system that has access to the same CAD data base in which the part model resides, is used to perform NC part programming, it makes little sense to recreate the geometry of the part during the programming procedure. Instead, the programmer generally retrieves the part geometry model from storage and uses that model to construct the appropriate cutter path. The significant advantage of using CAD/CAM in this way is that it eliminates one of the time-consuming steps in computer-assisted part programming: geometry definition. After the part geometry has been retrieved, the usual procedure is to label the geometric elements that will be used during part programming. These labels are the variable names (symbols) given to the lines, circles, and surfaces that comprise the part. Most systems have the capacity to automatically label the geometry elements of the part and to display the labels on the monitor. The programmer can then refer to those labeled elements during tool path construction.

An NC programmer who does not have access to the data base must define the geometry of the part, using similar interactive graphics techniques that the product designer would use to design the part. Points are defined in a coordinate system using the computer graphics system, lines and circles are defined from the points, surfaces are defined, and so forth, to construct a geometric model of the part. The advantage of the interactive graphics system over conventional computer-assisted part programming is that the programmer receives immediate visual verification of the geometric elements being created. This tends to improve the speed and accuracy of the geometry definition process.

Tool Path Generation Using CAD/CAM.   The second task of the NC programmer in computer-assisted part programming is tool path specification. First, the programmer selects the cutting tool for the operation. Most CAD/CAM systems have tool libraries that can be called by the programmer to identify the tools that are available in the tool crib. The programmer must decide which of the available tools is most appropriate for the operation under consideration and then specify it for the tool path. This permits the tool diameter and other dimensions to be entered automatically for tool offset calculations. If the desired cutting tool is not available in the library, the programmer can specify an appropriate tool. It then becomes part of the library for future use.

The next step is tool path definition. There are differences in capabilities of the various CAD/CAM systems, which result in different approaches for generating the tool path. The most basic approach involves the use of the interactive graphics system to enter the motion commands one by one, similar to computer-assisted part programming. Individual statements in APT or other part programming language are entered, and the CAD/CAM system provides an immediate graphic display of the action resulting from the command, thereby validating the statement.

A more advanced approach for generating tool path commands is to use one of the automatic software modules available on the CAD/CAM system. These modules have been developed to accomplish a number of common machining cycles for milling, drilling, and turning. They are subroutines in the NC programming package that can be called and the required parameters given to execute the machining cycle. Several of these modules are identified in Table 7.5 and Figure 7.19.

When the complete part program has been prepared, the CAD/CAM system can provide an animated simulation of the program for validation purposes.

Computer-Automated Part Programming.   In the CAD/CAM approach to NC part programming, several aspects of the procedure are automated. In the future, it should be possible to automate the complete NC part programming procedure. We are referring to this fully automated procedure as computer-automated part programming. Given the geometric model of a part that has been defined during product design, the computer-automated system would possess sufficient logic and decision-making capability to accomplish NC part programming for the entire part without human assistance.

This can most readily be done for certain NC processes that involve well-defined, relatively simple part geometries. Examples are point-to-point operations such as NC drilling and electronic component assembly machines. In these processes, the program

TABLE 7.5 Some Common NC Modules for Automatic Programming of Machining Cycles

Module Type	Brief Description
Profile milling	Generates cutter path around the periphery of a part, usually a two-dimensional contour where depth remains constant.
Pocket milling	Generates the tool path to machine a cavity, as in Figure 7.19(a). A series of cuts is usually required to complete the bottom of the cavity to the desired depth.
Lettering (engraving, milling)	Generates tool path to engrave (mill) alphanumeric characters and other symbols to specified font and size.
Contour turning	Generates tool path for a series of turning cuts to provide a defined contour on a rotational part, as in Figure 7.19(b).
Facing (turning)	Generates tool path for a series of facing cuts to remove excess stock from the part face or to create a shoulder on the part by a series of facing operations, as in Figure 7.19(c).
Threading (turning)	Generates tool path for a series of threading cuts to cut external, internal, or tapered threads on a rotational part, as in Figure 7.19(d) for external threads.

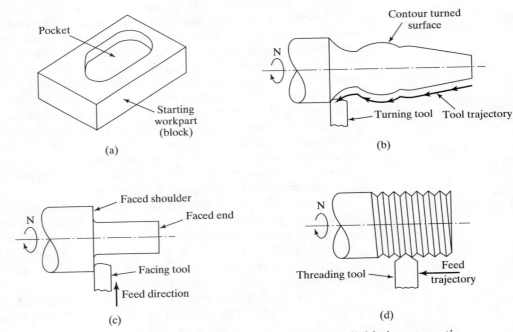

Figure 7.19 Examples of machining cycles available in automatic programming modules. (a) pocket milling, (b) contour turning, (c) facing and shoulder facing, and (d) threading (external).

TABLE 7.6 Typical Sequence of Steps in CNC Part Programming Using Mastercam for a Sequence of Milling and Drilling Operations

Step	Description
1	Develop a CAD model of the part to be machined using Mastercam, or import the CAD model from a compatible CAD package.
2	Orient the starting workpiece relative to the axis system of the machine.
3	Identify the workpiece material and specified grade (e.g., Aluminum 2024, for selection of cutting conditions).
4	Select the operation to be performed (e.g., drilling, pocket milling, contouring) and the surface to be machined.
5	Select the cutting tool (e.g., 0.250-inch drill) from the tool library.
6	Enter applicable cutting parameters such as hole depth.
7	Repeat steps 4 through 6 for each additional machining operation to be performed on the part.
8	Select appropriate postprocessor to generate the word address for the machine tool on which the machining job will be accomplished.
9	Verify the part program by animated simulation of the sequence of machining operations to be performed on the part.

consists basically of a series of locations in an *x-y* coordinate system where work is to be performed (e.g., holes are to be drilled or components are to be inserted). These locations are determined by data that are generated during product design. Special algorithms can be developed to process the design data and generate the NC program for the particular system. NC contouring systems will eventually be capable of a similar level of automation. Automatic programming of this type is closely related to computer-automated process planning (CAPP), discussed in Chapter 24.

Mastercam. Mastercam is the leading commercial CAD/CAM software package for CNC part programming. It is available from CNC Software, Inc. [16]. The package includes a CAD capability for designing parts in addition to its CAM features for part programming. If an alternative computer-aided design package is used for design (e.g., AutoCAD®, SolidWorks®), files from these other packages can be translated for use within Mastercam. Processes to which Mastercam can be applied include milling and drilling, turning, plasma cutting, and laser cutting. More information is available on the website of the company [16]. The typical steps that a programmer would use in Mastercam to accomplish a part-programming job are listed in Table 7.6. The output of the program would be a word address format program.

7.6.4 Manual Data Input

Manual and computer-assisted part programming require a relatively high degree of formal documentation and procedure. There is lead time required to write and validate the programs. CAD/CAM part programming automates a substantial portion of the procedure, but a significant commitment in equipment, software, and training is required. One method of simplifying the procedure is to have the machine operator perform the part programming task at the machine tool. This is called *manual data input* (abbreviated MDI) because the operator manually enters the part geometry data and motion commands directly into the MCU prior to running the job. MDI, also known as *conversational programming* [5], [11] is perceived as a way for the small machine shop to introduce NC into its operations without needing to acquire special NC part programming equipment and hiring a part programmer. MDI permits the shop to make a minimal initial investment to begin the transition to modern CNC technology. The limitation, of manual data input is the risk of programming errors as jobs become more complicated. For this reason, MDI is usually applied for relatively simple parts.

Communication between the machine operator-programmer and the MDI system is accomplished using a display monitor and alphanumeric keyboard. Entering the programming commands into the controller is typically done using a menu-driven procedure in which the operator responds to prompts and questions posed by the NC system about the job to be machined. The sequence of questions is designed so that the operator inputs the part geometry and machining commands in a logical and consistent manner. A computer graphics capability is included in modern MDI programming systems to permit the operator to visualize the machining operations and verify the program. Typical verification features include tool path display and animation of the tool path sequence.

A minimum of training in NC part programming is required of the machine operator. The operator must have the ability to read an engineering drawing of the part and must be familiar with the machining process. An important caveat in the use of MDI is to make certain that the NC system does not become an expensive toy that stands idle while the operator is entering the programming instructions. Efficient use of the system requires that programming for the next part be accomplished while the current part is being machined. Most MDI systems permit these two functions to be performed simultaneously to reduce changeover time between jobs.

REFERENCES

[1] CHANG, C.H., and M. MELKANOFF, *NC Machine Programming and Software Design*, Prentice Hall, Englewood Cliffs, NJ, 1989.

[2] GROOVER, M. P., and E. W. ZIMMERS, Jr., *CAD/CAM: Computer-Aided Design and Manufacturing*, Prentice-Hall, Englewood Cliffs, NJ, 1984.

[3] Illinois Institute of Technology Research Institute, *APT Part Programming*, McGraw-Hill Book Company, NY, 1967.

[4] LIN, S.C., *Computer Numerical Control: Essentials of Programming and Networking*, Delmar Publishers Inc., Albany, NY, 1994.

[5] LYNCH, M., *Computer Numerical Control for Machining*, McGraw-Hill, NY, 1992.

[6] NOAKER, P. M., "Down the Road with DNC," *Manufacturing Engineering*, November 1992, pp. 35–38.

[7] NOAKER, P. M., "The PC's CNC Transformation," *Manufacturing Engineering*, August 1995, pp. 49–53.

[8] NOBLE, D. F., *Forces of Production*, Alfred A. Knopf, NY, 1984.

[9] QUESADA, R., *Computer Numerical Control, Machining and Turning Centers,* Pearson/Prentice Hall, Upper Saddle River, NJ, 2005.

[10] REINTJES, J. F., *Numerical Control: Making a New Technology*, Oxford University Press, NY, 1991.

[11] STENERSON, J., and K. CURRAN, *Computer Numerical Control: Operation and Programming*, 3d ed., Pearson/Prentice Hall, Upper Saddle River, NJ, 2007.

[12] VALENTINO, J. V., and J. GOLDENBERG, *Introduction to Computer Numerical Control (CNC)*, 3d ed., Pearson/Prentice Hall, Upper Saddle River, NJ, 2003.

[13] WAURZYNIAK, P., "Machine Controllers: Smarter and Faster," *Manufacturing Engineering*, June 2005, pp 61–73.

[14] WAURZYNIAK, P., "Software Controls Productivity," *Manufacturing Engineering*, August 2005, pp 67–73.

[15] WAURZYNIAK, P., "Under Control," *Manufacturing Engineering*, June 2006, pp 51–58.

[16] Website of CNC Software, Inc.: *www. mastercam. com*

REVIEW QUESTIONS

7.1 What is numerical control?

7.2 What are the three basic components of an NC system?

7.3 What is the right-hand rule in NC and where is it used?

7.4 What is the difference between point-to-point and continuous path control in a motion control system?

7.5 What is linear interpolation, and why is it important in NC?

7.6 What is the difference between absolute positioning and incremental positioning?

7.7 How is computer numerical control (CNC) distinguished from conventional NC?

7.8 Name five of the ten features and capabilities of a modern CNC machine control unit listed in the text.

7.9 What is distributed numerical control (DNC)?

7.10 What are some of the machine tool types to which numerical control has been applied?

7.11 What is a machining center?

7.12 Name four of the six part characteristics that are most suited to the application of numerical control listed in the text.

7.13 Although NC technology is most closely associated with machine tool applications, it has been applied to other processes also. Name three of the six examples listed in the text.

7.14 What are four advantages of numerical control when properly applied in machine tool operations?

7.15 What are three disadvantages of implementing NC technology?

7.16 Briefly describe the differences between the two basic types of positioning control systems used in NC.

7.17 What is an optical encoder, and how does it work?

7.18 With reference to precision in an NC positioning system, what is control resolution?

7.19 What is the difference between manual part programming and computer-assisted part programming?

7.20 What is postprocessing in computer-assisted part programming?

7.21 What are some of the advantages of CAD/CAM-based NC part programming compared to computer-assisted part programming?

7.22 What is manual data input of the NC part program?

PROBLEMS

NC Applications

7.1 A machinable grade of aluminum is to be milled on an NC machine with a 20 mm diameter four tooth end milling cutter. Cutting speed = 120 m/min and feed = 0.008 mm/tooth. Convert these values to rev/min and mm/rev, respectively.

7.2 A cast iron workpiece is to be face milled on an NC machine using cemented carbide inserts. The cutter has 16 teeth and is 120 mm in diameter. Cutting speed = 200 m/min and feed = 0.005 mm/tooth. Convert these values to rev/min and mm/rev, respectively.

7.3 An end milling operation is performed on an NC machining center. The total length of travel is 625 mm along a straight line path to cut a particular workpiece. Cutting speed = 2.0 m/s and chip load (feed/tooth) = 0.075 mm. The end milling cutter has two teeth and its diameter = 15.0 mm. Determine the feed rate and time to complete the cut.

7.4 A turning operation is to be performed on an NC lathe. Cutting speed = 2.5 m/s, feed = 0.2 mm/rev, and depth = 4.0 mm. Workpiece diameter = 100 mm and length = 400 mm. Determine (a) rotational speed of the work bar, (b) feed rate, (c) metal removal rate, and (d) time to travel from one end of the part to the other.

7.5 A numerical control drill press drills 10.0 mm diameter holes at four locations on a flat aluminum plate in a production work cycle. Although the plate is only 12 mm thick, the drill must travel a full 20 mm vertically at each hole location to allow for clearance above the plate and breakthrough of the drill on the underside of the plate. Cutting conditions: speed = 0.4 m/s and feed = 0.10 mm/rev. Hole locations are indicated in the following table:

Hole number	x-coordinate (mm)	y-coordinate (mm)
1	25.0	25.0
2	25.0	100.0
3	100.0	100.0
4	100.0	25.0

The drill starts out at point (0,0) and returns to the same position after the work cycle is completed. Travel rate of the table in moving from one coordinate position to another is 500 mm/min. Owing to effects of acceleration and deceleration, and time required for the control system to achieve final positioning, a time loss of 3 s is experienced at each stopping position of the table. Assume that all moves are made so as to minimize the total cycle time. If loading and unloading the plate take 20 s (total handling time), determine the time required for the work cycle.

Analysis of Open Loop Positioning Systems

7.6 Two stepping motors are used in an open loop system to drive the leadscrews for x-y positioning. The range of each axis is 250 mm. The shaft of the motors are connected directly to the leadscrews. The pitch of each leadscrew is 3.0 mm, and the number of step angles on the stepping motor is 125. (a) How closely can the position of the table be controlled, assuming there are no mechanical errors in the positioning system? (b) What are the required pulse train frequencies and corresponding rotational speeds of each stepping motor in order to drive the table at 275 mm/min in a straight line from point $(x = 0, y = 0)$ to point $(x = 130$ mm$, y = 220$ mm$)$?

7.7 One axis of an NC positioning system is driven by a stepping motor. The motor is connected to a leadscrew whose pitch is 4.0 mm, and the leadscrew drives the table. Control resolution for the table is specified as 0.015 mm. Determine (a) the number of step angles required to achieve the specified control resolution, (b) size of each step angle in the motor, and (c) linear travel rate of the motor at a pulse frequency of 200 pulses per second.

7.8 The worktable in an NC positioning system is driven by a leadscrew with a 4 mm pitch. The leadscrew is powered by a stepping motor which has 250 step angles. The worktable is programmed to move a distance of 100 mm from its present position at a travel speed of 300 mm/min. (a) How many pulses are required to move the table the specified distance? (b) What is the required motor speed and pulse rate to achieve the desired table speed?

7.9 A stepping motor with 200 step angles is coupled to a leadscrew through a gear reduction of 5:1 (five rotations of the motor for each rotation of the leadscrew). The leadscrew has 2.4 threads/cm. The worktable driven by the leadscrew must move a distance = 25.0 cm at a feed rate = 75 cm/minute. Determine (a) the number of pulses required to move the table, (b) required motor speed, and (c) pulse rate to achieve the desired table speed.

7.10 A component insertion machine takes 2.0 sec to put a component into a printed circuit (PC) board, once the board has been positioned under the insertion head. The *x-y* table that positions the PC board uses a stepper motor directly linked to a leadscrew for each axis. The leadscrew has a pitch = 5.0 mm. The motor step angle = 7.2 degrees and the pulse train frequency = 400 Hz. Two components are placed on the PC board, one each at positions (25, 25) and (50, 150), where coordinates = mm. The sequence of positions is (0,0), (25, 25), (50, 150), (0,0). Time required to unload the completed board and load the next blank onto the machine table = 5.0 sec. Assume that 0.25 sec. is lost due to acceleration and deceleration on each move. What is the hourly production rate for this PC board?

7.11 The two axes of an *x-y* positioning table are each driven by a stepping motor connected to a leadscrew with a 4:1 gear reduction. The number of step angles on each stepping motor is 200. Each leadscrew has a pitch = 5.0 mm and provides an axis range = 400.0 mm. There are 16 bits in each binary register used by the controller to store position data for the two axes. (a) What is the control resolution of each axis? (b) What are the required rotational speeds and corresponding pulse train frequencies of each stepping motor in order to drive the table at 600 mm/min in a straight line from point (25,25) to point (300,150)? Ignore acceleration.

Analysis of Closed Loop Positioning Systems

7.12 A DC servomotor is used to drive one of the table axes of an NC milling machine. The motor is coupled directly to the leadscrew for the axis, and the leadscrew pitch = 5 mm. The optical encoder attached to the leadscrew emits 500 pulses per revolution of the lead-screw. The motor rotates at a normal speed of 300 rev/min. Determine (a) control resolution of the system, expressed in linear travel distance of the table axis, (b) frequency of the pulse train emitted by the optical encoder when the servomotor operates at full speed, and (c) travel rate of the table at normal rpm of the motor.

7.13 In Problem 7.3, the axis corresponding to the feed rate uses a DC servomotor as the drive unit and an optical encoder as the feedback sensing device. The motor is geared to the lead-screw with a 10:1 reduction (ten turns of the motor for each turn of the leadscrew). If the leadscrew pitch = 5 mm, and the optical encoder emits 400 pulses per revolution, determine the rotational speed of the motor and the pulse rate of the encoder in order to achieve the feed rate indicated.

7.14 The worktable of an NC machine is driven by a closed-loop positioning system which consists of a servomotor, leadscrew, and optical encoder. The leadscrew pitch = 4 mm and it is coupled directly to the motor shaft (gear ratio = 1:1). The optical encoder generates 225 pulses per leadscrew revolution. The table has been programmed to move a distance of 200 mm at a feed rate = 450 mm/min. (a) How many pulses are received by the control system to verify that the table has moved the programmed distance? What are (b) the pulse rate and (c) motor speed that correspond to the specified feed rate?

7.15 A NC machine tool table is powered by a servomotor, leadscrew, and optical encoder. The leadscrew has a pitch = 5.0 mm and is connected to the motor shaft with a gear ratio of 16:1 (16 turns of the motor for each turn of the leadscrew). The optical encoder is connected-directly to the lead screw and generates 200 pulses/rev of the leadscrew. The table must move a distance = 100 mm at a feed rate = 500 mm/min. Determine (a) the pulse count received by the control system to verify that the table has moved exactly 100 mm; and (b) the pulse rate and (c) motor speed that correspond to the feed rate of 500 mm/min.

7.16 Solve the previous problem assuming the optical encoder is directly coupled to the motor shaft rather than to the leadscrew.

7.17 A leadscrew coupled directly to a DC servomotor is used to drive one of the table axes of an NC milling machine. The leadscrew has 2.5 threads/cm. The optical encoder attached to the leadscrew emits 100 pulses/rev of the leadscrew. The motor rotates at a maximum speed of 800 rev/min. Determine (a) the control resolution of the system, expressed in linear travel distance of the table axis, (b) frequency of the pulse train emitted by the optical encoder when the servomotor operates at maximum speed; and (c) travel speed of the table at maximum motor speed.

7.18 Solve the previous problem only the servomotor is connected to the leadscrew through a gear box whose reduction ratio = 10:1 (ten revolutions of the motor for each revolution of the lead screw).

7.19 A milling operation is performed on a NC machining center. Total travel distance = 300 mm in a direction parallel to one of the axes of the worktable. Cutting speed = 1.25 m/s and chip load = 0.05 mm. The end milling cutter has four teeth and its diameter = 20.0 mm. The axis uses a DC servomotor whose output shaft is coupled to a leadscrew with pitch = 6.0 mm. The feedback sensing device is an optical encoder which emits 250 pulses per revolution. Determine (a) feed rate and time to complete the cut, (b) rotational speed of the motor and (c) pulse rate of the encoder at the feed rate indicated.

7.20 A DC servomotor drives the x-axis of a NC milling machine table. The motor is coupled directly to the table leadscrew, whose pitch = 6.25 mm. An optical encoder is connected to the leadscrew. The optical encoder emits 125 pulses per revolution. To execute a certain programmed instruction, the table must move from point ($x = 87.5$ mm, $y = 35.0$) to point ($x = 25.0$ mm, $y = 180.0$ mm) in a straight-line trajectory at a feed rate = 200 mm/min. Determine (a) the control resolution of the system for the x-axis, (b) rotational speed of the motor, and (c) frequency of the pulse train emitted by the optical encoder at the desired feed rate.

Resolution and Accuracy of Positioning Systems

7.21 A two-axis NC system used to control a machine tool table uses a bit storage capacity of 16 bits in its control memory for each axis. The range of the x-axis is 600 mm and the range of the y-axis is 500 mm. The mechanical accuracy of the machine table can be represented by a Normal distribution with standard deviation = 0.002 mm for both axes. For each axis of the NC system, determine (a) the control resolution, (b) accuracy, and (c) repeatability.

7.22 Stepping motors are used to drive the two axes of an insertion machine used for electronic assembly. A printed circuit board is mounted on the table which must be positioned accurately for reliable insertion of components into the board. Range of each axis = 700 mm. The leadscrew used to drive each of the two axes has a pitch of 3.0 mm. The inherent mechanical errors in the table positioning can be characterized by a Normal distribution with standard deviation = 0.005 mm. If the required accuracy for the table is 0.04 mm, determine (a) the number of step angles that the stepping motor must have, and (b) how many bits are required in the control memory for each axis to uniquely identify each control position.

7.23 Referring back to Problem 7.8, the mechanical inaccuracies in the open loop positioning system can be described by a normal distribution whose standard deviation = 0.005 mm. The range of the worktable axis is 500 mm, and there are 12 bits in the binary register used by the digital controller to store the programmed position. For the positioning system, determine (a) control resolution, (b) accuracy, and (c) repeatability. (d) What is the minimum number of bits that the binary register should have so that the mechanical drive system becomes the limiting component on control resolution?

7.24 The positioning table for a component insertion machine uses a stepping motor and leadscrew mechanism. The design specifications require a table speed of 0.4 m/s and an accuracy = 0.02 mm. The pitch of the leadscrew = 5.0 mm, and the gear ratio = 2:1 (two turns of the motor for each turn of the leadscrew). The mechanical errors in the motor, gear box, leadscrew, and table connection are characterized by a normal distribution with standard deviation = 0.0025 mm. Determine (a) the minimum number of step angles in the stepping motor and (b) frequency of the pulse train required to drive the table at the desired maximum speed.

7.25 The two axes of an x-y positioning table are each driven by a stepping motor connected to a leadscrew with a 10:1 gear reduction. The number of step angles on each stepping motor is 20. Each leadscrew has a pitch = 4.5 mm and provides an axis range = 300 mm. There are 16 bits in each binary register used by the controller to store position data for the two axes. (a) What is the control resolution of each axis? (b) What are the required rotational speeds and corresponding pulse train frequencies of each stepping motor in order to drive the table at 500 mm/min in a straight line from point (30,30) to point (100,200)? Ignore acceleration and deceleration.

NC Manual Part Programming

Note: Appendix A7 will be required to solve the problems in this group.

7.26 Write the part program to drill the holes in the part shown in Figure P7.26. The part is 12.0 mm thick. Cutting speed = 100 m/min and feed = 0.06 mm/rev. Use the lower left corner of the part as the origin in the x-y axis system. Write the part program in the word address format using absolute positioning. The program style should be similar to Example A7.1.

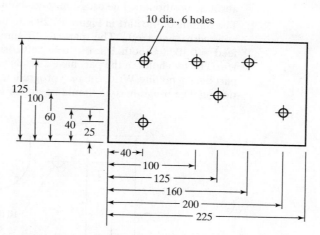

Figure P7.26 Part drawing for Problem 7.26. Dimensions are in millimeters.

7.27 The part in Figure P7.27 is to be drilled on a turret-type drill press. The part is 15.0 mm thick. There are three drill sizes to be used: 8 mm, 10 mm, and 12 mm. These drills are to be specified in the part program by tool turret positions T01, T02, and T03. All tooling is high speed steel. Cutting speed = 75 mm/min and feed = 0.08 mm/rev. Use the lower left corner of the part as the origin in the x-y axis system. Write the part program in the word address format using absolute positioning. The program style should be similar to Example A7.1.

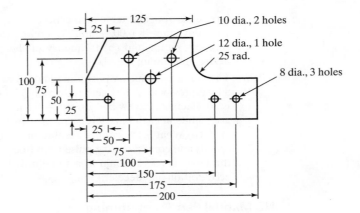

Figure P7.27 Part drawing for Problem 7.27. Dimensions are in millimeters.

7.28 The outline of the part in the previous problem is to be profile milled using a 30 mm diameter end mill with four teeth. The part is 15 mm thick. Cutting speed = 150 mm/min and feed = 0.085 mm/tooth. Use the lower left corner of the part as the origin in the x-y axis system. Two of the holes in the part have already been drilled and will be used for clamping the part during profile milling. Write the part program in the word address format. Use absolute positioning. The program style should be similar to Example A7.2.

7.29 The outline of the part in Figure P7.29 is to be profile milled, using a 20 mm diameter end mill with two teeth. The part is 10 mm thick. Cutting speed = 125 mm/min and feed = 0.10 mm/tooth. Use the lower left corner of the part as the origin in the x-y axis system. The two holes in the part have already been drilled and will be used for clamping the part during milling. Write the part program in the word address format. Use absolute positioning. The program style should be similar to Example A7.2.

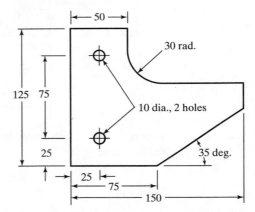

Figure P7.29 Part drawing for Problem 7.29. Dimensions are in millimeters.

NC Part Programming in APT

Note: Appendix B7 will be required to solve the problems in this group.

7.30 Write the APT geometry statements to define the hole positions of the part in Figure P7.26. Use the lower left corner of the part as the origin in the x-y axis system.

7.31 Write the complete APT part program to perform the drilling operations for the part drawing in Figure P7.26. Cutting speed = 0.4 m/s, feed = 0.10 mm/rev., and table travel speed between holes = 500 mm/min. Postprocessor call statement is MACHIN/DRILL, 04.

7.32 Write the APT geometry statements to define the hole positions of the part in Figure P7.27. Use the lower left corner of the part as the origin in the *x-y* axis system.

7.33 Write the APT part program to perform the drilling operations for the part drawing in Figure P7.27. Use the TURRET command to call the different drills required. Cutting speed = 0.4 m/s, feed = 0.10 mm/rev, and table travel speed between holes = 500 mm/min. Postprocessor call statement is MACHIN/TURDRL, 02.

7.34 Write the APT geometry statements to define the outline of the part in Figure P7.27. Use the lower left corner of the part as the origin in the *x-y* axis system.

7.35 Write the complete APT part program to profile mill the outside edges of the part in Figure P7.27. The part is 15 mm thick. Tooling = 30 mm diameter end mill with four teeth, cutting speed = 150 mm/min, and feed = 0.085 mm/tooth. Use the lower left corner of the part as the origin in the *x-y* axis system. Two of the holes in the part have already been drilled and will be used for clamping the part during profile milling. Postprocessor call statement is MACHIN/MILL, 06.

7.36 Write the APT geometry statements to define the part geometry shown in Figure P7.29. Use the lower left corner of the part as the origin in the *x-y* axis system.

7.37 Write the complete APT part program to perform the profile milling operation for the part drawing in Figure P7.29. Tooling = 20 mm diameter end mill with two teeth, cutting speed = 125 mm/min, and feed = 0.10 mm/tooth. The part is 10 mm thick. Use the lower left corner of the part as the origin in the *x-y* axis system. The two holes in the part have already been drilled and will be used for clamping the part during milling. Postprocessor call statement is MACHIN/MILL, 01.

7.38 Write the APT geometry statements to define the outline of the cam shown in Figure P7.38.

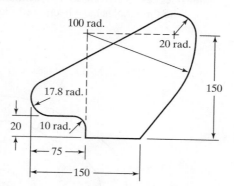

Figure P7.38   Part drawing for Problem 7.38. Dimensions are in millimeters.

7.39 The outline of the cam in Figure P7.38 is to be machined in an end milling operation, using a 12.5 mm diameter end mill with two teeth. The part is 7.5 mm thick. Write the complete APT program for this job, using a feed rate = 80 mm/min and a spindle speed = 500 rev/min. Postprocessor call statement is MACHIN/MILL, 03. Assume the rough outline for the part has been obtained in a band saw operation. Ignore clamping issues in the problem.

7.40 The part outline in Figure P7.40 is to be profile milled in several passes from a rectangular slab (outline of slab shown in dashed lines), using a 25 mm diameter end mill with four teeth. The initial passes are to remove no more than 5 mm of material from the periphery of the part, and the final pass should remove no more than 2 mm to cut the outline to final shape. Write the APT geometry and motion statements for this job. The final part thickness

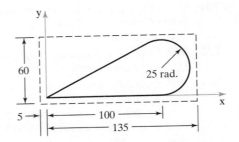

Figure P7.40 Part drawing for Problem 7.40. Dimensions are in millimeters.

is to be the same as the starting slab thickness, which is 10 mm, so no machining is required on the top and bottom of the part.

7.41 The top surface of a large cast iron plate is to be face-milled. The area to be machined is 400 mm wide and 700 mm long. The insert type face milling cutter has eight teeth and is 100 mm in diameter. Define the origin of the axis system at the lower left corner of the part with the long side parallel to the x-axis. Write the APT geometry and motion statements for this job.

7.42 Write the APT geometry statements to define the part geometry shown in Figure P7.42.

7.43 The part in Figure P7.42 is to be milled, using a 20 mm diameter end mill with four teeth. Write the APT geometry and motion statements for this job. Assume that preliminary passes have been completed so that only the final pass ("to size") is to be completed in this program. Cutting speed = 500 rev./min, and feed rate = 250 mm/min. The starting slab thickness is 15 mm, so no machining is required on the top or bottom surfaces of the part. The three holes have been predrilled for fixturing in this milling sequence.

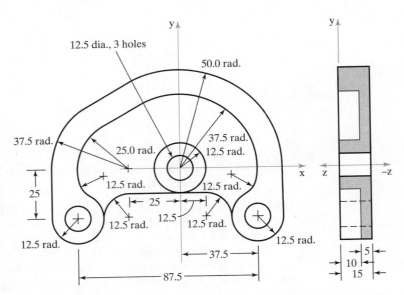

Figure P7.42 Part drawing for Problem 7.42. Dimensions are in millimeters.

APPENDIX A 7:   CODING FOR MANUAL PART PROGRAMMING

Instruction blocks in word address format consist of a series of words, each identified by a prefix label. The common prefixes are listed in Table A7.1 together with examples. As indicated in the text, the usual sequence of words in a block is (1) N-word, or sequence number, (2) G-word, or preparatory word, (3) X, Y, Z coordinates, (4) F-word, or feed

TABLE A7.1   Common Word Prefixes Used in Word Address Format

Word Prefix	Example	Function
N	N01	Sequence number; identifies block of instruction. One to four digits can be used.
G	G21	Preparatory word; prepares controller for instructions given in the block. See Table A7.2. There may be more than one G-word in a block. (Example specifies that numerical values are in millimeters.)
X, Y, Z	X75.0	Coordinate data for three linear axes. Can be specified in either inches or millimeters. (Example defines x-axis value as 75 mm.)
U, W	U25.0	Coordinate data for incremental moves in turning in the x- and z-directions, respectively. (Example specifies an incremental move of 25 mm in the x-direction.)
A, B, C	A90.0	Coordinate data for three rotational axes. A is the rotational axis about x-axis; B rotates about y-axis; and C rotates about z-axis. Specified in degrees of rotation. (Example defines 90° of rotation about x-axis.)
R	R100.0	Radius of arc; used in circular interpolation. (Example defines radius = 100 mm for circular interpolation.) The R-code can also be used to enter cutter radius data for defining the tool path offset distance from the part edge.
I, J, K	I32 J67	Coordinate values of arc center, corresponding to x-, y-, and z-axes, respectively; used in circular interpolation. (Example defines center of arc for circular interpolation to be at x = 32 mm and y = 67 mm.)
F	G94 F40	Feed rate per minute or per revolution in either inches or millimeters, as specified by G-words in Table A7.2. (Example specifies feed rate = 40 mm/min in milling or drilling operation.)
S	S0800	Spindle rotation speed in revolutions per minute, expressed in four digits. For some machines, spindle rotation speed is expressed as a percentage of maximum speed available on machine, expressed in two digits.
T	T14	Tool selection, used for machine tools with automatic tool changers or tool turrets. (Example specifies that the cutting tool to be used in the present instruction block is in position 14 in the tool drum.)
D	D05	Tool diameter word used in contouring moves for offsetting the tool from the workpart by a distance stored in the indicated register; usually the distance is the cutter radius. (Example indicates that the radius offset distance is stored in offset register number 05 in the controller.)
P	P05 R15.0	Used to store cutter radius data in offset register number 05. (Example indicates that a cutter radius value of 15.0 mm is to be stored in offset register 05.)
M	M03	Miscellaneous command. See Table A7.3. (Example commands the machine to start spindle rotation in clockwise direction.)

Note: Dimensional values in the examples are specified in millimeters.

rate, (5) S-word, or spindle speed, (6) T-word, for tool selection, if applicable, and (7) M-word, or miscellaneous command. Tables A7.2 and A7.3 list the common G-words and M-words, respectively.

In our coverage, statements are illustrated with dimensions given in millimeters. The values are expressed in four digits including one decimal place. For example, X020.0 means $x = 20.0$ mm. It should be noted that many CNC machines use formats that differ from ours, and so the instruction manual for each particular machine tool must be consulted to determine its own proper format. Our format is designed to convey principles and for easy reading.

TABLE A7.2 Common G-words (Preparatory Word)

G-word	Function
G00	Point-to-point movement (rapid traverse) between previous point and endpoint defined in current block. Block must include x-y-z coordinates of end position.
G01	Linear interpolation movement. Block must include x-y-z coordinates of end position. Feed rate must also be specified.
G02	Circular interpolation, clockwise. Block must include either arc radius or arc center; coordinates of end position must also be specified.
G03	Circular interpolation, counterclockwise. Block must include either arc radius or arc center; coordinates of end position must also be specified.
G04	Dwell for a specified time.
G10	Input of cutter offset data, followed by a P-code and an R-code.
G17	Selection of x-y plane in milling.
G18	Selection of x-z plane in milling.
G19	Selection of y-z plane in milling.
G20	Input values specified in inches.
G21	Input values specified in millimeters.
G28	Return to reference point.
G32	Thread cutting in turning.
G40	Cancel offset compensation for cutter radius (nose radius in turning).
G41	Cutter offset compensation, left of part surface. Cutter radius (nose radius in turning) must be specified in block.
G42	Cutter offset compensation, right of part surface. Cutter radius (nose radius in turning) must be specified in block.
G50	Specify location of coordinate axis system origin relative to starting location of cutting tool. Used in some lathes. Milling and drilling machines use G92.
G90	Programming in absolute coordinates.
G91	Programming in incremental coordinates.
G92	Specify location of coordinate axis system origin relative to starting location of cutting tool. Used in milling and drilling machines and some lathes. Other lathes use G50.
G94	Specify feed per minute in milling and drilling.
G95	Specify feed per revolution in milling and drilling.
G98	Specify feed per minute in turning.
G99	Specify feed per revolution in turning.

Note: Some G-words apply to milling and/or drilling only, whereas others apply to turning only.

TABLE A7.3 Common M-words Used in Word Address Format

M-Word	Function
M00	Program stop; used in middle of program. Operator must restart machine.
M01	Optional program stop; active only when optional stop button on control panel has been depressed.
M02	End of program. Machine stop.
M03	Start spindle in clockwise direction for milling machine (forward for turning machine).
M04	Start spindle in counterclockwise direction for milling machine (reverse for turning machine).
M05	Spindle stop.
M06	Execute tool change, either manually or automatically. If manually, operator must restart machine. Does not include selection of tool, which is done by T-word if automatic, by operator if manual.
M07	Turn cutting fluid on flood.
M08	Turn cutting fluid on mist.
M09	Turn cutting fluid off.
M10	Automatic clamping of fixture, machine slides, etc.
M11	Automatic unclamping.
M13	Start spindle in clockwise direction for milling machine (forward for turning machine) and turn on cutting fluid.
M14	Start spindle in counterclockwise direction for milling machine (reverse for turning machine) and turn on cutting fluid.
M17	Spindle and cutting fluid off.
M19	Turn spindle off at oriented position.
M30	End of program. Machine stop. Rewind tape (on tape-controlled machines).

In preparing the NC part program, the part programmer must initially define the origin of the coordinate axes and then reference the succeeding motion commands to this axis system. This is accomplished in the first statement of the part program. The directions of the x-, y-, and/or z-axes are predetermined by the machine tool configuration, but the origin of the coordinate system can be located at any desired position. The part programmer defines this position relative to some part feature that can be readily recognized by the machine operator. The operator is instructed to move the tool to this position at the beginning of the job. With the tool in position, the G92 code is used by the programmer to define the origin as

$$\text{G92 X0 Y-050.0 Z010.0}$$

where the x, y, and z values specify the coordinates of the tool location in the coordinate system; in effect, this defines the location of the origin. In some CNC lathes and turning centers, the code G50 is used instead of G92. Our x, y, and z values are specified in millimeters, and this must be explicitly stated. Thus, a more complete instruction block would be

$$\text{G21 G92 X0 Y-050.0 Z010.0}$$

where the G21 code indicates that the subsequent coordinate values are in millimeters.

Motions are programmed by the codes G00, G01, G02, and G03. G00 is used for a point-to-point rapid traverse movement of the tool to the coordinates specified in the command; for example,

$$\text{G00 X050.0 Y086.5 Z100.0}$$

specifies a rapid traverse motion from the current location to the location defined by the coordinates $x = 50.0$ mm, $y = 86.5$ mm, and $z = 100.0$ mm. This command would be appropriate for NC drilling machines in which a rapid move is desired to the next hole location, with no specification on the tool path. The velocity with which the move is achieved in rapid traverse mode is set by parameters in the MCU and is not specified numerically in the instruction block. The G00 code is not intended for contouring operations.

Linear interpolation is accomplished by the G01 code. This is used when it is desired for the tool to execute a contour cutting operation along a straight line path. For example, the command

G01 G94 X050.0 Y086.5 Z100.0 F40 S800

specifies that the tool is to move in a straight line from its current position to the location defined by $x = 50.0$ mm, $y = 86.5$ mm, and $z = 100.0$ mm, at a feed rate of 40 mm/min and spindle speed of 800 rev/min.

The G02 and G03 codes are used for circular interpolation, clockwise and counterclockwise, respectively. As indicated in Table 7.1, circular interpolation on a milling machine is limited to one of three planes, x-y, x-z, or y-z. The distinction between clockwise and counterclockwise is established by viewing the plane from the front view. Selection of the desired plane is accomplished by entering one of the codes, G17, G18, or G19, respectively. Thus, the instruction

G02 G17 X088.0 Y040.0 R028.0 F30

moves the tool along a clockwise circular trajectory in the x-y plane to the final coordinates defined by $x = 88$ mm and $y = 40$ mm at a feed rate of 30 mm/min. The radius of the circular arc is 28 mm. The path taken by the cutter from an assumed starting point ($x = 40$, $y = 60$) is illustrated in Figure A7.1.

In a point-to-point motion statement (G00), it is usually desirable to position the tool so that its center is located at the specified coordinates. This is appropriate for operations such as drilling, in which a hole is to be positioned at the coordinates indicated in

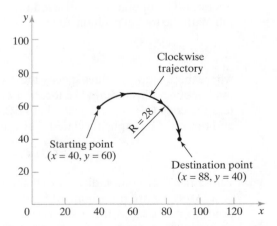

Figure A7.1 Tool path in circular interpolation for the statement: G02 G17 X088.0 Y040.0 R028.0. Units are millimeters.

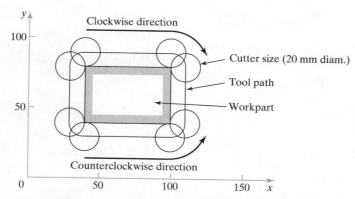

Figure A7.2 Cutter offset for a simple rectangular part. The tool path is separated from the part perimeter by a distance equal to the cutter radius. To invoke cutter offset compensation, the G41 code is used to follow the clockwise path, which keeps the tool on the left-hand side of the part. G42 is used to follow the counterclockwise path, which keeps the tool on the right-hand side of the part.

the statement. But in contouring motions, it is almost always desirable to separate the path followed by the center of the tool from the actual surface of the part by a distance equal to the cutter radius. This is shown in Figure A7.2 for profile milling the outside edges of a rectangular part in two dimensions. For a three-dimensional surface, the shape of the end of the cutter would also have to be considered in the offset computation. This tool path compensation is called the *cutter offset*, and the calculation of the correct coordinates of the endpoints of each move can be time consuming and tedious for the part programmer. Modern CNC machine tool controllers perform these cutter offset calculations automatically when the programmer uses the G40, G41, and G42 codes. The G40 code is used to cancel the cutter offset compensation. The G41 and G42 codes invoke the cutter offset compensation of the tool path on the left- or right-hand side of the part, respectively. The left- and right-hand sides are defined according to the tool path direction. To illustrate, in the rectangular part in Figure A7.2, a clockwise tool path around the part would always position the tool on the left-hand side of the edge being cut, so a G41 code would be used to compute the cutter offset compensation. By contrast, a counter-clockwise tool path would keep the tool on the right-hand side of the part, so G42 would be used. Accordingly, the instruction for profile milling the bottom edge of the part, assuming the cutter begins along the bottom left corner, would read

G42 G01 X100.0 Y040.0 D05

where D05 refers to the cutter radius value stored in MCU memory. Certain registers are reserved in the control unit for these cutter offset values. The D-code references the value contained in the identified register. D05 indicates that the radius offset distance is stored in the number 5 offset register in the controller. This data can be entered into the controller as either a manual input or an instruction in the part program. Manual input is more flexible because the tooling used to machine the part may change from one setup to the next. At the time the job is run, the operator knows which tool will be used, and the

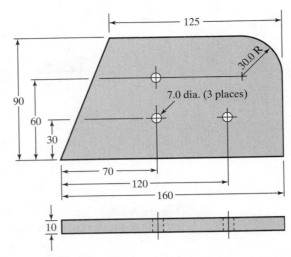

Figure A7.3 Sample part to illustrate NC part programming. Dimensions are in millimeters. General tolerance = ±0.1 mm. Work material is a machinable grade of aluminum.

data can be loaded into the proper register as one of the steps in the setup. When the offset data is entered as a part program instruction, the statement has the form

$$G10 \ P05 \ R10.0$$

where G10 is a preparatory word indicating that cutter offset data will be entered, P05 indicates that the data will be entered into offset register number 05, and R10.0 is the radius value, here 10.0 mm.

To demonstrate manual part programming, we present two examples using the sample part shown in Figure A7.3. The first example is a point-to-point program to drill the three holes in the part. The second example is a two-axis contouring program to accomplish profile milling around the periphery of the part.

EXAMPLE A7.1 Point-to-Point Drilling

This example presents the NC part program in word address format for drilling the three holes in the sample part shown in Figure A7.3. We assume that the outside edges of the starting workpart have been rough cut (by jig sawing) and are slightly oversized for subsequent profile milling. The three holes to be drilled in this example will be used to locate and fixture the part for profile milling in the following example. For the present drilling sequence, the part is gripped in place so that its top surface is 40 mm above the surface of the machine tool table to provide ample clearance beneath the part for hole drilling. We will define the x-, y-, and z-axes as shown in Figure A7.4. A 7.0-mm diameter drill, corresponding to the specified hole size, has been chucked in the CNC drill press. The drill will be operated at a feed of 0.05 mm/rev and a spindle speed of 1000 rev/min (corresponding to a surface speed of about 0.37 m/sec, which is slow for the aluminum work material). At the beginning of the job, the drill point will be positioned at a target point located at $x = 0$, $y = -50$, and $z = +10$ (axis units are millimeters). The program begins with the tool positioned at this target point.

NC Part Program Code	Comments
N001 G21 G90 G92 X0 Y-050.0 Z010.0;	Define origin of axes.
N002 G00 X070.0 Y030.0;	Rapid move to first hole location.
N003 G01 G95 Z-15.0 F0.05 S1000 M03;	Drill first hole.
N004 G01 Z010.0;	Retract drill from hole.
N005 G00 Y060.0;	Rapid move to second hole location.
N006 G01 G95 Z-15.0 F0.05;	Drill second hole.
N007 G01 Z010.0;	Retract drill from hole.
N008 G00 X120.0 Y030.0;	Rapid move to third hole location.
N009 G01 G95 Z-15.0 F0.05;	Drill third hole.
N010 G01 Z010.0;	Retract drill from hole.
N011 G00 X0 Y-050.0 M05;	Rapid move to target point.
N012 M30;	End of program, stop machine.

EXAMPLE A7.2 Two-Axis Milling

The three holes drilled in the previous example can be used for locating and holding the workpart to completely mill the outside edges without re-fixturing. The axis coordinates are shown in Figure A7.4 (same coordinates as in the

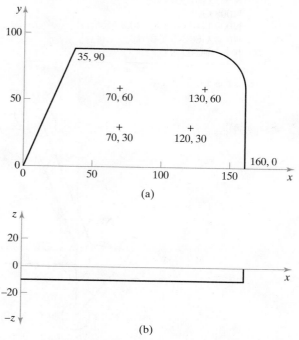

Figure A7.4 Sample part aligned relative to (a) x- and y-axes, and (b) z-axis. Coordinates are given for significant part features in (a).

previous drilling sequence). The part is fixtured so that its top surface is 40 mm above the surface of the machine tool table. Thus, the origin of the axis system will be 40 mm above the table surface. A 20-mm diameter end mill with four teeth will be used. The cutter has a side tooth engagement length of 40 mm. Throughout the machining sequence, the bottom tip of the cutter will be positioned 25 mm below the part top surface, which corresponds to $z = -25$ mm. Since the part is 10 mm thick, this z-position will allow the side cutting edges of the milling cutter to cut the full thickness of the part during profile milling. The cutter will be operated at a spindle speed = 1000 rev/min (which corresponds to a surface speed of about 1.0 m/sec) and a feed rate = 50 mm/min (which corresponds to 0.20 mm/tooth). The tool path to be followed by the cutter is shown in Figure A7.5, with numbering that corresponds to the sequence number in the program. Cutter diameter data has been manually entered into offset register 05. At the beginning of the job, the cutter will be positioned so that its center tip is at a target point located at $x = 0$, $y = -50$, and $z = +10$. The program begins with the tool positioned at this location.

NC Part Program Code	Comments
N001 G21 G90 G92 X0 Y-050.0 Z010.0;	Define origin of axes.
N002 G00 Z-025.0 S1000 M03;	Rapid move to cutter depth, turn spindle on.
N003 G01 G94 G42 Y0 D05 F40;	Engage part, start cutter offset.
N004 G01 X160.0;	Mill lower part edge.
N005 G01 Y060.0;	Mill right straight edge.
N006 G17 G03 X130.0 Y090.0 R030.0;	Circular interpolation around arc.
N007 G01 X035.0;	Mill upper part edge.
N008 G01 X0 Y0;	Mill left part edge.
N009 G40 G00 X-040.0 M05;	Rapid exit from part, cancel offset.
N010 G00 X0 Y-050.0;	Rapid move to target point.
N011 M30;	End of program, stop machine.

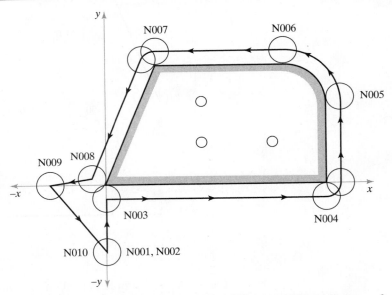

Figure A7.5 Cutter path for profile milling outside perimeter of sample part.

APPENDIX B 7:   PART PROGRAMMING WITH APT

APT is an acronym that stands for Automatically Programmed Tooling. It is a three-dimensional NC part programming system that was developed in the late 1950s and early 60s (Historical Note 7.1). Today it remains an important language in the United States and around the world, and most of the CAD/CAM approaches to part programming (Section 7.5.3) are based on APT. APT is also important because many of the concepts incorporated into it formed the basis for other subsequently developed systems in interactive graphics. APT was originally intended as a contouring language, but modern versions can be used for both point-to-point and contouring operations in up to five axes. Our discussion will be limited to the three linear axes, x, y, and z. APT can be used for a variety of machining operations. Our coverage will concentrate on drilling (point-to-point) and milling (contouring) operations. There are more than 500 words in the APT vocabulary. Only a small (but important) fraction of the total lexicon will be covered here.

APT is not only a language; it is also the computer program that processes the APT statements to calculate the corresponding cutter positions and generate the machine tool control commands. To program in APT, the programmer must first define the part geometry. Then the tool is directed to various point locations and along surfaces of the workpart to accomplish the required machining operations. The viewpoint of the programmer is that the workpiece remains stationary, and the tool is instructed to move relative to the part. To complete the program, speeds and feeds must be specified, tools must be called, tolerances must be given for circular interpolation, and so forth. Thus, there are four basic types of statements in the APT language:

1. *Geometry statements* are used to define the geometry elements that comprise the part.
2. *Motion commands* are used to specify the tool path.
3. *Postprocessor statements* control the machine tool operation, for example, to specify speeds and feeds, set tolerance values for circular interpolation, and actuate other capabilities of the machine tool.
4. *Auxiliary statements* are a group of miscellaneous statements used to name the part program, insert comments in the program, and accomplish similar functions.

These statements are constructed of APT vocabulary words, symbols, and numbers, all arranged using appropriate punctuation. APT vocabulary words consist of six or fewer characters. Such a restriction seems archaic today, but it must be remembered that APT was developed during the 1950s, when computer memory technology was extremely limited. Most APT statements include a slash (/) as part of the punctuation. APT vocabulary words that immediately precede the slash are called *major words*, whereas those that follow the slash are called *minor words*.

B7.1 APT Geometry Statements.

The geometry of the part must be defined to identify the surfaces and features that are to be machined. Accordingly, the points, lines, and surfaces must be defined in the program prior to specifying the motion statements. The general form of an APT geometry statement is the following:

SYMBOL = GEOMETRY TYPE/descriptive data

An example of such a statement is

P1 = POINT/20.0, 40.0, 60.0

An APT geometry statement consists of three sections. The first is the symbol used to identify the geometry element. A symbol can be any combination of six or fewer alphabetical and numerical characters, at least one of which must be alphabetical. Also, the symbol cannot be an APT vocabulary word. The second section of the APT geometry statement is an APT major word that identifies the type of geometry element. Examples are POINT, LINE, CIRCLE, and PLANE. The third section of the APT geometry statement provides the descriptive data that define the element precisely, completely, and uniquely. These data may include numerical values to specify dimensional and position data, previously defined geometry elements, and APT minor words.

Punctuation in an APT geometry statement is indicated in the preceding geometry statements. The geometry definition is written as an equation, the symbol being equated to the element type, followed by a slash with descriptive data to the right of the slash. Commas are used to separate the words and numerical values in the descriptive data. There are a variety of ways to specify geometry elements. In the following discussion, examples of APT statements will be presented for points, lines, planes, and circles.

Points. Specification of a point is most easily accomplished by designating its x-, y-, and z-coordinates:

P1 = POINT/20.0, 40.0, 60.0

where the descriptive data following the slash indicate the x-, y-, and z-coordinates. The specification can be done in either inches or millimeters (metric). We use metric values in our examples. As an alternative, a point can be defined as the intersection of two intersecting lines, as in the following:

P1 = POINT/INTOF, L1, L2

where the APT word INTOF in the descriptive data stands for "intersection of."

Other methods of defining points are also available. Several are illustrated in Figure B7.1. The associated points are identified in the following APT statements:

P2 = POINT/YLARGE, INTOF, L3, C2

P2 = POINT/XSMALL. INTOF, L3, C2

P3 = POINT/XLARGE, INTOF, L3, C2

P3 = POINT/YSMALL , INTOF, L3, C2

P4 = POINT/YLARGE, INTOF, C1, C2

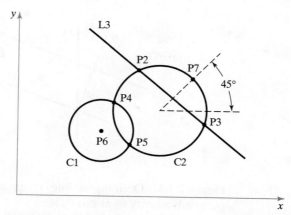

Figure B7.1 Defining a point using intersections of previously defined lines and circles.

P5 = POINT/YSMALL, INTOF, C1, C2

P6 = POINT/CENTER, C1

P7 = POINT/C2, ATANGL, 45

where the word ATANGL means "at angle" in the last statement.

Lines. A line defined in APT is considered to be of infinite length in both directions. Also, APT treats a line as a vertical plane that is perpendicular to the *x-y* plane. The easiest way to specify a line is by two points through which it passes, as in Figure B7.2:

L1 = LINE/P1, P2

The same line can be defined by indicating the coordinate positions of the two points by giving their *x-, y,-,* and *z*-coordinates in sequence; for example,

L1 = LINE/20, 30, 0, 70, 50, 0

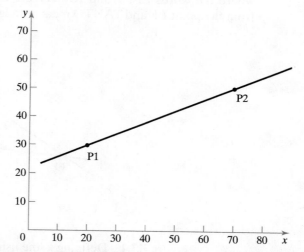

Figure B7.2 Defining a line using two previously defined points.

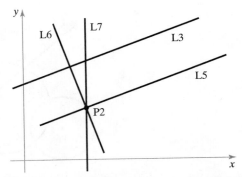

Figure B7.3 Defining a line using a point and parallelism or perpendicularity to another line.

In some situations, the part programmer may find it more convenient to define a new line as being parallel to or perpendicular to one of the axes or another line that has been previously defined; for example, with reference to Figure B7.3,

$$L5 = LINE/P2, PARLEL, L3$$

$$L6 = LINE/P2, PERPTO, L3$$

$$L7 = LINE/P2, PERPTO, XAXIS$$

where PARLEL and PERPTO are APT's way of spelling "parallel to" and "perpendicular to," respectively.

Lines can also be defined in relation to a point and a circle, as in Figure B7.4, as in the geometry statements

$$L1 = LINE/P1, LEFT, TANTO, C1$$

$$L2 = LINE/P1, RIGHT, TANTO, C1$$

where the words LEFT and RIGHT are used by looking in the direction of the circle from the point P1, and TANTO means "tangent to."

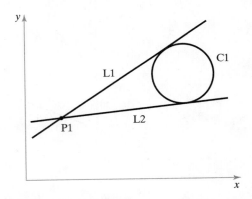

Figure B7.4 Defining a line using a point and a circle.

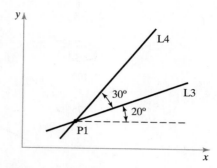

Figure B7.5 Defining a line using a point and the *x*-axis or another line.

Finally, lines can be defined using a point and the angle of the line relative to the *x*-axis or some other line, as in Figure B7.5. The following statements illustrate the definitions:

L3 = LINE/P1, ATANGL, 20, XAXIS

L4 = LINE/P1, ATANGL, 30, L3

Planes. A plane can be defined by specifying three points through which the plane passes, as in the following:

PL1 = PLANE/P1, P2, P3

Of course, the three points must be non-collinear. A plane can also be defined as being parallel to another plane that has been previously defined; for instance,

PL2 = PLANE/P2, PARLEL, PL1

which states that plane PL2 passes through point P2 and is parallel to plane PL1. In APT, a plane extends indefinitely.

Circles. In APT, a circle is considered to be a cylindrical surface that is perpendicular to the *x-y* plane and extends to infinity in the *z*-direction. The easiest way to define a circle is by its center and radius, as in the following two statements, illustrated in Figure B7.6:

C1 = CIRCLE/CENTER, P1, RADIUS, 32

C1 = CIRCLE/CENTER, 100, 50, 0, RADIUS, 32

Two additional ways of defining a circle utilize previously defined points P2, P3, and P4, or line L1 in the same figure:

C1 = CIRCLE/CENTER, P2, P3, P4

C1 = CIRCLE/CENTER, P1, TANTO, L1

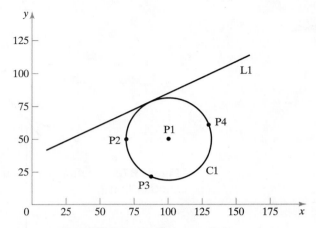

Figure B7.6 Defining a circle.

Obviously, the three points in the first statement must not be collinear.

Other ways to define circles make use of existing lines L2 and L3 in Figure B7.7. The statements for the four circles in the figure are the following:

$$C2 = CIRCLE/XSMALL, L2, YSMALL, L3, RADIUS, 25$$

$$C3 = CIRCLE/YLARGE, L2, YLARGE, L3, RADIUS, 25$$

$$C4 = CIRCLE/XLARGE, L2, YLARGE, L3, RADIUS, 25$$

$$C5 = CIRCLE/ YSMALL, L2, YSMALL, L3, RADIUS, 25$$

Ground Rules. Certain ground rules must be obeyed when formulating APT geometry statements. Following are four important rules in APT:

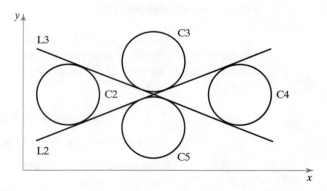

Figure B7.7 Defining a circle using two intersecting lines.

1. Coordinate data must be specified in the order x, then y, then z, because the statement

$$P1 = POINT/20.5, 40.0, 60.0$$

is interpreted to mean $x = 20.5$ mm, $y = 40.0$ mm, and $z = 60.0$ mm.

2. Any symbols used as descriptive data must have been previously defined; for example, in the statement

$$P1 = POINT/INTOF, L1, L2$$

the two lines L1 and L2 must have been previously defined. In setting up the list of geometry statements, the APT programmer must be sure to define symbols before using them in subsequent statements.

3. A symbol can be used to define only one geometry element. The same symbol cannot be used to define two different elements. For example, the following statements would be incorrect if they were included in the same program:

$$P1 = POINT/20, 40, 60$$

$$P1 = POINT/30, 50, 70$$

4. Only one symbol can be used to define any given element. For example, the following two statements in the same part program would be incorrect:

$$P1 = POINT/20, 40, 60$$

$$P2 = POINT/20, 40, 60$$

EXAMPLE B7.1   Part Geometry Using APT

Let us construct the geometry of our sample part in Figure A7.3 (Appendix A7). The geometry elements of the part to be defined in APT are labeled in Figure 7.17 in the text. Reference is also made to Figure A7.4 in Appendix A7, which shows the coordinate values of the points used to dimension the part. Only the geometry statements are given in the APT sequence that follows:

```
P1 = POINT/0, 0, 0
P2 = POINT/160.0, 0, 0
P3 = POINT/160.0, 60.0, 0
P4 = POINT/35.0, 90.0, 0
P5 = POINT/70.0, 30.0, 0
P6 = POINT/120.0, 30.0, 0
P7 = POINT/70.0, 60.0, 0
P8 = POINT/130.0, 60.0, 0
L1 = LINE/P1, P2
L2 = LINE/P2, P3
C1 = CIRCLE/CENTER, P8, RADIUS, 30.0
L3 = LINE/P4, PARLEL, L1
L4 = LINE/P4, P1
```

B7.2 APT Motion Commands

All APT motion statements follow a common format, just as geometry statements have their own format. The format of an APT motion command is

MOTION COMMAND/descriptive data

An example of an APT motion statement is

GOTO/P1

The statement consists of two sections separated by a slash. The first section is the basic command that indicates what move the tool should make. The descriptive data following the slash tell the tool where to go. In the above example, the tool is directed to go to (GOTO) point P1, which has been defined in a previous geometry statement.

At the beginning of the sequence of motion statements, the tool must be given a starting point. This is likely to be the target point, the location where the operator has positioned the tool at the start of the job. The part programmer keys in this starting position with the statement

FROM/PTARG

where FROM is an APT vocabulary word indicating that this is the initial point from which all other geometry elements will be referenced, and PTARG is the symbol assigned to the starting point. Another way to make this statement is

FROM/$-20.0, -20.0, 0$

where the descriptive data in this case are the x-, y-, and z-coordinates of the starting point. The FROM statement occurs only at the start of the motion sequence.

Point-To-Point Commands. In our discussion of APT motion statements, it is appropriate to distinguish between point-to-point motions and contouring motions. For point-to-point motions, there are only two commands: GOTO and GODLTA. The GOTO statement instructs the tool to go to a particular point location specified in the descriptive data. Two examples are

GOTO/P2

GOTO/25.0, 40.0, 0

In the first command, P2 is the destination of the tool point, and its location has been previously defined in a geometry statement. In the second command, the tool has been instructed to go to the location whose coordinates are $x = 25.0$, $y = 40.0$, and $z = 0$.

The GODLTA command specifies an incremental move for the tool. To illustrate, the following statement instructs the tool to move from its present position by a distance of 50.0 mm in the x-direction, 120.0 mm in the y-direction, and 40 mm in the z-direction:

GODLTA/50.0, 120.0, 40.0

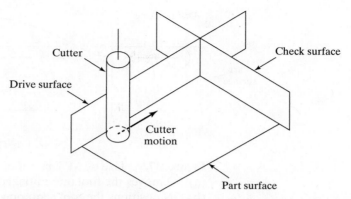

Figure B7.8 Three surfaces in APT contouring motions that guide the cutting tool.

The GODLTA statement is useful in drilling and related machining operations. The tool can be directed to go to a given hole location; then the GODLTA command can be used to drill the hole, as in the following sequence:

GOTO/P2
GODLTA/0, 0, −50.0
GODLTA/0, 0, 50.0

Contouring Motions. Contouring commands are more complicated than PTP commands because the tool's position must be continuously controlled throughout the move. To exercise this control, the tool is directed along two intersecting surfaces until it reaches a third surface, as shown in Figure B7.8. These three surfaces have specific names in APT:

1. *Drive surface.* This is the surface that guides the side of the cutter. It is pictured as a plane in our figure.
2. *Part surface.* This is the surface, again pictured as a plane, on which the bottom or nose of the tool is guided.
3. *Check surface.* This is the surface that stops the forward motion of the tool in the execution of the current command. One might say that this surface "checks" the advance of the tool.

It should be noted here that the "part surface" may or may not be an actual surface of the part. The part programmer may elect to use an actual part surface or some other previously defined surface for the purpose of maintaining continuous path control of the tool. The same qualification goes for the drive surface and check surface.

There are several ways in which the check surface can be used. This is determined by using any of four APT modifier words in the descriptive data of the motion statement. The four modifier words are TO, ON, PAST, and TANTO. As depicted in Figure B7.9, the word TO positions the leading edge of the tool in contact with the check surface,

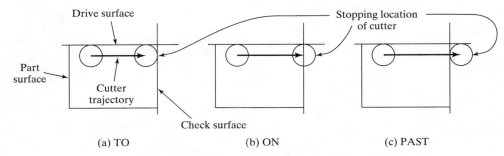

Figure B7.9 Use of APT modifier words in motion statements: (a) TO moves the tool into initial contact with the check surface; (b) ON positions the tool center on the check surface; and (c) PAST moves the tool just beyond the check surface.

ON positions the center of the tool on the check surface, and PAST puts the tool beyond the check surface so that its trailing edge is in contact with the check surface. The fourth modifier word TANTO is used when the drive surface is tangent to a circular check surface, as in Figure B7.10. TANTO moves the cutting tool to the point of tangency with the circular surface.

An APT contouring motion command causes the cutter to proceed along a trajectory defined by the drive surface and part surface; when the tool reaches the check surface it stops according to one of the modifier words TO, ON, PAST, or TANTO. In writing a motion statement, the part programmer must keep in mind the direction from which the tool is coming in the preceding motion command. The programmer must pretend to be riding on top of the tool, as if driving a car. After the tool reaches the check surface in the preceding move, does the next move involve a right turn or left turn or what? The answer to this question is determined by one of the following six motion words, whose interpretations are illustrated in Figure B7.11:

- GOLFT commands the tool to make a left turn relative to the last move.
- GORGT commands the tool to make a right turn relative to the last move.
- GOFWD commands the tool to move forward relative to the last move.
- GOBACK commands the tool to reverse direction relative to the last move.

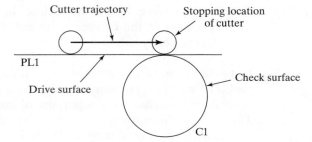

Figure B7.10 Use of the APT modifier word TANTO. TANTO moves the tool to the point of tangency between two surfaces, at least one of which is a circular surface.

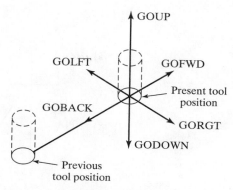

Figure B7.11 Use of the APT motion words. The tool has moved from a previous position to its present position. The direction of the next move is determined by one of the APT motion words GOLFT, GORGT, GOFWD, GOBACK, GOUP, or GODOWN.

- GOUP commands the tool to move upward relative to the last move.
- GODOWN commands the tool to move down relative to the last move.

In many cases, the next move will be in a direction that is a combination of two pure directions. For example, the direction might be somewhere between go forward and go right. In these cases, the proper motion command would designate the largest direction component among the choices available.

To begin the sequence of motion commands, the FROM statement is used in the same manner as for point-to-point moves. The statement following the FROM command defines the initial drive surface, part surface, and check surface. With reference to Figure B7.12, the sequence takes the following form:

FROM/PTARG

GO/TO, PL1, TO, PL2, TO, PL3

The symbol PTARG represents the target point where the operator has set up the tool. The GO command instructs the tool to move to the intersection of the drive surface (PL1), the part surface (PL2), and the check surface (PL3). Because the modifier word TO has been used for each of the three surfaces, the circumference of the cutter is tangent to PL1 and PL3, and the bottom of the cutter is on PL2. The three surfaces included in the GO statement must be specified in the order: (1) drive surface, (2) part surface, and (3) check surface.

Note that GO/TO is not the same as the GOTO command. GOTO is used only for PTP motions. The GO/command is used to initialize a sequence of contouring motions and may take alternative forms such as GO/ON, GO/TO, or GO/PAST.

After initialization, the tool is directed along its path by one of the six motion command words. It is not necessary to redefine the part surface in every motion command after it has been initially defined as long as it remains the same in subsequent commands. In the preceding motion command, the cutter has been directed from PTARG to the intersection of surfaces PL1, PL2, and PL3. Suppose we now wish to move the tool along

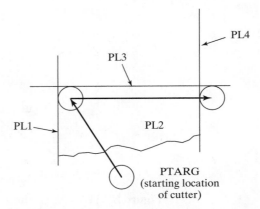

Figure B7.12 Initialization of APT contouring motion sequence.

plane PL3 in Figure B7.12, with PL2 remaining as the part surface. The following command would accomplish this motion:

GORGT/PL3, PAST, PL4

Note that PL2 is not mentioned in this new command. PL3, which was the check surface in the preceding command, is the drive surface in the new command, and the new check surface is PL4. Although the part surface may remain the same throughout the motion sequence, the drive surface and check surface must be redefined in each new contouring motion command.

There are many parts whose features can all be defined in two axes, x and y. Although such parts certainly possess a third dimension, there are no features to be machined in this direction. Our sample part is a case in point. In the engineering drawing, Figure A7.3 in Appendix A7, the sides of the part appear as lines, although they are three-dimensional surfaces on the physical part. In cases like this, it is more convenient for the programmer to define the part profile in terms of lines and circles rather than planes and cylinders. Fortunately, the APT language system allows this because in APT, lines are treated as planes and circles are treated as cylinders, which are both perpendicular to the x-y plane. Hence, the planes around the part outline in Figure A7.3 can be replaced by lines (call them L1, L2, L3, and L4), and the preceding APT commands can be replaced by the following:

FROM/PTARG

GO/TO, L1, TO, PL2, TO L3

GORGT/L3, PAST, L4

Substitution of lines and circles for planes and cylinders in APT is allowed only when the sides of the part are perpendicular to the x-y plane. Note that plane PL2 has not been converted to a line. As the "part surface" in the motion statement, it must maintain its status as a plane parallel to the x- and y-axes.

EXAMPLE B7.2   APT Contouring Motion Commands

Let us write the APT motion commands to profile mill the outside edges of our sample workpart. The geometry elements are labeled in Figure 7.17(b) in the text, and the tool path is shown in Figure A7.5 in Appendix A7. The tool begins its motion sequence from a target point PTARG located at $x = 0$, $y = -50$ mm and $z = 10$ mm. We also assume that "part surface" PL2 has been defined as a plane parallel to the x-y plane and located 25 mm below the top surface of the part (Figure A7.4). The reason for defining it this way is to ensure that the cutter will machine the entire thickness of the part:

```
FROM/PTARG
GO/TO, L1, TO, PL2, ON, L4
GORGT/L1, PAST, L2
GOLFT/L2, TANTO, C1
GOFWD/C1, PAST, L3
GOFWD/L3, PAST, L4
GOLFT/L4, PAST, L1
GOTO/P0
```

B7.3 Postprocessor and Auxiliary Statements

A complete APT part program must include functions not accomplished by geometry statements and motion commands. These additional functions are implemented by postprocessor statements and auxiliary statements.

Postprocessor statements control the operation of the machine tool and play a supporting role in generating the tool path. Such statements are used to define cutter size, specify speeds and feeds, turn coolant flow on and off, and control other features of the particular machine tool on which the machining job will be performed. The general form of a postprocessor statement is

POSTPROCESSOR COMMAND/descriptive data

where the POSTPROCESSOR COMMAND is an APT major word indicating the type of function or action to be accomplished, and the descriptive data consists of APT minor words and numerical values. In some commands, the descriptive data is omitted. Some examples of important postprocessor statements are the following:

- UNITS/MM indicates that the specified units used in the program are INCHES or MM.
- INTOL/0.02 specifies inward tolerance for circular interpolation.
- OUTTOL/0.02 specifies outward tolerance for circular interpolation.
- CUTTER/20.0 defines cutter diameter in Figure B7.13(a) for tool path offset calculations. The statement CUTTER/20, 5 indicates that the cutter has a corner radius of 5 mm [Figure B7.13(b)], for three-dimensional contouring. The length and other dimensions of the tool can also be specified, if necessary.
- SPINDL/1000, CLW specifies spindle rotation speed in revolutions per minute. Either CLW (clockwise) or CCLW (counterclockwise) can be specified.

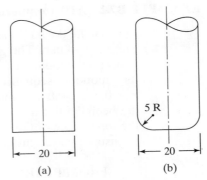

Figure B7.13 Cutter definition for a 20 mm diameter milling cutter. (a) where corner radius is zero and (b) where corner radius = 5 mm.

- SPINDL/OFF stops spindle rotation.
- FEDRAT/40, IPM specifies feed rate in millimeters per minute or inches per minute. Minor words IPM or IPR are used to indicate whether the feed rate is units per minute or units per revolution of the cutter, where the units are specified as inches or millimeters in a preceding UNITS statement.
- RAPID engages rapid traverse (high feed rate) for next move(s).
- COOLNT/FLOOD turns cutting fluid on.
- LOADTL/01 used with automatic tool-changers to identify which cutting tool should be loaded into the spindle.
- DELAY/30 temporarily stops the machine tool for a period specified in seconds.

Auxiliary statements are used to identify the part program, specify which postprocessor to use, insert remarks into the program, and so on. Auxiliary statements have no effect on the generation of tool path. The following are APT words used in auxiliary statements:

- PARTNO is the first statement in an APT program, used to identify the program; for example,

PARTNO SAMPLE PART NUMBER ONE

- MACHIN/ permits the part programmer to specify the postprocessor, which in effect specifies the machine tool.
- CLPRNT stands for "cutter location print," which is used to print out the cutter location sequence.
- REMARK is used to insert explanatory comments into the program that are not interpreted or processed by the APT processor.
- FINI indicates the end of an APT program.

The major word MACHIN requires a slash (/) as indicated in our list above, with descriptive data that identify the postprocessor to be used. Words such as CLPRNT and FINI are complete without descriptive data. PARTNO and REMARK have a format that is an exception to the normal APT statement structure. These are words that are followed by descriptive data, but without a slash separating the APT word from the descriptive data.

PARTNO is used at the very beginning of the part program and is followed by a series of alphanumeric characters that label the program. REMARK permits the programmer to insert comments that the APT processor does not process.

B7.4 Some APT Part Programming Examples

As examples of APT, we will prepare two part programs for our sample part, one to drill the three holes and the second to profile mill the outside edges. As in our example programs in Appendix A7, the starting workpiece is an aluminum plate of the desired thickness, and its perimeter has been rough cut slightly oversized in anticipation of the profile milling operation. In effect, these APT programs will accomplish the same operations as previous Examples A7.1 and A7.2 in which manual part programming was used.

EXAMPLE B7.3   Drilling Sequence in APT

Let us write the APT program to perform the drilling sequence for our sample part in Figure A7.3. We will show the APT geometry statements only for the three hole locations, saving the remaining elements of geometry for Example B7.4.

```
PARTNO SAMPLE PART DRILLING OPERATION
MACHIN/DRILL, 01
CLPRNT
UNITS/MM
REMARK Part geometry. Points are defined 10 mm above part surface.
PTARG = POINT/0, −50.0, 10.0
P5 = POINT/70.0, 30.0, 10.0
P6 = POINT/120.0, 30.0, 10.0
P7 = POINT/70.0, 60.0, 10.0
REMARK Drill bit motion statements.
FROM/PTARG
RAPID
GOTO/P5
SPINDL/1000, CLW
FEDRAT/0.05, IPR
GODLTA/0, 0, −25
GODLTA/0, 0, 25
RAPID
GOTO/P6
SPINDL/1000, CLW
FEDRAT/0.05, IPR
GODLTA/0, 0, −25
GODLTA/0, 0, 25
RAPID
GOTO/P7
SPINDL/1000, CLW
FEDRAT/0.05, IPR
GODLTA/0, 0, −25
GODLTA/0, 0, 25
```

RAPID
GOTO/PTARG
SPINDL/OFF
FINI

EXAMPLE B7.4 Two-Axis Profile Milling in APT

The three holes drilled in Example B7.3 will be used for locating and holding the workpart for milling the outside edges. Axis coordinates are given in Figure A7.4. The top surface of the part is 40 mm above the surface of the machine table. A 20-mm diameter end mill with four teeth and a side tooth engagement of 40 mm will be used. The bottom tip of the cutter will be positioned 25 mm below the top surface during machining, thus ensuring that the side cutting edges of the cutter will cut the full thickness of the part. Spindle speed = 1000 rev/min and feed rate = 50 mm/min. The tool path, shown in Figure A7.5, is the same as that followed in Example A7.2:

```
PARTNO SAMPLE PART MILLING OPERATION
MACHIN/MILLING, 02
CLPRNT
UNITS/MM
CUTTER/20.0
REMARK Part geometry. Points and lines are defined 25 mm below
part top surface.
PTARG = POINT/0, −50.0, 10.0
P1 = POINT/0, 0, −25
P2 = POINT/160, 0, −25
P3 = POINT/160, 60, −25
P4 = POINT/35, 90, −25
P8 = POINT/130, 60, −25
L1 = LINE/P1, P2
L2 = LINE/P2, P3
C1 = CIRCLE/CENTER, P8, RADIUS, 30
L3 = LINE/P4, LEFT, TANTO, C1
L4 = LINE/P4, P1
PL1 = PLANE/P1, P2, P4
REMARK Milling cutter motion statements.
FROM/PTARG
SPINDL/1000, CLW
FEDRAT/50, IPM
GO/TO, L1, TO, PL1, ON, L4
GORGT/L1, PAST, L2
GOLFT/L2, TANTO, C1
GOFWD/C1, PAST, L3
GOFWD/L3, PAST, L4
GOLFT/L4, PAST, L1
RAPID
GOTO/PTARG
SPINDL/OFF
FINI
```

Chapter 8

Industrial Robotics

CHAPTER CONTENTS

An *industrial robot* is a general-purpose, programmable machine possessing certain anthropomorphic characteristics. The most obvious anthropomorphic characteristic of an industrial robot is its mechanical arm, which is used to perform various industrial tasks. Other human-like characteristics are the robot's capabilities to respond to sensory inputs,

From Chapter 8 of *Automation, Production Systems, and Computer-Integrated Manufacturing*, Third Edition.
Mikell P. Groover. Copyright © 2008 by Pearson Education, Inc. Publishing as Prentice Hall. All rights reserved.

communicate with other machines, and make decisions. These capabilities permit robots to perform a variety of useful tasks. The development of robotics technology followed the development of numerical control (Historical Note 8.1), and the two technologies are quite similar. They both involve coordinated control of multiple axes (the axes are called *joints* in robotics), and they both use dedicated digital computers as controllers. Whereas NC machines are designed to perform specific processes (e.g., machining, sheetmetal hole punching, and thermal cutting), robots are designed for a wider variety of tasks. Typical production applications of industrial robots include spot welding, material transfer, machine loading, spray painting, and assembly.

Some of the qualities that make industrial robots commercially and technologically important are listed here.

- Robots can be substituted for humans in hazardous or uncomfortable work environments.
- A robot performs its work cycle with a consistency and repeatability that cannot be attained by humans.
- Robots can be reprogrammed. When the production run of the current task is completed, a robot can be reprogrammed and equipped with the necessary tooling to perform an altogether different task.
- Robots are controlled by computers and can therefore be connected to other computer systems to achieve computer integrated manufacturing.

Historical Note 8.1   A short history of industrial robots [5] [11]

The word "robot" entered the English language through a Czechoslovakian play titled *Rossum's Universal Robots*, written by Karel Capek in the early 1920s. The Czech word "robota" means forced worker. In the English translation, the word was converted to "robot." The story line of the play centers around a scientist named Rossum who invents a chemical substance similar to protoplasm and uses it to produce robots. The scientist's goal is for robots to serve humans and perform physical labor. Rossum continues to make improvements in his invention, ultimately perfecting it. These "perfect beings" begin to resent their subservient role in society and turn against their masters, killing off all human life.

Capek's play was pure science fiction. Our short history must include two real inventors who made original contributions to the technology of industrial robotics. The first was Cyril W. Kenward, a British inventor who devised a manipulator that moved on an *x-y-z* axis system. In 1954, Kenward applied for a British patent for his robotic device, and in 1957 the patent was issued.

The second inventor was an American named George C. Devol. Devol is credited with two inventions related to robotics. The first was a device for magnetically recording electrical signals so that the signals could be played back to control the operation of machinery. This device was invented around 1946, and a U.S. patent was issued in 1952. The second invention was a robotic device developed in the 1950s, which Devol called "Programmed Article Transfer." This device was intended for parts handling. The U.S. patent was finally issued in 1961. It was a prototype for the hydraulically driven robots that were later built by Unimation, Inc.

Although Kenward's robot was chronologically the first (at least in terms of patent date), Devol's proved ultimately to be far more important in the development and commercialization of robotics technology. The reason for this was a catalyst in the person of Joseph Engelberger. Engelberger had graduated with a degree in physics in 1949. As a student, he had read science fiction novels about robots. By the mid-1950s, he was working for a company that made control systems for jet engines. Hence, by the time a chance meeting occurred between Engelberger and Devol in 1956, Engelberger was "predisposed by education, avocation, and occupation toward the notion of robotics."[1] The meeting took place at a cocktail party in Fairfield, Connecticut. Devol described his programmed article transfer invention to Engelberger, and they subsequently began considering how to develop the device as a commercial product for industry. In 1962, Unimation, Inc. was founded, with Engelberger as president. The name of the company's first product was "Unimate," a polar configuration robot. The first application of a Unimate robot was unloading a die casting machine at a General Motors plant in New Jersey in 1961.

8.1 ROBOT ANATOMY AND RELATED ATTRIBUTES

The *manipulator* of an industrial robot consists of a series of joints and links. Robot anatomy is concerned with the types and sizes of these joints and links and other aspects of the manipulator's physical construction.

8.1.1 Joints and Links

A joint of an industrial robot is similar to a joint in the human body: It provides relative motion between two parts of the body. Each joint, or *axis* as it is sometimes called, provides the robot with a so-called *degree-of-freedom* (d.o.f.) of motion. In nearly all cases, only one degree-of-freedom is associated with each joint. Robots are often classified according to the total number of degrees-of-freedom they possess. Connected to each joint are two links, an input link and an output link. Links are the rigid components of the robot manipulator. The purpose of the joint is to provide controlled relative movement between the input link and the output link.

Most robots are mounted on a stationary base on the floor. Let us refer to this base and its connection to the first joint as link 0. It is the input link to joint 1, the first in the series of joints used in the construction of the robot. The output link of joint 1 is link 1. Link 1 is the input link to joint 2, whose output link is link 2, and so forth. This joint-link numbering scheme is illustrated in Figure 8.1.

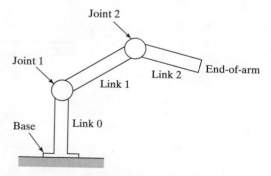

Figure 8.1 Diagram of robot construction showing how a robot is made up of a series of joint-link combinations.

[1]This quote was too good to resist. It was borrowed from Groover et al., *Industrial Robotics: Technology, Programming, and Applications* [5].

Nearly all industrial robots have mechanical joints that can be classified into one of five types: two types that provide translational motion and three types that provide rotary motion. These joint types are illustrated in Figure 8.2 and are based on a scheme described in [5]. The five joint types are

1. *Linear joint* (type L joint). The relative movement between the input link and the output link is a translational sliding motion, with the axes of the two links parallel.
2. *Orthogonal joint* (type O joint). This is also a translational sliding motion, but the input and output links are perpendicular to each other during the move.
3. *Rotational joint* (type R joint). This type provides rotational relative motion, with the axis of rotation perpendicular to the axes of the input and output links.
4. *Twisting joint* (type T joint). This joint also involves rotary motion, but the axis of rotation is parallel to the axes of the two links.
5. *Revolving joint* (type V joint, V from the "v" in revolving). In this joint type, the axis of the input link is parallel to the axis of rotation of the joint, and the axis of the output link is perpendicular to the axis of rotation.

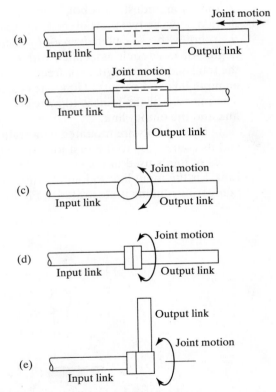

Figure 8.2 Five types of joints commonly used in industrial robot construction: (a) linear joint (type L joint), (b) orthogonal joint (type O joint), (c) rotational joint (type R joint), (d) twisting joint (type T joint), and (e) revolving joint (type V joint).

Each of these joint types has a range over which it can be moved. The range for a translational joint is usually less than a meter, but for large gantry robots, the range may be several meters. The three types of rotary joints may have a range as small as a few degrees or as large as several complete turns.

8.1.2 Common Robot Configurations

A robot manipulator can be divided into two sections: a body-and-arm assembly and a wrist assembly. There are usually three degrees-of-freedom associated with the body-and-arm, and either two or three degrees-of-freedom associated with the wrist. At the end of the manipulator's wrist is a device related to the task that must be accomplished by the robot. The device, called an *end effector* (Section 8.3), is usually either (1) a gripper for holding a workpart or (2) a tool for performing some process. The body-and-arm of the robot is used to position the end effector, and the robot's wrist is used to orient the end effector.

Body-and-Arm Configurations. Given the five types of joints defined above, there are $5 \times 5 \times 5 = 125$ different combinations of joints that can be used to design the body-and-arm assembly for a three-degree-of-freedom robot manipulator. In addition, there are design variations within the individual joint types (e.g., physical size of the joint and range of motion). It is somewhat remarkable, therefore, that there are only five basic configurations commonly available in commercial industrial robots.[2] These five configurations are:

1. *Polar configuration.* This configuration (Figure 8.3) consists of a sliding arm (L joint) actuated relative to the body, which can rotate about both a vertical axis (T joint) and a horizontal axis (R joint).
2. *Cylindrical configuration.* This robot configuration (Figure 8.4) consists of a vertical column, relative to which an arm assembly is moved up or down. The arm can be moved in and out relative to the axis of the column. Our figure shows one possible way in which this configuration can be constructed, using a T joint to rotate the column about its axis. An L joint is used to move the arm assembly vertically along the column, while an O joint is used to achieve radial movement of the arm.
3. *Cartesian coordinate robot.* Other names for this configuration include rectilinear robot and *x-y-z* robot. As shown in Figure 8.5, it is composed of three sliding joints, two of which are orthogonal.
4. *Jointed-arm robot.* This robot manipulator (Figure 8.6) has the general configuration of a human arm. The jointed arm consists of a vertical column that swivels about the base using a T joint. At the top of the column is a shoulder joint (shown as an R joint in our figure), whose output link connects to an elbow joint (another R joint).
5. *SCARA.* SCARA is an acronym for *Selective Compliance Assembly Robot Arm.* This configuration (Figure 8.7) is similar to the jointed arm robot except that the shoulder and elbow rotational axes are vertical, which means that the arm is very rigid in the vertical direction, but compliant in the horizontal direction. This permits the robot to perform insertion tasks (for assembly) in a vertical direction, where some side-to-side alignment may be needed to mate the two parts properly.

---

[2]There are possible variations in the joint types that can be used to construct the five basic configurations.

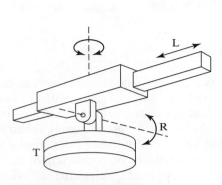

Figure 8.3 Polar coordinate body-and-arm assembly.

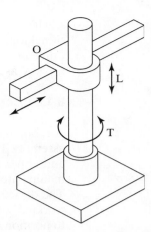

Figure 8.4 Cylindrical body-and-arm assembly.

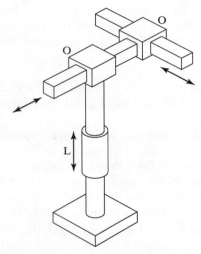

Figure 8.5 Cartesian coordinate body-and-arm assembly.

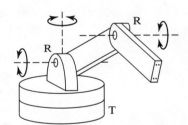

Figure 8.6 Jointed-arm body-and-arm assembly.

Wrist Configurations. The robot's wrist is used to establish the orientation of the end effector. Robot wrists usually consist of two or three degrees-of-freedom. Figure 8.8 illustrates one possible configuration for a three-degree-of-freedom wrist assembly. The three joints are defined as follows: (1) *roll*, using a T joint to accomplish rotation about the robot's arm axis; (2) *pitch*, which involves up-and-down rotation, typically using a R joint; and (3) *yaw*, which involves right-and-left rotation, also accomplished by means of an R-joint. A two-d.o.f wrist typically includes only roll and pitch joints (T and R joints).

To avoid confusion in the pitch and yaw definitions, the wrist roll should be assumed in its center position, as shown in our figure. To demonstrate the possible confusion, consider a two-jointed wrist assembly. With the roll joint in its center position, the second joint (R joint) provides up-and-down rotation (pitch). However, if the roll position were

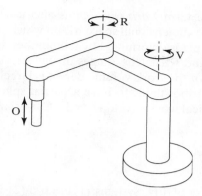

Figure 8.7 SCARA body-and-arm assembly.

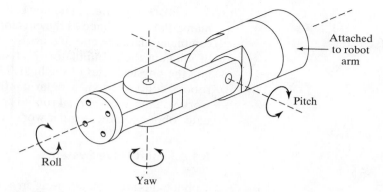

Figure 8.8 Typical configuration of a three-degree-of-freedom wrist assembly showing roll, pitch, and yaw.

90 degrees from center (either clockwise or counterclockwise), the second joint would provide a right-left rotation (yaw).

The SCARA robot configuration (Figure 8.7) is unique in that it typically does not have a separate wrist assembly. As indicated in our description, it is used for insertion type assembly operations in which the insertion is made from above. Accordingly, the orientation requirements are minimal, and the wrist is therefore not needed. Orientation of the object to be inserted is sometimes required, and an additional rotary joint can be provided for this purpose. The other four body-and-arm configurations possess wrist assemblies that almost always consist of combinations of rotary joints of types R and T.

Joint Notation System. The letter symbols for the five joint types (L, O, R, T, and V) can be used to define a joint notation system for the robot manipulator. In this notation system, the manipulator is described by the joint types that make up the body-and-arm assembly, followed by the joint symbols that make up the wrist. For example, the notation TLR : TR represents a five degrees-of-freedom manipulator whose body-and-arm is made up of a twisting joint (joint 1 = T), a linear joint (joint 2 = L), and a rotational joint (joint 3 = R). The wrist consists of two joints, a twisting joint (joint 4 = T) and a rotational joint (joint 5 = R). A colon separates the body-and-arm notation from the wrist notation. Typical joint notations for the five common body-and-arm configurations are presented in Table 8.1. Common wrist joint notations are TRR and TR.

TABLE 8.1 Joint Notations
for Five Common Robot Body-and-Arm Configurations

Body-and-Arm	Joint Notation	Alternative Configurations
Polar	TRL (Figure 8.3)	
Cylindrical	TLO (Figure 8.4)	LVL
Cartesian coordinate	LOO (Figure 8.5)	OOO
Jointed arm	TRR (Figure 8.6)	VVR
SCARA	VRO (Figure 8.7)	

Note: In some cases, more than one joint notation is given because the configuration can be constructed using more than one series of joint types.

Work Volume. The work volume (the term *work envelope* is also used) of the manipulator is defined as the envelope or three-dimensional space within which the robot can manipulate the end of its wrist. Work volume is determined by the number and types of joints in the manipulator (body-and-arm and wrist), the ranges of the various joints, and the physical sizes of the links. The shape of the work volume depends largely on the robot's configuration. A polar configuration robot tends to have a partial sphere as its work volume, a cylindrical robot has a cylindrical work envelope, and a Cartesian coordinate robot has a rectangular work volume.

8.1.3 Joint Drive Systems

Robot joints are actuated using any of three types of drive systems: (1) electric, (2) hydraulic, or (3) pneumatic. Electric drive systems use electric motors as joint actuators (e.g., servomotors or stepper motors, the same types of motors used in NC positioning systems, Chapter 7). Hydraulic and pneumatic drive systems use devices such as linear pistons and rotary vane actuators to accomplish the motion of the joint.

Pneumatic drive is typically limited to smaller robots used in simple material transfer applications. Electric drive and hydraulic drive are used on more sophisticated industrial robots. Electric drive has become the preferred drive system in commercially available robots, as electric motor technology has advanced in recent years. It is more readily adaptable to computer control, which is the dominant technology used today for robot controllers. Electric drive robots are relatively accurate compared with hydraulically powered robots. By contrast, the advantages of hydraulic drive include greater speed and strength.

The drive system, position sensors (and speed sensors if used), and feedback control systems for the joints determine the dynamic response characteristics of the manipulator. The speed with which the robot can achieve a programmed position and the stability of its motion are important characteristics of dynamic response in robotics. *Speed* refers to the absolute velocity of the manipulator at its end-of-arm. The maximum speed of a large robot is around 2 m/sec (6 ft/sec). Speed can be programmed into the work cycle so that different portions of the cycle are carried out at different velocities. What is sometimes more important than speed is the robot's capability to accelerate and decelerate in a controlled manner. In many work cycles, much of the robot's movement is performed in a confined region of the work volume, so the robot never achieves its top-rated velocity. In these cases, nearly all of the motion cycle is engaged in acceleration and deceleration rather than in constant speed. Other factors that influence speed of motion are the weight (mass) of the object that is being manipulated and the precision with which the object must be located at the end of a given move. A term that takes all of these factors into consideration is *speed of response*, which refers to the time required for the manipulator to move from one point in space to the next. Speed of response is important because it influences the robot's cycle time, which in turn affects the production rate in the application. *Stability* refers to the amount of overshoot and oscillation that occurs in the robot motion at the end-of-arm as it attempts to move to the next programmed location. More oscillation in the motion is an indication of less stability. The problem is that robots with greater stability are inherently slower in their response, whereas faster robots are generally less stable.

Load carrying capacity depends on the robot's physical size and construction as well as the force and power that can be transmitted to the end of the wrist. The weight carrying capacity of commercial robots ranges from less than 1 kg up to approximately 900 kg (2000 lb). Medium sized robots designed for typical industrial applications have capacities in the range of 10 to 45 kg (25 to 100 lb). One factor that should be kept in mind when considering load carrying capacity is that a robot usually works with a tool or gripper attached to its wrist. Grippers are designed to grasp and move objects about the work cell. The net load carrying capacity of the robot is obviously reduced by the weight of the gripper. If the robot is rated at a 10 kg (22 lb) capacity and the weight of the gripper is 4 kg (9 lbs), then the net weight carrying capacity is reduced to 6 kg (13 lb).

8.2 ROBOT CONTROL SYSTEMS

The actuations of the individual joints must be controlled in a coordinated fashion for the manipulator to perform a desired motion cycle. Microprocessor-based controllers are commonly used today in robotics as the control system hardware. The controller is organized in a hierarchical structure as indicated in Figure 8.9 so that each joint has its own feedback control system, and a supervisory controller coordinates the combined actuations of the joints according to the sequence of the robot program. Different types of control are required for different applications. Robot controllers can be classified into four categories [5]: (1) limited sequence control, (2) playback with point-to-point control, (3) playback with continuous path control, and (4) intelligent control.

Limited Sequence Control. This is the most elementary control type. It can be utilized only for simple motion cycles, such as pick-and-place operations (i.e., picking an object up at one location and placing it at another location). It is usually implemented by setting limits or mechanical stops for each joint and sequencing the actuation of the joints to accomplish the cycle. Feedback loops are sometimes used to indicate that the particular joint actuation has been accomplished so that the next step in the sequence can be initiated. However, there is no servo-control to accomplish precise positioning of the joint. Many pneumatically driven robots are limited sequence robots.

Playback with Point-to-Point Control. Playback robots represent a more sophisticated form of control than limited sequence robots. *Playback control* means that the controller has a memory to record the sequence of motions in a given work cycle as well as the locations and other parameters (such as speed) associated with each motion and then to subsequently play back the work cycle during execution of the program. In

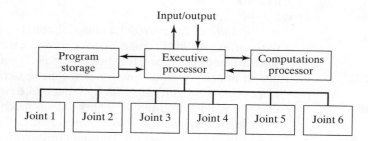

Figure 8.9 Hierarchical control structure of a robot microcomputer controller.

point-to-point (PTP) control, individual positions of the robot arm are recorded into memory. These positions are not limited to mechanical stops for each joint as in limited sequence robots. Instead, each position in the robot program consists of a set of values representing locations in the range of each joint of the manipulator. Thus, each "point" consists of five or six values corresponding to the positions of each of the five or six joints of the manipulator. For each position defined in the program, the joints are thus directed to actuate to their respective specified locations. Feedback control is used during the motion cycle to confirm that the individual joints achieve the specified locations in the program.

Playback with Continuous Path Control. Continuous path robots have the same playback capability as the previous type. The difference between continuous path and point-to-point is the same in robotics as it is in NC (Section 7.1.3). A playback robot with continuous path control is capable of one or both of the following:

1. *Greater storage capacity.* The controller has a far greater storage capacity than its point-to-point counterpart, so the number of locations that can be recorded into memory is far greater than for point-to-point. Thus, the points constituting the motion cycle can be spaced very closely together to permit the robot to accomplish a smooth continuous motion. In PTP, only the final location of the individual motion elements are controlled, so the path taken by the arm to reach the final location is not controlled. In a continuous path motion, the movement of the arm and wrist is controlled during the motion.
2. *Interpolation calculations.* The controller computes the path between the starting point and the ending point of each move using interpolation routines similar to those used in NC. These routines generally include linear and circular interpolation (Table 7.1).

The difference between PTP and continuous path control can be distinguished in the following mathematical way. Consider a three-axis Cartesian coordinate manipulator in which the end-of-arm is moved in x-y-z space. In point-to-point systems, the x, y, and z axes are controlled to achieve a specified point location within the robot's work volume. In continuous path systems, not only are the x, y, and z axes controlled, but the velocities dx/dt, dy/dt, and dz/dt are controlled simultaneously to achieve the specified linear or curvilinear path. Servo-control is used to continuously regulate the position and speed of the manipulator. It should be mentioned that a playback robot with continuous path control has the capacity for PTP control.

Intelligent Control. Industrial robots are becoming increasingly intelligent. In this context, an *intelligent robot* is one that exhibits behavior that makes it seem intelligent. Some of the characteristics that make a robot appear intelligent include the capacities to interact with its environment, make decisions when things go wrong during the work cycle, communicate with humans, make computations during the motion cycle, and respond to advanced sensor inputs such as machine vision.

In addition, robots with intelligent control possess playback capability for both PTP or continuous path control. These features require (1) a relatively high level of computer control and (2) an advanced programming language to input the decision-making logic and other "intelligence" into memory.

8.3 END EFFECTORS

In our discussion of robot configurations (Section 8.1.2), we mentioned that an end effector is usually attached to the robot's wrist. The end effector enables the robot to accomplish a specific task. Because there is a wide variety of tasks performed by industrial robots, the end effector is usually custom-engineered and fabricated for each different application. The two categories of end effectors are grippers and tools.

8.3.1 Grippers

Grippers are end effectors used to grasp and manipulate objects during the work cycle. The objects are usually workparts that are moved from one location to another in the cell. Machine loading and unloading applications fall into this category (Section 8.5.1). Owing to the variety of part shapes, sizes, and weights, most grippers must be custom designed. Types of grippers used in industrial robot applications include the following:

- *Mechanical grippers*, consisting of two or more fingers that can be actuated by the robot controller to open and close to grasp the workpart; Figure 8.10 shows a two-finger gripper
- *Vacuum grippers*, in which suction cups are used to hold flat objects
- *Magnetized devices*, for holding ferrous parts
- *Adhesive devices*, which use an adhesive substance to hold a flexible material such as a fabric
- *Simple mechanical devices* such as hooks and scoops.

Mechanical grippers are the most common gripper type. Some of the innovations and advances in mechanical gripper technology include:

- *Dual grippers*, consisting of two gripper devices in one end effector for machine loading and unloading. With a single gripper, the robot must reach into the production machine twice, once to unload the finished part from the machine and position it in a location external to the machine, and the second time to pick up the next part and load it into the machine. With a dual gripper, the robot picks up the next workpart

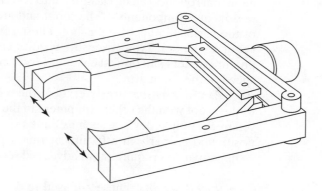

Figure 8.10   Robot mechanical gripper.

while the machine is still processing the preceding part. When the machine cycle is finished, the robot reaches into the machine only once: to remove the finished part and load the next part. This reduces the cycle time per part.

- *Interchangeable fingers* that can be used on one gripper mechanism. To accommodate different parts, different fingers are attached to the gripper.
- *Sensory feedback* in the fingers that provide the gripper with capabilities such as (1) sensing the presence of the workpart or (2) applying a specified limited force to the workpart during gripping (for fragile workparts).
- *Multiple fingered grippers* that possess the general anatomy of a human hand.
- *Standard gripper products* that are commercially available, thus reducing the need to custom-design a gripper for each separate robot application.

8.3.2 Tools

The robot uses tools to perform processing operations on the workpart. The robot manipulates the tool relative to a stationary or slowly moving object (e.g., workpart or subassembly). Examples of the tools used as end effectors by robots to perform processing applications include spot welding gun, arc welding tool; spray painting gun; rotating spindle for drilling, routing, grinding, and similar operations; assembly tool (e.g., automatic screwdriver); heating torch; ladle (for metal casting); and water jet cutting tool. In each case, the robot must not only control the relative position of the tool with respect to the work as a function of time, it must also control the operation of the tool. For this purpose, the robot must be able to transmit control signals to the tool for starting, stopping, and otherwise regulating its actions.

In some applications, the robot may use multiple tools during the work cycle. For example, several sizes of routing or drilling bits must be applied to the workpart. Thus, the robot must have a means of rapidly changing the tools. The end effector in this case takes the form of a fast-change tool holder for quickly fastening and unfastening the various tools used during the work cycle.

8.4 SENSORS IN ROBOTICS

The general topic of sensors as components in control systems was discussed in Chapter 6 (Section 6.1). Here we discuss sensors as they are applied in robotics. Sensors used in industrial robotics can be classified into two categories: (1) internal and (2) external. *Internal sensors* are components of the robot and are used to control the positions and velocities of the various joints of the robot. These sensors form a feedback control loop with the robot controller. Typical sensors used to control the position of the robot arm include potentiometers and optical encoders. Tachometers of various types are used to control the speed of the robot arm.

External sensors are external to the robot and are used to coordinate the operation of the robot with the other equipment in the cell. In many cases, these external sensors are relatively simple devices, such as limit switches that determine whether a part has been positioned properly in a fixture or that a part is ready to be picked up at a conveyor. Other situations require more advanced sensor technologies, including the following:

- *Tactile sensors.* These are used to determine whether contact is made between the sensor and another object. Tactile sensors can be divided into two types in robot applications: (1) touch sensors and (2) force sensors. *Touch sensors* indicate simply that contact has been made with the object. *Force sensors* indicate the magnitude of the force with the object. This might be useful in a gripper to measure and control the force being applied to grasp a delicate object.

- *Proximity sensors.* These indicate when an object is close to the sensor. When this type of sensor is used to indicate the actual distance of the object, it is called a *range sensor.*

- *Optical sensors.* Photocells and other photometric devices can be utilized to detect the presence or absence of objects and are often used for proximity detection.

- *Machine vision.* Machine vision is used in robotics for inspection, parts identification, guidance, and other uses. In Section 22.6, we provide a more complete discussion of machine vision in automated inspection. Improvements in programming of vision-guided robot (VGR) systems have made implementations of this technology easier and faster [12].

- *Other sensors.* A miscellaneous category includes other types of sensors that might be used in robotics, such as devices for measuring temperature, fluid pressure, fluid flow, electrical voltage, current, and various other physical properties.

8.5 INDUSTRIAL ROBOT APPLICATIONS

One of the earliest installations of an industrial robot was in 1961 in a die casting operation (Historical Note 8.1). The robot was used to unload castings from the die casting machine. The typical environment in die casting is not pleasant for humans due to the heat and fumes emitted by the casting process. It seemed quite logical to use a robot in this type of work environment in place of a human operator. Work environment is one of several characteristics that should be considered when selecting a robot application. The general characteristics of industrial work situations that tend to promote the substitution of robots for human labor are the following:

1. *Hazardous work for humans.* When the work and the environment in which it is performed are hazardous, unsafe, unhealthful, uncomfortable, or otherwise unpleasant for humans, it is desirable (also morally and socially imperative) to consider an industrial robot for the task. In addition to die casting, there are many other work situations that are hazardous or unpleasant for humans, including forging, spray painting, arc welding, and spot welding. Industrial robots are utilized in all of these processes.

2. *Repetitive work cycle.* A second characteristic that tends to promote the use of robotics is a repetitive work cycle. If the sequence of elements in the cycle is the same, and the elements consist of relatively simple motions, a robot is usually capable of performing the work cycle with greater consistency and repeatability than a human worker. Greater consistency and repeatability are usually manifested as higher product quality than can be achieved in a manual operation.

3. *Difficult handling for humans.* If the task involves the handling of parts or tools that are heavy or otherwise difficult to manipulate, an industrial robot may be available that can perform the operation. Parts or tools that are too heavy for humans to handle conveniently are well within the load carrying capacity of a large robot.

4. *Multishift operation.* In manual operations requiring second and third shifts, substitution of a robot provides a much faster financial payback than a single shift operation. Instead of replacing one worker, the robot replaces two or three workers.

5. *Infrequent changeovers.* Most batch or job shop operations require a changeover of the physical workplace between one job and the next. The time required to make the changeover is nonproductive time since parts are not being made. Consequently, robots have traditionally been easier to justify for relatively long production runs where changeovers are infrequent. Advances have been made in robot technology to reduce programming time, and shorter production runs have become more economical.

6. *Part position and orientation are established in the work cell.* Most robots in today's industrial applications are without vision capability. Their capacity to pick up an object during each work cycle relies on the part being in a known position and orientation. A means of presenting the part to the robot at the same location each cycle must be engineered.

Robots are being used in a wide field of applications in industry. Most of the current applications are in manufacturing. The applications can usually be classified into one of the following categories: (1) material handling, (2) processing operations, and (3) assembly and inspection. At least some of the work characteristics discussed above must be present in the application to make the installation of a robot technically and economically feasible.

8.5.1 Material Handling Applications

In material handling applications, the robot moves materials or parts from one place to another. To accomplish the transfer, the robot is equipped with a gripper type end effector. The gripper must be designed to handle the specific part or parts that are to be moved in the application. Included within this application category are (1) material transfer and (2) machine loading and/or unloading. In nearly all material handling applications, the parts must be presented to the robot in a known position and orientation. This requires some form of material handling device to deliver the parts into the work cell in this defined position and orientation.

Material Transfer. These applications are ones in which the primary purpose of the robot is to pick up parts at one location and place them at a new location. In many cases, reorientation of the part is accomplished during the relocation. The basic application in this category is the relatively simple *pick-and-place* operation, in which the robot picks up a part and deposits it at a new location. Transferring parts from one conveyor to another is an example. The requirements of the application are modest; a low-technology robot, (e.g., limited sequence type) is usually sufficient. Only two, three, or four joints are required for most of the applications. Pneumatically powered robots are often used.

A more complex example of material transfer is *palletizing*, in which the robot retrieves parts, cartons, or other objects from one location and deposits them onto a pallet or other container at multiple positions on the pallet. The problem is illustrated in Figure 8.11. Although the pickup point is the same for every cycle, the deposit location on the pallet is different for each carton. This adds to the degree of difficulty of the task. Either the robot must be taught each position on the pallet using the powered leadthrough

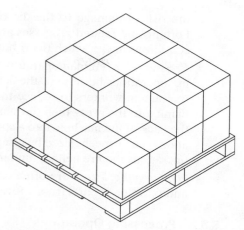

Figure 8.11 Typical part arrangement for a robot palletizing operation.

method (Section 8.6.1), or it must compute the location based on the dimensions of the pallet and the center distances between the cartons (in both x- and y-directions).

Other applications similar to palletizing include *depalletizing*, which consists of removing parts from an ordered arrangement in a pallet and placing them at another location (e.g., onto a moving conveyor), *stacking* operations, which involve placing flat parts on top of each other, such that the vertical location of the drop-off position is continuously changing with each cycle, and *insertion* operations, in which the robot inserts parts into the compartments of a divided carton.

Machine Loading and/or Unloading. In machine loading and/or unloading applications, the robot transfers parts into and/or from a production machine. The three possible cases are (1) *machine loading*, in which the robot loads parts into the production machine, but the parts are unloaded from the machine by some other means; (2) *machine unloading*, in which the raw materials are fed into the machine without using the robot, and the robot unloads the finished parts; and (3) *machine loading and unloading*, which involves both loading of the raw workpart and unloading of the finished part by the robot. Industrial robot applications of machine loading and/or unloading include the following processes:

- *Die casting.* The robot unloads parts from the die casting machine. Peripheral operations sometimes performed by the robot include dipping the parts into a water bath for cooling.
- *Plastic molding.* Plastic molding is similar to die casting. The robot unloads molded parts from the injection molding machine.
- *Metal machining operations.* The robot loads raw blanks into the machine tool and unloads finished parts from the machine. The change in shape and size of the part before and after machining often presents a problem in end effector design, and dual grippers (Section 8.3.1) are often used to deal with this issue.
- *Forging.* The robot typically loads the raw hot billet into the die, holds it during the forging blows, and removes it from the forge hammer. The hammering action and

the risk of damage to the die or end effector are significant technical problems. Forging and related processes are difficult robot applications because of the severe conditions under which the robot must operate.

- *Pressworking.* Human operators work at considerable risk in sheetmetal pressworking operations because of the action of the press. Robots are used as substitutes for the human workers to reduce the danger. In these applications, the robot loads the blank into the press, then the stamping operation is performed and the part falls out the back of the machine into a container. In high-production runs, pressworking operations can be mechanized by using sheetmetal coils instead of individual blanks. These operations require neither humans nor robots to participate directly in the process.

- *Heat treating.* These are often relatively simple operations in which the robot loads and/or unloads parts from a furnace.

8.5.2 Processing Operations

In processing applications, the robot performs some processing operation on a workpart, such as grinding or spray painting. A distinguishing feature of this category is that the robot is equipped with some type of tool as its end effector (Section 8.3.2). To perform the process, the robot must manipulate the tool relative to the part during the work cycle. In some processing applications, more than one tool must be used during the work cycle. In these instances, a fast-change tool holder is used to exchange tools during the cycle. Examples of industrial robot applications in the processing category include spot welding, arc welding, spray painting, and various machining and other rotating spindle processes.

Spot Welding. Spot welding is a metal joining process in which two sheet metal parts are fused together at localized points of contact. Two electrodes squeeze the metal parts together and then a large electrical current is applied across the contact point to cause fusion to occur. The electrodes, together with the mechanism that actuates them, constitute the welding gun in spot welding. Because of its widespread use in the automobile industry for car body fabrication, spot welding represents one of the most common applications of industrial robots today. The end effector is the spot welding gun used to pinch the car panels together and perform the resistance welding process. The welding gun used for automobile spot welding is typically heavy. Prior to the application of robots, human workers performed this operation, and the heavy welding tools were difficult for humans to manipulate accurately. As a consequence, there were many instances of missed welds, poorly located welds, and other defects, resulting in overall low quality of the finished product. The use of industrial robots in this application has dramatically improved the consistency of the welds.

Robots used for spot welding are usually large, with sufficient payload capacity to wield the heavy welding gun. Five or six axes are generally required to achieve the required position and orientation of the welding gun. Playback robots with point-to-point are used. Jointed arm coordinate robots are the most common type in automobile spot welding lines, which may consist of several dozen robots.

Arc Welding. Arc welding is used to provide continuous welds rather than individual spot welds at specific contact points. The resulting arc welded joint is substantially stronger than in spot welding. Since the weld is continuous, it can be used in airtight pressure vessels and other weldments in which strength and continuity are required. There are various forms of arc welding, but they all follow the general description given here.

The working conditions for humans who perform arc welding are not good. The welder must wear a face helmet for eye protection against the ultraviolet light emitted by the arc welding process. The helmet window must be dark enough to mask the UV radiation. High electrical current is used in the welding process, and this creates a hazard for the welder. Finally, there is the obvious danger from the high temperatures in the process, high enough to melt the steel, aluminum, or other metal that is being welded. A significant amount of hand-eye coordination is required by human welders to make sure that the arc follows the desired path with sufficient accuracy to make a good weld. This, together with the conditions described above, results in a high level of worker fatigue. Consequently, the welder is only accomplishing the welding process for perhaps 20–30% of the time. This percentage is called the *arc-on time*, defined as the proportion of time during the shift when the welding arc is on and performing the process. To assist the welder, a second worker is usually present at the work site, called the *fitter*, whose job is to set up the parts to be welded and to perform other similar chores in support of the welder.

Because of these conditions in manual arc welding, automation is used where technically and economically feasible. For welding jobs involving long continuous joints that are accomplished repetitively, mechanized welding machines have been designed to perform the process. These machines are used for long straight sections and regular round parts, such as pressure vessels, tanks, and pipes.

Industrial robots can also be used to automate the arc welding process. The cell consists of the robot, the welding apparatus (power unit, controller, welding tool, and wire feed mechanism), and a fixture that positions the components for the robot. The fixture might be mechanized with one or two degrees-of-freedom so that it can present different portions of the work to the robot for welding (the term *positioner* is used for this type of fixture). For greater productivity, a double fixture is often used so that a human helper can unload the completed job and load the components for the next work cycle while the robot is simultaneously welding the present job. Figure 8.12 illustrates this kind of workplace arrangement.

The robot used in arc welding jobs must be capable of continuous path control. Jointed arm robots consisting of six joints are frequently used. Some robot vendors provide manipulators that have hollow upper arms, so that the cables connected to the welding torch can be contained in the arm for protection, rather than attached to the exterior. Also, programming improvements for arc welding based on CAD/CAM have made it much easier and faster to implement a robot welding cell. The weld path can be developed directly from the CAD model of the assembly [8].

Spray Coating.   Spray coating directs a spray gun at the object to be coated. Fluid (e.g., paint) flows through the nozzle of the spray gun to be dispersed and applied over the surface of the object. Spray painting is the most common application in the category, but spray coating refers to a broader range of applications besides painting.

The work environment for humans who perform this process is filled with health hazards. These hazards include harmful and noxious fumes in the air, risk of flash fires, and noise from the spray gun nozzle. Largely because of these hazards, robots are being used more and more for spray coating tasks.

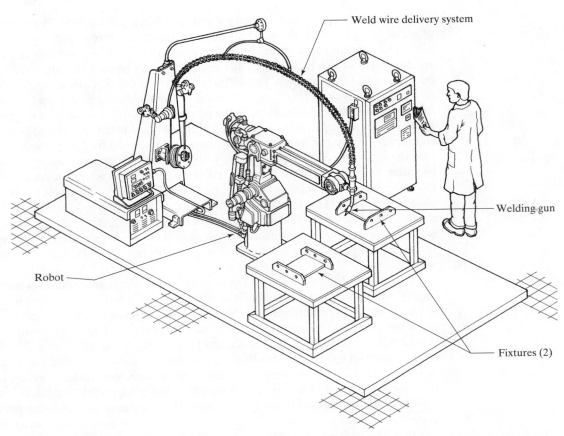

Figure 8.12 Robot arc welding cell (protective shields have been removed for clarity of illustration; in fact, there would be a protective barrier between the robot and the human worker).

Robot applications include spray coating of appliances, automobile car bodies, engines, and other parts; spray staining of wood products; and spraying of porcelain coatings on bathroom fixtures. The robot must be capable of continuous path control to accomplish the smooth motion sequences required in spray painting. The most convenient programming method is manual leadthrough (Section 8.6.1). Jointed arm robots seem to be the most common anatomy for this application. The robot must possess a work volume sufficient to access the areas of the workpart to be coated in the application.

The use of industrial robots for spray coating applications offers a number of benefits in addition to protecting workers from a hazardous environment. These other benefits include greater uniformity in applying the coating than humans can accomplish, reduced waste of paint, lower needs for ventilating the work area since humans are not present during the process, and greater productivity.

Other Processing Applications. Spot welding, arc welding, and spray coating are the most familiar processing applications of industrial robots. The list of industrial processes that are being performed by robots is continually growing. Among these are the following:

- *Drilling, routing, and other machining processes.* These applications use a rotating spindle as the end effector. The cutting tool is mounted in the spindle chuck. One of the problems with this application is the high cutting forces encountered in machining. The robot must be strong enough to withstand these cutting forces and maintain the required accuracy of the cut.
- *Grinding, wire brushing, and similar operations.* These operations also use a rotating spindle to drive the tool (grinding wheel, wire brush, polishing wheel, etc.) at high rotational speed to accomplish finishing and deburring operations on the work.
- *Waterjet cutting.* This is a process in which a high-pressure stream of water is forced through a small nozzle at high speed to cut plastic sheets, fabrics, cardboard, and other materials with precision. The end effector is the waterjet nozzle that is directed to follow the desired cutting path by the robot.
- *Laser cutting.* The function of the robot in this application is similar to its function in waterjet cutting. The laser tool is attached to the robot as its end effector. Laser beam welding is a similar application.

8.5.3 Assembly and Inspection

In some respects, assembly and inspection are hybrids of the previous two application categories: material handling and processing. Assembly and inspection applications can involve either the handling of materials or the manipulation of a tool. For example, assembly operations typically involve the addition of components to build a product. This requires the movement of parts from a supply location in the workplace to the product being assembled, which is material handling. In some cases, the fastening of the components requires a tool to be used by the robot (e.g., welding, driving a screw). Similarly, some robot inspection operations require that parts be manipulated, while other applications require that an inspection tool be manipulated.

Traditionally, assembly and inspection are labor-intensive activities. They are also highly repetitive and usually boring. For these reasons, they are logical candidates for robotic applications. However, assembly work typically involves diverse and sometimes difficult tasks, often requiring adjustments to be made in parts that don't quite fit together. A sense of feel is often required to achieve a close fitting of parts. Inspection work requires high precision and patience, and human judgment is often needed to determine whether a product is within quality specifications or not. Because of these complications in both types of work, the application of robots has not been easy. Nevertheless, the potential rewards are so great that substantial efforts are being made to develop the necessary technologies to achieve success in these applications.

Assembly. Assembly involves the combination of two or more parts to form a new entity, called a subassembly or assembly. The new entity is made secure by fastening the parts together using mechanical fastening techniques (e.g., screws, bolts and nuts, rivets) or joining processes (e.g., welding, brazing, soldering, or adhesive bonding). We have already discussed robot applications in welding.

Because of the economic importance of assembly, automated methods are often applied. Fixed automation is appropriate in mass production of relatively simple products, such as pens, mechanical pencils, cigarette lighters, and garden hose nozzles. Robots are usually at a disadvantage in these high-production situations because they cannot operate at the high speeds that fixed automated equipment can. The most appealing application of industrial robots for assembly is situations where a mixture of similar products or models are produced in the same work cell or assembly line. Examples of these kinds of products include electric motors, small appliances, and various other small mechanical and electrical products. In these instances, the basic configuration of the different models is the same, but there are variations in size, geometry, options, and other features. Such products are often made in batches on manual assembly lines. However, the pressure to reduce inventories makes mixed model assembly lines (Section 15.4) more attractive. Robots can be used to substitute for some or all of the manual stations on these lines. What makes robots viable in mixed model assembly is their capability to execute programmed variations in the work cycle to accommodate different product configurations.

Industrial robots used for the types of assembly operations described here are typically small, with light load capacities. The most common configurations are jointed arm, SCARA, and Cartesian coordinate. Accuracy requirements in assembly work are often more demanding than in other robot applications, and some of the more precise robots in this category have repeatabilities as close as ± 0.05 mm (± 0.002 in). In addition to the robot itself, the requirements of the end effector are often demanding. The end effector may have to perform multiple functions at a single workstation to reduce the number of robots required in the cell. These multiple functions can include handling more than one part geometry and performing both as a gripper and an automatic assembly tool.

Inspection. There is often a need in automated production and assembly systems to inspect the work that is done. Inspections accomplish the following functions: (1) making sure that a given process has been completed, (2) ensuring that parts have been added in assembly as specified, and (3) identifying flaws in raw materials and finished parts. The topic of automated inspection is considered in more detail in Chapter 21. Our purpose here is to identify the role played by industrial robots in inspection. Inspection tasks performed by robots can be divided into the following two cases:

1. The robot performs loading and unloading tasks to support an inspection or testing machine. This case is really machine loading and unloading, where the machine is an inspection machine. The robot picks parts (or assemblies) that enter the cell, loads and unloads them to carry out the inspection process, and places them at the cell output. In some cases, the inspection may result in sorting of parts that must be accomplished by the robot. Depending on the quality level of the parts, the robot places them in different containers or on different exit conveyors.
2. The robot manipulates an inspection device, such as a mechanical probe, to test the product. This case is similar to a processing operation in which the end effector attached to the robot's wrist is the inspection probe. To perform the process, the part must be presented at the workstation in the correct position and orientation, and the robot must manipulate the inspection device as required.

8.6 ROBOT PROGRAMMING

To do useful work, a robot must be programmed to perform its motion cycle. A *robot program* can be defined as a path in space to be followed by the manipulator, combined with peripheral actions that support the work cycle. Examples of the peripheral actions include opening and closing the gripper, performing logical decision making, and communicating with other pieces of equipment in the robot cell. A robot is programmed by entering the programming commands into its controller memory. Different robots use different methods of entering the commands.

In the case of limited sequence robots, programming is accomplished by setting limit switches and mechanical stops to control the endpoints of its motions. The sequence in which the motions occur is regulated by a sequencing device. This device determines the order in which each joint is actuated to form the complete motion cycle. Setting the stops and switches and wiring the sequencer is more akin to a manual setup than programming.

Today, nearly all industrial robots have digital computers as their controllers, and compatible storage devices as their memory units. For these robots, three programming methods can be distinguished: (1) leadthrough programming, (2) computer-like robot programming languages, and (3) off-line programming.

8.6.1 Leadthrough Programming

Leadthrough programming dates from the early 1960s before computer control was prevalent. The same basic methods are used today for many computer controlled robots. In leadthrough programming, the task is taught to the robot by moving the manipulator through the required motion cycle, simultaneously entering the program into the controller memory for subsequent playback.

Powered Leadthrough Versus Manual Leadthrough. There are two methods of performing the leadthrough teach procedure: (1) powered leadthrough and (2) manual leadthrough. The difference between the two is in the manner in which the manipulator is moved through the motion cycle during programming. Powered leadthrough is commonly used as the programming method for playback robots with point-to-point control. It involves the use of a teach pendant (hand-held control box) that has toggle switches and/or contact buttons for controlling the movement of the manipulator joints. Figure 8.13 illustrates the important components of a teach pendant. Using the toggle switches or buttons, the programmer power drives the robot arm to the desired positions, in sequence, and records the positions into memory. During subsequent playback, the robot moves through the sequence of positions under its own power.

Manual leadthrough is convenient for programming playback robots with continuous path control where the continuous path is an irregular motion pattern such as in spray painting. This programming method requires the operator to physically grasp the end-of-arm or the tool that is attached to the arm and move it through the motion sequence, recording the path into memory. Because the robot arm itself may have significant mass and would therefore be difficult to move, a special programming device often replaces the actual robot for the teach procedure. The programming device has the same joint configuration as the robot and is equipped with a trigger handle (or other control switch), which the operator activates when recording motions into memory. The motions are recorded as

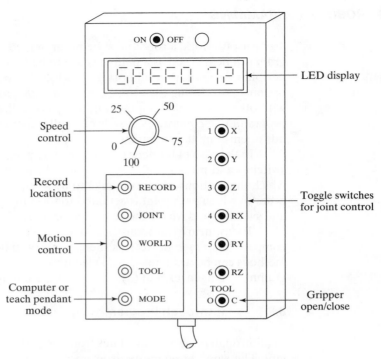

Figure 8.13 A typical robot teach pendant.

a series of closely spaced points. During playback, the path is recreated by controlling the actual robot arm through the same sequence of points.

Motion Programming.   The leadthrough methods provide a very natural way to program motion commands into the robot controller. In manual leadthrough, the operator simply moves the arm through the required path to create the program. In powered leadthrough, the operator uses a teach pendant to drive the manipulator. The teach pendant is equipped with a toggle switch or contact buttons for each joint. By activating these switches or buttons in a coordinated fashion for the various joints, the programmer moves the manipulator to the required positions in the work space.

Coordinating the individual joints with the teach pendant is an awkward and tedious way to enter motion commands to the robot. For example, it is difficult to coordinate the individual joints of a jointed-arm robot (TRR configuration) to drive the end-of-arm in a straight-line motion. Therefore, many of the robots using powered leadthrough provide two alternative methods for controlling movement of the entire manipulator during programming, in addition to controls for individual joints. With these methods, the programmer can move the robot's wrist end in straight line paths. The names given to these alternatives are (1) world-coordinate system and (2) tool-coordinate system. Both systems make use of a Cartesian coordinate system. In a *world-coordinate* system, the origin and axes are defined relative to the robot base, as illustrated in Figure 8.14(a). In a *tool-coordinate* system, shown in Figure 8.14(b), the alignment of the axis system is defined relative to the orientation of the wrist faceplate (to which the end effector is

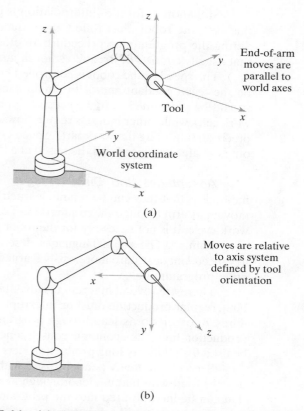

End-of-arm
moves are
parallel to
world axes

Tool

World coordinate
system

(a)

Moves are relative
to axis system
defined by tool
orientation

(b)

Figure 8.14 (a) World coordinate system. (b) Tool coordinate system.

attached). In this way, the programmer can orient the tool in a desired way and then control the robot to make linear moves in directions parallel or perpendicular to the tool.

The world- and tool-coordinate systems are useful only if the robot has the capacity to move its wrist end in a straight line motion, parallel to one of the axes of the coordinate system. Straight line motion is quite natural for a Cartesian coordinate robot (LOO configuration) but unnatural for robots with any combination of rotational joints (types R, T, and V). Accomplishing straight line motion requires manipulators with these types of joints to carry out a linear interpolation process. In *straight line interpolation*, the control computer calculates the sequence of addressable points in space through which the wrist end must move to achieve a straight line path between two points.

Other types of interpolation are available. More common than straight line interpolation is joint interpolation. When a robot is commanded to move its wrist end between two points using *joint interpolation*, it actuates each of the joints simultaneously at its own constant speed such that all of the joints start and stop at the same time. The advantage of joint interpolation over straight line interpolation is that usually less total motion energy is required to make the move. This may mean that the move could be made in slightly less time. It should be noted that in the case of a Cartesian coordinate robot, joint interpolation and straight line interpolation result in the same motion path.

Still another form of interpolation is used in manual leadthrough programming. In this case, the robot must follow the sequence of closely spaced points that are defined during the programming procedure. In effect, this is an interpolation process for a path that usually consists of irregular smooth motions, such as in spray painting.

The speed of the robot is controlled by means of a dial or other input device, located on the teach pendant and/or the main control panel. Certain motions in the work cycle should be performed at high speeds (e.g., moving parts over substantial distances in the work cell), while other motions require low speed operation (e.g., motions that require high precision in placing the workpart). Speed control also permits a given program to be tried out at a safe slow speed and then used at a higher speed during production.

Advantages and Disadvantages. The advantage offered by the leadthrough methods is that they can be readily learned by shop personnel. Programming the robot by moving its arm through the required motion path is a logical way for someone to teach the work cycle. It is not necessary for the robot programmer to possess knowledge of computer programming. The robot languages described in the next section, especially the more advanced languages, are more easily learned by someone whose background includes computer programming.

There are several inherent disadvantages of the leadthrough programming methods. First, regular production must be interrupted during the leadthrough programming procedures. In other words, leadthrough programming results in downtime of the robot cell or production line. The economic consequence of this is that the leadthrough methods must be used for relatively long production runs and are inappropriate for small batch sizes.

Second, the teach pendant used with powered leadthrough and the programming devices used with manual leadthrough are limited in terms of the decision-making logic that can be incorporated into the program. It is much easier to write logical instructions using the computer-like robot languages than the leadthrough methods.

Third, since the leadthrough methods were developed before computer control became common for robots, these methods are not readily compatible with modern computer-based technologies such as CAD/CAM, manufacturing data bases, and local communications networks. The capability to readily interface the various computer-automated subsystems in the factory for transfer of data is considered a requirement for achieving computer integrated manufacturing.

8.6.2 Robot Programming Languages

The use of textual programming languages became an appropriate programming method as digital computers took over the control function in robotics. Their use has been stimulated by the increasing complexity of the tasks that robots are called on to perform, with the concomitant need to imbed logical decisions into the robot work cycle. These computer-like programming languages are really on-line/off-line methods of programming, because the robot must still be taught its locations using the leadthrough method. Textual programming languages for robots provide the opportunity to perform the following functions that leadthrough programming cannot readily accomplish:

- enhanced sensor capabilities, including the use of analog as well as digital inputs and outputs,
- improved output capabilities for controlling external equipment,

- program logic that is beyond the capabilities of leadthrough methods,
- computations and data processing similar to computer programming languages,
- communications with other computer systems.

This section reviews some of the capabilities of the robot programming languages. Many of the language statements are taken from commercially available robot languages.

Motion Programming. Motion programming with robot languages usually requires a combination of textual statements and leadthrough techniques. Accordingly, this method of programming is sometimes referred to as *on-line/off-line programming.* The textual statements are used to describe the motion, and the leadthrough methods are used to define the position and orientation of the robot during and/or at the end of the motion. To illustrate, the basic motion statement is

MOVE P1

which commands the robot to move from its current position to a position and orientation defined by the variable name P1. The point P1 must be defined, and the most convenient way to define P1 is to use either powered leadthrough or manual leadthrough to place the robot at the desired point and record that point into memory. Statements such as

HERE P1

or

LEARN P1

are used in the leadthrough procedure to indicate the variable name for the point. What is recorded into the robot's control memory is the set of joint positions or coordinates used by the controller to define the point. For example, the aggregate

$$(236, 158, 65, 0, 0, 0)$$

could be utilized to represent the joint positions for a six-jointed manipulator. The first three values (236, 158, 65) give the joint positions of the body-and-arm, and the last three values (0, 0, 0) define the wrist joint positions. The values are specified in millimeters or degrees, depending on the joint types.

There are variants of the MOVE statement. These include the definition of straight line interpolation motions, incremental moves, approach and depart moves, and paths. For example, the statement

MOVES P1

denotes a move that is to be made using straight line interpolation. The suffix S on MOVE designates straight-line motion.

An incremental move is one whose endpoint is defined relative to the current position of the manipulator rather than to the absolute coordinate system of the robot. For example, suppose the robot is presently at a point defined by the joint coordinates (236, 158, 65, 0, 0, 0), and it is desired to move joint 4 (corresponding to a twisting motion of the wrist) from 0 to 125. The following form of statement might be used to accomplish this move:

$$\text{DMOVE } (4, 125)$$

The new joint coordinates of the robot would therefore be given by (236, 158, 65, 125, 0, 0). The prefix D is interpreted as delta, so DMOVE represents a delta move, or ncremental move.

Approach and depart statements are useful in material handling operations. The APPROACH statement moves the gripper from its current position to within a certain distance of the pickup (or drop-off) point, and then a MOVE statement positions the end effector at the pickup point. After the pickup is made, a DEPART statement moves the gripper away from the point. The following statements illustrate the sequence:

APPROACH P1, 40 MM

MOVE P1

(command to actuate gripper)

DEPART 40 MM

The destination is point P1, but the APPROACH command moves the gripper to a safe distance (40 mm) above the point. This might be useful to avoid obstacles such as other parts in a tote pan. The orientation of the gripper at the end of the APPROACH move is the same as that defined for the point P1, so that the final MOVE P1 is really a spatial translation of the gripper. This permits the gripper to be moved directly to the part for grasping.

A path in a robot program is a series of points connected together in a single move. The path is given a variable name, as illustrated in the following statement:

$$\text{DEFINE PATH123} = \text{PATH(P1, P2, P3)}$$

This is a path that consists of points P1, P2, and P3. The points are defined in the manner described above. A MOVE statement is used to drive the robot through the path.

MOVE PATH123

The speed of the robot is controlled by defining either a relative velocity or an absolute velocity. The following statement represents the case of relative velocity definition:

SPEED 75

When this statement appears within the program, it is typically interpreted to mean that the manipulator should operate at 75% of the initially commanded velocity in the statements that follow in the program. The initial speed is given in a command that precedes the execution of the robot program. For example,

SPEED 0.5 MPS

EXECUTE PROGRAM1

indicates that the program named PROGRAM1 is to be executed by the robot at a speed of 0.5 m/sec.

Interlock and Sensor Commands. The two basic interlock commands (Section 5.3.2) used for industrial robots are WAIT and SIGNAL. The WAIT command is used to implement an input interlock. For example,

WAIT 20, ON

would cause program execution to stop at this statement until the input signal coming into the robot controller at port 20 was in an "on" condition. This might be used in a situation where the robot needed to wait for the completion of an automatic machine cycle in a loading and unloading application.

The SIGNAL statement is used to implement an output interlock. This is used to communicate to some external piece of equipment. For example,

SIGNAL 20, ON

would switch on the signal at output port 20, perhaps to actuate the start of an automatic machine cycle.

Both of the above examples indicate on/off signals. Some robot controllers possess the capacity to control analog devices that operate at various levels. Suppose the programmer wanted the robot to turn on an external device that operates on variable voltages in the range 0 to 10 V. The command

SIGNAL 20, 6.0

is typical of a control statement that might be used to output a voltage level of 6.0 V to the device from controller output port 20.

All of the above interlock commands represent situations where the execution of the statement occurs at the point in the program where the statement appears. There are other situations in which it is desirable for an external device to be continuously monitored for any change that might occur in the device. This would be useful, for example, in safety monitoring where a sensor is set up to detect the presence of humans who might wander into the robot's work volume. The sensor reacts to the presence of the humans by signaling the robot controller. The following type of statement might be used for this case:

REACT 25, SAFESTOP

This command would be written to continuously monitor input port 25 for any changes in the incoming signal. If and when a change in the signal occurs, regular program execution is interrupted, and control is transferred to a subroutine called SAFESTOP. This subroutine would stop the robot from further motion and/or cause some other safety action to be taken.

Although end effectors are attached to the wrist of the manipulator, they are actuated very much like external devices. Special commands are usually written for controlling the end effector. In the case of grippers, the basic commands are

OPEN

and

CLOSE

which cause the gripper to actuate to fully open and fully closed positions, respectively, where fully closed is the position for grasping the object in the application. Greater control over the gripper is available in some sensored and servo-controlled hands. For grippers with force sensors that can be regulated through the robot controller, a command such as

CLOSE 2.0 N

controls the closing of the gripper until a 2.0-N force is encountered by the gripper fingers. A similar command used to close the gripper to a given opening width is

CLOSE 25 MM

A special set of statements is often required to control the operation of tool-type end effectors, such as spot welding guns, arc welding tools, spray painting guns, and powered spindles (e.g., for drilling or grinding). Spot welding and spray painting controls are typically simple binary commands (e.g., open/close and on/off), and these commands would be similar to those used for gripper control. In the case of arc welding and powered spindles, a greater variety of control statements is needed to control feed rates and other parameters of the operation.

Computations and Program Logic. Many robot languages possess capabilities for performing computations and data processing operations that are similar to computer programming languages. Most present-day robot applications do not require a high level of computational power. As the complexity of robot applications grows in the future, it is expected that these capabilities will be better utilized than at present.

Many of today's applications of robots require the use of branches and subroutines in the program. Statements such as

GO TO 150

and

IF (logical expression) GO TO 150

cause the program to branch to some other statement in the program (e.g., to statement number 150 in the above illustrations).

A subroutine in a robot program is a group of statements that are to be executed separately when called from the main program. In a preceding example, the subroutine SAFESTOP was named in the REACT statement for use in safety monitoring. Other uses of subroutines include making calculations or performing repetitive motion sequences at a number of different places in the program. Using a subroutine is more efficient than writing the same steps several times in the program.

8.6.3 Simulation and Off-Line Programming

The trouble with leadthrough methods and textual programming techniques is that the robot must be taken out of production for a certain length of time to accomplish the programming. Off-line programming permits the robot program to be prepared at a remote computer terminal and downloaded to the robot controller for execution without interrupting production. In true off-line programming, there is no need to physically locate the positions in the workspace for the robot as required with present textual programming languages. Some form of graphical computer simulation is required to validate the programs developed off-line, similar to off-line procedures used in NC part programming.

The off-line programming procedures being developed and commercially offered use graphical simulation to construct a three-dimensional model of a robot cell for evaluation and off-line programming. The cell might consist of the robot, machine tools, conveyors, and other hardware. The simulator displays these cell components on the graphics monitor and shows the robot performing its work cycle in animated computer graphics. After the program has been developed using the simulation procedure, it is then converted into the textual language corresponding to the particular robot employed in the cell. This is a step in the off-line programming procedure that is equivalent to postprocessing in NC part programming.

In the current commercial off-line programming packages, some adjustment must be performed to account for geometric differences between the three-dimensional model in the computer system and the actual physical cell. For example, the position of a machine tool in the physical layout might be slightly different than in the model used to do the off-line programming. For the robot to reliably load and unload the machine, it must have an accurate location of the load/unload point recorded in its control memory. This module is used to calibrate the three-dimensional computer model by substituting location data from the actual cell for the approximate values developed in the original model. The disadvantage of calibrating the cell is that production time is lost in performing this procedure.

8.7 ROBOT ACCURACY AND REPEATABILITY

The capacity of a robot to position and orient the end of its wrist with accuracy and repeatability is an important control attribute in nearly all industrial applications. Some assembly applications require that objects be located with a precision of 0.05 mm (0.002 in). Other applications, such as spot welding, usually require accuracies of 0.5–1.0 mm (0.020–0.040 in). Let us examine how a robot is able to move its various joints to achieve accurate and repeatable positioning. Several terms must be defined in the context of this discussion: (1) control resolution, (2) accuracy, and (3) repeatability. These terms have the same basic meanings in robotics that they have in NC. In robotics, the characteristics are defined at the end of the wrist and in the absence of any end effector attached to the wrist.

Control resolution refers to the capability of the robot's positioning system to divide the range of the joint into closely spaced points, called *addressable points*, to which the joint can be moved by the controller. Recall from Section 7.5.3 that the capability to divide the range into addressable points depends on two factors: (1) limitations of the electromechanical components that make up each joint-link combination and (2) the controller's bit storage capacity for that joint.

If the joint-link combination consists of a leadscrew drive mechanism, as in the case of an NC positioning system, then the methods of Section 7.5.3 can be used to determine the control resolution. We identified this electromechanical control resolution as CR_1. Unfortunately, from our viewpoint of attempting to analyze the control resolution of the robot manipulator, there is a much wider variety of joints used in robotics than in NC machine tools. And it is not possible to analyze the mechanical details of all of the types here. Let it suffice to recognize that there is a mechanical limit on the capacity to divide the range of each joint-link system into addressable points, and that this limit is given by CR_1.

The second limit on control resolution is the bit storage capacity of the controller. If B = the number of bits in the bit storage register devoted to a particular joint, then the number of addressable points in that joint's range of motion is given by 2^B. The control resolution is therefore defined as the distance between adjacent addressable points. This can be determined as

$$CR_2 = \frac{R}{2^B - 1} \tag{8.1}$$

where CR_2 = control resolution determined by the robot controller; and R = range of the joint-link combination, expressed in linear or angular units, depending on whether the joint provides a linear motion (joint types L or O) or a rotary motion (joint types R, T, or V). The control resolution of each joint-link mechanism will be the maximum of CR_1 and CR_2, that is,

$$CR = \text{Max}\{CR_1, CR_2\} \tag{8.2}$$

In our discussion of control resolution for NC (Section 7.5.3), we indicated that it is desirable for $CR_2 \leq CR_1$, which means that the limiting factor in determining control resolution is the mechanical system, not the computer control system. Because the mechanical structure of a robot manipulator is much less rigid than that of a machine tool, the control resolution for each joint of a robot will almost certainly be determined by mechanical factors (CR_1).

Similar to the case of an NC positioning system, the ability of a robot manipulator to position any given joint-link mechanism at the exact location defined by an addressable point is limited by mechanical errors in the joint and associated links. The mechanical errors arise from factors such as gear backlash, link deflection, hydraulic fluid leaks, and various other sources that depend on the mechanical construction of the given joint-link combination. If we characterize the mechanical errors by a normal distribution, as we did in Section 7.5.3, with mean μ at the addressable point and standard deviation σ characterizing the magnitude of the error dispersion, then we can determine accuracy and repeatability for the axis.

Repeatability is the easier term to define. *Repeatability* is a measure of the robot's ability to position its end-of-wrist at a previously taught point in the work volume. Each time the robot attempts to return to the programmed point it will return to a slightly different position. Repeatability variations have as their principal source the mechanical errors previously mentioned. Therefore, as in NC, for a single joint-link mechanism,

$$\text{Repeatability} = \pm 3\sigma \tag{8.3}$$

where σ = standard deviation of the error distribution.

Accuracy is the robot's ability to position the end of its wrist at a desired location in the work volume. For a single axis, using the same reasoning used to define accuracy in our discussion of NC, we have

$$\text{Accuracy} = \frac{CR}{2} + 3\sigma \tag{8.4}$$

where CR = control resolution from Eq. (8.2).

The terms control resolution, accuracy, and repeatability are illustrated in Figure 7.15 of the previous chapter for one axis that is linear. For a rotary joint, these parameters can be conceptualized as either an angular value of the joint itself or an arc length at the end of the joint's output link.

EXAMPLE 8.1   Control Resolution, Accuracy, and Repeatability in Robotic Arm Joint

One of the joints of a certain industrial robot has a type L joint with a range of 0.5 meters. The bit storage capacity of the robot controller is 10 bits for this joint. The mechanical errors form a normally distributed random variable about a given taught point. The mean of the distribution is zero and the standard deviation is 0.06 mm in the direction of the output link of the joint. Determine the control resolution (CR_2), accuracy, and repeatability for this robot joint.

Solution: The number of addressable points in the joint range is $2^{10} = 1024$. The control resolution is therefore

$$CR_2 = \frac{0.5}{1024 - 1} = 0.004888 \text{ m} = 0.4888 \text{ mm}$$

Accuracy is given by Eq. (8.4):

$$\text{Accuracy} = \frac{0.4888}{2} + 3(0.06) = 0.4244 \text{ mm}$$

Repeatability is defined as ±3 standard deviations

$$\text{Repeatability} = 3 \times 0.06 = 0.18 \text{ mm}$$

Our definitions of control resolution, accuracy, and repeatability have been depicted using a single joint or axis. To be of practical value, the accuracy and repeatability of a robot manipulator should include the effect of all of the joints, combined with the effect of their mechanical errors. For a multiple degrees-of-freedom robot, accuracy and repeatability will vary depending on where in the work volume the end-of-wrist is positioned. The reason for this is that certain joint combinations will tend to magnify the effect of the control resolution and mechanical errors. For example, for a polar configuration robot (TRL) with its linear joint fully extended, any errors in the R or T joints will be larger than when the linear joint is fully retracted.

Robots move in three-dimensional space, and the distribution of repeatability errors is therefore three-dimensional. In three dimensions, we can conceptualize the normal distribution as a sphere whose center (mean) is at the programmed point and whose radius is equal to three standard deviations of the repeatability error distribution. For conciseness, repeatability is usually expressed in terms of the radius of the sphere; for example, ±1.0 mm (±0.040 in). Some of today's small assembly robots have repeatability values as low as ±0.05 mm (±0.002 in).

In reality, the shape of the error distribution will not be a perfect sphere in three dimensions. In other words, the errors will not be isotropic. Instead, the radius will vary because the associated mechanical errors will be different in certain directions than in others. The mechanical arm of a robot is more rigid in certain directions, and this rigidity influences the errors. Also, the so-called sphere will not remain constant in size throughout the robot's work volume. As with control resolution, it will be affected by the particular combination of joint positions of the manipulator. In some regions of the work volume, the repeatability errors will be larger than in other regions.

Accuracy and repeatability have been defined above as static parameters of the manipulator. However, these precision parameters are affected by the dynamic operation of the robot. Characteristics such as speed, payload, and direction of approach will affect the robot's accuracy and repeatability.

REFERENCES

[1] COLESTOCK, H., *Industrial Robotics: Selection, Design, and Maintenance*, McGraw-Hill, NY, 2004.

[2] CRAIG, J. J., *Introduction to Robotics: Mechanics and Control*, 2d ed., Addison-Wesley Publishing Company, Reading, MA, 1989.

[3] CRAWFORD, K. R., "Designing Robot End Effectors," *Robotics Today*, October 1985, pp 27–29.

[4] ENGELBERGER, J. F., *Robotics in Practice*, AMACOM (American Management Association), NY, 1980.

[5] GROOVER, M. P., M. WEISS, R. N. NAGEL, and N. G. ODREY, *Industrial Robotics: Technology, Programming, and Applications*, McGraw-Hill Book Company, NY, 1986.

[6] NIEVES, E., "Robots: More Capable, Still Flexible," *Manufacturing Engineering*, May 2005, pp 131–143.

[7] SCHREIBER, R. R., "How to Teach a Robot," *Robotics Today*, June 1984, pp. 51–56.

[8] SPROVIERI, J., "Arc Welding with Robots," *Assembly*, July 2006, pp 26–31.

[9] TOEPPERWEIN, L. L., M. T. BLACKMAN, et al., "ICAM Robotics Application Guide," *Technical Report AFWAL-TR-80-4042*, Vol. II, Material Laboratory, Air Force Wright Aeronautical Laboratories, OH, April 1980.

[10] WAURZYNIAK, P., "Robotics Evolution," *Manufacturing Engineering*, February 1999, pp. 40–50.

[11] WAURZYNIAK, P., "Masters of Manufacturing: Joseph F. Engelberger," *Manufacturing Engineering*, July 2006, pp. 65–75.

[12] ZENS, R. G., Jr., "Guided by Vision," *Assembly*, September 2005, pp. 52–58.

REVIEW QUESTIONS

8.1 What is an industrial robot?

8.2 What was the first application of an industrial robot?

8.3 What are the five joint types used in robotic arms and wrists?

8.4 Name the five common body-and-arm configurations identified in the text.

8.5 What is the work volume of a robot manipulator?

8.6 What is a playback robot with point-to-point control?

8.7 What is an end effector?

8.8 In a machine loading and unloading application, what is the advantage of a dual gripper over a single gripper?

8.9 Robotic sensors are classified as internal and external. What is the distinction?

8.10 What are four of the six general characteristics of industrial work situations that tend to promote the substitution of robots for human workers?

8.11 What are the three categories of robot industrial applications, as identified in the text?

8.12 What is a palletizing operation?

8.13 What is a robot program?

8.14 What is the difference between powered leadthrough and manual leadthrough in robot programming?

8.15 What is control resolution in a robot positioning system?

8.16 What is the difference between repeatability and accuracy in a robotic manipulator?

PROBLEMS

Robot Anatomy

8.1 Using the notation scheme for defining manipulator configurations (Section 8.1.2), draw diagrams (similar to Figure 8.1) of the following robots: (a) TRT, (b) VVR, (c) VROT.

8.2 Using the notation scheme for defining manipulator configurations (Section 8.1.2), draw diagrams (similar to Figure 8.1) of the following robots: (a) TRL, (b) OLO, (c) LVL.

8.3 Using the notation scheme for defining manipulator configurations (Section 8.1.2), draw diagrams (similar to Figure 8.1) of the following robots: (a) TRT:R, (b) TVR:TR, (c) RR:T.

8.4 Using the robot configuration notation scheme discussed in Section 8.1, write the configuration notations for some of the robots in your laboratory or shop.

8.5 Describe the differences in orientation capabilities and work volumes for a :TR and a :RT wrist assembly. Use sketches as needed.

Robot Applications

8.6 A robot performs a loading and unloading operation for a machine tool. The work cycle consists of the following sequence of activities:

Seq.	Activity	Time
1	Robot reaches out, picks part from incoming conveyor, and loads part into fixture on machine tool.	5.5 seconds
2	Machining cycle (automatic)	33.0 seconds
3	Robot reaches in, retrieves part from machine tool, and deposits it onto outgoing conveyor.	4.8 seconds
4	Robot moves back to pickup position	1.7 seconds

The activities are performed sequentially as listed. Every 30 workparts, the cutting tools in the machine must be changed. This irregular cycle takes 3.0 minutes to accomplish. The uptime efficiency of the robot is 97% and the uptime efficiency of the machine tool is 98%, not including interruptions for tool changes. These two efficiencies are assumed not to overlap (i.e., if the robot breaks down, the cell will cease to operate, so the machine tool will not have the opportunity to break down, and vice versa). Downtime results from electrical and mechanical malfanctions of the robot, machine tool, and fixture. Determine the hourly production rate, taking into account the lost time due to tool changes and the uptime efficiency.

8.7 In the previous problem, suppose that a double gripper is used instead of a single gripper. The activities in the cycle would be changed as follows:

Seq.	Activity	Time
1	Robot reaches out, picks raw part from incoming conveyor in one gripper, and awaits completion of machining cycle. This activity is performed simultaneously with machining cycle.	3.3 seconds
2	At completion of previous machining cycle, robot reaches in, retrieves finished part from machine, loads raw part into fixture, and moves a safe distance from machine.	5.0 seconds
3	Machining cycle (automatic).	33.0 seconds
4	Robot moves to outgoing conveyor and deposits part. This activity is performed simultaneously with machining cycle.	3.0 seconds
5	Robot moves back to pickup position. This activity is performed simultaneously with machining cycle.	1.7 seconds

Steps 1, 4, and 5 are performed simultaneously with the automatic machining cycle. Steps 2 and 3 must be performed sequentially. The same tool change statistics and uptime efficiencies are applicable. Determine the hourly production rate when the double gripper is used, taking into account the lost time due to tool changes and the uptime efficiency.

8.8 Since the robot's portion of the work cycle requires much less time than the machine tool in Problem 8.6, the company is considering installing a cell with two machines. The robot would load and unload both machines from the same incoming and outgoing conveyors. The machines would be arranged so that distances between the fixture and the conveyors are the same for both machines. Thus, the activity times given in Problem 8.6 are valid for the two-machine cell. The machining cycles would be staggered so that the robot would be servicing only one machine at a time. The tool change statistics and uptime efficiencies in Problem 8.6 are applicable. Determine the hourly production rate for the two-machine cell. The lost time due to tool changes and the uptime efficiency should be accounted for. Assume that if one of the two machine tools is down, the other machine can continue to operate, but if the robot is down, the cell operation is stopped.

8.9 Determine the hourly production rate for a two-machine cell as in Problem 8.8, assuming the robot is equipped with a double gripper as in Problem 8.7. Assume the activity times from Problem 8.7 apply here.

8.10 The arc-on time is a measure of efficiency in an arc welding operation. As indicated in our discussion of arc welding in Section 8.5.2, typical arc-on times in manual welding range between 20% and 30%. Suppose that a certain welding operation is currently performed using a welder and a fitter. Production requirements are steady at 500 units per week. The fitter's job is to load the component parts into the fixture and clamp them in position for the welder. The welder then welds the components in two passes, stopping to reload the welding rod between the two passes. Some time is lost each cycle for repositioning the welding rod on the work. The fitter's and welder's activities are done sequentially, with times for the various elements as follows:

Seq.	Worker and Activity	Time
1	Fitter: load and clamp parts	4.2 minutes
2	Welder: weld first pass	2.5 minutes
3	Welder: reload weld rod	1.8 minutes
4	Welder: weld second pass	2.4 minutes
5	Welder: reposition	2.0 minutes
6	Delay time between work cycles	1.1 minutes

Because of fatigue, the welder must take a 20 minute rest at mid-morning and mid-afternoon, and a 40 minute lunch break around noon. The fitter joins the welder in these rest breaks. The nominal time of the work shift is eight hours, but the last 20 minutes of the shift is nonproductive time for clean-up at each workstation. A proposal has been made to install a robot welding cell to perform the operation. The cell would be set up with two fixtures, so that the robot could be welding one job (the set of parts to be welded) while the fitter is unloading the previous job and loading the next job. In this way, the welding robot and the human fitter could be working simultaneously rather than sequentially. Also, a continuous wire feed would be used rather than individual welding rods. It has been estimated that the continuous wire feed must be changed only once every 40 parts and the lost time will be 20 minutes to make the wire change. The times for the various activities in the regular work cycle are as follows:

Seq.	Fitter and Robot Activities	Times
1	Fitter: Load and clamp parts	4.2 minutes
2	Robot: Weld complete	4.0 minutes
3	Repositioning time	1.0 minutes
4	Delay time between work cycles	0.3 minutes

The fitter would take a 10 minute break in the morning, another in the afternoon, and 40 minutes for lunch. Clean-up time at the end of the shift is 20 minutes. In your calculations, assume that the proportion uptime of the robot will be 98%. Determine the following: (a) arc-on times (expressed as a percent, using the eight hour shift as the base) for the manual welding operation and the robot welding station, and (b) hourly production rate on average throughout the eight-hour shift for the manual welding operation and the robot welding station.

Programming Exercises

Note: Problems 8.11 through 8.17 are all programming exercises to be performed on robots available. The solutions depend on the particular programming methods or languages used.

8.11 The setup for this problem requires a felt-tipped pen mounted to the robot's end-of-arm (or held securely in the robot's gripper), and a thick cardboard, mounted on the surface of the worktable. Pieces of plain white paper will be pinned or taped to the cardboard surface. Program the robot to write your initials on the paper with the felt-tipped pen.

8.12 As an enhancement of the previous programming exercise, consider the problem of programming the robot to write any letter that is entered at the alphanumeric keyboard. Obviously, a textual programming language is required to accomplish this exercise.

8.13 Apparatus required for this exercise consists of two wood or plastic blocks of two different colors that can be grasped by the robot gripper. The blocks should be placed in specific positions (call the positions A and B on either side of a center location, position C). The robot should be programmed to do the following: (1) pick up the block at position A and place it at the central position C, (2) pick up the block at position B and place it at position A, and (3) pick up the block at position C and place it at position B. (4) Repeat steps (1), (2), and (3) continually.

8.14 Apparatus for this exercise consists of a cardboard box and a dowel about 4 inches long (any straight thin cylinder will suffice, e.g., pen, pencil, etc.). The dowel is attached to the robot's end-of-arm or held in its gripper. The dowel is intended to simulate an arc welding torch, and the edges of the cardboard box are intended to represent the seams that are to be welded. With the box oriented with one of its corners pointing towards the robot, program the robot to weld the three edges that lead into the corner. The dowel (welding torch) must be continuously oriented at a 45° angle with respect to the edge being welded. See Figure P8.14.

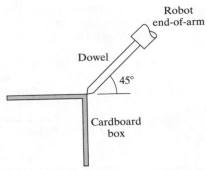

Figure P8.14 Orientation of arc welding torch for Problem 8.14.

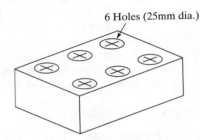

Figure P8.15 Approximate pallet dimensions for Problem 8.15.

8.15 This exercise is intended to simulate a palletizing operation. The apparatus includes six wooden (or plastic or metal) cylinders approximately 20 mm in diameter and 75 mm in length, and a 20 mm thick wooden block approximately 100 mm by 133 mm. The block is to have six holes of diameter 25 mm drilled in it as illustrated in Figure P8.15. The wooden cylinders represent workparts and the wooden block represents a pallet. (As an alternative to the wooden block, the layout of the pallet can be sketched on a plain piece of paper attached to the work table.) Using the powered leadthrough programming method, program the robot to pick up the parts from a fixed position on the worktable and place them into the six positions in the pallet. The fixed position on the table might be a stop point on a conveyor. (The student may have to manually place the parts at the position if a real conveyor is not available.)

8.16 Repeat the previous exercise using a robot programming language, and defining the positions of the pallet by calculating their x and y coordinates by whatever method is available in the particular programming language used.

8.17 Repeat Problem 8.16 in the reverse order to simulate a depalletizing operation.

Accuracy and Repeatability

8.18 The linear joint (type L) of a certain industrial robot is actuated by a piston mechanism. The length of the joint when fully retracted is 600 mm and when fully extended is 1000 mm. If the robot's controller has an eight-bit storage capacity, determine the control resolution for this robot.

8.19 In the previous problem, the mechanical errors associated with the linear joint form a normal distribution in the direction of the joint actuation with standard deviation = 0.08 mm. Determine (a) spatial resolution, (b) accuracy, and (c) repeatability for the robot.

8.20 The revolving joint (type V) of an industrial robot has a range of 240° rotation. The mechanical errors in the joint and the input/output links can be described by a normal distribution with its mean at any given addressable point, and a standard deviation of 0.25°. Determine the number of storage bits required in the controller memory so that the accuracy of the joint is as close as possible to, but less than, its repeatability. Use six standard deviations as the measure of repeatability.

8.21 A cylindrical robot has a T-type wrist axis that can be rotated a total of five rotations (each rotation is a full 360°). The robot's job requires it to position its wrist with a control resolution of 0.5° between adjacent addressable points. Determine the number of bits required in the binary register for that axis in the robot's control memory.

8.22 One axis of an RRL robot is a linear slide with a total range of 950 mm. The robot's control memory has an 10-bit capacity. It is assumed that the mechanical errors associated with the arm are normally distributed with a mean at the given taught point and an isotropic standard deviation of 0.10 mm. Determine (a) the control resolution for the axis under consideration, (b) the spatial resolution for the axis, (c) the defined accuracy, (d) the repeatability.

8.23 A TLR robot has a rotational joint (type R) whose output link is connected to the wrist assembly. Considering the design of this joint only, the output link is 600 mm long, and the total range of rotation of the joint is 40°. The spatial resolution of this joint is expressed as a linear measure at the wrist, and is specified to be ±0.5 mm. It is known that the mechanical inaccuracies in the joint result in an error of ±0.018° rotation, and it is assumed that the output link is perfectly rigid so as to cause no additional errors due to deflection. (a) Determine the minimum number of bits required in the robot's control memory in order to obtain the spatial resolution specified. (b) With the given level of mechanical error in the joint, show that it is possible to achieve the spatial resolution specified.

Chapter 9

Discrete Control Using Programmable Logic Controllers and Personal Computers

Numerical control (Chapter 7) and industrial robotics (Chapter 8) are primarily concerned with motion control, because the applications of machine tools and robots involve the movement of a cutting tool or end effector, respectively. A more general control category is discrete control, defined in Section 5.2.2. In the present chapter, we provide a more complete discussion of discrete control, and we examine the two principal industrial controllers used to implement discrete control: (1) programmable logic controllers (PLCs) and (2) personal computers (PCs).

9.1 DISCRETE PROCESS CONTROL

Discrete process control systems deal with parameters and variables that are discrete and that change values at discrete moments in time. The parameters and variables are typically binary; they can have either of two possible values, 1 or 0. The values mean ON or OFF, true or false, object present or not present, high voltage value or low voltage value, and so on, depending on the application. The binary variables in discrete process control are associated with input signals to the controller and output signals from the controller. Input signals are typically generated by binary sensors, such as limit switches or photosensors that are interfaced to the process. Output signals are generated by the controller to operate the process in response to the input signals and as a function of time. These output signals turn on and off switches, motors, valves, and other binary actuators related to the process. We have compiled a list of binary sensors and actuators, along with the interpretation of their 0 and 1 values, in Table 9.1. The purpose of the controller is to coordinate the various actions of the physical system, such as transferring parts into a workholder, feeding a machining workhead, and so on.

Discrete process control can be divided into two categories: (1) logic control, which is concerned with event-driven changes in the system, and (2) sequencing, which is concerned with time-driven changes in the system. Both are referred to as *switching systems* in the sense that they switch their output values on and off in response to changes in events or time.

9.1.1 Logic Control

A logic control system, also referred to as *combinational logic control*, is a switching system whose output at any moment is determined exclusively by the values of the current inputs. A logic control system has no memory and does not consider any previous values of input signals in determining the output signal. Neither does it have any operating characteristics that perform directly as a function of time.

Let us use an example from robotics to illustrate logic control. Suppose that in a machine-loading application, the robot is programmed to pick up a raw workpart from a known stopping point along a conveyor and place it in a forging press. Three conditions must be satisfied to initiate the loading cycle. First, the raw workpart must be at the stopping point; second, the forge press must have completed the process on the previous part; and third, the previous part must be removed from the die. The first condition can be indicated by means of a simple limit switch that senses the presence of the part at the conveyor stop and transmits an ON signal to the robot controller. The second condition can be indicated by the forge press, which sends an ON signal after it has completed the previous cycle. The third condition might be determined by a photodetector located so as to sense the presence or absence of the part in the forging die. When the finished part is removed from the die, an ON signal is transmitted by the photocell. All three of these ON signals must be received by the robot controller to initiate the next work cycle. When

TABLE 9.1 Binary Sensors and Actuators Used in Discrete Process Control

Sensor	One/Zero Interpretation	Actuator	One/Zero Interpretation
Limit switch	Contact/no contact	Motor	On/off
Photodetector	On/off	Control relay	Contact/no contact
Push-button switch	On/off	Light	On/off
Timer	On/off	Valve	Closed/open
Control relay	Contact/no contact	Clutch	Engaged/not engaged
Circuit breaker	Contact/no contact	Solenoid	Energized/not energized

these input signals have been received by the controller, the robot loading cycle is switched on. No previous conditions or past history are needed.

Elements of Logic Control. The basic elements of logic control are the logic gates AND, OR, and NOT. In each case, the logic gate is designed to provide a specified output value based on the values of the input(s). For both inputs and outputs, the values can be one of two levels, the binary values 0 or 1. For purposes of industrial control, we define 0 (zero) to mean OFF, and 1 (one) to mean ON.

The AND gate outputs a value of 1 if all of the inputs are 1, and 0 otherwise. Figure 9.1 illustrates the operation of a logical AND gate. If both switches X1 and X2 (representing inputs) in the circuit are closed, then the lamp Y (representing the output) is on. The AND gate might be used in an automated production system to indicate that two (or more) actions have been successfully completed, therefore signaling that the next step in the process should be initiated. The interlock system in our previous robot forging example illustrates the AND gate. All three conditions must be satisfied before loading of the forge press is allowed to occur.

The OR gate outputs a value of 1 if either of the inputs has a value of 1, and 0 otherwise. Figure 9.2 shows how the OR gate operates. In this case, the two input signals X1 and X2 are arranged in a parallel circuit, so that if either switch is closed, the lamp Y will be on. A possible use of the OR gate in a manufacturing system is for safety monitoring. Suppose that two sensors are utilized to monitor two different safety hazards. When either hazard is present, the respective sensor emits a positive signal that sounds an alarm buzzer.

Both the AND and OR gates can be used with two or more inputs. The NOT gate has a single input. The NOT gate reverses the input signal: If the input is 1, then the output is 0; if the input is 0, then the output is 1. Figure 9.3 shows a circuit in which the input switch X1 is arranged in parallel with the output so that the voltage flows through the lower path when the switch is closed (thus $Y = 0$), and through the upper path when the switch is open (thus $Y = 1$). The NOT gate can be used to open a circuit upon receipt of a control signal.

Boolean Algebra and Truth Tables. The logic elements form the foundation for a special algebra that was developed around 1847 by George Boole and that bears his name. Its original purpose was to provide a symbolic means of testing whether complex statements of logic were TRUE or FALSE. In fact, Boole called it *logical algebra*. It was not until about a century later that Boolean algebra was shown to be useful in digital logic systems. We briefly describe some of its fundamentals here.

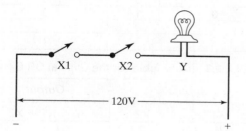

Figure 9.1 Electrical circuit illustrating the operation of the logical AND gate.

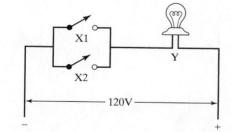

Figure 9.2 Electrical circuit illustrating the operation of the logical OR gate.

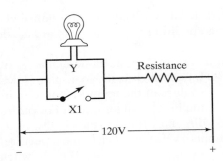

Figure 9.3 Electrical circuit illustrating the operation of the logical NOT gate.

In Boolean algebra, the AND function is expressed as

$$Y = X1 \cdot X2 \tag{9.1}$$

This is called the logical product of X1 and X2. As a logic statement, it means: Y is true if both X1 and X2 are true; otherwise, Y is false. The truth table is often used to present the operation of logic systems. A *truth table* is a tabulation of all of the combinations of input values to the corresponding logical output values. The truth table for the AND gate for four possible combinations of two input binary variables is presented in Table 9.2.

The OR function in Boolean algebra notation is given by

$$Y = X1 + X2 \tag{9.2}$$

This is called the logical sum of X1 and X2. In logic, the statement says: Y is true if either X1 or X2 is true; otherwise, Y is false. The outputs of the OR function for four possible combinations of two input binary variables are listed in the truth table of Table 9.3.

The NOT function is referred to as the negation or inversion of the variable. It is indicated by placing a bar above the variable (e.g., $\overline{X1}$). The truth table for the NOT function is listed in Table 9.4, and the corresponding Boolean equation is as follows:

$$Y = \overline{X1} \tag{9.3}$$

In addition to the three basic elements, there are two more elements that can be used in switching circuits: the NAND and NOR gates. The logical NAND gate is formed

TABLE 9.2 Truth Table for the Logical AND Gate

Inputs		Output
X1	X2	$Y = X1 \cdot X2$
0	0	0
0	1	0
1	0	0
1	1	1

TABLE 9.3 Truth Table for the Logical OR Gate

Inputs		Output
X1	X2	$Y = X1 + X2$
0	0	0
0	1	1
1	0	1
1	1	1

TABLE 9.4 Truth Table for the Logical NOT Gate

Inputs	Output
X1	$Y = \overline{X1}$
0	1
1	0

by combining an AND gate and a NOT gate in sequence, yielding the truth table shown in Table 9.5(a). In equation form,

$$Y = \overline{X1 \cdot X2} \tag{9.4}$$

The logical NOR gate is formed by combining an OR gate followed by a NOT gate, providing the truth table in Table 9.5(b). The Boolean algebra equation for the NOR gate is written as follows:

$$Y = \overline{X1 + X2} \tag{9.5}$$

Various diagramming techniques have been developed to represent the logic elements and their relationships in a given logic control system. The logic network diagram is one of the most common methods. Symbols used in the logic network diagram are illustrated in Figure 9.4. We demonstrate the use of the logic network diagram in several examples later in this section.

TABLE 9.5 Truth Tables for (a) the Logical NAND Gate and (b) Logical NOR Gate

(a) NAND			(b) NOR		
Inputs		Output	Inputs		Output
X1	X2	$Y = \overline{X1 \cdot X2}$	X1	X2	$Y = \overline{X1 + X2}$
0	0	1	0	0	1
0	1	1	0	1	0
1	0	1	1	0	0
1	1	0	1	1	0

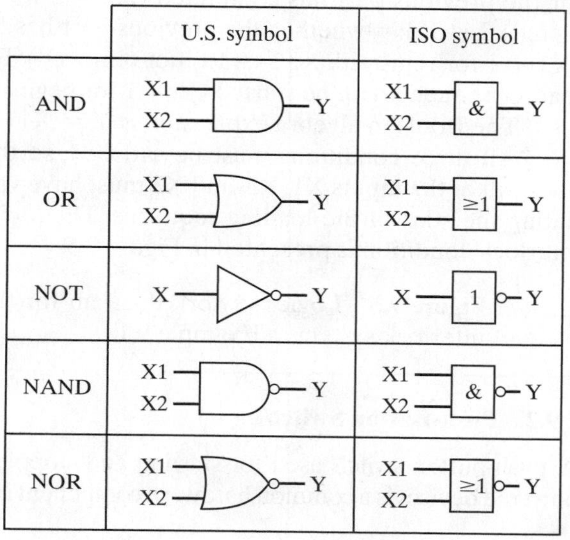

Figure 9.4 Symbols used for logical gates: U.S. and ISO.

TABLE 9.6 Laws and Theorems of Boolean Algebra

Commutative Law: $X + Y = Y + X$ $X \cdot Y = Y \cdot X$	**Law of Absorption:** $X \cdot (X + Y) = X + X \cdot Y = X$
Associative Law: $X + Y + Z = X + (Y + Z)$ $X + Y + Z = (X + Y) + Z$ $X \cdot Y \cdot Z = X \cdot (Y \cdot Z)$ $X \cdot Y \cdot Z = (X \cdot Y) \cdot Z$	**De Morgan's Laws:** $(\overline{X + Y}) = \overline{X} \cdot \overline{Y}$ $(\overline{X \cdot Y}) = \overline{X} + \overline{Y}$ **Consistency Theorem:** $X \cdot Y + X \cdot \overline{Y} = X$ $(X + Y) \cdot (X + \overline{Y}) = X$
Distributive Law: $X \cdot Y + X \cdot Z = X \cdot (Y + Z)$ $(X + Y) \cdot (Z + W) = X \cdot Z + X \cdot W + Y \cdot Z + Y \cdot W$	**Inclusion Theorem:** $X \cdot \overline{X} = 0$ $X + \overline{X} = 1$

There are certain laws and theorems of Boolean algebra. We cite them in Table 9.6. These laws and theorems can often be applied to simplify logic circuits and reduce the number of elements required to implement the logic, with resulting savings in hardware and/or programming time.

EXAMPLE 9.1 Robot Machine Loading

The robotic machine loading example described at the beginning of Section 9.1.1 required three conditions to be satisfied before the loading sequence was initiated. Determine the Boolean algebra expression and the logic network diagram for this interlock system.

Solution: Let $X1$ = whether the raw workpart is present at the conveyor stopping point ($X1 = 1$ for present, $X1 = 0$ for not present). Let $X2$ = whether the press cycle for the previous part has completed ($X2 = 1$ for completed, 0 for not completed). Let $X3$ = whether the previous part has been removed from the die ($X3 = 1$ for removed, $X3 = 0$ for not removed). Finally, let Y = whether the loading sequence can be started ($Y = 1$ for begin, $Y = 0$ for wait).

The Boolean algebra expression is $Y = X1 \cdot X2 \cdot X3$.

All three conditions must be satisfied, so the logical AND function is used. All of the inputs $X1$, $X2$, and $X3$ must have values of 1 before $Y = 1$, initiating the start of the loading sequence. The logic network diagram for this interlock condition is presented in Figure 9.5.

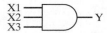

Figure 9.5 Logic network diagram for the robotic machine loading interlock system in Example 9.1.

EXAMPLE 9.2 Push-Button Switch

A push-button switch used for starting and stopping electric motors and other powered devices is a common hardware component in an industrial control system.

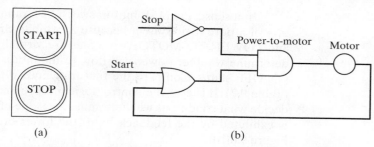

(a)   (b)

Figure 9.6   (a) Push-button switch of Example 9.2 and (b) its logic network diagram.

As shown in Figure 9.6(a), it consists of a box with two buttons, one for START and the other for STOP. When the START button is depressed momentarily by a human operator, power is supplied and maintained to the motor (or other load) until the STOP button is pressed. POWER-TO-MOTOR is the output of the push-button switch. The values of the variables can be defined as follows:

START = 0 is normally open contact status

START = 1 when the START button is pressed to contact

STOP = 0 is normally closed contact status

STOP = 1 when the STOP button is pressed to break contact

POWER-TO-MOTOR = 0 when the contacts are open

POWER-TO-MOTOR = 1 when the contacts are closed

MOTOR = 0 when off (not running)

MOTOR = 1 when on

The truth table for the push-button is presented in Table 9.7. From an initial motor off condition (MOTOR = 0), the motor is started by depressing the start button (START = 1). If the stop button is in its normally closed condition (STOP = 0), power will be supplied to the motor (POWER-TO-MOTOR = 1). While the motor is running (MOTOR = 1), it can be stopped by depressing the stop button (STOP = 1). The corresponding network logic diagram is shown in Figure 9.6(b).

TABLE 9.7   Truth Table for Push-Button Switch of Example 9.2

Start	Stop	Motor	Power-to-Motor
0	0	0	0
0	1	0	0
1	0	0	1
1	1	0	0
0	0	1	1
0	1	1	0
1	0	1	1
1	1	1	0

In a sense, the push-button switch of Example 9.2 goes slightly beyond our definition of a pure logic system because it exhibits characteristics of memory. The MOTOR and POWER-TO-MOTOR variables are virtually the same signal. The conditions that determine whether power will flow to the motor are different depending on the motor ON/OFF status. Compare the first four lines with the last four lines in the truth table (Table 9.7). It is as if the control logic must remember whether the motor is on or off to decide what conditions will determine the value of the output signal. This memory feature is exhibited by the feedback loop (the lower branch) in the logic network diagram of Figure 9.6(b).

9.1.2 Sequencing

A sequencing system uses internal timing devices to determine when to initiate changes in output variables. Washing machines, dryers, dishwashers, and similar appliances use sequencing systems to time the start and stop of cycle elements. There are many industrial applications of sequencing systems. For example, suppose an induction heating coil is used to heat the workpart in our previous example of a robotics forging application. Rather than using a temperature sensor, the coil could be set up with a timed heating cycle so that enough energy is provided to heat the workpart to the desired temperature. The heating process is sufficiently reliable and predictable that a certain duration of time in the induction coil will consistently heat the part to a certain temperature (with minimum variation).

Many applications in industrial automation require the controller to provide a prescheduled set of ON/OFF values for the output variables. The outputs are often generated in an open-loop fashion, meaning that there is no feedback verification that the control function has actually been executed. Another feature that typifies this mode of control is that the sequence of output signals is usually cyclical; the signals occur in the same repeated pattern within each regular cycle. Timers and counters illustrate this type of control component.

A *timer* is a device that switches its output ON or OFF at preset time intervals. Common timers used in industry and in homes switch on when activated and remain on for a programmed length of time. Some timers are activated by depressing a start button, for example, the water pump controls on whirlpool bath. Others operate off of a 24-hour clock. For example, some home security systems have timing features that turn lights on and off during the day to give the appearance of people at home.

Two additional types of timers used in discrete control systems can be distinguished: (1) delay-off timers and (2) delay-on timers. A *delay-off timer* switches power on immediately in response to a start signal, and then switches power off after a specified time delay. Many cars are equipped with this type of device. When you exit the car, the lights remain on for a certain length of time (e.g., 30 seconds), and then automatically turn off. A *delay-on timer* waits a specified length of time before switching power on when it receives a start signal. To program a timer, the user must specify the length of the time delay.

A *counter* is a component used to count electrical pulses and store the results of the counting procedure (Section 6.5.2). The instantaneous contents can be displayed and/or used in a process control algorithm. Counters are classified as up counters, down counters, and up/down counters. An *up counter* starts at zero and increments its contents (the count total) by one in response to each pulse. When a preset value has been reached, the up counter can be reset to zero. An application of such a device is counting the number of filled beer bottles moving along a conveyor for boxing into cases. Every set of 24 bottles adds up to one case, and the counter is then reset to zero. A *down counter* starts with a preset value and decrements the total by one for each pulse received. It could be used for the same bottling application as above, using a starting preset value of 24. An *up/down counter* combines the two counting operations, and might be useful for keeping track of the number of bottles remaining in a storage buffer. It adds the number of bottles entering and subtracts the number exiting the buffer to get a current tally of the buffer contents.

9.2 LADDER LOGIC DIAGRAMS

The logic network diagrams of the type shown in Figures 9.5 and 9.6(b) are useful for displaying the relationships between logic elements. Another diagramming technique that exhibits the logic and, to some extent, the timing and sequencing of the system is the ladder logic diagram. This graphical method has an important virtue in that it is analogous to the electrical circuits used to accomplish the logic and sequence control. In addition, ladder logic diagrams are familiar to shop personnel who must construct, test, maintain, and repair the discrete control system.

In a ladder logic diagram, the various logic elements and other components are displayed along horizontal lines or rungs connected on either end to two vertical rails, as illustrated in Figure 9.7. The diagram has the general configuration of a ladder, hence its name. The elements and components are *contacts* (representing logical inputs) and loads, also known as *coils* (representing outputs). Inputs include switches and relay contacts, and

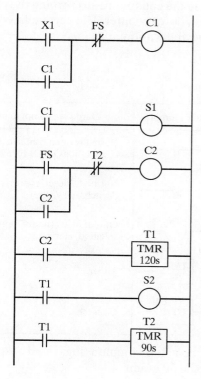

Figure 9.7 A ladder logic diagram.

263

loads include motors, lamps, and alarms. The power (e.g., 115 V alternating current) to the components is provided by the two vertical rails. It is customary in ladder diagrams to locate the inputs to the left of each rung and the outputs to the right.

Symbols used in ladder diagrams for the common logic and sequencing components are presented in Figure 9.8. There are two types of contacts: normally open and normally closed. A *normally open contact* remains open until activated. When activated, it closes to allow current to flow. A *normally closed contact* remains closed, allowing current to flow until activated. When activated, it opens, thereby turning off the flow of current. Normally open contacts of a switch or similar device are symbolized by two short vertical lines along a horizontal rung of the ladder, as in Figure 9.8(a). Normally closed contacts are shown as the same vertical lines only with a diagonal line across them as in Figure 9.8(b). Both types of contacts are used to represent ON/OFF inputs to the logic circuit. In addition to switches, inputs include relays, on/off sensors (e.g., limit switches and photodetectors), timers, and other binary contact devices.

Output loads such as motors, lights, alarms, solenoids, and other electrical components that are turned on and off by the logic control system are shown as nodes (circles) as in Figure 9.8(c). Timers and counters are symbolized by squares (or rectangles) with appropriate inputs and outputs to properly drive the device as shown in Figure 9.8(d) and (e). The simple timer requires the specification of the time delay and the identification of the input contact that activates the delay. When the input signal is received, the timer waits the specified delay time before switching on or off the output signal. The timer is reset (output is set back to its initial value) by turning off the input signal.

Counters require two inputs. The first is the pulse train (series of on/off signals) that is counted by the counter. The second is a signal to reset the counter and restart the counting procedure. Resetting the counter means zeroing the count for an up counter and setting the starting value for a down counter. The accumulated count is retained in memory for use if required for the application.

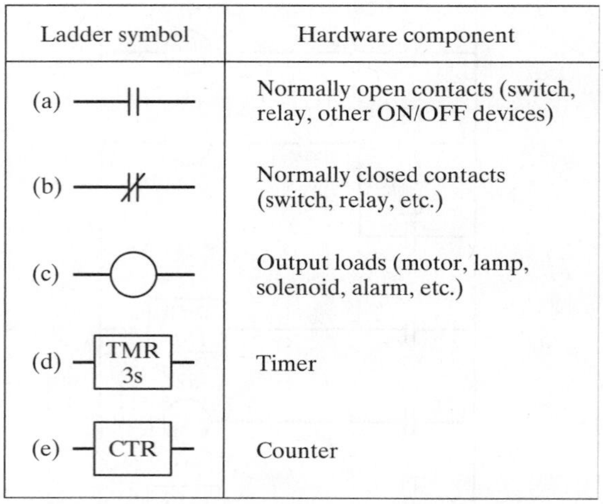

Ladder symbol	Hardware component
(a) —||—	Normally open contacts (switch, relay, other ON/OFF devices)
(b) —|/|—	Normally closed contacts (switch, relay, etc.)
(c) —O—	Output loads (motor, lamp, solenoid, alarm, etc.)
(d) —[TMR 3s]—	Timer
(e) —[CTR]—	Counter

Figure 9.8 Symbols for common logic and sequence elements used in ladder logic diagrams.

EXAMPLE 9.3   Three Simple Lamp Circuits

The three basic logic gates (AND, OR, and NOT) can be symbolized in ladder logic diagrams. Create diagrams for the three lamp circuits illustrated in Figures 9.1, 9.2, and 9.3.

Solution: The three ladder diagrams corresponding to these circuits are presented in Figure 9.9(a)-(c). Note the similarity between the original circuit diagrams and the ladder diagrams shown here. Notice that the NOT symbol is the same as a normally closed contact, which is the logical inverse of a normally open contact.

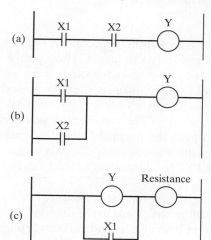

Figure 9.9   Three ladder logic diagrams for lamp circuits in (a) Figure 9.1, (b) Figure 9.2, and (c) Figure 9.3.

EXAMPLE 9.4   Push-Button Switch

The operation of the push-button switch of Example 9.2 can be depicted in a ladder logic diagram. From Figure 9.6, let START be represented by X1, STOP by X2, and MOTOR by Y, and create the diagram.

Solution: The ladder diagram is presented in Figure 9.10. X1 and X2 are input contacts, and Y is a load in the diagram. Note how Y also serves as an input contact to provide the POWER-TO-MOTOR connection.

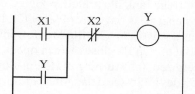

Figure 9.10   Ladder logic diagram for the push-button switch in Example 9.4.

EXAMPLE 9.5   Control Relay

The operation of a control relay can be demonstrated by means of the ladder logic diagram presented in Figure 9.11. A relay can be used to control on/off actuation of a powered device at some remote location. It can also be used to

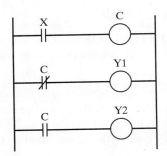

Figure 9.11 Ladder logic diagram for the control relay in Example 9.5.

define alternative decisions in logic control. Our diagram illustrates both uses. The relay is indicated by the load C (for control relay), which controls the on/off operation of two motors (or other types of output loads) Y1 and Y2. When the control switch X is open, the relay is deenergized, thereby connecting the load Y1 to the power lines. In effect, the open switch X turns on motor Y1 by means of the relay. When the control switch is closed, the relay becomes energized. This opens the normally closed contact of the second rung of the ladder and closes the normally open contact of the third rung. In effect, power is shut off to load Y1 and turned on to load Y2.

Example 9.5 illustrates several important features of a ladder logic diagram. First, the same input can be used more than once in the diagram. In our example, the relay contact C was used as an input on both the second and third rungs of the ladder. As we shall see in the following section, this feature of using a given relay contact in several different rungs of the ladder diagram to serve multiple logic functions provides a substantial advantage for the programmable controller over hardwired control units. With hardwired relays, separate contacts would have to be built into the controller for each logic function. A second feature of Example 9.5 is that it is possible for an output (load) on one rung of the diagram to be an input (contact) for another rung. The relay C was the output on the top rung in Figure 9.11, but that output was used as an input elsewhere in the diagram. This same feature was illustrated in the push-button ladder diagram of Example 9.4.

EXAMPLE 9.6

Consider the fluid storage tank illustrated in Figure 9.12. When the start button X1 is depressed, this energizes the control relay C1. In turn, this energizes solenoid S1, which opens a valve allowing fluid to flow into the tank. When the tank becomes full, the float switch FS closes, which opens relay C1, causing the solenoid S1 to be deenergized, thus turning off the in-flow. Switch FS also activates timer T1, which provides a 120-sec delay for a certain chemical reaction to occur in the tank. At the end of the delay time, the timer energizes a second relay C2, which controls two devices: (1) It energizes solenoid S2, which opens a valve to allow the fluid to flow out of the tank, and (2) it initiates timer T2, which waits 90 sec to allow the contents of the tank to be drained. At the end of the 90 sec, the timer breaks the current and deenergizes solenoid S2, thus closing the out-flow valve. Depressing the start button X1 resets the timers and opens their respective contacts. Construct the ladder logic diagram for the system.

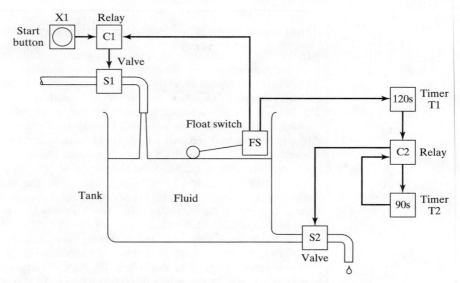

Figure 9.12 Fluid filling operation of Example 9.6.

Solution: The ladder logic diagram is constructed as shown in Figure 9.7.

The ladder logic diagram is an excellent way to represent the combinatorial logic control problems in which the output variables are based directly on the values of the inputs. As indicated by Example 9.6, it can also be used to display sequential control (timer) problems, although the diagram is somewhat more difficult to interpret and analyze for this purpose. The ladder diagram is the principal technique for setting up the control programs in PLCs.

9.3 PROGRAMMABLE LOGIC CONTROLLERS

A programmable logic controller (PLC) can be defined as a microcomputer-based controller that uses stored instructions in programmable memory to implement logic, sequencing, timing, counting, and arithmetic functions through digital or analog input/output (I/O) modules, for controlling machines and processes. PLC applications are found in both the process industries and discrete manufacturing. Examples of applications in process industries include chemical processing, paper mill operations, and food production. PLCs are primarily associated with discrete manufacturing industries to control individual machines, machine cells, transfer lines, material handling equipment, and automated storage systems. Before the PLC was introduced around 1970, hard-wired controllers composed of relays, coils, counters, timers, and similar components were used to implement this type of industrial control (Historical Note 9.1). Today, many older pieces of equipment are being retrofitted with PLCs to replace the original hard-wired controllers, often making the equipment more productive and reliable than when it was new.

Historical Note 9.1 Programmable logic controllers [2], [5], [7], [8].

In the mid-1960s, Richard Morley was a partner in Bedford Associates, a New England consulting firm specializing in control systems for machine tool companies. Most of the firm's work involved replacing relays with minicomputers in machine tool controls. In January 1968, Morley devised the notion and wrote the specifications for the first programmable controller.[1] It would overcome some of the limitations of conventional computers used for process control at the time; namely, it would be a real-time processor (Section 5.3.1), it would be predictable and reliable, and it would be modular and rugged. Programming would be based on ladder logic, which was widely used for industrial controls. The controller that emerged was named the Modicon Model 084. MODICON was an abbreviation of MOdular DIgital CONtroller. Model 084 was derived from the fact that this was the 84th product developed by Bedford Associates. Morley and his associates elected to start up a new company to produce the controllers, and Modicon was incorporated in October 1968. In 1977, Modicon was sold to Gould and became Gould's PLC division.

In the same year that Morley invented the PLC, the Hydramatic Division of General Motors Corporation developed a set of specifications for a PLC. The specifications were motivated by the high cost and lack of flexibility of electromechanical relay-based controllers used extensively in the automotive industry to control transfer lines and other mechanized and automated systems. The requirements for the device were that it must (1) be programmable and reprogrammable, (2) be designed to operate in an industrial environment, (3) accept 120 V ac signals from standard push-buttons and limit switches, (4) have outputs designed to switch and continuously operate loads such as motors and relays of 2-A rating, and (5) have a price and installation cost competitive with relay and solid-state logic devices then in use. In addition to Modicon, several other companies saw a commercial opportunity in the GM specifications and developed various versions of the PLC.

Capabilities of the first PLCs were similar to those of the relay controls they replaced. They were limited to on/off control. Within five years, product enhancements included better operator interfaces, arithmetic capability, data manipulation, and computer communications. Improvements over the next five years included larger memory, analog and positioning control, and remote I/O (permitting remote devices to be connected to a satellite I/O subsystem that was multiplexed to the PLC using twisted pair). Much of the progress was based on advancements taking place in microprocessor technology. By the mid-1980s, the micro PLC had been introduced. This was a down-sized PLC with much lower size (typical size = 75 mm by 75 mm by 125 mm) and cost (less than $500). By the mid-1990s, the nano PLC had arrived, which was even smaller and less expensive.

[1]Morley used the abbreviation PC to refer to the programmable controller. This term was used for many years until IBM began to call their personal computers by the same abbreviation in the early 1980s. The term PLC, widely used today for programmable logic controller, was coined by Allen-Bradley, a leading PLC supplier.

There are significant advantages to using a PLC rather than conventional relays, timers, counters, and other hard-wired control components. These advantages include (1) programming the PLC is easier than wiring the relay control panel; (2) the PLC can be reprogrammed, whereas conventional controls must be rewired and are often scrapped instead; (3) PLCs take less floor space than relay control panels; (4) reliability is greater, and maintenance is easier; (5) the PLC can be connected to computer systems more easily than relays; and (6) PLCs can perform a greater variety of control functions than can relay controls.

In this section, we describe the components, programming, and operation of the PLC. Although its principal applications are in logic control and sequencing (discrete control), many PLCs also perform additional functions, surveyed later in the section.

9.3.1 Components of the PLC

A schematic diagram of a PLC is presented in Figure 9.13. The basic components of the PLC are the following: (1) processor, (2) memory unit, (3) power supply, (4) I/O module, and (5) programming device. These components are housed in a suitable cabinet designed for the industrial environment.

The *processor* is the central processing unit (CPU) of the programmable controller. It executes the various logic and sequencing functions by operating on the PLC inputs to determine the appropriate output signals. The typical CPU operating cycle is described in Section 9.3.2. The CPU consists of one or more microprocessors similar to those used in PCs and other data processing equipment. The difference is that they have a real-time operating system and are programmed to facilitate I/O transactions and execute ladder logic functions. In addition, PLCs are hardened so that the CPU and other electronic components will operate in the electrically noisy environment of the factory.

Connected to the CPU is the PLC *memory unit*, which contains the programs of logic, sequencing, and I/O operations. It also holds data files associated with these programs, including I/O status bits, counter and timer constants, and other variable and parameter values. This memory unit is referred to as the user or application memory because its contents are entered by the user. In addition, the processor also contains the operating system memory, which directs the execution of the control program and coordinates I/O operations. The operating system is entered by the PLC manufacturer and cannot be accessed or altered by the user.

A *power supply* of 115 V ac is typically used to drive the PLC (some units operate on 230 V ac). The power supply converts the 115 V ac into direct current (dc) voltages of ± 5 V. These low voltages are used to operate equipment that may have much higher voltage and power ratings than the PLC itself. The power supply often includes a battery backup that switches on automatically in the event of an external power source failure.

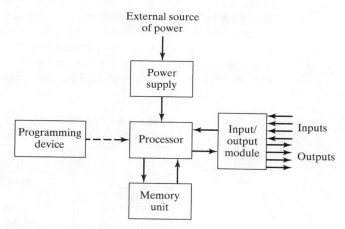

Figure 9.13   Components of a PLC.

The *input/output module* provides the connections to the industrial equipment or process that is to be controlled. Inputs to the controller are signals from limit switches, push-buttons, sensors, and other on/off devices. Outputs from the controller are on/off signals to operate motors, valves, and other devices required to actuate the process. In addition, many PLCs are capable of accepting continuous signals from analog sensors and generating signals suitable for analog actuators. The size of a PLC is usually rated in terms of the number of its I/O terminals, as indicated in Table 9.8.

The PLC is programmed by means of a programming device. The programming device is usually detachable from the PLC cabinet so that it can be shared among different controllers. Different PLC manufacturers provide different devices, ranging from simple teach-pendant type devices, similar to those used in robotics, to special PLC programming keyboards and displays. Personal computers can also be used to program PLCs. A PC used for this purpose sometimes remains connected to the PLC to serve a process monitoring or supervisory function and for conventional data processing applications related to the process.

9.3.2 PLC Operating Cycle

As far as the PLC user is concerned, the steps in the control program are executed simultaneously and continuously. In truth, a certain amount of time is required for the PLC processor to execute the user program during one cycle of operation. The typical operating cycle of the PLC, called a *scan*, consists of three parts: (1) input scan, (2) program scan, and (3) output scan. During the *input scan*, the inputs to the PLC are read by the processor and the status of these inputs is stored in memory. Next, the control program is executed during the *program scan*. The input values stored in memory are used in the control logic calculations to determine the values of the outputs. Finally, during the *output scan*, the outputs are updated to agree with the calculated values. The time to perform the scan is called the *scan time*, and this time depends on the number of inputs that must be read, the complexity of control functions to be performed, and the number of outputs that must be changed. Scan time also depends on the clock speed of the processor. Scan times typically vary between 1 and 25 msec [3].

One of the potential problems that can occur during the scan cycle is that the value of an input can change immediately after it has been sampled. Since the program uses the input value stored in memory, any output values that are dependent on that input are determined incorrectly. There is obviously a potential risk involved in this mode of operation. However, the risk is minimized because the time between updates is so short that it is unlikely that the output value being incorrect for such a short time will have a serious effect on process operation. The risk becomes most significant in processes in which

TABLE 9.8 Typical Classification of PLCs by Number of Input/Output Terminals

PLC Size	I/O Count
Large PLC	≥ 1024
Medium PLC	< 1024
Small PLC	< 256
Micro PLC	≤ 32
Nano PLC	≤ 16

the response times are very fast and where hazards can occur during the scan time. Some PLCs have special features for making "immediate" updates of output signals when input variables are known to cycle back and forth at frequencies faster than the scan time.

9.3.3 Additional Capabilities of the PLC

The logic control and sequencing functions described in Section 9.1 are likely to be the principal control operations accomplished by the PLC. These are the functions for which the programmable controller was originally designed. However, the PLC has evolved to include several capabilities in addition to logic control and sequencing. Some of these additional capabilities available on many commercial PLCs include

- *Analog control.* Proportional-integral-derivative (PID) control is available on some programmable controllers. These control algorithms have traditionally been implemented using analog controllers. Today the analog control schemes are approximated using the digital computer, with either a PLC or a computer process controller.
- *Arithmetic functions.* These functions are addition, subtraction, multiplication, and division. Use of these functions permits more complex control algorithms to be developed than what is possible with conventional logic and sequencing elements.
- *Matrix functions.* Some PLCs have the capability to perform matrix operations on stored values in memory. The capability can be used to compare the actual values of a set of inputs and outputs with the values stored in the PLC memory to determine if some error has occurred.
- *Data processing and reporting.* These functions are typically associated with business applications of PCs. PLC manufacturers have found it necessary to include these PC capabilities in their controller products, as the distinction between PCs and PLCs blurs.

9.3.4 Programming the PLC

Programming is the means by which the user enters the control instructions to the PLC through the programming device. The most basic control instructions consist of switching, logic, sequencing, counting, and timing. Virtually all PLC programming methods provide instruction sets that include these functions. Many control applications require additional instructions to accomplish analog control of continuous processes, complex control logic, data processing and reporting, and other advanced functions not readily performed by the basic instruction set. Owing to these differences in requirements, various PLC programming languages have been developed. A standard for PLC programming was published by the International Electrotechnical Commission in 1992, entitled *International Standard for Programmable Controllers* (IEC 1131–3). This standard specifies three graphical languages and two text-based languages for programming PLCs, respectively: (1) ladder logic diagrams, (2) function block diagrams, (3) sequential functions charts, (4) instruction list, and (5) structured text. Table 9.9 lists the five languages along with the most suitable application of each. IEC 1131–3 also states that the five languages must be able to interact with each other to allow for all possible levels of control sophistication in any given application.

TABLE 9.9 Features of the Five PLC Languages Specified in the IEC 1131–3 Standard

Language	Abbreviation	Type	Applications Best Suited for
Ladder logic diagram	(LD)	Graphical	Discrete control
Function block diagram	(FBD)	Graphical	Continuous control
Sequential function chart	(SFC)	Graphical	Sequencing
Instruction list	(IL)	Textual	Discrete control
Structured text	(ST)	Textual	Complex logic, computations, etc.

Ladder Logic Diagram. The most widely used PLC programming language today involves ladder diagrams (LDs), examples of which are shown in several previous figures. As indicated in Section 9.2, ladder diagrams are very convenient for shop personnel who are familiar with ladder and circuit diagrams but may not be familiar with computers and computer programming. To use ladder logic diagrams, they do not need to learn an entirely new programming language.

Direct entry of the ladder logic diagram into the PLC memory requires the use of a keyboard and monitor with graphics capability to display symbols representing the components and their interrelationships in the ladder logic diagram. The symbols are similar to those presented in Figure 9.8. The PLC keyboard is often designed with keys for each of the individual symbols. Programming is accomplished by inserting the appropriate components into the rungs of the ladder diagram. The components are of two basic types: contacts and coils, as described in Section 9.2. Contacts represent input switches, relay contacts, and similar elements. Coils represent loads such as motors, solenoids, relays, timers, and counters. In effect, the programmer inputs the ladder logic circuit diagram rung by rung into the PLC memory with the monitor displaying the results for verification.

Function Block Diagrams. A function block diagram (FBD) provides a means of inputting high-level instructions. Instructions are composed of operational blocks. Each block has one or more inputs and one or more outputs. Within a block, certain operations take place on the inputs to transform the signals into the desired outputs. The function blocks include operations such as timers and counters, control computations using equations (e.g., proportional-integral-derivative control), data manipulation, and data transfer to other computer-based systems. We leave further description of these function blocks to other references, such as Hughes [3] and the operating manuals for commercially available PLC products.

Sequential Function Charts. The sequential function chart (SFC, also called the *Grafcet* method) graphically displays the sequential functions of an automated system as a series of steps and transitions from one state of the system to the next. The sequential function chart is described in Boucher [1]. It has become a standard method for documenting logic control and sequencing in much of Europe. However, its use in the United States is more limited, and we refer the reader to the cited reference for more details on the method.

Instruction List. Instruction list (IL) programming also provides a way of entering the ladder logic diagram into PLC memory. In this method, the programmer uses a low-level computer language to construct the ladder logic diagram by entering statements

TABLE 9.10   Typical Low-Level Language Instruction Set for a PLC

STR	Store a new input and start a new rung of the ladder.
AND	Logical AND referenced with the previously entered element. This is interpreted as a series circuit relative to the previously entered element.
OR	Logical OR referenced with the previously entered element. This is interpreted as a parallel circuit relative to the previously entered element.
NOT	Logical NOT or inverse of entered element.
OUT	Output element for the rung of the ladder diagram.
TMR	Timer element. Requires one input signal to initiate timing sequence. Output is delayed relative to input by a duration specified by the programmer in seconds. Resetting the timer is accomplished by interrupting (stopping) the input signal.
CTR	Counter element. Requires two inputs: One is the incoming pulse train that is counted by the CTR element, the other is the reset signal indicating a restart of the counting procedure.

that specify the various components and their relationships for each rung of the ladder diagram. Let us explain this approach by introducing a hypothetical PLC instruction set. Our PLC "language" is a composite of various manufacturers' languages. It contains fewer features than most commercially available PLCs. We assume that the programming device consists of a suitable keyboard for entering the individual components on each rung of the ladder logic diagram. A monitor capable of displaying each ladder rung (and perhaps several rungs that precede it) is useful to verify the program. The instruction set for our PLC is presented in Table 9.10 with a concise explanation of each instruction. Let us examine the use of these commands with several examples.

EXAMPLE 9.7   Language Commands for AND, OR, and NOT Circuits

Using the command set in Table 9.10, write the PLC programs for the three ladder diagrams from Figure 9.10, depicting the AND, OR, and NOT circuits from Figures 9.1, 9.2, and 9.3.

Solution: Commands for the three circuits are listed below, with explanatory comments.

Command	Comment
(a) STR X1	Store input X1
AND X2	Input X2 in series with X1
OUT Y	Output Y
(b) STR X1	Store input X1
OR X2	Input X2 parallel with X1
OUT Y	Output Y
(c) STR NOT X1	Store inverse of X1
OUT Y	Output Y

EXAMPLE 9.8   Language Commands for Control Relay

Using the command set in Table 9.10, write the PLC program for the control relay depicted in the ladder logic diagram of Figure 9.11.

Solution: Commands for the control relay are listed below, with explanatory comments.

Command	Comment
STR X	Store input X
OUT C	Output contact relay C
STR NOT C	Store inverse of C output
OUT Y1	Output load Y1
STR C	Store C output
OUT Y2	Output load Y2

The low-level languages are generally limited to the kinds of logic and sequencing functions that can be defined in a ladder logic diagram. Although timers and counters have not been illustrated in the two preceding examples, some of the exercise problems at the end of the chapter require the reader to use them.

Structured Text. Structured text (ST) is a high-level computer-type language likely to become more common in the future to program PLCs and PCs for automation and control applications. The principal advantage of a high-level language is its capability to perform data processing and calculations on values other than binary. Ladder diagrams and low-level PLC languages are usually quite limited in their ability to operate on signals that are other than on/off types. The capability to perform data processing and computation permits the use of more complex control algorithms, communication with other computer-based systems, display of data on a monitor, and input of data by a human operator. Another advantage is the relative ease with which a complicated control program can be interpreted by a user. Explanatory comments can be inserted into the program to facilitate interpretation.

9.4 PERSONAL COMPUTERS USING SOFT LOGIC

In the early 1990s, PCs began to encroach into applications formerly dominated by PLCs. Previously, PLCs were favored for use in factories because they were designed to operate in harsh environments, while PCs were designed for office environments. In addition, with their built-in I/O interfaces and real-time operating systems, PLCs could be readily connected to external equipment for process control, whereas PCs required special I/O cards and programs to enable such functions. Finally, personal computers sometimes lock up for no apparent cause, and usually lockups cannot be tolerated in industrial control applications. PLCs are not prone to such malfunctions.

These PLC advantages notwithstanding, the technological evolution of programmable logic controllers has not kept pace with the development of personal computers, new generations of which are introduced with much greater frequency than PLCs. There is much more proprietary software and architecture in PLCs than in PCs, making it difficult to mix and match components from different vendors. Over time, these factors have resulted in a performance disadvantage for PLCs. Performance lags its PC counterpart by as much as two years, and the gap is increasing. PC speeds are typically doubling every 18 months or so, while improvements in PLC technology occur much more slowly and require that individual companies redesign their proprietary software and architectures for each new generation of microprocessors.

PCs are now available in more sturdy enclosures for the dirty and noisy plant environment. They can be equipped with membrane-type keyboards for protection against factory moisture, oil, and dirt. They can be ordered with I/O cards and related hardware to provide the necessary devices to connect to the plants' equipment and processes. Operating systems designed to implement real-time control applications can be installed in addition to traditional office software. PLC makers are responding to the PC challenge by including PC components and features in their controller products to distinguish them from conventional PLCs. Nevertheless, the future is likely to see increasing numbers of PCs used in factory control applications where PLCs would have formerly been used.

There are two basic approaches used in PC-based control systems [10]: soft logic and hard real-time control. In the *soft logic* configuration, the PC's operating system is Windows, and control algorithms are installed as high-priority programs under the operating system. However, it is possible to interrupt the control tasks in order to service certain system functions in Windows, such as network communications and disk access. When this happens, the control function is delayed, with possible negative consequences to the process. Thus, a soft logic control system cannot be considered a real-time controller in the sense of a PLC. In high-speed control applications or volatile processes, lack of real-time control is a potential hazard. In less critical processes, soft logic works well.

In a *hard real-time control* system, the PC's operating system is the real-time operating system, and the control software takes priority over all other software. Windows tasks are executed at a lower priority under the real-time operating system. Windows cannot interrupt the execution of the real-time controller. If Windows locks up, it does not affect the controller operation. Also, the real-time operating system resides in the PC's active memory, so a failure of the hard disk has no effect in a hard real-time control system.

REFERENCES

[1] BOUCHER, T. O., *Computer Automation in Manufacturing*, Chapman & Hall, London, UK, 1996.

[2] CLEVELAND, P., "PLCs get smaller, adapt to newest technology," *Instrumentation and Control Systems*, April 1997, pp. 23–32.

[3] HUGHES, T. A., *Programmable Controllers*, 2d ed., Instrument Society of America, Research Triangle Park, NC, 1997.

[4] *International Standard for Programmable Controllers*, Standard IEC 1131-3, International Electrotechnical Commission, Geneva, Switzerland, 1993.

[5] JONES, T., and L. A. BRYAN, *Programmable Controllers*, International Programmable Controls, Inc., An IPC/ASTEC Publication, Atlanta, GA, 1983.

[6] LAVALEE, R., "Soft Logic's New Challenge: Distributed Machine Control," *Control Engineering*, August, 1996, pp. 51–58.

[7] *MICROMENTOR: Understanding and Applying Micro Programmable Controllers*, Allen-Bradley Company, Inc., Milwaukee, WI, 1995.

[8] MORLEY, R., "The Techy History of Modicon," manuscript submitted to *Technology* magazine, 1989.

[9] RIZZONI, G., *Principles and Applications of Electrical Engineering*, 5th ed., McGraw-Hill, NY, 2007.

[10] STENEROSON, J., *Fundamentals of Programmable Logic Controllers, Sensors, and Communications*, 3d ed., Pearson/Prentice Hall, Upper Saddle River, NJ, 2004.

[11] WEBB, J. W., and R. A. REIS, *Programmable Logic Controllers: Principles and Applications*, 4th ed., Pearson/Prentice Hall, Upper Saddle River, NJ, 1999.

[12] WILHELM, E. *Programmable Controller Handbook*, Hayden Book Company, Hasbrouck Heights, NJ, 1985.

REVIEW QUESTIONS

9.1 Briefly define the two categories of discrete process control.

9.2 What is an AND gate? How does it operate on two binary inputs?

9.3 What is an OR gate? How does it operate on two binary inputs?

9.4 What is Boolean algebra? What was its original purpose?

9.5 What is the difference between a delay-off timer and a delay-on timer?

9.6 What is the difference between an up counter and a down counter?

9.7 What is a ladder logic diagram?

9.8 The two types of components in a ladder logic diagram are contacts and coils. Give two examples of each type.

9.9 What is a programmable logic controller?

9.10 What are the advantages of using a PLC rather than conventional relays, timers, counters, and other hard-wired control components?

9.11 What are the five basic components of a PLC?

9.12 The typical operating cycle of the PLC, called a *scan*, consists of three parts: (1) input scan, (2) program scan, and (3) output scan. Briefly describe what is accomplished in each part.

9.13 Name the five PLC programming methods identified in the International Standard for Programmable Controllers (IEC 1131–3).

9.14 What are three of the reasons and factors that explain why personal computers are being used with greater and greater frequency for industrial control applications?

9.15 Name the two basic approaches used in PC-based control systems.

PROBLEMS

9.1 Write the Boolean logic expression for the pushbutton switch of Example 9.2 using the following symbols: X1 = START, X2 = STOP, Y1 = MOTOR, and Y2 = POWER-TO-*MOTOR*.

9.2 Construct the ladder logic diagram for the robot interlock system in Example 9.1.

9.3 In the circuit of Figure 9.1, suppose a photodetector were used to determine whether the lamp worked. If the lamp does not light when both switches are closed, the photodetector causes a buzzer to sound. Construct the ladder logic diagram for this system.

9.4 Construct the ladder logic diagrams for (a) the NAND gate and (b) the NOR gate.

9.5 Construct the ladder logic diagrams for the following Boolean logic equations:
(a) $Y = (X1 + X2) \cdot X3$, (b) $Y = (X1 + X2) \cdot (X3 + X4)$,
(c) $Y = (X1 \cdot X2) + X3$.

9.6 Write the low level language statements for the robot interlock system in Example 9.1 using the instruction set in Table 9.10.

9.7 Write the low level language statements for the lamp and photodetector system in Problem 9.4 using the instruction set in Table 9.10.

9.8 Write the low level language statements for the fluid filling operation in Example 9.6 using the instruction set in Table 9.10.

9.9 Write the low level language statements for the four parts of Problem 9.5 using the instruction set in Table 9.10.

9.10 In the fluid filling operation of Example 9.6, suppose a sensor (e.g., a submerged float switch) is used to determine whether the contents of the tank have been evacuated, rather than relying on timer T2 to empty the tank. (a) Construct the ladder logic diagram for this revised system. (b) Write the low level language statements for the system using the PLC instruction set in Table 9.10.

9.11 In the manual operation of a sheet metal stamping press, a two button safety interlock system is often used to prevent the operator from inadvertently actuating the press while his hand is in the die. Both buttons must be depressed to actuate the stamping cycle. In this system, one button is located on one side of the press while the other button is located on the opposite side. During the work cycle the operator inserts the part into the die and depresses both buttons, using both hands. (a) Write the truth table for this interlock system. (b) Write the Boolean logic expression for the system. (c) Construct the logic network diagram for the system. (d) Construct the ladder logic diagram for the system.

9.12 An emergency stop system is to be designed for a certain automatic production machine. A single "start" button is used to turn on the power to the machine at the beginning of the day. In addition, there are three "stop" buttons located at different locations around the machine, any one of which can be pressed to immediately turn off power to the machine. (a) Write the truth table for this system. (b) Write the Boolean logic expression for the system. (c) Construct the logic network diagram for the system. (d) Construct the ladder logic diagram for the system.

9.13 An industrial robot performs a machine loading and unloading operation. A PLC is used as the robot cell controller. The cell operates as follows: (1) a human worker places a workpart into a nest, (2) the robot reaches over and picks up the part and places it into an induction heating coil, (3) the part is heated for 10 seconds, and (4) the robot reaches in, retrieves the part, and places it on an outgoing conveyor. A limit switch X1 (normally open) will be used in the nest to indicate part presence in step (1). Output contact Y1 will be used to signal the robot to execute step (2) of the work cycle. This is an output contact for the PLC, but an input interlock for the robot controller. Timer T1 will be used to provide the ten-second delay in step (3). Output contact Y2 will be used to signal the robot to execute step (4). (a) Construct the ladder logic diagram for the system. (b) Write the low level language statements for the system using the PLC instruction set in Table 9.10.

9.14 A PLC is used to control the sequence in an automatic drilling operation. A human operator loads and clamps a raw workpart into a fixture on the drill press table and presses a start button to initiate the automatic cycle. The drill spindle turns on, feeds down into the part to a certain depth (the depth is determined by limit switch), and then retracts. The fixture then indexes to a second drilling position, and the drill feed-and-retract is repeated. After the second drilling operation, the spindle turns off, and the fixture moves back to the first position. The worker then unloads the finished part and loads another raw part. (a) Specify the input/output variables for this system operation and define symbols for them (e.g., X1, X2, C1, Y1, etc.). (b) Construct the ladder logic diagram for the system. (c) Write the low level language statements for the system using the PLC instruction set in Table 9.10.

9.15 An industrial furnace is to be controlled as follows: The contacts of a bimetallic strip inside the furnace close if the temperature falls below the set point, and open when the temperature is above the set point. The contacts regulate a control relay which turns on and off the heating elements of the furnace. If the door to the furnace is opened, the heating elements are temporarily turned off until the door is closed. (a) Specify the input/output variables for this system operation and define symbols for them (e.g., X1, X2, C1, Y1, etc.). (b) Construct the ladder logic diagram for the system. (c) Write the low level language statements for the system using the PLC instruction set in Table 9.10.

Material Transport Systems

CHAPTER CONTENTS

Material handling is defined by the Material Handling Industry of America[1] as "the movement, storage, protection and control of materials throughout the manufacturing and distribution process including their consumption and disposal" [10]. The handling of materials must be performed safely, efficiently, at low cost, in a timely manner, accurately (the right materials in the right quantities to the right locations), and without damage to the materials. Material handling is an important yet often overlooked issue in production.

[1]The Material Handling Industry of America (MHIA) is the trade association for material handling companies that do business in North America. The definition is published in their Annual Report each year [10].

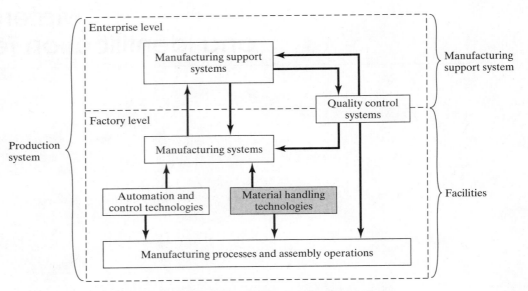

Figure 10.1 Material handling in the production system.

The cost of material handling is a significant portion of total production cost, estimates averaging around 20–25% of total manufacturing labor cost in the United States [3]. The proportion of total cost varies, depending on the type of production and degree of automation in the material handling function.

In this part of the book, we discuss the material handling and identification systems used in production. The position of material handling in the larger production system is shown in Figure 10.1. In our coverage, we divide the subject into three major categories: (1) material transport systems, discussed in the present chapter, (2) storage systems, described in Chapter 11, and (3) automatic identification and tracking systems, presented in Chapter 12. In addition, several kinds of material handling devices are discussed in other chapters of the text, including industrial robots used for material handling (Chapter 8), pallet shuttles in NC machining centers (Chapter 14), conveyors in manual assembly lines (Chapter 15), transfer mechanisms in automated transfer lines (Chapter 16), and parts feeding devices in automated assembly (Chapter 17).

10.1 INTRODUCTION TO MATERIAL HANDLING

Material handling is an important activity within the larger system by which materials are moved, stored, and tracked in our commercial infrastructure. The term commonly used for the larger system is *logistics*, which is concerned with the acquisition, movement, storage, and distribution of materials and products, as well as the planning and control of these operations in order to satisfy customer demand. Logistics operations can be divided into two basic categories: external logistics and internal logistics. *External logistics* is concerned with transportation and related activities that occur outside of a facility. In general, these activities involve the movement of materials between different geographical locations. The five traditional modes of transportation are rail, truck, air, ship, and pipeline. *Internal logistics*, more popularly known as material handling, involves the movement and storage of materials inside a given facility. Our interest in this book is on internal logistics. In this section, we first describe the various types of equipment used in material handling, and then identify some of the considerations required in the design of material handling systems.

10.1.1 Material Handling Equipment

A great variety of material handling equipment is available commercially. The equipment can be classified into four categories: (1) material transport equipment, (2) storage systems, (3) unitizing equipment, and (4) identification and tracking systems.

Material Transport Equipment. Material transport equipment is used to move materials inside a factory, warehouse, or other facility. The five main types of equipment are (1) industrial trucks, (2) automated guided vehicles, (3) rail-guided vehicles, (4) conveyors, and (5) hoists and cranes. These equipment types are described in Section 10.2.

Storage Systems. Although it is generally desirable to reduce the storage of materials in manufacturing, it seems unavoidable that raw materials and work-in-process will spend some time being stored, even if only temporarily. And finished products are likely to spend some time in a warehouse or distribution center before being delivered to the final customer. Accordingly, companies must give consideration to the most appropriate methods for storing materials and products prior to, during, and after manufacture. Storage methods and equipment can be classified into two major categories: (1) conventional storage methods and (2) automated storage systems. Conventional storage methods include bulk storage (storing items in an open floor area), rack systems, shelving and bins, and drawer storage. In general, conventional storage methods are labor intensive. Human workers put materials into storage and retrieve them from storage. Automated storage systems are designed to reduce or eliminate the manual labor involved in these functions. There are two major types of automated storage systems: (1) automated storage/retrieval systems and (2) carousel systems. These storage methods are described in greater detail in Chapter 11. In addition, mathematical models are developed to predict throughput and other performance characteristics of automated storage systems.

Unitizing Equipment. The term unitizing equipment refers to (1) containers used to hold individual items during handling and (2) equipment used to load and package the containers. Containers include pallets, boxes, baskets, barrels, pails, and drums, some of which are shown in Figure 10.2. Although seemingly mundane, containers are very important for moving materials efficiently as a unit load, rather than as individual items. Pallets and other containers that can be handled by forklift equipment are widely used in production and distribution operations. Most factories, warehouses, and distribution centers use forklift trucks to move unit loads on pallets. A given facility must often standardize on a specific type and size of container if it utilizes automatic transport and/or storage equipment to handle the loads.

The second category of unitizing equipment, loading and packaging equipment, includes *palletizers,* which are designed to automatically load cartons onto pallets and shrink-wrap plastic film around them for shipping, and *depalletizers*, which are designed to unload cartons from pallets. Other wrapping and packaging machines are also included in this equipment category.

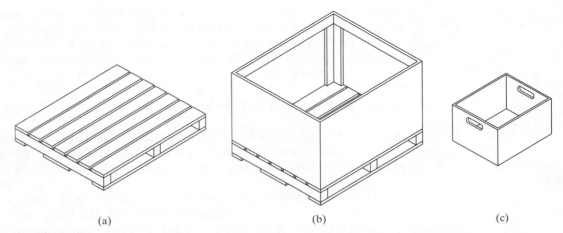

Figure 10.2 Examples of unit load containers for material handling:
(a) wooden pallet, (b) pallet box, and (c) tote box.

Identification and Tracking Systems. Material handling must include a means of keeping track of the materials being moved or stored. This is usually done by affixing some kind of label to the item, carton, or unit load that uniquely identifies it. The most common label used today is a bar code that can be read quickly and automatically by bar code readers. This is the same basic technology used by grocery stores and retail merchandisers. An alternative identification technology that is growing in importance is RFID (for radio frequency identification). Bar codes, RFID, and other automatic identification techniques are discussed in Chapter 12.

10.1.2 Design Considerations in Material Handling

Material handling equipment is usually assembled into a system. The system must be specified and configured to satisfy the requirements of a particular application. Design of the system depends on the materials to be handled, quantities and distances to be moved, type of production facility served by the handling system, and other factors, including available budget. In this section, we consider these factors that influence the design of the material handling system.

Material Characteristics. For handling purposes, materials can be classified by the physical characteristics presented in Table 10.1, suggested by a classification scheme

TABLE 10.1 Characteristics of Materials in Material Handling

Category	Measures or Descriptors
Physical state	Solid, liquid, or gas
Size	Volume, length, width, height
Weight	Weight per piece, weight per unit volume
Shape	Long and flat, round, square, etc.
Condition	Hot, cold, wet, dirty, sticky
Risk of damage	Fragile, brittle, sturdy
Safety risk	Explosive, flammable, toxic, corrosive, etc.

of Muther and Haganas [16]. Design of the material handling system must take these factors into account. For example, if the material is a liquid and is to be moved in this state over long distances in great volumes, then a pipeline is the appropriate transport means. But this handling method would be infeasible for moving a liquid contained in barrels or other containers. Materials in a factory usually consist of solid items: raw materials, parts, and finished or semifinished products.

Flow Rate, Routing, and Scheduling. In addition to material characteristics, other factors must be considered in determining which type of equipment is most appropriate for the application. These other factors include (1) quantities and flow rates of materials to be moved, (2) routing factors, and (3) scheduling of the moves.

The amount or quantity of material to be moved affects the type of handling system that should be installed. If large quantities of material must be handled, then a dedicated handling system is appropriate. If the quantity of a particular material type is small but there are many different material types to be moved, then the handling system must be designed to be shared by the various materials moved. The amount of material moved must be considered in the context of time, that is, how much material is moved within a given time period. We refer to the amount of material moved per unit time as the *flow rate*. Depending on the form of the material, flow rate is measured in pieces/hr, pallet loads/hr, tons/hr, ft^3/day, or similar units. Whether the material must be moved as individual units, in batches, or continuously has an effect on the selection of handling method.

Routing factors include pickup and drop-off locations, move distances, routing variations, and conditions that exist along the routes. Given that other factors remain constant, handling cost is directly related to the distance of the move: The longer the move distance, the greater the cost. Routing variations occur because different materials follow different flow patterns in the factory or warehouse. If these differences exist, the material handling system must be flexible enough to deal with them. Conditions along the route include floor surface condition, traffic congestion, whether a portion of the move is outdoors, whether the path is straight line or involves turns and changes in elevation, and the presence or absence of people along the path. All of these factors affect the design of the material transport system.

Scheduling relates to the timing of each individual delivery. In production as well as in many other material handling applications, the material must be picked up and delivered promptly to its proper destination to maintain peak performance and efficiency of the overall system. To the extent required by the application, the handling system must be responsive to this need for timely pickup and delivery of the items. Rush jobs increase material handling cost. Scheduling urgency is often mitigated by providing space for buffer stocks of materials at pickup and drop-off points. This allows a "float" of materials to exist in the system, thus reducing the pressure on the handling system for immediate response to a delivery request.

Plant Layout. Plant layout is an important factor in the design of a material handling system. When a new facility is being planned, the design of the handling system should be considered part of the layout. In this way, there is greater opportunity to create a layout that optimizes material flow in the building and utilizes the most appropriate type of handling system. In the case of an existing facility, there are more constraints on the design of the handling system. The present arrangement of departments and equipment in the building usually limits the attainment of optimum flow patterns.

The plant layout design should provide the following data for use in the design of the handling system: total area of the facility and areas within specific departments in the plant, relative locations of departments, arrangement of equipment in the layout, locations where materials must be picked up (load stations) and delivered (unload stations), possible routes between these locations, and distances traveled. Each of these factors affects flow patterns and selection of material handling equipment.

In Section 2.3, we described the conventional types of plant layout used in manufacturing: (1) process layout, (2) product layout, and (3) fixed-position layout. Different material handling systems are generally required for the three layout types. Table 10.2 summarizes the characteristics of the three conventional layout types and the kinds of material handling equipment usually associated with each layout type.

In process layouts, various different products are manufactured in small or medium batch sizes. The handling system must be flexible to deal with the variations. Considerable work-in-process is usually one of the characteristics of batch production, and the material handling system must be capable of accommodating this inventory. Hand trucks and forklift trucks (for moving pallet loads of parts) are commonly used in process layouts. Factory applications of automated guided vehicle systems are growing because they represent a versatile means of handling the different load configurations in medium and low volume production. Work-in-progress is often stored on the factory floor near the next scheduled machines. More systematic ways of managing in-process inventory include automated storage systems (Section 11.3).

A product layout involves production of a standard or nearly identical types of product in relatively high quantities. Final assembly plants for cars, trucks, and appliances are usually designed as product layouts. The transport system that moves the product is typically characterized as fixed route, mechanized, and capable of large flow rates. It sometimes serves as a storage area for work-in-process to reduce effects of downtime between production areas along the line of product flow. Conveyor systems are common in product layouts. Delivery of component parts to the various assembly workstations along the flow path is accomplished by trucks and similar unit load vehicles.

Finally, in a fixed-position layout, the product is large and heavy and therefore remains in a single location during most of its fabrication. Heavy components and subassemblies must be moved to the product. Handling systems used for these moves in fixed-position layouts are large and often mobile. Cranes, hoists, and trucks are common in this situation.

Unit Load Principle. The unit load principle stands as an important and widely applied principle in material handling. A *unit load* is simply the mass that is to be moved

TABLE 10.2   Types of Material Handling Equipment Associated with Three Layout Types

Layout Type	Characteristics	Typical Material Handling Equipment
Process	Variations in product and processing, low and medium production rates	Hand trucks, forklift trucks, automated guided vehicle systems
Product	Limited product variety, high production rate	Conveyors for product flow, industrial trucks and automated guided vehicles to deliver components to stations
Fixed-position	Large product size, low production rate	Cranes, hoists, industrial trucks

TABLE 10.3 Standard Pallet Sizes Commonly Used in Factories and Warehouses

Depth = x Dimension	Width = y Dimension
800 mm (32 in)	1000 mm (40 in)
900 mm (36 in)	1200 mm (48 in)
1000 mm (40 in)	1200 mm (48 in)
1060 mm (42 in)	1060 mm (42 in)
1200 mm (48 in)	1200 mm (48 in)

Sources: [6], [17].

or otherwise handled at one time. The unit load may consist of only one part, a container loaded with multiple parts, or a pallet loaded with multiple containers of parts. In general, the unit load should be designed to be as large as is practical for the material handling system that will move or store it, subject to considerations of safety, convenience, and access to the materials making up the unit load. This principle is widely applied in the truck, rail, and ship industries. Palletized unit loads are collected into truck loads, which then become larger unit loads themselves. Then these truck loads are aggregated once again on freight trains or ships, in effect becoming even larger unit loads.

There are good reasons for using unit loads in material handling, as described in Tompkins et al [17]: (1) multiple items can be handled simultaneously, (2) the required number of trips is reduced, (3) loading and unloading times are reduced, and (4) product damage is decreased. Using unit loads results in lower cost and higher operating efficiency.

Included in the definition of unit load is the container that holds or supports the materials to be moved. To the extent possible, these containers are standardized in size and configuration to be compatible with the material handling system. Examples of containers used to form unit loads in material handling are illustrated in Figure 10.2. Of the available containers, pallets are probably the most widely used, owing to their versatility, low cost, and compatibility with various types of material handling equipment. Most factories and warehouses use forklift trucks to move materials on pallets. Table 10.3 lists some of the most popular standard pallet sizes in use today. We use these standard pallet sizes in some of our analysis of automated storage/retrieval systems in Chapter 11.

10.2 MATERIAL TRANSPORT EQUIPMENT

In this section we examine the five categories of material transport equipment commonly used to move parts and other materials in manufacturing and warehouse facilities: (1) industrial trucks, manual and powered; (2) automated guided vehicles; (3) monorails and other rail-guided vehicles; (4) conveyors; and (5) cranes and hoists. Table 10.4 summarizes the principal features and kinds of applications for each equipment category. In Section 10.3, we consider quantitative techniques by which material transport systems consisting of this equipment can be analyzed.

10.2.1 Industrial Trucks

Industrial trucks are divided into two categories: nonpowered and powered. The nonpowered types are often referred to as hand trucks because they are pushed or pulled by human workers. Quantities of material moved and distances traveled are relatively low when this type of equipment is used to transport materials. Hand trucks are classified as

TABLE 10.4 Summary of Features and Applications of Five Categories of Material Handling Equipment

Material Handling Equipment	Features	Typical Applications
Industrial trucks, manual	Low cost Low rate of deliveries/hour	Moving light loads in a factory
Industrial trucks, powered	Medium cost	Movement of pallet loads and palletized containers in a factory or warehouse
Automated guided vehicle systems	High cost Battery-powered vehicles Flexible routing Nonobstructive pathways	Moving pallet loads in factory or warehouse Moving work-in-process along variable routes in low and medium production
Monorails and other rail-guided vehicles	High cost Flexible routing On-the-floor or overhead types	Moving single assemblies, products, or pallet loads along variable routes in factory or warehouse Moving large quantities of items over fixed routes in a factory or warehouse
Conveyors, powered	Great variety of equipment In-floor, on-the-floor, or overhead Mechanical power to move loads resides in pathway	Moving products along a manual assembly line Sortation of items in a distribution center
Cranes and hoists	Lift capacities of more than 100 tons	Moving large, heavy items in factories, mills, warehouses, etc.

either two-wheel or multiple-wheel. Two-wheel hand trucks, Figure 10.3(a), are generally easier to manipulate by the worker but are limited to lighter loads. Multiple-wheeled hand trucks are available in several types and sizes. Two common types are dollies and pallet trucks. Dollies are simple frames or platforms as shown in Figure 10.3(b). Various wheel configurations are possible, including fixed wheels and caster-type wheels. Pallet trucks, shown in Figure 10.3(c), have two forks that can be inserted through the openings in a pallet.

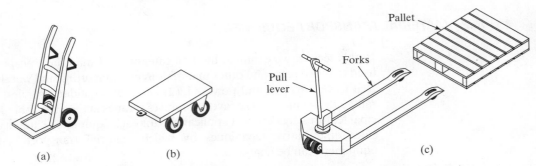

(a)   (b)   (c)

Figure 10.3 Examples of nonpowered industrial trucks (hand trucks): (a) two-wheel hand truck, (b) four-wheel dolly, and (c) hand-operated low-lift pallet truck.

A lift mechanism is actuated by the worker to lift and lower the pallet off the ground using small diameter wheels near the end of the forks. In operation, the worker inserts the forks into the pallet, elevates the load, pulls the truck to its destination, lowers the pallet, and removes the forks.

Powered trucks are self-propelled to relieve the worker of having to move the truck manually. Three common types are used in factories and warehouses: (a) walkie trucks, (b) forklift rider trucks, and (c) towing tractors. Walkie trucks, Figure 10.4(a), are battery-powered vehicles equipped with wheeled forks for insertion into pallet openings but with no provision for a worker to ride on the vehicle. The truck is steered by a worker using a control handle at the front of the vehicle. The forward speed of a walkie truck is limited to around 3 mi/hr (5 km/hr), about the normal walking speed of a human.

Forklift rider trucks, Figure 10.4(b), are distinguished from walkie trucks by the presence of a modest cab for the worker to sit in and drive the vehicle. Forklift trucks range in load carrying capacity from about 450 kg (1,000 lb) up to more than 4,500 kg (10,000 lb). Forklift trucks have been modified to suit various applications. Some trucks have high reach capacities for accessing pallet loads on high rack systems, while others are capable of operating in the narrow aisles of high-density storage racks. Power sources for forklift trucks are either internal combustion engines (gasoline, liquefied petroleum gas, or compressed natural gas) or electric motors (using on-board batteries).

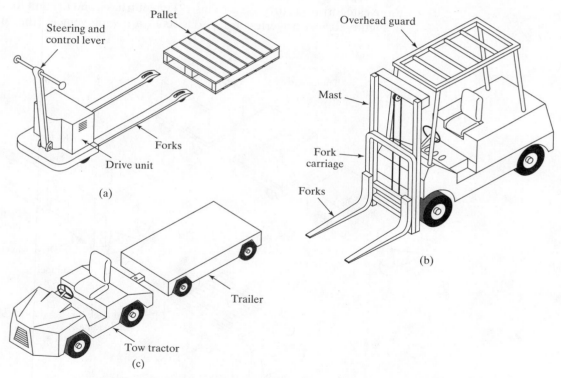

Figure 10.4 Three principal types of powered trucks: (a) walkie truck, (b) forklift truck, and (c) towing tractor.

Industrial towing tractors, Figure 10.4(c), are designed to pull one or more trailing carts over the relatively smooth surfaces found in factories and warehouses. They are generally used for moving large amounts of materials between major collection and distribution areas. The runs between origination and destination points are usually fairly long. Power is supplied either by electrical motor (battery-powered) or internal combustion engine. Tow tractors also find significant applications in air transport operations for moving baggage and air freight in airports.

10.2.2 Automated Guided Vehicles

An *automated guided vehicle system* (AGVS) is a material handling system that uses independently operated, self-propelled vehicles guided along defined pathways. The vehicles are powered by on-board batteries that allow many hours of operation (8–16 hr is typical) before needing to be recharged. A distinguishing feature of an AGVS, compared to rail-guided vehicle systems and most conveyor systems, is that the pathways are unobtrusive. An AGVS is appropriate where different materials are moved from various load points to various unload points. An AGVS is therefore suitable for automating material handling in batch production and mixed model production.

Types of Vehicles. Automated guided vehicles can be divided into the following three categories: (1) driverless trains, (2) pallet trucks, and (3) unit load carriers, illustrated in Figure 10.5. A driverless train consists of a towing vehicle (the AGV) pulling one or

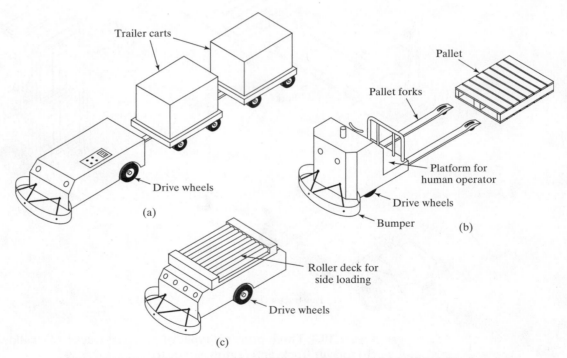

Figure 10.5 Three types of automated guided vehicles: (a) driverless automated guided train, (b) AGV pallet truck, and (c) unit load carrier.

more trailers to form a train, as in Figure 10.5(a). It was the first type of AGVS to be introduced and is still widely used today. A common application is moving heavy payloads over long distances in warehouses or factories with or without intermediate pickup and drop-off points along the route. For trains consisting of five to ten trailers, this is an efficient transport system.

Automated guided pallet trucks, Figure 10.5(b), are used to move palletized loads along predetermined routes. In the typical application the vehicle is backed into the loaded pallet by a human worker who steers the truck and uses its forks to elevate the load slightly. Then the worker drives the pallet truck to the guidepath and programs its destination, and the vehicle proceeds automatically to the destination for unloading. The capacity of an AGVS pallet truck ranges up to several thousand kilograms, and some trucks are capable of handling two pallets rather than one. A more recent introduction related to the pallet truck is the forklift AGV. This vehicle can achieve significant vertical movement of its forks to reach loads on racks and shelves.

AGV unit load carriers are used to move unit loads from one station to another. They are often equipped for automatic loading and unloading of pallets or tote pans by means of powered rollers, moving belts, mechanized lift platforms, or other devices built into the vehicle deck. A typical unit load AGV is illustrated in Figure 10.5(c). Variations of unit load carriers include light load AGVs and assembly line AGVs. The light load AGV is a relatively small vehicle with corresponding light load capacity (typically 250 kg or less). It does not require the same large aisle width as a conventional AGV. Light load guided vehicles are designed to move small loads (single parts, small baskets, or tote pans of parts.) through plants of limited size engaged in light manufacturing. An assembly line AGV is designed to carry a partially completed subassembly through a sequence of assembly workstations to build the product.

AGVS Applications. Automated guided vehicle systems are used in a growing number and variety of applications. The applications tend to correlate with the vehicle types previously described. The principal AGVS applications in production and logistics are (1) driverless train operations, (2) storage and distribution, (3) assembly line applications, and (4) flexible manufacturing systems. We have already described driverless train operations, which involve the movement of large quantities of material over relatively long distances.

A second application is storage and distribution operations. Unit load carriers and pallet trucks are typically used in these applications, which involve movement of material in unit loads. The applications often interface the AGVS with some other automated handling or storage system, such as an automated storage/retrieval system (AS/RS) in a distribution center. The AGVS delivers incoming unit loads contained on pallets from the receiving dock to the AS/RS, which places the items into storage, and the AS/RS retrieves individual pallet loads from storage and transfers them to vehicles for delivery to the shipping dock. Storage/distribution operations also include light manufacturing and assembly plants in which work-in-process is stored in a central storage area and distributed to individual workstations for processing. Electronics assembly is an example of these kinds of applications. Components are "kitted" at the storage area and delivered in tote pans or trays to the assembly workstations in the plant. Light load AGVs are the appropriate vehicles in these applications.

AGV systems are used in assembly line applications, based on a trend that began in Europe. Unit load carriers and light load guided vehicles are used in these lines. In the usual application, the production rate is relatively low (the product spending perhaps 4–10 min per station), and there are several different product models made on the line, each requiring a different processing time. Workstations are generally arranged in parallel to allow the line to deal with differences in assembly cycle time for different products. Between stations, components are kitted and placed on the vehicle for the assembly operations to be performed at the next station. The assembly tasks are usually performed with the work unit onboard the vehicle, thus avoiding the extra time required for unloading and reloading.

Another application area for AGVS technology is flexible manufacturing systems (FMSs, Chapter 19). In the typical operation, starting workparts are placed onto pallet fixtures by human workers in a staging area, and the AGVs deliver the parts to the individual workstations in the system. When the AGV arrives at the assigned station, the pallet is transferred from the vehicle platform to the station (such as the worktable of a machine tool) for processing. At the completion of processing, a vehicle returns to pick up the work and transport it to the next assigned station. An AGVS provides a versatile material handling system to complement the flexibility of the FMS.

AGVS technology is still developing, and the industry is continually working to design new systems to respond to new application requirements. An interesting example that combines two technologies involves the use of a robotic manipulator mounted on an automated guided vehicle to provide a mobile robot for performing complex handling tasks at various locations in a plant.

Vehicle Guidance Technology. The guidance system is the method by which AGVS pathways are defined and vehicles are controlled to follow the pathways. In this section, we discuss three technologies that are used in commercial systems for vehicle guidance: (1) imbedded guide wires, (2) paint strips, and (3) self-guided vehicles.

In the imbedded guide wire method, electrical wires are placed in a small channel cut into the surface of the floor. The channel is typically 3–12 mm (1/8–1/2 in) wide and 13–26 mm (1/2–1.0 in) deep. After the guide wire is installed, the channel is filled with cement to eliminate the discontinuity in the floor surface. The guide wire is connected to a frequency generator, which emits a low-voltage, low-current signal with a frequency in the range 1–15 kHz. This induces a magnetic field along the pathway that can be followed by sensors on board each vehicle. The operation of a typical system is illustrated in Figure 10.6. Two sensors (coils)

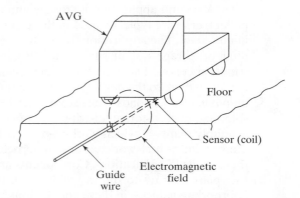

Figure 10.6 Operation of the on-board sensor system that uses two coils to track the magnetic field in the guide wire.

are mounted on the vehicle on either side of the guide wire. When the vehicle is located such that the guide wire is directly between the two coils, the intensity of the magnetic field measured by each coil is equal. If the vehicle strays to one side or the other, or if the guide wire path changes direction, the magnetic field intensity at the two sensors will become unequal. This difference is used to control the steering motor, which makes the required changes in vehicle direction to equalize the two sensor signals, thereby tracking the guide wire.

A typical AGVS layout contains multiple loops, branches, side tracks, and spurs, as well as pickup and drop-off stations. The most appropriate route must be selected from the alternative pathways available to a vehicle as it moves to a specified destination in the system. When a vehicle approaches a branching point where the guide path forks into two (or more) pathways, the vehicle must have a means of deciding which path to take. The two principal methods of making this decision in commercial wire guided systems are (1) the frequency select method and (2) the path switch select method. In the *frequency select method,* the guide wires leading into the two separate paths at the switch have different frequencies. As the vehicle enters the switch, it reads an identification code on the floor to determine its location. Depending on its programmed destination, the vehicle selects the correct guidepath by following only one of the frequencies. This method requires a separate frequency generator for each different frequency used in the guidepath layout.

The *path switch select method* operates with a single frequency throughout the guidepath layout. To control the path of a vehicle at a switch, the power is turned off in all other branches except the one that the vehicle is to travel on. To accomplish routing by the path switch select method, the guidepath layout is divided into blocks that are electrically insulated from each other. The blocks can be turned on and off either by the vehicles themselves or by a central control computer.

When paint strips are used to define the pathway, the vehicle uses an optical sensor system capable of tracking the paint. The strips can be taped, sprayed, or painted on the floor. One system uses a 1-in-wide paint strip containing fluorescent particles that reflect an ultraviolet (UV) light source from the vehicle. An on-board sensor detects the reflected light in the strip and controls the steering mechanism to follow it. Paint strip guidance is useful in environments where electrical noise renders the guide wire system unreliable or when the installation of guide wires in the floor surface is not practical. One problem with this guidance method is that the paint strip deteriorates with time. It must be kept clean and periodically replaced.

Self-guided vehicles (SGVs) represent the latest AGVS guidance technology. Unlike the previous two guidance methods, SGVs operate without continuously defined pathways. Instead, they use a combination of dead reckoning and beacons located throughout the plant that can be identified by on-board sensors. *Dead reckoning* refers to the capability of a vehicle to follow a given route in the absence of a defined pathway in the floor. Movement of the vehicle along the route is accomplished by computing the required number of wheel rotations in a sequence of specified steering angles. The computations are performed by the vehicle's on-board computer. As one would expect, positioning accuracy of dead reckoning decreases over long distances. Accordingly, the location of the self-guided vehicle must be periodically verified by comparing the calculated position with one or more known positions. These known positions are established using beacons located strategically throughout the plant. There are various types of beacons used in commercial SGV systems. One system uses bar-coded beacons mounted along the aisles. These beacons can be sensed by a rotating laser scanner on the vehicle. Based on the positions of the beacons, the on-board navigation computer uses triangulation to update the positions calculated by dead reckoning. Another guidance system uses magnetic beacons imbedded in the plant floor along the pathway. Dead reckoning is used to move the vehicle between beacons, and the actual locations of the beacons provide data to update the computer's dead reckoning map.

It should be noted that dead reckoning can be used by AGV systems that are normally guided by in-floor guide wires or paint strips. This capability allows the vehicle to cross steel plates in the factory floor where guide wires cannot be installed or to depart from the guidepath for positioning at a load/unload station. At the completion of the dead reckoning maneuver, the vehicle is programmed to return to the guidepath to resume normal guidance control.

The advantage of self-guided vehicle technology over fixed pathways (guide wires and paint strips) is its flexibility. The SGV pathways are defined in software. The path network can be changed by entering the required data into the navigation computer. New docking points can be defined. The pathway network can be expanded by installing new beacons. These changes can be made quickly and without major alterations to the plant facility.

Vehicle Management. For the AGVS to operate efficiently, the vehicles must be well managed. Delivery tasks must be allocated to vehicles to minimize waiting times at load/unload stations. Traffic congestion in the guidepath network must be minimized. In this discussion we consider two aspects of vehicle management: (1) traffic control and (2) vehicle dispatching.

The purpose of traffic control in an automated guided vehicle system is to minimize interference between vehicles and to prevent collisions. Two methods of traffic control used in commercial AGV systems are (1) on-board vehicle sensing and (2) zone control. The two techniques are often used in combination. *On-board vehicle sensing*, also called *forward sensing*, uses one or more sensors on each vehicle to detect the presence of other vehicles and obstacles ahead on the guide path. Sensor technologies include optical and ultrasonic devices. When the on-board sensor detects an obstacle in front of it, the vehicle stops. When the obstacle is removed, the vehicle proceeds. If the sensor system is 100% effective, collisions between vehicles are avoided. The effectiveness of forward sensing is limited by the capability of the sensor to detect obstacles that are in front of it on the guide path. These systems are most effective on straight pathways. They are less effective at turns and convergence points where forward vehicles may not be directly in front of the sensor.

In *zone control*, the AGVS layout is divided into separate zones, and the operating rule is that no vehicle is permitted to enter a zone that is already occupied by another vehicle. The length of a zone is at least sufficient to hold one vehicle plus allowances for safety and other considerations. Other considerations include number of vehicles in the system, size and complexity of the layout, and the objective of minimizing the number of separate zones. For these reasons, the zones are normally much longer than a vehicle length. Zone control is illustrated in Figure 10.7 in its simplest form. When one vehicle occupies a given zone, any trailing vehicle is not allowed to enter that zone. The leading vehicle must proceed into the next zone before the trailing vehicle can occupy the current zone. When the forward movement of vehicles in the separate zones is controlled, collisions are prevented, and traffic in the overall system is controlled.

One means of implementing zone control is to use separate control units mounted along the guide path. When a vehicle enters a given zone, it activates the block in that zone to prevent any trailing vehicle from moving forward and colliding with the present

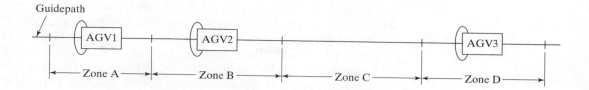

Figure 10.7 Zone control to implement blocking system. Zones A, B, and D are blocked. Zone C is free. Vehicle 2 is blocked from entering Zone A by Vehicle 1. Vehicle 3 is free to enter Zone C.

vehicle. As the present vehicle moves into the next (downstream) zone, it activates the block in that zone and deactivates the block in the previous zone. In effect, zones are turned on and off to control vehicle movement by the blocking system. Another method to implement zone control is to use a central computer, which monitors the location of each vehicle and attempts to optimize the movement of all vehicles in the system.

For an AGVS to serve its function, vehicles must be dispatched in a timely and efficient manner to the points in the system where they are needed. Several methods are used in AGV systems to dispatch vehicles: (1) on-board control panels, (2) remote call stations, and (3) central computer control. These dispatching methods are generally used in combination to maximize responsiveness and efficiency.

Each guided vehicle is equipped with some form of on-board control panel for the purpose of manual vehicle control, vehicle programming, and other functions. Most commercial vehicles can be dispatched by means of this control panel to a given station in the AGVS layout. Dispatching with an on-board control panel represents the lowest level of sophistication among the possible methods. It provides the AGVS with flexibility and timeliness in coping with changes and variations in delivery requirements.

Remote call stations represent another method for an AGVS to satisfy delivery requirements. The simplest call station is a push-button mounted at the load/unload station. This transmits a hailing signal for any available vehicle in the neighborhood to dock at the station and either pick up or drop off a load. The on-board control panel might then be used to dispatch the vehicle to the desired destination point. More sophisticated remote call stations permit the vehicle's destination to be programmed at the same time the vehicle is called. This is a more automated dispatching method that is useful in AGV systems capable of automatic loading and unloading operations.

In a large factory or warehouse involving a high degree of automation, the AGVS servicing the facility must also be highly automated to achieve efficient operation of the entire production-storage-handling system. Central computer control is used to accomplish automatic dispatching of vehicles according to a preplanned schedule of pickups and deliveries in the layout and/or in response to calls from the various load/unload stations. In this dispatching method, the central computer issues commands to the vehicles in the system concerning their destinations and the operations they must perform. To accomplish the dispatching function, the central computer must possess current information on the location of each vehicle in the system so that it can make appropriate decisions about which vehicles to dispatch to what locations. Hence, the vehicles must continually communicate their whereabouts to the central controller. Radio frequency (RF) is commonly used to achieve the required communication links.

Vehicle Safety. The safety of humans located along the pathway is an important objective in AGVS design. An inherent safety feature of an AGV is that its traveling speed is slower than the normal walking pace of a human. This minimizes the danger that it will overtake a human walking along the path in front of the vehicle.

In addition, AGVs are usually provided with several other features specifically for safety reasons. A safety feature included in most guidance systems is automatic stopping of the vehicle if it strays more than a short distance, typically 50–150 mm (2–6 in), from the guidepath; the distance is referred to as the vehicle's *acquisition distance*. This automatic stopping feature prevents a vehicle from running wild in the building. Alternatively, in the event that the vehicle is off the guidepath (e.g., for loading), its sensor system is capable of locking onto the guidepath when the vehicle is moved to within the acquisition distance.

Another safety device is an obstacle detection sensor located on each vehicle. This is the same on-board sensor used for traffic control. The sensor can detect obstacles along the path ahead, including humans. The vehicles are programmed either to stop when an obstacle is sensed ahead or to slow down. The reason for slowing down is that the sensed object may be located off to the side of the vehicle path or directly ahead but beyond a turn in the guide path, or the obstacle may be a person who will move out of the way as the AGV approaches. In any of these cases, the vehicle is permitted to proceed at a slower (safer) speed until it has passed the obstacle. The disadvantage of programming a vehicle to stop when it encounters an obstacle is that this delays the delivery and degrades system performance.

A safety device included on virtually all commercial AGVs is an emergency bumper. The bumpers are prominent in the illustrations shown in Figure 10.5. The bumper surrounds the front of the vehicle and protrudes ahead of it by a distance of 300 mm (12 in) or more. When the bumper makes contact with an object, the vehicle is programmed to brake immediately. Depending on the speed of the vehicle, its load, and other conditions, the distance the vehicle needs to come to a complete stop will vary from several inches to several feet. Most vehicles are programmed to require manual restarting after an obstacle has been encountered by the emergency bumper. Other safety devices on a typical vehicle include warning lights (blinking or rotating lights) and/or warning bells, which alert humans that the vehicle is present.

10.2.3 Monorails and Other Rail-Guided Vehicles

The third category of material transport equipment consists of motorized vehicles that are guided by a fixed rail system. The rail system consists of either one rail (called a monorail) or two parallel rails. Monorails in factories and warehouses are typically suspended overhead from the ceiling. In rail-guided vehicle systems using parallel fixed rails, the tracks generally protrude up from the floor. In either case, the presence of a fixed rail pathway distinguishes these systems from automated guided vehicle systems. As with AGVs, the vehicles operate asynchronously and are driven by an on-board electric motor. But unlike AGVs, which are powered by their own on-board batteries, rail guided vehicles pick up electrical power from an electrified rail (similar to an urban rapid transit rail system). This relieves the vehicle from periodic recharging of its battery; however, the electrified rail system introduces a safety hazard not present in an AGVS.

Routing variations are possible in rail-guided vehicle systems through the use of switches, turntables, and other specialized track sections. This permits different loads to travel different routes, similar to an AGVS. Rail-guided systems are generally considered to be more versatile than conveyor systems but less versatile than automated guided vehicle systems. One of the original applications of nonpowered monorails was in the meat processing industry before 1900. The slaughtered animals were hung from meat hooks attached to overhead monorail trolleys. The trolleys were moved through the different departments of the plant manually by the workers. It is likely that Henry Ford got the idea for the assembly line from observing these meat packing operations. Today, the automotive industry makes considerable use of electrified overhead monorails to move large components and subassemblies in its manufacturing operations.

10.2.4 Conveyors

A *conveyor* is a mechanical apparatus for moving items or bulk materials, usually inside a facility. Conveyors are used when material must be moved in relatively large quantities between specific locations over a fixed path, which may be in the floor, above the floor, or overhead. Conveyors are either powered or nonpowered. In *powered conveyors*, the power mechanism is contained in the fixed path, using chains, belts, rotating rolls, or other devices to propel loads along the path. Powered conveyors are commonly used in automated material transport systems in manufacturing plants, warehouses, and distribution centers. In *nonpowered conveyors*, materials are moved either manually by human workers who push the loads along the fixed path or by gravity from one elevation to a lower elevation.

Types of Conveyors. A variety of conveyor equipment is commercially available. Our primary interest here is in powered conveyors. Most of the major types of powered conveyors, organized according to the type of mechanical power provided in the fixed path, are briefly described in the following:

- *Roller conveyors.* In roller conveyors, the pathway consists of a series of tubes (rollers) that are perpendicular to the direction of travel, as in Figure 10.8(a). Loads must possess a flat bottom surface of sufficient area to span several adjacent rollers. Pallets, tote pans, or cartons serve this purpose well. The rollers are contained in a fixed frame that elevates the pathway above floor level from several inches to several feet. The loads are moved forward as the rollers rotate. Roller conveyors can either be powered or nonpowered. Powered roller conveyors are driven by belts or chains. Nonpowered roller conveyors are often driven by gravity so that the pathway has a downward slope sufficient to overcome rolling friction. Roller conveyors are used in a wide variety of applications, including manufacturing, assembly, packaging, sortation, and distribution.

- *Skate-wheel conveyors.* These are similar in operation to roller conveyors. Instead of rollers, they use skate wheels rotating on shafts connected to a frame to roll pallets, tote pans, or other containers along the pathway, as in Figure 10.8(b). Skate-wheel conveyors are lighter in weight than roller conveyors. Applications of skate-wheel conveyors are similar to those of roller conveyors, except that the loads must generally be lighter since the contacts between the loads and the conveyor are much more concentrated. Because of their light weight, skate-wheel conveyors are sometimes built as portable units that can be used for loading and unloading truck trailers at shipping and receiving docks at factories and warehouses.

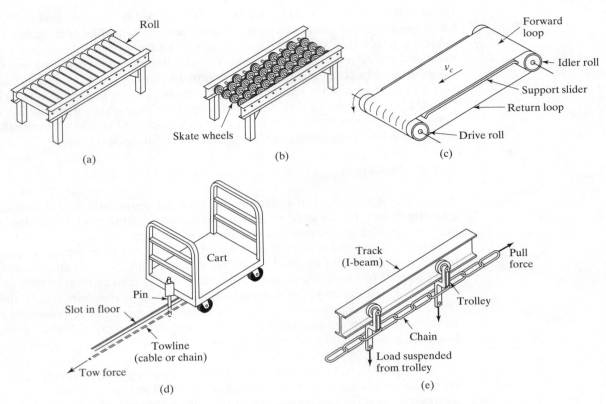

Figure 10.8 (a) Roller conveyor, (b) skate-wheel conveyor, (c) belt (flat) conveyor (support frame not shown), (d) in-floor towline conveyor, and (e) overhead trolley conveyor.

- *Belt Conveyors.* Belt conveyors consist of a continuous loop. Half its length is used for delivering materials, and the other half is the return run, as in Figure 10.8(c). The belt is made of reinforced elastomer (rubber), so that it possesses high flexibility but low extensibility. At one end of the conveyor is a drive roll that powers the belt. The flexible belt is supported by a frame that has rollers or support sliders along its forward loop. Belt conveyors are available in two common forms: (1) flat belts for pallets, individual parts, or even certain types of bulk materials; and (2) troughed belts for bulk materials. Materials placed on the belt surface travel along the moving pathway. In the case of troughed belt conveyors, the rollers and supports give the flexible belt a V shape on the forward (delivery) loop to contain bulk materials such as coal, gravel, grain, or similar particulate materials.

- *Chain conveyors.* The typical equipment in this category consists of chain loops in an over-and-under configuration around powered sprockets at the ends of the pathway. The conveyor may consist of one or more chains operating in parallel. The chains travel along channels in the floor that provide support for the flexible chain sections. Either the chains slide along the channel or they ride on rollers in the channel. The loads are generally dragged along the pathway using bars that project up from the moving chain.

- *In-floor towline conveyor.* These conveyors use four-wheel carts powered by moving chains or cables located in trenches in the floor, as in Figure 10.8(d). The chain or cable is called a towline. Pathways for the conveyor system are defined by the trench and cable, and the cable is driven as a powered pulley system. It is possible to switch between powered pathways to achieve flexibility in routing. The carts use steel pins that project below floor level into the trench to engage the chain for towing. (Gripper devices are substituted for pins when cable is used as the pulley system, as in the San Francisco trolleys.) The pin can be pulled out of the chain (or the gripper releases the cable) to disengage the cart for loading, unloading, switching, accumulating parts, and manually pushing a cart off the main pathway. Towline conveyor systems are used in manufacturing plants and warehouses.

- *Overhead trolley conveyor.* A *trolley* in material handling is a wheeled carriage running on an overhead rail from which loads can be suspended. An overhead trolley conveyor, Figure 10.8(e), consists of multiple trolleys, usually equally spaced along a fixed track. The trolleys are connected together and moved along the track by means of a chain or cable that forms a complete loop. Suspended from the trolleys are hooks, baskets, or other receptacles to carry loads. The chain (or cable) is attached to a drive pulley that pulls the chain at a constant velocity. The conveyor path is determined by the configuration of the track system, which has turns and possible changes in elevation. Overhead trolley conveyors are often used in factories to move parts and assemblies between major production departments. They can be used for both delivery and storage.

- *Power-and-free overhead trolley conveyor.* This conveyor is similar to the overhead trolley conveyor, except that the trolleys can be disconnected from the drive chain, providing the conveyor with an asynchronous capability. This is usually accomplished by using two tracks, one just above the other. The upper track contains the continuously moving endless chain, and the trolleys that carry loads ride on the lower track. Each trolley includes a mechanism by which it can be connected to the drive chain and disconnected from it. When connected, the trolley is pulled along its track by the moving chain in the upper track. When disconnected, the trolley is idle.

- *Cart-on-track conveyor.* This equipment consists of individual carts riding on a track a few feet above floor level. The carts are driven by means of a rotating shaft, as illustrated in Figure 10.9. A drive wheel, attached to the bottom of the cart and set at an angle to the rotating tube, rests against it and drives the cart forward. The cart speed is controlled by regulating the angle of contact between the drive wheel and the spinning tube. When the axis of the drive wheel is 45°, the cart is propelled forward. When the axis of the drive wheel is parallel to the tube, the cart does not move. Thus, control of the drive wheel angle on the cart allows power-and-free operation of the conveyor. One of the advantages of cart-on-track systems relative to many other conveyors is that the carts can be positioned with high accuracy. This permits their use for positioning work during production. Applications of cart-on-track systems include robotic spot welding lines in automobile body plants and mechanical assembly systems.

- *Other Conveyor Types.* Other powered conveyors include vibration-based systems and vertical lift conveyors. Screw conveyors are powered versions of the Archimedes screw, the water-raising device devised in ancient times, consisting of a large screw inside a tube, turned by hand to pump water uphill for irrigation purposes. Vibration-based conveyors use a flat track connected to an electromagnet that

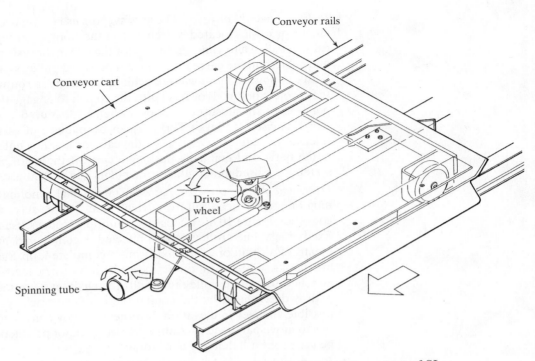

Figure 10.9 Cart-on-track conveyor. (Diagram courtesy of SI Division, Paragon Technologies, Inc.)

imparts an angular vibratory motion to the track to propel items in the desired direction. This same principle is used in vibratory bowl feeders to deliver components in automated assembly systems (Section 17.1.2). Vertical lift conveyors include a variety of mechanical elevators designed to provide vertical motion, such as between floors or to link floor-based conveyors with overhead conveyors. Other conveyor types include nonpowered chutes, ramps, and tubes, which are driven by gravity.

Conveyor Operations and Features. As indicated by our preceding discussion, conveyor equipment covers a wide variety of operations and features. Let us restrict our discussion here to powered conveyors. Conveyor systems divide into two basic types in terms of the characteristic motion of the materials moved by the system: (1) continuous and (2) asynchronous. Continuous motion conveyors move at a constant velocity v_c along the path. They include belt, roller, skate-wheel, and overhead trolley.

Asynchronous conveyors operate with a stop-and-go motion in which loads, usually contained in carriers (e.g., hooks, baskets, carts), move between stations and then stop and remain at the station until released. Asynchronous handling allows independent movement of each carrier in the system. Examples of this type include overhead power-and-free trolley, in-floor towline, and cart-on-track conveyors. Some roller and skate-wheel conveyors can also be operated asynchronously. Reasons for using asynchronous conveyors include (1) to accumulate loads, (2) to temporarily store items, (3) to allow for differences in production rates between adjacent processing areas, (4) to smooth production when cycle times are variable at stations along the conveyor, and (5) to accommodate different conveyor speeds along the pathway.

Conveyors can also be classified as (1) single direction, (2) continuous loop, and (3) recirculating. In Section 10.3.2, we present equations and techniques with which to analyze these conveyor systems. Single direction conveyors are used to transport loads one way from origination point to destination point, as depicted in Figure 10.10(a). These systems are appropriate when there is no need to move loads in both directions or to return containers or carriers from the unloading stations back to the loading stations. Single direction powered conveyors include roller, skate-wheel, belt, and chain-in-floor types. In addition, all gravity conveyors operate in one direction.

Continuous loop conveyors form a complete circuit, as in Figure 10.10(b). An overhead trolley conveyor is an example of this conveyor type. However, any conveyor type can be configured as a loop, even those previously identified as single direction conveyors, simply by connecting several single direction conveyor sections into a closed loop. A continuous loop system allows materials to be moved between any two stations along the pathway. Continuous loop conveyors are used when loads are moved in carriers (e.g., hooks, baskets) between load and unload stations and the carriers are affixed to the conveyor loop. In this design, the empty carriers are automatically returned from the unload station back to the load station.

The preceding description of a continuous loop conveyor assumes that items loaded at the load station are unloaded at the unload station. There are no loads in the return loop; the purpose of the return loop is simply to send the empty carriers back for reloading. This method of operation overlooks an important opportunity offered by a closed-loop conveyor: to store as well as deliver items. Conveyor systems that allow parts or products to remain on the return loop for one or more revolutions are called *recirculating conveyors*. In providing a storage function, the conveyor system can be used to accumulate parts to smooth out effects of loading and unloading variations at stations in the conveyor. There are two problems that can plague the operation of a recirculating conveyor system. One is that there may be times during the operation of the conveyor that no empty carriers are immediately available at the loading station when needed. The other problem is that no loaded carriers are immediately available at the unloading station when needed.

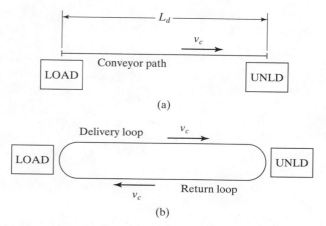

Figure 10.10 (a) Single direction conveyor and (b) continuous loop conveyor.

It is possible to construct branching and merging points into a conveyor track to permit different routing of different loads moving in the system. In nearly all conveyor systems, it is possible to build switches, shuttles, or other mechanisms to achieve these alternate routings. In some systems, a push-pull mechanism or lift-and-carry device is required to actively move the load from the current pathway onto the new pathway.

10.2.5 Cranes and Hoists

The fifth category of transport equipment in material handling is cranes and hoists. Cranes are used for horizontal movement of materials in a facility, and hoists are used for vertical lifting. A crane invariably includes a hoist; thus, the hoist component of the crane lifts the load, and the crane transports the load horizontally to the desired destination. This class of material handling equipment includes cranes capable of lifting and moving very heavy loads, in some cases over 100 tons.

A *hoist* is a mechanical device used to raise and lower loads. As seen in Figure 10.11, a hoist consists of one or more fixed pulleys, one or more moving pulleys, and a rope, cable, or chain strung between the pulleys. A hook or other means for attaching the load is connected to the moving pulley(s). The number of pulleys in the hoist determines its mechanical advantage, which is the ratio of the load weight to the driving force required to lift the weight. The mechanical advantage of the hoist in our illustration is 4.0. The driving force to operate the hoist is applied either manually or by electric or pneumatic motor.

Cranes include a variety of material handling equipment designed for lifting and moving heavy loads using one or more overhead beams for support. Principal types of cranes found in factories include (a) bridge cranes, (b) gantry cranes, and (c) jib cranes. In all three types, at least one hoist is mounted to a trolley that rides on the overhead beam

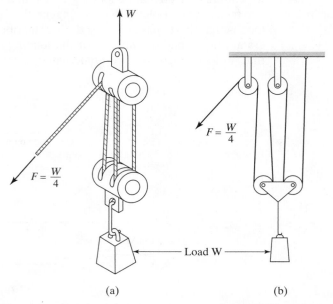

(a)   (b)

Figure 10.11 A hoist with a mechanical advantage of 4.0: (a) sketch of the hoist and (b) diagram to illustrate mechanical advantage.

of the crane. A *bridge crane* consists of one or two horizontal girders or beams suspended between fixed rails on either end which are connected to the structure of the building, as shown in Figure 10.12(a). The hoist trolley can be moved along the length of the bridge, and the bridge can be moved the length of the rails in the building. These two drive capabilities provide motion in the x- and y-axis of the building, and the hoist provides motion in the z-axis direction. Thus the bridge crane achieves vertical lifting due to its hoist and horizontal movement due to its orthogonal rail system. Large bridge cranes have girders that span up to 36.5 m (120 ft) and are capable of carrying loads up to 90,000 kg (100 tons). Large bridge cranes are controlled by operators riding in cabs on the bridge. Applications include heavy machinery fabrication, steel and other metal mills, and power-generating stations.

A *gantry crane* is distinguished from a bridge crane by the presence of one or two vertical legs that support the horizontal bridge. As with the bridge crane, a gantry crane includes one or more hoists that accomplish vertical lifting. Gantries are available in a variety of sizes and capacities, the largest possessing spans of about 46 m (150 ft) and load capacities of 136,000 kg (150 tons). A double gantry crane has two legs. A half gantry crane, Figure 10.12(b), has a single leg on one end of the bridge, and the other end is supported by a rail mounted on the wall or other structural member of a building. A cantilever gantry crane has a bridge that extends beyond the span created by the support legs.

A *jib crane* consists of a hoist supported on a horizontal beam that is cantilevered from a vertical column or wall support, as illustrated in Figure 10.12(c). The horizontal beam pivots about the vertical axis formed by the column or wall to provide a horizontal sweep for the crane. The beam also serves as the track for the hoist trolley to provide radial travel along the length of the beam. Thus, the horizontal area included by a jib crane is circular or semicircular. As with other cranes, the hoist provides vertical lift and lower motions. Standard capacities of jib cranes range up to about 5000 kg. Wall-mounted jib cranes can achieve a swing of about 180°, while a floor-mounted jib crane using a column or post as its vertical support can sweep a full 360°.

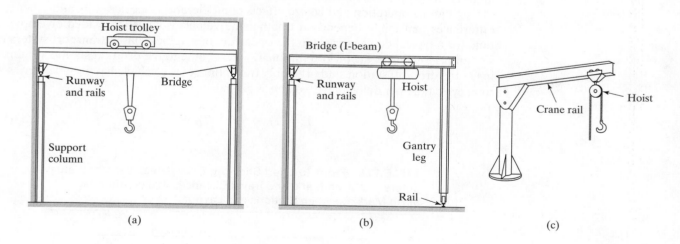

Figure 10.12 Three types of cranes: (a) bridge crane, (b) gantry crane (a half gantry crane is shown), and (c) jib crane.

10.3 ANALYSIS OF MATERIAL TRANSPORT SYSTEMS

Quantitative models are useful for analyzing material flow rates, delivery cycle times, and other aspects of system performance. The analysis may be useful in determining equipment requirements; for example, how many forklift trucks will be required to satisfy a given flow rate specification. Material transport systems can be classified as vehicle-based systems or conveyor systems. Our coverage of the quantitative models is organized along these lines.

10.3.1 Analysis of Vehicle-Based Systems

Equipment used in vehicle-based material transport systems includes industrial trucks (both hand trucks and powered trucks), automated guided vehicles, monorails and other rail-guided vehicles, and certain types of conveyor systems (e.g., in-floor towline conveyors). These systems are commonly used to deliver individual loads between several different origination and destination points. Two graphical tools that are useful for displaying and analyzing data in these deliveries are the from-to chart and the network diagram. The *from-to chart* is a table that can be used to indicate material flow data and distances between multiple locations. Table 10.5 illustrates a from-to chart that lists flow rates and distances between five workstations in a manufacturing system. The left-hand vertical column lists the origination points (loading stations), while the horizontal row at the top identifies the destination locations (unloading stations).

Network diagrams can also be used to indicate the same type of information. A *network diagram* consists of nodes and arrows, and the arrows indicate relationships among the nodes. In material handling, the nodes represent locations (e.g., load and unload stations), and the arrows represent material flows and/or distances between the stations. Figure 10.13 shows a network diagram that provides the same information as Table 10.5.

Mathematical equations can be developed to describe the operation of vehicle-based material transport systems. We assume that the vehicle operates at a constant velocity throughout its operation and ignore effects of acceleration, deceleration, and other speed differences that might depend on whether the vehicle is traveling loaded or empty. The time for a typical delivery cycle in the operation of a vehicle-based transport system consists of (1) loading at the pickup station, (2) travel time to the drop-off station, (3) unloading at the drop-off station, and (4) empty travel time of the vehicle between deliveries. The total cycle time per delivery per vehicle is given by

$$T_c = T_L + \frac{L_d}{v_c} + T_U + \frac{L_e}{v_o} \tag{10.1}$$

TABLE 10.5 From-To Chart Showing Flow Rates, loads/hr (Value Before the Slash Mark) and Travel Distances, (Value After the Slash Mark) Between Stations in a Layout

	To	1	2	3	4	5
From	1	0	9/50	5/120	6/205	0
	2	0	0	0	0	9/80
	3	0	0	0	2/85	3/170
	4	0	0	0	0	8/85
	5	0	0	0	0	0

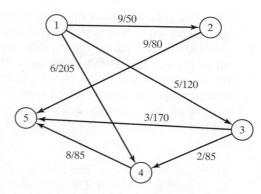

Figure 10.13 Network diagram showing material deliveries between load/unload stations. Nodes represent the load/unload stations, and arrows are labeled with flow rates and distances.

where T_c = delivery cycle time (min/del), T_L = time to load at load station (min), L_d = distance the vehicle travels between load and unload station (m, ft), v_c = carrier velocity (m/min, ft/min), T_U = time to unload at unload station (min), and L_e = distance the vehicle travels empty until the start of the next delivery cycle (m, ft).

The T_c calculated by Eq. (10.1) must be considered an ideal value, because it ignores any time losses due to reliability problems, traffic congestion, and other factors that may slow down a delivery. In addition, not all delivery cycles are the same. Originations and destinations may be different from one delivery to the next, which will affect the L_d and L_e terms in the equation. Accordingly, these terms are considered to be average values for the population of loaded and empty distances traveled by the vehicle during the course of a shift or other period of analysis.

The delivery cycle time can be used to determine certain parameters of interest in the vehicle-based transport system. Let us make use of T_c to determine two parameters: (1) rate of deliveries per vehicle and (2) number of vehicles required to satisfy a specified total delivery requirement. We will base our analysis on hourly rates and requirements; however, the equations can readily be adapted for other periods.

The hourly rate of deliveries per vehicle is 60 min divided by the delivery cycle time T_c, adjusting for any time losses during the hour. The possible time losses include (1) availability, (2) traffic congestion, and (3) efficiency of manual drivers in the case of manually operated trucks. *Availability* (symbolized A) is a reliability factor (Section 3.1.3) defined as the proportion of total shift time that the vehicle is operational and not broken down or being repaired.

To deal with the time losses due to traffic congestion, let us define the *traffic factor F_t* as a parameter for estimating the effect of these losses on system performance. Sources of inefficiency accounted for by the traffic factor include waiting at intersections, blocking of vehicles (as in an AGVS), and waiting in a queue at load/unload stations. If these situations do not occur, then $F_t = 1.0$. As blocking increases, the value of F_t decreases. Waiting at intersections, blocking, and waiting in line at load/unload stations are affected by the number of vehicles in the system relative to the size of the layout. If there is only one vehicle in the system, no blocking should occur, and the traffic factor will be 1.0. For systems with many vehicles, there will be more instances of blocking and congestion, and the traffic factor will take a lower value. Typical values of traffic factor for an AGVS range between 0.85 and 1.0 [4].

For systems based on industrial trucks, including both hand trucks and powered trucks that are operated by human workers, traffic congestion is probably not the main cause of low operating performance. Instead, performance depends primarily on the work efficiency of the operators who drive the trucks. Let us define *worker efficiency* here as the actual work rate of the human operator relative to work rate expected under standard or normal performance. Let E_w symbolize the worker efficiency.

With these factors defined, we can now express the available time per hour per vehicle as 60 min adjusted by A, F_t, and E_w. That is,

$$AT = 60AF_tE_w \tag{10.2}$$

where AT = available time (min/hr per vehicle), A = availability, F_t = traffic factor, and E_w = worker efficiency. The parameters A, F_t, and E_w do not take into account poor vehicle routing, poor guidepath layout, or poor management of the vehicles in the system. These factors should be minimized, but if present they are accounted for in the values of L_d and L_e.

We can now write equations for the two performance parameters of interest. The rate of deliveries per vehicle is given by

$$R_{dv} = \frac{AT}{T_c} \tag{10.3}$$

where R_{dv} = hourly delivery rate per vehicle (del/hr per vehicle), T_c = delivery cycle time computed by Eq. (10.1) (min/del), and AT = the available time in 1 hr with adjustments for time losses (min/hr).

The total number of vehicles (trucks, AGVs, trolleys, carts, etc.) needed to satisfy a specified total delivery schedule R_f in the system can be estimated by first calculating the total workload required and then dividing by the available time per vehicle. Workload is defined as the total amount of work, expressed in terms of time, that must be accomplished by the material transport system in 1 hr. This can be expressed as

$$WL = R_fT_c \tag{10.4}$$

where WL = workload (min/hr), R_f = specified flow rate of total deliveries per hour for the system (del/hr), and T_c = delivery cycle time (min/del). Now the number of vehicles required to accomplish this workload can be written as

$$n_c = \frac{WL}{AT} \tag{10.5}$$

where n_c = number of carriers required, WL = workload (min/hr), and AT = available time per vehicle (min/hr per vehicle). Substituting Eqs. (10.3) and (10.4) into Eq. (10.5) provides an alternative way to determine

$$n_c = \frac{R_f}{R_{dv}} \tag{10.6}$$

where n_c = number of carriers required, R_f = total delivery requirements in the system (del/hr), and R_{dv} = delivery rate per vehicle (del/hr per vehicle). Although the traffic factor accounts for delays experienced by the vehicles, it does not include delays encountered by a load/unload station that must wait for the arrival of a vehicle. Because of the random nature of the load/unload demands, workstations are likely to experience waiting time while vehicles are busy with other deliveries. The preceding equations do not consider this idle time or its impact on operating cost. If station idle time is to be minimized, then more vehicles may be needed than the number indicated by Eqs. (10.5) or (10.6). Mathematical models based on queueing theory are appropriate to analyze this more complex stochastic situation.

EXAMPLE 10.1 Determining Number of Vehicles in an AGVS

Consider the AGVS layout in Figure 10.14. Vehicles travel counterclockwise around the loop to deliver loads from the load station to the unload station. Loading time at the load station = 0.75 min, and unloading time at the unload station = 0.50 min. We are interested in determining how many vehicles are required to satisfy demand for this layout if a total of 40 del/hr must be completed by the AGVS. The following performance parameters are given: vehicle velocity = 50 m/min, availability = 0.95, traffic factor = 0.90, and operator efficiency does not apply, so $E_w = 1.0$. Determine (a) travel distances loaded and empty, (b) ideal delivery cycle time, and (c) number of vehicles required to satisfy the delivery demand.

Solution: (a) Ignoring effects of slightly shorter distances around the curves at corners of the loop, the values of L_d and L_e are readily determined from the layout to be 110 m and 80 m, respectively.

(b) Ideal cycle time per delivery per vehicle is given by Eq. (10.1):

$$T_c = 0.75 + \frac{110}{50} + 0.50 + \frac{80}{50} = 5.05 \text{ min}$$

(c) To determine the number of vehicles required to make 40 del/hr, we compute the workload of the AGVS and the available time per hour per vehicle:

$$WL = 40(5.05) = 202 \text{ min/hr}$$

$$AT = 60(0.95)(0.90)(1.0) = 51.3 \text{ min/hr per vehicle}$$

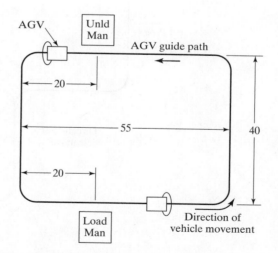

Figure 10.14 AGVS loop layout for Example 10.1. Key: Unld = unload, Man = manual operation, dimensions in meters (m).

Therefore, the number of vehicles required is

$$n_c = \frac{202}{51.3} = 3.94 \text{ vehicles}$$

This value should be rounded up to $n_c = 4$ vehicles, since the number of vehicles must be an integer.

Determining the average travel distances, L_d and L_e, requires analysis of the particular AGVS layout. For a simple loop layout such as in Figure 10.14, determining these values is straightforward. For a complex AGVS layout, the problem is more difficult. The following example illustrates the issue.

EXAMPLE 10.2   Determining L_d for a More-Complex AGVS Layout

The layout for this example is shown in Figure 10.15, and the from-to chart is presented in Table 10.5. The AGVS includes load station 1 where raw parts enter the system for delivery to any of three production stations 2, 3, and 4. Unload station 5 receives finished parts from the production stations. Load

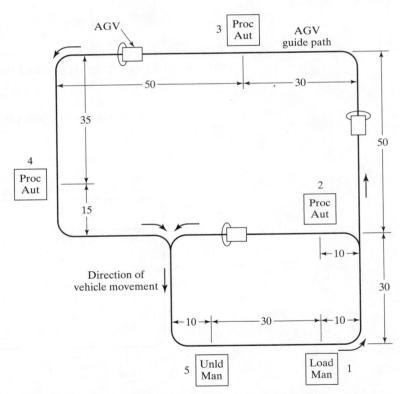

Figure 10.15   AGVS layout for production system of Example 10.2. Key: Proc = processing operation, Aut = automated, Unld = unload, Man = manual operation, dimensions in meters (m).

and unload times at stations 1 and 5 are each 0.5 min. Production rates for each workstation are indicated by the delivery requirements in Table 10.5. A complicating factor is that some parts must be transshipped between stations 3 and 4. Vehicles move in the direction indicated by the arrows in the figure. Determine the average delivery distance, L_d.

Solution. Table 10.5 shows the number of deliveries and corresponding distances between the stations. The distance values are taken from the layout drawing in Figure 10.15. To determine the value of L_d, a weighted average must be calculated based on the number of trips and corresponding distances shown in the from-to chart for the problem:

$$L_d = \frac{9(50) + 5(120) + 6(205) + 9(80) + 2(85) + 3(170) + 8(85)}{9 + 5 + 6 + 9 + 2 + 3 + 8} = \frac{4360}{42} = 103.8 \text{ m}$$

Determining L_e, the average distance a vehicle travels empty during a delivery cycle, is more complicated. It depends on the dispatching and scheduling methods used to decide how a vehicle should proceed from its last drop-off to its next pickup. In Figure 10.15, if each vehicle must travel back to station 1 after each drop-off at stations 2, 3, and 4, then the empty distance between pick-ups would be very long indeed. L_e would be greater than L_d. On the other hand, if a vehicle could exchange a raw workpart for a finished part while stopped at a given workstation, then empty travel time for the vehicle would be minimized. However, this would require a two-position platform at each station to enable the exchange. So this issue must be considered in the initial design of the AGVS. Ideally, L_e should be reduced to zero. It is highly desirable to minimize the average distance a vehicle travels empty through good AGVS design and good scheduling of the vehicles. Our mathematical model of vehicle-based systems indicates that the delivery cycle time will be reduced if L_e is minimized, and this will have a beneficial effect on the vehicle delivery rate and the number of vehicles required to operate the system. Two of our exercise problems at the end of the chapter ask the reader to determine L_e under different operating scenarios.

10.3.2 Conveyor Analysis

Conveyor operations have been analyzed in the research literature (see references [8], [9], [12], [13], [14], and [15]). In our discussion here, we consider the three basic types of conveyor operations discussed in Section 10.2.4: (1) single direction conveyors, (2) continuous loop conveyors, and (3) recirculating conveyors.

Single Direction Conveyors. Consider the case of a single direction powered conveyor with one load station at the upstream end and one unload station at the downstream end, as in Figure 10.10(a). Materials are loaded at one end and unloaded at the other. The materials may be parts, cartons, pallet loads, or other unit loads. Assuming the conveyor operates at a constant speed, the time required to move materials from load station to unload station is given by

$$T_d = \frac{L_d}{v_c} \tag{10.7}$$

where T_d = delivery time (min), L_d = length of conveyor between load and unload stations (m, ft), and v_c = conveyor velocity (m/min, ft/min).

The flow rate of materials on the conveyor is determined by the rate of loading at the load station. The loading rate is limited by the reciprocal of the time required to load the materials. Given the conveyor speed, the loading rate establishes the spacing of materials on the conveyor. Summarizing these relationships,

$$R_f = R_L = \frac{v_c}{s_c} \le \frac{1}{T_L} \tag{10.8}$$

where R_f = material flow rate (parts/min), R_L = loading rate (parts/min), s_c = center-to-center spacing of materials on the conveyor (m/part, ft/part), and T_L = loading time (min/part). One might be tempted to think that the loading rate R_L is the reciprocal of the loading time T_L. However, R_L is set by the flow rate requirement R_f, while T_L is determined by ergonomic factors. The worker who loads the conveyor may be capable of performing the loading task at a rate that is faster than the required flow rate. On the other hand, the flow rate requirement cannot be set faster than it is humanly possible to perform the loading task.

An additional requirement for loading and unloading is that the time required to unload the conveyor must be equal to or less than the reciprocal of material flow rate. That is,

$$T_U \le \frac{1}{R_f} \tag{10.9}$$

where T_U = unloading time (min/part). If unloading requires more time than the time interval between arriving loads, then loads may accumulate or be dumped onto the floor at the downstream end of the conveyor.

We are using parts as the material in Eqs. (10.8) and (10.9), but the relationships apply to other unit loads as well. The advantage of the unit load principle (Section 10.1.2) can be demonstrated by transporting n_p parts in a carrier rather than a single part. Recasting Eq. (10.8) to reflect this advantage, we have

$$R_f = \frac{n_p v_c}{s_c} \le \frac{1}{T_L} \tag{10.10}$$

where R_f = flow rate (parts/min), n_p = number of parts per carrier, s_c = center-to-center spacing of carriers on the conveyor (m/carrier, ft/carrier), and T_L = loading time per carrier (min/carrier). The flow rate of parts transported by the conveyor is potentially much greater in this case. However, loading time is still a limitation, and T_L may consist of not only the time to load the carrier onto the conveyor but also the time to load parts into the carrier. The preceding equations must be interpreted and perhaps adjusted for the given application.

EXAMPLE 10.3 Single Direction Conveyor

A roller conveyor follows a pathway 35 m long between a parts production department and an assembly department. Velocity of the conveyor is 40 m/min. Parts are loaded into large tote pans, which are placed onto the conveyor at the load station in the production department. Two operators work the loading station. The first worker loads parts into tote pans, which takes 25 sec. Each tote pan holds 20 parts. Parts enter the loading station from production at a rate that is in balance with this 25-sec cycle. The second worker loads tote pans onto the conveyor, which takes only 10 sec. Determine: (a) spacing between tote pans along the conveyor, (b) maximum possible flow rate in parts/min, and (c) the minimum time required to unload the tote pan in the assembly department.

Solution: (a) Spacing between tote pans on the conveyor is determined by the loading time. It takes only 10 sec to load a tote pan onto the conveyor, but 25 sec are required to load parts into the tote pan. Therefore, the loading cycle is limited by this 25 sec. At a conveyor speed of 40 m/min, the spacing will be

$$s_c = (25/60 \text{ min})(40 \text{ m/min}) = 16.67 \text{ m}$$

(b) Flow rate is given by Eq. (10.10):

$$R_f = \frac{20(40)}{16.67} = 48 \text{ parts/min}$$

This is consistent with the parts loading rate of 20 parts in 25 sec, which is 0.8 parts/sec or 48 parts/min.

(c) The minimum allowable time to unload a tote pan must be consistent with the flow rate of tote pans on the conveyor. This flow rate is one tote pan every 25 sec, so

$$T_U \leq 25 \text{ sec}$$

Continuous Loop Conveyors. Consider a continuous loop conveyor such as an overhead trolley in which the pathway is formed by an endless chain moving in a track loop, and carriers are suspended from the track and pulled by the chain. The conveyor moves parts in the carriers between a load station and an unload station. The complete loop is divided into two sections: a delivery (forward) loop in which the carriers are loaded and a return loop in which the carriers travel empty, as shown in Figure 10.10(b). The length of the delivery loop is L_d, and the length of the return loop is L_e. Total length of the conveyor is therefore $L = L_d + L_e$. The total time required to travel the complete loop is

$$T_c = \frac{L}{v_c} \tag{10.11}$$

where T_c = total cycle time (min), and v_c = speed of the conveyor chain (m/min, ft/min). The time a load spends in the forward loop is

$$T_d = \frac{L_d}{v_c} \tag{10.12}$$

where T_d = delivery time on the forward loop (min).

Carriers are equally spaced along the chain at a distance s_c apart. Thus, the total number of carriers in the loop is given by

$$n_c = \frac{L}{s_c} \tag{10.13}$$

where n_c = number of carriers, L = total length of the conveyor loop (m, ft), and s_c = center-to-center distance between carriers (m/carrier, ft/carrier). The value of n_c must be an integer, and so L and s_c must be consistent with that requirement.

Each carrier is capable of holding parts on the delivery loop, and it holds no parts on the return trip. Since only those carriers on the forward loop contain parts, the maximum number of parts in the system at any one time is given by

$$\text{Total parts in system} = \frac{n_p n_c L_d}{L} \qquad (10.14)$$

As in the single direction conveyor, the maximum flow rate between load and unload stations is

$$R_f = \frac{n_p v_c}{s_c}$$

where R_f = parts per minute. Again, this rate must be consistent with limitations on the time it takes to load and unload the conveyor, as defined in Eqs. (10.8)–(10.10).

Recirculating Conveyors. Recall the two problems complicating the operation of a recirculating conveyor system (Section 10.2.4): (1) the possibility that no empty carriers are immediately available at the loading station when needed and (2) the possibility that no loaded carriers are immediately available at the unloading station when needed. The case of a recirculating conveyor with one load station and one unload station was analyzed by Kwo [8], [9]. According to his analysis, three basic principles must be obeyed in designing such a conveyor system:

1. *Speed Rule.* The operating speed of the conveyor must be within a certain range. The lower limit of the range is determined by the required loading and unloading rates at the respective stations. These rates are dictated by the external systems served by the conveyor. Let R_L and R_U represent the required loading and unloading rates at the two stations, respectively. Then the conveyor speed must satisfy the relationship

$$\frac{n_p v_c}{s_c} \geq \text{Max}\{R_L, R_U\} \qquad (10.15)$$

   where R_L = required loading rate (parts/min), and R_U = the corresponding unloading rate. The upper speed limit is determined by the physical capabilities of the material handlers to perform the loading and unloading tasks. Their capabilities are defined by the time required to load and unload the carriers, so that

$$\frac{v_c}{s_c} \leq \text{Min}\left\{\frac{1}{T_L}, \frac{1}{T_U}\right\} \qquad (10.16)$$

   where T_L = time required to load a carrier (min/carrier), and T_U = time required to unload a carrier. In addition to Eqs. (10.15) and (10.16), another limitation is of course that the speed must not exceed the physical limits of the mechanical conveyor itself.

2. *Capacity Constraint.* The flow rate capacity of the conveyor system must be at least equal to the flow rate requirement to accommodate reserve stock and allow for the time elapsed between loading and unloading due to delivery distance. This can be expressed as follows:

$$\frac{n_p v_c}{s_c} \geq R_f \qquad (10.17)$$

   In this case, R_f must be interpreted as a system specification required of the recirculating conveyor.

3. *Uniformity Principle.* This principle states that parts (loads) should be uniformly distributed throughout the length of the conveyor, so that there will be no sections

of the conveyor in which every carrier is full while other sections are virtually empty. The reason for the uniformity principle is to avoid unusually long waiting times at the load or unload stations for empty or full carriers (respectively) to arrive.

EXAMPLE 10.4   Recirculating Conveyor Analysis: Kwo

A recirculating conveyor has a total length of 300 m. Its speed is 60 m/min, and the spacing of part carriers along its length is 12 m. Each carrier can hold two parts. The task time required to load two parts into each carrier is 0.20 min and the unload time is the same. The required loading and unloading rates are both defined by the specified flow rate, which is 4 parts/min. Evaluate the conveyor system design with respect to Kwo's three principles.

Solution: *Speed Rule:* The lower limit on speed is set by the required loading and unloading rates, which is 4 parts/min. Checking this against Eq. (10.15),

$$\frac{n_p v_c}{s_c} \geq \text{Max}\{R_L, R_U\}$$

$$\frac{(2 \text{ parts/carrier})(60 \text{ m/min})}{12 \text{ m/carrier}} = 10 \text{ parts/min} > 4 \text{ parts/min}$$

Checking the lower limit,

$$\frac{60 \text{ m/min}}{12 \text{ m/carrier}} = 5 \text{ carriers/min} \leq \text{Min}\left\{\frac{1}{0.2}, \frac{1}{0.2}\right\} = \text{Min}\{5, 5\} = 5$$

The Speed Rule is satisfied.

Capacity Constraint: The conveyor flow rate capacity = 10 parts/min as computed above. Since this is substantially greater than the required delivery rate of 4 parts/min, the capacity constraint is satisfied. Kwo provides guidelines for determining the flow rate requirement that should be compared to the conveyor capacity.

Uniformity Principle: The conveyor is assumed to be uniformly loaded throughout its length, since the loading and unloading rates are equal and the flow rate capacity is substantially greater than the load/unload rate. Conditions for checking the uniformity principle are available; the reader is referred to the original papers by Kwo [8], [9].

REFERENCES

[1]   BOSE, P. P., "Basics of AGV Systems," Special Report 784, *American Machinist and Automated Manufacturing*, March 1986, pp. 105–122.

[2]   CASTELBERRY, G., *The AGV Handbook*, AGV Decisions, Inc., published by Braun-Brumfield, Inc., Ann Arbor, MI, 1991.

[3]   EASTMAN, R. M., *Materials Handling*, Marcel Dekker, Inc., NY, 1987.

[4]   FITZGERALD, K. R., "How to Estimate the Number of AGVs You Need," *Modern Materials Handling*, October 1985, p. 79.

[5]   KULWIEC, R. A., *Basics of Material Handling*, Material Handling Institute, Pittsburgh, PA, 1981.

[6]   KULWIEC, R. A., Editor, *Materials Handling Handbook*, 2nd Edition, John Wiley & Sons, Inc., NY, 1985.

[7]   KULWEC, R., "Cranes for Overhead Handling," *Modern Materials Handling*, July 1998, pp. 43–47.

[8]   KWO, T. T., "A Theory of Conveyors," *Management Science*, Vol. 5, No. 1, 1958, pp. 51–71.

[9]   KWO, T. T., "A Method for Designing Irreversible Overhead Loop Conveyors," *Journal of Industrial Engineering*, Vol. 11, No. 6, 1960, pp. 459–466.

[10] Material Handling Industry, *Annual Report*, Charlotte, NC, 2006.

[11] MILLER, R. K., *Automated Guided Vehicle Systems*, Co-published by SEAI Institute, Madison, GA and Technical Insights, Fort Lee, NJ, 1983.

[12] MUTH, E. J., "Analysis of Closed-Loop Conveyor Systems," *AIIE Transactions*, Vol. 4, No. 2, 1972, pp. 134–143.

[13] MUTH, E. J., "Analysis of Closed-Loop Conveyor Systems: the Discrete Flow Case," *AIIE Transactions*, Vol. 6, No. 1, 1974, pp. 73–83.

[14] MUTH, E. J., "Modelling and Analysis of Multistation Closed-Loop Conveyors," *International Journal of Production Research*, Vol. 13, No. 6, 1975, pp. 559–566.

[15] MUTH, E. J., and J. A. WHITE, "Conveyor Theory: A Survey," *AIIE Transactions*, Vol. 11, No. 4, 1979, pp. 270–277.

[16] MUTHER, R., AND K. HAGANAS, *Systematic Handling Analysis*, Management and Industrial Research Publications, Kansas City, MO, 1969.

[17] TOMPKINS, J. A., J. A. WHITE, Y. A. BOZER, E. H. FRAZELLE, J. M. TANCHOCO, AND J. TREVINO, *Facilities Planning*, 3d ed., John Wiley & Sons, Inc., NY, 2003.

[18] WITT, C. E., "Palletizing Unit Loads: Many Options," *Material Handling Engineering*, January, 1999, pp 99–106.

[19] ZOLLINGER, H. A., "Methodology to Concept Horizontal Transportation Problem Solutions," paper presented at the *MHI 1994 International Research Colloquium*, Grand Rapids, MI, June 1994.

REVIEW QUESTIONS

10.1   Provide a definition of material handling.

10.2   How does material handling fit within the scope of logistics?

10.3   Name the four major categories of material handling equipment.

10.4   What is included within the term *unitizing equipment*?

10.5   What is the unit load principle?

10.6   What are the five categories of material transport equipment commonly used to move parts and materials inside a facility?

10.7   Give some examples of industrial trucks used in material handling.

10.8   What is an automated guided vehicle system (AGVS)?

10.9   Name three categories of automated guided vehicles.

10.10 What features distinguish self-guided vehicles from conventional AGVs?

10.11 What is forward sensing in AGVS terminology?

10.12 What are some of the differences between rail-guided vehicles and automated guided vehicles?

10.13 What is a conveyor?

10.14 Name some of the different types of conveyors used in industry.

10.15 What is a recirculating conveyor?

10.16 What is the difference between a hoist and a crane?

PROBLEMS

Analysis of Vehicle-based Systems

10.1 A flexible manufacturing system is being planned. It has a ladder layout as pictured in Figure P10.1 and uses a rail-guided vehicle system to move parts between stations in the layout. All workparts are loaded into the system at station 1, moved to one of three processing stations (2, 3, or 4), and then brought back to station 1 for unloading. Once loaded onto its RGV, each workpart stays on the vehicle throughout its time in the system. Load and unload times at station 1 are each 1.0 min. Processing times at other stations are 5.0 min at station 2, 7.0 min at station 3, and 9.0 min at station 4. Hourly production of parts through the system is seven parts through station 2, six parts through station 3, and five parts through station 4. (a) Develop the from-to chart for trips and distances using the same format as Table 10.5. (b) Develop the network diagram for this data similar to Figure 10.13. (c) Determine the number of rail-guided vehicles that are needed to meet the requirements of the flexible manufacturing system, if vehicle speed = 60 m/min and the anticipated traffic factor = 0.85. Assume reliability = 100%.

10.2 In Example 10.2 in the text, suppose that the vehicles operate according to the following scheduling rules: (1) vehicles delivering raw workparts from station 1 to stations 2, 3, and 4 must return empty to station 5, and (2) vehicles picking up finished parts at stations 2, 3, and 4 for delivery to station 5 must travel empty from station 1. (a) Determine the empty travel distances associated with each delivery and develop a from-to chart following the format of Table 10.5 in the text. (b) Suppose the AGVs travel at a speed of 50 m/min, and the traffic factor = 0.90. Assume reliability = 100%. As determined in Example 10.2, the delivery distance L_d = 103.8 m. Determine the value of L_e for the layout based on your table. (c) How many automated guided vehicles will be required to operate the system?

10.3 In Example 10.2 in the text, suppose that in order to minimize the distances the vehicles travel empty, vehicles delivering raw workparts from station 1 to stations 2, 3, and 4 must pick up finished parts at these respective stations for delivery to station 5. (a) Determine the empty travel distances associated with each delivery and develop a from-to chart following the format of Table 10.5 in the text. (b) Suppose the AGVs travel at a speed of 50 m/min, and the traffic factor = 0.90. Assume reliability = 100%. As determined in Example 10.2, the delivery distance L_d = 103.8 m. Determine the value of L_e for the layout based on your table. (c) How many automated guided vehicles will be required to operate the system?

10.4 A planned fleet of forklift trucks has an average travel distance per delivery = 500 ft loaded and an average empty travel distance = 350 ft. The fleet must make a total of 60

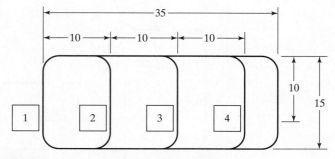

Figure P10.1 FMS layout for Problem 10.1.

deliveries per hour. Load and unload times are each 0.5 min and the speed of the vehicles = 300 ft/min. The traffic factor for the system = 0.85. Availability = 0.95, and worker efficiency = 90%. Determine (a) ideal cycle time per delivery, (b) the resulting average number of deliveries per hour that a forklift truck can make, and (c) the number of trucks required to accomplish the 60 deliveries per hour.

10.5 An automated guided vehicle system has an average travel distance per delivery = 200 m and an average empty travel distance = 150 m. Load and unload times are each 24 s and the speed of the AGV = 1 m/s. Traffic factor = 0.9. How many vehicles are needed to satisfy a delivery requirement of 30 deliveries/hour? Assume that availability = 0.95.

10.6 Four forklift trucks are used to deliver pallet loads of parts between work cells in a factory. Average travel distance loaded is 350 ft and the travel distance empty is estimated to be the same. The trucks are driven at an average speed of 3 miles/hr when loaded and 4 miles/hr when empty. Terminal time per delivery averages 1.0 min (load = 0.5 min and unload = 0.5 min). If the traffic factor is assumed to be 0.90, availability = 100%, and worker efficiency = 0.95, what is the maximum hourly delivery rate of the four trucks?

10.7 An AGVS has an average loaded travel distance per delivery = 400 ft. The average empty travel distance is not known. Required number of deliveries per hour = 60. Load and unload times are each 0.6 min and the AGV speed = 125 ft/min. Anticipated traffic factor = 0.85 and availability = 0.95. Develop an equation that relates the number of vehicles required to operate the system as a function of the average empty travel distance L_e.

10.8 A rail-guided vehicle system is being planned as part of an assembly cell. The system consists of two parallel lines, as in Figure P10.8. In operation, a base part is loaded at station 1 and delivered to either station 2 or 4, where components are added to the base part. The RGV then goes to either station 3 or 5, respectively, where further assembly of components is accomplished. From stations 3 or 5, the product moves to station 6 for removal from the system. Vehicles remain with the products as they move through the station sequence; thus, there is no loading and unloading of parts at stations 2, 3, 4, and 5. After unloading parts at station 6, the vehicles then travel empty back to station 1 for reloading. The hourly moves (parts/hr) and distances (ft) are listed in the table below. RGV speed = 100 ft/min. Assembly cycle times at stations 2 and 3 = 4.0 min each, and at stations 4 and 5 = 6.0 min each. Load and unload times at stations 1 and 6 respectively are each 0.75 min. Traffic factor = 1.0 and availability = 1.0. How many vehicles are required to operate the system?

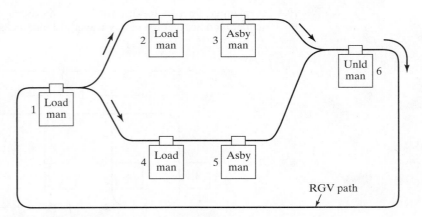

Figure P10.8 Layout for Problem 10.8.

	To	1	2	3	4	5	6
From	1	0/0	14L/200	0/NA	9L/150	0/NA	0/NA
	2	0/NA	0/0	14L/50	0/NA	0/NA	0/NA
	3	0/NA	0/NA	0/0	0/NA	0/NA	14L/50
	4	0/NA	0/NA	0/NA	0/0	9L/50	0/NA
	5	0/NA	0/NA	0/NA	0/NA	0/0	9L/100
	6	23E/400	0/NA	0/NA	0/NA	0/NA	0/0

10.9 An AGVS will be used to satisfy material flows indicated in the from-to chart in the table below, which shows deliveries per hour between stations (before the slash) and distances in meters between stations (after the slash). Moves indicated by "L" are trips in which the vehicle is loaded, while "E" indicates moves in which the vehicle is empty. It is assumed that availability = 0.90, traffic factor = 0.85, and efficiency = 1.0. Speed of an AGV = 0.9 m/s. If load handling time per delivery cycle = 1.0 min, determine the number of vehicles needed to satisfy the indicated deliveries per hour. Assume that availability = 0.90.

	To	1	2	3	4
From	1	0/0	9L/90	7L/120	5L/75
	2	5E/90	0/0	0/NA	4L/80
	3	7E/120	0/NA	0/0	0/NA
	4	9E/75	0/NA	0/NA	0/0

10.10 An automated guided vehicle system is being proposed to deliver parts between 40 workstations in a factory. Loads must be moved from each station about once every hour; thus, the delivery rate = 40 loads per hour. Average travel distance loaded is estimated to be 250 ft and travel distance empty is estimated to be 300 ft. Vehicles move at a speed = 200 ft/min. Total handling time per delivery = 1.5 min (load = 0.75 min and unload = 0.75 min). Traffic factor F_t becomes increasingly significant as the number of vehicles n_c increases; this can be modeled as:

$$F_t = 1.0 - 0.05 \, (n_c - 1) \qquad \text{for } n_c = \text{Integer} > 0$$

Determine the minimum number of vehicles needed in the factory to meet the flow rate requirement. Assume that availability = 1.0 and worker efficiency = 1.0.

10.11 An automated guided vehicle system is being planned for a warehouse complex. The AGVS will be a driverless train system, and each train will consist of the towing vehicle plus four carts. Speed of the trains will be 160 ft/min. Only the pulled carts carry loads. The average loaded travel distance per delivery cycle is 2000 ft and empty travel distance is the same. Anticipated travel factor = 0.95. Assume reliability = 1.0. The load handling time per train per delivery is expected to be 10 min. If the requirements on the AGVS are 25 cart loads per hour, determine the number of trains required.

10.12 The from-to chart in the table below indicates the number of loads moved per 8-hour day (before the slash) and the distances in ft (after the slash) between departments in a particular factory. Forklift trucks are used to transport materials between departments. They move at an average speed = 275 ft/min (loaded) and 350 ft/min (empty). Load handling time per delivery is 1.5 min, and anticipated traffic factor = 0.9. Assume reliability = 1.0 and worker efficiency = 110%. Use an availability factor = 95% and worker efficiency = 110%. Determine the number of trucks required under each of the following assumptions: (a) the trucks never travel empty; and (b) the trucks travel empty a distance equal to their loaded distance.

To Dept.	A	B	C	D	E
From Dept A	–	62/500	51/450	45/350	0
B	0	–	0	22/400	0
C	0	0	–	0	76/200
D	0	0	0	–	65/150
E	0	0	0	0	–

10.13 A warehouse consists of five aisles of racks (racks on both sides of each aisle) and a loading dock. The rack system is four levels high. Forklift trucks are used to transport loads between the loading dock and the storage compartments of the rack system in each aisle. The trucks move at an average speed = 140 m/min (loaded) and 180 m/min (empty). Load handling time (loading plus unloading) per delivery totals 1.0 min per storage/retrieval delivery on average, and the anticipated traffic factor = 0.90. Worker efficiency = 100% and vehicle reliability (availability) = 96%. The average distance between the loading dock and the centers of aisles 1 through 5 are 200 m, 300 m, 400 m, 500 m, and 600 m, respectively. These values are to be used to compute travel times. The desired rate of storage/retrieval deliveries is 100 per hour, distributed evenly among the five aisles, and the trucks perform either storage or retrieval deliveries, but not both in one delivery cycle. Determine the number of forklift trucks required to achieve the 100 deliveries per hour.

10.14 Suppose the warehouse in the preceding problem were organized according to a class-based dedicated storage strategy based on activity level of the pallet loads in storage, so that aisles 1 and 2 accounted for 70% of the deliveries (class A) and aisles 3, 4, and 5 accounted for the remaining 30% (class B). Assume that deliveries in class A are evenly divided between aisles 1 and 2, and that deliveries in class B are evenly divided between aisles 3, 4, and 5. How many trucks would be required to achieve 100 storage/retrieval deliveries per hour?

10.15 Major appliances are assembled on a production line at the rate of 55 per hour. The products are moved along the line on work pallets (one product per pallet). At the final workstation the finished products are removed from the pallets. The pallets are then removed from the line and delivered back to the front of the line for reuse. Automated guided vehicles are used to transport the pallets to the front of the line, a distance of 600 ft. Return trip distance (empty) to the end of the line is also 600 ft. Each AGV carries four pallets and travels at a speed of 150 ft/min (either loaded or empty). The pallets form queues at each end of the line, so that neither the production line nor an AGV is ever starved for pallets. Time required to load each pallet onto an AGV = 15 sec; time to release a loaded AGV and move an empty AGV into position for loading at the end of the line = 12 sec. The same times apply for pallet handling and release/positioning at the unload station located at the front of the production line. Assume the traffic factor is 1.0 since the route is a simple loop. How many vehicles are needed to operate the AGV system?

10.16 For the production line in the previous problem, assume that a single AGV train consisting of a tractor and multiple trailers is used to make deliveries rather than separate vehicles. Time required to load a pallet onto a trailer = 15 sec; and the time to release a loaded train and move an empty train into position for loading at the end of the production line = 30 sec. The same times apply for pallet handling and release/positioning at the unload station located at the front of the production line. If each trailer is capable of carrying four pallets, how many trailers should be included in the train?

10.17 An AGVS will be implemented to deliver loads between four workstations: A, B, C, and D. The hourly flow rates (loads/hr) and distances (m) within the system are given in the table below (travel loaded denoted by "L" and travel empty denoted by "E"). Load and unload times are each 0.45 min, and travel speed of each vehicle is 1.4 m/sec. A total of 43 loads enter the system at station A, and 30 loads exit the system at station A. In addition, six loads exit the system from workstation B each hour and seven loads exit the system from station

D. This is why there are a total of 13 empty trips made by the vehicles within the AGVS. How many vehicles are required to satisfy these delivery requirements, assuming the traffic factor is 0.85 and the reliability (availability) is 95%?

		Hourly rate (loads/hr)			
	To	A	B	C	D
From	A	–	18L	10L	15L
	B	6E		12L	
	C			–	22L
	D	30L, 7E			–

	Distances (m)			
	A	B	C	D
A	–	95	80	150
B		–	65	75
C			–	80
D				–

Analysis of Conveyor Systems

10.18 An overhead trolley conveyor is configured as a continuous closed loop. The delivery loop has a length of 120 m and the return loop = 80 m. All parts loaded at the load station are unloaded at the unload station. Each hook on the conveyor can hold one part and the hooks are separated by 4 m. Conveyor speed = 1.25 m/s. Determine (a) maximum number of parts in the conveyor system, (b) parts flow rate, and (c) maximum loading and unloading times that are compatible with the operation of the conveyor system.

10.19 A 300 ft long roller conveyor, which operates at a velocity = 80 ft/min, is used to move pallets between load and unload stations. Each pallet carries 12 parts. Cycle time to load a pallet is 15 sec and one worker at the load station is able to load pallets at the rate of 4 per min. It takes 12 sec to unload at the unload station. Determine (a) center-to-center distance between pallets, (b) the number of pallets on the conveyor at one time, and (c) hourly flow rate of parts. (d) By how much must conveyor speed be increased to increase flow rate to 3000 parts/hour.

10.20 A roller conveyor moves tote pans in one direction at 150 ft/min between a load station and an unload station, a distance of 200 ft. With one worker, the time to load parts into a tote pan at the load station is 3 sec per part. Each tote pan holds eight parts. In addition, it takes 9 sec to load a tote pan onto the conveyor. Determine (a) spacing between tote pan centers flowing in the conveyor system and (b) flow rate of parts on the conveyor system. (c) Consider the effect of the unit load principle. Suppose the tote pans were smaller and could hold only one part rather than eight. Determine the flow rate in this case if it takes 7 sec to load a tote pan onto the conveyor (instead of 9 sec for the larger tote pan), and it takes the same 3 sec to load the part into the tote pan.

10.21 A closed loop overhead conveyor must be designed to deliver parts from one load station to one unload station. The specified flow rate of parts that must be delivered between the two stations is 300 parts per hour. The conveyor has carriers, each holding one part. Forward and return loops will each be 90 m long. Conveyor speed = 0.5 m/s. Times to load and unload parts at the respective stations are each = 12 s. Is the system feasible and if so, what is the appropriate number of carriers and center-to-center spacing between carriers that will achieve the specified flow rate?

10.22 Consider the previous problem, only the carriers are larger and capable of holding up to four parts (n_p = 1, 2, 3, or 4). The loading time $T_L = 9 + 3n_p$, where T_L is in seconds. With other parameters defined as in the previous problem, determine which of the four values of n_p are feasible. For those values that are feasible, specify the appropriate design parameters for (a) spacing between carriers and (b) number of carriers that will achieve this flow rate.

10.23 A recirculating conveyor has a total length of 700 ft and a speed of 90 ft/min. Spacing of part carriers = 14 ft. Each carrier can hold one part. Automatic machines load and unload the conveyor at the load and unload stations. Time to load a part is 0.10 min and unload time is the same. To satisfy production requirements, the loading and unloading rates are each 2.0 parts per min. Evaluate the conveyor system design with respect to the three principles developed by Kwo.

10.24 A recirculating conveyor has a total length of 200 m and a speed of 50 m/min. Spacing of part carriers = 5 m. Each carrier holds two parts. Time needed to load a part carrier = 0.15 min. Unloading time is the same. The required loading and unloading rates are 6 parts per min. Evaluate the conveyor system design with respect to the three Kwo principles.

10.25 There is a plan to install a continuous loop conveyor system with a total length of 1000 ft and a speed of 50 ft/min. The conveyor will have carriers that are separated by 25 ft. Each carrier will be capable of holding one part. A load station and an unload station are to be located 500 ft apart along the conveyor loop. Each day, starting empty, the load station will load parts at the rate of one part every 30 seconds, continuing this loading operation for 10 min, then resting for 10 min during which no loading occurs. It will repeat this 20 min cycle of loading and then resting throughout the 8-hour shift. The unload station will wait until loaded carriers begin to arrive, then will unload parts at the rate of one part every minute during the eight hours, continuing until all carriers are empty. Will the planned conveyor system work? Present calculations and arguments to justify your answer.

Storage Systems

CHAPTER CONTENTS

The function of a material storage system is to store materials for a period of time and to permit access to those materials when required. Materials stored by manufacturing firms include a variety of types, as indicated in Table 11.1. Categories (1)–(5) relate directly to the product, (6)–(8) relate to the process, and (9) and (10) relate to overall support of factory operations. The different categories of materials require different storage methods and controls. Many production plants use manual methods for storing and retrieving items. The storage function is often accomplished inefficiently, in terms of human resources, factory floor space, and material control. Automated methods are available to improve the efficiency of the storage function.

In this chapter, we describe the types of storage equipment and methods, dividing them into conventional and automated types. The final section presents a quantitative analysis of automated storage systems, with emphasis on two important performance measures: storage capacity and throughput.

From Chapter 11 of *Automation, Production Systems, and Computer-Integrated Manufacturing*, Third Edition.
Mikell P. Groover. Copyright © 2008 by Pearson Education, Inc. Publishing as Prentice Hall. All rights reserved.

TABLE 11.1 Types of Materials Typically Stored in a Factory

Type	Description
1. Raw materials	Raw stock to be processed (e.g., bar stock, sheet metal, plastic molding compound)
2. Purchased parts	Parts from vendors to be processed or assembled (e.g., castings, purchased components)
3. Work-in-process	Partially completed parts between processing operations or parts awaiting assembly
4. Finished product	Completed product ready for shipment
5. Rework and scrap	Parts that do not meet specifications, either to be reworked or scrapped
6. Refuse	Chips, swarf, oils, other waste products left over after processing; these materials must be disposed of, sometimes using special precautions
7. Tooling	Cutting tools, jigs, fixtures, molds, dies, welding wire, and other tooling used in manufacturing and assembly; supplies such as helmets, gloves, etc.
8. Spare parts	Parts needed for maintenance and repair of factory equipment
9. Office supplies	Paper, paper forms, writing instruments, and other items used in support of plant office
10. Plant records	Records on product, equipment, and personnel

11.1 STORAGE SYSTEM PERFORMANCE AND LOCATION STRATEGIES

Before describing the storage methods and equipment, let us describe certain terms and operating characteristics related to storage systems. Our coverage is organized into the following topics: (1) storage system performance and (2) storage location strategies.

11.1.1 Storage System Performance

The performance of a storage system in accomplishing its function must be sufficient to justify its investment and operating expense. Various measures used to assess the performance of a storage system include (1) storage capacity, (2) storage density, (3) accessibility, and (4) throughput. In addition, standard measures used for mechanized and automated systems include (5) utilization and (6) reliability.

Storage capacity can be defined and measured in two ways: (1) as the total volumetric space available or (2) as the total number of storage compartments in the system available to hold items or loads. In many storage systems, materials are stored in unit loads that are held in standard size containers (pallets, tote pans, or other containers). The standard container can readily be handled, transported, and stored by the storage system and by the material transport system that may be connected to it. Hence, storage capacity is conveniently measured as the number of unit loads that can be stored in the system. The physical capacity of the storage system should be greater than the maximum number of loads anticipated to be stored, to provide available empty spaces for materials entering the system and to allow for variations in maximum storage requirements.

Storage density is defined as the volumetric space available for actual storage relative to the total volumetric space in the storage facility. In many warehouses, aisle space and wasted overhead space account for more volume than the volume available for actual storage of materials. Floor area is sometimes used to assess storage density, because it is convenient to measure this on a floor plan of the facility. However, volumetric density is usually a more appropriate measure than area density.

For efficient use of space, the storage system should be designed to achieve a high density. However, as storage density is increased, accessibility, another important measure of storage performance, is adversely affected. *Accessibility* refers to the capability to access any desired item or load stored in the system. In the design of a given storage system, appropriate tradeoffs must be made between storage density and accessibility.

System throughput is defined as the hourly rate at which the storage system (1) receives and puts loads into storage and/or (2) retrieves and delivers loads to the output station. In many factory and warehouse operations, there are certain periods of the day when the required rate of storage and/or retrieval transactions is greater than at other times. The storage system must be designed for the maximum throughput that will be required during the day.

System throughput is limited by the time to perform a storage or retrieval (S/R) transaction. A typical storage transaction consists of the following elements: (1) pick up load at input station, (2) travel to storage location, (3) place load in storage location, and (4) travel back to input station. A retrieval transaction consists of: (1) travel to storage location, (2) pick up item from storage, (3) travel to output station, and (4) unload at output station. Each element takes time. The sum of the element times is the transaction time that determines throughput of the storage system. Throughput can sometimes be increased by combining storage and retrieval transactions in one cycle, thus reducing travel time; this is called a *dual command cycle*. When either a storage or a retrieval transaction alone is performed in the cycle, it is called a *single command cycle*. The ability to perform dual command cycles rather than single command cycles depends on demand and scheduling issues. If, during a certain portion of the day, there is demand for only storage transactions and no retrievals, then it is not possible to include both types of transactions in the same cycle. If both transaction types are required, then greater throughput will be achieved by scheduling dual command cycles. This scheduling is more readily done by a computerized (automated) storage system than by one controlled manually.

There are variations in the way a storage/retrieval cycle is performed, depending on the type of storage system. In manually operated systems, time is often lost looking up the storage location of the item being stored or retrieved. On the other hand, manual systems can achieve greater efficiency by combining multiple storage and/or retrieval transactions in one cycle, thus reducing time traveling to and from the input/output station. Element times are subject to the variations and motivations of human workers, and there is a lack of control over the operations.

Two additional performance measures applicable to mechanized and automated storage systems are utilization and availability. *Utilization* is defined as the proportion of time that the system is actually being used for performing S/R operations compared with the time it is available. Utilization varies throughout the day, as requirements change from hour to hour. It is desirable to design an automated storage system for relatively high utilization, in the range 80–90%. If utilization is too low, then the system is probably overdesigned. If utilization is too high, then there is no allowance for rush periods or system breakdowns.

Availability is a measure of system reliability, defined as the proportion of time that the system is capable of operating (not broken down) compared with the normally scheduled shift hours (Section 3.1.3). Malfunctions and failures of the equipment cause downtime. Reasons for downtime include computer failures, mechanical breakdowns, load jams, improper maintenance, and incorrect procedures by personnel using the system. The reliability of an existing system can be improved by following good preventive maintenance procedures and by having repair parts on hand for critical components. Backup procedures should be devised to mitigate the effects of system downtime.

11.1.2 Storage Location Strategies

Several strategies can be used to organize stock in a storage system. These storage location strategies affect the performance measures discussed above. The two basic strategies are (1) randomized storage and (2) dedicated storage. Let us explain these strategies as they are commonly applied in warehousing operations. Each item type stored in a warehouse is known as a *stock-keeping-unit* (SKU). The SKU uniquely identifies that item type. The inventory records of the storage facility maintain a count of the quantities of each SKU that are in storage.

In randomized storage, items are stored in any available location in the storage system. In the usual implementation of randomized storage, incoming items are placed into storage in the nearest available open location. When an order is received for a given SKU, the stock is retrieved from storage according to a first-in-first-out policy so that the items held in storage the longest are used to make up the order.

In dedicated storage, SKUs are assigned to specific locations in the storage facility. This means that locations are reserved for all SKUs stored in the system, and so the number of storage locations for each SKU must be sufficient to accommodate its maximum inventory level. The basis for specifying the storage locations is usually one of the following: (1) items are stored in part number or product number sequence; (2) items are stored according to activity level, the more active SKUs being located closer to the input/output station; or (3) items are stored according to their activity-to-space ratios, the higher ratios being located closer to the input/output station.

When comparing the benefits of the two strategies, it is generally found that less total space is required in a storage system that uses randomized storage, but higher throughput rates can usually be achieved when a dedicated storage strategy is implemented based on activity level. Example 11.1 illustrates the storage density advantage of randomized storage.

EXAMPLE 11.1   Comparison of Storage Strategies

Suppose that a total of 50 SKUs must be stored in a storage system. For each SKU, average order quantity = 100 cartons, average depletion rate = 2 cartons/day, and safety stock level = 10 cartons. Each carton requires one storage location in the system. Based on this data, each SKU has an inventory cycle that lasts 50 days. Since there are 50 SKUs in all, management has scheduled incoming orders so that a different SKU arrives each day. Determine the number of storage locations required in the system under two alternative strategies: (a) randomized storage and (b) dedicated storage.

Solution: Our estimates of space requirements are based on average order quantities and other values in the problem statement. Let us first calculate the maximum inventory level and average inventory level for each SKU. The inventory for each SKU varies over time as shown in Figure 11.1. The maximum inventory

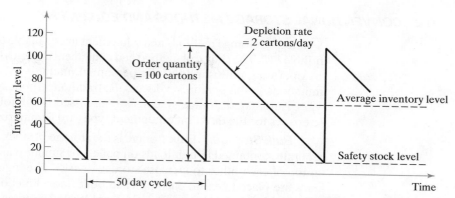

Figure 11.1 Inventory level as a function of time for each SKU in Example 11.1.

level, which occurs just after an order has been received, is the sum of the order quantity and safety stock level:

$$\text{Maximum inventory level} = 100 + 10 = 110 \text{ cartons}$$

The average inventory is the average of the maximum and minimum inventory levels under the assumption of uniform depletion rate. The minimum value occurs just before an order is received when the inventory is depleted to the safety stock level:

$$\text{Minimum inventory level} = 10 \text{ cartons}$$

$$\text{Average inventory level} = (110 + 10)/2 = 60 \text{ cartons}$$

(a) Under a randomized storage strategy, the number of locations required for each SKU is equal to the average inventory level of the item, since incoming orders are scheduled each day throughout the 50-day cycle. This means that when the inventory level of one SKU near the beginning of its cycle is high, the level for another SKU near the end of its cycle is low. Thus, the number of storage locations required in the system is

$$\text{Number of storage locations} = (50 \text{ SKUs})(60 \text{ cartons}) = 3{,}000 \text{ locations}$$

(b) Under a dedicated storage strategy, the number of locations required for each SKU must equal its maximum inventory level. Thus, the number of storage locations required in the system is

$$\text{Number of storage locations} = (50 \text{ SKUs})(110 \text{ cartons}) = 5{,}500 \text{ locations}$$

Some of the advantages of both storage strategies can be obtained in a class-based dedicated storage allocation, in which the storage system is divided into several classes according to activity level, and a randomized storage strategy is used within each class. The classes containing more active SKUs are located closer to the input/output point of the storage system for increased throughput, and the randomized locations within the classes reduce the total number of storage compartments required. We examine the effect of class-based dedicated storage on throughput in Example 11.4 and several of our end-of-chapter problems.

11.2 CONVENTIONAL STORAGE METHODS AND EQUIPMENT

A variety of storage methods and equipment are available to store the various materials listed in Table 11.1. The choice of method and equipment depends largely on the material to be stored, the operating philosophy of the personnel managing the storage facility, and budgetary limitations. In this section, we discuss the traditional (nonautomated) methods and equipment types. Automated storage systems are discussed in the following section. Application characteristics for the different equipment types are summarized in Table 11.2.

Bulk Storage. Bulk storage is the storage of stock in an open floor area. The stock is generally contained in unit loads on pallets or similar containers, and unit loads are stacked on top of each other to increase storage density. The highest density is achieved when unit loads are placed next to each other in both floor directions, as in Figure 11.2(a). However, this provides very poor access to internal loads. To increase accessibility, bulk storage loads can be organized into rows and blocks, so that natural aisles are created between pallet loads, as in Figure 11.2(b). The block widths can be designed to provide an appropriate balance between density and accessibility. Depending on the shape and physical support provided by the items stored, there may be a restriction on how high the unit loads can be stacked. In some cases, loads cannot be stacked on top of each other, either because of the physical shape or limited compressive strength of the individual loads. The inability to stack loads in bulk storage reduces storage density, removing one of its principal benefits.

Although bulk storage is characterized by the absence of specific storage equipment, material handling equipment must be used to put materials into storage and to retrieve them. Industrial trucks such as pallet trucks and powered forklifts (Section 10.2.1) are typically used for this purpose.

Rack Systems. Rack systems provide a method of stacking unit loads vertically without the need for the loads themselves to provide support. One of the most common rack systems is the *pallet rack*, consisting of a frame that includes horizontal load-supporting beams, as illustrated in Figure 11.3. Pallet loads are stored on these horizontal beams. Alternative storage rack systems include

TABLE 11.2 Application Characteristics of the Types of Storage Equipment and Methods

Storage Equipment	Advantages and Disadvantages	Typical Applications
Bulk storage	Highest density is possible Low accessibility Low possible cost per square foot	Storage of low turnover, large stock, or large unit loads
Rack systems	Low cost Good storage density Good accessibility	Palletized loads in warehouses
Shelves and bins	Some stock items not clearly visible	Storage of individual items on shelves Storage of commodity items in bins
Drawer storage	Contents of drawer easily visible Good accessibility Relatively high cost	Small tools Small stock items Repair parts
Automated storage systems	High throughput rates Facilitate use of computerized inventory control system Highest cost equipment Facilitate interface to automated material handling systems	Work-in-process storage Final product warehousing and distribution center Order picking Kitting of parts for electronic assembly

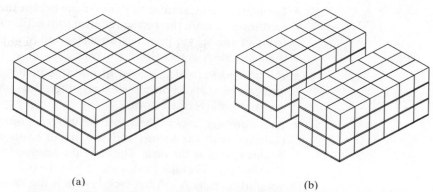

(a)   (b)

Figure 11.2 Various bulk storage arrangements: (a) high-density bulk storage provides low accessibility; (b) bulk storage with loads arranged to form rows and blocks for improved accessibility.

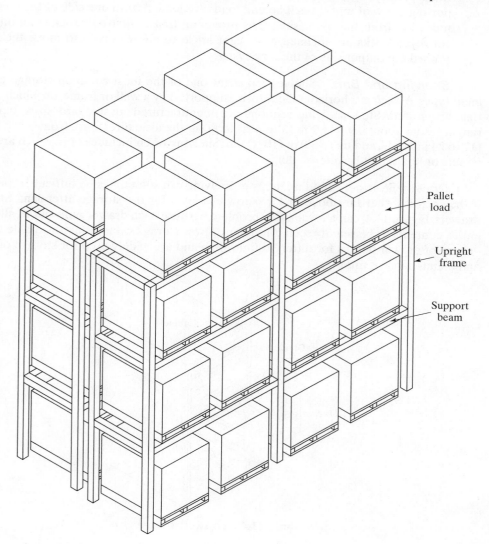

Pallet load

Upright frame

Support beam

Figure 11.3 Pallet rack system for storage of unit loads on pallets.

- *Cantilever racks*, similar to pallet racks except the supporting horizontal beams are cantilevered from the vertical central frame. Elimination of the vertical beams at the front of the frame provides unobstructed spans, which facilitates storage of long materials such as rods, bars, and pipes.
- *Portable racks*, which consist of portable box-frames that hold a single pallet load and can be stacked on top of each other, thus preventing load crushing that might occur in bulk vertical storage.
- *Drive-through racks*. These consist of aisles, open at each end, having two vertical columns with supporting rails for pallet loads on either side but no obstructing beams spanning the aisle. The rails are designed to support pallets of specific widths (Table 10.3). Forklift trucks are driven into the aisle to place the pallets onto the supporting rails. A related rack system is the *drive-in rack*, which is open at one end, permitting forklifts to access loads from one direction only.
- *Flow-through racks*. In place of the horizontal load-supporting beams in a conventional rack system, the flow-through rack uses long conveyor tracks capable of supporting a row of unit loads. The unit loads are loaded from one side of the rack and unloaded from the other side, thus providing first-in-first-out stock rotation. The conveyor tracks are inclined at a slight angle to allow gravity to move the loads toward the output side of the rack system.

Shelving and Bins.   Shelves represent one of the most common storage equipment types. A *shelf* is a horizontal platform, supported by a wall or frame, on which materials are stored. Steel shelving sections are manufactured in standard sizes, typically ranging from about 0.9 to 1.2 m (3 to 4 ft) long (in the aisle direction), from 0.3 to 0.6 m (12 to 24 in) wide, and up to 3.0 m (10 ft) tall. Shelving often includes *bins*, which are containers or boxes that hold loose items.

Drawer Storage.   Finding items in shelving can sometimes be difficult, especially if the shelf is either far above or far below eye level for the storage attendant. Storage drawers, Figure 11.4, can alleviate this problem because each drawer pulls out to allow its entire contents to be readily seen. Modular drawer storage cabinets are available with a variety of drawer depths for different item sizes and are widely used for storage of tools and maintenance items.

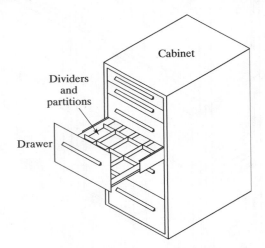

Figure 11.4   Drawer storage.

11.3 AUTOMATED STORAGE SYSTEMS

The storage equipment described in the preceding section requires a human worker to access the items in storage. The storage system itself is static. Mechanized and automated storage systems are available that reduce or eliminate the amount of human intervention required to operate the system. The level of automation varies. In less automated systems, a human operator is required to handle each storage/retrieval transaction. In highly automated systems, loads are entered or retrieved under computer control, with no human participation except to input data to the computer. Table 11.2 lists the advantages and disadvantages as well as typical applications of automated storage systems.

An automated storage system represents a significant investment, and it often requires a new and different way of doing business. Companies have different reasons for automating the storage function. Table 11.3 provides a list of possible objectives and reasons behind company decisions to automate their storage operations. Automated storage systems divide into two general types: (1) automated storage/retrieval systems and (2) carousel storage systems. These two types are discussed in the following sections.

11.3.1 Automated Storage/Retrieval Systems

An automated storage/retrieval system (AS/RS) is a storage system that performs storage and retrieval operations with speed and accuracy under a defined degree of automation. Figure 11.5 shows one aisle of an AS/RS that handles and stores unit loads on pallets. A wide range of automation is found in commercially available AS/RSs. At the most sophisticated level, the operations are totally automated, computer controlled, and fully integrated with factory and/or warehouse operations. At the other extreme, human workers control the equipment and perform the storage/retrieval transactions. Automated storage/retrieval systems are custom designed for each application, although the designs are based on standard modular components available from each respective AS/RS supplier.

Our definition of AS/RS might be interpreted to include carousel storage systems. However, in the material handling industry, carousel-based systems are distinguished from AS/RSs. The biggest difference is in the construction of the equipment. The basic AS/RS consists of a rack structure for storing loads and a storage/retrieval mechanism whose motions are linear (x-y-z motions), as pictured in Figure 11.5. By contrast, the carousel system uses storage baskets suspended from an overhead conveyor that revolves

TABLE 11.3 Possible Objectives and Reasons
for Automating a Company's Storage Operations

- To increase storage capacity
- To increase storage density
- To recover factory floor space presently used for storing work-in-process
- To improve security and reduce pilferage
- To improve safety in the storage function
- To reduce labor cost and/or increase labor productivity in storage operations
- To improve control over inventories
- To improve stock rotation
- To improve customer service
- To increase throughput

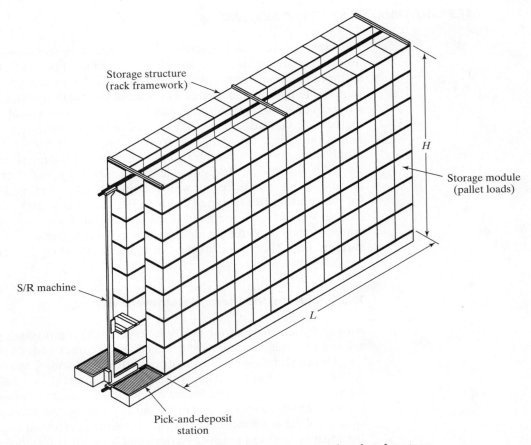

Storage structure
(rack framework)

Storage module
(pallet loads)

H

S/R machine

L

Pick-and-deposit
station

Figure 11.5   A unit load automated storage/retrieval system.

around an oval track loop to deliver the baskets to a load/unload station, as pictured in Figure 11.6. The differences between an AS/RS and a carousel storage system are summarized in Table 11.4.

An AS/RS consists of one or more storage aisles that are each serviced by a storage/retrieval (S/R) machine. (The S/R machines are sometimes referred to as cranes.) The aisles have storage racks for holding the stored materials. The S/R machines are used to deliver materials to the storage racks and to retrieve materials from the racks. Each AS/RS aisle has one or more input/output stations where materials are delivered into the storage system or moved out of the system. The input/output stations are called pickup-and-deposit (P&D) stations in AS/RS terminology. P&D stations can be manually operated or interfaced to some form of automated transport system such as a conveyor or an AGVS.

AS/RS Types.   Several important categories of automated storage/retrieval system can be distinguished. The following are the principal types:

- *Unit load AS/RS.* The unit load AS/RS is typically a large automated system designed to handle unit loads stored on pallets or in other standard containers. The

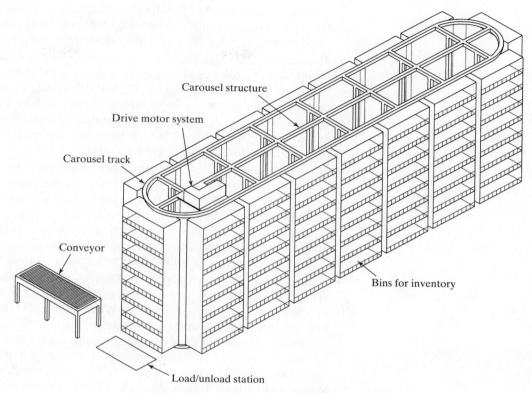

Figure 11.6 A horizontal storage carousel.

system is computer controlled, and the S/R machines are automated and designed to handle the unit load containers. The AS/RS pictured in Figure 11.5 is a unit load system. Other systems described below represent variations of the unit load AS/RS.

- *Deep-lane AS/RS.* The deep-lane AS/RS is a high-density unit load storage system that is appropriate when large quantities of stock are stored, but the number of separate stock types (SKUs) is relatively small. Instead of storing each unit load so that it can be accessed directly from the aisle (as in a conventional unit load system), the deep-lane system stores ten or more loads in a single rack, one load behind the next.

TABLE 11.4 Differences Between an AS/RS and a Carousel Storage System

Feature	Basic AS/RS	Basic Carousel Storage System
Storage structure	Rack system to support pallets or shelf system to support tote bins	Baskets suspended from overhead convey or trolleys
Motions	Linear motions of S/R machine	Revolution of overhead conveyor trolleys around oval track
Storage/retrieval operation	S/R machine travels to compartments in rack structure	Conveyor revolves to bring baskets to load/unload station
Replication of storage capacity	Multiple aisles, each consisting of rack structure and S/R machine	Multiple carousels, each consisting of oval track and suspended bins

Each rack is designed for "flow-through," with input on one side and output on the other side. Loads are picked up from one side of the rack by an S/R-type machine designed for retrieval, and another machine inputs loads on the entry side of the rack.

- *Miniload AS/RS.* This storage system is used to handle small loads (individual parts or supplies) that are contained in bins or drawers in the storage system. The S/R machine is designed to retrieve the bin and deliver it to a P&D station at the end of the aisle so that individual items can be withdrawn from the bins. The P&D station is usually operated by a human worker. The bin or drawer must then be returned to its location in the system. A miniload AS/RS is generally smaller than a unit load AS/RS and is often enclosed for security of the items stored.

- *Man-on-board AS/RS.* A man-on-board (also called man-aboard) storage/retrieval system represents an alternative approach to the problem of retrieving individual items from storage. In this system, a human operator rides on the carriage of the S/R machine. Whereas the miniload system delivers an entire bin to the end-of-aisle pick station and must return it subsequently to its proper storage compartment, with the man-on-board system the worker picks individual items directly at their storage locations. This offers an opportunity to increase system throughput.

- *Automated item retrieval system.* These storage systems are also designed for retrieval of individual items or small product cartons; however, the items are stored in lanes rather than bins or drawers. When an item is retrieved, it is pushed from its lane and drops onto a conveyor for delivery to the pickup station. The operation is somewhat similar to a candy vending machine, except that an item retrieval system has more storage lanes and a conveyor to transport items to a central location. The supply of items in each lane is periodically replenished, usually from the rear of the system so that there is flow-through of items, thus permitting first-in/first-out inventory rotation.

- *Vertical lift storage modules* (VLSM) [10]. These are also called vertical lift automated storage/retrieval systems (VL-AS/RS) [7]. All of the preceding AS/RS types are designed around a horizontal aisle. The same principle of using a center aisle to access loads is used except that the aisle is vertical. Vertical lift storage modules, some with heights of 10 m (30 ft) or more, are capable of holding large inventories while saving valuable floor space in the factory.

AS/RS Applications. Most applications of AS/RS technology have been associated with warehousing and distribution operations. An AS/RS can also be used to store raw materials and work-in-process in manufacturing. Three application areas can be distinguished for automated storage/retrieval systems: (1) unit load storage and handling, (2) order picking, and (3) work-in-process storage. Unit load storage and retrieval applications are represented by the unit load AS/RS and deep-lane storage systems. These kinds of applications are commonly found in warehousing for finished goods in a distribution center, rarely in manufacturing. Deep-lane systems are used in the food industry. As described above, order picking involves retrieving materials in less than full unit load quantities. Miniload, man-on-board, and item retrieval systems are used for this second application area.

Work-in-process (WIP) storage is a more recent application of automated storage technology. While it is desirable to minimize the amount of work-in-process, WIP is unavoidable and must be effectively managed. Automated storage systems, either automated storage/retrieval systems or carousel systems, represent an efficient way to store materials between processing steps, particularly in batch and job shop production. In high production, work-in-process is often carried between operations by conveyor systems, which thus serve both storage and transport functions.

The merits of an automated WIP storage system for batch and job shop production can best be seen be comparing it with the traditional way of dealing with work-in-process. The typical factory contains multiple work cells, each performing its own processing operations on different parts. At each cell, orders consisting of one or more parts are waiting on the plant floor to be processed, while other completed orders are waiting to be moved to the next cell in the sequence. It is not unusual for a plant engaged in batch production to have hundreds of orders in progress simultaneously, all of which represent work-in-process. The disadvantages of keeping all of this inventory in the plant include (1) time spent searching for orders, (2) parts or even entire orders becoming temporarily or permanently lost, sometimes resulting in repeat orders to reproduce the lost parts, (3) orders not being processed according to their relative priorities at each cell, and (4) orders spending too much time in the factory, causing customer deliveries to be late. These problems indicate poor control of work-in-process.

Automated storage/retrieval systems are also used in high-production operations. In the automobile industry, some final assembly plants use large capacity AS/RSs to temporarily store car and small truck bodies between major assembly steps. The AS/RS can be used for staging and sequencing the work units according to the most efficient production schedule [1].

Automated storage systems help to regain control over WIP. Reasons that justify the installation of automated storage systems for work-in-process include

- *Buffer storage in production.* A storage system can be used as a buffer storage zone between two processes whose production rates are significantly different. A simple example is a two-process sequence in which the first processing operation feeds a second process, which operates at a slower production rate. The first operation requires only one shift to meet production requirements, while the second step requires two shifts to produce the same number of units. An in-process buffer is needed between these operations to temporarily store the output of the first process.

- *Support of just-in-time delivery.* Just-in-time (JIT) is a manufacturing strategy in which parts required in production and/or assembly are received immediately before they are needed in the plant (Section 26.2). This results in a significant dependency of the factory on its suppliers to deliver the parts on time for use in production. To reduce the chance of stock-outs due to late supplier deliveries, some plants have installed automated storage systems as storage buffers for incoming materials. Although this approach subverts the objectives of JIT, it also reduces some of its risks.

- *Kitting of parts for assembly.* The storage system is used to store components for assembly of products or subassemblies. When an order is received, the required components are retrieved, collected into kits (tote pans), and delivered to the production floor for assembly.

- *Compatible with automatic identification systems.* Automated storage systems can be readily interfaced with automatic identification devices such as bar code readers. This allows loads to be stored and retrieved without needing human operators to identify the loads.

- *Computer control and tracking of materials*. Combined with automatic identification, an automated WIP storage system permits the location and status of work-in-process to be known.
- *Support of factory-wide automation*. Given the need for some storage of work-in-process in batch production, an appropriately sized automated storage system is an important subsystem in a fully automated factory.

Components and Operating Features of an AS/RS. Virtually all of the automated storage/retrieval systems described above consist of the following components, shown in Figure 11.5: (1) storage structure, (2) S/R machine, (3) storage modules (e.g., pallets for unit loads), and (4) one or more pickup-and-deposit stations. In addition, a control system is required to operate the AS/RS.

The storage structure is the rack framework, made of fabricated steel, which supports the loads contained in the AS/RS. The rack structure must possess sufficient strength and rigidity that it does not deflect significantly due to the loads in storage or other forces on the framework. The individual storage compartments in the structure must be designed to hold the storage modules used to contain the stored materials. The rack structure may also be used to support the roof and siding of the building in which the AS/RS resides. Another function of the storage structure is to support the aisle hardware required to align the S/R machines with respect to the storage compartments of the AS/RS. This hardware includes guide rails at the top and bottom of the structure as well as end stops and other features required to provide safe operation.

The S/R machine is used to accomplish storage transactions, delivering loads from the input station into storage, and retrieving loads from storage and delivering them to the output station. To perform these transactions, the storage/retrieval machine must be capable of horizontal and vertical travel to align its carriage (which carries the load) with the storage compartment in the rack structure. The S/R machine consists of a rigid mast on which is mounted an elevator system for vertical motion of the carriage. Wheels are attached at the base of the mast to permit horizontal travel along a rail system that runs the length of the aisle. A parallel rail at the top of the storage structure is used to maintain alignment of the mast and carriage with respect to the rack structure.

The carriage includes a shuttle mechanism to move loads into and from their storage compartments. The design of the shuttle system must also permit loads to be transferred from the S/R machine to the P&D station or other material-handling interface with the AS/RS. The carriage and shuttle are positioned and actuated automatically in the usual AS/RS. Man-on-board S/R machines are equipped for a human operator to ride on the carriage.

To accomplish the desired motions of the S/R machine, three drive systems are required: horizontal movement of the mast, vertical movement of the carriage, and shuttle transfer between the carriage and a storage compartment. Modern S/R machines are available with horizontal speeds up to 200 m/min (600 ft/min) along the aisle and vertical or lift speeds up to around 50 m/min(150 ft/min). These speeds determine the time required for the carriage to travel from the P&D station to a particular location in the storage aisle. Acceleration and deceleration have a more significant effect on travel time over short distances. The shuttle transfer is accomplished by any of several mechanisms, including forks (for pallet loads) and friction devices for flat bottom tote pans.

The storage modules are the unit load containers of the stored material. These include pallets, steel wire baskets and containers, plastic tote pans, and special drawers (used in miniload systems). The storage modules are generally made to a standard base size that can be handled automatically by the carriage shuttle of the S/R machine. The standard size is also designed to fit in the storage compartments of the rack structure.

The pickup-and-deposit station is where loads are transferred into and out of the AS/RS. It is generally located at the end of the aisle for access by the external handling system that brings loads to the AS/RS and takes loads away. Pickup stations and deposit stations may be located at opposite ends of the storage aisle or combined at the same location. This depends on the origin of incoming loads and the destination of output loads. A P&D station must be compatible with both the S/R machine shuttle and the external handling system. Common methods to handle loads at the P&D station include manual load/unload, forklift truck, conveyor (e.g., roller), and AGVS.

The principal AS/RS controls problem is positioning the S/R machine within an acceptable tolerance at a storage compartment in the rack structure to deposit or retrieve a load. The locations of materials stored in the system must be determined to direct the S/R machine to a particular storage compartment. Within a given aisle in the AS/RS, each compartment is identified by its horizontal and vertical positions and whether it is on the right side or left side of the aisle. A scheme based on alphanumeric codes can be used for this purpose. Using this location identification scheme, each unit of material stored in the system can be referenced to a particular location in the aisle. The record of these locations is called the "item location file." Each time a storage transaction is completed, the transaction must be recorded in the item location file.

Given a specified storage compartment to go to, the S/R machine must be controlled to move to that location and position the shuttle for load transfer. One positioning method uses a counting procedure in which the number of bays and levels are counted in the direction of travel (horizontally and vertically) to determine position. An alternative method is a numerical identification procedure in which each compartment has a reflective target with binary-coded location identifications on its face. The S/R machine uses optical scanners to read the target and position the shuttle for depositing or retrieving a load.

Computer controls and programmable logic controllers are used to determine the required location and guide the S/R machine to its destination. Computer control permits the physical operation of the AS/RS to be integrated with the supporting information and record-keeping system. It allows storage transactions to be entered in real-time, inventory records to be accurately maintained, system performance to be monitored, and communications to be facilitated with other factory computer systems. These automatic controls can be superseded or supplemented by manual controls when required under emergency conditions or for man-on-board operation of the machine.

11.3.2 Carousel Storage Systems

A carousel storage system consists of a series of bins or baskets suspended from an overhead chain conveyor that revolves around a long oval rail system, as depicted in Figure 11.6. The purpose of the chain conveyor is to position bins at a load/unload station at the end of the oval. The operation is similar to the powered overhead rack system used by dry cleaners to deliver finished garments to the front of the store. Most carousels are operated by a human worker at the load/unload station. The worker activates the powered carousel to deliver a desired bin to the station. One or more parts are removed from or added to the bin, and then the cycle is repeated. Some carousels are automated by using transfer mechanisms at the load/unload station to move loads into and from the carousel.

Carousel Technology. Carousels can be classified as horizontal or vertical. The more common horizontal configuration, shown in Figure 11.6, comes in a variety of sizes, ranging between 3 m (10 ft) and 30 m (100 ft) in length. Carousels at the upper end of the range have higher storage density, but the average access cycle time is greater. Accordingly, most carousels are 10–16 m (30–50 ft) long to achieve a proper balance between these competing factors.

A horizontal carousel storage system consists of welded steel framework that supports the oval rail system. The carousel can be either an overhead system (called a top-driven unit) or a floor-mounted system (called a bottom-driven unit). In the top-driven unit, a motorized pulley system is mounted at the top of the framework and drives an overhead trolley system. The bins are suspended from the trolleys. In the bottom-driven unit, the pulley drive system is mounted at the base of the frame, and the trolley system rides on a rail in the base. This provides more load-carrying capacity for the carousel storage system. It also eliminates the problem of dirt and oil dripping from the overhead trolley system onto the storage contents in top-driven systems.

The design of the individual bins and baskets of the carousel must be consistent with the loads to be stored. Bin widths range from about 50 to 75 cm (20 to 30 in), and depths are up to about 55 cm (22 in). Heights of horizontal carousels are typically 1.8–2.4 m (6–8 ft). Standard bins are made of steel wire to increase operator visibility.

Vertical carousels are constructed to operate around a vertical conveyor loop. They occupy much less floor space than the horizontal configuration, but require sufficient overhead space. The ceiling of the building limits the height of vertical carousels, and therefore their storage capacity is typically lower than for the average horizontal carousel.

Controls for carousel storage systems range from manual call controls to computer control. Manual controls include foot pedals, hand switches, and specialized keyboards. Foot pedal control allows the operator at the pick station to rotate the carousel in either direction to the desired bin position. Hand control involves use of a hand-operated switch that is mounted on an arm projecting from the carousel frame within easy reach of the operator. Again, bidirectional control is the usual mode of operation. Keyboard control permits a greater variety of control features than the previous control types. When the operator enters the desired bin position, the carousel is programmed to deliver the bin to the pick station by the shortest route (i.e., clockwise or counterclockwise motion of the carousel).

Computer control increases opportunities for automation of the mechanical carousel and for management of the inventory records. On the mechanical side, automatic loading and unloading is available on modern carousel storage systems. This allows the carousel to be interfaced with automated handling systems without the need for human participation in the load/unload operations. Data management features provided by computer control include the capability to maintain data on bin locations, items in each bin, and other inventory control records.

Carousel Applications. Carousel storage systems provide a relatively high throughput and are often an attractive alternative to a miniload AS/RS in manufacturing operations where their relatively low cost, versatility, and high reliability are recognized. Typical applications of carousel storage systems include (1) storage and retrieval operations, (2) transport and accumulation, (3) work-in-process, and (4) specialized uses.

Storage and retrieval operations can be efficiently accomplished using carousels when individual items must be selected from groups of items in storage. Sometimes called "pick and load" operations, these procedures are common in order-picking of tools in a toolroom, raw materials in a stockroom, service parts or other items in a wholesale firm, and work-in-process in a factory. In small electronics assembly, carousels are used for kitting of parts to be transported to assembly workstations.

In transport and accumulation applications, the carousel is used to transport and/or sort materials as they are stored. One example of this is in progressive assembly operations where the workstations are located around the periphery of a continuously moving carousel, and the workers have access to the individual storage bins of the carousel. They remove work from the bins to complete their own respective assembly tasks, then place their work into another bin for the next operation at some other workstation. Another example of transport and accumulation applications is sorting and consolidation of items. Each bin is defined for collecting the items of a particular type or customer. When the bin is full, the collected load is removed for shipment or other disposition.

Carousel storage systems often compete with automated storage and retrieval systems for applications where work-in-process is to be temporarily stored. Applications of carousel systems in the electronics industry are common.

One example of specialized use of carousel systems is electrical testing of products or components, where the carousel is used to store the item during testing for a specified period of time. The carousel is programmed to deliver the items to the load/unload station at the conclusion of the test period.

11.4 ENGINEERING ANALYSIS OF STORAGE SYSTEMS

Several aspects of the design and operation of a storage system are susceptible to quantitative engineering analysis. In this section, we examine capacity sizing and throughput performance for the two types of automated storage systems.

11.4.1 Automated Storage/Retrieval Systems

While the methods developed here are specifically for automated storage/retrieval systems, similar approaches can be used for analyzing traditional storage facilities, such as warehouses consisting of pallet racks and bulk storage.

Sizing the AS/RS Rack Structure. The total storage capacity of one storage aisle depends on how many storage compartments are arranged horizontally and vertically in the aisle, as indicated in our diagram in Figure 11.7. This can be expressed as

$$\text{Capacity per aisle} = 2n_y n_z \tag{11.1}$$

where n_y = number of load compartments along the length of the aisle, and n_z = number of load compartments that make up the height of the aisle. The constant, 2, accounts for the fact that loads are contained on both sides of the aisle.

If we assume a standard size compartment (to accept a standard size unit load), then the compartment dimensions facing the aisle must be larger than the unit load dimensions. Let x and y = the depth and width dimensions of a unit load (e.g., a standard pallet size as given in Table 10.3), and z = the height of the unit load. The width, length, and height of the rack structure of the AS/RS aisle are related to the unit load dimensions and number of compartments as follows [6]:

$$W = 3(x + a) \tag{11.2a}$$

$$L = n_y(y + b) \tag{11.2b}$$

$$H = n_z(z + c) \tag{11.2c}$$

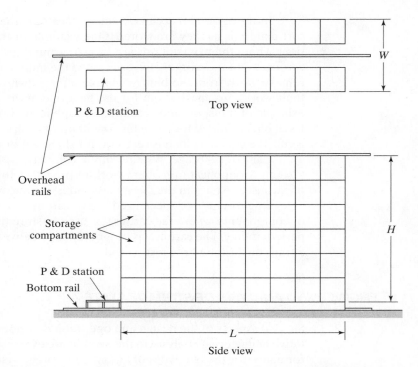

Figure 11.7 Top and side views of a unit load AS/RS, with nine storage compartments horizontally ($n_y = 9$) and six compartments vertically ($n_z = 6$).

where W, L, and H are the width, length, and height of one aisle of the AS/RS rack structure (mm, in); x, y, and z are the dimensions of the unit load (mm, in); and a, b, and c are allowances designed into each storage compartment to provide clearance for the unit load and to account for the size of the supporting beams in the rack structure (mm, in). For the case of unit loads contained on standard pallets, recommended values for the allowances [6] are: $a = 150$ mm (6 in), $b = 200$ mm (8 in), and $c = 250$ mm (10 in). For an AS/RS with multiple aisles, W is simply multiplied by the number of aisles to obtain the overall width of the storage system. The rack structure is built above floor level by 300–600 mm (12–24 in), and the length of the AS/RS extends beyond the rack structure to provide space for the P&D station.

EXAMPLE 11.2 Sizing an AS/RS System

Each aisle of a four-aisle AS/RS contains 60 storage compartments in the length direction and 12 compartments vertically. All storage compartments are the same size to accommodate standard size pallets of dimensions: $x = 42$ in and $y = 48$ in. The height of a unit load $z = 36$ in. Using the allowances, $a = 6$ in, $b = 8$ in, and $c = 10$ in, determine (a) how many unit loads can be stored in the AS/RS and (b) the width, length, and height of the AS/RS.

Solution: (a) The storage capacity is given by Eq. (11.1):

Capacity per aisle = 2(60)(12) = 1,440 unit loads. With four aisles, the total capacity is

$$\text{AS/RS capacity} = 4(1440) = 5760 \text{ unit loads}$$

(b) From Eqs. (11.2), we can compute the dimensions of the storage rack structure:

$$W = 3(42 + 6) = 144 \text{ in} = 12 \text{ ft/aisle}$$

$$\text{Overall width of the AS/RS} = 4(12) = 48 \text{ ft}$$

$$L = 60(48 + 8) = 3,360 \text{ in} = 280 \text{ ft}$$

$$H = 12(36 + 10) = 552 \text{ in} = 46 \text{ ft}$$

AS/RS Throughput. System throughput is defined as the hourly rate of S/R transactions that the automated storage system can perform (Section 11.1). A transaction involves depositing a load into storage or retrieving a load from storage. Either of these transactions alone is accomplished in a single command cycle. A dual command cycle accomplishes both transaction types in one cycle; since this reduces travel time per transaction, throughput is increased by using dual command cycles.

Several methods are available to compute AS/RS cycle times to estimate throughput performance. The method presented here is recommended by the Material Handling Institute [2]. It assumes (1) randomized storage of loads in the AS/RS (i.e., any compartment in the storage aisle is equally likely to be selected for a transaction), (2) storage compartments of equal size, (3) the P&D station located at the base and end of the aisle, (4) constant horizontal and vertical speeds of the S/R machine, and (5) simultaneous horizontal and vertical travel. For a single command cycle, the load to be entered or retrieved is assumed to be located at the center of the rack structure, as in Figure 11.8(a). Thus, the S/R machine must travel half the length and half the height of the AS/RS, and it must return the same distance. The single command cycle time can therefore be expressed by

$$T_{cs} = 2 \, \text{Max} \left\{ \frac{0.5L}{v_y}, \frac{0.5H}{v_z} \right\} + 2T_{pd} = \text{Max} \left\{ \frac{L}{v_y}, \frac{H}{v_z} \right\} + 2T_{pd} \qquad (11.3a)$$

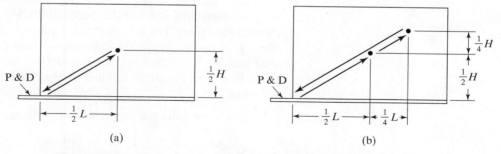

(a)   (b)

Figure 11.8 Assumed travel trajectory of the S/R machine for (a) single command cycle and (b) dual command cycle.

where T_{cs} = cycle time of a single command cycle (min/cycle), L = length of the AS/RS rack structure (m, ft), v_y = velocity of the S/R machine along the length of the AS/RS (m/min, ft/min), H = height of the rack structure (m, ft), v_z = velocity of the S/R machine in the vertical direction of the AS/RS (m/min, ft/min), and T_{pd} = pickup-and-deposit time (min). Two P&D times are required per cycle, representing load transfers to and from the S/R machine.

For a dual command cycle, the S/R machine is assumed to travel to the center of the rack structure to deposit a load, and then it travels to 3/4 the length and height of the AS/RS to retrieve a load, as in Figure 11.8(b). Thus, the total distance traveled by the S/R machine is 3/4 the length and 3/4 the height of the rack structure, and back. In this case, cycle time is given by

$$T_{cd} = 2 \, \text{Max} \left\{ \frac{0.75L}{v_y}, \frac{0.75H}{v_z} \right\} + 4T_{pd} = \text{Max} \left\{ \frac{1.5L}{v_y}, \frac{1.5H}{v_z} \right\} + 4T_{pd} \quad (11.3b)$$

where T_{cd} = cycle time for a dual command cycle (min/cycle), and the other terms are defined above.

System throughput depends on the relative numbers of single and dual command cycles performed by the system. Let R_{cs} = number of single command cycles performed per hour, and R_{cd} = number of dual command cycles per hour at a specified or assumed utilization level. We can formulate the following equation for the amounts of time spent in performing single command and dual command cycles each hour:

$$R_{cs}T_{cs} + R_{cd}T_{cd} = 60U \quad (11.4)$$

where U = system utilization during the hour. The right-hand side of the equation gives the total number of minutes of operation per hour. To solve Eq. (11.4), the relative proportions of R_{cs} and R_{cd} must be determined, or assumptions about these proportions must be made. When solved, the total hourly cycle rate is given by

$$R_c = R_{cs} + R_{cd} \quad (11.5)$$

where R_c = total S/R cycle rate (cycles/hr). Note that the total number of storage and retrieval transactions per hour will be greater than this value unless $R_{cd} = 0$, since there are two transactions accomplished in each dual command cycle. Let R_t = the total number of transactions performed per hour; then

$$R_t = R_{cs} + 2R_{cd} \quad (11.6)$$

EXAMPLE 11.3   AS/RS Throughput Analysis

Consider the AS/RS from previous Example 11.2, in which an S/R machine is used for each aisle. The length of the storage aisle = 280 ft and its height = 46 ft. Suppose horizontal and vertical speeds of the S/R machine are 200 ft/min and 75 ft/min, respectively. The S/R machine requires 20 sec to accomplish a P&D operation. Determine (a) the single command and dual command cycle times per aisle and (b) throughput per aisle under the assumptions that storage system utilization = 90% and the number of single command and dual command cycles are equal.

Solution: (a) We first compute the single and dual command cycle times by Eqs. (11.3):

$$T_{cs} = \text{Max}\{280/200, 46/75\} + 2(20/60) = 2.066 \text{ min/cycle}$$

$$T_{cd} = \text{Max}\{1.5 \times 280/200, 1.5 \times 46/75\} + 4(20/60) = 3.432 \text{ min/cycle}$$

(b) From Eq. (11.4), we can establish the single command and dual command activity levels each hour as follows:

$$2.066 R_{cs} + 3.432 R_{cd} = 60(0.90) = 54.0 \text{ min}$$

According to the problem statement, the number of single command cycles is equal to the number of dual command cycles. Thus, $R_{cs} = R_{cd}$.

Substituting this relation into the above equation, we have

$$2.066 R_{cs} + 3.432 R_{cs} = 54$$

$$5.498 R_{cs} = 54$$

$$R_{cs} = 9.822 \text{ single command cycles/hr}$$

$$R_{cd} = R_{cs} = 9.822 \text{ dual command cycles/hr}$$

System throughput is equal to the total number of S/R transactions per hour from Eq. (11.6):

$$R_t = R_{cs} + 2R_{cd} = 29.46 \text{ transactions/hr}$$

With four aisle, R_t for the AS/RS = 117.84 transactions/hr

EXAMPLE 11.4   AS/RS Throughput Using a Class-Based Dedicated Storage Strategy

The aisles in the AS/RS of the previous example will be organized following a class-based dedicated storage strategy. There will be two classes, according to activity level. The more active stock is stored in the half of the rack system that is located closest to the input/output station, and the less active stock is stored in the other half of the rack system farther away from the input/output station. Within each half of the rack system, random storage is used. The more active stock accounts for 80% of the transactions, and the less active stock accounts for the remaining 20%. As before, assume that system utilization = 90%, and the number of single command cycles = the number of dual command cycles. Determine the throughput of the AS/RS, basing the computation of cycle times on the same kinds of assumptions used in the MHI method.

Solution: With a total length of 280 ft, each half of the rack system will be 140 ft long and 46 ft high. Let us identify the stock nearest the input/output station (accounting for 80% of the transactions) as Class A, and the other half of the stock (accounting for 20% of the transactions) as Class B. The cycle times are computed as follows:

For Class A stock:

$$T_{scA} = \text{Max} \left\{ \frac{140}{200}, \frac{46}{75} \right\} + 2(0.333) = 1.366 \text{ min}$$

$$T_{dcA} = \text{Max} \left\{ \frac{1.5 \times 140}{200}, \frac{1.5 \times 46}{75} \right\} + 4(0.333) = 2.382 \text{ min}$$

For Class B stock:

$$T_{scB} = 2 \, \text{Max} \left\{ \frac{140 + 0.5(140)}{200}, \frac{0.5(46)}{75} \right\} + 2(0.333) = 2.766 \text{ min}$$

$$T_{dcB} = 2\,\text{Max}\left\{\frac{140 + 0.75(140)}{200}, \frac{0.75(46)}{75}\right\} + 4(0.333) = 3.782\;\text{min}$$

Consistent with the previous problem, let us conclude that

$$R_{csA} = R_{cdA} \text{ and } R_{csB} = R_{cdB} \tag{a}$$

We are also given that 80% of the transactions are Class A and 20% are Class B. Accordingly,

$$R_{csA} = 4R_{csB} \text{ and } R_{cdA} = 4R_{cdB} \tag{b}$$

We can establish the following equation for how each aisle spends its time during 1 hr:

$$R_{csA}T_{csA} + R_{cdA}T_{cdA} + R_{csB}T_{csB} + R_{cdB}T_{cdB} = 60(.90)$$

Based on Eqs. (a),

$$R_{csA}T_{csA} + R_{csA}T_{cdA} + R_{csB}T_{csB} + R_{csB}T_{cdB} = 60(.90)$$

Based on Eqs. (b),

$$4R_{csB}T_{csA} + 4R_{csB}T_{cdA} + R_{csB}T_{csB} + R_{csB}T_{cdB} = 60(.90)$$

$$4(1.366)R_{csB} + 4(2.382)R_{csB} + 2.766R_{csB} + 3.782R_{csB} = 54$$

$$21.54\,R_{csB} = 54$$
$$R_{csB} = 2.507$$
$$R_{csA} = 4R_{csB} = 10.028$$
$$R_{cdB} = R_{csB} = 2.507$$
$$R_{cdA} = 4R_{cdB} = 10.028$$

For one aisle,

$$R_t = R_{csA} + R_{csB} + 2(R_{cdA} + R_{cdB})$$

$$= 10.028 + 2.507 + 2(10.028 + 2.507) = 37.605\;\text{transactions/hr}$$

For four aisles, $R_t = 150.42$ transactions/hr

This represents almost a 28% improvement over the randomized storage strategy in Example 11.3.

11.4.2 Carousel Storage Systems

Let us develop the corresponding capacity and throughput relationships for a carousel storage system. Because of their construction, carousel systems do not possess nearly the volumetric capacity of an AS/RS. However, according to our calculations, a typical carousel system is likely to have higher throughput rates than an AS/RS.

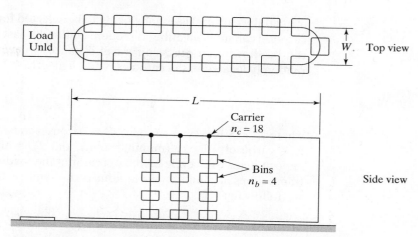

Figure 11.9 Top and side views of horizontal storage carousel with 18 carriers ($n_c = 18$) and four bins/carrier ($n_b = 4$).

Storage Capacity. The size and capacity of a carousel can be determined with reference to Figure 11.9. Individual bins or baskets are suspended from carriers that revolve around an oval rail with circumference given by

$$C = 2(L - W) + \pi W \tag{11.7}$$

where C = circumference of oval conveyor track (m, ft), and L and W are the length and width of the track oval (m, ft).

The capacity of the carousel system depends on the number and size of the bins (or baskets) in the system. Assuming standard size bins are used, each of a certain volumetric capacity, then the number of bins can be used as our measure of capacity. As illustrated in Figure 11.9, the number of bins hanging vertically from each carrier is n_b, and n_c = the number of carriers around the periphery of the rail. Thus,

$$\text{Total number of bins} = n_c n_b \tag{11.8}$$

The carriers are separated by a certain distance so that they do not interfere with each other while traveling around the ends of the carousel. Let s_c = the center-to-center spacing of carriers along the oval track. Then the following relationship must be satisfied by the values of s_c and n_c:

$$s_c n_c = C \tag{11.9}$$

where C = circumference (m, ft), s_c = carrier spacing (m/carrier, ft/carrier), and n_c = number of carriers, which must be an integer value.

Throughput Analysis. The storage/retrieval cycle time can be derived based on the following assumptions. First, only single command cycles are performed; a bin is accessed in the carousel either to put items into storage or to retrieve one or more items from storage. Second, the carousel operates with a constant speed v_c; acceleration and deceleration effects are ignored. Third, random storage is assumed; that is, any location around the carousel is equally likely to be selected for an S/R transaction. And fourth, the carousel can move in either direction. Under this last assumption of bidirectional travel, it can be shown that the mean travel distance between the load/unload station and a bin randomly located in the carousel is $C/4$. Thus, the S/R cycle time is given by

$$T_c = \frac{C}{4v_c} + T_{pd} \tag{11.10}$$

where T_c = S/R cycle time (min), C = carousel circumference as given by Eq. (11.7) (m, ft), v_c = carousel velocity (m/min, ft/min), and T_{pd} = the average time required to pick or deposit items each cycle by the operator at the load/unload station (min). The number of transactions accomplished per hour is the same as the number of cycles and is given by the following:

$$R_t = R_c = \frac{60}{T_c} \tag{11.11}$$

EXAMPLE 11.5 Carousel Operation

The oval rail of a carousel storage system has length = 12 m and width = 1 m. There are 75 carriers equally spaced around the oval. Suspended from each carrier are six bins. Each bin has volumetric capacity = 0.026 m^3. Carousel speed = 20 m/min. Average P&D time for a retrieval = 20 sec. Determine (a) volumetric capacity of the storage system and (b) hourly retrieval rate of the storage system.

Solution: (a) Total number of bins in the carousel is

$$n_c n_b = 75 \times 6 = 450 \text{ bins}$$

Total volumetric capacity = $450(0.026) = 11.7 \text{ m}^3$

(b) The circumference of the carousel rail is determined by Eq. (11.7):

$$C = 2(12 - 1) + 1\pi = 25.14 \text{ m}$$

Cycle time per retrieval is given by Eq. (11.10):

$$T_c = \frac{25.14}{4(20)} + 20/60 = 0.647 \text{ min}$$

Expressing throughput as an hourly rate, we have

$$R_t = 60/0.647 = 92.7 \text{ retrieval transactions/hr}$$

REFERENCES

[1] FEARE, T., "GM Runs in Top Gear with AS/RS Sequencing," *Modern Materials Handling*, August 1998, pp. 50–52.

[2] KULWIEC, R. A., Editor, *Materials Handling Handbook*, 2nd Edition, John Wiley & Sons, Inc., NY, 1985.

[3] Material Handling Institute, *AS/RS in the Automated Factory*, Pittsburgh, PA, 1983.

[4]   Material Handling Institute, *Consideration for Planning and Installing an Automated Storage/Retrieval System*, Pittsburgh, PA, 1977.

[5]   MULCAHY, D. E., *Materials Handling Handbook*, McGraw-Hill, NY, 1999.

[6]   TOMPKINS, J. A., J. A. WHITE, Y. A. BOZER, E. H. FRAZELLE, J. M. TANCHOCO, and J. TREVINO, *Facilities Planning*, 3d ed., John Wiley & Sons, Inc., NY, 2003.

[7]   TRUNK, C., "The Sky's the Limit for Vertical Lifts," *Material Handling Engineering*, August 1998, pp. 36–40.

[8]   TRUNK, C., "Pick-To-Light: Choices, Choices, Choices," *Material Handling Engineering*, September 1998, pp. 44–48.

[9]   TRUNK, C., "ProMat Report: New Ideas for Carousels," *Material Handling Engineering*, April 1999, pp. 69–74.

[10] "Vertical Lift Storage Modules: Advances Drive Growth," *Modern Materials Handling*, October 1998, pp. 42–43.

[11] WEISS, D. J., "Carousel Systems Capabilities and Design Considerations," *Automated Material Handling and Storage* (J. A. Tompkins and J. D. Smith, Editors), Auerbach Publishers, Inc., Pennsauken, NJ, 1983.

REVIEW QUESTIONS

11.1   Materials stored in manufacturing include a variety of types. Name six of the ten categories listed in Table 11.1.

11.2   Name and briefly describe four of the six measures used to assess the performance of a storage system.

11.3   Briefly describe the two basic storage location strategies.

11.4   What is a class-based dedicated storage strategy?

11.5   Name the four traditional (non automated) methods for storing materials.

11.6   Which of the four traditional storage methods is capable of the highest storage density?

11.7   What are some of the objectives and reasons behind company decisions to automate their storage operations? Name six of the ten objectives and reasons listed in Table 11.3.

11.8   What are the two basic categories of automated storage systems?

11.9   What are the differences between the two basic types of automated storage systems?

11.10 Identify the three application areas of automated storage/retrieval systems.

11.11 What are the four basic components of nearly all automated storage/retrieval systems?

11.12 What is the advantage of a vertical storage carousel over a horizontal storage carousel?

PROBLEMS

Sizing the AS/RS Rack Structure

11.1   Each aisle of a six-aisle Automated Storage/Retrieval System is to contain 50 storage compartments in the length direction and 8 compartments in the vertical direction. All storage compartments will be the same size to accommodate standard size pallets of dimensions: $x = 36$ in and $y = 48$ in. The height of a unit load $z = 30$ in. Using the allowances $a = 6$ in, $b = 8$ in, and $c = 10$ in, determine (a) how many unit loads can be stored in the

AS/RS, and (b) the width, length, and height of the AS/RS. The rack structure will be built 18 in above floor level.

11.2 A unit load AS/RS is being designed to store 1000 pallet loads in a distribution center located next to the factory. Pallet dimensions are $x = 1000$ mm, $y = 1200$ mm; and the maximum height of a unit load = 1300 mm. The AS/RS will consist of two aisles with one S/R machine per aisle, the length of the structure should be approximately five times its height, and the rack structure will be built 500 mm above floor level. Using the allowances $a = 150$ mm, $b = 200$ mm, and $c = 250$ mm, determine the width, length, and height of the AS/RS rack structure.

11.3 Consider the rack structure whose dimensions were computed in Problem 11.2. Assuming that only 80% of the storage compartments are occupied on average, and that the average volume of a unit load per pallet in storage = 0.75 m^3, compute the ratio of the total volume of unit loads in storage relative to the total volume occupied by the storage rack structure.

11.4 A unit load AS/RS for work-in-process storage in a factory must be designed to store 2,000 pallet loads, with an allowance of no less than 20% additional storage compartments for peak periods and flexibility. The unit load pallet dimensions are: depth (x) = 36 in and width (y) = 48 in. Maximum height of a unit load = 42 in. It has been determined that the AS/RS will consist of four aisles with one S/R machine per aisle. The maximum ceiling height (interior) of the building permitted by local ordinance is 60 ft, so the AS/RS must fit within this height limitation. The rack structure will be built 2 ft above floor level, and the clearance between the rack structure and the ceiling of the building must be at least 18 in. Determine the dimensions (height, length, and width) of the rack structure.

AS/RS Throughput Analysis

11.5 The length of the storage aisle in an AS/RS = 240 ft and its height = 60 ft. Suppose horizontal and vertical speeds of the S/R machine are 400 ft/min and 60 ft/min, respectively. The S/R machine requires 18 sec to accomplish a pickup-and-deposit operation. Find: (a) the single-command and dual-command cycle times per aisle, and (b) throughput for the aisle under the assumptions that storage system utilization = 85% and the number of single command and dual command cycles are equal.

11.6 Solve the previous problem using a ratio of single-command to dual-command cycles of 3:1 instead of 1:1.

11.7 An AS/RS is used for work-in-process storage in a manufacturing facility. The AS/RS has five aisles, each 120 ft long and 40 ft high. The horizontal and vertical speeds of the S/R machine are 400 ft/min and 50 ft/min, respectively. The S/R machine requires 12 sec to accomplish a pickup and deposit operation. The number of single command cycles equals the number of dual command cycles. If the throughput rate must be 200 S/R transactions per hour during periods of peak activity, will the AS/RS satisfy this requirement? If so, what is the utilization of the AS/RS during peak hours?

11.8 An automated storage/retrieval system installed in a warehouse has five aisles. The storage racks in each aisle are 30 ft high and 150 ft long. The S/R machine for each aisle travels at a horizontal speed of 350 ft/min and a vertical speed of 60 ft/min. The pickup-and-deposit time = 0.25 min. Assume that the number of single command cycles per hour is equal to the number of dual command cycles per hour and that the system operates at 75% utilization. Determine the throughput rate (loads moved/hour) of the AS/RS.

11.9 A 10-aisle automated storage/retrieval system is located in an integrated factory-warehouse facility. The storage racks in each aisle are 18 m high and 95 m long. The S/R machine for each aisle travels at a horizontal speed of 2.5 m/s and a vertical speed of 0.5 m/s. Pickup and deposit time = 20 s. Assume that the number of single command cycles per hour is half the number of dual command cycles per hour and that the system operates at 80% utilization. Determine the throughput rate (loads moved/hour) of the AS/RS.

11.10 An automated storage/retrieval system for work-in-process has five aisles. The storage racks in each aisle are 10 m high and 50 m long. The S/R machine for each aisle travels at a horizontal speed of 2.0 m/s and a vertical speed of 0.4 m/s. Pickup and-deposit time = 15 s. Assume that the number of single command cycles per hour is three times the number of dual command cycles per hour and that the system operates at 90% utilization. Determine the throughput rate (loads moved/hour) of the AS/RS.

11.11 The length of one aisle in an AS/RS is 100 m and its height is 20 m. Horizontal travel speed is 4.0 m/s. The vertical speed is specified so that the storage system is "square in time," which means that $L/v_y = H/v_z$. The pickup-and deposit time is 12 s. Determine the expected throughput rate (transactions per hour) for the aisle if the expected ratio of the number of transactions performed under single-command cycles to the number of transactions performed under dual-command cycles is 2:1. The system operates continuously during the hour.

11.12 An automated storage/retrieval system has four aisles. The storage racks in each aisle are 40 ft high and 200 ft long. The S/R machine for each aisle travels at a horizontal speed of 400 ft/min and a vertical speed of 60 ft/min. If the pickup-and-deposit time = 0.3 min, determine the throughput rate (loads moved/hour) of the AS/RS, under the assumption that time spent each hour performing single command cycles is twice the time spent performing dual command cycles, and that the AS/RS operates at 90% utilization.

11.13 An AS/RS with one aisle is 300 ft long and 60 ft high. The S/R machine has a maximum speed of 300 ft/min in the horizontal direction. It accelerates from zero to 300 ft/min over a distance of 15 ft. On approaching its target position (where the S/R machine will transfer a load onto or off its platform), it decelerates from 300 ft/min to a full stop in 15 ft. The maximum vertical speed is 60 ft/min, and the acceleration and deceleration distances are each 3 ft. Assume simultaneous horizontal and vertical movement, and constant rates of acceleration and deceleration in both directions. The pickup-and-deposit time = 0.3 min. Using the general approach of the MHI method for computing cycle time but adding considerations for acceleration and deceleration, determine the single command and dual command cycle times.

11.14 An AS/RS with four aisles is 80 m long and 18 m high. The S/R machine has a maximum speed of 1.6 m/s in the horizontal direction. It accelerates from zero to 1.6 m/s in a distance of 2.0 m. On approaching its target position (where the S/R machine will transfer a load onto or off its platform), it decelerates from 1.6 m/s to a full stop in 2.0 m. The maximum vertical speed is 0.5 m/s, and the acceleration and deceleration distances are each 0.3 m. Rates of acceleration and deceleration are constant in both directions. Pickup-and-deposit time = 12 s. Utilization of the AS/RS is assumed to be 90%, and the number of dual command cycles = the number of single command cycles. (a) Calculate the single command and dual command cycle times, including considerations for acceleration and deceleration. (b) Determine the throughput rate for the system.

11.15 Your company is seeking proposals for an automated storage/retrieval system that will have a throughput rate of 300 storage/retrieval transactions/hour during the one 8-hour shift per day. The request for proposals indicates that the number of single command cycles is expected to be four times the number of dual command cycles. The first proposal received is from a vendor who specifies the following: ten aisles, each aisle 150 ft long and 50 ft high; horizontal and vertical speeds of the S/R machine = 200 ft/min and 66.67 ft/min, respectively; and pickup-and-deposit time = 0.3 min. As the responsible engineer for the project, you must analyze the proposal and make recommendations accordingly. One of the difficulties you see in the proposed AS/RS is the large number of S/R machines that would be required: one for each of the 10 aisles. This makes the proposed system very expensive. Your recommendation is to reduce the number of aisles from 10 to 6 and to select a S/R machine with horizontal and vertical speeds of 300 ft/min and 100 ft/min, respectively. Although each high speed S/R machine is slightly more expensive than the slower model, reducing the number of machines from 10 to 6 will significantly reduce total cost. Also, fewer aisles will reduce the cost of the rack structure even though each aisle will be somewhat larger since total storage capacity must remain the same. The problem is that throughput rate will be adversely affected by the larger rack system. (a) Determine the throughput rate of the proposed 10-aisle AS/RS and calculate its utilization relative to the specified 300 transactions/hour. (b) Determine the length and height of a 6-aisle AS/RS whose storage capacity would be the same as the proposed 10-aisle system. (c) Determine the throughput rate of the 6-aisle AS/RS and calculate its utilization relative to the specified 300 transactions/hour. (d) Given the dilemma now confronting you, what other alternatives would you analyze and what recommendations would you make to improve the design of the system?

11.16 A unit load automated storage/retrieval system has five aisles. The storage racks are 60 ft high and 280 ft long. The S/R machine travels at a horizontal speed of 200 ft/min and a vertical speed of 80 ft/min. The pickup-and-deposit time = 0.30 min. Assume that the number of single command cycles per hour is four times the number of dual command cycles per hour and that the system operates at 80% utilization. A class-based dedicated storage strategy is used for organizing the stock, in which unit loads are separated into two classes, according to activity level. The more active stock is stored in the half of the rack system located closest to the input/output station, and the less active stock is stored in the other half of the rack system (farther away from the input/output station). Within each half of the rack system, random storage is used. The more active stock accounts for 75% of the transactions, and the less active stock accounts for the remaining 25% of the transactions. Determine the throughput rate (loads moved/hour into and out of storage) of the AS/RS, basing your computation of cycle times on the same types of assumptions used in the MHI method. Assume that when dual command cycles are performed the two transactions per cycle are both in the same class.

11.17 The AS/RS aisle of Problem 11.5 will be organized following a class-based dedicated storage strategy. There will be two classes, according to activity level. The more active stock is stored in the half of the rack system that is closest to the input/output station, and the less active stock is stored in the other half of the rack system farther away from the input/output station. Within each half of the rack system, random storage is used. The more active stock accounts for 80% of the transactions, and the less active stock accounts for the remaining 20%. Assume that system utilization = 85% and the number of single command cycles = the number of dual command cycles in each half of the AS/RS. (a) Determine the throughput of the AS/RS, basing the computation of cycle times on the same kinds of assumptions used in the MHI method. (b) A class-based dedicated storage strategy is supposed to increase throughput. Why is throughput less here than in Problem 11.5?

Carousel Storage Systems

11.18 A single carousel storage system is located in a factory making small assemblies. It is 20 m long and 1.0 m wide. The pickup-and-deposit time is 0.25 min. The speed at which the carousel operates is 0.5 m/s. The storage system has a 90% utilization. Determine the hourly throughput rate.

11.19 A storage system serving an electronics assembly plant has three storage carousels, each with its own manually operated pickup-and-deposit station. The pickup-and-deposit time is 0.30 min. Each carousel is 60 ft long and 2.5 ft wide. The speed at which the system revolves is 85 ft/min. Determine the throughput rate of the storage system.

11.20 A single carousel storage system has an oval rail loop that is 30 ft long and 3 ft wide. Sixty carriers are equally spaced around the oval. Suspended from each carrier are five bins.

Each bin has a volumetric capacity = 0.75 ft³. Carousel speed = 100 ft/min. Average pickup-and-deposit time for a retrieval = 20 sec. Determine (a) volumetric capacity of the storage system and (b) hourly retrieval rate of the storage system.

11.21 A carousel storage system is to be designed to serve a mechanical assembly plant. The system must have a total of 400 storage bins and a throughput of at least 125 storage and retrieval transactions per hour. Two alternative configurations are being considered: (1) a one-carousel system and (2) a two-carousel system. In either case, the width of the carousel is to be 4.0 ft and the spacing between carriers = 2.5 ft. One picker-operator will be required for the one-carousel system and two picker-operators will be required for the two-carousel system. In either system v_c = 75 ft/min. For the convenience of the picker-operator, the height of the carousel will be limited to five bins. The standard time for a pickup-and-deposit operation at the load/unload station = 0.4 min if one part is picked or stored per bin and 0.6 min if more than one part is picked or stored. Assume that 50% of the transactions will involve more than one component. Determine (a) the required length of the one-carousel system, (b) the corresponding throughput rate, (c) the required length of the two-carousel system, and (d) the corresponding throughput rate. (e) Which system better satisfies the design specifications?

11.22 Given your answers to Problem 11.21, compare the costs of the two carousel systems. The one-carousel system has an installed cost of $50,000, and the comparable cost of the two-carousel system is $75,000. Labor cost for a picker-operator is $20/hour, including fringe benefits and applicable overhead. The storage systems will be operated 250 days per year for seven hours per day, although the operators will be paid for eight hours. Using a three-year period in your analysis, and a 25% rate of return, determine (a) the equivalent annual cost for the two design alternatives, assuming no salvage value at the end of three years; and (b) the average cost per storage/retrieval transaction.

Chapter 12

Automatic Identification and Data Capture

CHAPTER CONTENTS

Automatic identification and data capture (AIDC) refers to the technologies that provide direct entry of data into a computer or other microprocessor controlled system without using a keyboard. Many of these technologies require no human involvement in the data capture and entry process. Automatic identification systems are being used increasingly to collect data in material handling and manufacturing applications. In material handling, the applications include shipping and receiving, storage, sortation, order picking, and kitting of parts for assembly. In manufacturing, the applications include monitoring the status of order processing, work-in-process, machine utilization, worker attendance, and other measures of factory operations and performance. Of course, AIDC has many important applications outside the factory, including retail sales and inventory control, warehousing and distribution center operations, mail and parcel handling, patient identification in hospitals, check processing in banks, and security systems. Our interest in this chapter emphasizes material handling and manufacturing applications.

The alternative to automatic data capture is manual collection and entry of data. This typically involves a worker recording the data on paper and later entering them into the computer by means of a keyboard. There are several drawbacks to this method:

1. *Errors* occur in both data collection and keyboard entry of the data when it is accomplished manually. The average error rate of manual keyboard entry is one error per 300 characters.

2. *Time factor*. Manual methods are inherently more time consuming than automated methods. Also, when manual methods are used, there is a time delay between when the activities and events occur and when the data on status are entered into the computer.

3. *Labor cost*. The full attention of human workers is required in manual data collection and entry, with the associated labor cost.

These drawbacks are virtually eliminated when automatic identification and data capture are used. With AIDC, the data on activities, events, and conditions are acquired at the location and time of their occurrence and entered into the computer immediately or shortly thereafter.

Automatic data capture is often associated with the material handling industry. The AIDC industry trade association, the Automatic Identification Manufacturers Association (AIM), started as an affiliate of the Material Handling Institute, Inc. Many of the applications of this technology relate to material handling. But automatic identification and data capture has also become important in shop floor control in manufacturing plants (Chapter 25). In the present chapter, we examine the important AIDC technologies, with emphasis on manufacturing.

12.1 OVERVIEW OF AUTOMATIC IDENTIFICATION METHODS

Nearly all of the automatic identification technologies consist of three principal components, which also comprise the sequential steps in AIDC [8]:

1. *Data encoder*. A *code* is a set of symbols or signals that usually represent alphanumeric characters. When data are encoded, the characters are translated into a machine-readable code. (For most AIDC techniques, the encoded data are not readable by humans.) A label or tag containing the encoded data is attached to the item that is to be identified.

2. *Machine reader* or *scanner*. This device reads the encoded data, converting them to alternative form, usually an electrical analog signal.

3. *Data decoder*. This component transforms the electrical signal into digital data and finally back into the original alphanumeric characters.

Many different technologies are used to implement automated identification and data collection. Within the category of bar codes alone (currently the leading AIDC technology), more than 250 different bar code schemes have been devised. AIDC technologies can be divided into the following six categories [18]:

1. *Optical.* Most of these technologies use high-contrast graphical symbols that can be interpreted by an optical scanner. They include linear (one-dimensional) and two-dimensional bar codes, optical character recognition, and machine vision.

2. *Electromagnetic.* The important AIDC technology in this group is radio frequency identification (RFID), which uses a small electronic tag capable of holding more data than a bar code. Its applications are gaining on bar codes due to several mandates from companies like Wal-Mart and from the U.S. Department of Defense.

3. *Magnetic.* These technologies encode data magnetically, similar to recording tape. The two important techniques in this category are (a) magnetic stripe, widely used in plastic credit cards and bank access cards, and (b) magnetic ink character recognition, widely used in the banking industry for check processing.

4. *Smart card.* This term refers to small plastic cards (the size of a credit card) imbedded with microchips capable of containing large amounts of information. Other terms used for this technology include *chip card* and *integrated circuit card.*

5. *Touch techniques.* These include touch screens and button memory.

6. *Biometric.* These technologies are utilized to identify humans or to interpret vocal commands of humans. They include voice recognition, fingerprint analysis, and retinal eye scans.

The most widely used AIDC technologies in production and distribution are bar codes and radio frequency methods. The common applications of AIDC technologies are (1) receiving, (2) shipping, (3) order picking, (4) finished goods storage, (5) manufacturing processing, (6) work-in-process storage, (7) assembly, and (8) sortation. Some of the identification applications require workers to be involved in the data collection procedure, usually to operate the identification equipment in the application. These techniques are therefore semiautomated rather than automated methods. Other applications accomplish the identification with no human participation. The same basic sensor technologies may be used in both cases. For example, certain types of bar code readers are operated by humans, whereas other types operate automatically.

As indicated in our chapter introduction, there are good reasons for using automatic identification and data collection techniques: (1) data accuracy, (2) timeliness, and (3) labor reduction. First and foremost, the accuracy of the data collected is improved with AIDC, in many cases by a significant margin. The error rate in bar code technology is approximately 10,000 times lower than in manual keyboard data entry. The error rates of most of the other technologies are not as good as for bar codes but are still better than manual methods. The second reason for using automatic identification techniques is to reduce the time required by human workers to make the data entry. The speed of data entry for handwritten documents is approximately 5–7 characters/sec and it is 10–15 characters/sec (at best) for keyboard entry [16]. Automatic identification methods are capable of reading hundreds of characters per second. The time savings from using automatic identification techniques can mean substantial labor cost benefits for large plants with many workers.

Although the error rate in automatic identification and data collection technologies is much lower than for manual data collection and entry, errors do occur in AIDC. The industry has adopted two parameters to measure the errors:

1. *First Read Rate* (FRR). This is the probability of a successful (correct) reading by the scanner in its initial attempt.
2. *Substitution Error Rate* (SER). This is the probability or frequency with which the scanner incorrectly reads the encoded character as some other character. In a given set of encoded data containing n characters, the expected number of errors = SER multiplied by n.

Obviously, it is desirable for the AIDC system to possess a high first read rate and a low substitution error rate. A subjective comparison of substitution error rates for several AIDC technologies is presented in Table 12.1.

TABLE 12.1 Comparison of AIDC Techniques and Manual Keyboard Data Entry

Technique	Time to Enter*	Error Rate**	Equipment Cost	Advantages/(Disadvantages)
Manual entry	Slow	High	Low	Low initial cost (Requires human operator) (Slow speed) (High error rate)
Bar codes: 1-D	Medium	Low	Low	High speed Good flexibility (Low data density)
Bar codes: 2-D	Medium	Low	High	High speed High data density
Radio frequency	Fast	Low	High	Label need not be visible to reader Read-write capability available High data density (Expensive labeling)
Magnetic stripe	Medium	Low	Medium	Much data can be encoded Data can be changed (Vulnerable to magnetic fields) (Contact required for reading)
OCR	Medium	Medium	Medium	Can be read by humans (Low data density) (High error rate)
Machine vision	Fast	***	Very high	High speed (Equipment expensive) (Not suited to general AIDC applications)

Source: Based on data from Palmer [14].

*Time to enter data is based on a 20-character field. All techniques except machine vision use a human worker to either enter the data (manual entry) or to operate the AIDC equipment (bar codes, RFID, magnetic stripe, OCR). Key: Slow = 5–10 sec, Medium = 2–5 sec, Fast =< 2 sec

**Substitution error rate (SER); see definition (Section 12.1).

***Application dependent.

12.2 BAR CODE TECHNOLOGY

As indicated earlier, bar codes divide into two basic types: (1) linear, in which the encoded data are read using a linear sweep of the scanner, and (2) two-dimensional, in which the encoded data must be read in both directions.

12.2.1 Linear (One-Dimensional) Bar Codes

As mentioned previously, linear bar codes are the most widely used automatic identification and data collection technique. There are actually two forms of linear bar code symbologies, illustrated in Figure 12.1: (a) *width-modulated,* in which the symbol consists of bars and spaces of varying width; and (b) *height-modulated,* in which the symbol consists of evenly spaced bars of varying height. The only significant application of the height-modulated bar code symbologies is in the U.S. Postal Service for ZIP code identification, so our discussion will focus on the width-modulated bar codes, which are used widely in retailing and manufacturing.

In linear width-modulated bar code technology, the symbol consists of a sequence of wide and narrow colored bars separated by wide and narrow spaces (the colored bars are usually black and the spaces are white for high contrast). The pattern of bars and spaces is coded to represent numeric or alphanumeric characters. Palmer [14] uses the interesting analogy that bar codes might be thought of as a printed version of the Morse code, where narrow bands represent dots and wide bands represent dashes. Using this scheme, the bar code for the familiar SOS distress signal would be as shown in Figure 12.2. Bar codes do not follow Morse code, however; the difficulties with a "Morse" bar code symbology are (1) only the dark bars are used, thus increasing the length of the coded symbol, and (2) the number of bars making up the alphanumeric characters differs, making decoding more difficult [14].

Bar code readers interpret the code by scanning and decoding the sequence of bars. The reader consists of the scanner and decoder. The scanner emits a beam of light that is swept past the bar code (either manually or automatically) and senses light reflections to distinguish between the bars and spaces. The light reflections are sensed by a photodetector, which converts the spaces into an electrical signal and the bars into absence of an electrical signal. The width of the bars and spaces is indicated by the duration of the corresponding signals. The procedure is depicted in Figure 12.3. The decoder analyzes the pulse train to validate and interpret the corresponding data.

(a)   (b)

Figure 12.1 Two forms of linear bar codes are (a) width-modulated, exemplified here by the Universal Product Code, and (b) height-modulated, exemplified here by Postnet, used by the U.S. Postal Service.

Figure 12.2 The SOS distress signal in "Morse" bar codes.

Certainly a major reason for the acceptance of bar codes is their widespread use in grocery markets and other retail stores. In 1973, the grocery industry adopted the Universal Product Code (UPC) as its standard for item identification. This is a 12-digit bar code that uses six digits to identify the manufacturer and five digits to identify the product. The final digit is a check character. The U.S. Department of Defense provided another major endorsement in 1982 by adopting a bar code standard (Code 39) that must be applied by vendors on product cartons supplied to the various agencies of DOD. The UPC is a numerical code (0–9), while Code 39 provides the full set of alphanumeric characters plus other symbols (44 characters in all). These two linear bar codes and several others are compared in Table 12.2.

The Bar Code Symbol. The bar code standard adopted by the automotive industry, the Department of Defense, the General Services Administration, and many other manufacturing industries is Code 39, also known as AIM USD–2 (Automatic Identification Manufacturers Uniform Symbol Description-2), although this is actually a subset of Code 39. We describe this format as an example of linear bar code symbols [3], [6], [14]. Code 39 uses a series of wide and narrow elements (bars and spaces) to represent alphanumeric and other characters. The wide elements are equivalent to a binary value of one and the narrow elements are equal to zero. The width of the wide bars and spaces is between two and three times the width of the narrow bars and spaces. Whatever the wide-to-narrow ratio, the width must be uniform throughout the code for the reader to be able to consistently interpret the resulting pulse train. Figure 12.4 presents the character structure for USD–2, and Figure 12.5 illustrates how the character set might be developed in a typical bar code.

The reason for the name Code 39 is that nine elements (bars and spaces) are used in each character and three of the elements are wide. The placement of the wide spaces

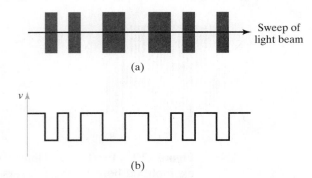

(a)

(b)

Figure 12.3 Conversion of bar code into a pulse train of electrical signals, (a) bar code and (b) corresponding electrical signal.

TABLE 12.2 Some Widely Used Linear Bar Codes

Bar Code	Date	Description	Applications
Codabar	1972	Only 16 characters: 0–9, $, :, /, ., +, −	Used in libraries, blood banks, and some parcel freight applications
UPC*	1973	Numeric only, length = 12 digits	Widely used in U.S. and Canada, in grocery and other retail stores
Code 39	1974	Alphanumeric (see text for description)	Adopted by Department of Defense, automotive, and other manufacturing industries
Postnet	1980	Numeric only**	U.S. Postal Service code for ZIP code numbers
Code 128	1981	Alphanumeric, but higher density	Substitutes in some Code 39 applications
Code 93	1982	Similar to Code 39 but higher density	Same applications as Code 39

Sources: Nelson [13], Palmer [14].

*UPC= Universal Product Code, adopted by the grocery industry in 1973 and based on a symbol developed by IBM Corporation in early grocery tests. A similar standard bar code system was developed in Europe, called the European Article Numbering system (EAN), in 1978.

**This is the only height-modulated bar code in the table. All others are width-modulated.

and bars in the code uniquely designates the character. Each code begins and ends with either a wide or narrow bar. The code is sometimes referred to as code three-of-nine. In addition to the character set in the bar code, there must also be a so-called "quiet zone" both preceding and following the bar code, in which there is no printing that might confuse the decoder. This quiet zone is shown in Figure 12.5.

Bar Code Readers. Bar code readers come in a variety of configurations; some require a human to operate them and others are stand-alone automatic units. They are usually classified as contact or noncontact readers. Contact bar code readers are hand-held wands or light pens operated by moving the tip of the wand quickly past the bar code on the object or document. The wand tip must be in contact with the bar code surface or in very close proximity during the reading procedure. In a factory data collection application, they are usually part of a keyboard entry terminal. The terminal is sometimes referred to as a stationary terminal in the sense that it is placed in a fixed location in the shop. When a transaction is entered in the factory, the data are usually communicated to the computer system immediately. In addition to their use in factory data collection systems, stationary contact bar code readers are widely used in retail stores to enter the item in a sales transaction.

Contact bar code readers are also available as portable units that can be carried around the factory or warehouse by a worker. They are battery-powered and include a solid-state memory device capable of storing data acquired during operation. The data can be transferred to the computer system subsequently. Portable bar code readers often include a keypad that can be used by the operator to input data that cannot be entered via bar code. These portable units are used for order picking in a warehouse and similar applications that require a worker to move significant distances in a building.

Noncontact bar code readers focus a light beam on the bar code, and a photodetector reads the reflected signal to interpret the code. The reader probe is located a certain distance from the bar code (several inches to several feet) during the read procedure. Noncontact readers are classified as fixed beam and moving beam scanners.

Char.	Bar pattern	9 bits	Char.	Bar pattern	9 bits
1		100100001	K		100000011
2		001100001	L		001000011
3		101100000	M		101000010
4		000110001	N		000010011
5		100110000	O		100010010
6		001110000	P		001010010
7		000100101	Q		000000111
8		100100100	R		100000110
9		001100100	S		001000110
0		000110100	T		000010110
A		100001001	U		110000001
B		001001001	V		011000001
C		101001000	W		111000000
D		000011001	X		010010001
E		100011000	Y		110010000
F		001011000	Z		011010000
G		000001101	-		010000101
H		100001100	.		110000100
I		001001100	space		011000100
J		000011100	*		010010100

*Denotes a start/stop code that must be placed at the beginning and end of every bar code message.

Figure 12.4 Character set in USD-2 bar code, a subset of Code 39 [6].

Fixed beam readers are stationary units that use a fixed beam of light. They are usually mounted beside a conveyor and depend on the movement of the bar code past the light beam for their operation. Applications of fixed beam bar code readers are typically in warehousing and material handling operations where large quantities of materials must be identified as they flow past the scanner on conveyors. Fixed beam scanners in these kinds of operations represent some of the first applications of bar codes in industry.

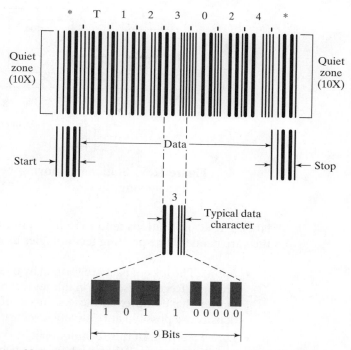

Figure 12.5 A typical grouping of characters to form a bar code in Code 39. (Reprinted from Automatic Identification Manufacturers, Inc [6] by permission.)

Moving beam scanners use a highly focused beam of light, often a laser, actuated by a rotating mirror to traverse an angular sweep in search of the bar code on the object. A scan is defined as a single sweep of the light beam through the angular path. The high rotational speed of the mirror allows for very high scan rates—up to 1440 scans/sec [1]. This means that many scans of a single bar code can be made during a typical reading procedure, thus permitting verification of the reading. Moving beam scanners can be either stationary or portable units. Stationary scanners are located in a fixed position to read bar codes on objects as they move past on a conveyor or other material handling equipment. They are used in warehouses and distribution centers to automate the product identification and sortation operations. A typical setup using a stationary scanner is illustrated in Figure 12.6. Portable scanners are hand-held devices that the user points at the bar code like a pistol. The vast majority of bar code scanners used in factories and warehouses are of this type.

Bar Code Printers. In many bar code applications, the labels are printed in medium-to-large quantities for product packages and the cartons used to ship the packaged products. These preprinted bar codes are usually produced off-site by companies specializing in these operations. The labels are printed in either identical or sequenced symbols. Printing technologies include traditional techniques such as letterpress, offset lithography, and flexographic printing.

Bar codes can also be printed on-site by methods in which the process is controlled by microprocessor to achieve individualized printing of the bar coded document or item

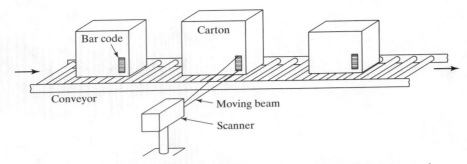

Figure 12.6 Stationary moving beam bar code scanner located along a moving conveyor.

label. These applications tend to require multiple printers distributed at locations where they are needed. The printing technologies used in these applications include [7], [9], [14]

- *Ink-jet.* The ink-jet bars are formed by overlapping dots, and the dots are made by ink droplets. Recent advances in ink-jet technology, motivated by the personal computer market, have improved the resolution of ink-jet printing, and so bar codes of high density are possible at relatively low cost.

- *Direct thermal.* In this technique, light-colored paper labels are coated with a heat-sensitive chemical that darkens when heated. The printing head of the thermal printer consists of a linear array of small heating elements that heat localized areas of the label as it moves past the head, causing the desired bar code image to be formed. Bar codes by direct thermal printing are of good quality, and the cost is low. Care must be taken with the printed label to avoid prolonged exposure to elevated temperatures and ultraviolet light.

- *Thermal transfer.* This technology is similar to direct thermal printing, except that the thermal printing head is in contact with a special ink ribbon that transfers its ink to the moving label in localized areas when heated. Unlike direct thermal printing, this technique can use plain (uncoated) paper, and so the concerns about ambient temperature and ultraviolet light do not apply. The disadvantage is that the thermally activated ink ribbon is consumed in the printing process and must be periodically replaced.

- *Laser printing.* Laser printing is widely used in printers for personal computers. In laser printing, the bar code image is written onto a photosensitive surface (usually a rotating drum) by a controllable light source (the laser), forming an electrostatic image on the surface. The surface is then brought into contact with toner particles that are attracted to selected regions of the image. The toner image is then transferred to plain paper (the label) and cured by heat and pressure. High-quality bar codes can be printed by this technique.

- *Laser etching.* A laser etching process can mark bar codes onto metal parts, providing a permanent identification mark on the item that is not susceptible to damage in the harsh environments encountered in many manufacturing operations. Other processes also used to form permanent three-dimensional bar codes on parts include molding, casting, engraving, and embossing [7]. Special scanners are required to read these codes.

Examples of applications of these individualized bar code printing methods include keyboard entry of data for inclusion in the bar code of each item that is labeled, unique identification of production lots for pharmaceutical products, and preparation of route sheets and other documents included in a shop packet traveling with a production order, as in Figure 12.7. Production workers use bar code readers to indicate order number and completion of each step in the operation sequence.

12.2.2 Two-Dimensional Bar Codes

The first two-dimensional (2-D) bar code was introduced in 1987. Since then, more than a dozen 2-D symbol schemes have been developed, and the number is expected to increase. The advantage of 2-D codes is their capacity to store much greater amounts of data at higher area densities. Their disadvantage is that special scanning equipment is required to read the codes, and the equipment is more expensive than scanners used for conventional bar codes. Two-dimensional symbologies divide into two basic types: (1) stacked bar codes and (2) matrix symbologies.

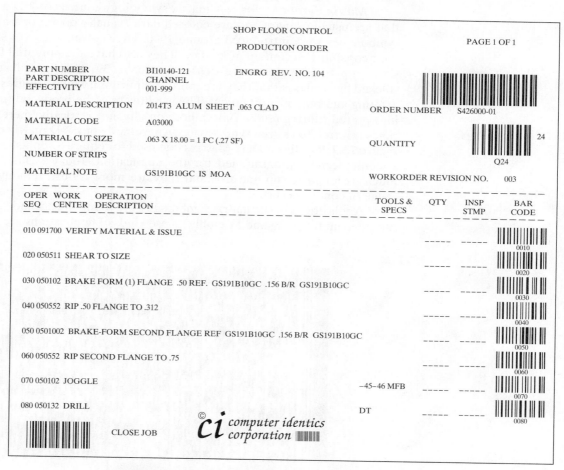

Figure 12.7 Bar-coded production order and route sheet. (Courtesy of Computer Identics Corporation.)

Stacked Bar Codes. The first 2-D bar code to be introduced was a stacked symbology. It was developed in an effort to reduce the area required for a conventional bar code. But its real advantage is that it can contain significantly greater amounts of data. A stacked bar code consists of multiple rows of conventional linear bar codes stacked on top of each other. Several stacking schemes have been devised over the years, nearly all of which allow for multiple rows and variations in the numbers of encoded characters possible. An example of a 2-D stacked bar code is illustrated in Figure 12.8. The data density of stacked bar codes is typically five to seven times that of the linear bar code 39.

The encoded data in a stacked bar code are decoded using laser-type scanners that read the lines sequentially. The technical problems encountered in reading a stacked bar code include (1) keeping track of the different rows during scanning, (2) dealing with scanning swaths that cross between rows, and (3) detecting and correcting localized errors [14]. As in linear bar codes, printing defects in the 2-D bar codes are also a problem.

Matrix Symbologies. A matrix symbology consists of 2-D patterns of data cells that are usually square and are colored dark (usually black) or white. The 2-D matrix symbologies were introduced around 1990. Their advantage over stacked bar codes is their capability to contain more data. They also have the potential for higher data densities—up to 30 times more dense than code 39. Their disadvantage compared to stacked bar codes is that they are more complicated, which requires more sophisticated printing and reading equipment. The symbols must be produced (during printing) and interpreted (during reading) both horizontally and vertically; therefore they are sometimes referred to as area symbologies. An example of a 2-D matrix code is illustrated in Figure 12.9. Reading a Data Matrix code used to require a sophisticated machine vision system specially programmed for the application. Today's Data Matrix readers are much easier to set up and use, and they are more robust, reliably operating under a range of conditions [4].

Applications of the matrix symbologies are found in part and product identification during manufacturing and assembly. These kinds of applications are expected to grow as

Figure 12.8 A 2-D stacked bar code. Shown is an example of a PDF417 symbol.

Figure 12.9 A 2-D matrix bar code. Shown is an example of the Data Matrix symbol.

computer-integrated manufacturing becomes more pervasive throughout industry. The semiconductor industry has adopted Data Matrix ECC200 (a variation of the Data Matrix code shown in Figure 12.9) as its standard for marking and identifying wafers and other electronic components [12].

12.3 RADIO FREQUENCY IDENTIFICATION

Radio frequency identification technology (RFID) represents the biggest challenge to the dominance of bar codes. Companies including Wal-Mart, Target, and Metro AG (in Germany), as well as the U.S. Department of Defense, have mandated that their suppliers use RFID on incoming materials. In fact the Department of Defense requires a combination of Data Matrix and RFID on all "mission-critical" parts, assemblies, and equipment [4]. These requirements have provided a significant impetus for the implementation of RFID in industry. To use Wal-Mart as an example, nearly 1000 of its stores and 600 of its suppliers are currently using RFID at time of writing (end of 2006). According to a study of Wal-Mart cited by Weber [17], "RFID stores are 63 percent more effective in replenishing out-of-stock items than traditional stores."

In radio frequency identification, an "identification tag" containing electronically encoded data is attached to the subject item, which can be a part, product, or container (e.g., carton, tote pan, pallet). The identification tag consists of an integrated circuit chip and a small antenna. These components are usually enclosed in a protective plastic container, but they can also be imbedded in labels that are attached to containers. The tag is designed to satisfy the Electronic Product Code (EPC) standard, which is the RFID counterpart to the Universal Product Code (UPC) used in bar codes. The tag communicates the encoded data by RF to a reader as the item is brought into the reader's proximity. The reader can be portable or stationary. It decodes and confirms the RF signal before transmitting the associated data to a collection computer system.

Although the RF signals are similar to those used in wireless radio and television transmission, there are differences in how RF technology is used in product identification.

One difference is that the communication is in two directions rather than in one direction as in commercial radio and TV. The identification tag is a *transponder*, a device that emits a signal of its own when it receives a signal from an external source. To activate it, the reader transmits a low-level RF magnetic field that serves as the power source for the transponder when they are in close proximity. Another difference between RFID and commercial radio and TV is that the signal power is substantially lower in RFID applications (milliwatts to several watts), and the communication distances usually range between several millimeters and several meters. Finally, there are differences in the allowable frequencies that can be used for RFID applications versus radio, TV, and other commercial and military users.

RF identification tags are available in two general types: (1) passive and (2) active. *Passive tags* have no internal power source; they derive their electrical power for transmitting a signal from radio waves generated by the reader when in close proximity. *Active tags* include their own battery power packs. Passive tags are smaller, less expensive, longer lasting, and have a shorter radio communication range. Active tags generally possess a larger memory capacity and a longer communication range (typically 10 m and more). Applications of active tags tend to be associated with higher value items due to the higher cost per tag.

One of the initial uses of RFID was in Britain in World War II to distinguish between enemy and allied airplanes flying across the English Channel. Commercial and military aircraft still use transponders for identification purposes. Another early application was tracking railway cargo. In this application, the term "tag" may be misleading, because a brick-sized container was used to house the electronics for data storage and RF communications. Subsequent applications use tags available in a variety of different forms, such as credit-card-sized plastic labels for product identification and very small glass capsules injected into wild animals for tracking and research purposes. The principal applications of RFID in industry (in approximate descending order of frequency) are (1) inventory management, (2) supply chain management, (3) tracking systems, (4) warehouse control, (5) location identification, and (6) work-in-progress [17].

Identification tags in RFID have traditionally been read-only devices that contain up to 20 characters of data identifying the item and representing other information that is to be communicated. Advances in the technology have provided much higher data storage capacity and the ability to change the data in the tag (read/write tags). This has opened opportunities for incorporating much more status and historical information into the automatic identification tag rather than using a central database. Table 12.3 compares the two major AIDC technologies, bar codes, and RFID.

TABLE 12.3 Bar Codes versus Radio Frequency Identification

Comparison	Bar Codes	RFID
Technology	Optical	Radio frequency
Read-write capability	Read only	Read-write available
Memory capacity	14 to 16 digits (linear bar codes)	96 to 256 digits
Line-of-sight reading	Required	Not required
Reusability	One-time use	Reusable
Cost per label	Very low cost per label	Approx.10 times cost of bar code
Durability	Susceptible to dirt and scratches	More durable in plant environment

Source : Based mostly on Weber [17].

Advantages of RFID include the following: (1) identification does not depend on physical contact or direct line of sight observation by the reader, (2) much more data can be contained in the identification tag than with most AIDC technologies, and (3) data in the read/write tags can be altered for historical usage purposes or reuse of the tag. The disadvantage of RFID is that the labels and hardware are more expensive than for most other AIDC technologies. For this reason, RFID systems have traditionally been appropriate only for data collection situations in which environmental factors preclude the use of optical techniques such as bar codes, for example, to identify products in manufacturing processes that would obscure any optically coded data (e.g., spray painting). The applications are now expanding beyond these limits due to the mandates set forth by Wal-Mart, the Department of Defense, and others.

In addition to RF identification, radio frequencies are also widely used to augment bar code and other AIDC techniques by providing the communication link between remote bar code readers and some central terminal. This latter application is called radio frequency data communication (RFDC), as distinguished from RFID.

12.4 OTHER AIDC TECHNOLOGIES

The other automated identification and data collection techniques are either used in special applications in factory operations, or they are widely applied outside the factory.

12.4.1 Magnetic Stripes

Magnetic stripes attached to the product or container are used for item identification in factory and warehouse applications. A magnetic stripe is a thin plastic film containing small magnetic particles whose pole orientations can be used to encode bits of data into the film. The film can be encased in or attached to a plastic card or paper ticket for automatic identification. These are the same kinds of magnetic stripes used to encode data onto plastic credit cards and bank access cards. Two advantages of magnetic stripes are their large data storage capacity and the ability to alter the data contained in them. Although they are widely used in the financial community, their use seems to be declining in shop floor control applications for the following reasons: (1) the magnetic stripe must be in contact with the scanning equipment for reading to be accomplished, (2) there are no convenient shop floor encoding methods to write data into the stripe, and (3) the magnetic stripe labels are more expensive than bar code labels.

12.4.2 Optical Character Recognition

Optical character recognition (OCR) is the use of specially designed alphanumeric characters that are machine readable by an optical reading device. Optical character recognition is a 2-D symbology, and scanning involves interpretation of both the vertical and horizontal features of each character during decoding. Accordingly, when manually operated scanners are used, a certain level of skill is required by the human operator, and first read rates are relatively low (often less than 50% [14]). The substantial benefit of OCR technology is that the characters and associated text can be read by humans as well as by machines.

As an interesting historical note, OCR was selected as the standard automatic identification technology by the National Retail Merchants Association (NRMA) shortly after the UPC bar code was adopted by the grocery industry. Many retail establishments made the investment in OCR equipment at that time. However, the problems with the technology became apparent by the mid-1980s [14]: (1) low first read rate and high substitution error rate when hand-held scanners were used, (2) lack of an omnidirectional scanner for automatic checkout, and (3) widespread and growing adoption of bar code technology. NRMA was subsequently forced to revise its recommended standard from OCR technology to bar codes.

For factory and warehouse applications, the list of disadvantages includes (1) the requirement for near-contact scanning, (2) lower scanning rates, and (3) higher error rates compared to bar code scanning.

12.4.3 Machine Vision

The principal application of machine vision is for automated inspection tasks (Section 22.6). For AIDC applications, machine vision systems are used to read 2-D matrix symbols, such as Data Matrix (Figure 12.9), and they can also be used for stacked bar codes, such as PDF-417 (Figure 12.8) [11]. Applications of machine vision also include other types of automatic identification problems, and these applications may grow in number as the technology advances. For example, machine vision systems are capable of distinguishing among a limited variety of products moving down a conveyor so that the products can be sorted. The recognition task is accomplished without using special identification codes on the products and is instead based on the inherent geometric features of the object.

REFERENCES

[1] Accu-Sort Systems, Inc., *Bar Code Technology—Present State, Future Promise*, 2d ed., Telford, PA (no date).

[2] "AIDC Technologies—Who Uses Them and Why," *Modern Materials Handling*, March 1993, pp. 12–13.

[3] AGAPAKIS, J., and A. STUEBLER, "Data Matrix and RFID – Partnership in Productivity," *Assembly*, October 2006, pp. 56–59.

[4] ALLAIS, D. C., *Bar Code Symbology*, Intermec Corporation, 1984.

[5] ATTARAN, M., "RFID Pays Off," *Industrial Engineer*, September 2006, pp. 46–50.

[6] Automatic Identification Manufacturers, *Automatic Identification Manufacturers Manual*, Pittsburgh, PA.

[7] "Bar Codes Move into the Next Dimension," *Modern Materials Handling/AIDC News & Solutions*, June 1998, p. A11.

[8] COHEN, J., *Automatic Identification and Data Collection Systems*, McGraw-Hill Book Company Europe, Berkshire, UK, 1994.

[9] FORCINO, H., "Bar Code Revolution Conquers Manufacturing," *Managing Automation*, July 1998, pp. 59–61.

[10] KINSELLA, B., "Delivering the Goods," *Industrial Engineer*, March 2005, pp. 24–30.

[11] MOORE, B., "New Scanners for 2D Symbols," *Material Handling Engineering*, March 1998, pp. 73–77.

[12] NAVAS, D., "Vertical Industry Overview: Electronics '98," *ID Systems*, February 1998, pp. 16–26.

[13]   NELSON, B., *Punched Cards to Bar Codes*, Helmers Publishing, Inc., Peterborough, NH, 1997.

[14]   PALMER, R. C., *The Bar Code Book*, 3rd ed., Helmers Publishing, Inc., Peterborough, NH, 1995.

[15]   "RFID: Wal-Mart Has Spoken. Will You Comply?" *Material Handling Management*, December 2003, pp. 24–30.

[16]   SOLTIS, D. J., "Automatic Identification System: Strengths, Weaknesses, and Future Trends," *Industrial Engineering*, November 1985, pp. 55–59.

[17]   WEBER, A., "RFID on the Line," *Assembly*, January 2006, pp. 78–92.

[18]   Website of AIM USA: *www.aimusa.org/techinfo/aidc.html*.

REVIEW QUESTIONS

12.1   What is automatic identification and data capture?

12.2   What are the drawbacks of manual collection and entry of data?

12.3   What are the three principal components in automatic identification technologies?

12.4   Name four of the six categories of AIDC technologies that are identified in the text.

12.5   Name five common applications of AIDC technologies in production and distribution.

12.6   There are two forms of linear bar codes. Name them, and indicate what the difference is.

12.7   What was the major industry to first use the Universal Product Code (UPC)?

12.8   What are the two basic types of two-dimensional bar codes?

12.9   What does RFID stand for?

12.10   What is a transponder in RFID?

12.11   What is the difference between a passive tag and an active tag?

12.12   What are the relative advantages of RFID over bar codes?

12.13   What are the relative advantages of bar codes over RFID?

12.14   Why are magnetic stripes not widely used in factory floor operations?

12.15   What is the advantage of optical character recognition technology over bar code technology?

12.16   What is the principal application of machine vision in industry?

Introduction to Manufacturing Systems

In this part of the book, we consider how automation and material handling technologies, as well as human workers, are synthesized to create manufacturing systems. We define a *manufacturing system* to be a collection of integrated equipment and human resources, whose function is to perform one or more processing and/or assembly operations on a starting raw material, part, or set of parts. The integrated equipment includes production machines and tools, material handling and work positioning devices, and computer systems. Human resources are required either full-time or periodically to keep the system

From Chapter 13 of *Automation, Production Systems, and Computer-Integrated Manufacturing*, Third Edition.
Mikell P. Groover. Copyright © 2008 by Pearson Education, Inc. Publishing as Prentice Hall. All rights reserved.

running. The manufacturing system is where the value-added work is accomplished on the parts and products. The position of the manufacturing system in the larger production system is seen in Figure 13.1. The following are examples of manufacturing systems described in this part of the book.

- *Single station cell.* A common situation is one worker tending one production machine that operates on semi-automatic cycle.
- *Machine cluster.* One worker tends a group of semi-automatic machines.
- *Manual assembly line.* This is a production line consisting of a series of workstations at which assembly operations are performed to gradually build a product such as an automobile. Human workers perform the assembly tasks as the product is moved along the line, usually by mechanized conveyor.
- *Automated transfer line.* This is a production line consisting of a series of automated workstations that perform processing operations such as machining. The transfer of workparts between stations is also automated.
- *Automated assembly system.* This system performs a sequence of automated or mechanized assembly operations. The products are generally simpler than those made on a manual assembly line, for example, ballpoint pens, light bulbs, and small electric motors.
- *Machine cell.* This series of manually operated production machines and workstations is often laid out in a U-shaped configuration. It performs a sequence of operations on a family of parts or products that are similar but not identical. The term cellular manufacturing is often applied to this form of manufacturing system.
- *Flexible manufacturing system (FMS).* This is a highly automated machine cell that produces part or product families. The most common form of FMS consists of workstations that are CNC machine tools.

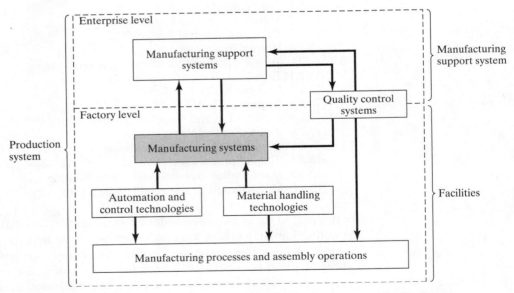

Figure 13.1 The position of the manufacturing system in the larger production system.

In the present chapter, we provide an overview of these manufacturing systems by describing their common components and features. We then develop a framework for how the components are combined and organized into systems to achieve various capabilities in production.

13.1 COMPONENTS OF A MANUFACTURING SYSTEM

A manufacturing system consists of several components. In a given system these components usually include (1) production machines plus tools, fixtures, and other related hardware, (2) a material handling system, (3) a computer system to coordinate and/or control the preceding components, and (4) human workers to operate and manage the system.

13.1.1 Production Machines

In virtually all modern manufacturing systems, most of the actual processing or assembly work is accomplished by machines or with the aid of tools. In terms of worker participation, the machines can be classified as (a) manually operated, (b) semi-automated or (c) fully automated. The three types are depicted graphically in Figure 13.2.

Manually operated machines are controlled or supervised by a human worker. The machine provides the power for the operation and the worker provides the control. Conventional machine tools (such as lathes, milling machines, drill presses) fit into this category. The worker must be at the machine continuously to engage the feed, position the tool, load and unload workparts, and perform other tasks related to the operation.

A *semi-automated machine* performs a portion of the work cycle under some form of program control, and a worker tends to the machine for the remainder of the cycle, as indicated in Figure 13.2(b). An example of this category is a CNC lathe or other programmable production machine that is controlled for most of the work cycle by the part program, but requires a worker to unload the finished part and load the next workpiece at the end of each cycle of the part program. In these cases, the worker must attend to the machine every cycle, but need not be continuously present during the cycle. If the automatic machine cycle takes, say, 10 minutes while the part unloading and loading portion of the work cycle only takes one minute, then the worker may be able to tend several machines. We analyze this possibility in Chapter 14 (Section 14.4.2).

What distinguishes a *fully automated machine* from the two previous types is the capability to operate with no human attention for periods of time that are longer than one work cycle. Although a worker's attention is not required during each cycle, some form of machine tending may be needed periodically. For example, after a certain number of cycles, a new supply of raw material may need to be loaded into the automated machine.

In manufacturing systems, we use the term *workstation* to refer to a location in the factory where some well-defined task or operation is accomplished by an automated machine, a worker-and-machine combination, or a worker using hand tools and/or portable powered tools. In this last case, there is no definable production machine at the location. Many assembly tasks are in this category. A given manufacturing system may consist of one or more workstations. A system with multiple stations is called a

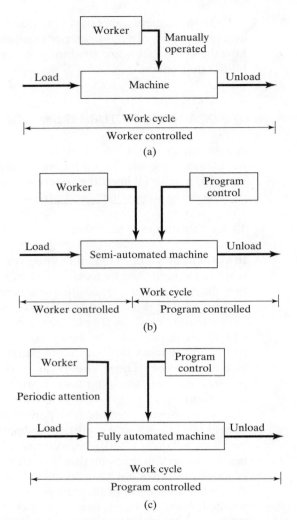

Figure 13.2 Three types of production machines are (a) manually operated, (b) semi-automated, and (c) fully automated.

production line, assembly line, machine cell, or other name, depending on its configuration and function.

13.1.2 Material Handling System

In most processing and assembly operations performed on discrete parts and products, the following material handling functions must be performed: (1) *loading* work units at each station, (2) *positioning* the work units at the station, and (3) *unloading* the work units from the station. In manufacturing systems composed of multiple workstations, (4) *transporting* work units between stations is also required. In many cases, workers perform these functions, but more often some form of mechanized or automated material transport system (Chapter 10) is used to reduce the human effort. Most material transport systems used in production also provide (5) a *temporary storage* function as well. The purpose of storage in these systems is usually to make sure that work is always present for the stations, so that the stations are not starved (meaning that they have nothing to work on).

Some of the issues related to the material handling system are unique to the particular type of manufacturing system, so it makes sense to discuss the details of each handling system when we discuss the manufacturing system itself in later chapters. Our discussion here is concerned with more general material handling issues.

Loading, Positioning, and Unloading. These three material handling functions occur at each workstation. Loading involves moving the work units into the production machine or processing equipment from a source inside the station. For example, starting parts in batch processing operations are often stored in containers (pallets, tote bins, etc.) in the immediate vicinity of the station. For most processing operations, especially those requiring accuracy and precision, the work unit must be positioned in the production machine. Positioning requires the part to be in a known location and orientation relative to the workhead or tooling that performs the operation. Positioning in the production equipment is often accomplished by means of a workholder. A *workholder* is a device that accurately locates, orients, and clamps the part for the operation, and resists any forces that may occur during processing. Common workholders include jigs, fixtures, and chucks. When the production operation has been completed, the work unit must be unloaded, that is, removed from the production machine and either placed in a container at the workstation or prepared for transport to the next workstation in the processing sequence. "Prepared for transport" may simply mean the part is loaded onto a conveyor leading to the next station.

When the production machine is manually operated or semi-automatic, loading, positioning, and unloading are performed by the worker either by hand or with the aid of a hoist. In fully automated stations, a mechanized device such as an industrial robot, parts feeder, coil feeder (in sheet metal stamping), or automatic pallet changer is used to accomplish these material handling functions.

Work Transport between Stations. In the context of manufacturing systems, work *transport* means moving parts between workstations in a multi-station system. The transport function can be accomplished manually or by appropriate material transport equipment.

In some manufacturing systems, work units are passed from station to station by hand, either one at a time or in batches. Moving parts in batches is generally more efficient according to the Unit Load Principle (Section 10.1.2). Manual work transport is limited to cases in which the parts are small and light, so that the manual labor is ergonomically acceptable. When the load to be moved exceeds certain weight standards, powered hoists (Section 10.2.5) and similar lift equipment are used. Manufacturing systems that utilize manual work transport include manual assembly lines and group technology machine cells.

Various types of mechanized and automated material handling equipment are widely used to transport work units in manufacturing systems. We distinguish two general categories of work transport, according to the type of routing between stations: (1) fixed routing and (2) variable routing. In *fixed routing*, the work units always flow through the same sequence of workstations. This means that the work units are identical, or similar enough that the processing sequence is identical. Fixed routing transport is commonly used on production lines. In *variable routing*, work units are transported through a variety of different station sequences. This means that the manufacturing system processes or assembles different types of work units. Variable routing transport is associated with job shop production and many batch production operations. Manufacturing systems that use variable routing include machine cells and flexible manufacturing systems. The difference between fixed and variable routing is portrayed in Figure 13.3. Table 13.1 lists some of the typical material transport equipment used for the two types of part routing.

Pallet Fixtures and Work Carriers in Transport Systems. Depending on the geometry of the work units and the nature of the processing and/or assembly operations performed, the transport system may be designed to accommodate some form of pallet fixture. A *pallet fixture* is a workholder that is designed to be transported by the material handling system. The part is accurately attached to the fixture on the upper face of the pallet, and the under portion of the pallet is designed to be moved, located, and clamped in position at each workstation in the system. Since the part is accurately located in the fixture, and the pallet is accurately clamped at the station, the part is accurately located at each station for processing or assembly. Use of pallet fixtures is common in automated manufacturing systems, such as single machine cells with automatic pallet changers, transfer lines, and automated assembly systems.

The fixtures can be designed with modular features that allow them to be used for different workpart geometries. With different components and a few adjustments, the fixture can accommodate variations in part sizes and shapes. These *modular pallet fixtures* are ideal for use in flexible manufacturing systems.

Alternative methods of workpart transport avoid the use of pallet fixtures. Instead, parts are moved by the handling system either with or without work carriers. A *work*

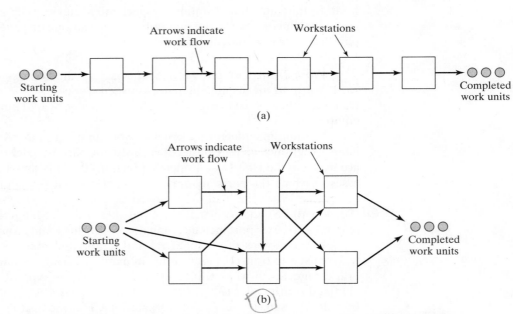

Figure 13.3 Two types of routing in multi-station manufacturing systems are (a) fixed routing and (b) variable routing.

TABLE 13.1 Common Material Transport Equipment Used for Fixed and Variable Routing in Multiple Station Manufacturing Systems

Type of Part Routing	Fixed Routing	Variable Routing
Material handling equipment (described in Chapters 10 and 16)	Powered roller conveyor Belt conveyor Drag chain conveyor Overhead trolley conveyor Rotary indexing mechanisms Walking beam transfer equipment	Automated guided vehicle system Power-and-free overhead conveyor Monorail system Cart-on-track conveyor

carrier is a container (for example, tote pan, flat pallet, or wire basket) that holds one or more parts and can be moved in the system. Work carriers do not fixture the part(s) in an exact position. Their role is simply to contain parts during transport. When the parts arrive at the desired destination, any locating requirements for the next operation must be satisfied at that station (this is usually done manually).

An alternative to using pallet fixtures or work carriers is *direct transport*, in which the transport system is designed to move the work unit itself. The obvious benefit of this arrangement is that it avoids the expense of purchasing pallet fixtures or work carriers, as well as the ongoing costs of returning them to the starting point in the system for reuse. In manually operated manufacturing systems, direct transport is quite feasible, since any positioning required at workstations can be accomplished by workers. In automated manufacturing systems, in particular systems that require accurate positioning at workstations, the feasibility of direct transport depends on the part's geometry and whether an automated handling method can be devised that is capable of moving, locating, and clamping the part with sufficient precision and accuracy. Not all part shapes allow for direct handling by a mechanized or automated system.

13.1.3 Computer Control System

In modern automated manufacturing systems, a computer system is required to control the automated and semi-automated equipment and to participate in the overall coordination and management of the manufacturing system. Even in manually driven manufacturing systems, such as manual assembly lines, a computer system is useful to support production. Typical computer system functions include the following:

- *Communicate instructions to workers.* In manually operated workstations that perform different tasks on different work units, operators must receive processing or assembly instructions for the specific work unit.
- *Download part programs.* The computer sends these instructions to computer-controlled machines.
- *Control material handling system.* This function also coordinates the activities of the material handling system with those of the workstations.
- *Schedule production.* Certain production scheduling functions may be accomplished at the site of the manufacturing system.

- *Diagnose failures.* This involves diagnosing equipment malfunctions, preparing preventive maintenance schedules, and maintaining the spare parts inventory.
- *Monitor safety.* This function ensures that the system does not operate in an unsafe manner. The goal of safety monitoring is to protect both the human workers and the equipment comprising the system.
- *Maintain quality control.* The purpose of this control function is to detect and reject defective work units produced by the system.
- *Manage operations.* This consists of managing the overall operations of the manufacturing system, either directly (by supervisory computer control) or indirectly (by preparing the necessary reports for management personnel).

13.1.4 Human Resources

In many manufacturing systems, humans perform some or all of the value-added work that is accomplished on the parts or products. In these cases, the human workers are referred to as *direct labor*. Through their physical labor, they directly add to the value of the work unit by performing manual work on it or by controlling the machines that perform the work. In systems that are fully automated, direct labor is still needed to perform activities such as loading and unloading parts, changing tools, and resharpening tools. Human workers are also needed in automated manufacturing systems to manage or support the system as computer programmers, computer operators, part programmers for CNC machine tools (Chapter 7), maintenance and repair personnel, and similar indirect roles. In automated systems, the distinction between direct and indirect labor is not always precise.

13.2 A CLASSIFICATION SCHEME FOR MANUFACTURING SYSTEMS

In this section, we explore the various types of manufacturing systems and develop a classification scheme based on the factors that define and distinguish them. The factors are (1) types of operations performed, (2) number of workstations, (3) system layout, (4) automation and manning level, and (5) part or product variety. These five factors are briefly identified in Table 13.2 and discussed below.

TABLE 13.2 Factors in Manufacturing Systems Classification Scheme

Factor	Alternatives
Types of operations performed	Processing operations versus assembly operations Types of processing or assembly operations
Number of workstations	Single-station cell versus multi-station system
System layout	For more than one station, fixed routing versus variable routing
Automation and manning level	Manual or semi-automated workstations that require full-time operator attention versus fully automated stations that require only periodic worker attention
Part or product variety	Identical work units versus variations in work units that require differences in processing

13.2.1 Types of Operations Performed

First of all, manufacturing systems are distinguished by the types of operations they perform. At the highest level, the distinction is between (1) processing operations on individual work units, and (2) assembly operations to combine individual parts into assembled entities. Beyond this distinction, there are the technologies of the individual processing and assembly operations (Section 2.2.1).

Additional parameters of the product that play a role in determining the design of the manufacturing system include the following:

- *Type of material processed.* Different engineering materials require different types of processes. Processing operations used for metals are usually different from those used for plastics or ceramics. These differences affect the type of equipment and handling method in the manufacturing system.
- *Size and weight of the part or product.* Larger and heavier work units require bigger equipment with greater power capacity. Safety hazards increase with the size and weight of parts and products.
- *Part or product complexity.* In general, part complexity correlates with the number of processing operations required, and product complexity correlates with the number of components that must be assembled.
- *Part geometry.* Machined parts can be classified as rotational or non-rotational. Rotational parts are cylindrical or disk-shaped and require turning and related rotational operations. Non-rotational parts are rectangular or cube-like and require milling and related machining operations to shape them. Manufacturing systems that perform machining operations must be distinguished according to whether they make rotational or non-rotational parts. The distinction is important not only because of differences in the machining processes and machine tools required, but because the material handling system must be engineered differently for the two cases.

13.2.2 Number of Workstations

The number of workstations is a key factor in our manufacturing systems classification scheme. It exerts a strong influence on the performance of the manufacturing system in terms of performance factors such as workload capacity, production rate, and reliability. Let us denote the number of workstations in the system by the symbol n. The individual stations in a manufacturing system can be identified by the subscript i, where $i = 1, 2, \ldots, n$. This might be useful in identifying parameters of the individual workstations, such as operation time or number of workers at a station. In our classification scheme, we distinguish between single-station cells ($n = 1$) and multi-station systems ($n > 1$).

The number of workstations in the manufacturing system is a convenient measure of its size. As the number of stations increases, the amount of work that can be accomplished by the system increases. This may translate into a higher production rate, certainly compared to the output of a single station, but also compared to the same number of single stations working independently. There must be a synergistic benefit obtained from multiple stations working together rather than independently; otherwise, it makes more sense for the stations to work as independent entities. The synergistic benefit is usually derived from the fact that the total amount of work performed on the part or product is too complex to accomplish at a single workstation. There are too many tasks to perform at one station. When separate tasks are assigned to individual stations, the task performed at each station is simplified. We build on this notion in Section 13.2.3.

More stations also mean that the system is more complex, and therefore more difficult to manage and maintain. The system consists of more workers, more machines, and more parts being handled. The material handling system is more complex in a multi-station system. It becomes increasingly complex as n increases. The logistics and coordination of the system is more involved. Reliability and maintenance problems occur more frequently.

13.2.3 System Layout

Closely related to the number of workstations is the configuration of the workstations, that is, the way the system is laid out. This, of course, applies mainly to systems with multiple stations. Workstation layouts organized for fixed routing are usually arranged linearly, as in a production line, while layouts organized for variable routing can have a variety of possible configurations. The layout of stations is an important factor in determining the most appropriate material handling system.

The relationship between the two factors, number of workstations and system layout, is depicted in Table 13.3. This relationship applies to manufacturing systems that perform either processing or assembly operations. Although these operations are different, the manufacturing systems to perform them possess similar configurations. For example, some production lines perform processing operations, while others perform assembly operations.

Let us consider the relationship between the two factors, number of stations and system layout. The most obvious relationship deals with the workload capacity of the system. The *workload* is the amount of processing or assembly work accomplished by the system, expressed in terms of the time required to perform the work. It is the sum of the cycle times of all the work units completed by the system in a given period of interest. It stands to reason that two workstations can accomplish twice the workload of one station. Thus, one obvious relationship is that the workload capacity of a manufacturing system increases in proportion to the number of workstations in it.

The question remains why a multi-system manufacturing system with n stations would have any advantage over n single stations. If workload capacity is proportional to the number of stations, then why is one n-station system not equivalent to n single stations? The answer is that in manufacturing systems with multiple workstations ($n > 1$), the total work content required to process or assemble one work unit is divided among the stations so that different tasks are performed by different stations. The different stations are designed to

TABLE 13.3 Relationship Between Number of Workstations and System Layout in Manufacturing Systems

Number of Workstations	$n = 1$	$n \geq 2$
System layout	Single-station cell	Multi-station system with fixed routing (e.g., production line) Multi-station system with variable routing (various layouts possible)

specialize in their own assigned tasks. The total work content to produce one work unit would be too much to complete at one station, because the sum of the tasks involves a scope and complexity that is beyond the capability of one workstation. By breaking the total work content down into tasks, and assigning different tasks to different stations, the work at each station is simplified. This is what provides a multi-station system with its synergistic benefit, referred to in Section 13.2.2. Because of the specialization that is designed into each station in a multi-station system, such a system is able to deal with product complexity better than the same number of single-stations that each performs the total work content on the part or product. The result is a higher production rate for complex parts and products. Automobile final assembly plants are a good example of this advantage. The total work content required to assemble each car in the plant is typically 15 to 20 hours – too much time and too much complexity for one workstation to cope with. However, when the total work content is divided into simple tasks of about one-minute duration, and these tasks are assigned to individual workers at stations along the line of flow, cars are produced at the rate of about 60 per hour.

13.2.4 Automation and Manning Levels

The level of automation is another factor that characterizes the manufacturing system. As defined above, the workstation machines in a manufacturing system can be manually operated, semi-automated, or automated. Inversely correlated with the level of automation is the proportion of time that a worker must be in attendance at each station. The *manning level* of a workstation, symbolized M_i, is the proportion of time that a worker is at the station. If $M_i = 1$ for station i, it means that one worker must be at the station continuously. If one worker tends four automatic machines, then $M_i = 0.25$ for each of the four machines, assuming each machine requires the same amount of attention. On sections of an automobile final assembly line, there are stations each tended by multiple workers, in which case $M_i = 2$ or 3 or more. In general, high values of M_i ($M_i \geq 1$) indicate manual operations at the workstation, while low values ($M_i < 1$) denote some form of automation.

The average manning level of a multi-station manufacturing system is a useful indicator of the direct labor content of the system. Let us define it as

$$M = \frac{w_u + \sum_{i=1}^{n} w_i}{n} = \frac{w}{n} \tag{13.1}$$

where M = average manning level for the system; w_u = number of utility workers assigned to the system; w_i = number of workers assigned specifically to station i, for $i = 1, 2, \ldots, n$; and w = total number of workers assigned to the system. *Utility workers* are workers who are not specifically assigned to individual processing or assembly stations; instead they perform functions such as (1) relieving workers at stations for personal breaks, (2) maintenance and repair of the system, (3) material handling, and (4) tool changing. Even a fully automated multi-station manufacturing system is likely to have one or more utility workers who are responsible for keeping it running.

Including automation and manning level in our classification scheme, we have two possible levels for single stations and three possible levels for multi-station systems. The two levels for single stations are manned and fully automated. The manned station is identified by the fact that one or more workers must be at the station every cycle. This means that any machine at the station is manually operated or semi-automatic and that manning is equal to or greater than one ($M \geq 1$). However, in some cases, one worker may be able to attend more than one machine (e.g., a machine cluster) if the semi-automatic cycle is long relative to the service required each cycle of the worker (thus, $M < 1$). We discuss machine clusters in Section 14.4.2. A fully automated station requires less than full-time attention of a worker ($M < 1$). For multi-station systems, the same two levels are applicable (manned and fully automated), but a third level is also possible for the system. This is of a hybrid system, in which some stations are manned while others are fully automated. Expanding the information portrayed in previous Table 13.3 to include automation and manning level, we have Table 13.4.

13.2.5 Part or Product Variety

A fifth factor that characterizes a manufacturing system is the degree to which it is capable of dealing with variations in the parts or products it produces. Examples of possible variations that a manufacturing system may have to cope with include

- Variations in type and/or color of plastic of molded parts in injection molding
- Variations in electronic components placed on a standard size printed circuit board
- Variations in the size of printed circuit boards handled by a component placement machine
- Variations in geometry of machined parts
- Variations in parts and options in an assembled product.

In this section, we identify three types of manufacturing systems, distinguished by their capability to cope with part or product variety. We then discuss two ways in which manufacturing systems can be endowed with this capability.

Part or Product Variety: Three Cases. Borrowing from the terminology used in manual assembly line technology (Section 15.1.4), the three cases of part or product variety in manufacturing systems are (1) single model, (2) batch model, and (3) mixed model.

TABLE 13.4 Manufacturing Systems Framework That Includes Automation and Manning Level in Addition to Number of Workstations and System Layout

Number of Workstations	$n = 1$	$n \geq 2$
System layout	Single-station cell Manual Fully automated	Multi-station system with fixed routing (e.g., production line) Manual ($M_i \geq 1$ for all stations) Fully automated ($M_i < 1$ for all stations) Hybrid (some manual and some automated stations) Multi-station system with variable routing (various layouts possible) Manual ($M_i \geq 1$ for all stations) Fully automated ($M_i < 1$ for all stations) Hybrid (some manual and some automated stations)

The three cases are depicted in Figure 13.4, in which the differences in shading of the work units represent the relative degree of part or product variety.

In the *single model case*, all parts or products made by the manufacturing system are identical. There are no variations. In this case, demand for the item must be sufficient to justify dedicating the system to production of that item for an extended period of time, perhaps several years. Equipment associated with the system is specialized and designed for maximum efficiency. Fixed automation (Section 1.2.1) is common in single model systems.

In the *batch model case*, different parts or products are made by the system, but they are made in batches because a changeover in physical setup and/or equipment programming is required between models. Changeover of the manufacturing system is required because the differences in part or product style are significant enough that the system cannot cope unless changes in tooling and programming are made. It is a case of hard product variety (Section 2.3). The time needed to accomplish the changeover requires the system to be operated in a batch mode, in which a batch of one product style is followed by a batch of another, and so on. The changeover time between batches is lost time on the manufacturing system.

In the *mixed model case*, different parts or products are made by the manufacturing system, but the differences are not significant (soft product variety, Section 2.3). Thus, the system is able to handle the differences without the need for time-consuming changeovers in setup or program. This means that the mixture of different styles can be produced continuously rather than in batches. In effect, continuous production of different part or product styles is achieved by designing the system so that whatever changes need to be made from one style to the next can be made quickly enough to economically produce the units in batch sizes of one.

Flexibility in Mixed Model Manufacturing Systems. Flexibility allows a mixed model manufacturing system to cope with a certain level of variation in part or product style without interruptions in production for changeovers between models. Flexibility is

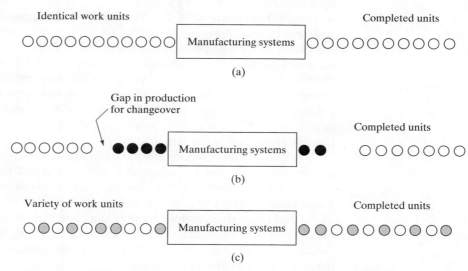

Figure 13.4 Three cases of part or product variety in manufacturing systems: (a) single model case, (b) batch model case, and (c) mixed model case.

generally a desirable feature of a manufacturing system. Systems that possess it are called *flexible manufacturing systems*, or *flexible assembly systems*, or similar names. They can produce different part styles, or they can readily adapt to new part styles when the previous ones become obsolete. In order to be flexible, a manufacturing system must possess the following capabilities:

- *Identification of the different work units.* Different part or product styles require different operations. The manufacturing system must identify the work unit in order to perform the correct operation. In a manually operated or semi-automatic system, this task is usually an easy one for the worker(s). In an automated system, some means of automatic work unit identification must be devised.
- *Quick changeover of operating instructions.* The instructions, or part program in the case of computer controlled production machines, must correspond to the correct operation for the given part. In the case of a manually operated system, this generally means workers who (1) are skilled in the variety of operations needed to process or assemble the different work unit styles, and (2) know which operations to perform on each work unit style. In semi-automatic and fully automated systems, it means that the required part programs are readily available to the control unit.
- *Quick changeover of physical setup.* Flexibility in manufacturing means that the different work units are not produced in batches. To enable different work unit styles to be produced with no time lost between one unit and the next, the flexible manufacturing system must be capable of making any necessary changes in fixturing and tooling in a very short time (the changeover time should correspond approximately to the time required to exchange the completed work unit for the next unit to be processed).

These capabilities are often difficult to engineer. In manually operated manufacturing systems, human errors can cause problems—operators not performing the correct operations on the different work unit styles. In automated systems, sensor systems must be designed to enable work unit identification. Part program changeover is accomplished with relative ease using today's computer technology. Changing the physical setup is often the most challenging problem, and it becomes more difficult as part or product variety increases. Endowing a manufacturing system with flexibility increases its complexity. The material handling system and/or pallet fixtures must be designed to hold a variety of part shapes. The required number of different tools increases. Inspection becomes more complicated because of part variety. The logistics of supplying the system with the correct quantities of different starting workparts is more involved. Scheduling and coordinating the system become more difficult.

Single-station manned cells inherently possess the greatest flexibility. Human workers are dexterous and can adapt to a great variety of tasks requiring a range of skills. Given the proper tools, a worker can change over his or her workstation to accommodate a significant diversity of jobs and work units. However, single stations are limited in terms of the part or product complexity they can cope with. If the work unit is simple, requiring only one or a limited number of processing or assembly operations, then a single-station cell can be justified for high as well as low annual production quantities. Higher quantities make automated cells more attractive.

As the complexity of the work unit increases, the advantage shifts toward multi-station systems. The larger number of tasks and additional tooling required for more complex parts or products begins to overwhelm a single station. Dividing the work among multiple stations is a way to reduce complexity at each station. If there is no product variety or soft product variety, and the product is made in high quantities, then a multi-station system with fixed routing is appropriate. As product variety increases for production quantities in the medium range, a multi-station system with variable routing becomes more appropriate. Variable routing allows different work units to follow their own individual sequence of stations and operations in the system. Finally, in cases of significant product variety and low production quantities, the greatest flexibility is achieved by using a collection of single-station cells, each organized to perform a limited set of tasks but integrated to complete the total work content on each work unit. Of course, what we are describing is a job shop, the most flexible but least efficient of the factory organizations. Much of the discussion in the preceding sections is summarized in the Figure 13.5(a) and (b).

Flexibility itself is a complex issue, certainly more complex than it appears in this introductory treatment of it. It is recognized as an important attribute for a system to possess. We provide a more in-depth discussion of the issue in Chapter 19.

Reconfigurable Manufacturing Systems. In an era when product styles have ever shortening life cycles, the cost of designing, building, and installing a new manufacturing system every time a new part or product must be produced is becoming prohibitive, both in terms of time and money. One alternative is to reuse and reconfigure components of the original system in a new manufacturing system. In modern manufacturing engineering practice, even single model manufacturing systems are being built with features that enable them to change over to new product styles when necessary. These kinds of features include [1]

- *Ease of mobility.* Machine tools and other production machines may be designed with three-point bases that allow then to be lifted readily and moved by a crane or forklift truck. The three-point base facilitates leveling of the machine after moving.
- *Modular design of system components.* This permits hardware components from different machine builders to be connected together.
- *Open architecture in computer controls.* This permits data interchange between software packages from different vendors.

13.3 OVERVIEW OF THE CLASSIFICATION SCHEME

In this final section, we provide an overview of the three basic categories of manufacturing systems: (1) single-station cells, (2) multi-station systems with fixed routing, and (3) multi-station systems with variable routing. These systems are described more completely in the following chapters.[1]

13.3.1 Single-Station Cells

Applications of single workstations are widespread. The typical case is a worker-machine cell. Our classification scheme distinguishes two categories: (1) *manned workstations*, in which a worker must be in attendance either continuously or for a portion of each work cycle, and (2) *automated stations*, in which periodic attention is required less frequently

---

[1]One of the examples used here is the job shop, which is described in Section 2.3. No elaboration of the job shop is provided in the following chapters.

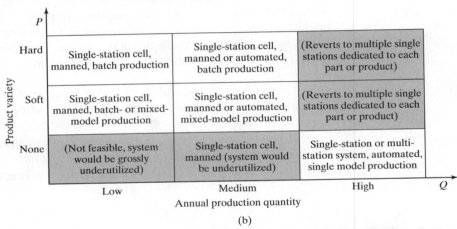

Medium or high part or product complexity (high total work content)

(a)

Low part or product complexity (low total work content)

(b)

Figure 13.5 Types of manufacturing systems that are appropriate for various combinations of part or product complexity, part or product variety, and annual production quantities: (a) high product complexity (high total work content) and (b) low product complexity (low total work content). Gray blocks indicate cases that are infeasible, are unlikely to occur, or revert to some other case.

than every cycle. In either case, these systems are used for processing as well as assembly operations, and their applications include single model, batch model or mixed model production. Several examples of these systems are listed in Table 13.5.

The single model workstation is popular because (1) it is the easiest and least expensive manufacturing system to implement, especially the manned version; (2) it is the most adaptable, adjustable, and flexible manufacturing system; and (3) a manned single workstation can be converted to an automated station if demand for the parts or products made in the station justify this conversion.

TABLE 13.5 Examples of Single-Station Manufacturing Cells

Example	Operation	Automation	Typical Part or Product Variety
Worker at CNC lathe (semi-automated)	Processing	Manned	Batch or mixed model
Worker at stamping press (manual)	Processing	Manned	Single or batch model
Welder and fitter at arc welding setup	Assembly	Manned	Single, batch, or mixed model
CNC turning center with parts carousel operating unattended using a robot to load and unload parts	Processing	Automated	Batch or mixed model
Assembly system in which one robot performs multiple assembly tasks to complete a product	Assembly	Automated	Single or batch model

13.3.2 Multi-Station Systems with Fixed Routing

A multi-station manufacturing system with fixed routing is a production line. A *production line* consists of a series of workstations laid out so that the part or product moves from one station to the next, and a portion of the total work content is performed on it at each station. Transfer of work units from one station to the next is usually accomplished by a conveyor or other mechanical transport system. However, in some cases the work is simply pushed between stations by hand. Production lines are generally associated with mass production, although they can also be applied in batch production. Conditions that favor the use of a production line are the following:

- The quantity of parts or products to be made is very high (up to millions of units)
- The work units are identical or very similar (thus, they require the same or similar operations to be performed in the same sequence)
- The total work content can be divided into separate tasks of approximately equal duration that can be assigned to individual workstations.

Table 13.6 lists some examples of multi-station manufacturing systems with fixed routing, most of which would be called production lines. Production lines are used for either processing or assembly operations, and they can be either manually operated or automated.

TABLE 13.6 Examples of Multi-Station Manufacturing Systems with Fixed Routing

Example	Operation	Automation	Typical Part or Product Variety
Manual assembly line that produces small power tools	Assembly	Manned	Single, batch, or mixed model
Machining transfer line	Processing	Automated	Single model
Automated assembly machine with a carousel system for work transport	Assembly	Automated	Single model
Automobile final assembly plant, in which many of the spot welding and spray painting operations are automated while general assembly is manual	Assembly and processing	Hybrid	Mixed model

Manual production lines usually perform assembly operations, and we discuss manual assembly lines in Chapter 15. Automated lines perform either processing or assembly operations, and we discuss these two system types in Chapters 16 and 17. There are also hybrid systems, in which both manual and automated stations exist in the same line. This case is analyzed in Section 17.2.4.

13.3.3 Multi-Station Systems with Variable Routing

A multiple-station system with variable routing is a group of workstations organized to achieve some special purpose. It is typically intended for production quantities in the medium range (annual production $= 10^2$ to 10^4 parts or products), although its applications sometimes extend beyond these boundaries. The special purpose may be any of the following:

- Production of a family of parts having similar processing operations
- Assembly of a family of products having similar assembly operations
- Production of the complete set of components that are used in the assembly of one unit of final product. Producing all of the parts in one product, rather than performing batch production of the parts, reduces work-in-process inventory.

As this list indicates, multi-station systems with variable routing are applicable to either processing or assembly operations. The list also indicates that the applications usually involve part or product variety, which means differences in operations and sequences of operations that must be performed. The machine groups must possess flexibility in order to cope with this variety. The most flexible machine group for coping with product variety is the job shop, included in the list of examples in Table 13.7. It is really a collection of single-station cells organized to accomplish the particular mission of the shop.

The machines in a multi-station system with variable routing may be manually operated, semi-automatic, or fully automated. When manually operated or semi-automatic, the machine groups are often called *machine cells*, and the use of these cells in a factory is called *cellular manufacturing*. Cellular manufacturing and its companion topic, group technology, are discussed in Chapter 18. When the machines in the group are fully automated, with automated material handling between workstations, the system is referred to as a *flexible manufacturing system* or *flexible manufacturing cell*. We discuss flexibility and flexible manufacturing systems in Chapter 19.

TABLE 13.7 Examples in Multi-Station Manufacturing Systems with Variable Routing

Example	Operation	Automation	Typical Part or Product Variety
Job shop with a process layout consisting of a variety of machine tools that each can be equipped for a variety of machining operations	Processing and assembly	Manned	Batch or mixed model
Group technology machine cell	Processing	Manned	Mixed model
Flexible manufacturing system	Processing	Automated	Mixed model

REFERENCES

[1] ARONSON, R. B., "Operation Plug-and-Play is On the Way," *Manufacturing Engineering*, March 1997, pp. 108–112.

[2] GROOVER, M. P., *Fundamentals of Modern Manufacturing: Materials, Processes, and Systems*, 3d ed., John Wiley & Sons, Inc., Hoboken, NJ, 2007.

[3] GROOVER, M. P., and O. MEJABI, "Trends in Manufacturing System Design," *Proceedings*, IIE Fall Conference, Nashville, TN, November 1987.

REVIEW QUESTIONS

13.1 What is a manufacturing system?

13.2 Name the four components of a manufacturing system.

13.3 What are the three classifications of production machines, in terms of worker participation?

13.4 What are the five material handling functions that must be provided in a manufacturing system?

13.5 What is the difference between fixed routing and variable routing in manufacturing systems consisting of multiple workstations?

13.6 What is a pallet fixture in work transport in a manufacturing system?

13.7 A computer system is an integral component in a modern manufacturing system. Name four of the eight functions of the computer system listed in the text.

13.8 What are the five factors that can be used to distinguish manufacturing systems in the classification scheme proposed in the chapter?

13.9 Why is manning level inversely correlated with automation level in a manufacturing system?

13.10 Name and briefly define the three cases of part or product variety in manufacturing systems.

13.11 What is flexibility in a manufacturing system?

13.12 What are the three capabilities that a manufacturing system must possess in order to be flexible?

Chapter 14

Single-Station Manufacturing Cells

CHAPTER CONTENTS

Single stations constitute the most common manufacturing system in industry. They operate independently of other workstations in the factory, although their activities are coordinated within the larger production system. Single-station manufacturing cells can be manually operated or automated. They are used for either processing or assembly operations. They can be designed for single model production (where all parts or products made by the system are identical), for batch production (where different part styles are made in batches), or for mixed-model production (where different parts are made sequentially, not in batches).

This chapter describes the features and operations of single-station manufacturing cells. Figure 14.1 provides a roadmap of the discussion. We also examine two analysis issues that must be considered in the planning of single-station systems: (1) how many

From Chapter 14 of *Automation, Production Systems, and Computer-Integrated Manufacturing*, Third Edition.
Mikell P. Groover. Copyright © 2008 by Pearson Education, Inc. Publishing as Prentice Hall. All rights reserved.

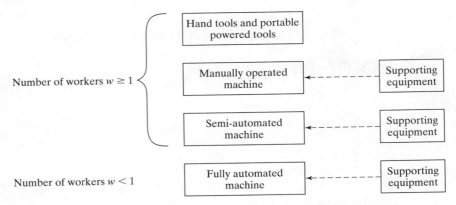

Figure 14.1 Classification of single-station manufacturing cells.

workstations are required to satisfy production requirements, and (2) how many machines can be assigned to one worker in a machine cluster. A *machine cluster* is a collection of two or more identical or similar machines that are serviced by one worker.

14.1 SINGLE-STATION MANNED CELLS

The single-station manned cell, the standard model for which consists of one worker tending one machine, is probably the most widely used production method today. It dominates job shop production and batch production, and it is not uncommon even in high production. There are many reasons for its widespread adoption.

- It requires the shortest amount of time to implement. The user company can quickly launch production of a new part or product using one or more manual workstations, while it plans and designs a more automated production method.
- It requires the least capital investment of all manufacturing systems.
- Technologically, it is the easiest system to install and operate. Its maintenance requirements are usually minimal.
- For many situations, particularly for low quantities of production, it results in the lowest cost per unit produced.
- In general, it is the most flexible manufacturing system with regard to changeovers from one part or product style to the next.

In a one-machine/one-worker station ($n = 1, w = 1$), the machine is manually operated or semi-automated. In a *manually operated station*, the operator controls the machine and loads and unloads the work. A typical processing example is a worker operating a standard machine tool such as an engine lathe, drill press, or forge hammer. The work cycle requires the attention of the worker either continuously or for most of the cycle (for example, the operator might relax temporarily during the cycle when the machine feed is engaged on the lathe or drill press).

The manually operated workstation also includes the case of a worker using hand tools (for example, screwdriver and wrench in mechanical assembly) or portable powered tools (for example, powered hand-held drill, soldering iron, or arc welding gun). The key factor is that the worker performs the task at one location (one workstation) in the factory.

In a *semi-automated station*, the machine is controlled by some form of program, such as a part program that controls a CNC machine tool during a portion of the work cycle, and the worker's function is to load and unload the machine each cycle, and periodically to change cutting tools. In this case, the worker's attendance at the station is required every work cycle, although the worker's attention may not be continuously occupied throughout the cycle.

There are several variations from the standard model of a one-machine/one-worker station. First, the single-station manned cell classification includes the case where two or more workers are needed full-time to operate the machine or accomplish the task at the workplace ($n = 1, w > 1$). Examples include

- Two workers required to manipulate heavy forgings in a forge press
- A welder and fitter working in an arc welding setup
- Multiple workers combining their efforts to assemble one large piece of machinery at a single assembly station.

Another variation from the standard case occurs when there is a principal production machine, plus other equipment in the station that supports the principal machine. The other equipment is clearly subordinate to the main machine. Examples of clearly subordinate equipment include

- Drying equipment used to dry plastic molding powder prior to molding in a manually operated injection molding machine
- A grinder used at an injection molding machine to grind the sprues and runners from plastic moldings for recycling
- Trimming shears used in conjunction with a forge hammer to trim flash from the forgings.

14.2 SINGLE-STATION AUTOMATED CELLS

The single-station automated cell consists of a fully automated machine capable of unattended operation for a time period longer than one machine cycle. A worker is not required to be at the machine except periodically to load and unload parts or otherwise tend it. Advantages of this system include the following:

- Labor cost is reduced compared to the single-station manned cell.
- Among automated manufacturing systems, the single-station automated cell is the easiest and least expensive system to implement.
- Production rates are generally higher than for a comparable manned machine.
- It often represents the first step toward implementing an integrated multi-station automated system. The user company can install and debug the single automated machines individually, and subsequently integrate them (1) electronically by means of a supervisory computer system and/or (2) physically by means of an automated

material handling system. Recall the automation migration strategy from Chapter 1 (Section 1.4.3).

The issue of supporting equipment arises in single-station automated cells, just as it does in manned single-station cells. In the case of a fully automated injection molding machine that uses drying equipment for the incoming plastic molding compound, the drying equipment clearly plays a supporting role to the molding machine. Other examples of supporting equipment in automated cells include

- A robot loading and unloading an automated production machine. The production machine is the principal machine in the cell, and the robot plays a supporting role.
- Bowl feeders and other parts feeding devices used to deliver components in a single robot assembly cell. In this case, the assembly robot is the principal production machine in the cell, and the parts feeders are subordinate.

Let us consider some of the technological features of this type of manufacturing system, beginning with the enablers that make it possible.

14.2.1 Enablers for Unattended Cell Operation

A key feature of a single-station automated cell is its capability to operate unattended for extended periods of time. The enablers required for unattended operation in single and batch model production must be distinguished from those required for mixed model production.

Enablers for Unattended Single Model and Batch Model Production. The technical attributes required for unattended operation of a single model or batch model cell are the following:

- *Programmed cycle* that allows the machine to perform every step of the processing or assembly cycle automatically.
- *Parts storage subsystem* and a supply of parts that permit continuous operation beyond one machine cycle. The storage system must be capable of holding both raw workparts as well as completed work units. This sometimes means that two storage units are required, one for the starting workparts and the second for the completed parts.
- *Automatic transfer of workparts* between the storage system and the machine (automatic unloading of finished parts from the machine and loading of raw workparts to the machine). This transfer is a step in the regular work cycle. The parts storage subsystem and automatic transfer of parts are discussed in more detail in Section 14.2.2.
- *Periodic attention of a worker* who resupplies raw workparts, takes away finished parts, changes tools as they wear out (depending on the process), and performs other machine tending functions that are necessary for the particular processing or assembly operation.
- *Built-in safeguards* that protect the system against operating under conditions that may be (1) unsafe to workers, (2) self-destructive, or (3) destructive to the work units being processed or assembled. Some of these safeguards may simply

be in the form of very high process and equipment reliability. In other cases, the cell must be furnished with the capability for error detection and recovery (Section 4.2.3).

Enablers for Mixed Model Production.   The preceding list of enablers applies to single model and batch model production. In cases when the system is designed to process or assemble a variety of part or product styles in sequence (that is, a flexible manufacturing workstation), then the following additional enablers must be provided:

- *Work identification subsystem* that can distinguish the different starting work units entering the station, so that the correct processing sequence can be used for that part or product style. This may take the form of sensors that can recognize the features of the work unit, or the identification subsystem may consist of automatic identification methods such as bar codes (Chapter 12). In some cases, identical starting work units are subjected to different processing operations according to a specified production schedule. If the starting units are identical, a workpart identification subsystem is unnecessary.
- *Program downloading capability* to transfer the machine cycle program corresponding to the identified part or product style. This assumes that programs have been prepared in advance for all part styles, and that these programs are stored in the machine control unit or that the control unit has access to them.
- *Quick setup changeover capability* so that the necessary workholding devices and other tools for each part are available on demand.

The same enablers that we have described here are required for the unattended operation of workstations in multi-station flexible manufacturing systems discussed in later chapters.

14.2.2   Parts Storage Subsystem and Automatic Parts Transfer

The parts storage subsystem and automatic transfer of parts between the storage subsystem and the processing station are necessary conditions for an automated cell that operates unattended for extended periods of time. The storage subsystem has a designed parts storage capacity n_p. Accordingly, the cell can theoretically operate unattended for a length of time given by

$$UT = \sum_{j=1}^{n_p} T_{cj} \tag{14.1}$$

where UT = unattended time of operation of the manufacturing cell, min; T_{cj} = cycle time for part j that is held in the parts storage subsystem, for $j = 1, 2, \ldots, n_p$, where n_p = parts storage capacity of the storage subsystem, pc. This equation assumes that one work unit is processed each cycle. If all of the parts are identical and require the same machine cycle, then the equation simplifies to the following:

$$UT = n_p T_c \tag{14.2}$$

In reality, the unattended time of operation will be somewhat less than this amount (by one or more cycle times), because the worker needs time to unload all of the finished pieces and load starting work units into the storage subsystem.

Capacities of parts storage subsystems range from one part to hundreds. As Eq. (14.2) indicates, the time of unattended operation increases directly with storage capacity, so there is an advantage to designing the storage subsystem with sufficient capacity to satisfy the plant's operational objectives. Typical objectives for storage capacity include the following, expressed in terms of the time periods of unattended operation:

- A fixed time interval that allows a worker to tend multiple machines
- The time between scheduled tool changes, so that tools and parts can be changed during the same machine downtime
- One complete shift
- Overnight operation, sometimes referred to as *lights out operation*. The objective is to keep the machines running with no workers in the plant during the middle and/or night shifts.

Storage Capacity of One Part. The minimum storage capacity of a parts storage subsystem is one workpart. This case is represented by an automatic parts transfer mechanism operating with manual loading/unloading rather than a parts storage subsystem. An example of this arrangement in machining is a two-position *automatic pallet changer* (APC), used as the parts input/output interface for a CNC machining center. The APC is used to exchange pallet fixtures between the machine tool worktable and the load/unload position. The workparts are clamped and located on the pallet fixtures, so that when the pallet fixture is accurately positioned in front of the spindle, the part itself is accurately located. Figure 14.2 shows an APC set up for manual unloading and loading of parts.

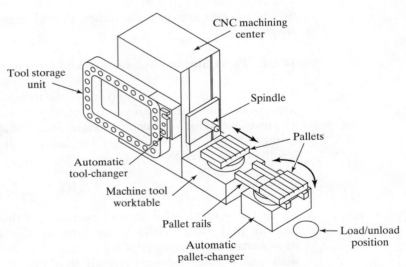

Figure 14.2 Automatic pallet changer integrated with a CNC machining center, set up for manual unloading and loading of workparts. At the completion of the machining cycle, the pallet currently at the spindle is moved onto the automatic pallet changer (APC), and the APC table is rotated 180° to move the other pallet into position for transfer to the machine tool worktable.

When the storage capacity is only one part, this usually means that the worker must be in attendance at the machine full-time. While the machine is processing one workpart, the worker is unloading the piece just finished and loading the next workpart to be processed. This is an improvement over no storage capacity, in which case the processing machine is not being utilized during unloading and loading. If T_m = machine processing time and T_s = worker service time (to perform unloading and loading or other tending duties), then the overall cycle time of the single station with no storage is

$$T_c = T_m + T_s \tag{14.3}$$

By contrast, the overall cycle time for a single station with one part storage capacity, such as the case in Figure 14.2, is

$$T_c = \text{Max}\{T_m, T_s\} + T_r \tag{14.4}$$

where T_r = the *repositioning time* to move the completed part away from the processing head and move the raw workpart into position in front of the workhead. In most instances, the worker service time is less than the machine processing time, and machine utilization is high. If $T_s > T_m$, the machine experiences forced idle time during each work cycle, and this is undesirable. Methods analysis should be applied to reduce T_s so that $T_s < T_m$.

Storage Capacities Greater than One. Larger storage capacities allow unattended operation, so long as loading and unloading of all parts can be accomplished in less time than the machine processing time. Figure 14.3 shows several possible designs of parts storage subsystems for CNC machining centers. The parts storage unit is interfaced with an automatic pallet changer, shuttle cart, or other mechanism that is interfaced directly with the machine tool. Comparable arrangements are available for turning centers, in which an industrial robot is commonly used to perform loading and unloading between the machine tool and the parts storage subsystem. Pallet fixtures are not employed; instead, the robot uses a specially designed dual gripper (Section 7.3.1) to handle the raw parts and finished parts during the unloading/loading portion of the work cycle.

In processes other than machining, a variety of techniques can achieve parts storage. In many cases, the starting material is not a discrete workpart. The following examples illustrate some of the methods:

- *Sheet metal stamping.* In sheet metal pressworking, automated operation of the press is accomplished using a starting sheet metal coil, whose length is enough for hundreds or even thousands of stampings. The stampings either remain attached to the remainder of the coil or are collected in a container. Periodic attention is required by a worker to change the starting coil and remove the completed stampings.

- *Plastic injection molding.* The starting molding compound is in the form of small pellets, which are loaded into a hopper above the heating barrel of the molding machine. The hopper contains enough material for dozens or hundreds of molded parts. Prior to being loaded into the hopper, the molding compound is often subjected to a drying process to remove moisture and this represents another material storage unit. The molded parts drop by gravity after each molding cycle and are stored temporarily in a container beneath the mold. A worker must periodically attend the machine to load molding compound into the dryer or hopper and to collect the molded parts.

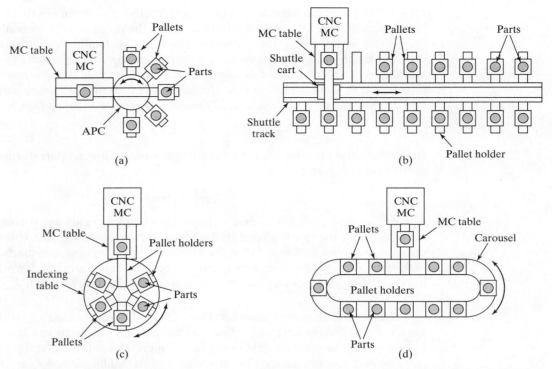

Figure 14.3 Alternative designs of parts storage subsystems that might be used with CNC machining centers: (a) automatic pallet changer with pallet holders arranged radially, parts storage capacity = 5; (b), in-line shuttle cart system with pallet holders along its length, parts storage capacity = 16; (c) pallets held on indexing table, parts storage capacity = 6; and (d) parts storage carousel, parts storage capacity = 12.

- *Plastic extrusion.* Plastic extrusion operations are similar to injection molding except that the product is continuous rather than discrete. The starting material and the methods for loading into the extrusion machine are basically the same as for injection molding. A pliable product can be collected in a coil, while a rigid one is usually cut to standard lengths. Either method can be automated to allow unattended operation of the extrusion machine.

In single-station automated assembly systems, parts storage must be provided for each component as well as the assembled work unit. Various parts storage and delivery systems are used in practice. We discuss these systems in Chapter 17 on automated assembly.

14.3 APPLICATIONS OF SINGLE-STATION CELLS

Single-station cells are abundant. Most industrial production operations are based on the use of single-station manned and automated cells. Let us distinguish the applications between manned and automated single stations.

14.3.1 Applications of Single-Station Manned Cells

Our examples in Section 14.1 illustrate the variety of possible manually operated and semi-automatic work cells. Some additional applications include

- A CNC machining center producing identical parts. The machine executes a part program for each part. A worker is required to be at the machine at the end of each program execution to unload the part just completed and load a raw workpart onto the machine table.
- A CNC machining center that produces nonidentical parts. In this case, the machine operator must call the appropriate part program and load it into the CNC control unit for each consecutive workpart.
- A cluster of two CNC turning centers, each producing the same part but operating independently from its own machine control unit. A single worker attends to the loading and unloading of both machines. The part programs are long enough relative to the load/unload portion of the work cycle that the worker can service both machines without forced machine idle time.
- A plastic injection molding machine on semi-automatic cycle, with a worker present to remove the molding, sprue, and runner system when the mold opens each molding cycle, and to place parts in a box. Another worker must periodically exchange the tote box and resupply molding compound to the machine.
- An electronics assembly workstation where a worker places components onto printed circuit boards in a batch operation. The worker must periodically delay production and replace the supply of components stored in tote bins at the station. Starting and finished boards are stored in magazines that must be periodically replaced by another worker.
- An assembly workstation where a worker performs mechanical assembly of a simple product (or subassembly of a product) from components located in tote bins at the station.
- A stamping press that punches and forms sheet metal parts from flat blanks in a stack near the press. A worker is required to load the blank into the press, actuate the press, and then remove the stamping each cycle. Completed stampings are stored in four-wheel trucks that have been specially designed for the part.

14.3.2 Applications of Single-Station Automated Cells

Following are examples of single-station automated cells. We have taken each of the preceding examples of manned cells and converted it into an automated cell.

- A computer numerical control (CNC) machining center with parts carousel and automatic pallet changer, as in the layout of Figure 14.3(d). The parts are identical, and the machining cycle is controlled by a part program. Each part is held on a pallet fixture. The machine cuts the parts one by one. When all of the parts in the carousel have been machined, a worker removes the finished pieces from the carousel and loads starting workparts. Loading and unloading of the carousel can be performed concurrently while the machine is operating.
- A CNC machining center producing nonidentical parts. In this case, the appropriate part program is automatically downloaded to the CNC control unit for each

consecutive workpart, based on either a given production schedule or an automatic part recognition system that identifies the raw part.

- A cluster of ten CNC turning centers, each producing a different part. Each workstation has its own parts carousel and robotic arm for loading and unloading between the machine and the carousel. A single worker must attend all ten machines by periodically unloading and loading the storage carousels. The time required to service a carousel is short relative to the time each machine can run unattended, so all ten machines can be serviced by the worker with no machine idle time.

- A plastic injection-molding machine on automatic cycle, with a mechanical arm to ensure removal of the molding, sprue, and runner system each molding cycle. Parts are collected in a tote box beneath the mold. A worker must periodically exchange the tote box and resupply molding compound to the machine.

- An automated insertion machine assembling electronic components onto printed circuit boards in a batch operation. Starting boards and finished boards are stored in magazines for periodic replacement by a human worker. The worker must also periodically replace the supply of components that are stored in long magazines.

- A robotic assembly cell consisting of one robot that assembles a simple product (or subassembly of a product) from components presented by several parts delivery systems (for example, bowl feeders).

- A stamping press that punches and forms small sheet metal parts from a long coil, as depicted in Figure 14.4. The press operates at a rate of 180 cycles per minute, and 9,000 parts can be stamped from each coil. The stampings are collected in a tote box on the output side of the press. When the coil runs out, it must be replaced with a new coil, and the tote box is replaced at the same time.

14.3.3 CNC Machining Centers and Related Machine Tools

Several of our application examples of single station manufacturing cells consisted of CNC machining centers. Let us discuss this important class of machine tool, which was identified in Section 7.4.1. The *machining center*, developed in the late 1950s before the advent of CNC, is a machine tool capable of performing multiple machining operations

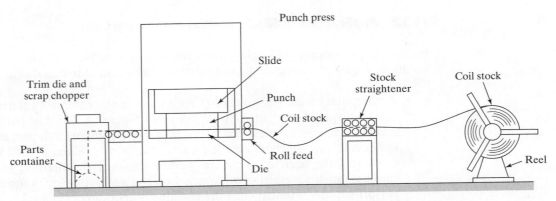

Figure 14.4 Stamping press on automatic cycle producing stampings from a sheet metal coil.

on a workpart in one setup under NC program control. Today's machining centers are CNC. Typical cutting operations performed on a machining center use a rotating cutting tool, such as milling, drilling, reaming, and tapping.

Machining centers are classified as vertical, horizontal, or universal. The designation refers to the orientation of the machine spindle. A vertical machining center has its spindle on a vertical axis relative to the worktable, and a horizontal machining center has its spindle on a horizontal axis. This distinction generally results in a difference in the type of work that is performed on the machine. A vertical machining center (VMC) is typically used for flat work that requires tool access from the top. Examples include mold and die cavities, and large components of aircraft. A horizontal machining center (HMC) is used for cube-shaped parts where tool access can best be achieved on the sides of the cube. A universal machining center (UMC) has a workhead that swivels its spindle axis to any angle between horizontal and vertical, making this a very flexible machine tool. Airfoil shapes and other curvilinear geometries often require the capabilities of a UMC.

Numerical control machining centers are usually designed with features to reduce nonproductive time. These features include the following:

- *Automatic tool-changer.* A variety of machining operations means that a variety of cutting tools is required. The tools are contained in a tool storage unit that is integrated with the machine tool. When a cutter needs to be changed, the tool drum rotates to the proper position, and an automatic tool changer (ATC), operating under part program control, exchanges the tool in the spindle for the tool in the tool storage unit. Capacities of the tool storage unit commonly range from 16 to 80 cutting tools.
- *Automatic workpart positioner.* Many horizontal and universal machining centers have the capability to orient the workpart relative to the spindle. This is accomplished by means of a rotary table on which the workpart is fixtured. The table can be oriented at any angle about a vertical axis to permit the cutting tool to access almost the entire surface of the part in a single setup.
- *Automatic pallet changer.* Machining centers are often equipped with two (or more) separate pallets that can be presented to the cutting tool using an automatic pallet changer (Section 14.2.2). While machining is being performed with one pallet in position at the machine, the other pallet is in a safe location away from the spindle. In this safe location, the operator can unload the finished part from the prior cycle and then fixture the raw workpart for the next cycle.

A numerically controlled horizontal machining center, with many of the features described above, is shown in Figure 14.2. Machining centers are being used increasingly by the automotive industry for high-volume production of transmission components, engine blocks, and engine heads [7].

The success of NC machining centers motivated the development of NC turning centers. A modern *NC turning center*, Figure 14.5, is capable of performing various turning and related operations, contour turning, and automatic tool indexing, all under computer control. In addition, the most sophisticated turning centers can accomplish (1) workpart gaging (checking key dimensions after machining), (2) tool monitoring (sensing when the tools are worn), (3) automatic tool changing when tools become worn, and (4) automatic workpart changing at the completion of the work cycle.

Another development in NC machine tool technology is the *mill-turn center*, which has the general configuration of a turning center, but also the capability to position a cylindrical

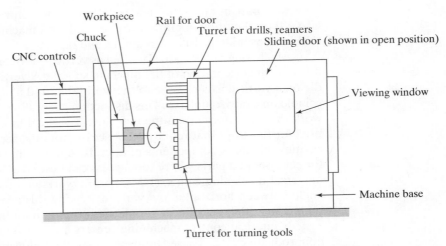

Figure 14.5 Front view of CNC turning center showing two tool turrets, one for single point turning tools and the other for drills and similar tools. Turrets can be positioned under NC control to cut the workpiece.

workpart at a specified angle so that a rotating cutter can machine features into the outside surface of the part, as illustrated in Figure 14.6. The mill-turn center has the traditional x- and z-axes of a NC lathe. In addition, orientation of the work provides a third axis, while manipulation of the rotational tool with respect to the work provides two more axes. A conventional NC turning center does not have the capability to stop the rotation of the workpart at a defined angular position, and it does not possess rotating tool spindles.

The trend in the machine tool industry is in the direction of designing machines that perform multiple operations in one setup, continuing the advancements in machining centers, turning centers, and mill-turn centers. The term *multitasking machine* is used to include all of these machine tools that accomplish multiple and often quite different types of operations [4], [6]. The processes that might be available on a single multitasking machine include milling, drilling, tapping, turning, grinding, and welding. To fully automate the work cycle, industrial robots are often used to perform loading and unloading of the workparts. Advantages of this new class of highly versatile machine, compared to more conventional CNC machine tools, include (1) fewer setups, (2) reduced part handling, (3) increased accuracy and repeatability because the parts utilize the same fixture throughout their processing, and (4) faster delivery of parts in small lot sizes. The availability of computer-aided manufacturing software for part programming, simulation, and selection of cutting conditions has become an essential prerequisite for successful implementation of these technologically advanced machines [1].

CNC machining centers, turning centers, mill-turn centers, and other multitasking machines can be operated either as manned or automated manufacturing systems. Whether it operates with a worker in continuous attendance or as an automated single station depends on the existence of an integrated parts storage subsystem with automatic transfer of workparts between the machine tool and the storage unit. These machine tools can also be used in flexible machine cells (Chapters 18 and 19).

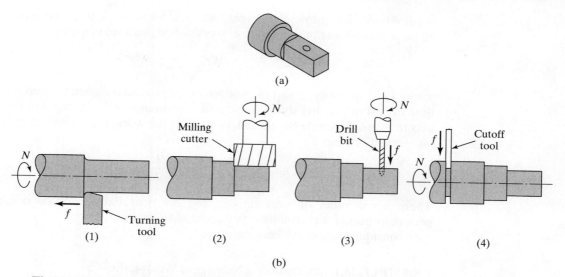

Figure 14.6 Operation of a mill-turn center: (a) example part with turned, milled, and drilled surfaces; and (b) sequence of cutting operations: (1) turn smaller diameter; (2) mill flat with part in programmed angular position, four positions for square cross section; (3) drill hole with part in programmed angular position, and (4) cutoff of the machined piece.

14.4 ANALYSIS OF SINGLE-STATION SYSTEMS

Two analysis issues related to single-station manufacturing systems are the determination of (1) the number of single stations required to satisfy specified production requirements and (2) the number of machines to assign to a worker in a machine cluster.

14.4.1 Number of Workstations Required

Any manufacturing system must be designed to produce a specified quantity of parts or products at a specified production rate. In the case of single-station manufacturing systems, this may mean that more than one single-station cell is required to achieve the specifications. The problem is to determine the number of workstations required to achieve a given production rate or to produce a given quantity of work units. The basic approach is the following: (1) determine the total workload that must be accomplished in a certain period (hour, week, month, year), where *workload* is defined as the total hours required to complete a given amount of work or to produce a given number of work units scheduled during the period, and then (2) divide the workload by the hours available on one workstation in the same period.

Workload is figured as the quantity of work units to be produced during the period of interest multiplied by the time (hours) required for each work unit. The time required for each work unit is the cycle time on the machine in most cases, so workload is given by the following:

$$WL = QT_c \tag{14.5}$$

where WL = workload scheduled for a given period, hr of work/hr or hr of work/wk; Q = quantity to be produced during the period, pc/hr or pc/wk, etc.; and T_c = cycle time required per piece, hr/pc. If the workload includes multiple part or product styles that can all be produced on the same type of workstation, then we can use

$$WL = \sum_j Q_j T_{cj} \qquad (14.6)$$

where Q_j = quantity of part or product style j produced during the period, pc; T_{cj} = cycle time of part or product style j, hr/pc; and the summation includes all of the parts or products to be made during the period. In step (2) the workload is divided by hours available on one station; that is,

$$n = \frac{WL}{AT} \qquad (14.7)$$

where n = number of workstations; and AT = available time on one station in the period, hr/period. Let us illustrate the use of these equations with a simple example, and then consider some of the complications.

EXAMPLE 14.1   Determining Number of Workstations

A total of 800 shafts must be produced in the lathe section of the machine shop during a particular week. Each shaft is identical and requires a machine cycle time T_c = 11.5 min. All of the lathes in the department are equivalent in terms of their capability to produce the shaft in the specified cycle time. How many lathes must be devoted to shaft production during the given week, if there are 40 hours of available time on each lathe?

Solution: The workload consists of 800 shafts at 11.5 min per shaft.

$$WL = 800(11.5 \text{ min}) = 9{,}200 \text{ min} = 153.33 \text{ hr}$$

Time available per lathe during the week AT = 40 hr

$$n = \frac{153.33}{40} = 3.83 \text{ lathes}$$

This calculated value would be rounded up to four lathes that are assigned to the production of shafts during the given week.

Several factors present in most real life manufacturing systems complicate the computation of the number of workstations. These factors include

- *Setup time in batch production.* During setup, the workstation is not producing, but time marches on.
- *Availability.* This is a reliability factor that reduces the available production time.
- *Utilization.* Workstations may not be fully utilized due to scheduling problems, lack of work for a given machine type, workload imbalance among workstations, and other reasons.
- *Defect rate.* The output of the manufacturing system may not be 100% good quality. Defective units are produced at a certain fraction defect rate. To compensate, the system must increase the total number of units processed.

These factors affect how many workstations and/or workers are required to accomplish a given workload. They influence either the workload or the amount of time available at the workstation during the period of interest.

Setup time in batch production occurs between batches because the tooling and fixturing must be changed over from the current part style to the next part style, and the equipment controller must be reprogrammed. It is lost time when no parts are produced (except perhaps trial parts to check out the new setup and program). Yet it consumes available time at a workstation. The following examples illustrate two possible ways of dealing with the issue, depending on the information given.

EXAMPLE 14.2   Number of Setups is Known

A total of 800 shafts must be produced in the lathe section of the machine shop during a particular week. The shafts are of 20 different types, and each type is produced in its own batch. Average batch size is 40 parts. Each batch requires a setup and the average setup time is 3.5 hr. The average machine cycle time to produce a shaft $T_c = 11.5$ min. How many lathes are required during the week?

Solution: In this case we know how many setups are required during the week, because we know how many batches will be produced: 20. We can determine the workload for the 20 setups and the workload for 20 production batches:

$$WL = 20(3.5) + 20(40)(11.5/60) = 70 + 153.33 = 223.33 \text{ hr}$$

Given that each lathe is available 40 hr/wk (since setup is included in the workload calculation),

$$n = \frac{223.33}{40} = 5.58 \text{ lathes}$$

Again, rounding up, the shop would have to dedicate six lathes to the shaft work.

EXAMPLE 14.3   Number of Setups Not Known

This is similar to Example 14.2, but the number of setups is equal to the number of machines that will be required, n and we do not yet know what that number is. The setup takes 3.5 hr. How many lathes are required during the week?

Solution: In this problem formulation, the number of hours available on any lathe used for the shaft order is reduced by the setup time. The workload to actually produce the parts remains the same, 153.33 hr. Adding the setup workload,

$$WL = 153.33 + 3.5n$$

Now dividing by the available time of 40 hours per lathe, we have

$$n = \frac{153.33 + 3.5\text{n}}{40} = 3.83325 + 0.0875\text{n}$$

Solving for n,

$$n - 0.0875n = 0.9125n = 3.83325$$

$$n = 4.2$$

This would round up to five lathes that must be devoted to the shaft job. That's a shame, because the lathes will not be fully utilized. With five lathes, utilization will be

$$U = \frac{4.20}{5} = 0.840 \quad (84.0\%)$$

Given this unfortunate result, it might be preferable to offer overtime to the workers on four of the lathes. How much overtime above the regular 40 hours will be required?

$$OT = \left(3.5 + \frac{153.33}{4}\right) - 40 = (3.5 + 38.33) - 40 = 1.83 \text{ hr}$$

This is a total of $4(1.83 \text{ hr}) = 7.33$ hr for the four machine operators.

Availability and utilization (Section 3.1.3) tend to reduce the available time on the workstation. The available time becomes the actual shift time in the period multiplied by availability and utilization. In equation form,

$$AT = H_{sh}AU \tag{14.8}$$

where AT = available time, hr; H_{sh} = shift hours during the period, hr; A = availability; and U = utilization.

Defect rate is the fraction of parts produced that are defective. We discuss the issue of fraction defect rate in more detail later (Section 21.5). A defect rate greater than zero increases the quantity of work units that must be processed in order to yield the desired quantity. If a process is known to produce parts at a certain average scrap rate, then the starting batch size is increased by a scrap allowance to compensate for the defective parts that will be made. The relationship between the starting quantity and the quantity produced is

$$Q = Q_o(1 - q) \tag{14.9}$$

where Q = quantity of good units made in the process; Q_o = original or starting quantity; and q = fraction defect rate. Thus, if we want to produce Q good units, we must process a total of Q_o starting units, which is

$$Q_o = \frac{Q}{(1 - q)} \tag{14.10}$$

The combined effect of worker efficiency and fraction defect rate is given in the following equation, which amends the workload formula, Eq. (14.5):

$$WL = \frac{QT_c}{(1 - q)} \tag{14.11}$$

EXAMPLE 14.4 Including Availability, Utilization, and Defect Rate

Suppose that in Example 14.2 the anticipated availability of the lathes is 100% during setup and 92% during the production run, and the expected utilization for calculation purposes is 100%. The fraction defect rate for lathe work of this type is 5%. Other data from Example 14.1 are applicable. How many lathes are required during the week, given this additional information?

Solution: When there is a separation of tasks between two or more types of work (in this case, setup and run are two separate types of work), we must be careful to use the various factors only where they are applicable. For example, fraction defect rate does not apply to the setup time. Availability is also assumed not to apply to setup (how can the machine break down if it's not running?). Accordingly, it is appropriate to compute the number of equivalent workstations for setup separately from the number for running production.

For setup, the workload is simply the time spent performing the 20 setups:

$$WL = 20(3.5) = 70.0 \text{ hr}$$

The available hours during the week are

$$AT = 40(1.0)(1.0) = 40$$

Thus, the number of lathes required just for setup is determined as follows:

$$n_{su} = \frac{70}{40} = 1.75 \text{ lathes}$$

The total workload for the 20 production runs is computed as follows:

$$WL = \frac{20(40)(11.5/60)}{(1 - 0.05)} = 161.4 \text{ hr}$$

The available time is affected by the 92% availability:

$$AT = 40(0.92) = 36.8 \text{ hr/machine}$$

$$n_{pr} = \frac{161.4}{36.8} = 4.39$$

Total machines required = 1.75 + 4.39 = 6.14 lathes

This should be rounded up to seven lathes, unless the remaining time on the seventh lathe can be used for other production.

Note that the rounding up should occur after adding the machine fractions; otherwise we risk overestimating machine requirements (not in this problem, however).

14.4.2 Machine Clusters

When the machine in a single workstation does not require the continuous attention of a worker during its semi-automatic machine cycle, an opportunity exists to assign more than one machine to the worker. The workstation still requires operator attention every work cycle. However, the manning level of the workstation is reduced from $M = 1$ to $M < 1$. This kind of machine organization has sometimes been referred to as a "machine cell"; however, let us use the term machine cluster. A *machine cluster* is defined here as collection of two or more machines producing parts or products with identical cycle times and serviced (usually loaded and unloaded) by one worker. By contrast, a *machine cell* consists of one or more machines organized to produce a family of parts or products. We discuss machine cells in Chapters 18 and 19.

Several conditions must be satisfied in order to organize a collection of machines into a machine cluster: (1) the semi-automatic machine cycle must be long relative to the service portion of the cycle that requires the worker's attention; (2) the semi-automatic machine cycle time must be the same for all machines; (3) the machines that the worker would service must be located in close enough proximity to allow time to walk between them; and (4) the work rules of the plant must permit a worker to service more than one machine.

Consider a collection of single workstations, all producing the same parts and operating on the same semi-automatic machine cycle time. Each machine operates for a certain portion of the total cycle under its own control T_m (machine time), and then it requires servicing by the worker, which takes time T_s. Thus, assuming the worker is always available when servicing is needed, so that the machine is never idle, the total cycle time of a machine is $T_c = T_m + T_s$. If more than one machine is assigned to the worker, a certain amount of time will be lost while the worker walks from one machine to the next, referred to here as the *repositioning time*, T_r. The time required for the operator to service one machine is therefore $T_s + T_r$, and the time to service n machines is $n(T_s + T_r)$. For the system to be perfectly balanced in terms of worker time and machine cycle time,

$$n(T_s + T_r) = T_m + T_s$$

We can determine from this the number of machines that should be assigned to one worker by solving for

$$n = \frac{T_m + T_s}{T_s + T_r} \tag{14.12}$$

where n = number of machines; T_m = machine semi-automatic cycle time, min; T_s = worker service time per machine, min; T_r = worker repositioning time between machines, min.

It is likely that the calculated value of n will not be an integer, which means that the worker time in the cycle, that is, $n(T_s + T_r)$, cannot be perfectly balanced with the cycle time T_c of the machines. However, the actual number of machines in the manufacturing system must be an integer, so either the worker or the machines will experience some idle time. The number of machines will either be the integer that is greater than n from Eq. (14.12) or it will be the integer that is less than n. Let us identify these two integers as n_1 and n_2. We can determine which of the alternatives is preferable by introducing cost factors into the analysis. Let C_L = the labor cost rate and C_m = machine cost rate. Certain overheads may be applicable to these rates (see Section 3.2.2). The decision will be based on the cost per work unit produced by the system.

Case 1: If we use n_1 = maximum integer $\leq n$, then the worker will have idle time, and the cycle time of the machine cluster will be the cycle time of the machines $T_c = T_m + T_s$. Assuming one work unit is produced by each machine during a cycle, we have

$$C_{pc}(n_1) = \left(\frac{C_L}{n_1} + C_m \right)(T_m + T_s) \tag{14.13}$$

where $C_{pc}(n_1)$ = cost per work unit, \$/pc; C_L = labor cost rate, \$/min; C_m = cost rate per machine, \$/min; and $(T_m + T_s)$ is expressed in min.

Case 2: If we use n_2 = minimum integer $> n$, then the machines will have idle time, and the cycle time of the machine cluster will be the time it takes for the worker to service the n_2 machines, which is $n_2(T_s + T_r)$. The corresponding cost per piece is given by

$$C_{pc}(n_2) = (C_L + C_m n_2)(T_s + T_r) \qquad (14.14)$$

The selection of n_1 or n_2 is based on whichever case results in the lower cost per work unit.

In the absence of cost data needed to make these calculations, we feel that it is generally preferable to assign machines to a worker such that the worker has some idle time and the machines are utilized 100%. The reason for this is that the total hourly cost rate of n production machines is usually greater than the labor rate of one worker. Therefore, machine idle time costs more than worker idle time. The corresponding number of machines to assign the worker is therefore given by

$$n_1 = \text{maximum integer} \le \frac{T_m + T_s}{T_s + T_r} \qquad (14.15)$$

EXAMPLE 14.5   How Many Machines For One Worker?

A machine shop contains many CNC lathes that operate on a semi-automatic machining cycle under part program control. A significant number of these machines produce the same part, whose machining cycle time = 2.75 min. One worker is required to perform unloading and loading of parts at the end of each machining cycle. This takes 25 sec. Determine how many machines one worker can service if it takes an average of 20 sec to walk between the machines and no machine idle time is allowed.

Solution: Given that T_m = 2.75 min, T_s = 25 sec = 0.4167 min, and T_r = 20 sec = 0.3333 min, Eq. (14.15) can be used to obtain

$$n_1 = \text{maximum integer} \le \left(\frac{2.75 + 0.4167}{0.4167 + 0.3333} = \frac{3.1667}{0.75} = 4.22 \right) = 4 \text{ machines}$$

Each worker can be assigned four machines. With a machine cycle T_c = 3.1667 min, the worker will spend 4(0.4167) = 1.667 min servicing the machines, 4(0.3333) = 1.333 min walking between machines, and the worker's idle time during the cycle will be 0.167 min (10 sec).

Note the regularity that exists in the worker's schedule in this example. If we imagine the four machines laid out on the four corners of a square, the worker services each machine and then proceeds clockwise to the machine in the next corner. Each cycle of servicing and walking takes 3.0 min, with a slack time of 10 sec left over for the worker.

If this kind of regularity characterizes the operations of a cluster of single-station automated cells, then the same kind of analysis can be applied to determine the number of cells to assign to one worker. If, on the other hand, servicing of each cell is required at random and unpredictable intervals, then there will be periods when several cells require servicing simultaneously, overloading the capabilities of the human worker, while during other periods the worker will have no cells to service. Queueing analysis is appropriate in this case of random service requirements.

REFERENCES

[1]   ABRAMS, M., "Simply Complex," *Mechanical Engineering*, January 2006, pp. 28–31.

[2]   ARONSON, R., "Turning's Just the Beginning," *Manufacturing Engineering*, June 1999, pp 42–53.

[3]   ARONSON, R., "Cells and Centers," *Manufacturing Engineering*, February 1999, pp. 52–60.

[4]   ARONSON, R., "Multitalented Machine Tools," *Manufacturing Engineering*, January 2005, pp. 65–75.

[5]   DROZDA, T. J., and Wick, C., (eds.), *Tool and Manufacturing Engineers Handbook*, 4th ed., *Volume I: Machining*, Society of Manufacturing Engineers, Dearborn, MI, 1983.

[6]   LORINCZ, J., "Multitasking Machining," *Manufacturing Engineering*, February 2006, pp. 45–54.

[7]   LORINCZ, J., "Just Say VMC," *Manufacturing Engineering*, June 2006, pp. 61–67.

[8]   WAURZYNIAK, P., "Programming for MTM," *Manufacturing Engineering*, April 2005, pp. 83–91.

REVIEW QUESTIONS

14.1   Name three reasons why single-station manned cells are so widely used in industry.

14.2   What does the term *semi-automated station* mean?

14.3   What is a single-station automated cell?

14.4   What are the five enablers required for unattended operation of a single model or batch model automated production cell?

14.5   What are the additional three enablers required for unattended operation of a mixed model automated production cell?

14.6   What is an automatic pallet changer?

14.7   What is a machining center?

14.8   What are some of the features on a NC machining center used to reduce nonproductive time in the work cycle?

14.9   What is a machine cluster?

PROBLEMS

Unattended Operation

14.1   A CNC machining center has a programmed cycle time = 25.0 min for a certain part. The time to unload the finished part and load a starting work unit = 5.0 min. (a) If loading and unloading are done directly onto the machine tool table and no automatic storage capacity exists at the machine, what are the cycle time and hourly production rate? (b) If the machine tool has an automatic pallet changer so that unloading and loading can be accomplished while the machine is cutting another part, and the repositioning time = 30 sec, what are the total cycle time and hourly production rate? (c) If the machine tool has an automatic pallet changer that interfaces with a parts storage unit whose capacity is 12 parts, and the repositioning time = 30 sec, what are the total cycle time and hourly production rate? Also, how long does it take to perform the loading and unloading of the 12 parts by the human worker, and what is the time the machine can operate unattended between parts changes?

Determining Workstation Requirements

14.2 A total of 7,000 stampings must be produced in the press department during the next three days. Manually operated presses will be used to complete the job and the cycle time is 27 sec. Each press must be set up before production starts. Setup time for this job is 2.0 hr. How many presses and operators must be devoted to this production during the three days, if there are 7.5 hours of available time per day?

14.3 A stamping plant must be designed to supply an automotive engine plant with sheet metal stampings. The plant will operate one 8-hour shift for 250 days per year and must produce 15,000,000 good quality stampings annually. Batch size = 10,000 good stampings produced per batch. Scrap rate = 5%. On average it takes 3.0 sec to produce each stamping when the presses are running. Before each batch, the press must be set up, and it takes 4 hr to accomplish each setup. Presses are 90% reliable during production and 100% reliable during setup. How many stamping presses will be required to accomplish the specified production?

14.4 A new forging plant must supply parts to the automotive industry. Because forging is a hot operation, the plant will operate 24 hr per day, five days per week, 50 weeks per year. Total output from the plant must be 10,000,000 forgings per year in batches of 1250 parts per batch. Anticipated scrap rate = 3%. Each forging cell will consist of a furnace to heat the parts, a forging press, and a trim press. Parts are placed in the furnace an hour prior to forging; they are then removed and forged and trimmed one at a time. On average the forging and trimming cycle takes 0.6 min to complete one part. Each time a new batch is started, the forging cell must be changed over, which consists of changing the forging and trimming dies for the next part style. It takes 2.0 hr on average to complete a changeover between batches. Each cell is considered to be 96% reliable during operation and 100% reliable during changeover. Determine the number of forging cells that will be required in the new plant.

14.5 A plastic injection molding plant will be built to produce 6 million molded parts per year. The plant will run three 8-hour shifts per day, five days per week, 50 weeks per year. For planning purposes, the average batch size = 6000 moldings, average changeover time between batches = 6 hrs, and average molding cycle time per part = 30 sec. Assume scrap rate = 2%, and average uptime proportion (reliability) per molding machine = 97%, which applies to both run time and changeover time. How many molding machines are required in the new plant?

14.6 A plastic extrusion plant will be built to produce 30 million meters of plastic extrusions per year. The plant will run three 8-hour shifts per day, 360 days per year. For planning purposes, the average run length = 3000 meters of extruded plastic. The average changeover time between runs = 2.5 hr, and average extrusion speed = 15 m/min. Assume scrap rate = 1%, and average uptime proportion per extrusion machine = 95% during run time. Uptime proportion during changeover is assumed to be 100%. If each extrusion machine requires 500 sq. ft of floor space, and there is an allowance of 40% for aisles and office space, what is the total area of the extrusion plant?

14.7 Future production requirements in a machine shop call for several automatic bar machines to be acquired to produce three new parts (A, B, and C) that have been added to the shop's product line. Annual quantities and cycle times for the three parts are given in the table below. The machine shop operates one 8-hour shift for 250 days per year. The machines are expected to be 95% reliable, and the scrap rate is 3%. How many automatic bar machines will be required to meet the specified annual demand for the three new parts?

Part	Annual Demand	Machining Cycle Time
A	25,000	5.0 min
B	40,000	7.0 min
C	50,000	10.0 min

14.8 A certain type of machine will be used to produce three products: A, B, and C. Sales forecasts for these products are 52,000, 65,000, and 70,000 units per year, respectively. Production rates for the three products are, 12, 15, and 10 pc/hr; and scrap rates are 5%, 7%, and 9%. The plant will operate 50 weeks per year, 10 shifts per week, and 8 hr per shift. It is anticipated that production machines of this type will be down for repairs on average 10% of the time. How many machines will be required to meet demand?

14.9 An emergency situation has occurred in the milling department, because the ship carrying a certain quantity of a required part from an overseas supplier sank on Friday evening. A certain number of machines in the department must therefore be dedicated to the production of this part during the next week. A total of 1000 of these parts must be produced, and the production cycle time per part = 16.0 min. Each milling machine used for this emergency production job must first be set up, which takes 5.0 hr. A scrap rate of 2% can be expected. (a) If the production week consists of 10 shifts at 8.0 hr per shift, how many machines will be required? (b) It so happens that only two milling machines can be spared for this emergency job, due to other priority jobs in the department. To cope with the emergency situation, plant management has authorized a three-shift operation for six days next week. Can the 1000 replacement parts be completed within these constraints?

14.10 A machine shop has dedicated one CNC machining center to the production of two parts (A and B) used in the final assembly of the company's main product. The machining center is equipped with an automatic pallet changer and a parts carousel that holds ten parts. One thousand units of the product are produced per year, and one of each part is used in the product. Part A has a machining cycle time of 50 min. Part B has a machining cycle time of 80 min. These cycle times include the operation of the automatic pallet changer. No other changeover time is lost between parts. The anticipated scrap rate is zero. The machining center is 95% reliable. The machine shop operates 250 days per year. How many hours must the CNC machining center operate each day on average to supply parts for the product?

Machine Clusters

14.11 The CNC grinding section has a large number of machines devoted to grinding shafts for the automotive industry. The grinding machine cycle takes 3.6 min. At the end of this cycle an operator must be present to unload and load parts, which takes 40 sec. (a) Determine how many grinding machines the worker can service if it takes 20 sec to walk between the machines and no machine idle time is allowed. (b) How many seconds during the work cycle is the worker idle? (c) What is the hourly production rate of this machine cluster?

14.12 A worker is currently responsible for tending two machines in a machine cluster. The service time per machine is 0.35 min and the time to walk between machines is 0.15 min. The machine automatic cycle time is 1.90 min. If the worker's hourly rate = $12/hr and the hourly rate for each machine = $18/hr, determine (a) the current hourly rate for the cluster, and (b) the current cost per unit of product, given that two units are produced by each machine during each machine cycle. (c) What is the percentage of time that the worker is idle? (d) What is the optimum number of machines that should be used in the machine cluster, if minimum cost per unit of product is the decision criterion?

14.13 Let n = the number of machines in a machine cluster. Each production machine is identical and has an automatic processing time $T_m = 4.0$ min. The servicing time $T_s = 12$ sec for each machine. The full cycle time for each machine in the cell is $T_c = T_s + T_m$. The repositioning time for the worker is given by $T_r = 5 + 3n$, where T_r is in sec. T_r increases with n because the distance between machines increases with more machines. (a) Determine the maximum number of machines in the cell if no machine idle time is allowed. For your answer, compute (b) the cycle time and (c) the worker idle time expressed as a percent of the cycle time.

14.14 An industrial robot will service n production machines in a machine cluster. Each production machine is identical and has an automatic processing time $T_m = 130$ sec. The robot servicing and repositioning time for each machine is given by the equation $(T_s + T_r) = 15 + 4n$, where T_s is the servicing time (sec), T_r is the repositioning time (sec), and n = number of machines that the robot services. $(T_s + T_r)$ increases with n because more time is needed to reposition the robot arm as n increases. The full cycle time for each machine in the cell is $T_c = T_s + T_m$. (a) Determine the maximum number of machines in the cell such that machines are not kept waiting. For your answer, (b) what is the machine cycle time, and (c) what is the robot idle time expressed as a percent of the cycle time T_c?

14.15 A factory production department consists of a large number of work cells. Each cell consists of one human worker performing electronics assembly tasks. The cells are organized into sections within the department, and one foreman supervises each section. The foreman's job consists of two tasks: (1) provide each cell with a sufficient supply of parts that it can work for 4.0 hr before it needs to be resupplied and (2) prepare production reports for each work-cell. Task (1) takes 18.0 min on average per workcell and must be done twice per day. The foreman must schedule the resupply of parts to every cell so that no idle time occurs in any cell. Task (2) takes 9.0 min per workcell and must be done once per day. The plant operates one shift which is 8.0 working hr, and neither the workers nor the foreman may work more than 8.0 hr per day. Each day, the cells continue production from where they stopped the day before. (a) What is the maximum number of work cells that should be assigned to a foreman, with the proviso that the work cells are never idle? (b) With the number of work cells from part (a), how many idle minutes does the foreman have each day?

410

Chapter 15

Manual Assembly Lines

CHAPTER CONTENTS

Most manufactured consumer products are assembled. Each product consists of multiple components joined together by various assembly processes (Section 2.2.1). These kinds of

From Chapter 15 of *Automation, Production Systems, and Computer-Integrated Manufacturing*, Third Edition.
Mikell P. Groover. Copyright © 2008 by Pearson Education, Inc. Publishing as Prentice Hall. All rights reserved.

products are usually made on a manual assembly line. Factors favoring the use of manual assembly lines include the following:

- Demand for the product is high or medium
- The products made on the line are identical or similar
- The total work required to assemble the product can be divided into small work elements
- It is technologically impossible or economically infeasible to automate the assembly operations.

Products characterized by these factors that are usually made on a manual assembly line are listed in Table 15.1.

There are several reasons why manual assembly lines are so productive compared to alternative methods in which multiple workers each perform all of the tasks to assemble the products.

- *Specialization of labor.* Called "division of labor" by Adam Smith (Historical Note 15.1), this principle asserts that when a large job is divided into small tasks and each task is assigned to one worker, the worker becomes highly proficient at performing the single task. Each worker becomes a specialist.
- *Interchangeable parts*, in which each component is manufactured to sufficiently close tolerances that any part of a certain type can be selected for assembly with its mating component. Without interchangeable parts, assembly would require filing and fitting of mating components, rendering assembly line methods impractical.
- *Work flow principle*, which involves moving the work to the worker rather than vice versa. Each work unit flows smoothly through the production line, traveling the minimum distance between stations.
- *Line pacing.* Workers on an assembly line are usually required to complete their assigned tasks on each product unit within a certain cycle time, which paces the line to maintain a specified production rate. Pacing is generally implemented by means of a mechanized conveyor.

In the present chapter we discuss the engineering and technology of manual assembly lines. Automated assembly systems are covered in Chapter 17.

TABLE 15.1 Products Usually Made on Manual Assembly Lines

Audio equipment	Furniture	Pumps
Automobiles	Lamps	Refrigerators
Cameras	Luggage	Stoves
Cooking ranges	Microwave ovens	Telephones
Dishwashers	Personal computers and	Toasters and toaster ovens
Dryers (laundry)	peripherals (printers,	Trucks, light and heavy
DVD players	monitors, etc.)	Video game consoles
Electric motors	Power tools (drills, saws, etc.)	Washing machines (laundry)

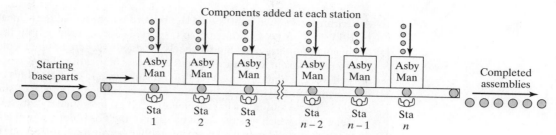

Figure 15.1 Configuration of a manual assembly line. Key: Asby = assembly, Man = manual, Sta = workstation, n = number of stations on the line.

15.1 FUNDAMENTALS OF MANUAL ASSEMBLY LINES

A *manual assembly line* is a production line that consists of a sequence of workstations where assembly tasks are performed by human workers, as depicted in Figure 15.1. Products are assembled as they move along the line. At each station, a worker performs a portion of the total work on the unit. The common practice is to "launch" base parts onto the beginning of the line at regular intervals. Each base part travels through successive stations and workers add components that progressively build the product. A mechanized material transport system is typically used to move the base parts along the line as they are gradually transformed into final products. The production rate of an assembly line is determined by its slowest station. Stations capable of working faster are ultimately limited by the slowest station.

Manual assembly line technology has made a significant contribution to the development of American industry in the 20th century, as indicated in our Historical Note 15.1. It remains an important production system throughout the world in the manufacture of automobiles, consumer appliances, and other assembled products listed in Table 15.1.

Historical Note 15.1   Origins of the manual assembly line.

Manual assembly lines are based largely on two fundamental work principles. The first is *division of labor*, argued by Adam Smith (1723-1790) in his book *The Wealth of Nations*, which was published in England in 1776. Using a pin factory to illustrate the division of labor, the book describes how 10 workers, specializing in the various distinct tasks required to make a pin, produce 48,000 pins per day, compared to a system in which each worker performs all of the tasks on each pin, that produces only a few pins per day. Smith did not invent division of labor; there had been other examples of its use in Europe for centuries, but he was the first to note its significance in production.

The second work principle is *interchangeable parts*, based on the efforts of Eli Whitney (1765–1825) and others at the beginning of the 19th century [15]. The origins of the interchangeable parts principle were previously described in Historical Note 1.1. Without interchangeable parts, assembly line technology would not be possible.

The origins of modern production lines can be traced to the meat industry in Chicago, Illinois and Cincinnati, Ohio. In the mid and late 1800s, meat packing plants used unpowered overhead conveyors to move the slaughtered stock from one worker to the next. These unpowered conveyors were later replaced by power-driven chain conveyors to create "disassembly lines," which were the predecessor of the assembly line. The work organization permitted each meat cutter to concentrate on a single task (division of labor).

American automotive industrialist Henry Ford had observed these meat packing operations. In 1913, he and his engineering colleagues designed an assembly line in Highland Park, Michigan to produce magneto flywheels. Productivity increased four-fold. Flushed by success, Ford applied assembly line techniques to chassis fabrication. The use of chain-driven conveyors and workstations arranged for the convenience and comfort of his assembly line workers increased productivity by a factor of eight, compared to previous single-station assembly methods. These and other improvements resulted in dramatic reductions in the price of the Model T Ford, which was the main product of the Ford Motor Company at the time. Masses of Americans could now afford an automobile because of Ford's achievement in cost reduction. This stimulated further development and use of production line techniques, including automated transfer lines. It also forced Ford's competitors and suppliers to imitate his methods, and the manual assembly line became intrinsic to American industry.

15.1.1 Assembly Workstations

A *workstation* on a manual assembly line is a designated location along the work flow path at which one or more work elements are performed by one or more workers. The work elements represent small portions of the total work that must be accomplished to assemble the product. Typical assembly operations performed at stations on a manual assembly line are listed in Table 15.2. A given workstation also includes the tools (hand tools or powered tools) required to perform the task assigned to the station.

Some workstations are designed for workers to stand, while others allow the workers to sit. When the workers stand, they can move about the station area to perform their assigned task. This is common for assembly of large products such as cars, trucks, and major appliances. The product is typically moved by a conveyor at constant velocity through the station. The worker begins the assembly task near the upstream side of the station and moves along with the work unit until the task is completed, then walks back to the next work unit and repeats the cycle. For smaller assembled products (such as small appliances, electronic devices, and subassemblies used on larger products), the workstations are usually designed to allow the workers to sit while they perform their tasks. This is more comfortable and less fatiguing for the workers and is generally more conducive to precision and accuracy in the assembly task.

We have previously defined manning level in Chapter 13 (Section 13.2.4) for various types of manufacturing systems. For a manual assembly line, the *manning level* of workstation i, symbolized M_i, is the number of workers assigned to that station, where $i = 1, 2, \ldots, n$

TABLE 15.2 Typical Assembly Operations Performed on a Manual Assembly Line

Application of adhesive	Expansion fitting applications	Snap fitting of two parts
Application of sealant	Insertion of components	Soldering
Arc welding	Press fitting	Spotwelding
Brazing	Printed circuit board assembly	Stapling
Cotter pin applications	Riveting and eyelet applications	Stitching
Crimping	Shrink fitting applications	Threaded fastener applications

Source: See Groover [12] for definitions.

and n = number of workstations on the line. The generic case is one worker: $M_i = 1$. In cases where the product is large, such as a car or a truck, multiple workers are often assigned to one station, so that $M_i > 1$. Multiple manning conserves valuable floor space in the factory and reduces line length and throughput time because fewer stations are required. The average manning level of a manual assembly line is simply the total number of workers on the line divided by the number of stations, that is,

$$M = \frac{w}{n}$$

(15.1)

where M = average manning level of the line, workers/station; w = number of workers on the line; and n = number of stations on the line. This seemingly simple ratio is complicated by the fact that manual assembly lines often include more workers than those assigned to stations, so M is not a simple average of M_i values. These additional workers, called *utility workers*, are not assigned to specific workstations; instead they are responsible for functions such as (1) helping workers who fall behind, (2) relieving workers for personal breaks, (3) material handling tasks, and (4) maintenance and repair duties. Including the utility workers in the worker count, we have

$$M = \frac{w_u + \sum_{i=1}^{n} w_i}{n}$$

(15.2)

where w_u = number of utility workers assigned to the system and w_i = number of workers assigned specifically to station i for $i = 1, 2, \ldots, n$. The parameter w_i is almost always an integer, except for the unusual case where a worker is shared between two adjacent stations.

15.1.2   Work Transport Systems

There are two basic ways to accomplish the movement of work units along a manual assembly line: (1) manually or (2) by a mechanized system. Both methods provide the fixed routing (all work units proceed through the same sequence of stations) that is characteristic of production lines.

Manual Methods of Work Transport.   In manual work transport, the units of product are passed from station to station by the workers themselves. Two problems result from this mode of operation: starving and blocking. *Starving* is the situation in which the assembly operator has completed the assigned task on the current work unit, but the next unit has not yet arrived at the station. The worker is thus starved for work. When a station is *blocked*, it means that the operator has completed the assigned task on the current work unit but cannot pass the unit to the downstream station because that worker is not yet ready to receive it. The operator is therefore blocked from working.

To mitigate the effects of these problems, storage buffers are sometimes used between stations. In some cases, the work units made at each station are collected in batches and then moved to the next station. In other cases, work units are moved individually along a flat table or nonpowered conveyor. When the task is finished at each station, the worker simply pushes the unit toward the downstream station. Space is often allowed for one or more work units in front of each workstation. This provides an available supply of work for the station, as well as room for completed units from the upstream station. Hence, starving and blocking are minimized. The trouble with this method of operation is that it can result in significant work-in-process, which is economically undesirable. Also, workers are unpaced in lines that rely on manual transport methods, and production rates tend to be lower.

Mechanized Work Transport. Powered conveyors and other types of mechanized material handling equipment are widely used to move units along a manual assembly line. These systems can be designed to provide paced or unpaced operation of the line. Three major categories of work transport systems in production lines are (a) continuous transport, (b) synchronous transport, and (c) asynchronous transport. These are illustrated schematically in Figure 15.2. Table 15.3 identifies some of the material transport equipment commonly associated with each of the categories.

A *continuous transport system* uses a continuously moving conveyor that operates at constant velocity, as in Figure 15.2(a). This method is common on manual assembly lines. The conveyor usually runs the entire length of the line. However, if the line is very long, such as the case of an automobile final assembly plant, it is divided into segments with a separate conveyor for each segment.

Continuous transport can be implemented in two ways: (1) work units are fixed to the conveyor, and (2) work units are removable from the conveyor. In the first case, the product is large and heavy (e.g., automobile, washing machine) and cannot be removed

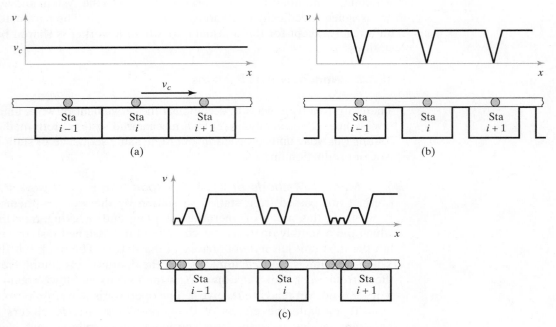

Figure 15.2 Velocity-distance diagram and physical layout for three types of mechanized transport systems used in production lines; (a) continuous transport, (b) synchronous transport, and (c) asynchronous transport. Key: v = velocity, v_c = constant velocity of continuous transport conveyor, x = distance in conveyor direction, Sta. = workstation, i = workstation identifier.

TABLE 15.3 Material Handling Equipment Used to Obtain the Three Types of Fixed Routing Work Transport Depicted in Figure 15.2

Work Transport System	Material Handling Equipment (Text Reference)
Continuous transport	Overhead trolley conveyor (Section 10.2.4)
	Belt conveyor (Section 10.2.4)
	Roller conveyor (Section 10.2.4)
	Drag chain conveyor (Section 10.2.4)
Synchronous transport	Walking beam transport equipment (Section 16.1.2)
	Rotary indexing mechanisms (Section 16.1.2)
Asynchronous transport	Power-and-free overhead conveyor (Section 10.2.4)
	Cart-on-track conveyor (Section 10.2.4)
	Powered roller conveyors (Section 10.2.4)
	Automated guided vehicle system (Section 10.2.2)
	Monorail systems (Section 10.2.3)
	Chain-driven carousel systems (Section 11.3.2)

from the conveyor. The worker must therefore walk along with the product at the speed of the conveyor in order to accomplish the assigned task.

In the case where work units are small and lightweight, they can be removed from the conveyor for the physical convenience of the operator at each station. Another convenience for the worker is that the assigned task at the station does not need to be completed within a fixed cycle time. Each worker has flexibility to deal with technical problems that may be encountered with a particular work unit. However, on average, each worker must maintain a production rate equal to that of the rest of the line. Otherwise, the line will produce *incomplete units,* which occurs when parts that were supposed to be added at a station are not added because the worker runs out of time.

In *synchronous transport systems*, all work units are moved simultaneously between stations with a quick, discontinuous motion, and then positioned at their respective stations. Depicted in Figure 15.2(b), this type of system is also known as *intermittent transport*, which describes the motion experienced by the work units. Synchronous transport is not common for manual lines, due to the requirement that the task must be completed within a certain time limit. This can cause undue stress on the assembly workers and result in incomplete products. Despite its disadvantages for manual assembly lines, synchronous transport is often ideal for automated production lines, in which mechanized workstations operate on a constant cycle time.

In an *asynchronous transport system*, a work unit leaves a given station when the assigned task has been completed and the worker releases the unit. Work units move independently, rather than synchronously. At any moment, some units are moving between workstations while others are positioned at stations, as in Figure 15.2(c). With asynchronous transport systems, small queues of work units are permitted to form in front of each station. This system tends to be forgiving of variations in worker task times.

15.1.3 Line Pacing

A manual assembly line operates at a certain cycle time that is established to achieve the required production rate of the line. We explain how this cycle time is determined in Section 15.2. On average each worker must complete the assigned task at his/her station within the cycle time, or else the required production rate will not be achieved. This pacing of the workers is one of the reasons why a manual assembly line is successful. Pacing provides a discipline for the assembly line workers that more or less guarantees a certain production rate. From the viewpoint of management, this is desirable.

Manual assembly lines can be designed with three alternative levels of pacing: (1) rigid pacing, (2) pacing with margin, and (3) no pacing. In *rigid pacing,* each worker is allowed only a certain fixed time each cycle to complete the assigned task. The allowed time is implemented by a synchronous work transport system and is (usually) equal to the cycle time of the line. Rigid pacing has two undesirable aspects, as mentioned above. First, rigid pacing is emotionally and physically stressful to human workers. Although some level of stress is conducive to improved human performance, fast pacing on an assembly line throughout an eight-hour shift (or longer) can have harmful effects on workers. Second, in a rigidly paced operation, if the task has not been completed within the fixed cycle time, the work unit exits the station incomplete. This may inhibit completion of subsequent tasks at downstream stations. Whatever tasks are left undone on the work unit at the regular workstations must later be completed by some other worker in order to yield an acceptable product.

In *pacing with margin*, the worker is allowed to complete the task at the station within a specified time range. The maximum time of the range is longer than the cycle time, so that a worker is permitted to take more time if a problem occurs or if the task time required for a particular work unit is longer than the average (this occurs when different product styles are produced on the same assembly line). There are several ways in which pacing with margin can be achieved: (1) allowing queues of work units to form between stations, (2) designing the line so that the time a work unit spends inside each station is longer than the cycle time, and (3) allowing the worker to move beyond the boundaries of his/her own station. In method (1), implemented using an asynchronous transport system, work units are allowed to form queues in front of each station, thus guaranteeing that the workers are never starved for work, but also providing extra time for some work units as long as other units take less time. Method (2) applies to lines in which work units are fixed to a continuously moving conveyor and cannot be removed. Because the conveyor speed is constant, when the station length is longer than the distance needed by the worker to complete the assigned task, the time spent by the work unit inside the station boundaries (called the *tolerance time*) is longer than the cycle time. In method (3), the worker is simply allowed to either move upstream beyond the immediate station to get an early start on the next work unit, or move downstream past the current station boundary to finish the task on the current work unit. In either case, there are usually practical limits on how far the worker can move upstream or downstream, making this a case of pacing with margin. The terms *upstream allowance* and *downstream allowance* are sometimes used to designate these limits in movement. In all of these methods, as long as the worker maintains an average pace that matches the cycle time, the required cycle rate of the line will be achieved.

The third level of pacing is when there is *no pacing*, meaning that no time limit exists within which the task at the station must be finished. In effect, each assembly operator works at his/her own pace. This case can occur when (1) manual work transport is used on the line, (2) work units can be removed from the conveyor, allowing the worker to take as much time as desired to complete a given unit, or (3) an asynchronous conveyor is used and the worker controls the release of each work unit from the station. In each of these cases, there is no mechanical means of achieving a pacing discipline on the line. To reach the required production rate, the workers are motivated to achieve a certain pace either by their own collective work ethic or by an incentive system sponsored by the company.

15.1.4 Coping with Product Variety

Owing to the versatility of human workers, manual assembly lines can be designed to deal with differences in assembled products. In general, the product variety must be relatively soft (Sections 2.3). Three types of assembly line can be distinguished: (1) single model, (2) batch model, and (3) mixed model.

A *single model line* produces only one product in large quantities. Every work unit is identical, so the task performed at each station is the same for all products. This line type is intended for products with high demand.

Batch model and mixed model lines are designed to produce two or more models, but different approaches are used to cope with the model variations. A *batch model line* produces each model in batches. Workstations are set up to produce the required quantity of the first model, then the stations are reconfigured to produce the next model, and so on. Products are often assembled in batches when demand for each product is medium. It is generally more economical to use one assembly line to produce several products in batches than to build a separate line for each different model.

When we state that the workstations are set up, we are referring to the assignment of tasks to each station on the line, including the special tools needed to perform the tasks, and the physical layout of the station. The models made on the line are usually similar, and the tasks to make them are therefore similar. However, differences exist among models so that a different sequence of tasks is usually required, and the tools used at a given workstation for the last model might not be the same as those required for the next model. One model may take more total time than another, requiring the line to be operated at a slower pace. Worker retraining or new equipment may be needed to produce each new model. For these kinds of reasons, changes in the station setup must be made before production of the next model can begin. These changeovers result in lost production time on a batch model line.

A *mixed model line* also produces more than one model; however, the models are not produced in batches. Instead, they are made simultaneously on the same line. While one station is working on one model, the next station is processing a different model. Each station is equipped to perform the variety of tasks needed to produce any model that moves through it. Many consumer products are assembled on mixed model lines. Examples are automobiles and major appliances, which are characterized by model variations, differences in available options, and even brand name differences in some cases.

Advantages of a mixed model line over a batch model line include (1) no lost production time changing over between models, (2) avoidance of the high inventories typical of batch production, and (3) the ability to alter production rates of different models as product demand changes. On the other hand, the problem of assigning tasks to workstations so that they all share an equal workload is more complex on a mixed model line. Scheduling (determining the sequence of models) and logistics (getting the right parts to each workstation for the model currently at that station) are more difficult in this type of line. And in general, a batch model line can accommodate wider variations in model configurations.

As a summary of this discussion, Figure 15.3 indicates the position of each of the three assembly line types on a scale of product variety.

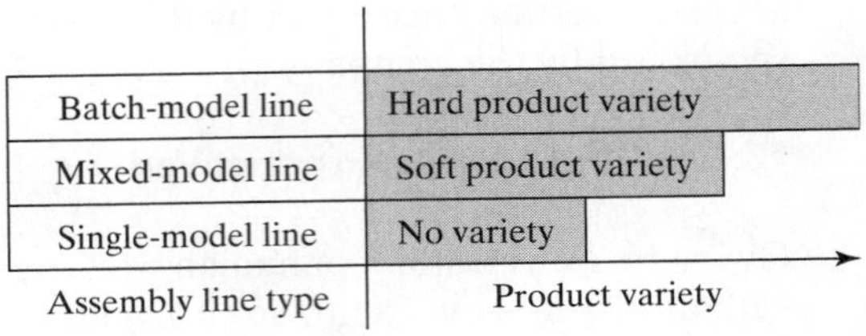

Batch-model line	Hard product variety	
Mixed-model line	Soft product variety	
Single-model line	No variety	
Assembly line type	Product variety	

Figure 15.3   Three types of manual assembly line related to product variety.

15.2   ANALYSIS OF SINGLE MODEL ASSEMBLY LINES

The relationships developed in this and the following section are applicable to single model assembly lines. With a little modification the same relationships apply to batch model lines. We consider mixed model assembly lines in Section 15.4.

The assembly line must be designed to achieve a production rate, R_p, sufficient to satisfy demand for the product. Product demand is often expressed as an annual quantity, which can be reduced to an hourly rate. Management must decide on the number of shifts per week that the line will operate, and the number of hours per shift. Assuming the plant operates 50 weeks per year, then the required hourly production rate is given by

$$R_p = \frac{D_a}{50 S_w H_{sh}} \tag{15.3}$$

where R_p = average hourly production rate, units/hr; D_a = annual demand for the single product to be made on the line, units/yr; S_w = number of shifts/wk; and H_{sh} = hr/shift. If the line operates 52 weeks rather than 50, then $R_p = D_a/52 S_w H_{sh}$. If a time period other than a year is used for product demand, then the equation can be adjusted by using consistent time units in the numerator and denominator.

This production rate must be converted to a cycle time T_c, which is the time interval at which the line will be operated. The cycle time must take into account the reality that some production time will be lost due to occasional equipment failures, power outages, lack of a certain component needed in assembly, quality problems, labor problems, and other reasons. As a consequence of these losses, the line will be up and operating only a certain proportion of time out of the total shift time available; this uptime proportion is referred to as the *line efficiency*. The cycle time can be determined as

$$T_c = \frac{60E}{R_p} \tag{15.4}$$

where T_c = cycle time of the line, min/cycle; R_p = required production rate, as determined from Eq. (15.3), units/hr; the constant 60 converts the hourly production rate to a cycle time in minutes; and E = line efficiency. Typical values of E for a manual assembly line are in the range 0.90 to 0.98. The cycle time T_c establishes the ideal cycle rate for the line

$$R_c = \frac{60}{T_c} \tag{15.5}$$

where R_c = cycle rate for the line, cycles/hr; and T_c is in min/cycle as in Eq. (15.4). This rate R_c must be greater than the required production rate R_p because the line efficiency E is less than 100%. Line efficiency E is therefore defined as

$$E = \frac{R_p}{R_c} = \frac{T_c}{T_p} \tag{15.6}$$

where T_p = average production cycle time ($T_p = 60/R_p$).

An assembled product requires a certain total amount of time to build, called the *work content time* (T_{wc}.) This is the total time of all work elements that must be performed on the line to make one unit of the product. It represents the total amount of work that is to be accomplished on the product by the assembly line. It is useful to compute a theoretical minimum number of workers that will be required on the assembly line to produce a product with known T_{wc} and specified production rate R_p. The approach is basically the same as we used in Section 14.4.1 to compute the number of workstations required to achieve a specified production workload. Let us use Eq. (14.7) in that section to determine the number of workers on the production line:

$$w = \frac{WL}{AT} \tag{15.7}$$

where w = number of workers on the line; WL = workload to be accomplished in a given time period; and AT = available time in the period. The time period of interest will be 60 min. The *workload* in that period is the hourly production rate multiplied by the work content time of the product, that is,

$$WL = R_p T_{wc} \tag{15.8}$$

where R_p = production rate, pc/hr; and T_{wc} = work content time, min/pc.

Eq. (15.4) can be rearranged to the form $R_p = 60E/T_c$. Substituting this into Eq. (15.8), we have

$$WL = \frac{60 E T_{wc}}{T_c}$$

The available time AT = one hour (60 min) multiplied by the proportion uptime on the line; that is,

$$AT = 60E$$

Substituting these terms for WL and AT into Eq. (15.7), the equation reduces to the ratio T_{wc}/T_c. Since the number of workers must be an integer, we can state

$$w^* = \text{Minimum Integer} \geq \frac{T_{wc}}{T_c} \tag{15.9}$$

where w^* = theoretical minimum number of workers. If we assume one worker per station ($M_i = 1$ for all i, $i = 1, 2, \ldots, n$; and the number utility workers $w_u = 0$), then this ratio also gives the theoretical minimum number of workstations on the line.

Achieving this minimum theoretical value in practice is very unlikely. Eq. (15.9) ignores two critical factors that exist in a real assembly line and tend to increase the number of workers above the theoretical minimum.

1. *Repositioning losses.* Some time will be lost at each station for repositioning of the work unit or the worker. Thus, the time available per worker to perform assembly is less than T_c.
2. *The line balancing problem.* It is virtually impossible to divide the work content time evenly amongst all workstations. Some stations are bound to have an amount of work that requires less time than T_c. This tends to increase the number of workers.

Let us consider repositioning losses and imperfect balancing in the following discussion. For simplicity, we limit the discussion to the case where one worker is assigned to each station ($M_i = 1$). Thus, when we refer to a certain station, we are referring to the worker at that station, and when we refer to a certain worker, we are referring to the station where that worker is assigned.

15.2.1 Repositioning Losses

Repositioning losses on a production line occur because some time is required each cycle to reposition the worker, or the work unit, or both. For example, on a continuous transport line with work units attached to the conveyor and moving at a constant speed, time is required for the worker to walk from the unit just completed to the upstream unit entering the station. In other conveyorized systems, time is required to remove the work unit from the conveyor and position it at the station for the worker to perform his or her task on it. In all manual assembly lines, there is some lost time for repositioning. Let us define T_r as the time required each cycle to reposition the worker, the work unit, or both. In our subsequent analysis, we assume that T_r is the same for all workers, although repositioning times may actually vary among stations.

The repositioning time T_r must be subtracted from the cycle time T_c to obtain the available time remaining to perform the actual assembly task at each workstation. Let us refer to the time to perform the assigned task at each station as the *service time.* It is symbolized T_{si}, where i is used to identify station i, $i = 1, 2, \ldots, n$. Service times will vary amongst stations because the total work content cannot be allocated evenly amongst stations. Some stations will have more work than others. There will be at least one station at which T_{si} is maximum. This is referred to as the *bottleneck station* because it establishes the cycle time for the entire line. This maximum service time must be no greater than the difference between the cycle time T_c and the repositioning time T_r; that is,

$$\text{Max}\{T_{si}\} \leq T_c - T_r \qquad \text{for } i = 1, 2, \ldots n \qquad (15.10)$$

where $\text{Max}\{T_{si}\}$ = maximum service time amongst all stations, min/cycle; T_c = cycle time for the assembly line from Eq. (15.4), min/cycle; and T_r = repositioning time (assumed the same for all stations), min/cycle. For simplicity of notation, let us use T_s to denote this maximum allowable service time, that is,

$$T_s = \text{Max}\{T_{si}\} \leq T_c - T_r \qquad (15.11)$$

At all stations where T_{si} is less than T_s, workers will be idle for a portion of the cycle, as portrayed in Figure 15.4. When the maximum service time does not consume the entire available time $T_c - T_r$ (that is, when $T_s < T_c - T_r$), this means that the line could be operated at a faster pace than T_c from Eq. (15.4). In this case, the cycle time T_c is usually reduced so that $T_c = T_s + T_r$; this allows the production rate to be increased slightly.

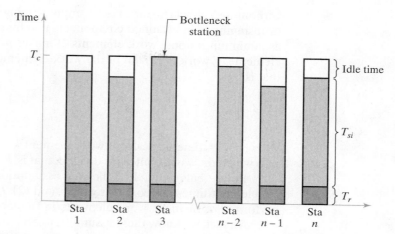

Figure 15.4 Components of cycle time at several stations on a manual assembly line. At the slowest station, the bottleneck station, idle time = zero; at other stations idle time exists. Key: Sta. = workstation, n = number of workstations on the line, T_r = repositioning time, T_{si} = service time, T_c = cycle time.

Repositioning losses reduce the amount of time that can be devoted to productive assembly work on the line. These losses can be expressed in terms of an efficiency factor as

$$E_r = \frac{T_s}{T_c} = \frac{T_c - T_r}{T_c} \tag{15.12}$$

where E_r = *repositioning efficiency* and the other terms are defined above.

15.2.2 The Line Balancing Problem

The work content performed on an assembly line consists of many separate and distinct work elements. Invariably, the sequence in which these elements can be performed is restricted, at least to some extent, and the line must operate at a specified production rate, which reduces to a required cycle time as defined by Eq. (15.4). Given these conditions, the line balancing problem is concerned with assigning the individual work elements to workstations so that all workers have an equal amount of work. Let us discuss the terminology of the line balancing problem in this section. We present some of the algorithms to solve it in Section 15.3.

Two important concepts in line balancing are the separation of the total work content into minimum rational work elements and the precedence constraints that must be satisfied by these elements. Based on these concepts we can define performance measures for solutions to the line balancing problem.

Minimum Rational Work Elements. A minimum rational work element is a small amount of work that has a specific limited objective, such as adding a component to the base part, joining two components, or performing some other small portion of the total work content. A minimum rational work element cannot be subdivided any further without loss of practicality. For example, drilling a through-hole in a piece of sheet metal or fastening two machined components together with a bolt and screw would be defined as minimum rational work elements. It makes no sense to divide these tasks into smaller elements of work. The sum of the work element times is equal to the work content time; that is,

$$T_{wc} = \sum_{k=1}^{n_e} T_{ek} \qquad (15.13)$$

where T_{ek} = time to perform work element k, min; and n_e = number of work elements into which the work content is divided, that is, $k = 1, 2, \ldots, n_e$.

In line balancing, the following assumptions are made about work element times: (1) element times are constant values, and (2) T_{ek} values are additive; that is, the time to perform two or more work elements in sequence is the sum of the individual element times. In fact, we know these assumptions are not quite true. Work element times are variable, leading to the problem of task time variability. And there is often motion economy that can be achieved by combining two or more work elements, violating the additivity assumption. Nevertheless, these assumptions are made to allow solution of the line balancing problem.

The task time at station i, or service time as we are calling it, T_{si}, is composed of the work element times that have been assigned to that station, that is,

$$T_{si} = \sum_{k \in i} T_{ek} \qquad (15.14)$$

An underlying assumption in this equation is that all T_{ek} are less than the maximum service time T_s.

Different work elements require different times, and when the elements are grouped into logical tasks and assigned to workers, the station service times T_{si} are likely not to be equal. Thus, simply because of the variation among work element times, some workers will be assigned more work, while others will be assigned less. Although service times vary from station to station, they must add up to the work content time:

$$T_{wc} = \sum_{i=1}^{n} T_{si} \qquad (15.15)$$

Precedence Constraints.   In addition to the variation in element times that make it difficult to obtain equal service times for all stations, there are restrictions on the order in which the work elements can be performed. Some elements must be done before others. For example, to create a threaded hole, the hole must be drilled before it can be tapped. A machine screw that will use the tapped hole to attach a mating component cannot be fastened before the hole has been drilled and tapped. These technological requirements on the work sequence are called *precedence constraints*. As we shall see later, they complicate the line balancing problem.

Precedence constraints can be presented graphically in the form of a *precedence diagram,* which is a network diagram that indicates the sequence in which the work elements must be performed. Work elements are symbolized by nodes, and the precedence requirements are indicated by arrows connecting the nodes. The sequence proceeds from left to right. Figure 15.5 presents the precedence diagram for the following example, which illustrates the terminology and some of the equations presented here.

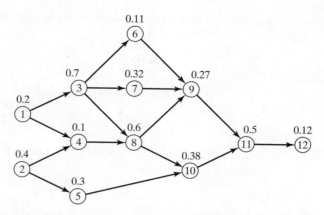

Figure 15.5 Precedence diagram for Example 15.1. Nodes represent work elements, and arrows indicate the sequence in which the elements must be done. Element times are shown above each node.

EXAMPLE 15.1   A Problem for Line Balancing.

A small electrical appliance is to be produced on a single model assembly line. The work content of assembling the product has been reduced to the work elements listed in Table 15.4. The table also lists the times for each element and the precedence order in which they must be performed. The line is to be balanced for an annual demand of 100,000 units per year. The line will operate 50 weeks/yr, 5 shifts/week, and 7.5 hr per shift. Manning level will be one worker per station. Previous experience suggests that the uptime efficiency for the line will be 96%, and repositioning time lost per cycle will be 0.08 min. Determine (a) total work content time T_{wc}, (b) required hourly production rate R_p to achieve the annual demand, (c) cycle time T_c, (d) theoretical minimum number of workers required on the line, and (e) service time T_s to which the line must be balanced.

TABLE 15.4   Work Elements for Example 15.1.

No.	Work Element Description	T_{ek} (Minutes)	Must be Preceded By
1	Place frame in workholder and clamp	0.2	—
2	Assemble plug, grommet to power cord	0.4	—
3	Assemble brackets to frame	0.7	1
4	Wire power cord to motor	0.1	1, 2
5	Wire power cord to switch	0.3	2
6	Assemble mechanism plate to bracket	0.11	3
7	Assemble blade to bracket	0.32	3
8	Assemble motor to brackets	0.6	3,4
9	Align blade and attach to motor	0.27	6, 7, 8
10	Assemble switch to motor bracket	0.38	5, 8
11	Attach cover, inspect, and test	0.5	9, 10
12	Place in tote pan for packing	0.12	11

Solution: (a) The total work content time is the sum of the work element times in Table 15.4.

$$T_{wc} = 4.0 \text{ min}$$

(b) Given the annual demand, the hourly production rate is

$$R_p = \frac{100,000}{50(5)(7.5)} = 53.33 \text{ units/hr}$$

(c) The corresponding cycle time T_c with an uptime efficiency of 96% is

$$T_c = \frac{60(0.96)}{53.33} = 1.08 \text{ min}$$

(d) The theoretical minimum number of workers is given by Eq. (15.9):

$$w^* = \text{Min Int} \geq \frac{4.0}{1.08} (= 3.7) = 4 \text{ workers}$$

(e) The available service time against which the line must be balanced is

$$T_s = 1.08 - 0.08 = 1.00 \text{ min}$$

Measures of Line Balance Efficiency. Owing to the differences in minimum rational work element times and the precedence constraints among the elements, it is virtually impossible to obtain a perfect line balance. Measures must be defined to indicate how good a given line balancing solution is. One possible measure is *balance efficiency*, which is the work content time divided by the total available service time on the line:

$$E_b = \frac{T_{wc}}{wT_s} \tag{15.16}$$

where E_b = balance efficiency, often expressed as a percent; T_s = the maximum available service time on the line $(\text{Max}\{T_{si}\})$, min/cycle; and w = number of workers. The denominator in Eq. (15.16) gives the total service time available on the line to devote to the assembly of one product unit. The closer the values of T_{wc} and wT_s, the less idle time on the line. E_b is therefore a measure of how good the line balancing solution is. A perfect line balance yields a value of $E_b = 1.00$. Typical line balancing efficiencies in industry range between 0.90 and 0.95.

The complement of balance efficiency is *balance delay*, which indicates the amount of time lost due to imperfect balancing as a ratio to the total time available, that is,

$$d = \frac{(wT_s - T_{wc})}{wT_s} \tag{15.17}$$

where d = balance delay and the other terms have the same meaning as before. A balance delay of zero indicates a perfect balance. Note that $E_b + d = 1$.

Worker Requirements. In our discussion of the relationships in this section, we have identified three factors that reduce the productivity of a manual assembly line. They can all be expressed as efficiencies:

1. *Line efficiency*, the proportion of uptime on the line E, as defined in Eq. (15.6)
2. *Repositioning efficiency*, E_r, as defined in Eq. (15.12)
3. *Line balancing efficiency*, E_b, as defined in Eq. (15.16).

Together, they constitute the overall labor efficiency on the assembly line:

$$\text{Labor efficiency on the assembly line} = EE_rE_b \qquad (15.18)$$

Using this measure of labor efficiency, we can calculate a more realistic value for the number of workers on the assembly line, based on previous Eq. (15.9):

$$w = \text{Minimum Integer} \geq \frac{R_pT_{wc}}{60EE_rE_b} = \frac{T_{wc}}{E_rE_bT_c} = \frac{T_{wc}}{E_bT_s} \quad (15.19)$$

where w = actual number of workers required on the line; R_p = hourly production rate, units/hr; and T_{wc} = work content time per product to be accomplished on the line, min/unit. The trouble with this relationship is that it is difficult to determine values for E, E_r, and E_b before the line is built and operated. Nevertheless, the equation provides an accurate model of the parameters that affect the number of workers required to accomplish a given workload on a single model assembly line.

15.3 LINE BALANCING ALGORITHMS

The objective in line balancing is to distribute the total workload on the assembly line as evenly as possible among the workers. This objective can be expressed mathematically in two alternative but equivalent forms:

$$\text{Minimize } (wT_s - T_{wc}) \text{ or Minimize } \sum_{i=1}^{w}(T_s - T_{si}) \qquad (15.20)$$

subject to:

$$(1) \sum_{k \in i} T_{ek} \leq T_s$$

and

(2) all precedence requirements are obeyed.

In this section we consider several methods to solve the line balancing problem, using the data of Example 15.1 to illustrate. The algorithms are (1) largest candidate rule, (2) Kilbridge and Wester method, and (3) ranked positional weights method. These methods are heuristic, meaning they are based on common sense and experimentation rather than mathematical optimization. In each of the algorithms, we assume that the manning level is one, so when we identify station i, we are also identifying the worker at station i.

15.3.1 Largest Candidate Rule

In this method, work elements are arranged in descending order according to their T_{ek} values, as in Table 15.5. Given this list, the algorithm consists of the following steps: (1) assign elements to the worker at the first workstation by starting at the top of the list and selecting the first element that satisfies precedence requirements and does not cause the total sum of T_{ek} at that station to exceed the allowable T_s; when an element is selected for assignment to the station, start back at the top of the list for subsequent assignments; (2) when no more elements can be assigned without exceeding T_s, then proceed to the next

TABLE 15.5 Work Elements Arranged According to T_{ek} Value for the Largest Candidate Rule

Work Element	T_{ek} (min)	Preceded By
3	0.7	1
8	0.6	3, 4
11	0.5	9, 10
2	0.4	–
10	0.38	5, 8
7	0.32	3
5	0.3	2
9	0.27	6, 7, 8
1	0.2	–
12	0.12	11
6	0.11	3
4	0.1	1, 2

station; (3) repeat steps 1 and 2 for as many additional stations as necessary until all elements have been assigned.

EXAMPLE 15.2 Largest Candidate Rule

Apply the largest candidate rule to Example Problem 15.1.

Solution: Work elements are arranged in descending order in Table 15.5, and the algorithm is carried out as presented in Table 15.6. Five workers and stations are required in the solution. Balance efficiency is computed as

$$E_b = \frac{4.0}{5(1.0)} = 0.80$$

Balance delay $d = 0.20$. The line balancing solution is presented in Figure 15.6.

TABLE 15.6 Work Elements Assigned to Stations According to the Largest Candidate Rule

Station	Work Element	T_{ek} (min)	Station Time (min)
1	2	0.4	
	5	0.3	
	1	0.2	
	4	0.1	1.0
2	3	0.7	
	6	0.11	0.81
3	8	0.6	
	10	0.38	0.98
4	7	0.32	
	9	0.27	0.59
5	11	0.5	
	12	0.12	0.62

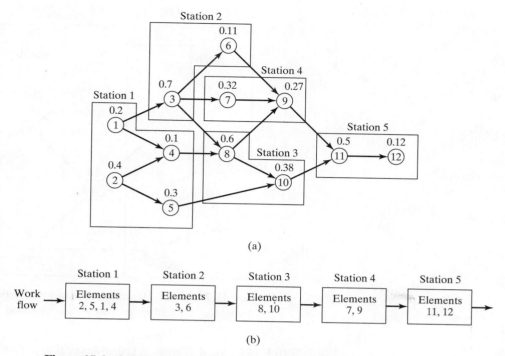

(a)

| Station 1 | Station 2 | Station 3 | Station 4 | Station 5 |

Work flow → Elements 2, 5, 1, 4 → Elements 3, 6 → Elements 8, 10 → Elements 7, 9 → Elements 11, 12 →

(b)

Figure 15.6 Solution for Example 15.2, which indicates: (a) assignment of elements according to the largest candidate rule, and (b) physical sequence of stations with assigned work elements.

15.3.2 Kilbridge and Wester Method

This method has received considerable attention since its introduction in 1961 [17] and has been applied with apparent success to several complicated line balancing problems in industry [21]. It is a heuristic procedure that selects work elements for assignment to stations according to their position in the precedence diagram. This overcomes one of the difficulties with the largest candidate rule in which an element may be selected because of a high T_e value but irrespective of its position in the precedence diagram. In general, the Kilbridge and Wester method provides a superior line balance solution to that provided by the largest candidate rule (although this is not the case for our example problem).

In the Kilbridge and Wester method, work elements in the precedence diagram are arranged into columns, as shown in Figure 15.7. The elements can then be organized into a list according to their columns, with the elements in the first column listed first. We have developed such a list of elements for our example problem in Table 15.7. If a given element can be located in more than one column, then all of the columns for that element should be listed, as we have done in the case of element 5. In our list, we have added the feature that elements in a given column are presented in the order of their T_{ek} value; that is, we have applied the largest candidate rule within each column. This is helpful when assigning elements to stations, because it ensures that the larger elements are selected

Column

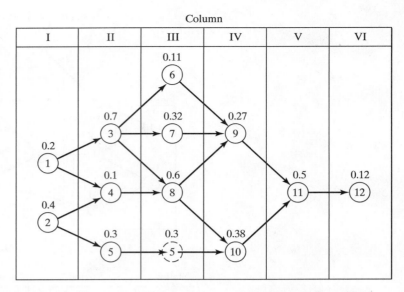

Figure 15.7 Work elements in example problem arranged into columns for the Kilbridge and Wester method.

TABLE 15.7 Work Elements Listed According to Columns from Figure 15.7 for the Kilbridge and Wester method

Work Element	Column	T_{ek} (min)	Preceded By
2	I	0.4	–
1	I	0.2	–
3	II	0.7	1
5	II, III	0.3	2
4	II	0.1	1, 2
8	III	0.6	3, 4
7	III	0.32	3
6	III	0.11	3
10	IV	0.38	5, 8
9	IV	0.27	6, 7, 8
11	V	0.5	9, 10
12	VI	0.12	11

first, thus increasing our chances of making the sum of T_{ek} in each station closer to the allowable T_s limit. Once the list is established, the same three-step procedure is used as before.

EXAMPLE 15.3 Kilbridge and Wester Method

Apply the Kilbridge and Wester method to Example Problem 15.1.

Solution: Work elements are arranged in order of columns in Table 15.7. The Kilbridge and Wester solution is presented in Table 15.8. Five workers are required and the balance efficiency is $E_b = 0.80$. Note that although the balance efficiency is the same as in the largest candidate rule, the allocation of work elements to stations is different.

TABLE 15.8 Work Elements Assigned to Stations According to the Kilbridge and Wester Method

Station	Work Element	Column	T_{ek} (min)	Station Time (min)
1	2	I	0.4	
	1	I	0.2	
	5	II	0.3	
2	4	II	0.1	1.0
	3	II	0.7	
3	6	III	0.11	0.81
	8	III	0.6	
4	7	III	0.32	0.92
	10	IV	0.38	
5	9	IV	0.27	0.65
	11	V	0.5	
	12	VI	0.12	0.62

15.3.3 Ranked Positional Weights Method

The ranked positional weights method was introduced by Helgeson and Birnie [13]. In this method, a ranked positional weight value (call it *RPW* for short) is computed for each element. The *RPW* takes into account both the T_{ek} value and its position in the precedence diagram. Specifically, RPW_k is calculated by summing T_{ek} and all other times for elements that follow T_{ek} in the arrow chain of the precedence diagram. Elements are compiled into a list according to their *RPW* value, and the algorithm proceeds using the same three steps as before.

EXAMPLE 15.4 Ranked Positional Weights Method

Apply the ranked positional weights method to Example Problem 15.1.

Solution: The *RPW* must be calculated for each element. To illustrate,

$$RPW_{11} = 0.5 + 0.12 = 0.62$$

$$RPW_8 = 0.6 + 0.27 + 0.38 + 0.5 + 0.12 = 1.87$$

Work elements are listed according to *RPW* value in Table 15.9. Assignment of elements to stations proceeds with the solution presented in Table 15.10. Note that the largest T_s value is 0.92 min. This can be exploited by operating the line at this faster rate, with the result that line balance efficiency is improved and production rate is increased:

$$E_b = \frac{4.0}{5(.92)} = 0.87$$

The cycle time is $T_c = T_s + T_r = 0.92 + 0.08 = 1.00$; therefore,

$$R_c = \frac{60}{1.0} = 60 \text{ cycles/hr}$$

And given that line efficiency $E = 0.96$, $R_p = 60(0.96) = 57.6$ units/hr

TABLE 15.9 Elements Ranked According to Their Ranked Positional Weights (*RPW*)

Work Element	RPW	T_{ek} (min)	Preceded By
1	3.30	0.2	–
3	3.00	0.7	1
2	2.67	0.4	–
4	1.97	0.1	1, 2
8	1.87	0.6	3, 4
5	1.30	0.3	2
7	1.21	0.32	3
6	1.00	0.11	3
10	1.00	0.38	5, 8
9	0.89	0.27	6, 7, 8
11	0.62	0.5	9, 10
12	0.12	0.12	11

TABLE 15.10 Work Elements Assigned to Stations According to the Ranked Positional Weights (*RPW*) method

Station	Work Element	T_{ek} (min)	Station Time (min)
1	1	0.2	
	3	0.7	0.90
2	2	0.4	
	4	0.1	
	5	0.3	
	6	0.11	0.91
3	8	0.6	
	7	0.32	0.92
4	10	0.38	
	9	0.27	0.65
5	11	0.5	
	12	0.12	0.62

This is a better solution than the ones that the previous line balancing methods provided. It turns out that the performance of a given line balancing algorithm depends on the problem to be solved. Some line balancing methods work better on some problems, while other methods work better on other problems.

15.4 MIXED MODEL ASSEMBLY LINES

A mixed model assembly line is a manual production line capable of producing a variety of different product models simultaneously and continuously (not in batches). Each workstation specializes in a certain set of assembly work elements, but the stations are sufficiently flexible that they can perform their respective tasks on different models. Mixed model lines are typically used to accomplish the final assembly of automobiles, small and large trucks, and major and small appliances. In this section, we discuss some of the technical issues related to mixed model assembly lines, specifically (1) determining the number of workers and other operating parameters, (2) line balancing, and (3) model launching.

15.4.1 Determining the Number of Workers on the Line

To determine the number of workers required for a mixed model assembly line, we again start with Eq. (15.7):

$$w = \frac{WL}{AT}$$

where w = number of workers; WL = workload to be accomplished by the workers in the scheduled time period, min/hr; and AT = available time per worker in the same time period, min/hr per worker. The time period used here is an hour, but units could be min/shift, or min/week, depending on the information available and the analyst's preference.

The *workload* consists of the work content time of each model multiplied by its respective production rate during the period, that is,

$$WL = \sum_{j=1}^{P} R_{pj} T_{wcj} \tag{15.21}$$

where WL = workload min/hr; R_{pj} = Production rate of model j, pc/hr; T_{wcj} = work content time of model j, min/pc; p = the number of models to be produced during the period; and j is used to identify the model, $j = 1, 2, \ldots, P$.

Available time per worker is the number of minutes available to accomplish assembly work on the product during the hour. In the ideal case, where repositioning and line balance efficiencies are 100%, then $AT = 60E$, where E = proportion uptime on the line. This allows us to determine the theoretical minimum number of workers:

$$w^* = \text{Minimum Integer} \geq \frac{WL}{60E} \tag{15.22}$$

More realistically, repositioning and balance efficiencies will be less than 100%, and this fact should be factored into the available time:

$$AT = 60EE_r E_b \tag{15.23}$$

where AT = available time per worker, min/hr; 60 = number of minutes in an hour, min/hr; E = line efficiency; E_r = repositioning efficiency; and E_b = balance efficiency.

EXAMPLE 15.5 Number of Workers Required in a Mixed Model Line

The hourly production rate and work content time for two models to be produced on a mixed model assembly line are given in this table:

Model j	R_{pj}	T_{wcj}
A	4/hr	27.0 min
B	6/hr	25.0 min

Also given is that line efficiency $E = 0.96$ and manning level $M = 1$. Determine (a) the theoretical minimum number of workers required on the assembly line and (b) the actual number of workers if it is known that repositioning efficiency $E_r = 0.974$ and line balancing efficiency $E_b = 0.921$.

Solution: (a) Workload per hour is computed by using Eq. (15.21):

$$WL = 4(27) + 6(25) = 258 \text{ min/hr}$$

Available time per hour is 60 min corrected for E, but not for E_r and E_b:

$$AT = 60(0.96) = 57.6 \text{ min}$$

Using Eq. (15.22), the theoretical minimum number of workers is therefore

$$w^* = \text{Minimum Integer} \geq \frac{258}{57.6} = 4.48 \rightarrow 5 \text{ workers}$$

(b) Now using Eq. (15.23), $AT = 60(0.96)(0.974)(0.921) = 51.67 \text{ min}$

$$w = \text{Minimum Integer} \geq \frac{258}{51.67} = 4.99 \rightarrow 5 \text{ workers}$$

15.4.2 Mixed Model Line Balancing

The objective in mixed model line balancing is the same as for single model lines: to spread the workload amongst stations as evenly as possible. Algorithms used to solve the mixed model line balancing problem are usually adaptations of methods developed for single model lines. Our treatment of this topic is admittedly limited. The interested reader can pursue mixed model line balancing and its companion problem, model sequencing, in several of our references, including [7], [21], [22], [23], [26]. A literature review of these topics is presented in [11].

In single model line balancing, work element times are utilized to balance the line, as in Section 15.3. In mixed model assembly line balancing, total work element times per hour (or per shift) are used. The objective function can be expressed as

$$\text{Minimize } (wAT - WL) \text{ or Minimize } \sum_{i=1}^{w} (AT - TT_{si}) \qquad (15.24)$$

where w = number of workers or stations (we are again assuming the $M_i = 1$, so that $n = w$); AT = available time in the period of interest (e.g., hour, shift), min; WL = workload to be accomplished during the same period, min; and TT_{si} = total service time at station i to perform its assigned portion of the workload, min.

The two statements in Eq. (15.24) are equivalent. Workload can be calculated as before, using Eq. (15.21):

$$WL = \sum_{j=1}^{P} R_{pj} T_{wcj}$$

To determine total service time at station i, we must first compute the total time to perform each element in the workload. Let T_{ejk} = time to perform work element k on product j. The total time per element is given by

$$TT_k = \sum_{j=1}^{P} R_{pj} T_{ejk} \qquad (15.25)$$

where TT_k = total time within the workload that must be allocated to element k for all products, min. Based on these TT_k values, element assignments can be made to each station according to one of the line balancing algorithms. Total service times at each station are computed:

$$TT_{si} = \sum_{k \in i} TT_k \qquad (15.26)$$

where TT_{si} = total service time at station i which equals the sum of the times of the elements that have been assigned to that station, min.

Measures of balance efficiency for mixed model assembly line balancing correspond to those in single model line balancing,

$$E_b = \frac{WL}{w(Max\{TT_{si}\})} \qquad (15.27)$$

where E_b = balance efficiency; WL = workload from Eq. (15.21), min; w = number of workers (stations); and $Max\{TT_{si}\}$ = maximum value of total service time among all stations in the solution. It is possible that the line balancing solution will yield a value of $Max\{TT_{si}\}$ that is less than the available total time AT. This situation occurs in the following example.

EXAMPLE 15.6 Mixed Model Assembly Line Balancing

This is a continuation of Example 15.5. For the two models A and B, hourly production rates are 4 units/hr for A and 6 units/hr for B. Most of the work elements are common to the two models, but in some cases the elements take longer for one model than for the other. The elements, times, and precedence requirements are given in Table 15.11. Also given: $E = 0.96$, repositioning time $T_r = 0.15$ min, and $M_i = 1$. (a) Construct the precedence diagram for each model and for both models combined into one diagram. (b) Use the Kilbridge and Wester method to solve the line balancing problem. (c) Determine the balance efficiency for the solution in (b).

Solution: (a) The precedence diagrams are shown in Figure 15.8.

(b) To use the Kilbridge and Wester method, we must (1) calculate total production time requirements for each work element, TT_k, according to Eq. (15.25) — this is done in Table 15.12; (2) arrange the elements according to columns in the precedence diagram, as in Table 15.13 (within columns, we have listed the

TABLE 15.11 Work Elements for Models A and B in Example 15.6

Work Element k	T_{eAK} (min)	Preceded By	T_{eBK} (min)	Preceded By
1	3	–	3	–
2	4	1	4	1
3	2	1	3	1
4	6	1	5	1
5	3	2	–	–
6	4	3	2	3
7	–	–	4	4
8	5	5,6	4	7
T_{wc}	27		25	

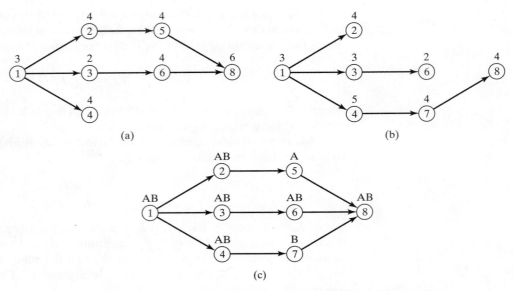

Figure 15.8 Precedence diagrams for Example 15.6: (a) for model A, (b) for model B, and (c) for both models combined.

TABLE 15.12 Total Times Required for Each Element in Each Model to Meet Respective Production Rates and for Both Models in Example 15.6

Element k	$R_{pA}T_{eAk}$ (min)	$R_{pB}T_{eBk}$ (min)	$\sum_{j=A,B} R_{pj}T_{ejk}$ (min)
1	12	18	30
2	16	24	40
3	8	18	26
4	24	30	54
5	12	0	12
6	16	12	28
7	0	24	24
8	20	24	44
			258

TABLE 15.13 Elements Arranged in Columns in Example 15.6

Element	Column	TT_k	Preceded By
1	I	30	–
4	II	54	1
2	II	40	1
3	II	26	1
6	III	28	3
7	III	24	4
5	III	12	2
8	IV	44	5,6,7

elements according to the largest candidate rule); and (3) allocate elements to workstations by using the three-step procedure defined in Section 15.5.1. To accomplish this third step, we need to compute the available time per worker, given proportion uptime $E = 0.96$ and repositioning efficiency E_r. To determine E_r, we note that total production rate is

$$R_p = 4 + 6 = 10 \text{ units/hr}$$

The corresponding cycle time is found by multiplying the reciprocal of this rate by proportion uptime E and accounting for the difference in time units, as follows:

$$T_c = \frac{60(0.96)}{10} = 5.76 \text{ min}$$

The service time each cycle is the cycle time less the repositioning time T_r:

$$T_s = 5.76 - 0.15 = 5.61 \text{ min}$$

Now repositioning efficiency can be determined as follows:

$$E_r = 5.61/5.76 = 0.974$$

Hence, we have the available time against which the line is to be balanced:

$$AT = 60(0.96)(0.974) = 56.1 \text{ min}$$

Allocating elements to stations against this limit, we have the final solution presented in Table 15.14.

(c) Balance efficiency is determined by Eq. (15.27). $\text{Max}\{TT_{si}\} = 56$ min. Note that this is slightly less than the available time of 56.1 min, so our line will operate slightly faster than we originally designed it to operate.

$$E_b = \frac{258}{5(56)} = 0.921 = 92.1 \%$$

TABLE 15.14 Allocation of Work Elements to Stations in Example 15.6 by Using the Kilbridge and Wester Method

Station	Element	TT_k (min)	TT_{si} (min)
1	1	30	
	3	26	56
2	4	54	54
3	2	40	
	5	12	52
4	6	28	
	7	24	52
5	8	44	44
			258

15.4.3 Model Launching in Mixed Model Lines

We previously noted that production on a manual assembly line typically involves launching of base parts onto the beginning of the line at regular time intervals. In a single model line, this time interval is constant and set equal to the cycle time T_c. The same applies for a batch model line, but T_c is likely to differ for each batch because the models are different and their production requirements are probably different. In a mixed model line, model launching is more complicated because each model is likely to have a different work content time, which translates into different station service times. Thus, the time interval between launches and the selection of which model to launch are interdependent. For example, if a series of models with high work content times are launched at short time intervals, the assembly line will quickly become congested (overwhelmed with too much work). On the other hand, if a series of models with low work content times are launched at long time intervals, then stations will be starved for work (with resulting idleness). Neither congestion nor idleness is desirable.

Determining the time interval between successive launches is referred to as the *launching discipline*. Two alternative launching disciplines are available in mixed model assembly lines: (1) variable-rate launching and (2) fixed-rate launching.

Variable-Rate Launching. In *variable-rate launching*, the time interval between the launching of the current base part and the next is set equal to the cycle time of the current unit. Since different models have different work content times and thus different task times per station, their cycle times and launch time intervals vary. The time interval in variable-rate launching can be expressed as

$$T_{cv}(j) = \frac{T_{wcj}}{w E_r E_b} \tag{15.28}$$

where $T_{cv}(j)$ = time interval before the next launch in variable-rate launching, min; T_{wcj} = work content time of the product just launched (model j), min; w = number of workers on the line; E_r = repositioning efficiency; and E_b = balance efficiency. If manning level $M_i = 1$ for all i, then the number of stations n can be substituted for w. With variable-rate launching, as long as the launching interval is determined by this formula, then models can be launched in any sequence desired.

EXAMPLE 15.7   Variable-Rate Launching in a Mixed Model Assembly Line

Determine the variable-rate launching intervals for models A and B in Examples 15.5 and 15.6. From the results of Example 15.6, we have $E_r = 0.974$ and $E_b = 0.921$.

Solution: Applying Eq. (15.28) for model A, we have

$$T_{cv}(A) = \frac{27.0}{5(.974)(.921)} = 6.020 \text{ min}$$

And for model B,

$$T_{cv}(B) = \frac{25.0}{5(.974)(.921)} = 5.574 \text{ min}$$

When a unit of model A is launched onto the front of the line, 6.020 min must elapse before the next launch. When a unit of model B is launched onto the front of the line, 5.574 min must elapse before the next launch.

The advantage of variable-rate launching is that units can be launched in any order without causing idle time or congestion at workstations, as long as the specified model mix is achieved by the end of the shift. The model mix can be adjusted at a moment's notice to adapt to changes in demand for the various products made on the line. However, certain technical and logistical issues must be addressed when variable-rate launching is used. One technical issue is that the work carriers on a moving conveyor are usually located at constant intervals along its length and so the work units must be attached only at these positions. This is not compatible with variable-rate launching which presumes that work units can be attached at any location along the conveyor corresponding to the variable-rate launching interval T_{cv} for the preceding model. One of the logistical issues in variable-rate launching is the problem of supplying the correct components and subassemblies to the individual stations for the models being assembled on the line at any given moment. Because of these kinds of issues, industry seems to prefer fixed-rate launching.

Fixed-Rate Launching for Two Models. In *fixed-rate launching*, the time interval between two consecutive launches is constant. This launching discipline is usually set by the speed of the conveyor and the spacing between work carriers (for example, hooks on a chain conveyor that occur at regular spacings in the chain). The time interval in fixed-rate launching depends on the product mix and production rates of models on the line. Of course, the schedule must be consistent with the available time and manpower available on the line, so repositioning efficiency and line balance efficiency must be figured in. Given the hourly production schedule, as well as values of E_r and E_b, the launching time interval is determined as

$$T_{cf} = \frac{\dfrac{1}{R_p} \displaystyle\sum_{j=1}^{P} R_{pj} T_{wcj}}{w E_r E_b} \tag{15.29}$$

where T_{cf} = time interval between launches in fixed-rate launching, min; R_{pj} = production rate of model j, units/hr; T_{wcj} = work content time of model j, min/unit; R_p = total production rate of all models in the schedule or simply the sum of R_{pj} values; P = the number of models produced in the scheduled period, $j = 1, 2, \ldots, P$; and w, E_r, and E_b have the same meaning as before. If manning level $M_i = 1$ for all i, then n can be used in place of w in the equation.

In fixed-rate launching, the models must be launched in a specific sequence; otherwise, station congestion and/or idle time (starving) will occur. Several algorithms, each with advantages and disadvantages, have been developed to select the model sequence [6], [10], [21], [23], [26]. In our present coverage, we attempt to synthesize the findings of this previous research to provide two approaches to the fixed-rate launching problem, one that works for the case of two models and another that works for three or more models.

Congestion and idle time can be identified in each successive launch as the difference between the cumulative fixed-rate launching interval and the sum of the launching intervals for the individual models that have been launched onto the line. This difference can be expressed mathematically as

$$\text{Congestion time or idle time} = \sum_{h=1}^{m} T_{cjh} - m T_{cf} \tag{15.30}$$

where T_{cf} = fixed-rate launching interval determined by Eq. (15.29), min; m = launch sequence during the period of interest; h = launch index number for summation purposes; and T_{cjh} = the cycle time associated with model j in launch position h, min, calculated as

$$T_{cjh} = \frac{T_{wcj}}{wE_rE_b} \tag{15.31}$$

where the symbols on the right-hand side of the equation are the same as for Eq. (15.28).

Congestion is recognized when Eq. (15.30) yields a positive difference, indicating that the actual sum of task times for the models thus far launched (m) exceeds the planned cumulative task time. *Idle time* is identified when Eq. (15.30) yields a negative value, indicating that the actual sum of task times is less than the planned time for the current launch m. It is desirable to minimize both congestion and idle time. Accordingly, let us propose the following procedure, in which the model sequence is selected so that the square of the difference between the cumulative fixed-rate launching interval and the cumulative individual model-launching interval is minimized for each launch. Expressing this procedure in equation form, we have

$$\text{For each launch } m, \text{ select } j \text{ so as to minimize } \left(\sum_{h=1}^{m} T_{cjh} - mT_{cf} \right)^2 \tag{15.32}$$

where all terms have been defined above.

EXAMPLE 15.8   Fixed-Rate Launching in a Mixed Model Assembly Line for Two Models

Determine: (a) the fixed-rate launching interval for the production schedule in Example 15.5, and (b) the launch sequence of models A and B during the hour. Use E_r and E_b from Example 15.5(b).

Solution: (a) The combined production rate of models A and B is $R_p = 4 + 6 = 10$ units/hr. The fixed time interval is computed by using Eq. (15.29):

$$T_{cf} = \frac{\dfrac{1}{10}\left(4(27) + 6(25) \right)}{5(.974)(.921)} = 5.752 \text{ min}$$

(b) To use the sequencing rule in Eq. (15.32), we need to compute T_{cjh} for each model by Eq. (15.31). The values are the same as those computed in previous Example 15.7 for the variable launching case: for model A, $T_{cAh} = 6.020$ min; and for model B, $T_{cBh} = 5.574$ min.

To select the first launch, compare

$$\text{For model A, } \left(6.020 - 1(5.752) \right)^2 = 0.072$$

$$\text{For model B, } \left(5.574 - 1(5.752) \right)^2 = 0.032$$

The value is minimized for model B; therefore a base part for model B is launched first ($m = 1$). To select the second launch, compare

$$\text{For model A, } \left(5.574 + 6.020 - 2(5.752) \right)^2 = 0.008$$

$$\text{For model B, } \left(5.574 + 5.574 - 2(5.752)\right)^2 = 0.127$$

The value is minimized for model A; therefore a base part for model A is launched second ($m = 2$). The procedure continues in this way, with the results displayed in Table 15.15.

TABLE 15.15 Fixed-Rate Launching Sequence-Obtained for Example 15.8

Launch m	mT_{cf}	$\left(\sum_{h=1}^{m-1} T_{cjh} + T_{cAm} - mT_{cf}\right)^2$	$\left(\sum_{h=1}^{m-1} T_{cjh} + T_{cBm} - mT_{cf}\right)^2$	Model
1	5.752	0.072	**0.032**	B
2	11.504	**0.008**	0.127	A
3	17.256	0.128	**0.008**	B
4	23.008	**0.032**	0.071	A
5	28.760	0.201	**0.000**	B
6	34.512	0.073	**0.031**	B
7	40.264	**0.008**	0.125	A
8	46.016	0.130	**0.007**	B
9	51.768	**0.033**	0.070	A
10	57.520	0.202	**0.000**	B

Fixed-Rate Launching for Three or More Models. The reader will note that four units of A and six units of B are scheduled in the sequence in Table 15.15, which is consistent with the production rate data given in the original example. This schedule is repeated each successive hour. When only two models are being launched in a mixed model assembly line, Eq. (15.32) yields a sequence that matches the desired schedule used to calculated T_{cf} and T_{cjh}. However, when three or more models are being launched onto the line, Eq. (15.32) is likely to yield a schedule that does not provide the desired model mix during the period. What happens is that models whose T_{cjh} values are close to T_{cf} are overproduced, while models with T_{cjh} values significantly different from T_{cf} are underproduced or even omitted from the schedule. Our sequencing procedure can be adapted for the case of three or more models by adding a term to the equation that forces the desired schedule to be met. The additional term is the ratio of the quantity of model j to be produced during the period divided by the quantity of model j units that have yet to be launched in the period, that is,

$$\text{Additional term for three or more models} = \frac{R_{pj}}{Q_{jm}}$$

where R_{pj} = quantity of model j to be produced during the period, that is, the production rate of model j, units/hr; and Q_{jm} = quantity of model j units remaining to be launched during the period as m (number of launches) increases, units/hr. Accordingly, the fixed-rate launching procedure for three or more models can be expressed as

$$\text{For each launch } m, \text{ select } j \text{ so as to minimize } \left(\sum_{h=1}^{m} T_{cjh} - mT_{cf}\right)^2 + \frac{R_{pj}}{Q_{jm}} \quad (15.33)$$

where all terms have been previously defined. The effect of the additional term is to reduce the chances that a unit of any model j will be selected for launching as the number of units of that model already launched during the period increases. When the last unit of model j scheduled during the period has been launched, the chance of launching another unit of model j becomes zero.

Selecting the sequence in fixed-rate launching can sometimes be simplified by dividing all R_{pj} values in the schedule by the largest common denominator (if one exists) that results in a set of new values all of which are integers. For instance, in Example 15.8 the hourly schedule consists of four units of model A and six units of model B. Both number are divisible by two, reducing the schedule to two units of A and three units of B each half hour. These values can then be used in the ratio in Eq. (15.33). The model sequence obtained from Eq. (15.33) is then repeated as necessary to fill out the hour or shift.

EXAMPLE 15.9 Fixed-Rate Launching in a Mixed Model Assembly Line for Three Models

Let us add a third model, C, to the production schedule in previous Example 15.8. Two units of model C will be produced each hour, and its work content time = 30 min. The proportion of uptime $E = 0.96$, as before.

Solution: Let us begin by calculating the total hourly production rate

$$R_p = 4 + 6 + 2 = 12 \text{ units/hr.}$$

Cycle time is determined based on this rate and the given value of proportion uptime E:

$$T_c = \frac{60(0.96)}{12} = 4.80 \text{ min}$$

Then

$$T_s = 4.80 - 0.15 = 4.65 \text{ min}$$

Using these values, we can determine repositioning efficiency.

$$E_r = 4.65/4.80 = 0.96875$$

To determine balance efficiency, we need to divide the workload by the available time on the line, where available time is adjusted for line efficiency E and repositioning efficiency E_r. Workload is computed as follows:

$$WL = 4(27) + 6(25) + 2(30) = 318 \text{ min}$$

Available time to be used in line balancing is thus

$$AT = 60(.96)(.96875) = 55.80 \text{ min}$$

The number of workers (and stations, since $M_i = 1$) required is given by

$$w = \text{Minimum Integer} \geq \frac{318}{55.8} = 5.7 \longrightarrow 6 \text{ workers}$$

For our example, let us assume that the line can be balanced with six workers, leading to the following balance efficiency:

$$E_b = \frac{318}{6(55.8)} = 0.94982$$

Using the values of E_r and E_b in Eq. (15.29), the fixed-rate launching interval is calculated:

$$T_{cf} = \frac{\frac{1}{12}(318)}{6(0.96875)(0.94982)} = 4.80 \text{ min}$$

The T_{cjh} values for each model are, respectively,

$$T_{cAh} = \frac{27}{6(0.96875)(0.94982)} = 4.891 \text{ min}$$

$$T_{cBh} = \frac{25}{6(0.96875)(0.94982)} = 4.528 \text{ min}$$

$$T_{cCh} = \frac{30}{6(0.96875)(0.94982)} = 5.434 \text{ min}$$

Let us note that the models A, B, and C are produced at rates of four, six, and two units per hour. Dividing by two these rates can be reduced to 2, 3, and 1 per half hour, respectively. These are the values we will use in the additional term of Eq. (15.33). The starting values of Q_{jm} for $m = 1$ are $Q_{A1} = 2$, $Q_{B1} = 3$, and $Q_{C1} = 1$. According to our procedure, we have

$$\text{For model A, } (4.891 - 4.80)^2 + \frac{2}{2} = 1.008$$

$$\text{For model B, } (4.528 - 4.80)^2 + \frac{3}{3} = 1.074$$

$$\text{For model C, } (5.434 - 4.80)^2 + \frac{1}{1} = 1.402$$

The minimum value occurs if a unit of model A is launched. Thus, the first launch ($m = 1$) is model A. The value of Q_{A1} is decremented by the one unit already launched, so that $Q_{A2} = 1$. For the second launch, we have

$$\text{For model A, } \left(4.891 + 4.891 - 2(4.80)\right)^2 + \frac{2}{1} = 2.033$$

$$\text{For model B, } \left(4.891 + 4.528 - 2(4.80)\right)^2 + \frac{3}{3} = 1.033$$

$$\text{For model C, } \left(4.891 + 5.434 - 2(4.80)\right)^2 + \frac{1}{1} = 1.526$$

The minimum occurs when a model B unit is launched. Thus, for $m = 2$, a unit of model B is launched, and $Q_{B3} = 2$. The procedure continues in this way, with the results displayed in Table 15.16.

TABLE 15.16 Fixed-Rate Launching Sequence Obtained for Example 15.9

m	mT_{cf}	$\left(\sum\limits_{h=1}^{m-1} T_{cjh} + T_{cAm} - mT_{cf}\right)^2 + \dfrac{R_{pA}}{Q_{Am}}$	$\left(\sum\limits_{h=1}^{m-1} T_{cjh} + T_{cBm} - mT_{cf}\right)^2 + \dfrac{R_{pB}}{Q_{Bm}}$	$\left(\sum\limits_{h=1}^{m-1} T_{cjh} + T_{cCm} - mT_{cf}\right)^2 + \dfrac{R_{pC}}{Q_{Cm}}$	Model
1	4.80	**1.008**	1.074	1.402	A
2	9.60	**2.033**	**1.033**	1.526	B
3	14.40	2.008	1.705	**1.205**	C
4	19.20	2.296	**1.526**	∞	B
5	24.00	**2.074**	3.008	∞	A
6	28.80	∞	**3.000**	∞	B

15.5 WORKSTATION CONSIDERATIONS

Let us attach a quantitative definition to some of the assembly line parameters discussed in Section 15.1.1. A workstation is a position along the assembly line where one or more workers perform assembly tasks. If the manning level is one for all stations ($M_i = 1.0$ for $i = 1, 2, \ldots, n$) then the number of stations is equal to the number of workers. In general, for any value of M for the line,

$$n = \frac{w}{M} \tag{15.34}$$

A workstation has a length dimension L_{si}, where i denotes station i. The total length of the assembly line is the sum of the station lengths:

$$L = \sum_{i=1}^{n} L_{si} \tag{15.35}$$

Here L = length of the assembly line, m (ft); and L_{si} = length of station i, m (ft). In the case when all L_{si} are equal,

$$L = nL_s \tag{15.36}$$

where L_s = station length, m (ft).

A common transport system used on manual assembly lines is a constant speed conveyor. Let us consider this case in developing the following relationships. Base parts are launched onto the beginning of the line at constant time intervals equal to the cycle time T_c. This provides a constant feed rate of base parts, and if the base parts remain fixed to the conveyor during their assembly, this feed rate will be maintained throughout the line. The feed rate is simply the reciprocal of the cycle time,

$$f_p = \frac{1}{T_c} \tag{15.37}$$

where f_p = feed rate on the line, products/min. A constant feed rate on a constant speed conveyor provides a center-to-center distance between base parts given by

$$s_p = \frac{v_c}{f_p} = v_c T_c \tag{15.38}$$

where s_p = center-to-center spacing between base parts, m/part (ft/part); and v_c = velocity of the conveyor, m/min (ft/min).

As we discussed in Section 15.1.3, pacing with margin is a desirable way to operate the line so as to achieve the desired production rate and at the same time allow for some product-to-product variation in task times at workstations. One way to achieve pacing with margin in a continuous transport system is to provide a tolerance time that is greater than the cycle time. *Tolerance time* is defined as the time a work unit spends inside the boundaries of the workstation. It is determined as the length of the station divided by the conveyor velocity, that is,

$$T_t = \frac{L_s}{v_c} \tag{15.39}$$

where T_t = tolerance time, min/part, assuming that all station lengths are equal. If stations have different lengths, identified by L_{si}, then the tolerance times will differ proportionally, since v_c is constant.

The total elapsed time a work unit spends on the assembly line can be determined simply as the length of the line divided by the conveyor velocity. It is also equal to the tolerance time multiplied by the number of stations. Expressing these relationships in equation form, we have

$$ET = \frac{L}{v_c} = \sum_{i=1}^{n} T_{ti} \tag{15.40}$$

where ET = elapsed time a work unit (specifically, the base part) spends on the conveyor during its assembly, min. If all tolerance times are equal, then $ET = nT_t$.

15.6 OTHER CONSIDERATIONS IN ASSEMBLY LINE DESIGN

The line balancing algorithms described in Section 15.3 are precise computational procedures that allocate work elements to stations based on deterministic quantitative data. However, the designer of a manual assembly line should not overlook certain other factors, some of which may improve line performance beyond what the balancing algorithms provide. Following are some of the considerations.

- *Line efficiency.* The uptime proportion E is a critical parameter in assembly line operation. When the entire line goes down, all workers are idled. It is the responsibility of management to maintain a value of E as close to 100% as possible. Steps that can be taken include (1) implementing a preventive maintenance program to minimize downtime occurrences, (2) employing well-trained repair crews to quickly fix breakdowns when they occur, (3) managing incoming components so that parts shortages do not cause line stoppages, and (4) insisting on the highest quality of incoming parts from suppliers so that downtime is not caused by poor quality components.
- *Methods Analysis.* Methods analysis involves the study of human work activity to seek out ways in which the activity can be done with less effort, in less time, and with greater effect. This kind of analysis is an obvious step in the design of a manual assembly line, since the work elements need to be defined in order to balance the line. In addition, methods analysis can be used after the line is in operation to examine

workstations that turn out to be bottlenecks. The analysis may result in improved efficiency of workers' hand and body motions, better workplace layout, design of special tools and/or fixtures to facilitate manual work elements, or even changes in the product design for easier assembly (we discuss design for assembly in Chapter 24).

- *Subdividing Work Elements.* Minimum rational work elements are defined as small tasks that cannot be subdivided further. It is reasonable to define such tasks in the assembly of a given product, even though in some cases it may be technically possible to further subdivide the element. For example, suppose a hole is to be drilled through a rather thick cross-section in one of the parts to be assembled. It would normally make sense to define this drilling operation as a minimum rational work element. However, what if this drilling process were the bottleneck station? It might then be argued that the drilling operation should be subdivided into two separate steps, which might be performed at two adjacent stations. This would not only relieve the bottleneck; it would probably increase the tool lives of the drill bits, thus reducing downtime for tool changes.

- *Sharing work elements between two adjacent stations.* If a particular work element results in a bottleneck operation at one station, while the adjacent station has ample idle time, it might be possible for the element to be shared between the two stations, perhaps alternating every other cycle.

- *Utility workers.* We have previously mentioned utility workers in our discussion of manning levels. Utility workers can be used to relieve congestion at stations that are temporarily overloaded.

- *Changing Workhead Speeds at Mechanized Stations.* At stations where a mechanized operation is performed, such as the drilling step mentioned previously, the power feed or speed of the process may be increased or decreased to alter the time required to perform the task. If the mechanized operation takes too long, then an increase in speed or feed is indicated. On the other hand, if the mechanized process is of relatively short duration, so that idle time is associated with the station, then a reduction in speed and/or feed may be appropriate. The advantage of reducing the speed/feed combination is that tool life is increased. The opposite occurs when speed or feed is increased. Whether speeds and/or feeds are increased or decreased, procedures must be devised for efficiently changing the tools without causing undue downtime on the line.

- *Preassembly of Components.* To reduce the total amount of work done on the regular assembly line, certain subassemblies can be prepared offline, either by another assembly cell in the plant, or by purchasing them from an outside vendor that specializes in the type of processes required. Although it may seem like the work is simply being moved from one location to another, there are some good reasons for organizing assembly operations in this manner: (1) the required process may be difficult to implement on the regular assembly line, (2) task time variability (e.g., for adjustments or fitting) for the associated assembly operations may result in a longer overall cycle time if done on the regular line, and (3) an assembly cell set up in the plant or by a vendor with certain special capabilities to perform the work may be able to achieve higher quality.

- *Storage Buffers Between Stations.* A storage buffer is a location in the production line where work units are temporarily stored. Reasons to include one or more storage buffers in a production line include (1) to accumulate work units between two stages of the line when their production rates are different, (2) to smooth production between stations with large task time variations, and (3) to permit continued operation of certain sections of the line when other sections are temporarily down for service or repair. The use of storage buffers generally improves the performance of the line operation by increasing line efficiency E (discussed in the context of transfer lines in Chapter 16).

- *Zoning and Other Constraints.* In addition to precedence constraints, there may be other restrictions on the line balancing solution. *Zoning constraints* impose limitations on the grouping of work elements and/or their allocation to workstations. Zoning constraints may be positive or negative. A *positive zoning constraint* means that certain elements should be grouped together at the same workstation if possible. For example, spray painting elements should all be grouped together due to the need for special enclosures. A *negative zoning constraint* indicates that certain work elements might interfere with each other and therefore should not be located near each other. For example, a work element requiring delicate adjustments should not be located near an assembly operation in which loud sudden noises occur, such as hammering. Another limitation on the allocation of work to stations is a *position constraint*; it is encountered in the assembly of large products such as trucks and automobiles, when it is difficult for one worker to perform tasks on both sides of the work. To facilitate the work, operators are positioned on both sides of the assembly line.
- *Parallel Workstations.* Parallel stations are sometimes used to balance a production line. Their most obvious application is where a particular station has an unusually long task time that would cause the production rate of the line to be less than that required to satisfy product demand. In this case, two stations operating in parallel and both performing the same long task may eliminate the bottleneck. In other situations, the advantage of using parallel stations is not as obvious. Conventional line balancing methods, such as the largest candidate rule, the Kilbridge and Wester method, and ranked positional weights method, do not consider the use of parallel workstations. It turns out that the only way to achieve a perfect balance in our earlier example problem is by using parallel stations.

EXAMPLE 15.10 Parallel Work Stations for Better Line Balance

Can a perfect line balance be achieved in our Example 15.1 using parallel stations?

Solution: The answer is yes. Using a parallel station configuration to replace positions 1 and 2, and reallocating the elements as indicated in Table 15.17, will achieve a perfect balance. The solution is illustrated in Figure 15.9.

TABLE 15.17 Assignment of Work Elements to Stations for Example 15.11 Using Parallel Workstations

Station	Work Element	T_{ek} (min)	Station Time (min)
1, 2*	1	0.2	
	2	0.4	
	3	0.7	
	4	0.1	
3	8	0.6	2.00/2 = 1.00
	5	0.3	
	6	0.11	
	7	0.32	
4	9	0.27	1.00
	10	0.38	
	11	0.5	
	12	0.12	1.00

*Stations 1 and 2 are in parallel.

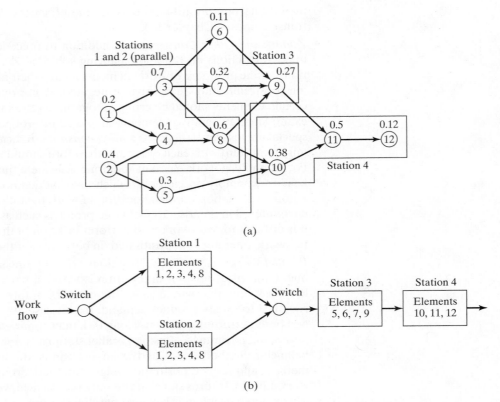

Figure 15.9 Solution for Example 15.10 using parallel workstations: (a) precedence diagram and (b) workstation layout showing
element assignments.

Work content time T_{wc} = 4.0 min as before. To figure the available service
time, we note that there are two conventional stations (3 and 4) with
T_s = 1.0 min each. The parallel stations (1 and 2) each have service times of
2.0 min, but each is working on its own unit of product so the effective
throughput of the two stations is one work unit every minute. Using this reasoning, we can compute the balance efficiency as follows:

$$E_b = \frac{4.0}{2(1.0) + 2.0} = 1.00 = 100\,\%$$

15.7 ALTERNATIVE ASSEMBLY SYSTEMS

The well-defined pace of a manual assembly line has merit from the viewpoint of
maximizing production rate. However, assembly line workers often complain about the
monotony of the repetitive tasks they must perform and the unrelenting pace they must
maintain when a moving conveyor is used. Poor quality workmanship, sabotage of the line
equipment, and other problems have occurred on high production assembly lines. To address

these issues, alternative assembly systems are available in which either the work is made less monotonous and repetitious by enlarging the scope of the tasks performed, or the work is automated. In this section, we identify the following alternative assembly systems: (1) single-station manual assembly cells, (2) assembly cells based on worker teams, and (3) automated assembly systems.

A *single-station manual assembly cell* consists of a single workplace in which assembly is accomplished on the product or some major subassembly of the product. This method is generally used on products that are complex and produced in small quantities, sometimes one-of-a-kind. The workplace may utilize one or more workers, depending on the size of the product and the required production rate. Custom-engineered products such as machine tools, industrial equipment, and prototype models of complex products (e.g., aircraft, appliances, cars) are assembled on single manual stations.

Assembly by worker teams involves the use of multiple workers assigned to a common assembly task. The pace of the work is controlled largely by the workers themselves rather than by a pacing mechanism such as a powered conveyor moving at a constant speed. Team assembly can be implemented in several ways. A single-station manual assembly cell in which there are multiple workers is a form of worker team. The assembly tasks performed by each worker are generally far less repetitious and broader in scope than the corresponding work on an assembly line.

Other ways of organizing assembly work by teams include moving the product through multiple workstations, but having the same worker team follow the product from station to station. This form of team assembly was pioneered by Volvo, the Swedish car maker. It uses independently operated automated guided vehicles (Section 10.2.2) that hold major components and/or subassemblies of the automobile and deliver them to manual assembly workstations along the line. At each station, the guided vehicle stops at the station and is not released to proceed until the assembly task at that station has been completed by the worker team. Thus, production rate is determined by the pace of the team, rather than by a moving conveyor. The reason for moving the work unit through multiple stations, rather than performing all the assembly at one station, is because the many component parts assembled to the car must be located at more than one station. As the car moves through each station, parts from that station are added. The difference between this and the conventional assembly line is that all work is done by one worker team moving with the car. Accordingly, the members of the team achieve greater personal satisfaction at having accomplished a major portion of the car assembly. Workers on a conventional line who perform a very small portion of the total car assembly do not usually have this level of job satisfaction.

The use of automated guided vehicles allows the assembly system to be configured with parallel paths, queues of parts between stations, and other features not typically found on a conventional assembly line. In addition, these team assembly systems can be designed to be highly flexible and capable of dealing with variations in product and corresponding variations in assembly cycle times at the different workstations. Accordingly, this type of team assembly is generally used when there are many different models to be produced, and the variations in the models result in significant differences in station service times.

Reported benefits of worker team assembly systems compared to conventional assembly line include greater worker satisfaction, better product quality, increased capability to accommodate model variations, and greater ability to cope with problems that require more time without stopping the entire production line. The principal disadvantage is that these team systems are not capable of the high production rates characteristic of a conventional assembly line.

Automated assembly systems use automated methods at workstations rather than humans. We defer discussion of automated assembly systems until Chapter 17, where we also discuss hybrid assembly systems consisting of both automated workstations and human assembly operators.

REFERENCES

[1] ANDREASEN, M., KAHLER, S., and LUND, T., *Design for Assembly*, IFS (Publications) Ltd., U.K., and Springer-Verlag, Berlin, 1983.

[2] BARD, J. F., E. M. DAR-EL, and A. SHTUB, "An Analytical Framework for Sequencing Mixed-Model Assembly Lines," *Int. J. Production Research*, Vol. 30, No. 1, 1992, pp. 35–48.

[3] BOOTHROYD, G., P. DEWHURST, and W. KNIGHT, *Product Design for Manufacture and Assembly*, Marcel Dekker, Inc., NY, 1994.

[4] BRALLA, J. G., *Handbook of Product Design for Manufacturing*, McGraw-Hill Book Company, NY, 1986, Chapter 7.

[5] CHOW, W-M., *Assembly Line Design: Methodology and Applications*, Marcel Dekker, Inc., NY, 1990.

[6] DAR-EL, E. M. and F. COTHER, "Assembly Line Sequencing for Model Mix," *Int. J. Production Research*, Vol. 13, No. 5, 1975, pp. 463–477.

[7] DAR-EL, E. M., "Mixed-Model Assembly Line Sequencing Problems," *OMEGA*, Vol. 6, No. 4, 1978, pp. 313–323.

[8] DAR-EL, E. M. and A. NAVIDI, "A Mixed-Model Sequencing Application," *Int. J. Production Research*, Vol. 19, No. 1, 1981, pp. 69–84.

[9] DEUTSCH, D. F., "A Branch and Bound Technique for Mixed-Model Assembly Line Balancing," *Ph.D. Dissertation*, (Unpublished), Arizona State University, 1971.

[10] FERNANDES, C. J. L., "Heuristic Methods for Mixed-Model Assembly Line Balancing and Sequencing," *M.S. Thesis*, (Unpublished), Lehigh University, 1992.

[11] FERNANDES, C. J. L., and M. P. GROOVER, "Mixed Model Assembly Line Balancing and Sequencing: A Survey," *Engineering Design and Automation*, Vol. 1, No. 1, 1995, pp. 33–42.

[12] GROOVER, M. P., *Fundamentals of Modern Manufacturing: Materials, Processes, and Systems*, 3d ed., John Wiley & Sons, Inc., Hoboken, NJ, 2007.

[13] HELGESON, W. B., and D. P. BIRNIE, "Assembly Line Balancing Using Ranked Positional Weight Technique," *Journal of Industrial Engineering*, Vol. 12, No. 6, 1961, pp. 394–398.

[14] HOFFMAN, T., "Assembly Line Balancing: A Set of Challenging Problems," *Int. J. Production Research*, Vol. 28, No. 10, 1990, pp. 1807–1815.

[15] HOUNSHELL, D. A., *From the American System to Mass Production, 1800–1932*, The Johns Hopkins University Press, Baltimore, MD, 1984.

[16] IGNALL, E. J., "A Review of Assembly Line Balancing," *Journal of Industrial Engineering*, Vol. 16, No. 4, 1965, pp. 244–252.

[17] KILBRIDGE, M. and WESTER, L., "A Heuristic Method of Assembly Line Balancing," *Journal of Industrial Engineering*, Vol. 12, No. 6, 1961, pp. 292–298.

[18] MACASKILL, J. L. C., "Production Line Balances for Mixed-Model Lines," *Management Science*, Vol. 19, No. 4, 1972, pp. 423–434.

[19] Moodie, C. L. and Young, H. H., "A Heuristic Method of Assembly Line Balancing for Assumptions of Constant or Variable Work Element Times," *J. Industrial Engineering*, Vol. 16, No. 1, 1965, pp. 23–29.

[20] Nof, S. Y., W. E. Wilhelm, and H.-J. Warnecke, *Industrial Assembly*, Chapman & Hall, London, UK, 1997.

[21] Prenting, T. O. and Thomopoulos, N. T., *Humanism and Technology in Assembly Systems*, Hayden Book Company, Inc., Rochelle Park, NJ, 1974.

[22] Rekiek, B., and A. Delchambre, *Assembly Line Design: The Balancing of Mixed-Model Hybrid Assembly Lines with Generic Algorithms*, Springer Verlag London Limited, UK, 2006.

[23] Sumichrast, R. T., Russel, R. R., and Taylor, B. W., "A Comparative Analysis of Sequencing for Mixed-Model Assembly Lines in a Just-In-Time Production System," *Int. J. Production Research*, Vol. 30, No. 1, 1992, pp. 199–214.

[24] Villa, C., "Multi Product Assembly Line Balancing," *Ph.D. Dissertation* (Unpublished), University of Florida, 1970.

[25] Whitney, D., *Mechanical Assemblies*, Oxford University Press, NY, 2004.

[26] Wild, R., *Mass Production Management*, John Wiley & Sons, London, UK, 1972.

REVIEW QUESTIONS

15.1 Name three of the four factors that favor the use of manual assembly lines.

15.2 What are the four reasons given in the text that explain why manual assembly lines are so productive compared to alternative methods in which multiple workers each perform all of the tasks to assemble the product?

15.3 What is a manual assembly line?

15.4 What is meant by the term *manning level* in the context of a manual assembly line?

15.5 What do the terms *starving* and *blocking* mean?

15.6 Identify and briefly describe the three major categories of mechanized work transport systems used in production lines.

15.7 The text describes, three types of assembly line that were developed to cove with product variety. Name the three types and explain the differences between them.

15.8 What does the term *line efficiency* mean in production line terminology?

15.9 The theoretical minimum number of workers on an assembly line $w*$ is the minimum integer that is greater than the ratio of the work content time T_{wc} divided by the cycle time T_c. Name the two factors identified in the text that make it difficult to achieve this minimum value in practice.

15.10 What is the difference between the cycle time T_c and the service time T_s?

15.11 What is a minimum rational work element?

15.12 What is meant by the term *precedence constraint*?

15.13 What is meant by the term *balance efficiency*?

15.14 What is the difference between the largest candidate rule and the Kilbridge and Wester method?

15.15 In a mixed model assembly line, what is the difference between variable-rate launching and fixed-rate launching?

15.16 What are storage buffers and why are they sometimes used on a manual assembly line?

PROBLEMS

Single Model Assembly Lines

15.1 A product whose work content time = 47.5 min is to be assembled on a manual production line. The required production rate is 30 units per hour. From previous experience, it is estimated that the manning level will be 1.25, proportion uptime = 0.95, and repositioning time = 6 sec. Determine (a) cycle time, and (b) ideal minimum number of workers required on the line. (c) If the ideal number in part (b) could be achieved, how many workstations would be needed?

15.2 A manual assembly line has 17 workstations with one operator per station. Work content time to assemble the product = 28.0 min. Production rate of the line = 30 units per hour. The proportion uptime = 0.94, and repositioning time = 6 sec. Determine the balance delay.

15.3 A manual assembly line must be designed for a product with annual demand = 100,000 units. The line will operate 50 wks/year, 5 shifts/wk, and 7.5 hr/shift. Work units will be attached to a continuously moving conveyor. Work content time = 42.0 min. Assume line efficiency $E = 0.97$, balancing efficiency $E_b = 0.92$, and repositioning time $T_r = 6$ sec. Determine (a) hourly production rate to meet demand, and (b) number of workers required.

15.4 A single model assembly line is being planned to produce a consumer appliance at the rate of 200,000 units per year. The line will be operated 8 hours per shift, two shifts per day, five days per week, 50 weeks per year. Work content time = 35.0 min. For planning purposes, it is anticipated that the proportion uptime on the line will be 95%. Determine (a) average hourly production rate R_p, (b) cycle time T_c, and (c) theoretical minimum number of workers required on the line. (d) If the balance efficiency is 0.93 and the repositioning time = 6 sec, how many workers will actually be required?

15.5 The required production rate = 50 units per hour for a certain product whose assembly work content time = 1.2 hours. The product will be produced on a production line that includes four workstations that are automated. Because the automated stations are not completely reliable, the line will have an expected uptime efficiency = 90%. The remaining manual stations will each have one worker. It is anticipated that 8% of the cycle time will be lost due to repositioning at the bottleneck station. If the balance delay is expected to be 0.07, determine (a) the cycle time, (b) number of workers, (c) number of workstations needed for the line, (d) average manning level on the line, including the automated stations, and (e) labor efficiency on the line.

15.6 A final assembly plant for a certain automobile model is to have a capacity of 225,000 units annually. The plant will operate 50 weeks/yr, two shifts/day, 5 days/week, and 7.5 hours/shift. It will be divided into three departments: (1) body shop, (2) paint shop, and (3) general assembly department. The body shop welds the car bodies using robots, and the paint shop coats the bodies. Both of these departments are highly automated. General assembly has no automation. There are 15.0 hours of work content time on each car in this third department, where cars are moved by a continuous conveyor. Determine (a) hourly production rate of the plant, (b) number of workers and workstations required in trim-chassis-final if no automated stations are used, the average manning level is 2.5, balancing efficiency = 90%, proportion uptime = 95%, and a repositioning time of 0.15 min is allowed for each worker.

15.7 Production rate for a certain assembled product is 47.5 units per hour. The assembly work content time = 32 min of direct manual labor. The line operates at 95% uptime. Ten workstations have two workers on opposite sides of the line so that both sides of the product can receive attention on simultaneously. The remaining stations have one worker. Repositioning time lost by each worker is 0.2 min/cycle. It is known that the number of workers on the line is two more than the number required for perfect balance. Determine (a) number of workers, (b) number of workstations, (c) balance efficiency, and (d) average manning level.

15.8 The work content for a product assembled on a manual production line is 48 min. The work is transported using a continuous overhead conveyor that operates at a speed of 5 ft/min. There are 24 workstations on the line, one-third of which have two workers while the remaining stations each have one worker. Repositioning time per worker is 9 sec, and uptime efficiency of the line is 95%. (a) What is the maximum possible hourly production rate if the line is assumed to be perfectly balanced? (b) If the actual production rate is only 92% of the maximum possible rate determined in part (a), what is the balance delay on the line?

15.9 Work content time for a product assembled on a manual production line is 45.0 min. Production rate of the line must be 40 units/hr. Work units are attached to a moving conveyor whose speed = 8 ft/min. Repositioning time per worker is 8 sec, line efficiency is 93%, and manning level = 1.25. Owing to imperfect line balancing, it is expected that the number of workers needed on the line will be about 10% more workers than the number required for perfect balance. If the workstations are arranged in a line, and the length of each station is 12 ft, (a) how long is the entire production line, and (b) what is the elapsed time a work unit spends on the line?

Line Balancing (Single Model Lines)

15.10 Show that the two statements of the objective function in single model line balancing in Eq. (15.20) are equivalent.

15.11 The table below defines the precedence relationships and element times for a new model toy. (a) Construct the precedence diagram for this job. (b) If the ideal cycle time = 1.1 min. repositioning time = 0.1 min, and uptime proportion is assumed to be 1.0, what is the theoretical minimum number of workstations required to minimize the balance delay under the assumption that there will be one worker per station? (c) Use the largest candidate rule to assign work elements to stations. (d) Compute the balance delay for your solution .

Work Element	T_e (Min)	Immediate Predecessors
1	0.5	–
2	0.3	1
3	0.8	1
4	0.2	2
5	0.1	2
6	0.6	3
7	0.4	4,5
8	0.5	3,5
9	0.3	7,8
10	0.6	6,9

15.12 Solve the previous problem by using the Kilbridge and Wester method in part (c).

15.13 Solve Problem 15.11 using the ranked positional weights method in part (c).

15.14 A manual assembly line is to be designed to make a small consumer product. The work elements, their times, and the precedence constraints are given in the table below. The workers will operate the line for 400 min per day and must produce 300 products per day. A mechanized belt, moving at a speed of 1.25 m/min, will transport the products between stations. Because of the variability in the time required to perform the assembly operations, it has been determined that the tolerance time should be 1.5 times the cycle time of the line. (a) Determine the ideal minimum number of workers on the line. (b) Use the Kilbridge and Wester method to balance the line. (c) Compute the balance delay for your solution in part (b).

Element	T_e (Min)	Preceded By	Element	T_e (Min)	Preceded By
1	0.4	–	6	0.2	3
2	0.7	1	7	0.3	4
3	0.5	1	8	0.9	4, 9
4	0.8	2	9	0.3	5, 6
5	1.0	2, 3	10	0.5	7, 8

15.15 Solve the previous problem by using the ranked positional weights method in part (b).

15.16 A manual assembly line operates with a mechanized conveyor. The conveyor moves at a speed of 5 ft/min, and the spacing between base parts launched onto the line is 4 ft. It has been determined that the line operates best when there is one worker per station and each station is 6 ft long. There are 14 work elements that must be accomplished to complete the assembly, and the element times and precedence requirements are listed in the table below. Determine (a) feed rate and corresponding cycle time, (b) tolerance time for each worker, and (c) ideal minimum number of workers on the line. (d) Draw the precedence diagram for the problem. (e) Determine an efficient line balancing solution. (f) For your solution, determine the balance delay.

Element	T_e (Min)	Preceded By	Element	T_e (Min)	Preceded By
1	0.2	–	8	0.2	5
2	0.5	–	9	0.4	5
3	0.2	1	10	0.3	6, 7
4	0.6	1	11	0.1	9
5	0.1	2	12	0.2	8, 10
6	0.2	3, 4	13	0.1	11
7	0.3	4	14	0.3	12, 13

15.17 A small electrical appliance is to be assembled on a single model assembly line. The line will be operated 250 days per year, 15 hours per day. The work content has been divided into work elements as defined in the table below. Also given are the element times and precedence requirements. Annual production is to be 200,000 units. It is anticipated that the line efficiency will be 0.96. Repositioning time for each worker is 0.08 min. Determine (a) average hourly production rate, (b) cycle time, and (c) theoretical minimum number of workers required to meet annual production requirements. (d) Use one of the line balancing algorithms to balance the line. For your solution, determine (e) balance efficiency and (f) overall labor efficiency on the line.

Element	Element Description	T_e (Min)	Preceded By
1	Place frame on workholder and clamp	0.15	–
2	Assemble fan to motor	0.37	–
3	Assemble bracket A to frame	0.21	1
4	Assemble bracket B to frame	0.21	1
5	Assemble motor to frame	0.58	1, 2
6	Affix insulation to bracket A	0.12	3
7	Assemble angle plate to bracket A	0.29	3
8	Affix insulation to bracket B	0.12	4
9	Attach link bar to motor and bracket B	0.30	4, 5

10	Assemble three wires to motor	0.45	5
11	Assemble nameplate to housing	0.18	–
12	Assemble light fixture to housing	0.20	11
13	Assemble blade mechanism to frame	0.65	6, 7, 8, 9
14	Wire switch, motor, and light	0.72	10, 12
15	Wire blade mechanism to switch	0.25	13
16	Attach housing over motor	0.35	14
17	Test blade mechanism, light, etc.	0.16	15, 16
18	Affix instruction label to cover plate	0.12	–
19	Assemble grommet to power cord	0.10	–
20	Assemble cord and grommet to cover plate	0.23	18, 19
21	Assemble power cord leads to switch	0.40	17, 20
22	Assemble cover plate to frame	0.33	21
23	Perform final inspection and remove from workholder	0.25	22
24	Package	1.75	23

Mixed Model Assembly Lines

15.18 Two product models, A and B, are to be produced on a mixed model assembly line. Hourly production rate and work content time for model A are 12 units/hr and 32.0 min, respectively; and for model B are 20 units/hr and 21.0 min. Line efficiency = 0.95, balance efficiency = 0.93, repositioning time = 0.10 min, and manning level = 1. Determine how many workers and workstations must be on the production line in order to produce this workload.

15.19 Three models, A, B, and C, will be produced on a mixed model assembly line. Hourly production rate and work content time for model A are 10 units/hr and 45.0 min; for model B are 20 units/hr and 35.0 min; and for model C are 30 units/hr and 25.0 min. Line efficiency is 95%, balance efficiency is 0.94, repositioning efficiency = 0.93, and manning level = 1.3. Determine how many workers and workstations must be on the production line in order to produce this workload.

15.20 For Problem 15.18, determine the variable rate launching intervals for models A and B.

15.21 For Problem 15.19, determine the variable rate launching intervals for models A, B, and C.

15.22 For Problem 15.18, determine (a) the fixed rate launching interval, and (b) the launch sequence of models A and B during one hour of production.

15.23 For Problem 15.19, determine (a) the fixed rate launching interval, and (b) the launch sequence of models A, B, and C during one hour or production.

15.24 Two models A and B are to be assembled on a mixed model line. Hourly production rates for the two models are: A, 25 units/hr; and B, 18 units/hr. The work elements, element times, and precedence requirements are given in the table that follows. Elements 6 and 8 are not required for model A, and elements 4 and 7 are not required for model B. Assume $E = 1.0$, $E_r = 1.0$, and $M_i = 1$. (a) Construct the precedence diagram for each model and for both models combined into one diagram. (b) Find the theoretical minimum number of workstations required to achieve the required production rate. (c) Use the Kilbridge and Wester method to solve the line balancing problem. (d) Determine the balance efficiency for your solution in (c).

Work Element k	T_{eAk} (Min)	Preceded By	T_{eBk} (Min)	Preceded By
1	0.5	–	0.5	–
2	0.3	1	0.3	1
3	0.7	1	0.8	1
4	0.4	2	–	–
5	1.2	2, 3	1.3	2, 3
6	–	–	0.4	3
7	0.6	4, 5	–	–
8	–	–	0.7	5, 6
9	0.5	7	0.5	8
T_{wc}	4.2		4.5	

15.25 For the data given in Problem 15.24, solve the mixed model line balancing problem by using the ranked positional weights method to determine the order of entry of work elements.

15.26 Three models A, B, and C are to be assembled on a mixed model line. Hourly production rates for the three models are A, 15 units/hr; B, 10 units/hr; and C, 5 units/hr. The work elements, element times, and precedence requirements are given in the table below. Assume $E = 1.0$, $E_r = 1.0$, and $M_i = 1$. (a) Construct the precedence diagram for each model and for all three models combined into one diagram. (b) Find the theoretical minimum number of workstations required to achieve the required production rate. (c) Use the Kilbridge and Wester method to solve the line balancing problem. (d) Determine the balance efficiency for the solution in (c).

Element	T_{eAk} (Min)	Preceded By	T_{eBk} (Min)	Preceded By	T_{eCk} (Min)	Preceded By
1	0.6	–	0.6	–	0.6	–
2	0.5	1	0.5	1	0.5	1
3	0.9	1	0.9	1	0.9	1
4	–		0.5	1	–	
5	–		–		0.6	1
6	0.7	2	0.7	2	0.7	2
7	1.3	3	1.3	3	1.3	3
8	–		0.9	4	–	
9	–		–		1.2	5
10	0.8	6, 7	0.8	6, 7, 8	0.8	6, 7, 9
T_{wc}	4.8		6.2		6.6	

15.27 For the data given in Problem 15.26, (a) solve the mixed model line balancing problem using line efficiency = 0.96 and repositioning efficiency = 0.95. (b) Determine the balance efficiency for your solution.

15.28 For Problem 15.26, determine (a) the fixed rate launching interval and (b) the launch sequence of models A, B, and C during one hour of production.

15.29 Two similar models, A and B, are to be produced on a mixed model assembly line. There are four workers and four stations on the line ($M_i = 1$ for i = 1, 2, 3, 4). Hourly production rates for the two models are for A, 7 units/hr; and for B, 5 units/hr. The work elements, element times, and precedence requirements for the two models are given in the table below. As the table indicates, most elements are common to both models. Element 5 is unique to model A, while elements 8 and 9 are unique to model B. Assume $E = 1.0$ and $E_r = 1.0$. (a) Develop the mixed model precedence diagram for the two models and for both models combined. (b) Determine a line balancing solution that allows the two models to be produced on the four stations at the specified rates. (c) Using your solution from (b), solve the fixed-rate model launching problem by determining the fixed-rate launching interval and constructing a table to show the sequence of model launchings during the hour.

Work element k	T_{eAk} (Min)	Preceded By	T_{eBk} (Min)	Preceded By
1	1	–	1	–
2	3	1	3	1
3	4	1	4	1, 8
4	2	–	2	8
5	1	2	–	–
6	2	2, 3, 4	2	2, 3, 4
7	3	5, 6	3	6, 9
8	–	–	4	–
9	–	–	2	4
T_{wc}	16		21	

Workstation Considerations

15.30 An overhead continuous conveyor is used to carry dishwasher base parts along a manual assembly line. The spacing between appliances = 2.2 m and the speed of the conveyor = 1.2 m/min. The length of each workstation is 3.5 m. There are 25 stations and 30 workers on the line. Determine (a) elapsed time a dishwasher base part spends on the line, (b) feed rate, and (c) tolerance time.

15.31 A moving belt line is used to assemble a product whose work content = 20 min. Production rate = 48 units/hour, and the proportion uptime = 0.96. The length of each station = 5 ft and manning level = 1.0. The belt speed can be set at any value between 1.0 and 6.0 ft/min. It is expected that the balance delay will be about 0.08 or slightly higher. Time lost for repositioning each cycle is 3 seconds. (a) Determine the number of stations needed on the line. (b) Using a tolerance time that is 50% greater than the cycle time, what would be an appropriate belt speed and spacing between parts?

15.32 In the general assembly department of an automobile final assembly plant, there are 500 workstations, and the cycle time = 0.95 min. If each workstation is 6.2 m long, and the tolerance time = the cycle time, determine the following: (a) speed of the conveyor, (b) center-to-center spacing between units on the line, (c) total length of the general assembly line, assuming no vacant space between stations, and (d) elapsed time a work unit spends in the general assembly department.

15.33 Total work content for a product assembled on a manual production line is 33.0 min. Production rate of the line must be 47 units/hr. Work units are attached to a moving conveyor whose speed = 7.5 ft/min. Repositioning time per worker is 6 sec, and uptime efficiency of the line is 94%. Due to imperfect line balancing, the number of workers needed on the line must be four more workers than the number required for perfect balance. Manning level = 1.6. Determine (a) the number of workers and (b) the number of workstations are on the line. (c) What is the balance efficiency for this line? (d) If the workstations are arranged in a line, and the length of each station is 11 ft, what is the tolerance time in each station? (e) What is the elapsed time a work unit spends on the line?

15.34 A manual assembly line is to be designed for a certain major appliance whose assembly work content time = 2.0 hours. The line will be designed for an annual production rate of 150,000 units. The plant will operate one 10-hour shift per day, 250 days per year. A continuous conveyor system will be used and it will operate at a speed = 1.6 m/min. The line must be designed under the following assumptions: balance delay = 6.5%, uptime efficiency = 96%, repositioning time = 6 sec for each worker, and average manning level = 1.25. (a) How many workers will be required to operate the assembly line? If each station is 2.0 m long, (b) how long will the production line be, and (c) what is the elapsed time a work unit spends on the line?

Chapter 16

Automated Production Lines

CHAPTER CONTENTS

The manufacturing systems considered in this chapter are used for high production of parts that require multiple processing operations. Each processing operation is performed at a workstation, and the stations are physically integrated by means of a mechanized work transport system to form an automated production line. Machining (milling, drilling, and similar rotating cutter operations) is commonly performed on these production lines, in which case the term *transfer line* or *transfer machine* is used. Other applications of automated production lines include robotic spot welding in automobile final assembly plants, sheet-metal pressworking, and electroplating of metals. Similar automated lines are used for assembly operations; however, the technology of automated assembly is sufficiently different that we postpone coverage of this topic until the following chapter.

From Chapter 16 of *Automation, Production Systems, and Computer-Integrated Manufacturing*, Third Edition.
Mikell P. Groover. Copyright © 2008 by Pearson Education, Inc. Publishing as Prentice Hall. All rights reserved.

Automated production lines require a significant capital investment. They are examples of fixed automation (Section 1.2.1), and it is generally difficult to alter the sequence and content of the processing operations once the line is built. Their application is therefore appropriate only under the following conditions:

- *High demand*, requiring high production quantities
- *Stable product design*, because frequent design changes are difficult to accommodate on an automated production line
- *Long product life*, at least several years in most cases
- *Multiple operations* performed on the product during its manufacture.

When the application satisfies these conditions, automated production lines provide the following benefits:

- Low amount of direct labor.
- Low product cost, because cost of fixed equipment is spread over many units
- High production rate
- Minimal work-in-progress and production lead time (the time between beginning of production and completion of a finished unit)
- Minimal use of factory floor space

In this chapter, we examine the technology of automated production lines and develop several mathematical models that can be used to analyze their operation.

16.1 FUNDAMENTALS OF AUTOMATED PRODUCTION LINES

An *automated production line* consists of multiple workstations that are automated and linked together by a work handling system that transfers parts from one station to the next, as depicted in Figure 16.1. A raw workpart enters one end of the line, and the processing steps are performed sequentially as the part progresses forward (from left to right in our drawing). The line may include inspection stations to perform intermediate quality checks. Also, manual stations may be located along the line to perform certain operations that are difficult or uneconomical to automate. Each station performs a different operation, so all the operations are required to complete one work unit. Multiple parts are processed simultaneously on the line, one part at each workstation. In the simplest form

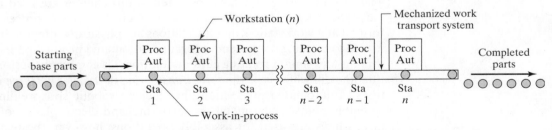

Figure 16.1 General configuration of an automated production line. Key: Proc = processing operation, Aut = automated workstation.

of production line, the number of parts on the line at any moment is equal to the number of workstations, as in our figure. In more complicated lines, provision is made for temporary parts storage between stations, in which case there is more than one part per station.

An automated production line operates in cycles, similar to a manual assembly line (Chapter 15). Each cycle consists of processing time plus the time to transfer parts to the next workstation. The slowest workstation on the line sets the pace of the line, just as in an assembly line. In Section 16.3, we develop equations to describe the cycle time performance of the transfer line and similar automated manufacturing systems.

Depending on workpart geometry, a transfer line may utilize pallet fixtures for part handling. A *pallet fixture* is a workholding device that is designed to (1) fixture the part in a precise location relative to its base and (2) be moved, located, and accurately clamped in position at successive workstations by the transfer system. With the parts accurately located on the pallet fixture, and the pallet accurately registered at a given workstation, the part is itself accurately positioned relative to the processing operation performed at the station. The location requirement is especially critical in machining operations, where tolerances are typically specified in hundredths of a millimeter or thousands of an inch. The term *palletized transfer line* is sometimes used to identify a transfer line that uses pallet fixtures or similar workholding devices. The alternative method of workpart location is to simply index the parts themselves from station to station. This is called a *free transfer line*, and it has the obvious benefit that it avoids the cost of the pallet fixtures. However, certain part geometries require the use of pallet fixtures to facilitate handling and ensure accurate location at a workstation. When pallet fixtures are used, a means must be provided for them to be delivered back to the front of the line for reuse.

16.1.1 System Configurations

Although Figure 16.1 shows the flow of work to be in a straight line, the work flow can actually take several different forms. We classify them as (1) in-line, (2) segmented in-line, and (3) rotary. The *in-line* configuration consists of a sequence of stations in a straight-line arrangement, as in Figure 16.1. This configuration is common for machining big workpieces, such as automotive engine blocks, engine heads, and transmission cases. Because these parts require a large number of operations, a production line with many stations is needed. The in-line configuration can accommodate a large number of stations. In-line systems can also be designed with integrated storage buffers along the flow path (Section 16.1.3).

The *segmented in-line* configuration consists of two or more straight-line transfer sections, where the segments are usually perpendicular to each other. Figure 16.2 shows several possible layouts of the segmented in-line category. There are a number of reasons for designing a production line in these configurations rather than in a pure straight line: (1) available floor space may limit the length of the line, (2) a workpiece in a segmented in-line configuration may be reoriented to present different surfaces for machining, and (3) the rectangular layout provides for swift return of workholding fixtures to the front of the line for reuse.

Figure 16.3 shows two transfer lines that perform metal machining operations on a truck rear axle housing. The first line, on the bottom right hand side, is a segmented in-line configuration in the shape of a rectangle. Pallet fixtures are used in this line to position the starting castings at the workstations for machining. The second line, in the upper left corner, is a conventional in-line configuration consisting of seven stations. When processing

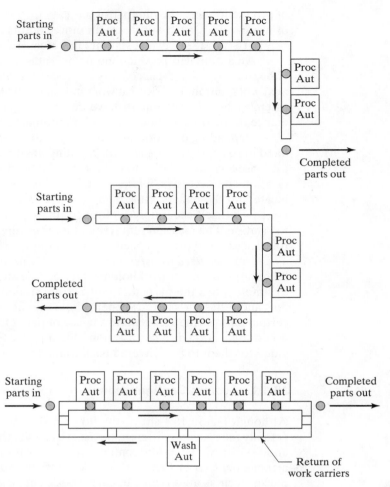

Figure 16.2 Several possible layouts of the segmented in-line configuration of an automated production line: (a) L-shaped, (b) U-shaped, and (c) rectangular. Key: Proc = processing operation, Aut = automated workstation, Wash = work carrier washing station.

on the first line is completed, the parts are manually transferred to the second line, where they are reoriented to present different surfaces for machining. In this line the parts are moved individually by the transfer mechanism, using no pallet fixtures.

In the *rotary* configuration, the workparts are attached to fixtures around the periphery of a circular worktable, and the table is indexed (rotated in fixed angular amounts) to present the parts to workstations for processing. A typical arrangement is illustrated in Figure 16.4. The worktable is often referred to as a dial, and the equipment is called a *dial indexing machine*, or simply, *indexing machine*. Although the rotary configuration does not seem to belong to the class of production systems called "lines," its operation is nevertheless very similar. By comparison with the in-line and segmented in-line configurations, rotary indexing systems are commonly limited to smaller workparts and

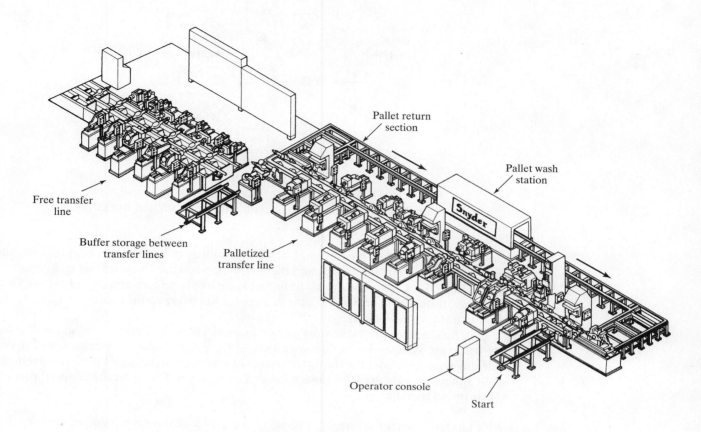

Figure 16.3 Two machining transfer lines. At bottom right, the first is a 12 station segmented in-line configuration that uses pallet fixtures to locate the workparts. The return loop brings the pallets back to the front of the line. At upper left, the second transfer line is a seven-station in-line configuration. The manual station between the lines is used to reorient the parts. (Courtesy of Snyder Corporation)

fewer workstations, and they cannot readily accommodate buffer storage capacity. On the positive side, the rotary system usually involves a less expensive piece of equipment and typically requires less floor space.

16.1.2 Workpart Transfer Mechanisms

The workpart transfer system moves parts between stations on the production line. Transfer mechanisms used on automated production lines are usually either synchronous or asynchronous (Section 15.1.2). Synchronous transfer has been the traditional means of moving parts in a transfer line. However, applications of asynchronous transfer systems are increasing because they provide certain advantages over synchronous parts

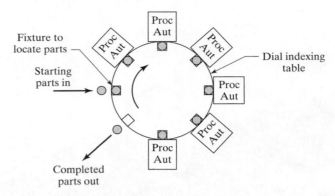

Figure 16.4 Rotary indexing machine (dial indexing machine). Key: Proc = processing operation, Aut = automated workstation.

movement [10]: (1) they have greater flexibility, (2) they require fewer pallet fixtures, and (3) it is easier to rearrange or expand the production system. These advantages come at a higher first cost. Continuous work transport systems are uncommon on automated lines due to the difficulty in providing accurate registration between the station workheads and the continuously moving parts.

In this section we divide workpart transfer mechanisms into two categories: (1) linear transport systems for in-line systems, and (2) rotary indexing mechanisms for dial indexing machines. Some of the linear transport systems provide synchronous movement, while others provide asynchronous motion. The rotary indexing mechanisms all provide synchronous motion.

Linear Transfer Systems. Most of the material transport systems described in Chapter 10 provide a linear motion, and some of these are used for workpart transfer in automated production systems. These include powered roller conveyors, belt conveyors, chain driven conveyors, and cart-on-track conveyors (Section 10.2.4). Figure 16.5 illustrates the possible application of a chain driven or belt conveyor to provide continuous or intermittent movement of parts between stations. Either a chain or flexible steel belt is used to transport parts using work carriers attached to the conveyor. The chain is driven by pulleys in either an "over-and-under" configuration, in which the pulleys turn about a

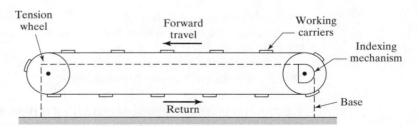

Figure 16.5 Side view of chain or steel belt driven conveyor ("over-and-under" type) for linear workpart transfer by using work carriers.

horizontal axis, or an "around-the-corner" configuration, in which the pulleys rotate about a vertical axis.

The belt conveyor can also be adapted for asynchronous movement of work units using friction between the belt and the part to move parts between stations. The forward motion of the parts is stopped at each station using pop-up pins or other stopping mechanisms.

Cart-on-track conveyors provide asynchronous parts movement and are designed to position their carts within about ± 0.12 mm (± 0.005 inch), which is adequate for many processing situations. In the other types, provision must be made to stop the workparts and locate them within the required tolerance at each workstation. Pin-in-hole mechanisms and detente devices can be used for this purpose.

Many machining type transfer lines utilize *walking beam transfer* systems, in which the parts are synchronously lifted up from their respective stations by a transfer beam and moved one position ahead, to the next station. The transfer beam lowers the parts into nests that position them for processing at their stations. The beam then retracts to make ready for the next transfer cycle. The action sequence is depicted in Figure 16.6.

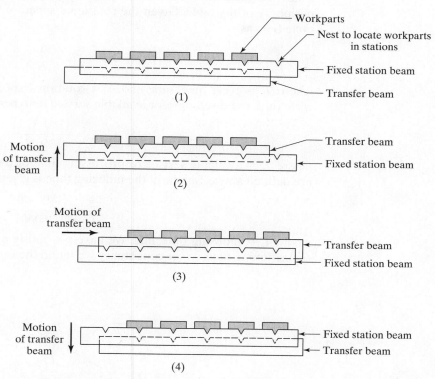

Figure 16.6 Operation of walking beam transfer system: (1) workparts at station positions on fixed station beam, (2) transfer beam is raised to lift workparts from nests, (3) elevated transfer beam moves parts to next station positions, and (4) transfer beam lowers to drop workparts into nests at new station positions. Transfer beam then retracts to original position shown in (1).

Rotary Indexing Mechanisms. Several mechanisms are available to provide the rotational indexing motion required in a dial indexing machine. Two representative types are explained here, Geneva mechanism and cam drive.

The *Geneva mechanism* uses a continuously rotating driver to index the table through a partial rotation, as illustrated in Figure 16.7. If the driven member has six slots for a six station dial indexing table, each turn of the driver results in a 1/6-th rotation of the worktable, or 60°. The driver only causes motion of the table through a portion of its own rotation. For a six-slotted Geneva, 120° of driver rotation is used to index the table. The remaining 240° of driver rotation is dwell time for the table, during which the processing operation must be completed on the work unit. In general,

$$\theta = \frac{360}{n_s} \tag{16.1}$$

where θ = angle of rotation of worktable during indexing (degrees of rotation), and n_s = number of slots in the Geneva. The angle of driver rotation during indexing = 2θ, and the angle of driver rotation during which the worktable experiences dwell time is $(360 - 2\theta)$. Geneva mechanisms usually have four, five, six, or eight slots, which establishes the maximum number of workstation positions that can be placed around the periphery of the table. Given the rotational speed of the driver, we can determine total cycle time as

$$T_c = \frac{1}{N} \tag{16.2}$$

where T_c = cycle time, min; and N = rotational speed of driver, rev/min. Of the total cycle time, the dwell time, or available service time per cycle, is given by

$$T_s = \frac{(180 + \theta)}{360N} \tag{16.3}$$

where T_s = available service or processing time or dwell time, min; and the other terms are defined above. Similarly, the indexing time is given by

$$T_r = \frac{(180 - \theta)}{360N} \tag{16.4}$$

where T_r = indexing time, min. We have previously referred to this indexing time as the repositioning time, so for consistency we retain the same notation.

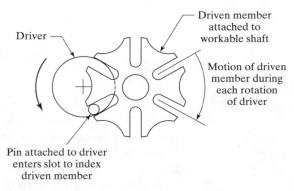

Figure 16.7   Geneva mechanism with six slots.

EXAMPLE 16.1 Geneva Mechanism for a Rotary Indexing Table

A rotary worktable is driven by a Geneva mechanism with six slots, as in Figure 16.7. The driver rotates at 30 rev/min. Determine the cycle time, available processing time, and the lost time each cycle to index the table.

Solution: With a driver rotational speed of 30 rev/min, the total cycle time is given by Eq. (16.2):

$$T_c = (30)^{-1} = 0.0333 \text{ min} = 2.0 \text{ sec}$$

The angle of rotation of the worktable during indexing for a six-slotted Geneva is given by Eq. (16.1):

$$\theta = \frac{360}{6} = 60°$$

Eqs. (16.3) and (16.4) give the available service time and indexing time, respectively, as

$$T_s = \frac{(180 + 60)}{360(30)} = 0.0222 \text{ min} = 1.333 \text{ sec}$$

$$T_r = \frac{(180 - 60)}{360(30)} = 0.0111 \text{ min} = 0.667 \text{ sec}$$

Various forms of *cam drive* mechanisms, one of which is illustrated in Figure 16.8, are used to provide an accurate and reliable method of indexing a rotary dial table. Although it is a relatively expensive drive mechanism, its advantage is that the cam can be designed to provide a variety of velocity and dwell characteristics.

16.1.3 Storage Buffers

Automated production lines can be designed with storage buffers. A *storage buffer* is a location in the production line where parts can be collected and temporarily stored before proceeding to subsequent (downstream) workstations. The storage buffers can be

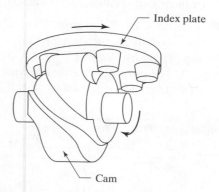

Figure 16.8 Cam mechanism to drive dial indexing table (reprinted from Boothroyd, Poli, and Murch [1]).

manually operated or automated. When it is automated, a storage buffer consists of a mechanism to accept parts from the upstream workstation, a place to store the parts, and a mechanism to supply parts to the downstream station. A key parameter of a storage buffer is its storage capacity, that is, the number of workparts it is capable of holding. Storage buffers may be located between every pair of adjacent stations, or between line stages containing multiple stations. We illustrate the case of one storage buffer between two stages in Figure 16.9.

There are several reasons why storage buffers are used on automated production lines:

- *To reduce the impact of station breakdowns.* Storage buffers between stages on a production line permit one stage to continue operation while the other stage is down for repairs. We analyze this situation in Section 16.3.2.
- *To provide a bank of parts to supply the line.* Parts can be collected into a storage unit and automatically fed to a downstream manufacturing system. This permits untended operation of the system between refills.
- *To provide a place to put the output of the line.*
- *To allow for curing time or other process delay.* A curing time is required for some processes such as painting or adhesive application. The storage buffer is designed to provide sufficient time for curing to occur before supplying the parts to the downstream station.
- *To smooth cycle time variations.* Although this is generally not an issue in an automated line, it is relevant in manual production lines, where cycle time variations are an inherent feature of human performance.

Storage buffers are more readily accommodated in the design of an in-line transfer machine than a rotary indexing machine. In the latter case, buffers are sometimes located (1) before a dial indexing system to provide a bank of raw starting workparts, (2) following the dial indexing machine to accept the output of the system, or (3) between pairs of adjacent dial indexing machines.

16.1.4 Control of the Production Line

Controlling an automated production line is complex, owing to the sheer number of sequential and simultaneous activities that occur during its operation. In this section we discuss (1) the basic control functions that are accomplished to run the line and (2) controllers used on automated lines.

Control Functions. Three basic control functions can be distinguished in the operation of an automatic transfer machine: (1) sequence control, (2) safety monitoring, and (3) quality control.

The purpose of sequence control is to coordinate the sequence of actions of the transfer system and associated workstations. The various activities of the production line must be

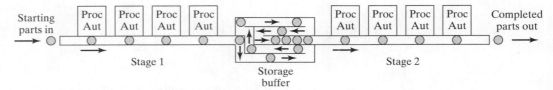

Figure 16.9   Storage buffer between two stages of a production line.

carried out with split-second timing and accuracy. On a transfer line, for example, the parts must be released from their current workstations, transported, located, and clamped into position at their respective next stations. Then the workheads must be actuated to begin their feed cycles, and so on. The sequence control function in automated production line operation includes both logic control and sequencing, as discussed in Chapter 9.

The safety monitoring function ensures that the production line does not operate in an unsafe manner. Safety applies to both the human workers in the area and the equipment itself. Additional sensors must be incorporated into the line beyond those required for sequence control, in order to complete the safety feedback loop and avoid hazardous operation. For example, interlocks must be installed to prevent the equipment from operating when workers are performing maintenance or other duties on the line. In the case of machining transfer lines, cutting tools must be monitored for breakage and/or excessive wear to prevent feeding a defective cutter into the work. A more complete treatment of safety monitoring in manufacturing systems is presented in Section 4.2.1.

In the quality control function, certain quality attributes of the workparts are monitored. The purpose is to detect and possibly reject defective work units produced on the line. The inspection devices required to accomplish quality control are sometimes incorporated into existing processing stations. In other cases, separate inspection stations are included in the line for the sole purpose of checking the desired quality characteristic. We discuss quality inspection principles and practices, as well as the associated inspection technologies, in Chapters 21 and 22, respectively.

Line Controllers. Programmable logic controllers (PLCs, Chapter 9) are the conventional controllers used on automated production lines today. Personal computers (PCs) equipped with control software and designed for the factory environment are also widely used. Computer control offers the following benefits:

- Opportunity to improve and upgrade the control software, such as adding specific control functions not anticipated in the original system design
- Recording of data on process performance, equipment reliability, and product quality for subsequent analysis. In some cases product quality records must be maintained for legal reasons.
- Diagnostic routines to expedite maintenance and repair when line breakdowns occur and to reduce the duration of downtime incidents
- Automatic generation of preventive maintenance schedules indicating when certain preventive maintenance activities should be performed. This helps to reduce the frequency of downtime occurrences.
- A more convenient human-machine interface between the operator and the automated line.

16.2 APPLICATIONS OF AUTOMATED PRODUCTION LINES

Automated production lines are applied in processing operations as well as assembly. We discuss automated assembly systems in Chapter 17. Machining is one of the most common processing applications and is the focus of most of our discussion in this section. Other processes performed on automated production lines and similar systems include sheet-metal forming and cutting, rolling mill operations, spot welding of automobile car bodies, painting, and plating operations.

16.2.1 Machining Systems

Many applications of machining transfer machines, both in-line and rotary configurations, are found in the automotive industry to produce engine and drive train components. In fact, the first transfer lines can be traced to the automobile industry (Historical Note 16.1). Machining operations commonly performed on transfer lines include milling, drilling, reaming, tapping, grinding, and similar rotational cutting tool operations. It is possible to perform turning and boring on transfer lines, but these applications are less common. In this section we discuss the various multiple station machining systems.

Transfer Lines. In a *transfer line*, the workstations containing machining workheads are arranged in an in-line or segmented in-line configuration and the parts are moved between stations by transfer mechanisms such as the walking beam system (Section 16.1.2). It is the most highly automated and productive system in terms of the number of operations that can be performed to accommodate complex work geometries and the rates of production that can be achieved. It is also the most expensive of the systems discussed in this section. Machining type transfer lines are pictured in Figure 16.3. The transfer line can include a large number of workstations, but the reliability of the system decreases as the number of stations is increased (we discuss this issue in Section 16.3). Among the variations in features and options found in transfer lines are the following:

- Workpart transport can be synchronous or asynchronous
- Workparts can be transported with or without pallet fixtures, depending on part geometry and ease of handling
- A variety of monitoring and control features can be included to manage the line.

Historical Note 16.1 Transfer lines [15]

Development of automated transfer lines originated in the automobile industry, which had become the largest mass production industry in the United States by the early 1920s, and was also a major industry in Europe. The Ford Motor Company had pioneered the development of the moving assembly line, but the operations performed on these lines were manual. The next step was to extend the principle of manual assembly lines by building lines capable of automatic or semiautomatic operation. The first fully automatic production line is credited to L. R. Smith of Milwaukee, Wisconsin during 1919 and 1920. This line produced automobile chassis frames out of sheet-metal, using air-powered riveting heads that rotated into position at each station to engage the workpart. The line performed a total of 550 operations on each frame and was capable of producing over a million chassis frames per year.

The first metal machining multi-station line was developed by Archdale Company in England for Morris Engines, Limited in 1923 to machine automobile engine blocks. It had 53 stations, performed 224 minutes of machining on each part, and had a production rate of 15 blocks/hour. It was not a true automatic line because it required manual transfer of work between stations. Yet it stands as an important forerunner of the automated transfer line.

The first machining line to use automatic work transfer between stations was built by Archdale Company for Morris Engines in 1924. The two companies had obviously benefited from their previous collaboration. This line performed 45 machining operations on gearboxes and produced at the rate of 17 units per hour. Reliability problems limited the success of this first transfer line.

In recent years, transfer lines have been designed for ease of changeover to allow different but similar workparts to be produced on the same line [10], [11], [13]. The workstations on these lines consist of a combination of fixed tooling and CNC machines, so that differences in workparts can be accommodated by the CNC stations while the common operations are performed by the stations with fixed tooling. Thus, we see a trend in transfer lines in the direction of flexible manufacturing systems (Chapter 19).

Rotary Transfer Machines and Related Systems. A rotary transfer machine consists of a horizontal circular worktable, upon which are fixtured the workparts to be processed, and around whose periphery are located stationary workheads. The worktable is indexed to present each workpart to each workhead to accomplish the sequence of machining operations. An example is shown in Figure 16.10. By comparison with a transfer line, the rotary indexing machine is limited to smaller, lighter workparts and fewer workstations.

Two variants of the rotary transfer machine are the center column machine and the trunnion machine. In the *center column machine*, vertical machining heads are mounted on a center column in addition to the stationary machining heads located on the outside of the horizontal worktable, thereby increasing the number of machining operations that can be performed. The center column machine, pictured in Figure 16.11, is considered to be a high production machine that makes efficient use of floor space. The *trunnion machine* gets its name from a vertically oriented worktable, or *trunnion*, to which are attached workholders to fixture the parts for machining. Since the trunnion indexes around a horizontal axis, this provides the opportunity to perform machining operations on opposite sides of the workpart. Additional workheads can be located around the periphery of the trunnion to increase the number of machining directions. Trunnion machines are more suitable for smaller workparts than the other rotary machines discussed here.

16.2.2 System Design Considerations

Most companies that use automated production lines and related systems turn the design of the system over to a machine tool builder who specializes in this type of equipment. The customer (company purchasing the equipment) must develop specifications that include design drawings of the part and the required production rate. Typically, several

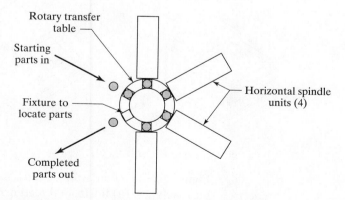

Figure 16.10 Plan view of a rotary transfer machine.

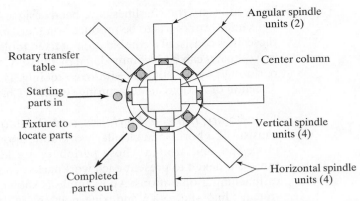

Figure 16.11 Plan view of the center column machine.

machine tool builders are invited to submit proposals. Each proposal is based on the machinery components comprising the builder's product line and depends on the ingenuity of the engineer preparing the proposal. The proposed line consists of standard workheads, spindles, feed units, drive motors, transfer mechanisms, bases, and other standard modules, all assembled into a special configuration to match the machining requirements of the particular part. Examples of these standard modules are illustrated in Figures 16.12 and 16.13. The controls for the system are either designed by the machine builder or sublet as a separate contract to a controls specialist. Transfer lines and indexing machines constructed using this building-block approach are sometimes referred to as *unitized production lines*.

An alternative approach in designing an automated line is to use standard machine tools and to connect them with standard or special material handling devices. The material handling hardware serves as the transfer system that moves work between the standard

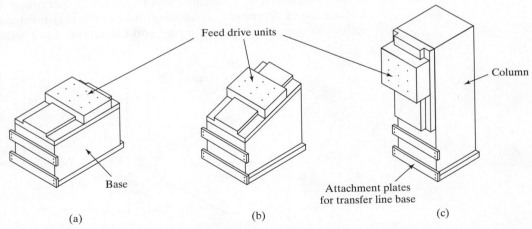

Figure 16.12 Standard feed units used with in-line or rotary transfer machines: (a) horizontal feed drive unit, (b) angular feed drive unit, and (c) vertical column unit.

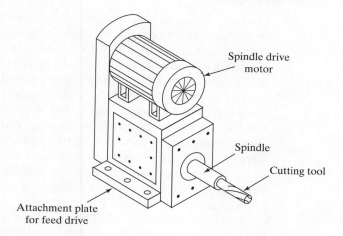

Figure 16.13   Standard milling head unit. This unit attaches to the feed drive units in Figure 16.12.

machines. The term *link line* is sometimes used in connection with this type of construction. In some cases, the individual machines are manually operated if there are fixturing and location problems at the stations that are difficult to solve without human assistance.

A company often prefers to develop a link line rather than a unitized production line because it can utilize existing equipment in the plant. This usually means the production line can be installed sooner and at lower cost. Since the machine tools in the system are standard, they can be reused when the production run is finished. Also, the lines can be engineered by personnel within the company rather than outside contractors. The limitation of the link line is that it tends to favor simpler part shapes and therefore fewer operations and workstations. Unitized lines are generally capable of higher production rates and require less floor space. However, their high cost makes them suitable only for very long production runs on products that are not subject to frequent design changes.

16.3   ANALYSIS OF TRANSFER LINES

In the analysis and design of automated production lines, three problem areas must be considered: (1) line balancing, (2) processing technology, and (3) system reliability.

The line-balancing problem is most closely associated with manual assembly lines (Section 15.2.2), but it is also a problem on automated production lines. Somehow, the total processing work that is to be accomplished on the automated line must be divided as evenly as possible among the workstations. In a manual assembly line, the total work content can be divided into much smaller work elements, and the elements can then be allocated to workstations to determine the task that is performed at each station, as detailed in the previous chapter. Each task has a corresponding service time. In an automated production line, the tasks consist of processing steps whose sequence and service times are limited by technological considerations. For example, in a machining transfer line, certain operations must be performed before others. Drilling must precede tapping to create a threaded hole. Locating surfaces must be machined before the features that will use those locating surfaces are machined. These precedence constraints, as we called them in

Chapter 15, impose a significant restriction on the order in which the processing steps can be carried out. Once the sequence of operations is established, then the service time at a given station depends on how long it takes to accomplish the operation at that station.

Process technology refers to the body of knowledge about the theory and principles of the particular manufacturing processes used on the production line. For example, in the machining process, process technology includes the metallurgy and machinability of the work material, the proper application of cutting tools, selection of speeds and feeds, chip control, and a host of other problem areas and issues. Many of the problems encountered in machining can be solved by direct application of good machining principles. The same is true of other processes. In each process, a technology has been developed over many years of research and practice. By making use of this technology, each individual workstation in the production line can be designed to operate at or near its maximum performance.

The third problem area in the analysis and design of automated production lines is the reliability problem. In a highly complex and integrated system such as an automated production line, failure of any one component can stop the entire system. This reliability problem is the primary focus of this section. Our coverage is divided into two parts: (1) analysis of transfer lines with no internal parts storage and (2) analysis of transfer lines with internal storage buffers.

16.3.1 Transfer Lines with No Internal Parts Storage

Figure 16.1 illustrates the configuration of a transfer line with no internal parts storage. The mathematical models developed in this section are also applicable to rotary indexing machines, shown in Figure 16.4. We make the following assumptions about the operation of these systems: (1) the workstations perform processing operations such as machining, not assembly; (2) processing times at each station are constant, though not necessarily equal; (3) workpart transport is synchronous; and (4) there are no internal storage buffers. In Section 16.3.2, we consider automated production lines with internal storage buffers.

Cycle Time Analysis. In the operation of an automated production line, parts are introduced into the first workstation and are processed and transported at regular intervals to succeeding stations. This interval defines the ideal cycle time T_c of the production line. T_c is the processing time for the slowest station on the line plus the transfer time, that is,

$$T_c = \text{Max}\{T_{si}\} + T_r \tag{16.5}$$

where T_c = ideal cycle time on the line, min; T_{si} = the processing time at station i, min; and T_r = repositioning time, called the transfer time here, min. We use the $\text{Max}\{T_{si}\}$ in Eq. (16.5) because this longest service time establishes the pace of the production line. The remaining stations with smaller service times must wait for the slowest station. Therefore, these other stations will experience idle time. The situation is the same as for a manual assembly line depicted in Figure 15.4.

In the operation of a transfer line, random breakdowns and planned stoppages cause downtime on the line. Common reasons for downtime on an automated production line are listed in Table 16.1. Although the breakdowns and line stoppages occur randomly, their frequency can be measured over the long run. When the line stops, it is down a certain amount of time for each downtime occurrence. Downtime occurrences cause the actual average

TABLE 16.1 Common Reasons for Downtime on an Automated Production Line

Mechanical failure of a workstation	Power outages
Mechanical failure of the transfer system	Stockouts of starting work units
Tool failure at a workstation	Insufficient space for completed parts
Tool adjustment at a workstation	Preventive maintenance on the line
Scheduled tool change at a station	Worker breaks
Electrical malfunctions	Poor quality starting workparts

production cycle time of the line to be longer than the ideal cycle time given by Eq. (16.5). We can formulate the following expression for the actual average production time T_p:

$$T_p = T_c + FT_d \qquad (16.6)$$

where F = downtime frequency, line stops/cycle; and T_d = average downtime per line stop, min. The downtime T_d includes the time for the repair crew to swing into action, diagnose the cause of the failure, fix it, and restart the line. Thus, FT_d = downtime averaged on a per cycle basis.

Line downtime is usually associated with failures at individual workstations. Many of the reasons for downtime listed in Table 16.1 represent malfunctions that cause a single station to stop production. Since all workstations on an automated production line without internal storage are interdependent, the failure of one station causes the entire line to stop. Let p_i = probability or frequency of a failure at station i, where $i = 1, 2, \ldots, n$, and where n = the number of workstations on the production line. The frequency of line stops per cycle is obtained by merely summing the frequencies p_i over the n stations, that is,

$$F = \sum_{i=1}^{n} p_i \qquad (16.7)$$

where F = expected frequency of line stops per cycle, first encountered in Eq. (16.6); p_i = frequency of station breakdown per cycle, causing a line stop; and n = number of workstations on the line. If all p_i are assumed equal, which is unlikely but useful for approximation and computation purposes, then

$$F = np \qquad (16.8)$$

where all p_i are equal, $p_1 = p_2 = \cdots = p_n = p$.

Performance Measures. One of the important measures of performance on an automated transfer line is production rate, which is the reciprocal of T_p,

$$R_p = \frac{1}{T_p} \qquad (16.9)$$

where R_p = actual average production rate, pc/min; and T_p is the actual average production time from Eq. (16.6), min. It is of interest to compare this rate with the ideal production rate given by

$$R_c = \frac{1}{T_c} \qquad (16.10)$$

where R_c = Ideal production rate, pc/min. It is customary to express production rates on automated production lines as hourly rates, so we must multiply the rates in Eqs. (16.9) and (16.10) by 60.

The machine tool builder uses the ideal production R_c in its proposal for the automated transfer line and speaks of it as the production ra... R_c in its proposal for the tunately, because of downtime, the line will not operate at 100% efficiency. Unfortunately seem deceptive for the machine tool builder to ignore the effect... efficiency. While it may tion rate, the amount of downtime experienced on the line is mo...wntime on production the company using the production line. In practice, most of the ...responsibility of occurrences in Table 16.1 represent factors that must be controlled for downtime user company.

In the context of automated production systems, line efficiency re...ged by the tion of uptime on the line and is really a measure of reliability more than ertheless, this is the terminology of production lines. Line efficiency can b...opor-

$$E = \frac{T_c}{T_p} = \frac{T_c}{T_c + FT_d}$$

where E = the proportion of uptime on the production line, and the other term... previously defined.

An alternative measure of performance is the proportion of downtime on the which is given by

$$D = \frac{FT_d}{T_p} = \frac{FT_d}{T_c + FT_d} \tag{16.12}$$

where D = proportion of downtime on the line. It is obvious that

$$E + D = 1.0 \tag{16.13}$$

An important economic measure of performance of an automated production line is the cost per unit produced. This piece cost includes the cost of the starting material that is to be processed on the line, the cost of time on the production line, and the cost of any tooling that is consumed (for example, cutting tools on a machining line). The piece cost can be expressed as the sum of the three factors

$$C_{pc} = C_m + C_o T_p + C_t \tag{16.14}$$

where C_{pc} = cost per piece, \$/pc; C_m = cost of starting material, \$/pc; C_o = cost per minute to operate the line, \$/min; T_p = average production time per piece, min/pc; and C_t = cost of tooling per piece, \$/pc. C_o includes the allocation of the capital cost of the equipment over its expected service life, labor to operate the line, applicable overheads, maintenance and other relevant costs, all reduced to a cost per minute. (See Section 3.2.3.)

Eq. (16.14) does not include factors such as scrap rates, inspection costs, and rework costs associated with fixing defective work units. These factors can usually be incorporated into the unit piece cost in a fairly straightforward way.

EXAMPLE 16.2   Transfer Line Performance

A machine tool builder submits a proposal for a 20-station transfer line to machine a certain component currently produced by conventional methods. The proposal states that the line will operate at a production rate of 50 pieces per hour at 100% efficiency. On similar transfer lines, the probability of station breakdowns per cycle is equal for all stations and $p = 0.005$ breakdowns/cycle. It is also estimated that the average downtime per line stop will be 8.0 min. The starting casting that is machined on the line costs \$3.00 per part. The line operates at a cost of $75.00/hr. The 20 cutting tools (one tool per station) last for 50 parts each, and the average cost per tool = $2.00 per cutting edge. Based on this data, compute (a) production rate, (b) line efficiency, and (c) cost per unit piece produced on the line.

Solution: (a) At 100% efficiency, the line produces 50 pieces per hour. The reciprocal gives the unit time, or ideal cycle time per piece:

$$T_c = \frac{1}{50} = 0.02 \text{ hr/pc} = 1.2 \text{ min}$$

With a station breakdown frequency $p = 0.005$, the frequency of line stops is

$$F = 20(0.005) = 0.10 \text{ breakdowns per cycle}$$

Given an average downtime of 8.0 min, the average production time per piece is

$$T_p = T_c + FT_d = 1.2 + 0.10(8.0) = 1.2 + 0.8 = 2.0 \text{ min/piece}$$

Actual average production rate is the reciprocal of average production time per piece:

$$R_p = \frac{1}{2.0} = 0.500 \text{ pc/min} = 30.0 \text{ pc/hr.}$$

Line efficiency is the ratio of ideal cycle time to actual average production time:

$$E = \frac{1.2}{2.0} = 0.60 = 60\%$$

Finally, for the cost per piece produced, we need the tooling cost per piece, which is computed as follows:

$$C_t = (20 \text{ tools})(\$2/\text{tool})/(50 \text{ parts}) = \$0.80/\text{pc}$$

Now the unit cost can be calculated by Eq. (16.14). The hourly rate of $75/hr to operate the line is equivalent to $1.25/min

$$C_{pc} = 3.00 + 1.25(2.0) + 0.80 = \$6.30/\text{pc.}$$

What the Equations Tell Us. Two general truths are revealed by the equations in this section about the operation of automated transfer lines with no internal parts storage:

- As the number of workstations on an automated production line increases, line efficiency and production rate are adversely affected.
- As the reliability of individual workstations decreases, line efficiency and production rate are adversely affected.

Perhaps the biggest difficulty in the practical use of the equations is determining the values of p_i for the various workstations. We want to use the equations to predict performance for a proposed transfer line, yet we do not know the critical reliability factors for the individual stations on the line. The most reasonable approach is to base the values of p_i on previous experience and historical data for similar workstations.

16.3.2 Transfer Lines with Internal Storage Buffers

As described in the previous section, in an automated production line with no internal parts storage, the workstations are interdependent. When one station breaks down, all other stations on the line are affected, either immediately or by the end of a few cycles of operation, due to starving or blocking. These terms have the same meanings as in the operation of manual assembly lines (Section 15.1.2). *Starving* on an automated production line means that a workstation is prevented from performing its cycle because it has no part to work on. When a breakdown occurs at any workstation on the line, the downstream stations will either immediately or eventually become starved for parts. *Blocking* means that a station is prevented from performing its work cycle because it cannot pass the part just completed to the neighboring downstream station. When a breakdown occurs at a station on the line, the upstream stations become blocked because the broken-down station cannot accept the next part for processing from its upstream neighbor. Therefore, none of the upstream stations can pass its completed part forward.

Downtime on an automated line due to starving and blocking can be reduced by adding one or more parts storage buffers between workstations. Storage buffers divide the line into stages that can operate independently for a number of cycles, the number depending on the storage capacity of the buffer. If one storage buffer is used, the line is divided into two stages. If two buffers are used at two different locations along the line, then a three-stage line is formed. And so forth. The upper limit on the number is to have storage buffers between every pair of adjacent stations. The number of stages will then equal the number of workstations. For an *n*-stage line, there will be $n - 1$ storage buffers, not including the raw parts inventory at the front of the line or the finished parts inventory at the end of the line.

Consider a two-stage transfer line, with a storage buffer separating the stages. Let us suppose that, on average, the storage buffer is half full. If the first stage breaks down, the second stage can continue to operate (avoid starving) using parts that have been collected in the buffer. And if the second stage breaks down, the first stage can continue to operate (avoid blocking) because it has the buffer to receive its output. The reasoning for a two-stage line can be extended to production lines with more than two stages. For any number of stages in an automated production line, the storage buffers allow each stage to operate somewhat independently, the degree of independence depending on the capacity of the upstream and downstream buffers.

Limits of Storage Buffer Effectiveness. Two extreme cases of storage buffer effectiveness can be identified: (1) no buffer storage capacity at all, and (2) infinite capacity storage buffers. In the analysis that follows, let us assume that the ideal cycle time T_c is the same for all stages considered. This is generally desirable in practice because it helps to balance production rates among stages.

In the case of no storage capacity, the production line acts as one stage. When a station breaks down, the entire line stops. This is the case of a production line with no internal storage analyzed in Section 16.3.1. The efficiency of the line is given by Eq. (16.11). We rewrite it here as the line efficiency of a zero capacity storage buffer,

$$E_0 = \frac{T_c}{T_c + FT_d} \tag{16.15}$$

where the subscript 0 identifies E_0 as the efficiency of a line with zero storage buffer capacity, and the other terms have the same meanings as before.

The opposite extreme is the theoretical case where buffer zones of infinite capacity are installed between every pair of stages. If we assume that each buffer zone is half full (in other words, each buffer zone has an infinite supply of parts as well as the capacity to accept an infinite number of additional parts), then each stage is independent of the rest. The presence of infinite storage buffers means that no stage will ever be blocked or starved because of a breakdown at some other stage. Of course, an infinite capacity storage buffer cannot be realized in practice.

For all transfer lines with storage buffers, the overall line efficiency is limited by the bottleneck stage. That is, production on all other stages is ultimately restricted by the slowest stage. The downstream stages can only process parts at the output rate of the bottleneck stage. And it makes no sense to run the upstream stages at higher production rates because this will only accumulate inventory in the storage buffer ahead of the bottleneck. As a practical matter, therefore, the upper limit on the efficiency of the entire line is determined by the efficiency of the bottleneck stage. Given that the cycle time T_c is the same for all stages, the efficiency of any stage k is given by

$$E_k = \frac{T_c}{T_c + F_k T_{dk}} \tag{16.16}$$

where the subscript k is used to identify the stage. According to our argument above, the overall line efficiency would be given by

$$E_\infty = \text{Minimum}\{E_k\} \quad \text{for } k = 1, 2, \ldots, K \tag{16.17}$$

where the subscript ∞ identifies E_∞ as the efficiency of a line whose storage buffers all have infinite capacity.

By including one or more storage buffers in an automated production line, we expect to improve the line efficiency above E_0 but we cannot expect to achieve E_∞, simply because buffer zones of infinite capacity are not possible. Hence, the actual value of line efficiency for a given buffer capacity b will fall somewhere between these extremes:

$$E_0 < E_b < E_\infty \tag{16.18}$$

Next, let us consider the problem of evaluating E_b for realistic levels of buffer capacity for a two-stage automated production line ($K = 2$).

Analysis of a Two-Stage Transfer Line. Most of the discussion in this section is based on the work of Buzacott, who pioneered the analytical research on production lines with buffer stocks. Several of his publications are listed in our references [2], [3], [4], [5], [6], and [7]. Our presentation in this section will follow Buzacott's analysis in [2].

The two-stage line is divided by a storage buffer of capacity b, expressed in terms of the number of workparts that it can store. The buffer receives the output of stage 1 and forwards it to stage 2, temporarily storing any parts not immediately needed by stage 2 up to its capacity b. The ideal cycle time T_c is the same for both stages. We assume the downtime distributions of each stage to be the same with mean downtime $= T_d$. Let F_1 and $F_2 =$ the breakdown rates of stages 1 and 2, respectively; F_1 and F_2 are not necessarily equal.

Over the long run, both stages must have equal efficiencies. If the efficiency of stage 1 was greater than that of stage 2, then inventory would build up in the storage buffer until its capacity b is reached. Thereafter, stage 1 would be blocked when it out-produced stage 2. Similarly, if the efficiency of stage 2 were greater than that of stage 1, the inventory in the buffer would become depleted, thus starving stage 2. Accordingly, the efficiencies in the two stages would tend to equalize over time. The overall line efficiency for the two-stage line can be expressed as

$$E_b = E_0 + D'_1 h(b) E_2 \qquad (16.19)$$

where E_b = overall line efficiency for a two-stage line with buffer capacity b; E_0 = line efficiency for the same line with no internal storage; and the second term on the right hand side $(D'_1 h(b) E_2)$ represents the improvement in efficiency that results from having a storage buffer with $b > 0$. Let us examine the RHS terms in Eq. (16.19). The value of E_0 was given by Eq. (16.15), but we write it next to explicitly define the two stage efficiency when $b = 0$:

$$E_0 = \frac{T_c}{T_c + (F_1 + F_2) T_d} \qquad (16.20)$$

The term D'_1 can be thought of as the proportion of total time that stage 1 is down, defined as follows:

$$D'_1 = \frac{F_1 T_d}{T_c + (F_1 + F_2) T_d} \qquad (16.21)$$

The term $h(b)$ is the proportion of the downtime D'_1 (when stage 1 is down) that stage 2 could be up and operating within the limits of storage buffer capacity b. Buzacott presents equations for evaluating $h(b)$ using Markov chain analysis. The equations cover several different downtime distributions based on the assumption that both stages are never down at the same time. Four of these equations are presented in Table 16.2.

TABLE 16.2 Formulas for Computing $h(b)$ in Eq. (16.19) for a Two-Stage Automated Production Line Under Several Downtime Distributions

Assumptions and definitions: Assume that the two stages have equal downtime distributions ($T_{d1} = T_{d2} = T_d$) and equal cycle times ($T_{c1} = T_{c2} = T_c$). Let F_1 = downtime frequency for stage 1 and F_2 = downtime frequency for stage 2. Define r to be the ratio of breakdown frequencies as follows:

$$r = \frac{F_1}{F_2} \qquad (16.22)$$

Equations for h(b): With these definitions and assumptions, we can express the relationships for $h(b)$ for two theoretical downtime distributions as derived by Buzacott [2]:

Constant downtime: Each downtime occurrence is assumed to be of constant duration T_d. This is a case of no downtime variation. Given buffer capacity b, define B and L as

$$b = B \frac{T_d}{T_c} + L \qquad (16.23)$$

where B = Maximum Integer $\le b \frac{T_c}{T_d}$ and L represents the leftover units, the amount by which b exceeds $B \frac{T_d}{T_c}$. There are two cases:

$$\text{Case 1: } r = 1.0. \ h(b) = \frac{B}{B + 1} + L \frac{T_c}{T_d} \frac{1}{(B + 1)(B + 2)} \qquad (16.24)$$

$$\text{Case 2: } r \ne 1.0. \ h(b) = r \frac{1 - r^B}{1 - r^{B+1}} + L \frac{T_c}{T_d} \frac{r^{B+1}(1 - r)^2}{(1 - r^{B+1})(1 - r^{B+2})} \qquad (16.25)$$

Geometric downtime distribution: In this downtime distribution, the probability that repairs are completed during any cycle duration T_c is independent of the time since repairs began. This is a case of maximum downtime variation. There are two cases:

$$\text{Case 1: } r = 1.0. \quad h(b) = \frac{b\dfrac{T_c}{T_d}}{2 + (b-1)\dfrac{T_c}{T_d}} \tag{16.26}$$

$$\text{Case 2: } r \neq 1.0. \text{ Define } K = \frac{1 + r - \dfrac{T_c}{T_d}}{1 + r - r\dfrac{T_c}{T_d}} \quad \text{then } h(b) = \frac{r(1 - K^b)}{1 - rK^b} \tag{16.27}$$

Finally, E_2 corrects for the unrealistic assumption in the calculation of $h(b)$ that both stages are never down at the same time. What is more realistic is that when stage 1 is down but stage 2 is producing using parts stored in the buffer, occasionally stage 2 itself will break down. E_2 is calculated as

$$E_2 = \frac{T_c}{T_c + F_2 T_d} \tag{16.28}$$

It should be mentioned that Buzacott's derivation of Eq. (16.19) in [2] omitted the E_2 term, relying on the assumption that stages 1 and 2 will not share downtimes. However, without E_2, the author has found that the equation tends to overestimate line efficiency. With E_2 included, as in our Eq. (16.19), the calculated values are much more realistic. In subsequent research, Buzacott developed other equations that agree closely with results given by Eq. (16.19).

EXAMPLE 16.3 Two-Stage Automated Production Line

A 20-station transfer line is divided into two stages of 10 stations each. The ideal cycle time of each stage is $T_c = 1.2$ min. All of the stations in the line have the same probability of stopping, $p = 0.005$. We assume that the downtime is constant when a breakdown occurs, $T_d = 8.0$ min. Compute the line efficiency for the following buffer capacities: (a) $b = 0$, (b) $b = \infty$, (c) $b = 10$, and (d) $b = 100$.

Solution: (a) A two-stage line with 20 stations and $b = 0$ turns out to be the same case as our previous Example 16.2. To review,

$$F = np = 20(0.005) = 0.10 \text{ and } T_p = T_c + FT_d = 1.2 + 0.1(8) = 2.0 \text{ min}$$

$$E_0 = \frac{1.2}{2.0} = 0.60$$

(b) For a two-stage line with 20 stations (each stage = 10 stations) and $b = \infty$,

$$F_1 = F_2 = 10(0.005) = 0.05 \text{ and } T_p = 1.2 + 0.05(8) = 1.6 \text{ min}$$

$$E_\infty = E_1 = E_2 = \frac{1.2}{1.6} = 0.75$$

(c) For a two-stage line with $b = 10$, we must determine each of the terms in Eq. (16.19). We have E_0 from part (a): $E_0 = 0.60$, and we have E_2 from part (b): $E_2 = 0.75$. Hence,

$$D_1' = \frac{0.05(8)}{1.2 + (0.05 + 0.05)(8)} = \frac{0.40}{2.0} = 0.20$$

Evaluation of $h(b)$ is from Eq. (16.24) for a constant repair distribution. In Eq. (16.23), the ratio

$$\frac{T_d}{T_c} = \frac{8.0}{1.2} = 6.667. \qquad \text{For } b = 10, B = 1 \text{ and } L = 3.333. \text{ Thus,}$$

$$h(b) = h(10) = \frac{1}{1 + 1} + 3.333\left(\frac{1.2}{8.0}\right)\frac{1}{(1 + 1)(1 + 2)} = 0.50 + 0.0833 = 0.5833$$

We can now use Eq. (16.19):

$$E_{10} = 0.600 + 0.20(0.5833)(0.75) = 0.600 + 0.0875 = 0.6875$$

(d) For $b = 100$, the only parameter in Eq. (16.19) that is different from part (c) is $h(b)$. For $b = 100$, $B = 15$ and $L = 0$ in Eq. (16.23). Thus, we have

$$h(b) = h(100) = \frac{15}{15 + 1} = 0.9375$$

Using this value,

$$E_{100} = 0.600 + 0.20(0.9375)(0.75) = 0.600 + 0.1406 = 0.7406$$

The value of $h(b)$ not only serves its role in Eq. (16.19), it also provides information on how much improvement in efficiency we get from any given value of b. Note in Example 16.3 that the difference between E_∞ and $E_0 = 0.75 - 0.60 = 0.15$. For $b = 10$, $h(b) = h(10) = 0.5833$, which means we get 58.33% of the maximum possible improvement in line efficiency by using a buffer capacity of 10 ($E_{10} = 0.6875 = 0.60 + 0.5833(0.75 - 0.60)$). For $b = 100$, $h(b) = h(100) = 0.9375$, which means we get 93.75% of the maximum improvement with $b = 100$ ($E_{100} = 0.7406 = 0.60 + 0.9375(0.75 - 0.60)$).

We are not only interested in the line efficiencies of a two-stage production line. We also want to know the corresponding production rates. These can be evaluated based on knowledge of the ideal cycle time T_c and the definition of line efficiency. According to Eq. (16.11), $E = T_c/T_p$. Since $R_p = $ the reciprocal of T_p, then $E = T_c R_p$. Rearranging, we have

$$R_p = \frac{E}{T_c} \tag{16.29}$$

EXAMPLE 16.4   Production Rates on the Two-Stage Line of Example 16.3

Compute the production rates for the four cases in Example 16.3. The value of $T_c = 1.2$ minutes as before.

Solution: (a) For $b = 0$, $E_0 = 0.60$. Applying Eq. (16.29), we have

$$R_p = 0.60/1.2 = 0.5 \text{ pc/min} = 30 \text{ pc/hr}$$

This is the same value calculated in Example 16.2.

(b) For $b = \infty$, $E_\infty = 0.75$, and $R_p = 0.75/1.2 = 0.625$ pc/min = 37.5 pc/hr

(c) For $b = 10$, $E_{10} = 0.6875$, and

$$R_p = 0.6875/1.2 = 0.5729 \text{ pc/min} = 34.375 \text{ pc/hr}$$

(d) For $b = 100$, $E_{100} = 0.7406$, and

$$R_p = 0.7406/1.2 = 0.6172 \text{ pc/min} = 37.03 \text{ pc/hr}$$

In Example 16.3, a constant repair distribution was assumed. Every breakdown had the same constant repair time of 8.0 min. It is more realistic to expect that there will be some variation in the repair time distribution. Table 16.2 provides two possible distributions, representing extremes in variability. We have already used the constant repair distribution in Examples 16.3 and 16.4, which represent the case of no downtime variation. This is covered in Table 16.2 by Eqs. (16.24) and (16.25). Let us consider the opposite extreme, the case of very high variation. This is presented in Table 16.2 as the geometric repair distribution, where $h(b)$ is computed by Eqs. (16.26) and (16.27).

EXAMPLE 16.5   Effect of High Variability in Downtime

Evaluate the line efficiencies and production rates for the two-stage line in Examples 16.3 and 16.4, using the geometric repair distribution instead of the constant downtime distribution.

Solution: For parts (a) and (b), the values of E_0 and E_∞ will be the same as in previous Example 16.4.

(a) $E_0 = 0.600$ and $R_p = 30$ pc/hr

(b) $E_\infty = 0.750$ and $R_p = 37.5$ pc/hr

(c) For $b = 10$, all of the parameters in Eq. (16.19) remain the same except $h(b)$.

Using Eq. (16.26) from Table 16.2, we have

$$h(b) = h(10) = \frac{10(1.2/8.0)}{2 + (10-1)(1.2/8.0)} = 0.4478$$

Now using Eqs. (16.19) and (16.29), we have

$$E_{10} = 0.600 + 0.20(0.4478)(0.75) = 0.6672 \text{ and}$$
$$R_p = 0.6672(60)/1.2 = 33.36 \text{ pc/hr}$$

(d) For $b = 100$, again the only change is in $h(b)$.

$$h(b) = h(100) = \frac{100(1.2/8.0)}{2 + (100-1)(1.2/8.0)} = 0.8902$$

$$E_{100} = 0.600 + 0.20(0.8902)(0.75) = 0.7333 \text{ and}$$
$$R_p = 0.7333(60)/1.2 = 36.67 \text{ pc/hr}$$

Note that when we compare the values of line efficiency and production rates for $b = 10$ and $b = 100$ in this example with the corresponding values in Examples 16.3 and 16.4, both values are lower here. We must conclude that increased downtime variability degrades line performance.

Transfer Lines with More than Two Stages. If the line efficiency of an automated production line can be increased by dividing it into two stages with a storage buffer between, then one might infer that further improvements in performance can be achieved by adding additional storage buffers. Although we do not present exact formulas for computing line efficiencies for the general case of any capacity b for multiple storage buffers, efficiency improvements can readily be determined for the case of infinite buffer capacity. In Examples 16.4 and 16.5 we saw the relative improvement in efficiency that results from intermediate buffer sizes between $b = 0$ and $b = \infty$.

EXAMPLE 16.6 Transfer Lines with More Than One Storage Buffer

For the same 20-station transfer line we have been considering in previous examples, compare line efficiencies and production rates for the following cases, assuming an infinite buffer capacity: (a) no storage buffers, (b) one buffer, (c) three buffers, and (d) 19 buffers. Assume in cases (b) and (c) that the buffers are located in the line so as to equalize the downtime frequencies, that is, all F_i are equal.

Solution: We have already computed the answer for (a) and (b) in Example 16.4.

(a) For the case of no storage buffer, $E_\infty = 0.60$ and $R_p = 0.60(60)/1.2 = 30$ pc/hr

(b) For one storage buffer (a two-stage line), $E_\infty = 0.75$ and $R_p = 0.75(60)/1.2 = 37.5$ pc/hr

(c) For the case of three storage buffers (a four-stage line), we have
$F_1 = F_2 = F_3 = F_4 = 5(.005) = 0.025$ and
$T_p = 1.2 + 0.025(8) = 1.4$ min/pc

$E_\infty = 1.2/1.4 = 0.8571$ and $R_p = 0.8571(60)/1.2 = 42.86$ pc/hr.

(d) For the case of 19 storage buffers (each stage is one station), we have
$F_1 = F_2 = \cdots = F_{20} = 1(0.005) = 0.005$ and
$T_p = 1.2 + 0.005(8) = 1.24$ min/pc

$E_\infty = 1.2/1.24 = 0.9677$ and $R_p = 0.9677(60)/1.2 = 48.39$ pc/hr.

This last value is very close to the ideal production rate of $R_c = 50$ pc/hr.

What the Equations Tell Us. The equations and analysis in this section provide some practical guidelines in the design and operation of automated production lines with internal storage buffers. The guidelines can be expressed as follows:

- If E_0 and E_∞ are nearly equal in value, little advantage is gained by adding a storage buffer to the line. If E_∞ is significantly greater than E_0, then storage buffers offer the possibility of significantly improving line performance.
- In considering a multi-stage automated production line, workstations should be divided into stages so as to make the efficiencies of all stages as equal as possible. In this way, the maximum difference between E_0 and E_∞ is achieved, and no single stage will stand out as a significant bottleneck.
- In the operation of an automated production line with storage buffers, if any of the buffers are nearly always empty or nearly always full, this indicates that the production rates of the stages on either side of the buffer are out of balance and that the storage buffer is serving little useful purpose.

- The maximum possible line efficiency is achieved by (1) setting the number of stages equal to the number of stations—that is, by providing a storage buffer between every pair of stations and (2) by using large capacity buffers.
- The "law of diminishing returns" operates in multi-stage automated lines. It is manifested in two ways: (1) as the number of storage buffers is increased, line efficiency improves at an ever decreasing rate, and (2) as the storage buffer capacity is increased, line efficiency improves at an ever decreasing rate.

REFERENCES

[1] BOOTHROYD, G., POLI, C., and MURCH, L. E., *Automatic Assembly*, Marcel Dekker, Inc., NY, 1982.

[2] BUZACOTT, J. A., "Automatic Transfer Lines with Buffer Stocks," *International Journal of Production Research*, Vol. 5, No. 3, 1967, pp. 183–200.

[3] BUZACOTT, J. A., "Prediction of the Efficiency of Production Systems without Internal Storage," *International Journal of Production Research*, Vol. 6, No. 3, 1968, pp. 173–188.

[4] BUZACOTT, J. A., "The Role of Inventory Banks in Flow-Line Production Systems," *International Journal of Production Research*, Vol. 9, No. 4, 1971, pp. 425–436.

[5] BUZACOTT, J. A., and L. E. HANIFIN, "Models of Automatic Transfer Lines with Inventory Banks—A Review and Comparison," *AIIE Transactions*, Vol. 10, No. 2, 1978, pp. 197–207.

[6] BUZACOTT, J. A., and L. E. HANIFIN, "Transfer Line Design and Analysis—An Overview," *Proceedings*, 1978 Fall Industrial Engineering Conference of AIIE, Atlanta, GA. December 1978.

[7] BUZACOTT, J. A., and SHANTHIKUMAR, J. G., *Stochastic Models of Manufacturing Systems*, Prentice Hall, Englewood Cliffs, NJ, 1993, Chapters 5 and 6.

[8] CHOW, W-M., *Assembly Line Design*, Marcel Dekker, Inc., NY, 1990.

[9] GROOVER, M. P., "Analyzing Automatic Transfer Lines," *Industrial Engineering*, Vol. 7, No. 11, 1975, pp. 26–31.

[10] KOELSCH, J. R., "A New Look to Transfer Lines," *Manufacturing Engineering*, April 1994, pp. 73–78.

[11] LAVALLEE, R. J., "Using a PC to Control a Transfer Line," *Control Engineering*, /2nd February 1991, pp. 43–56.

[12] MASON, F., "High Volume Learns to Flex," *Manufacturing Engineering*, April 1995, pp. 53–59.

[13] OWEN, J. V., "Transfer Lines Get Flexible," *Manufacturing Engineering*, January, 1999, pp. 42–50.

[14] RILEY, F. J., *Assembly Automation*, Industrial Press, NY, 1983.

[15] WILEY, R., *Mass-Production Management*, John Wiley & Sons, London, UK, 1972.

REVIEW QUESTIONS

16.1 Name three of the four conditions under which automated production lines are appropriate.

16.2 What is an automated production line?

16.3 What is a pallet fixture, as the term is used in the context of an automated production line?

16.4 What is a dial indexing machine?

16.5 Why are continuous work transport systems uncommon on automated production lines?

16.6 Is a Geneva mechanism used to provide linear motion or rotary motion?

16.7 What is a storage buffer as the term is used for an automated production line?

16.8 Name three reasons for including a storage buffer in an automated production line.

16.9 What are the three basic control functions that must be accomplished to operate an automated production line?

16.10 Name some of the industrial applications of automated production lines.

16.11 What is the difference between a unitized production line and a link line?

16.12 What are the three problem areas that must be considered in the analysis and design of an automated production line?

16.13 As the number of workstation on an automated production line increases, does line efficiency (a) decrease, (b) increase, or (c) remain unaffected?

16.14 What is starving on an automated production line?

16.15 In the operation of an automated production line with storage buffers, what does it mean if a buffer is nearly always empty or nearly always full?

PROBLEMS

Transfer Mechanisms

16.1 A rotary worktable is driven by a Geneva mechanism with five slots. The driver rotates at 48 rev/min. Determine (a) the cycle time, (b) available process time, and (c) indexing time each cycle.

16.2 A Geneva with six slots is used to operate the worktable of a dial indexing machine. The slowest workstation on the dial indexing machine has an operation time of 2.5 sec, so the table must be in a dwell position for this length of time. (a) At what rotational speed must the driven member of the Geneva mechanism be turned to provide this dwell time? (b) What is the indexing time each cycle?

16.3 Solve the previous problem using a Geneva with eight slots.

Automated Production Lines with No Internal Storage

16.4 A ten-station transfer machine has an ideal cycle time of 30 sec. The frequency of line stops is 0.075 stops per cycle. When a line stop occurs, the average downtime is 4.0 min. Determine (a) average production rate in pc/hr, (b) line efficiency, and (c) proportion downtime.

16.5 Cost elements associated with the operation of the ten-station transfer line in Problem 16.4 are as follows: raw workpart cost = $0.55/pc, line operating cost = $42.00/hr, and cost of disposable tooling = $0.27/pc. Compute the average cost of a workpiece produced.

16.6 In Problem 16.4, the line stop occurrences are due to random mechanical and electrical failures on the line. Suppose that in addition to these reasons for downtime, that the tools at each workstation on the line must be changed and/or reset every 150 cycles. This procedure takes a total of 12.0 min for all ten stations. Include this additional data to determine (a) average production rate in pc/hr, (b) line efficiency, and (c) proportion downtime.

16.7 The dial indexing machine of Problem 16.2 experiences a breakdown frequency of 0.06 stops/cycle. The average downtime per breakdown is 3.5 min. Determine (a) average production rate in pc/hr and (b) line efficiency.

16.8 In the operation of a certain 15-station transfer line, the ideal cycle time = 0.58 min. Breakdowns occur at a rate of once every 20 cycles, and the average downtime per breakdown is 9.2 min. The transfer line is located in a plant that works an 8-hr day, 5 days per week. Determine (a) line efficiency, and (b) how many parts the transfer line will produce in a week.

16.9 A 22-station in-line transfer machine has an ideal cycle time of 0.35 min. Station breakdowns occur with a probability of 0.01. Assume that station breakdowns are the only reason for line stops. Average downtime = 8.0 min per line stop. Determine (a) ideal production rate, (b) frequency of line stops, (c) average actual production rate, and (d) line efficiency.

16.10 A ten-station rotary indexing machine performs machining operations at nine workstations, while the tenth station is used for loading and unloading parts. The longest process time on the line is 1.30 min and the loading/unloading operation can be accomplished in less time than this. It takes 9.0 sec to index the machine between workstations. Stations break down with a frequency of 0.007, equal for all ten stations. When these stops occur, it takes an average of 10.0 min to diagnose the problem and make repairs. Determine (a) line efficiency and (b) average actual production rate.

16.11 A transfer machine has six stations that function as follows:

Station	Operation	Process Time (min)	p_i
1	Load part	0.78	0
2	Drill three holes	1.25	0.02
3	Ream two holes	0.90	0.01
4	Tap two holes	0.85	0.04
5	Mill flats	1.32	0.01
6	Unload parts	0.45	0

In addition, transfer time = 0.18 min. Average downtime per occurrence = 8.0 min. A total of 20,000 parts must be processed through the transfer machine. Determine (a) proportion downtime, (b) average actual production rate, and (c) how many hours of operation are required to produce the 20,000 parts.

16.12 The cost to operate a certain 20-station transfer line is $72/hr. The line operates with an ideal cycle time of 0.85 min. Downtime occurrences happen on average once per 14 cycles. Average downtime per occurrence is 9.5 min. Management proposes to install a new computer system and associated sensors to monitor the line and diagnose downtime occurrences when they happen. This new system is expected to reduce downtime from its present value to 7.5 min. If the cost of purchasing and installing the new system is $15,000, how many units must the system produce in order for the savings to pay for the computer system?

16.13 A 23-station transfer line has been logged for 5 days (total time = 2400 min). During this time there were a total of 158 downtime occurrences on the line. The accompanying table identifies the types of downtime occurrences, how many occurrences there were of each type, and how much total time was lost for each type. The transfer line performs a sequence of machining operations, the longest of which takes 0.42 min. The transfer mechanism takes 0.08 min to index the parts from one station to the next each cycle. Assuming no parts removal when the line jams, determine the following based on the five-day observation period: (a) the number of parts produced, (b) the downtime proportion, (c) the production rate, and (d) the frequency rate associated with the transfer mechanism failures.

Type of Downtime	Number of Occurrences	Total Time Lost (min)
Tool-related causes	104	520
Mechanical failures	21	189
Miscellaneous	7	84
Subtotal	132	793
Transfer mechanism	26	78
Total	158	871

16.14 An eight-station rotary indexing machine performs the machining operations shown in the accompanying table, together with processing times and breakdown frequencies for each station. The transfer time for the machine is 0.15 min per cycle. A study of the system was undertaken, during which time 2000 parts were completed. It was determined in this study that when breakdowns occur, it takes an average of 7.0 min to make repairs and get the system operating again. For the study period, determine (a) average actual production rate, (b) line uptime efficiency, and (c) the number of hours required to produce the 2000 parts.

Station	Process	Process Time (min)	Breakdowns
1	Load part	0.50	0
2	Mill top	0.85	22
3	Mill sides	1.10	31
4	Drill two holes	0.60	47
5	Ream two holes	0.43	8
6	Drill six holes	0.92	58
7	Tap six holes	0.75	84
8	Unload part	0.40	0

16.15 A 14-station transfer line has been logged for 2400 min to identify type of downtime occurrence, number of occurrences, and time lost. The results are presented in the table below. The ideal cycle time for the line is 0.50 min, including transfer time between stations. Determine (a) how many parts were produced during the 2400 min, (b) line uptime efficiency, (c) average actual production rate of acceptable parts per hour, and (d) frequency p associated with transfer system failures.

Type of Occurrence	Number	Time Lost (min)
Tool changes and failures	70	400
Station failures (mechanical and electrical)	45	300
Transfer system failures	25	150

16.16 A transfer machine has a mean time between failures (MTBF) = 50 minutes and a mean time to repair (MTTR) = 9 minutes. If the ideal cycle rate = 1/min (when the machine is running), what is the average hourly production rate?

16.17 A part is to be produced on an automated transfer line. The total work content time to make the part is 36 minutes, and this work will be divided evenly amongst the workstations, so that the processing time at each station is $36/n$, where n = the number of stations. In addition, the time required to transfer parts between workstations is 6 seconds. Thus, the cycle time = $0.1 + 36/n$ minutes. In addition, it is known that the station breakdown frequency will be 0.005, and that the average downtime per breakdown = 8.0 minutes. (a) Given these data, determine the number of workstations that should be included in the line to maximize production rate. Also, what is the (b) production rate and (c) line efficiency for this number of stations?

Automated Production Lines with Storage Buffers

16.18   A 30-station transfer line has an ideal cycle time of 0.75 min, an average downtime of 6.0 min per line stop occurrence, and a station failure frequency of 0.01 for all stations. A proposal has been submitted to locate a storage buffer between stations 15 and 16 to improve line efficiency. Determine (a) the current line efficiency and production rate, and (b) the maximum possible line efficiency and production rate that would result from installing the storage buffer.

16.19   Given the data in Problem 16.18, solve the problem except that (a) the proposal is to divide the line into three stages, that is, with two storage buffers located between stations 10 and 11, and between stations 20 and 21, respectively; and (b) the proposal is to use an asynchronous line with large storage buffers between every pair of stations on the line, that is, a total of 29 storage buffers.

16.20   In Problem 16.18, if the capacity of the proposed storage buffer is to be 20 parts, determine (a) line efficiency, and (b) production rate of the line. Assume that the downtime ($T_d = 6.0$ min) is a constant.

16.21   Solve Problem 16.20 but assume that the downtime ($T_d = 6.0$ min) follows the geometric repair distribution.

16.22   In the transfer line of Problems 16.18 and 16.20, suppose it is more technically feasible to locate the storage buffer between stations 11 and 12, rather than between stations 15 and 16. Determine (a) the maximum possible line efficiency and production rate that would result from installing the storage buffer, and (b) the line efficiency and production rate for a storage buffer with a capacity of 20 parts. Assume that downtime ($T_d = 6.0$ min) is a constant.

16.23   A proposed synchronous transfer line will have 20 stations and will operate with an ideal cycle time of 0.5 min. All stations are expected to have an equal probability of breakdown, $p = 0.01$. The average downtime per breakdown is expected to be 5.0. An option under consideration is to divide the line into two stages, each stage having 10 stations, with a buffer storage zone between the stages. It has been decided that the storage capacity should be 20 units. The cost to operate the line is $96.00/hr. Installing the storage buffer would increase the line operating cost by $12.00/hr. Ignoring material and tooling costs, determine (a) line efficiency, production rate, and unit cost for the one-stage configuration, and (b) line efficiency, production rate, and unit cost for the optional two-stage configuration.

16.24   A two-week study has been performed on a 12-station transfer line that is used to partially machine engine heads for a major automotive company. During the 80 hours of observation, the line was down a total of 42 hours, and a total of 1689 parts were completed. The accompanying table lists the machining operation performed at each station, the process times, and the downtime occurrences for each station. Transfer time between stations is 6 sec. To address the downtime problem, it has been proposed to divide the line into two stages, each consisting of six stations. The storage buffer between the stages would have a storage capacity of 20 parts. Determine (a) line efficiency and production rate of the current one-stage configuration, and (b) line efficiency and production rate of the proposed two-stage configuration. (c) Given that the line is to be divided into two stages, should each stage consist of six stations as proposed, or is there a better way to divide the stations into stages? Support your answer.

Station	Operation	Process Time (min)	Downtime Occurrences
1	Load part (manual)	0.50	0
2	Rough mill top	1.10	15
3	Finish mill top	1.25	18
4	Rough mill sides	0.75	23
5	Finish mill sides	1.05	31
6	Mill surfaces for drill	0.80	9
7	Drill two holes	0.75	22
8	Tap two holes	0.40	47
9	Drill three holes	1.10	30
10	Ream three holes	0.70	21
11	Tap three holes	0.45	30
12	Unload and inspect part (manual)	0.90	0
	Total	9.40	246

16.25 In Problem 16.24, the current line has an operating cost of $66.00/hr. The starting workpart is a casting that costs $4.50/pc. Disposable tooling costs $1.25/pc. The proposed storage buffer will add $6.00/hr to the operating cost of the line. Does the improvement in production rate justify this $20 increase?

16.26 A 16-station transfer line can be divided into two stages by installing a storage buffer between stations 8 and 9. The probability of failure at any station is 0.01. The ideal cycle time is 1.0 min and the downtime per line stop is 10.0 min. These values are applicable for both the one-stage and two-stage configurations. The downtime should be considered a constant value. The cost of installing the storage buffer is a function of its capacity. This cost function is $C_b = \$0.60b/hr = \$0.01b/min$, where $b =$ the buffer capacity. However, the buffer can only be constructed to store increments of 10 (in other words, b can take on values of 10, 20, 30, etc.). The cost to operate the line itself is $120/hr. Ignore material and tooling costs. Based on cost per unit of product, determine the buffer capacity b that will minimize unit product cost.

16.27 The uptime efficiency of a 20-station automated production line is only 40%. The ideal cycle time is 48 sec, and the average downtime per line stop occurrence is 3.0 min. Assume the frequency of breakdowns for all stations is equal ($p_i = p$ for all stations) and the downtime is constant. To improve uptime efficiency, it is proposed to install a storage buffer with a 15-part capacity for $14,000. The present production cost is $4.00 per unit, ignoring material and tooling costs. How many units would have to be produced in order for the $14,000 investment to pay for itself?

16.28 An automated transfer line is divided into two stages with a storage buffer between them. Each stage consists of nine stations. The ideal cycle time of each stage = 1.0 minute, and frequency of failure for each station is 0.01. The average downtime per stop is 8.0 minutes, and a constant downtime distribution should be assumed. Determine the required capacity of the storage buffer such that the improvement in line efficiency E compared to a zero buffer capacity would be 80% of the improvement yielded by a buffer with infinite capacity.

16.29 In Problem 16.17, suppose that a two-stage line were to be designed, with an equal number of stations in each stage. Work content time will be divided evenly between the two stages. The storage buffer between the stages will have a capacity $= 3\,T_d/T_c$. Assume a constant repair distribution. (a) For this two-stage line, determine the number of workstations that should be included in each stage of the line to maximize production rate. (b) What is the production rate and line efficiency for this line configuration? (c) What is the buffer storage capacity?

16.30 A 20-station transfer line presently operates with a line efficiency $E = 1/3$. The ideal cycle time $= 1.0$ min. The repair distribution is geometric with an average downtime per occurrence $= 8$ min, and each station has an equal probability of failure. It is possible to divide the line into two stages with 10 stations each, separating the stages by a storage buffer of capacity b. With the information given, determine the required value of b that will increase the efficiency from $E = 1/3$ to $E = 2/5$.

Automated Assembly Systems

CHAPTER CONTENTS

The term *automated assembly* refers to the use of mechanized and automated devices to perform the various assembly tasks in an assembly line or cell. Much progress has been made in the technology of assembly automation in recent years. Some of this progress has been motivated by advances in the field of robotics. Industrial robots are sometimes used as components in automated assembly systems (Chapter 8). In this chapter, we discuss automated assembly as a distinct field of automation. Although the manual assembly methods described in Chapter 15 will be used for many years into the future, there are significant opportunities for productivity gains in the use of automated methods.

Like the transfer lines discussed in the preceding chapter, automated assembly systems are usually included in the category of fixed automation. Most automated assembly systems are designed to perform a fixed sequence of assembly steps on a specific product. Automated assembly technology should be considered when the following conditions exist:

- *High product demand.* Automated assembly systems should be considered for products made in millions of units (or close to this range).
- *Stable product design.* In general, any change in the product design means a change in workstation tooling and possibly the sequence of assembly operations. Such changes can be very costly.
- *A limited number of components in the assembly.* Riley [13] recommends a maximum of around a dozen parts.
- *Product designed for automated assembly.* In Chapter 24, we examine the product design factors that allow for automated assembly.

Automated assembly systems involve a significant capital expense. However, the investments are generally less than for the automated transfer lines because (1) work units produced on automated assembly systems are usually smaller than those made on transfer lines and (2) assembly operations do not have the large mechanical force and power requirements of processing operations such as machining. Accordingly, in comparing an automated assembly system and a transfer line both having the same number of stations, the assembly system would tend to be physically smaller. This usually reduces the cost of the system.

17.1 FUNDAMENTALS OF AUTOMATED ASSEMBLY SYSTEMS

An automated assembly system performs a sequence of automated assembly operations to combine multiple components into a single entity. The single entity can be a final product or a subassembly in a larger product. In many cases, the assembled entity consists of a base part to which other components are attached. The components are usually joined one at a time, so the assembly is completed progressively.

A typical automated assembly system consists of the following subsystems: (1) one or more workstations at which the assembly steps are accomplished, (2) parts feeding devices that deliver the individual components to the workstations, and (3) a work handling system for the assembled entity. In assembly systems with one workstation, the work handling system moves the base part into and out of the station. In systems with multiple stations, the handling system transfers the partially assembled base part between stations.

Control functions required in automated assembly machines are the same as in the automated production lines of Chapter 16: (1) sequence control, (2) safety monitoring, and (3) quality control. These functions are described in Section 16.1.4.

17.1.1 System Configurations

Automated assembly systems can be classified according to physical configuration. The principal configurations, illustrated in Figure 17.1, are (a) in-line assembly machine, (b) dial-type assembly machine, (c) carousel assembly system, and (d) single-station assembly machine.

The *in-line assembly machine*, Figure 17.1(a), is a series of automatic workstations located along an in-line transfer system. It is the assembly version of the machining transfer

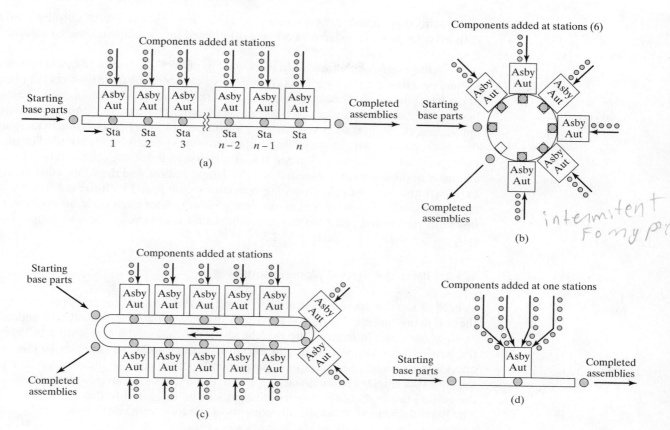

Figure 17.1 Types of automated assembly systems: (a) in-line, (b) dial-type, (c) carousel, and (d) single-station.

line. Synchronous and asynchronous transfer systems are the common means of transporting base parts from station to station with the in-line configuration.

In the typical application of the *dial-type machine*, Figure 17.1(b), base parts are loaded onto fixtures or nests attached to the circular dial. Components are added and/or joined to the base part at the various workstations located around the periphery of the dial. The dial indexing machine operates with a synchronous or intermittent motion, in which the cycle consists of the service time plus indexing time. Dial-type assembly machines are sometimes designed to use a continuous rather than intermittent motion. This is common in beverage bottling and canning plants, but not in mechanical and electronics assembly.

The operation of dial-type and in-line assembly systems is similar to the operation of their counterparts for processing operations described in Section 16.1.2, except that assembly operations are performed. For synchronous transfer of work between stations, the ideal cycle time equals the operation time at the slowest station plus the transfer time between stations. The production rate, at 100% uptime, is the reciprocal of the ideal cycle time. Owing to part jams at the workstations and other malfunctions, the system will always operate at less than 100% uptime. We analyze the performance of these systems in Section 17.2.2.

As seen in Figure 17.1(c), the *carousel assembly system* represents a hybrid between the circular work flow of the dial-type assembly machine and the straight work flow of the in-line system. The carousel configuration can be operated with continuous, synchronous, or asynchronous transfer mechanisms to move the work around the carousel. Carousels with asynchronous transfer of work are often used in partially automated assembly systems (Section 17.2.4).

In the *single-station assembly machine*, Figure 17.1(d), assembly operations are performed on a base part at a single location. The typical operating cycle involves the placement of the base part at a stationary position in the workstation, the addition of components to the base, and finally the removal of the completed assembly from the station. An important application of single-station assembly is the component insertion machine, widely used in the electronics industry to populate components onto printed circuit boards. For mechanical assemblies, the single-station cell is sometimes selected as the configuration for robotic assembly applications. Parts are fed into the single station, and the robot adds them to the base part and performs the fastening operations. Compared with the other three system types, the single-station system is inherently slower, since each cycle all of the assembly tasks are performed and only one assembled unit is completed. Single-station assembly systems are analyzed in Section 17.2.3.

17.1.2 Parts Delivery at Workstations

In each of the configurations described above, a workstation accomplishes one or both of the following tasks: (1) a part is delivered to the assembly workhead and added to the existing base part in front of the workhead (in the case of the first station in the system, the base part is often deposited into the work carrier), and (2) a fastening or joining operation is performed at the station to permanently attach parts to the existing base part. In the case of a single-station assembly system, these tasks are carried out multiple times at the single station. Task (1) requires the parts to be delivered to the assembly workhead. The parts delivery system typically consists of the following hardware:

1. *Hopper.* This is the container into which the components are loaded at the workstation. A separate hopper is used for each component type. The components are usually loaded into the hopper in bulk. This means that the parts are randomly oriented in the hopper.

2. *Parts feeder.* This is a mechanism that removes the components from the hopper one at a time for delivery to the assembly workhead. The hopper and parts feeder are often combined into one operating mechanism. A vibratory bowl feeder, pictured in Figure 17.2, is a very common example of the hopper-feeder combination.

3. *Selector* and/or *orientor*. These elements of the delivery system establish the proper orientation of the components for the assembly workhead. A *selector* is a device that acts as a filter, permitting only parts in the correct orientation to pass through. Incorrectly oriented parts are rejected back into the hopper. An *orientor* is a device that allows properly oriented parts to pass through, and reorients parts that are not properly oriented initially. Several selector and orientor schemes are illustrated in Figure 17.3. Selector and orientor devices are often combined and incorporated into one hopper-feeder system.

4. *Feed track.* The preceding elements of the delivery system are usually separated from the assembly workhead by a certain distance. A feed track moves the components from the hopper and parts feeder to the location of the assembly workhead, maintaining proper orientation of the parts during the transfer. There are two general categories of

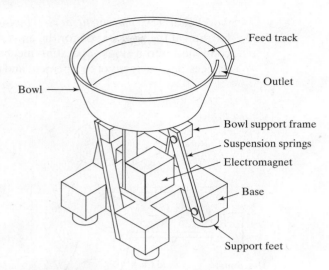

Figure 17.2 Vibratory bowl feeder.

feed tracks: gravity and powered. Gravity feed tracks are most common. In this type the hopper and parts feeder are located at an elevation above that of the workhead. The force of gravity is used to deliver the components to the workhead. The powered feed track uses vibratory action, air pressure, or other means to force the parts to travel along the feed track towards the assembly workhead.

5. *Escapement* and *placement device*. The *escapement device* removes components from the feed track at time intervals that are consistent with the cycle time of the assembly

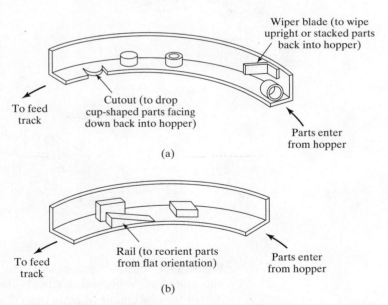

Figure 17.3 (a) Selector and (b) orientor devices used with parts feeders in automated assembly systems.

workhead. The *placement device* physically places the component in the correct location at the workstation for the assembly operation. These elements are sometimes combined into a single operating mechanism. In other cases, they are two separate devices. Several types of escapement and placement devices are pictured in Figure 17.4.

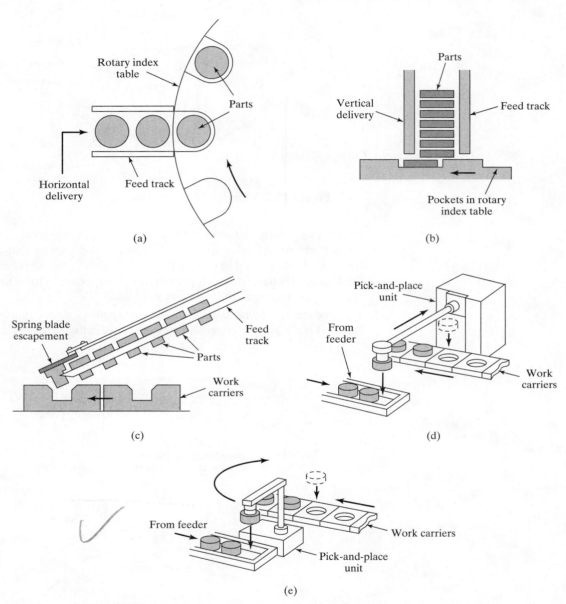

Figure 17.4 Various escapement and placement devices used in automated assembly systems: (a) horizontal device and (b) vertical device for placement of parts onto dial indexing table; (c) escapement of rivet-shaped parts actuated by work carriers, (d) and (e) two types of pick-and-place mechanisms (reprinted from Gay [6]).

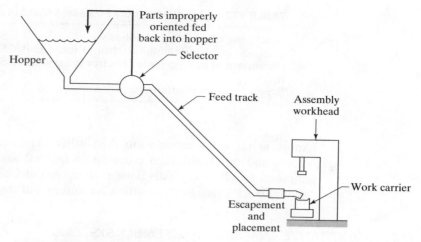

Figure 17.5 Hardware elements of the parts delivery system at an assembly workstation.

The hardware elements of the parts delivery system are illustrated schematically in Figure 17.5. A parts selector is illustrated in the diagram. Improperly oriented parts are returned to the hopper. In the case of a parts orientor, improperly oriented parts are reoriented and proceed to the feed track. A more detailed description of the various elements of the delivery system is provided in Boothroyd, Poli, and Murch [3].

One of the developments in the technology of parts feeding and delivery systems is the programmable parts feeder [7], [10]. A *programmable parts feeder* is capable of feeding components of varying geometries and needs only a few minutes to make the adjustments (change the program) for the differences. The flexibility of this type of feeder permits it to be used in batch production or when product design changes occur. Most parts feeders are designed as fixed automated systems for high production assembly of stable product designs.

17.1.3 Applications

Automated assembly systems are used to produce a wide variety of products and subassemblies. Table 17.1 presents a list of typical products made by automated assembly.

The kinds of operations performed on automated assembly machines cover a wide range. We provide a representative list of processes in Table 17.2. These processes are described in Groover [9]. It should be noted that certain assembly processes are

TABLE 17.1 Typical Products Made by Automated Assembly

Alarm clocks	Electrical plugs and sockets	Printed circuit board assemblies
Audio tape cassettes	Fuel injectors	Pumps for household appliances
Ball bearings	Gear boxes	Small electric motors
Ball point pens	Light bulbs	Spark plugs
Cigarette lighters	Locks	Wrist watches
Computer diskettes	Mechanical pens and pencils	

TABLE 17.2 Typical Assembly Processes Used in Automated Assembly Systems

Automatic dispensing of adhesive	Snap fitting
Insertion of components (electronic assembly)	Soldering
Placement of components (electronic assembly)	Spot welding
Riveting	Stapling
Screw fastening (automatic screwdriver)	Stitching

more suitable for automation than others. For example, threaded fasteners (screws, bolts, and nuts), although common in manual assembly, are a challenging assembly method to automate. This issue is discussed in Chapter 24, which also provides some guidelines for designing products for automated assembly.

17.2 QUANTITATIVE ANALYSIS OF ASSEMBLY SYSTEMS

Certain performance aspects of automated assembly systems can be studied using mathematical models. In this section, we develop models to analyze the following issues in automated assembly: (1) the parts delivery system at workstations, (2) multi-station automated assembly systems, (3) single-station automated assembly systems, and (4) partial automation.

17.2.1 Parts Delivery System at Workstations

In the parts delivery system, Figure 17.5, the parts feeding mechanism is capable of removing parts from the hopper at a certain rate f. These parts are assumed to be randomly oriented initially, and must be presented to the selector or orientor to establish the correct orientation. In the case of a selector, a certain proportion of the parts will be correctly oriented initially and these will be allowed to pass through. The remaining proportion that are incorrectly oriented will be rejected back to the hopper. In the case of an orientor, incorrectly oriented parts will be reoriented, resulting ideally in a 100% rate of parts passing through the device. In many delivery system designs, the functions of the selector and the orientor are combined. Let us define θ to be the proportion of components that pass through the selector-orientor process and are correctly oriented for delivery into the feed track. Hence the effective rate of delivery of components from the hopper into the feed track is $f\theta$. The remaining proportion, $(1 - \theta)$ is recirculated back into the hopper. Obviously, the delivery rate $f\theta$ of components to the workhead must be sufficient to keep up with the cycle rate of the assembly machine.

Assuming the delivery rate of components $f\theta$ is greater than the cycle rate R_c of the assembly machine, the system needs to have a means of limiting the size of the queue in the feed track. The usual solution is to place a sensor (for example, limit switch or optical sensor) near the top of the feed track to turn off the feeding mechanism when the feed track is full. This sensor is referred to as the high level sensor, and its location defines the active length L_{f2} of the feed track. If the length of a component in the feed track is L_c, then the number of parts that can be held in the feed track is $n_{f2} = L_{f2}/L_c$. The length of the components must be measured from a point on a given component to the corresponding point on the next component in the queue to allow for possible overlap of parts. The value of n_{f2} is the capacity of the feed track.

Another sensor placed along the feed track at some distance from the first sensor is used to restart the feeding mechanism. If we define the location of this low level sensor as L_{f1}, then the number of components in the feed track at this point is $n_{f1} = L_{f1}/L_c$.

The rate at which parts in the feed track are reduced when the high level sensor is actuated (turns off the feeder) $= R_c$, which is the cycle rate of the automated assembly workhead. On average, the rate at which the quantity of parts will increase upon actuation of the low level sensor (which turns on the feeder) is $f\theta - R_c$. However, the rate of increase will not be uniform due to the random nature of the feeder-selector operation. Accordingly, the value of n_{f1} must be large enough to virtually eliminate the probability of a stockout after the low level sensor has turned on the feeder.

EXAMPLE 17.1 Parts Delivery System in Automatic Assembly

The cycle time for a given assembly workhead $= 6$ sec. The parts feeder has a feed rate $= 50$ components/min. The probability that a given component fed by the feeder will pass through the selector is $\theta = 0.25$. The number of parts in the feed track corresponding to the low level sensor is $n_{f1} = 6$. The capacity of the feed track is $n_{f2} = 18$ parts. Determine (a) how long it will take for the supply of parts in the feed track to go from n_{f2} to n_{f1}, and (b) how long it will take on average for the supply of parts to go from n_{f1} to n_{f2}.

Solution: (a) $T_c = 6$ sec $= 0.1$ min. The rate of depletion of parts in the feed track starting from n_{f2} will be $R_c = 1/0.1 = 10$ parts/min

$$\text{Time to deplete feed track (time to go from } n_{f2} \text{ to } n_{f1}) = \frac{18 - 6}{10} = 1.2 \text{ min}$$

(b) The rate of parts increase in the feed track when the low level sensor is reached is $f\theta - R_c = (50)(0.25) - 10 = 12.5 - 10 = 2.5$ parts/min

$$\text{Time to replenish feed track (time go from } n_{f1} \text{ to } n_{f2}) = \frac{18 - 6}{2.5} = 4.8 \text{ min}$$

17.2.2 Multi-Station Assembly Machines

In this section, we analyze the operation and performance of automated assembly machines that have several workstations and use a synchronous transfer system. The types include the dial indexing machine, many in-line assembly systems, and certain carousel systems. Assumptions underlying the analysis are similar to those in our analysis of transfer lines: (1) assembly operations at the stations have constant element times, although the times are not necessarily equal at all stations; (2) synchronous parts transfer is used; and (3) there is no internal storage.

The analysis of an automated assembly machine with multiple stations shares much in common with the approach used for transfer lines in Section 16.3.1. Some modifications in the analysis must be made to account for the fact that components are being added at the various workstations in the assembly system. The general operations of the assembly systems are pictured in Figures 17.1(a), (b), and (c). In presenting the equations that describe these operations, we follow the approach developed by Boothroyd and Redford [2].

We assume that the typical operation at a workstation of an assembly machine consists of a component being added and/or joined in some fashion to an existing assembly.

The existing assembly consists of a base part plus the components assembled to it at previous stations. The base part is launched onto the line either at or before the first workstation. The components that are added must be clean, uniform in size and shape, of high quality, and consistently oriented. When the feed mechanism and assembly workhead attempt to join a component that does not satisfy this technical description, the station can jam. When a jam occurs, it results in the shutdown of the entire system until the fault is corrected. Thus, in addition to the other mechanical and electrical failures that interrupt the operation of a production line, the problem of defective components is one that specifically plagues the operation of an automatic assembly system.

The Assembly Machine as a Game of Chance. Defective parts occur in manufacturing with a certain fraction defect rate q $(0 \leq q \leq 1.0)$. In the operation of an assembly workstation, q is the probability that the component to be added during the current cycle is defective. When an attempt is made to feed and assemble a defective component, the defect might or might not cause the station to jam. Let m = probability that a defect results in a jam at the station and consequential stoppage of the line. Since the values of q and m may be different for different stations, we subscript these terms as q_i and m_i, where $i = 1, 2, \ldots n$, where n is the number of workstations on the assembly machine.

At a particular workstation, say station i, there are three possible events that might occur when the feed mechanism attempts to feed the next component and the assembly device attempts to join it to the existing assembly at the station.

1. *The component is defective and causes a station jam.* The probability of this event is the fraction defect rate of the parts at the station (q_i) multiplied by the probability that a defect will cause the station to jam (m_i). This product is the same term p_i in our previous analysis of transfer machines in Section 16.3.1. For an assembly machine, $p_i = m_i q_i$. When the station jams, the component must be cleared and the next component be allowed to feed and be assembled. We assume that if the next component in the feed track is defective, the operator who cleared the previous jam would notice and remove this next defect as well. Anyway, the probability of two consecutive defects is very small, equal to q_i^2.

2. *The component is defective but does not cause a station jam.* This has a probability $(1 - m_i)q_i$. With this outcome, a bad part is joined to the existing assembly, perhaps rendering the entire assembly defective.

3. *The component is not defective.* This is the most desirable outcome and the most likely by far (we hope). The probability that a part added at the station is not defective is equal to the proportion of good parts $(1 - q_i)$.

The probabilities of the three possible events must sum to unity for any workstation; that is,

$$m_i q_i + (1 - m_i)q_i + (1 - q_i) = 1 \tag{17.1}$$

For the special case where $m_i = m$ and $q_i = q$ for all i, this equation reduces to the following:

$$mq + (1 - m)q + (1 - q) = 1 \tag{17.2}$$

Although it is unlikely that all m_i are equal and all q_i are equal, the equation is nevertheless useful for computation and approximation purposes.

To determine the complete distribution of possible outcomes that can occur on an n-station assembly machine, the terms of Eq. (17.1) are multiplied together for all n stations:

$$\prod_{i=1}^{n} [m_i q_i + (1 - m_i)q_i + (1 - q_i)] = 1 \qquad (17.3)$$

In the special case where $m_i = m$ and $q_i = q$ for all i, this reduces to

$$[mq + (1 - m)q + (1 - q)]^n = 1 \qquad (17.4)$$

Expansion of Eq. (17.3) reveals the probabilities for all possible sequences of events that can take place on the n-station assembly machine. Regrettably, the number of terms in the expansion becomes very large for a machine with more than two or three stations. The exact number of terms is equal to 3^n, where n = number of stations. For example, for an eight-station line, the number of terms = $3^8 = 6561$, each term representing the probability of one of the 6561 possible outcome sequences on the assembly machine.

Measures of Performance. Fortunately, it is not necessary to calculate every term to use the description of assembly machine operation provided by Eq. (17.3). One of the characteristics of performance that we want to know is the proportion of assemblies that contain one or more defective components. Two of the three terms in Eq. (17.3) represent events in which a defective component is not added at the given station. The first term is $m_i q_i$, which indicates that a station jam has occurred, preventing a defective component from being added to the existing assembly. The other term is $(1 - q_i)$, which means that a good component has been added at the station. The sum of these two terms represents the probability that a defective component is not added at station i. Multiplying these probabilities for all stations, we get the proportion of acceptable product coming off the line

$$P_{ap} = \prod_{i=1}^{n} (1 - q_i + m_i q_i) \qquad (17.5)$$

where P_{ap} can be thought of as the *yield* of good assemblies produced by the assembly machine. If P_{ap} = the proportion of good assemblies, then the proportion of assemblies containing at least one defective component P_{qp} is given by

$$P_{qp} = 1 - P_{ap} = 1 - \prod_{i=1}^{n} (1 - q_i + m_i q_i) \qquad (17.6)$$

In the case of equal m_i and equal q_i, these two equations become, respectively,

$$P_{ap} = (1 - q + mq)^n \qquad (17.7)$$

$$P_{qp} = 1 - (1 - q + mq)^n \qquad (17.8)$$

The yield P_{ap} is certainly one of the important performance measures of an assembly machine. To have a certain proportion of assemblies with one or more defective components in the final output is a significant disadvantage. These assemblies must be identified through an inspection process and repaired, or they will become mixed in with the good assemblies, which would lead to undesirable consequences when the assemblies are placed in service.

Other performance measures of interest are the machine's production rate, the proportion of uptime and downtime, and the average cost per unit produced. To calculate production rate, we first determine the frequency of downtime occurrences per cycle F. If each station jam results in a machine downtime occurrence, F can be determined by taking the expected number of station jams per cycle; that is,

$$F = \sum_{i=1}^{n} p_i = \sum_{i=1}^{n} m_i q_i \qquad (17.9)$$

In the case of a station performing only a joining or fastening operation and not adding a part at the station, then the contribution to F made by that station is p_i, the probability of a station breakdown, where p_i does not depend on m_i and q_i.

If $m_i = m$ and $q_i = q$ for all stations, $i = 1, 2, \ldots, n$, then the above equation for F reduces to the following:

$$F = nmq \qquad (17.10)$$

The average actual production time per assembly is given by

$$T_p = T_c + \sum_{i=1}^{n} m_i q_i T_d \qquad (17.11)$$

where T_c = ideal cycle time of the assembly machine, which is the longest assembly task time on the machine plus the indexing or transfer time, min; and T_d = average downtime per occurrence, min. For the case of equal m_i and q_i,

$$T_p = T_c + nmq T_d \qquad (17.12)$$

From the average actual production time, we obtain the production rate, which is the reciprocal of production time:

$$R_p = \frac{1}{T_p} \qquad (17.13)$$

This is the same relationship as Eq. (16.9) in our previous chapter on transfer lines. However, the operation of assembly machines is different from processing machines. In an assembly machine, unless $m_i = 1.0$ for all stations, the production output will include some assemblies with one or more defective components. Accordingly, the production rate should be corrected to give the rate of acceptable product, that is, those that contain no defects. This is simply the yield P_{ap} multiplied by the production rate

$$R_{ap} = P_{ap} R_p = \frac{P_{ap}}{T_p} = \frac{\prod_{i=1}^{n}(1 - q_i + m_i q_i)}{T_p} \qquad (17.14)$$

where R_{ap} = production rate of acceptable product, units/min. When all m_i are equal and all q_i are equal, the corresponding equation is

$$R_{ap} = P_{ap} R_p = \frac{P_{ap}}{T_p} = \frac{(1 - q + mq)^n}{T_p} \qquad (17.15)$$

Eq. (17.13) gives the production rate of all assemblies made on the system, including those that contain one or more defective parts. Eqs. (17.14) and (17.15) give production rates for good product only. The problem still remains that the defective products are mixed in with the good units. We take up this issue of inspection and sortation in Chapter 21 (Section 21.5).

Line efficiency is calculated as the ratio of ideal cycle time to average actual production time. This is the same ratio as defined in Chapter 16, Eq. (16.11),

$$E = \frac{R_p}{R_c} = \frac{T_c}{T_p} \qquad (17.16)$$

where T_p is calculated from Eq. (17.11) or Eq. (17.12). The proportion downtime $D = 1 - E$, as before. No attempt has been made to correct line efficiency E for the yield of good assemblies. We are treating assembly machine efficiency and the quality of units produced on it as separate issues.

On the other hand, the cost per assembled product must take account of the output quality. Therefore, the general cost formula given in Eq. (16.14) in the previous chapter must be corrected for yield, as

$$C_{pc} = \frac{C_m + C_o T_p + C_t}{P_{ap}} \qquad (17.17)$$

where C_{pc} = cost per good assembly, $/pc; C_m = cost of materials, which includes the cost of the base part plus components added to it, $/pc; C_o = operating cost of the assembly system, $/min; T_p = average actual production time, min/pc; C_t = cost of disposable tooling, $/pc; and P_{ap} = yield from Eq. (17.5). The effect of the denominator is to increase the cost per assembly; as the quality of the individual components deteriorates, the average cost per good quality assembly increases.

In addition to the traditional ways of indicating line performance (production rate, line efficiency, cost per unit), we see an additional dimension of importance in the form of yield. While the yield of good product is an important issue in any automated production line, we see that it can be explicitly included in the formulas for assembly machine performance by means of q and m.

EXAMPLE 17.2 Multi-Station Automated Assembly System

A ten-station in-line assembly machine has an ideal cycle time = 6 sec. The base part is automatically loaded prior to the first station, and components are added at each of the stations. The fraction defect rate at each of the ten stations is $q = 0.01$, and the probability that a defect will jam is $m = 0.5$. When a jam occurs, the average downtime is 2 min. Cost to operate the assembly machine is $42.00/hr. Other costs are ignored. Determine (a) average production rate of all assemblies, (b) yield of good assemblies, (c) average production rate of good product, (d) uptime efficiency of the assembly machine, and (e) cost per unit.

Solution: (a) T_c = 6 sec = 0.1 min. The average production cycle time is

$$T_p = 0.1 + (10)(.5)(.01)(2.0) = 0.2 \text{ min}$$

The production rate is therefore

$$R_p = \frac{60}{0.2} = 300 \text{ total assemblies / hr}$$

(b) The yield is given by Eq. (17.7):

$$P_{ap} = \{1 - .01 + 0.5(0.01)\}^{10} = 0.9511$$

(c) Average actual production rate of good assemblies is determined by Eq. (17.15):

$$R_{ap} = 300(0.9511) = 285.3 \text{ good asbys/hr}$$

(d) The efficiency of the assembly machine is

$$E = 0.1/0.2 = 0.50 = 50\%$$

(e) Cost to operate the assembly machine $C_o = \$42/\text{hr} = \$0.70/\text{min}$

$$C_{pc} = (\$0.70/\text{min})(0.2 \text{ min/pc})/0.9511 = \$0.147/\text{pc}$$

EXAMPLE 17.3   Effect of Variations in q and m on Assembly System Performance

Let us examine how the performance measures in Example 17.2 are affected by variations in q and m. First, for $m = 0.5$, determine the production rate, yield, and efficiency for $q = 0$, $q = 0.01$, and $q = 0.02$. Second, for $q = 0.01$, determine the production rate, yield, and efficiency for $m = 0$, $m = 0.5$, and $m = 1.0$.

Solution: Computations similar to those in Example 17.2 provide the following results:

q	m	R_p (pc/hr)	Yield	R_{ap} (pc/hr)	E	C_{pc}
0	0.5	600	1.0	600	100%	$0.07
0.01	0.5	300	0.951	285	50%	$0.15
0.02	0.5	200	0.904	181	33.3%	$0.23
0.01	0	600	0.904	543	100%	$0.08
0.01	0.5	300	0.951	285	50%	$0.15
0.01	1.0	200	1.0	200	33.3%	$0.21

Let us discuss the results of Example 17.3. The effect of component quality, as indicated in the value of q, is predictable. As fraction defect rate increases, meaning that component quality gets worse, all five measures of performance suffer. Production rate drops, yield of good product is reduced, proportion uptime decreases, and cost per unit increases.

The effect of m (probability that a defect will jam the workhead and cause the assembly machine to stop) is less obvious. At low values of m ($m = 0$) for the same component quality level ($q = 0.01$), production rate and machine efficiency are high, but yield of good product is low. Instead of interrupting the assembly machine operation and causing downtime, all defective components pass through the assembly process to become part of the final product. At $m = 1.0$, all defective components are removed before they become part of the product. Therefore, yield is 100%, but removing the defects takes time, adversely affecting production rate, efficiency, and cost per unit.

17.2.3   Single-Station Assembly Machines

The single-station assembly system is depicted in Figure 17.1(d). We assume a single workhead, with several components feeding into the station to be assembled to a base part. Let n_e = the number of distinct assembly elements that are performed on the machine. Each element has an element time, T_{ej}, where $j = 1, 2, \ldots, n_e$. The ideal cycle time for the single-station assembly machine is the sum of the individual element times of the assembly operations to be performed on the machine, plus the handling time to load the base part into position and unload the completed assembly. We can express this ideal cycle time as

$$T_c = T_h + \sum_{j=1}^{n_e} T_{ej} \qquad (17.18)$$

where T_h = handling time, min.

Many of the assembly elements involve the addition of a component to the existing subassembly. As in our analysis of multiple station assembly, each component type has a certain fraction defect rate, q_j, and there is a certain probability that a defective component will jam the workstation, m_j. When a jam occurs, the assembly machine stops, and it takes an average T_d to clear the jam and restart the system. The inclusion of downtime resulting from jams in the machine cycle time gives

$$T_p = T_c + \sum_{j=1}^{n_e} q_j m_j T_d \qquad (17.19)$$

For elements that do not include the addition of a component, the value of $q_j = 0$ and m_j is irrelevant. This might occur, for example, when a fastening operation is performed with no part added during element j. In this type of operation, a term $p_j T_d$ would be included in the above expression to allow for a downtime during that element, where p_j = the probability of a station failure during element j. For the special case of equal q and equal m values for all components added, Eq. (17.19) becomes

$$T_p = T_c + nmq T_d \qquad (17.20)$$

Determining yield (proportion of assemblies that contain no defective components) for the single-station assembly machine makes use of the same equations as for the multiple station systems, Eqs. (17.5) or (17.7). Uptime efficiency is computed as $E = T_c / T_p$ using the values of T_c and T_p from Eqs. (17.18) and (17.19) or (17.20).

EXAMPLE 17.4 Single-Station Automatic Assembly System

A single-station assembly machine performs five work elements to assemble four components to a base part. The elements are listed in the table below, together with the fraction defect rate (q) and probability of a station jam (m) for each of the components added (NA means not applicable).

Element	Operation	Time	q	m	p
1	Add gear	4	0.02	1.0	
2	Add spacer	3	0.01	0.6	
3	Add gear	4	0.015	0.8	
4	Add gear and mesh	7	0.02	1.0	
5	Fasten	5	0	NA	0.012

Time to load the base part is 3 sec and time to unload the completed assembly is 4 sec, giving a total load/unload time of $T_h = 7$ sec. When a jam occurs, it takes an average of 1.5 minutes to clear the jam and restart the machine. Determine

(a) production rate of all product, (b) yield of good product, (c) production rate of good product, and (d) uptime efficiency of the assembly machine.

Solution: (a) The ideal cycle time of the assembly machine is

$$T_c = 7 + (4 + 3 + 4 + 7 + 5) = 30 \text{ sec} = 0.5 \text{ min}$$

Frequency of downtime occurrences is

$$F = 0.02(1.0) + 0.01(0.6) + 0.015(0.8) + 0.02(1.0) + 0.012 = 0.07$$

Adding the average downtime due to jams,

$$T_p = 0.5 + 0.07(1.5) = 0.5 + 0.105 = 0.605 \text{ min}$$

Production rate is therefore

$$R_p = 60/0.605 = 99.2 \text{ total assemblies/hr}$$

(b) Yield of good product is the following, from Eq. (17.5):

$$P_{ap} = \{1 - 0.02 + 1.0(0.02)\}\{1 - 0.01 + 0.6(0.01)\}$$
$$\{1 - 0.015 + 0.8(0.015)\}\{1 - 0.02 + 1.0(0.02)\}$$
$$= (1.0)(0.996)(0.997)(1.0) = 0.993$$

(c) Production rate of only good assemblies is

$$R_{ap} = 99.2(0.993) = 98.5 \text{ good assemblies/hr}$$

(d) Uptime efficiency is

$$E = 0.5/0.605 = 0.8264 = 82.64\%$$

As our analysis suggests, increasing the number of elements in the assembly machine cycle results in a longer cycle time, decreasing the production rate of the machine. Accordingly, applications of a single-station assembly machine are limited to lower volume, lower production rate situations. For higher production rates, one of the multi-station assembly systems is generally preferred.

17.2.4 Partial Automation

Many assembly lines in industry contain a combination of automated and manual work-stations. These cases of partially automated production lines occur for two main reasons:

1. *Automation is introduced gradually on an existing manual line.* Suppose that demand for the product made on a manually operated line increases, and the company decides to increase production and reduce labor costs by automating some or all of the stations. The simpler operations are automated first, and the transition toward a fully automated line is accomplished over a long period of time. Until then, the line operates as a partially automated system. (See Automation Migration Strategy, Section 1.4.3.)

2. *Certain manual operations are too difficult or too costly to automate.* Therefore, when the sequence of workstations is planned for the line, certain stations are designed to be automated while the others are designed as manual stations.

Examples of operations that might be too difficult to automate are assembly procedures or processing steps involving alignment, adjustment, or fine-tuning of the work unit. These operations often require special human skills and/or senses to carry out. Many inspection procedures also fall into this category. Defects in a product or part that can be easily perceived by a human inspector are sometimes extremely difficult to identify by an automated inspection device. Another problem is that the automated inspection device can only check for the defects for which it was designed, whereas a human inspector is capable of sensing a variety of unanticipated imperfections and problems.

To analyze the performance of a partially automated production line, we build on our previous analysis and make the following assumptions: (1) workstations perform either processing or assembly operations, (2) processing and assembly times at automated stations are constant, though not necessarily equal at all stations, (3) the system uses synchronous transfer of parts, (4) the system has no internal buffer storage, and (5) station breakdowns occur only at automated stations. Breakdowns do not occur at manual stations because the human workers are flexible enough, we assume, to adapt to the kinds of disruptions and malfunctions that would interrupt the operation of an automated workstation. For example, if a human operator were to retrieve a defective part from the parts bin at the station, the worker would immediately discard the part and select another without much lost time. Of course, this assumption of human adaptability is not always correct, but our analysis is based on it.

The ideal cycle time T_c is determined by the slowest station on the line, which is generally one of the manual stations. If the cycle time is in fact determined by a manual station, then T_c will exhibit a certain degree of variability, simply because there is random variation in any repetitive human activity. However, we assume that the average T_c remains constant over time. Given our assumption that breakdowns occur only at automated stations, let n_a = the number of automated stations and T_d = average downtime per occurrence. For the automated stations that perform processing operations, let p_i = the probability (frequency) of breakdowns per cycle, and for automated stations that perform assembly operations, let q_i and m_i equal, respectively, the defect rate and probability that the defect will cause station i to stop. We are now in a position to define the average actual production time:

$$T_p = T_c + \sum_{i \in n_a} p_i T_d \tag{17.21}$$

where the summation applies to the n_a automated stations only. For those automated stations that perform assembly operations in which a part is added,

$$p_i = m_i q_i$$

If all p_i, m_i, and q_i are equal, respectively to p, m, and q, then the preceding equations reduce to

$$T_p = T_c + n_a p T_d \tag{17.22}$$

and $p = mq$ for those stations that perform assembly consisting of the addition of a part.

Given that n_a is the number of automated stations, then n_w = the number of stations operated by workers, and $n_a + n_w = n$, where n = the total station count. Let C_{asi} = cost to operate automatic workstation i, \$/min; C_{wi} = cost to operate manual workstation i, \$/min; and C_{at} = cost to operate the automatic transfer mechanism. Then the total cost to operate the line is given by

$$C_o = C_{at} + \sum_{i \in n_a} C_{asi} + \sum_{i \in n_w} C_{wi} \tag{17.23}$$

where C_o = cost of operating the partially automated production system, \$/min. For all $C_{asi} = C_{as}$, and all $C_{wi} = C_w$, then

$$C_o = C_{at} + n_a C_{as} + n_w C_w \qquad (17.24)$$

Now the total cost per unit produced on the line can be calculated as

$$C_{pc} = \frac{C_m + C_o T_p + C_t}{P_{ap}} \qquad (17.25)$$

where C_{pc} = cost per good assembly, \$/pc; C_m = cost of materials and components being processed and assembled on the line, \$/pc; C_o = cost of operating the partially automated production system by either of Eqs. (17.23) or (17.24), \$/min; T_p = average actual production time, min/pc; C_t = any cost of disposable tooling, \$/pc; and P_{ap} = proportion of good assemblies by Eqs. (17.5) or (17.7).

EXAMPLE 17.5 Partial Automation

The company is considering replacing one of the current manual workstations with an automatic workhead on a 10-station production line. The current line has six automatic stations and four manual stations. Current cycle time is 30 sec. The limiting process time is at the manual station that is proposed for replacement. Implementing the proposal would allow the cycle time to be reduced to 24 sec. The new station would cost \$0.20/min. Other cost data: C_w = \$0.15/min, C_{as} = \$0.10/min, and C_{at} = \$0.12/min. Breakdowns occur at each automated station with a probability $p = 0.01$. The new automated station is expected to have the same frequency of breakdowns. Average downtime per occurrence $T_d = 3.0$ min, which will be unaffected by the new station. Material costs and tooling costs will be neglected in the analysis. It is desired to compare the current line with the proposed change on the basis of production rate and cost per piece. Assume a yield of 100% good product.

Solution: For the current line, $T_c = 30$ sec = 0.50 min

$$T_p = 0.50 + 6(0.01)(3.0) = 0.68 \text{ min}$$

$$R_p = 1/0.68 = 1.47 \text{ pc/min} = 88.2 \text{ pc/hr}$$

$$C_o = 0.12 + 4(0.15) + 6(0.10) = \$1.32/\text{min}$$

$$C_{pc} = 1.32(0.68) = \$0.898/\text{pc}$$

For the proposed line, $T_c = 24$ sec = 0.4 min

$$T_p = 0.40 + 7(0.01)(3.0) = 0.61 \text{ min}$$

$$R_p = 1/0.61 = 1.64 \text{ pc/min} = 98.4 \text{ pc/hr}$$

$$C_o = 0.12 + 3(0.15) + 6(0.10) + 1(0.20) = \$1.37/\text{min}$$

$$C_{pc} = 1.37(0.61) = \$0.836/\text{pc}$$

Even though the line would be more expensive to operate per unit time, the proposed change would increase production rate and reduce piece cost.

The preceding analysis assumes no buffer storage between stations. When the automated portion of the line breaks down, the manual stations must also stop due to either starving or blocking, depending on where the manual stations are located relative to the automated stations. Performance would be improved if the manual stations could continue to operate even when the automated stations stop for a temporary downtime incident. Storage buffers located before and after the manual stations would reduce forced downtime at these stations.

EXAMPLE 17.6   Storage Buffers on a Partially Automated Line

Considering the current line in previous Example 17.5, suppose that the ideal cycle time for the automated stations on the current line $T_c = 18$ sec. The longest manual time is 30 sec. Under the method of operation assumed in Example 17.5, both manual and automated stations are out of action when a breakdown occurs at an automated station. Suppose that storage buffers could be provided for every operator to insulate them from breakdowns at automated stations. What effect would this have on production rate and cost per piece?

Solution: Given $T_c = 18$ sec $= 0.3$ min, the average actual production time on the automated stations is computed as follows:

$$T_p = 0.30 + 6(0.01)(3.0) = 0.48 \text{ min}$$

Since this is less than the longest manual time of 0.50, the manual operations could work independently of the automated stations if storage buffers of sufficient capacity were placed before and after each manual station. Thus, the limiting cycle time on the line would be $T_c = 30$ sec $= 0.50$ min, and the corresponding production rate would be

$$R_p = R_c = 1/0.50 = 2.0 \text{ pc/min} = 120.0 \text{ pc/hr}$$

Using the line operating cost from Example 17.5, $C_o = \$1.32/\text{min}$, we have a piece cost of

$$C_{pc} = 1.32(0.50) = \$0.66/\text{pc}$$

When we compare this result with that in Example 17.5, we can see that storage buffers provide a dramatic improvement in production rate and unit cost.

17.2.5   What the Equations Tell Us

The equations derived in this section and their application in our examples reveal several practical guidelines for the design and operation of automated assembly systems and the products made on such systems.

- The parts delivery system at each station must be designed to deliver components to the assembly operation at a net rate (parts feeder multiplied by pass-through proportion of the selector/orientor) that is greater than or equal to the cycle rate of the assembly workhead. Otherwise, assembly system performance is limited by the parts delivery system rather than the assembly process technology.
- The quality of components added in an automated assembly system has a significant effect on system performance. Poor quality, as represented by the fraction defect rate, can result in
  1. Jams at stations that stop the entire assembly system, which has adverse effects on production rate, uptime proportion, and cost per unit produced; or

2. Assembly of defective parts in the product, which has adverse effects on yield of good assemblies and product cost.

- As the number of workstations increases in an automated assembly system, uptime efficiency and production rate tend to decrease due to parts quality and station reliability effects. This reinforces the need to use only the highest quality components on automated assembly systems.

- The cycle time of a multi-station assembly system is determined by the slowest station (longest assembly task) in the system. The number of assembly tasks to be performed is important only insofar as it affects the reliability of the assembly system. By comparison, the cycle time of a single-station assembly system is determined by the sum of the assembly element times rather than by the longest assembly element.

- Compared with a multi-station assembly machine, a single-station assembly system with the same number of assembly tasks has a lower production rate but a higher uptime efficiency.

- Multi-station assembly systems are appropriate for high production applications and long production runs. In comparison, single-station assembly systems have a longer cycle time and are more appropriate for mid-range quantities of product.

- Storage buffers should be used on partially automated production lines to isolate the manual stations from breakdowns of the automated stations. Use of storage buffers will increase production rates and reduce unit product cost.

- An automated station should be substituted for a manual station only if it reduces cycle time sufficiently to offset any negative effects of lower reliability.

REFERENCES

[1] ANDREASEN, M. M., S. KAHLER, and T. LUND, *Design for Assembly*, IFS (Publications) Ltd., U.K., and Springer-Verlag, Berlin, FRG, 1983.

[2] BOOTHROYD, G., and A. H. REDFORD, *Mechanized Assembly*, McGraw-Hill Publishing Company, Ltd., London, 1968.

[3] BOOTHROYD, G., C. POLI, and L. E. MURCH, *Automatic Assembly*, Marcel Dekker, Inc., NY, 1982.

[4] BOOTHROYD, G., P. Dewhurst, and W. Knight, *Product Design for Manufacture and Assembly*, Marcel Dekker, Inc., NY, 1994.

[5] DELCHAMBRE, A., *Computer-Aided Assembly Planning*, Chapman & Hall, London, UK, 1992.

[6] GAY, D. S., "Ways to Place and Transport Parts," *Automation*, June 1973.

[7] GOODRICH, J. L., and G. P. Maul, "Programmable Parts Feeders," *Industrial Engineering*, May 1983, pp. 28–33.

[8] GROOVER, M. P., M. WEISS, R. N. NAGEL, and N. G. ODREY, *Industrial Robotics: Technology, Programming, and Applications*, McGraw-Hill Book Company, NY, 1986, Chapter 15.

[9] GROOVER, M. P., *Fundamentals of Modern Manufacturing: Materials, Processes, and Systems*, 3d ed., John Wiley & Sons, Inc., Hoboken, NJ, 2007.

[10] MACKZKA, W. J., "Feeding the Assembly System," *Assembly Engineering*, April 1985, pp. 32–34.

[11] MURCH, L. E., and G. BOOTHROYD, "On-off Control of Parts Feeding," *Automation*, August 1970, pp. 32–34.

[12] NOF, S. Y., W. E. WILHELM, and H.-J. WARNECKE, *Industrial Assembly*, Chapman & Hall, London, UK, 1997.

[13] RILEY, F. J., *Assembly Automation*, Industrial Press Inc., NY, 1983.

[14] SCHWARTZ, W. H., "Robots Called to Assembly," *Assembly Engineering*, August 1985, pp. 20–23.

[15] SYNTRON (FMC Corporation), *Vibratory Parts Feeders*, FMC Corporation (Materials Handling Equipment Division), Homer City, PA.

[16] *Syntron Parts Handling Equipment*, Catalog No. PHE-10, FMC Corporation (Materials Handling Equipment Division), Homer City, PA.

[17] WARNECKE, H. J., M. SCHWEIZER, K. TAMAKI, and S. NOF, "Assembly," *Handbook of Industrial Engineering*, Institute of Industrial Engineers, John Wiley & Sons, Inc., NY, 1992, pp. 505–562.

REVIEW QUESTIONS

17.1 Name three of the four conditions under which automated assembly technology should be considered.

17.2 What are the four automated assembly system configurations listed in the text?

17.3 What are the typical hardware components of a workstation parts delivery system?

17.4 What is a programmable parts feeder?

17.5 Name six typical products that are made by automated assembly.

17.6 Considering the assembly machine as a game of chance, what are the three possible events that might occur when the feed mechanism attempts to feed the next component to the assembly workhead at a given workstation in a multi-station system?

17.7 Name some of the important performance measures for an automated assembly system.

17.8 Why is the production rate inherently lower on a single-station assembly system than on a multi-station assembly system?

17.9 What are two reasons for the existence of partially automated production lines?

17.10 What are the effects of poor quality parts, as represented by the fraction defect rate, on the performance of an automated assembly system?

17.11 Why are storage buffers used on partially automated production lines?

PROBLEMS

Parts Feeding

17.1 A feeder-selector device at one of the stations of an automated assembly machine has a feed rate of 25 parts per minute and provides a throughput of one part in four. The ideal cycle time of the assembly machine is 10 sec. The low level sensor on the feed track is set at 10 parts, and the high level sensor is set at 20 parts. (a) How long will it take for the supply of parts to be depleted from the high level sensor to the low level sensor once the feeder-selector device is turned off? (b) How long will it take for the parts to be resupplied from the low level sensor to the high level sensor, on average, after the feeder-selector device is turned on? (c) What proportion of the time that the assembly machine is operating will the feeder-selector device be turned on? Turned off?

17.2 Solve Problem 17.1 using a feed rate of 32 parts per minute. Note the importance of tuning the feeder-selector rate to the cycle rate of the assembly machine.

17.3 A synchronous assembly machine has eight stations and must produce at an average rate of 400 completed assemblies per hour. Average downtime per jam is 2.5 min. When a breakdown occurs, all subsystems (including the feeder) stop. The frequency of breakdowns of the machine is once every 50 parts. One of the eight stations is an automatic assembly operation that uses a feeder-selector. The components fed into the selector can have any of five possible orientations, each with equal probability, but only one of which is correct for passage into the feed track to the assembly workhead. Parts rejected by the selector are fed back into the hopper. What is the minimum rate at which the feeder must deliver components to the selector during system uptime in order to keep up with the assembly machine?

Multi-Station Assembly Systems

17.4 A dial indexing machine has six stations that perform assembly operations on a base part. The operations, element times, and q and m values for components added are given in the table below (NA means q and m are not applicable to the operation). The indexing time for the dial table is 2 sec. When a jam occurs, it requires 1.5 min to release the jam and put the machine back in operation. Determine (a) production rate for the assembly machine, (b) yield of good product (final assemblies containing no defective components), and (c) proportion uptime of the system.

Station	Operation	Element Time (sec)	q	m
1	Add part *A*	4	0.015	0.6
2	Fasten part *A*	3	NA	NA
3	Assemble part *B*	5	0.01	0.8
4	Add part *C*	4	0.02	1.0
5	Fasten part *C*	3	NA	NA
6	Assemble part *D*	6	0.01	0.5

17.5 An eight-station assembly machine has an ideal cycle time of 6 sec. The fraction defect rate at each of the eight stations is $q = 0.015$ and a defect always jams the affected station. When a breakdown occurs, it takes 1 minute, on average, for the system to be put back into operation. Determine the production rate for the assembly machine, the yield of good product (final assemblies containing no defective components), and proportion uptime of the system.

17.6 Solve Problem 17.5 assuming that defects never jam the workstations. Other data are the same.

17.7 Solve Problem 17.5 assuming that $m = 0.6$ for all stations. Other data are the same.

17.8 A six-station automatic assembly line has an ideal cycle time of 12 sec. Downtime occurs for two reasons. First, mechanical and electrical failures cause line stops that occur with a frequency of once per 50 cycles. Average downtime for these causes is 3 min. Second, defective components also result in downtime. The fraction defect rate of each of the six components added to the base part at the six stations is 2%. The probability that a defective component will cause a station jam is 0.5 for all stations. Downtime per occurrence for defective parts is 2 min. Determine (a) yield of assemblies that are free of defective components, (b) proportion of assemblies that contain at least one defective component, (c) average production rate of good product, and (d) uptime efficiency.

17.9 An eight-station automatic assembly machine has an ideal cycle time of 10 sec. Downtime is caused by defective parts jamming at the individual assembly stations. The average downtime per occurrence is 3.0 min. The fraction defect rate is 1.0% and the probability that a defective part will jam at a given station is 0.6 for all stations. The cost to operate the assembly machine is $90.00 per hour and the cost of components being assembled is $.60 per unit assembly. Ignore other costs. Determine (a) yield of good assemblies, (b) average production rate of good assemblies, (c) proportion of assemblies with at least one defective component, and (d) unit cost of the assembled product.

17.10 An automated assembly machine has four workstations. The first station presents the base part, and the other three stations add parts to the base. The ideal cycle time for the machine is 3 sec, and the average downtime when a jam results from a defective part is 1.5 min. The fraction defective rates (q) and probabilities that a defective part will jam the station (m) are given in the table below. Quantities of 100,000 for each of the bases, brackets, pins, and retainers are used to stock the assembly line for operation. Determine (a) proportion of good products to total products coming off the line, (b) production rate of good products coming off the line, and (c) total number of final assemblies produced, given the starting component quantities. Of the total, how many are good products, and how many are products that contain at least one defective component? (d) Of the number of defective assemblies determined in part (c), how many will have defective base parts? How many will have defective brackets? How many will have defective pins? How many will have defective retainers?

Station	Part Identification	q	m
1	Base	0.01	1.0
2	Bracket	0.02	1.0
3	Pin	0.03	1.0
4	Retainer	0.04	0.5

17.11 A six-station automatic assembly machine has an ideal cycle time of 6 sec. At stations 2 through 6, which are identical, parts feeders deliver identical components to be assembled to a base part that is added at the first station. That is, the completed product consists of the base part plus the five components. The base parts have zero defects, but the other components are defective at a rate q. When an attempt is made to assemble a defective component to the base part, the machine stops ($m = 1.0$). It takes an average of 2.0 min to make repairs and start the machine up after each stoppage. Since all components are identical, they are purchased from a supplier who can control the fraction defect rate very closely. However, the supplier charges a premium for better quality. The cost per component is determined by the equation

$$\text{Cost per component} = 0.1 + \frac{0.0012}{q}$$

where q = the fraction defect rate. Cost of the base part is 20 cents. Accordingly, the total cost of the base part and the five components is

$$\text{Product material cost} = 0.70 + \frac{0.006}{q}$$

The cost to operate the automatic assembly machine is $150.00 per hour. The problem facing the production manager is this: As the component quality decreases (q increases), the downtime increases which drives production costs up. As the quality improves (q decreases), the material cost increases because of the price formula used by the supplier. To minimize total cost, the optimum value of q must be determined. Determine by analytical methods (rather than trial-and-error) the value of q that minimizes the total cost per assembly. Also, determine the associated cost per assembly and production rate. Ignore other costs.

17.12 A six-station dial indexing machine is designed to perform four assembly operations at stations 2 through 5 after a base part has been manually loaded at station 1. Station 6 is the unload station. Each assembly operation involves the attachment of a component to the existing base. At each of the four assembly stations, a hopper-feeder is used to deliver components to a selector device that separates components that are improperly oriented and drops them back into the hopper. The system was designed with the operating parameters for stations 2 through 5 as given in the table below. It takes 2 sec to index the dial from one station position to the next. When a component jam occurs, it takes an average of 2 min to release the jam and restart the system. Line stops due to mechanical and electrical failures of the assembly machine are not significant and can be neglected. The foreman says the system was designed to produce at a certain hourly rate, which takes into account the jams resulting from defective components. However, the actual delivery of finished assemblies is far below that designed production rate. Analyze the problem and determine (a) the designed average production rate that the foreman alluded to, (b) the proportion of assemblies coming off the system that contain one or more defective components, (c) the problem that limits the assembly system from achieving the expected production rate, and (d) the production rate that the system is actually achieving. State any assumptions that you make in determining your answer.

Station	Assembly Time (sec)	Feed Rate f (pc/min)	Selector θ	q	m
2	4	32	0.25	0.01	1.0
3	7	20	0.50	0.005	0.6
4	5	20	0.20	0.02	1.0
5	3	15	1.0	0.01	0.7

17.13 For Example 17.4 in the text, dealing with a single-station assembly system, suppose that the sequence of assembly elements was to be accomplished on a seven-station assembly system with synchronous parts transfer. Each element is performed at a separate station (stations 2 through 6) and the assembly time at each respective station is the same as the element time given in Example 17.4. Assume that the handling time is divided evenly (3.5 sec each) between a load station (station 1) and an unload station (station 7). The transfer time is 2 sec, and the average downtime per downtime occurrence is 2.0 min. Determine (a) production rate of all completed units, (b) yield, (c) production rate of good quality completed units, and (d) uptime efficiency.

Single-Station Assembly Systems

17.14 A single-station assembly machine is considered as an alternative to the dial indexing machine in Problem 17.4. Use the data given in the table for that problem to determine (a) production rate, (b) yield of good product (final assemblies containing no defective components), and (c) proportion uptime of the system. Handling time to load the base part and unload the finished assembly is 7 sec and the downtime averages 1.5 min every time a component jams. Why is the proportion uptime so much higher than in the case of the dial indexing machine in Problem 17.4?

17.15 A single station robotic assembly system performs a series of five assembly elements, each of which adds a different component to a base part. Each element takes 4.5 sec. In addition, the handling time needed to move the base part into and out of position is 4 sec. For identification, the components and the elements that assemble them are numbered 1, 2, 3, 4, and 5. The fraction defect rate is 0.005 for all components, and the probability of a jam by a defective component is 0.7. Average downtime per occurrence = 2.5 min. Determine (a) production rate, (b) yield of good product in the output, (c) uptime efficiency, and (d) proportion of the output that contains a defective type 3 component.

17.16 A robotic assembly cell uses an industrial robot to perform a series of assembly operations. The base part and parts 2 and 3 are delivered by vibratory bowl feeders that use selectors to ensure that only properly oriented parts are delivered to the robot for assembly. The robot cell performs the elements in the table below (also given are feeder rates, selector proportion θ, element times, fraction defect rate q, and probability of jam m, and, for the last element, the frequency of downtime incidents p). In addition to the times given in the table, the time required to unload the completed subassembly is 4 sec. When a linestop occurs, it takes an average of 1.8 min to make repairs and restart the cell. Determine (a) yield of good product, (b) average production rate of good product, and (c) uptime efficiency for the cell. State any assumptions you must make about the operation of the cell in order to solve the problem.

Element	Feed Rate f (pc/min)	Selector θ	Element	Time T_e (sec)	q	m	p
1	15	0.30	Load base part	4	0.01	0.6	
2	12	0.25	Add part 2	3	0.02	0.3	
3	25	0.10	Add part 3	4	0.03	0.8	
4			Fasten	3			0.02

Partial Automation

17.17 A partially automated production line has three mechanized and three manual workstations. The ideal cycle time is 1.0 min, which includes a transfer time of 6 sec. Data on the six stations are listed in the accompanying table. Cost of the transfer mechanism C_{at} = $0.10/min, cost to run each automated station C_{as} = $0.12/min, and labor cost to operate each manual station C_w = $0.17/min. The company is considering substituting an automated station in place of station 5. The cost of this station is estimated at C_{as5} = $0.25/min and its breakdown rate p_5 = 0.02, but its process time would be only 30 sec, thus reducing the overall cycle time of the line from 1.0 min to 36 sec. Average downtime per breakdown of the current line, as well as for the proposed configuration, is 3.5 min. Determine the following for the current line and the proposed line: (a) production rate, (b) proportion uptime, and (c) cost per unit. Assume the line operates without storage buffers, so when an automated station stops, the whole line stops, including the manual stations. When computing costs, neglect material and tooling costs.

Station	Type	Process Time (sec)	p_i
1	Manual	36	0
2	Automatic	15	0.01
3	Automatic	20	0.02
4	Automatic	25	0.01
5	Manual	54	0
6	Manual	33	0

17.18 Reconsider Problem 17.17 assuming that both the current line and the proposed line will have storage buffers before and after the manual stations. The storage buffers will be of sufficient capacity to allow these manual stations to operate independently of the automated portions of the line. Determine (a) production rate, (b) proportion uptime, and (c) cost per unit for the current line and the proposed line.

17.19 A manual assembly line has six stations. The assembly time at each manual station is 60 sec. Parts are transferred by hand from one station to the next, and the lack of discipline in this method adds 12 sec ($T_r = 12$ sec) to the cycle time. Hence, the current cycle time is 72 sec. The following two proposals are made: (1) install a mechanized transfer system to pace the line, and (2) automate one or more of the manual stations using robots that would perform the tasks faster than humans. The second proposal requires the mechanized transfer system of the first proposal to be installed and would result in a partially or fully automated assembly line. The transfer system would have a transfer time of 6 sec, thus reducing the cycle time on the manual line to 66 sec. Regarding the second proposal, all six stations are candidates for automation. Each automated station would have an assembly time of 30 sec. Thus if all six stations were automated the cycle time for the line would be 36 sec. There are differences in the quality of parts added at the stations; these data are given in the accompanying table for each station (q = fraction defect rate, m = probability that a defect will jam the station). Average downtime per station jam at the automated stations is 3.0 min. Assume that the manual stations do not experience line stops due to defective components. Cost data: $C_{at} = \$0.10/\text{min}$; $C_w = \$0.20/\text{min}$; and $C_{as} = \$0.15/\text{min}$. Determine if either or both of the proposals should be accepted. If the second proposal is accepted, how many stations should be automated and which ones? Use cost per piece as the criterion of your decision. Assume for all cases considered that the line operates without storage buffers, so when an automated station stops, the whole line stops, including the manual stations.

Station	q_i	m_i	Station	q_i	m_i
1	0.005	1.0	4	0.020	1.0
2	0.010	1.0	5	0.025	1.0
3	0.015	1.0	6	0.030	1.0

17.20 Solve Problem 17.19, assuming that the probability that a defective part will jam the automated station is $m = 0.5$ for all stations.

Chapter 18

Cellular Manufacturing

CHAPTER CONTENTS

Batch manufacturing is estimated to be the most common form of production in the United States, constituting more than 50% of total manufacturing activity. It is important to make batch manufacturing as efficient and productive as possible. In addition, there has been a trend to integrate the design and manufacturing functions in a firm. An approach directed at both of these objectives is group technology (GT).

Group technology is a manufacturing philosophy in which similar parts are identified and grouped together to take advantage of their similarities in design and production. Similar parts are arranged into part families, where each *part family* possesses similar design and/or manufacturing characteristics. For example, a plant producing 10,000 different part numbers may be able to group the vast majority of these parts into 30 or 40 distinct families.

From Chapter 18 of *Automation, Production Systems, and Computer-Integrated Manufacturing*, Third Edition.
Mikell P. Groover. Copyright © 2008 by Pearson Education, Inc. Publishing as Prentice Hall. All rights reserved.

It is reasonable to believe that the processing of each member of a given family is similar, and this should result in manufacturing efficiencies. The efficiencies are generally achieved by arranging the production equipment into machine groups, or cells, to facilitate work flow. Organizing the production equipment into machine cells, where each cell specializes in the production of a part family, is called *cellular manufacturing*. Cellular manufacturing is an example of mixed model production (Section 13.2.5). The origins of group technology and cellular production can be traced to around 1925 (Historical Note 18.1).

Group technology and cellular manufacturing are applicable to a wide variety of manufacturing situations. GT is most appropriate under the following conditions:

- *The plant currently uses traditional batch production and a process type layout*, which results in much material handling effort, high in-process inventory, and long manufacturing lead times.

- *The parts can be grouped into part families*. This is a necessary condition. Each machine cell is designed to produce a given part family, or a limited collection of part families, so it must be possible to group parts made in the plant into families. Fortunately, in the typical mid-volume production plant, most of the parts can be grouped into part families.

There are two major tasks that a company must undertake when it implements group technology. These two tasks represent significant obstacles to the application of GT.

1. *Identifying the part families*. If the plant makes 10,000 different parts, reviewing all of the part drawings and grouping the parts into families is a substantial and time-consuming task.

Historical Note 18.1 Group technology

In 1925, R. Flanders of the United States presented a paper before the American Society of Mechanical Engineers that described a way of organizing manufacturing at Jones and Lamson Machine Company that would today be called group technology. In 1937, A. Sokolovskiy of the (former) Soviet Union described the essential features of group technology by proposing that parts of similar configuration be produced by a standard process sequence, thus permitting flow line techniques to be used for work normally accomplished by batch production. In 1949, A. Korling of Sweden presented a paper in Paris on "group production," whose principles are an adaptation of production line techniques to batch manufacturing. In the paper, he described how to decentralize work into independent groups, each containing the machines and tooling to produce "a special category of parts."

In 1959, researcher S. Mitrofanov of the Soviet Union published a book titled *Scientific Principles of Group Technology*. The book was widely read and is considered responsible for over 800 plants in the Soviet Union using group technology by 1965. Another researcher, H. Opitz in Germany, studied workparts manufactured by the German machine tool industry and developed the well-known parts classification and coding system for machined parts that bears his name (Section 18.2.2).

In the United States, the first application of group technology was at the Langston Division of Harris-Intertype in New Jersey around 1969. Traditionally a machine shop arranged as a process type layout, the company reorganized into "family of parts" lines, each of which specialized in producing a given part configuration. Part families were identified by taking photos of about 15% of the parts made in the plant and grouping them into families. When the changes were implemented, they improved productivity by 50% and reduced lead times from weeks to days.

2. *Rearranging production machines into machine cells.* It is time-consuming and costly to plan and accomplish this rearrangement, and the machines are not producing during the changeover.

Group technology offers substantial benefits to companies that have the perseverance to implement it.

- GT promotes standardization of tooling, fixturing, and setups
- Material handling is reduced because the distances within a machine cell are much shorter than within the entire factory
- Process planning and production scheduling are simplified
- Setup times are reduced, resulting in lower manufacturing lead times
- Work-in-process is reduced
- Worker satisfaction usually improves when workers collaborate in a GT cell
- Higher quality work is accomplished using group technology.

In this chapter, we discuss group technology, cellular manufacturing, and several related topics. Let us begin by defining an underlying concept in group technology: part families.

18.1 PART FAMILIES

A part family is a collection of parts that are similar either in geometric shape and size or in the processing steps required in their manufacture. The parts within a family are different, but their similarities are close enough to merit their inclusion as members of the part family. Figures 18.1 and 18.2 show two different part families. The two parts in Figure 18.1 are very similar in terms of geometric design, but quite different in terms of manufacturing because of differences in tolerances, production quantities, and materials. The parts shown in Figure 18.2 constitute a part family in manufacturing, but their different geometries make them appear quite different from a design viewpoint.

One of the important manufacturing advantages of grouping workparts into families can be explained with reference to Figures 18.3 and 18.4. Figure 18.3 shows a process type plant layout for batch production in a machine shop. The various machine tools are arranged by function. There is a lathe department, milling machine department, drill press department, and so on. To machine a given part, the workpiece

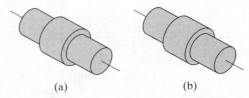

(a)   (b)

Figure 18.1 Two parts of identical shape and size but different manufacturing requirements: (a) 1,000,000 pc/yr, tolerance = ±0.010 in., material = 1015 CR steel, nickel plate; and (b) 100 pc/yr, tolerance = ±0.001 in., material = 18–8 stainless steel.

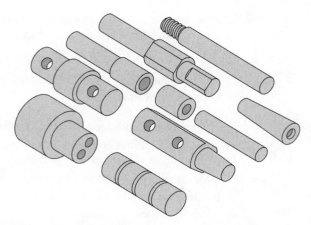

Figure 18.2 A family of parts with similar manufacturing process requirements but different design attributes. All parts are machined from cylindrical stock by turning; some parts require drilling and/or milling.

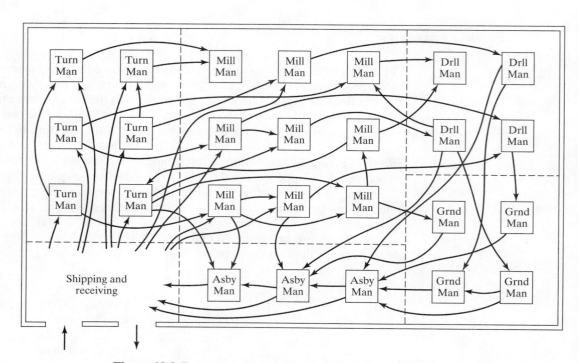

Figure 18.3 Process type plant layout. (Key: "Turn" = turning, "Mill" = milling, "Drll" = drilling, "Grnd" = grinding, "Asby" = assembly, "Man" = manual operation; arrows indicate work flow through plant, dashed lines indicate separation of machines into departments.)

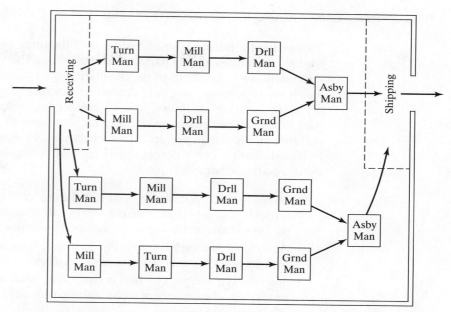

Figure 18.4 Group technology layout. (Key: "Turn" = turning, "Mill" = milling, "Drll" = drilling, "Grnd" = grinding, "Asby" = assembly, "Man" = manual operation; arrows indicate work flow in machine cells.)

must be transported between departments, perhaps visiting the same department several times. This results in much material handling, large in-process inventories, many machine setups, long manufacturing lead times, and high cost. Figure 18.4 shows a production shop of equivalent capacity, that has its machines arranged into cells. Each cell is organized to specialize in the production of a particular part family. Advantages are gained in the form of reduced workpiece handling, lower setup times, fewer setups (in some cases, no setup changes are necessary), less in-process inventory, and shorter lead times.

The biggest single obstacle in changing over to group technology from a conventional production shop is the problem of grouping the parts into families. There are three general methods for solving this problem. All three are time consuming and involve the analysis of much data by properly trained personnel. The three methods are (1) visual inspection, (2) parts classification and coding, and (3) production flow analysis. Let us provide a brief description of the visual inspection method and then examine the second and third methods in more detail.

The *visual inspection* method is the least sophisticated and least expensive method. It involves the classification of parts into families by looking at either the physical parts or their photographs and arranging them into groups having similar features. Although this method is generally considered the least accurate of the three, it was the method used in one of the first major success stories of GT in the United States, the Langston Division of Harris Intertype in Cherry Hill, New Jersey [20] (Historical Note 18.1).

18.2 PARTS CLASSIFICATION AND CODING

This method is the most time consuming of the three methods. In *parts classification and coding*, similarities among parts are identified and these similarities are related in a coding system. Two categories of part similarities can be distinguished: (1) design attributes, which are concerned with part characteristics such as geometry, size, and material, and (2) manufacturing attributes, which consider the processing steps required to make a part. While the design and manufacturing attributes of a part are usually correlated, the correlation is less than perfect. Accordingly, classification and coding systems are devised to include both a part's design attributes and its manufacturing attributes. Reasons for using a coding scheme include

- *Design retrieval.* A designer faced with the task of developing a new part can use a design retrieval system to determine if a similar part already exists. Simply changing an existing part would take much less time than designing a whole new part from scratch.
- *Automated process planning.* The part code for a new part can be used to search for process plans for existing parts with identical or similar codes.
- *Machine cell design.* The part codes can be used to design machine cells capable of producing all members of a particular part family, using the composite part concept (Section 18.4.1).

To accomplish parts classification and coding, an analyst must examine the design and/or manufacturing attributes of each part. The examination is sometimes done by looking in tables to match the subject part against the features described and diagrammed in the tables. An alternative and more productive approach involves using a computerized classification and coding system, in which the user responds to questions asked by the computer. On the basis of the responses, the computer assigns the code number to the part. Whichever method is used, the classification results in a code number that uniquely identifies the part's attributes.

The classification and coding procedure may be carried out on the entire list of active parts produced by the firm, or some sort of sampling procedure may be used to establish part families. For example, parts produced in the shop during a certain time period could be examined to identify part family categories. The trouble with any sampling procedure is the risk that the sample may not be representative of the population.

A number of classification and coding systems are described in the literature [15], [18], and [30], and a number of commercial coding packages have been developed. However, none of the systems has been universally adopted. One reason is that a classification and coding system must be customized for each company, because each company's products are unique. A system that is best for one company may not be best for another company.

18.2.1 Features of Parts Classification and Coding Systems

The principal functional areas that utilize a parts classification and coding system are design and manufacturing. Accordingly, parts classification systems fall into one of three categories:

1. Systems based on part design attributes
2. Systems based on part manufacturing attributes
3. Systems based on both design and manufacturing attributes.

Table 18.1 presents a list of the common design and manufacturing attributes typically included in classification schemes. A certain amount of overlap exists between design and manufacturing attributes, since a part's geometry is largely determined by the sequence of manufacturing processes that are performed on it.

In terms of the meaning of the symbols in the code, there are three structures used in classification and coding schemes:

1. *Hierarchical structure*, also known as a *monocode*, in which the interpretation of each successive symbol depends on the value of the preceding symbols

2. *Chain-type structure*, also known as a *polycode*, in which the interpretation of each symbol in the sequence is always the same; it does not depend on the value of preceding symbols

3. *Mixed-mode structure*, a hybrid of the two previous coding schemes.

To distinguish the hierarchical and chain-type structures, consider a two-digit code number for a part, such as 15 or 25. Suppose the first digit stands for the general shape of the part: 1 means the part is cylindrical (rotational), and 2 means the geometry is rectangular. In a hierarchical structure, the interpretation of the second digit depends on the value of the first digit. If preceded by 1, the 5 might indicate a length-to-diameter ratio; and if preceded by 2, the 5 might indicate an aspect ratio between the length and width dimensions of the part. In the chain-type structure, the symbol 5 would have the same meaning whether preceded by 1 or 2. For example, it might indicate the overall length of the part. The advantage of the hierarchical structure is that in general more information can be included in a code of a given number of digits. The mixed-mode structure uses a combination of hierarchical and chain-type structures. It is the most common structure found in GT parts classification and coding systems.

The number of digits in the code can range between 6 and 30. Coding schemes that contain only design data require fewer digits, perhaps 12 or fewer. Most modern classification and coding systems include both design and manufacturing data, and this usually requires 20 to 30 digits. This might seem like too many digits for a human reader to easily comprehend, but most of the data processing of the codes is accomplished by computer, for which a large number of digits is of minor concern.

TABLE 18.1 Design and Manufacturing Attributes Typically Included in a Group Technology Classification and Coding System

Part Design Attributes	Part Manufacturing Attributes
Basic external shape	Major processes
Basic internal shape	Minor operations
Rotational or rectangular shape	Operation sequence
Length-to-diameter ratio (rotational parts)	Major dimension
Aspect ratio (rectangular parts)	Surface finish
Material types	Machine tool
Part function	Production cycle time
Major dimensions	Batch size
Minor dimensions	Annual production
Tolerances	Fixtures required
Surface finish	Cutting tools used in manufacture

18.2.2 The Opitz Parts Classification and Coding System

The Opitz system is of interest because it was one of the first published classification and coding schemes for mechanical parts [29] (Historical Note 18.1) and is still widely used. It was developed by H. Opitz of the University of Aachen in Germany and represents one of the pioneering efforts in group technology. It is probably the best known, if not the most frequently used, of the parts classification and coding systems. It is intended for machined parts. The Opitz coding scheme uses the following digit sequence:

$$12345 \ 6789 \ ABCD$$

The basic code consists of nine digits, which can be extended by adding four more digits. The first nine are intended to convey both design and manufacturing data. The interpretation of the first nine digits is defined in Figure 18.5. The first five digits, 12345, are called the *form code*. This describes the primary design attributes of the part, such as external shape (for example, rotational versus rectangular) and machined features (for example, holes, threads, gear teeth, and so forth). The next four digits, 6789, constitute the *supplementary code*, which indicates some of the attributes that would be useful in manufacturing (for example, dimensions, work material, starting shape, and accuracy). The extra four digits, ABCD, are referred to as the *secondary code* and are intended to identify the production operation type and sequence. The secondary code can be designed by the user firm to serve its own particular needs.

The complete coding system is too complex to provide a comprehensive description here. Opitz wrote an entire book on his system [29]. However, to obtain a general

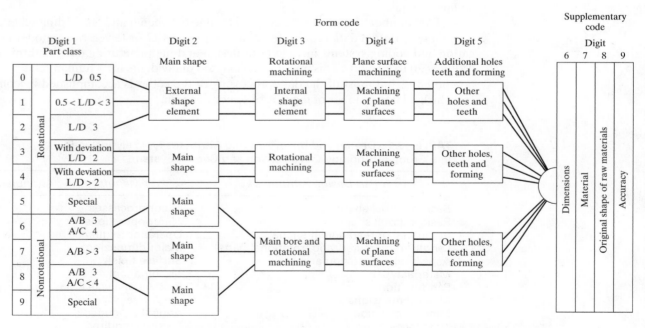

Figure 18.5 Basic structure of the Opitz system of parts classification and coding.

	Digit 1		Digit 2		Digit 3		Digit 4		Digit 5
	Part class		**External shape, external shape elements**		**Internal shape, internal shape elements**		**Plane surface machining**		**Auxiliary holes and gear teeth**
0	L/D 0.5	0	Smooth, no shape elements	0	No hole, no breakthrough	0	No surface machining	0	No auxiliary hole
1	0.5 < L/D < 3	1	No shape elements	1	No shape elements	1	Surface plane and/or curved in one direction, external	1	Axial, not on pitch circle diameter
2	L/D 3	2	Thread	2	Thread	2	External plane surface related by graduation around the circle	2	Axial on pitch circle diameter
3		3	Functional groove	3	Functional groove	3	External groove and/or slot	3	Radial, not on pitch circle diameter
4		4	No shape elements	4	No shape elements	4	External spline (polygon)	4	Axial and/or radial and/or other direction
5		5	Thread	5	Thread	5	External plane surface and/or slot, external spline	5	Axial and/or radial on PCD and/or other directions
6		6	Functional groove	6	Functional groove	6	Internal plane surface and/or slot	6	Spur gear teeth
7		7	Functional cone	7	Functional cone	7	Internal spline (polygon)	7	Bevel gear teeth
8		8	Operating thread	8	Operating thread	8	Internal and external polygon, groove and/or slot	8	Other gear teeth
9		9	All others	9	All others	9	All others	9	All others

Digit 1 labels: Rotational parts (0–6), Nonrotational parts (7–9).
Digit 2 labels: Stepped to one end or smooth (1–3), Stepped to both ends (4–6).
Digit 3 labels: Smooth or stepped to one end (1–3), Stepped to both ends (4–6).
Digit 5 labels: No gear teeth (0–5), With gear teeth (6–9).

Figure 18.6 Form code (digits 1 through 5) for rotational parts in the Opitz coding system. The first digit of the code is limited to values of 0, 1, or 2.

idea of how it works, let us examine the form code consisting of the first five digits, defined generally in Figure 18.5. The first digit identifies whether the part is rotational or nonrotational. It also describes the general shape and proportions of the part. We limit our survey here to rotational parts that do not possess any unusual features, those with first digit values of 0, 1, or 2. For this class of workparts, the coding of the first five digits is defined in Figure 18.6. Consider the following example to demonstrate the coding of a given part.

EXAMPLE 18.1   Opitz Part Coding System

Given the rotational part design in Figure 18.7, determine the form code in the Opitz parts classification and coding system.

Solution: With reference to Figure 18.6, the five-digit code is developed as follows:

Length-to-diameter ratio, L/D=1.5	Digit 1=1
External shape: stepped on both ends with screw thread on one end	Digit 2=5
Internal shape: part contains a through-hole	Digit 3=1
Plane surface machining: none	Digit 4=0
Auxiliary holes, gear teeth, etc.: none	Digit 5=0

The form code in the Opitz system is 15100.

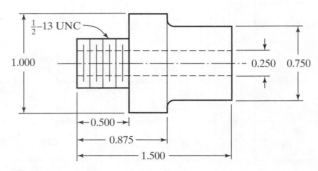

Figure 18.7 Part design for Example 18.1.

18.3 PRODUCTION FLOW ANALYSIS

Production flow analysis (PFA) is an approach to part family identification and machine cell formation that was pioneered by J. Burbidge [7], [8], [9]. It is a method for identifying part families and associated machine groupings that uses the information contained on production route sheets rather than part drawings. Workparts with identical or similar routings are classified into part families. These families can then be used to form logical machine cells in a group technology layout. Since PFA uses manufacturing data rather than design data to identify part families, it can overcome two possible anomalies that can occur in parts classification and coding. First, parts whose basic geometries are quite different may nevertheless require similar or even identical process routings. Second, parts whose geometries are quite similar may nevertheless require process routings that are quite different.

The procedure in production flow analysis must begin by defining the scope of the study, which means deciding on the population of parts to be analyzed. Should all of the parts in the shop be included in the study, or should a representative sample be selected for analysis? Once this decision is made, then the procedure in PFA consists of the following steps:

1. *Data collection.* The minimum data needed in the analysis are the part number and operation sequence, which is contained in shop documents called route sheets or operation sheets or some similar name. Each operation is usually associated with a particular machine, and so determining the operation sequence also determines the machine sequence.

2. *Sortation of process routings.* In this step, the parts are arranged into groups according to the similarity of their process routings. To facilitate this step, all operations or

TABLE 18.2 Possible Code Numbers Indicating Operations and/or Machines for Sortation in Production Flow Analysis (Highly Simplified)

Operation or Machine	Code
Cutoff	01
Lathe	02
Turret lathe	03
Mill	04
Drill—manual	05
NC drill	06
Grind	07

machines included in the shop are reduced to code numbers, such as those shown in Table 18.2. For each part, the operation codes are listed in the order in which they are performed. A sortation procedure is then used to arrange parts into "packs," which are groups of parts with identical routings. Some packs may contain only one part number, indicating the uniqueness of the processing of that part. Other packs will contain many parts, and these will constitute a part family.

3. *PFA chart.* The processes used for each pack are then displayed in a PFA chart, a simplified example of which is illustrated in Table 18.3.[1] The chart is a tabulation of the process or machine code numbers for all of the part packs. In recent GT literature [28], the PFA chart has been referred to by the term *part-machine incidence matrix.* In this matrix, the entries have a value $x_{ij} = 1$ or 0: a value of $x_{ij} = 1$ indicates that the corresponding part i requires processing on machine j, and $x_{ij} = 0$ indicates that no processing of component i is accomplished on machine j. For clarity in presenting the matrix, the 0's are often indicated as blank (empty) entries, as in our table.

4. *Cluster analysis.* From the pattern of data in the PFA chart, related groupings are identified and rearranged into a new pattern that brings together packs with similar machine sequences. One possible rearrangement of the original PFA chart is shown

TABLE 18.3 PFA Chart, Also Known as a Part-Machine Incidence Matrix

Machines (j)	Parts (i) A	B	C	D	E	F	G	H	I
1	1			1				1	
2					1				1
3			1		1				1
4		1				1			
5	1							1	
6			1						1
7		1				1	1		

[1]For clarity in the part-machine incidence matrices and related discussion, we identify parts by alphabetic character and machines by number. In practice, numbers would be used for both.

TABLE 18.4 Rearranged PFA Chart, Indicating Possible Machine Groupings

	Parts (*i*)								
Machines (*j*)	C	E	I	A	D	H	F	G	B
3	1	1	1						
2		1	1						
6	1		1						
1				1	1	1			
5				1		1			
7							1	1	1
4							1		1

in Table 18.4, where different machine groupings are indicated within blocks. The blocks might be considered as possible machine cells. It is often the case (but not in Table 18.4) that some packs do not fit into logical groupings. These parts might be analyzed to see if a revised process sequence can be developed that fits into one of the groups. If not, these parts must continue to be fabricated through a conventional process layout. In Section 18.6.1, we examine a systematic technique called *rank order clustering* that can be used to perform the cluster analysis.

The weakness of production flow analysis is that the data used in the technique are derived from existing production route sheets. In all likelihood, these route sheets have been prepared by different process planners, and the routings may contain operations that are nonoptimal, illogical, or unnecessary. Consequently, the final machine groupings obtained in the analysis may be suboptimal. Notwithstanding this weakness, PFA has the virtue of requiring less time than a complete parts classification and coding procedure. This is attractive to many firms wishing to introduce group technology into their plant operations.

18.4 CELLULAR MANUFACTURING

Whether part families have been determined by visual inspection, parts classification and coding, or production flow analysis, there is advantage in producing those parts using GT machine cells rather than a traditional process-type machine layout. When the machines are grouped, the term cellular manufacturing is used to describe this work organization. *Cellular manufacturing* is an application of group technology in which dissimilar machines or processes have been aggregated into cells, each of which is dedicated to the production of a part, product family, or limited group of families. The typical objectives in cellular manufacturing are similar to those of group technology:

- *To shorten manufacturing lead times* by reducing setup, workpart handling, waiting times, and batch sizes.
- *To reduce work-in-process inventory.* Smaller batch sizes and shorter lead times reduce work-in-process.
- *To improve quality.* This is accomplished by allowing each cell to specialize in producing a smaller number of different parts. This reduces process variability.

- *To simplify production scheduling.* The similarity among parts in the family reduces the complexity of production scheduling. Instead of scheduling parts through a sequence of machines in a process-type shop layout, the system simply schedules the parts though the cell.

- *To reduce setup times.* This is accomplished by using *group tooling* (cutting tools, jigs, and fixtures) that have been designed to process the part family, rather than part tooling, which is designed for an individual part. This reduces the number of individual tools required as well as the time to change tooling between parts.

In this section, we consider several aspects of cellular manufacturing: (1) the composite part concept and (2) machine cell design.

18.4.1 Composite Part Concept

Part families are defined by the fact that their members have similar design and/or manufacturing features. The composite part concept takes this part family definition to its logical conclusion. The *composite part* for a given family is a hypothetical part that includes all of the design and manufacturing attributes of the family. In general, an individual part in the family will have some of the features that characterize the family, but not all of them.

There is always a correlation between part design features and the production operations required to generate those features. Round holes are made by drilling, cylindrical shapes are made by turning, flat surfaces by milling, and so on. A production cell designed for the part family would include those machines required to make the composite part. Such a cell would be capable of producing any member of the family, simply by omitting those operations corresponding to features not possessed by the particular part. The cell would be designed to allow for size variations within the family as well as feature variations.

To illustrate, consider the composite part in Figure 18.8(a). It represents a family of rotational parts with features defined in part (b) of the figure. Associated with each feature is

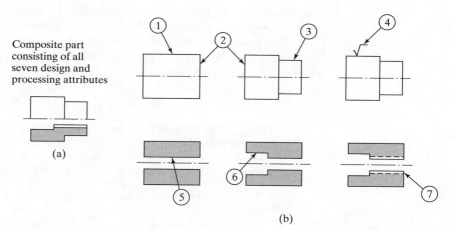

Composite part consisting of all seven design and processing attributes

(a)

(b)

Figure 18.8 Composite part concept: (a) the composite part for a family of machined rotational parts, and (b) the individual features of the composite part. See Table 18.5 for key to individual features and corresponding manufacturing operations.

a certain machining operation, as summarized in Table 18.5. A machine cell to produce this part family would be designed with the capability to accomplish all seven operations required to produce the composite part (the last column in the table). To produce a specific member of the family, operations would be included to fabricate the required features of the part. For parts without all seven features, unnecessary operations would simply be omitted. Machines, fixtures, and tools would be organized for efficient flow of workparts through the cell.

In practice, the number of design and manufacturing attributes is greater than seven, and allowances must be made for variations in overall size and shape of the parts in the family. Nevertheless, the composite part concept is useful for visualizing the machine cell design problem.

18.4.2 Machine Cell Design

Design of the machine cell is critical in cellular manufacturing. The cell design determines to a great degree the performance of the cell. In this section we discuss types of machine cells, cell layouts, and the key machine concept.

Types of Machine Cells and Layouts. GT manufacturing cells can be classified according to the number of machines and the degree to which the material flow is mechanized between machines. Here we identify four common GT cell configurations:

1. Single machine cell
2. Group machine cell with manual handling
3. Group machine cell with semi-integrated handling
4. Flexible manufacturing cell or flexible manufacturing system.

As its name indicates, the *single machine cell* consists of one machine plus supporting fixtures and tooling. This type of cell can be applied to workparts whose attributes allow them to be made on one basic type of process, such as turning or milling. For example, the composite part of Figure 18.8 could be produced on a conventional turret lathe with the possible exception of the cylindrical grinding operation (step 4).

The *group machine cell with manual handling* is an arrangement of more than one machine used collectively to produce one or more part families. There is no provision for mechanized parts movement between the machines in the cell. Instead, the human operators who run the cell perform the material handling function. The cell is often organized into a U-shaped layout, as shown in Figure 18.9. This layout is considered appropriate

TABLE 18.5 Design Features of the Composite Part in Figure 18.8 and the Manufacturing Operations Required to Shape Those Features

Label	Design Feature	Corresponding Manufacturing Operation
1	External cylinder	Turning
2	Cylinder face	Facing
3	Cylindrical step	Turning
4	Smooth surface	External cylindrical grinding
5	Axial hole	Drilling
6	Counterbore	Counterboring
7	Internal threads	Tapping

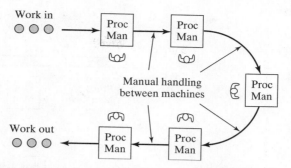

Figure 18.9 Machine cell with manual handling between machines. A U-shaped machine layout is shown. (Key: "Proc" = processing operation (mill, turn, etc.), "Man" = manual operation; arrows indicate work flow.)

when there is variation in the work flow among the parts made in the cell. It also allows the multifunctional workers in the cell to move easily between machines [27]. Other advantages of U-shaped cells in batch model assembly applications, compared to a conventional paced assembly line, include (1) easier changeover from one model to the next, (2) improved quality, (3) visual control of work-in-process, (4) lower initial investment because the cells are simpler and no powered conveyor is required, (5) greater worker satisfaction due to job enlargement and absence of pacing, and (6) more flexibility to adjust to increased demand simply by adding more cells [14].

The group machine cell with manual handling is sometimes achieved in a conventional process type layout without rearranging the equipment. This is done by simply assigning certain machines to be included in the machine group, and restricting their work to specified part families. This allows many of the benefits of cellular manufacturing to be achieved without the expense of rearranging equipment in the shop. Obviously, the material handling benefits of GT are minimized with this organization.

The *group machine cell with semi-integrated handling* uses a mechanized handling system, such as a conveyor, to move parts between machines in the cell. The *flexible manufacturing system* (FMS) combines a fully integrated material handling system with automated processing stations. The FMS is the most highly automated of the group technology machine cells. The following chapter is devoted to this form of automation, and we defer discussion of it until then.

Various layouts are used in GT cells. The U-shape in Figure 18.9 is a popular configuration in cellular manufacturing. Other GT layouts include in-line, loop, and rectangular, shown in Figure 18.10 for the case of semi-integrated handling.

Determining the most appropriate cell layout depends on the routings of parts produced in the cell. Four types of part movement can be distinguished in a mixed model part production system. They are illustrated in Figure 18.11 and defined as follows, where the forward direction of work flow is from left to right in the figure: (1) *repeat operation*, in which a consecutive operation is carried out on the same machine, so that the part does not actually move; (2) *in-sequence move*, in which the part moves forward from the current machine to an immediate neighbor; (3) *bypassing move*, in which the part moves forward from the current machine to another machine that is two or more machines ahead; and (4) *backtracking move*, in which the part moves backward from the current machine to another machine.

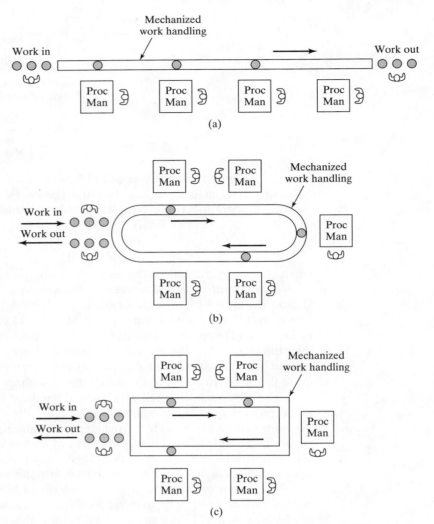

Figure 18.10 Machine cells with semi-integrated handling: (a) in-line layout, (b) loop layout, and (c) rectangular layout. (Key: "Proc" = processing operation (mill, turn, etc.), "Man" = manual operation; arrows indicate work flow.)

When the application consists exclusively of in-sequence moves, then an in-line layout is appropriate. A U-shaped layout also works well here and has the advantage of closer interaction among the workers in the cell. When the application includes repeated operations, then multiple stations (machines) are often required. For cells requiring by-passing moves, the U-shape layout is appropriate. When backtracking moves are needed, a loop or rectangular layout allows recirculation of parts within the cell. Additional factors that must be accommodated by the cell design include

- *Quantity of work to be done by the cell.* This includes the number of parts per year and the processing (or assembly) time per part at each station. These factors determine the

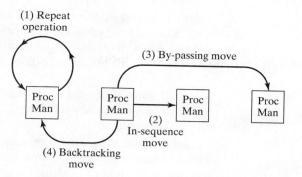

Figure 18.11 Four types of part moves in a mixed model production system. The forward flow of work is from left to right.

workload that must be accomplished by the cell and therefore the number of machines that must be included, as well as total operating cost of the cell and the investment that can be justified.

- *Part size, shape, weight, and other physical attributes.* These factors determine the size and type of material handling and processing equipment that must be used.

Key Machine Concept. In some respects, a GT machine cell operates like a manual assembly line, and it is desirable to spread the workload as evenly as possible among the machines in the cell. On the other hand, there is typically a certain machine in a cell (or perhaps more than one machine in a large cell) that is more expensive to operate than the other machines or that performs certain critical operations in the plant. This machine is referred to as the *key machine.* It is important that the utilization of this key machine be high, even if it means that the other machines in the cell have relatively low utilizations. The other machines are referred to as *supporting machines*, and they should be organized in the cell to keep the key machine busy. In a sense, the cell is designed so that the key machine becomes the bottleneck in the system.

The key machine concept is sometimes used to plan the GT machine cell. The approach is to decide what parts should be processed through the key machine and then determine what supporting machines are required to complete the processing of those parts.

There are generally two measures of utilization that are of interest in a GT cell: the utilization of the key machine and the utilization of the overall cell. The utilization of the key machine can be measured using the usual definition (Section 3.1.3). The utilization of each of the other machines can be evaluated similarly. The cell utilization is obtained by taking a simple arithmetic average of all the machines in the cell. One of the exercise problems at the end of the chapter illustrates the key machine concept and the determination of utilization.

18.5 APPLICATIONS OF GROUP TECHNOLOGY

In our chapter introduction, we defined group technology as a "manufacturing philosophy." GT is not a particular technique, although various tools and techniques, such as parts classification and coding and production flow analysis, have been developed to implement it. The group technology philosophy can be applied in a number of areas. Our discussion focuses on the two main areas of manufacturing and product design.

Manufacturing Applications. The most common applications of GT are in manufacturing, and the most common application in manufacturing involves the formation of cells of one kind or another. Not all companies rearrange machines to form cells. There are three ways in which group technology principles can be applied in manufacturing [21]:

1. *Informal scheduling and routing of similar parts through selected machines.* This approach achieves setup advantages, but no formal part families are defined and no physical rearrangement of equipment is undertaken.
2. *Virtual machine cells.* This approach involves the creation of part families and dedication of equipment to the manufacture of these part families, but without the physical rearrangement of machines into cells. The machines in the virtual cell remain in their original locations in the factory. Use of virtual cells seems to facilitate the sharing of machines with other virtual cells producing other part families [23].
3. *Formal machine cells.* This is the conventional GT approach in which a group of dissimilar machines are physically relocated into a cell that is dedicated to the production of one or a limited set of part families (Section 18.4.2). The machines in a formal machine cell are located in close proximity to one another in order to minimize part handling, throughput time, and work-in-process.

Other GT applications in manufacturing include process planning, family tooling, and numerical control part programs. Process planning of new parts can be facilitated by identifying part families. The new part is associated with an existing part family, and generation of the process plan for the new part follows the routing of the other members of the part family. This is done in a formalized way through the use of parts classification and coding. The approach is discussed in the context of automated process planning (Section 24.2.1).

Ideally, all members of the same part family require similar setups, tooling, and fixturing. This generally results in a reduction in the amount of tooling and fixturing needed. Instead of using a special tool kit developed for each part, a GT system uses a tool kit developed for each part family. The concept of a *modular fixture* can often be exploited, in which a common base fixture that can accommodate adaptations to rapidly switch between different parts in the family.

A similar approach can be applied in NC part programming. *Parametric programming* [26] involves the preparation of a common NC program that covers the entire part family. The program is then adapted for individual members of the family by inserting dimensions and other parameters applicable to the particular part. Parametric programming reduces both programming time and setup time.

Product Design Applications. The application of group technology in product design is principally for design retrieval systems that reduce part proliferation. It has been estimated that the cost of releasing a new part design ranges between $2,000 and $12,000 [33]. In a survey of industry reported in Wemmerlov and Myer [32], it was concluded that in about 20% of new part situations, an existing part design could have been used. In about 40% of the cases, an existing part design could have been used with modifications. The remaining cases required new part designs. If the cost savings for a company generating 1000 new part designs per year was 75% when an existing part design could be used (assuming that there would still be some cost of time associated with the new part for engineering analysis and design retrieval) and 50% when an existing design could be modified, then the total annual savings to the company would be $700,000 to $4,200,000, or 35%

of the company's total design expense due to part releases. The level of design savings described here require an efficient design retrieval procedure. Most part design retrieval procedures are based on parts classification and coding systems (Section 18.2).

Other design applications of group technology involve simplification and standardization of design parameters such as tolerances, inside radii on corners, chamfer sizes on outside edges, hole sizes, thread sizes, etc. These measures simplify design procedures and reduce part proliferation. Design standardization also pays dividends in manufacturing by reducing the required number of distinct lathe tool nose radii, drill sizes, and fastener sizes. There is also a benefit in reducing the amount of data and information that the company must handle. Fewer part designs, design attributes, tools, fasteners, and so on mean fewer and simpler design documents, process plans, and other data records.

18.6 QUANTITATIVE ANALYSIS IN CELLULAR MANUFACTURING

Many quantitative techniques have been developed to deal with problems in group technology and cellular manufacturing. In this section, we consider two problem areas: (1) grouping parts and machines into families, and (2) arranging machines in a GT cell. The first problem area has been and remains an active research area, and several of the more significant research publications are listed in our references [2], [3], [12], [13], [24], [25]. The technique we describe in the current section for solving the part and machine grouping problem is rank order clustering [24]. The second problem area has also been the subject of research, and several reports are listed in the references [1], [7], [9], [19]. In Section 18.6.2, we describe a heuristic approach introduced by Hollier [19].

18.6.1 Grouping Parts and Machines by Rank Order Clustering

The problem addressed here is determining how machines in an existing plant should be grouped into machine cells. The problem is the same whether the cells are virtual or formal (Section 18.5). It is basically the problem of identifying part families. After part families have been identified, the machines to produce a given part family can be selected and grouped together. As previously discussed, the three basic methods to identify part families are (1) visual inspection, (2) parts classification and coding, and (3) production flow analysis.

The rank order clustering technique, first proposed by King [24], is specifically applicable in production flow analysis. It is an efficient and easy-to-use algorithm for grouping machines into cells. In a starting part-machine incidence matrix that might be compiled to document the part routings in a machine shop (or other job shop), the occupied locations in the matrix are organized in a seemingly random fashion. Rank order clustering works by reducing the part-machine incidence matrix to a set of diagonalized blocks that represent part families and associated machine groups. Starting with the initial part-machine incidence matrix, the algorithm consists of the following steps:

1. In each row of the matrix, read the series of 1's and 0's (blank entries = 0's) from left to right as a binary number. Rank the rows in order of decreasing value. In case of a tie, rank the rows in the same order as they appear in the current matrix.

2. Numbering from top to bottom, is the current order of rows the same as the rank order determined in the previous step? If yes, go to step 7. If no, go to the following step.

3. Re-order the rows in the part-machine incidence matrix by listing them in decreasing rank order, starting from the top.

4. In each column of the matrix, read the series of 1's and 0's (blank entries = 0's) from top to bottom as a binary number. Rank the columns in order of decreasing value. In case of a tie, rank the columns in the same order as they appear in the current matrix.

5. Numbering from left to right, is the current order of columns the same as the rank order determined in the previous step? If yes, go to step 7. If no, go to the following step.

6. Re-order the columns in the part-machine incidence matrix by listing them in decreasing rank order, starting with the left column. Go to step 1.

7. Stop.

For readers unaccustomed to evaluating binary numbers in steps 1 and 4, it might be helpful to convert each binary value into its decimal equivalent. For example, the entries in the first row of the matrix in Table 18.3 are read as 100100010. This converts to its decimal equivalent as follows: $(1 \times 2^8) + (0 \times 2^7) + (0 \times 2^6) + (1 \times 2^5) + (0 \times 2^4) + (0 \times 2^3) + (0 \times 2^2) + (1 \times 2^1) + (0 \times 2^0) = 256 + 32 + 2 = 290$. Decimal conversion becomes impractical for the large numbers of parts found in practice, so it is preferable to compare the binary numbers.

EXAMPLE 18.2 Rank Order Clustering Technique

Apply the rank order clustering technique to the part-machine incidence matrix in Table 18.3.

Solution: Step 1 consists of reading the series of 1's and 0's in each row as a binary number. We have done this in Table 18.6(a), converting the binary value for each row to its decimal equivalent. The values are then rank ordered in the far right-hand column. In step 2, we see that the row order is different from the starting matrix. We therefore reorder the rows in step 3. In step 4, we read the series of 1's and 0's in each column from top to bottom as a binary number (again we have converted to the decimal equivalent) and rank the columns in order of decreasing value, as shown in Table 18.6(b). In step 5, we see that the column

TABLE 18.6(a) First Iteration (Step 1) in the Rank Order Clustering Technique Applied to Example 18.2

Binary values	2^8	2^7	2^6	2^5	2^4	2^3	2^2	2^1	2^0	Decimal Equivalent	Rank
Machines	A	B	C	D	E	F	G	H	I		
1	1			1				1		290	1
2					1				1	17	7
3			1		1				1	81	5
4		1				1				136	4
5	1							1		258	2
6			1						1	65	6
7		1				1	1			140	3

TABLE 18.6(b) Second Iteration (Steps 3 and 4) in the Rank Order Clustering Technique Applied to Example 18.2

Machines	Parts									Binary values
	A	B	C	D	E	F	G	H	I	
1	1			1				1		2^6
5	1							1		2^5
7		1				1	1			2^4
4		1				1				2^3
3			1		1				1	2^2
6			1						1	2^1
2					1				1	2^0
Decimal equivalent	96	24	6	64	5	24	16	96	7	
Rank	1	4	8	3	9	5	6	2	7	

TABLE 18.6(c) Solution of Example 18.2

Machines	Parts								
	A	H	D	B	F	G	I	C	E
1	1	1	1						
5	1	1							
7				1	1	1			
4				1	1				
3							1	1	1
6							1	1	
2							1		1

order is different from the preceding matrix. Proceeding from step 6 back to steps 1 and 2, we see that a reordering of the columns provides a row order that is in descending value and the algorithm is concluded (step 7). The final solution is shown in Table 18.6(c). A close comparison of this solution with Table 18.4 reveals that they are the same part-machine groupings.

In the example problem, it was possible to divide the parts and machines into three mutually exclusive part-machine groups. This represents the ideal case because the part families and associated machine cells are completely segregated. However, it is not uncommon for an overlap in processing requirements to exist between machine groups. That is, a given part type needs to be processed by more than one machine group. One way of dealing with the overlap is simply to duplicate the machine that is used by more than one part family, placing the same machine type in both cells. Other approaches, attributed to Burbidge [24], include (1) changing the routing so that all processing can be accomplished in the primary machine group, (2) redesigning the part to eliminate the processing requirement outside the primary machine group, and (3) purchasing the parts from an outside supplier.

18.6.2 Arranging Machines in a GT Cell

After part-machine groupings have been identified, the next problem is to organize the machines into the most logical sequence. Let us describe a simple yet effective method suggested by Hollier [19][2] that uses data contained in From/To charts (Section 10.3.1) and is intended to place the machines in an order that maximizes the proportion of in-sequence moves within the cell. The method is based on the use of From/To ratios determined by summing the total flow from and to each machine in the cell. The algorithm can be reduced to three steps:

1. *Develop the From-To chart.* The data contained in the chart indicate numbers of part moves between the machines (or workstations) in the cell. Moves into and out of the cell are not included in the chart.

2. *Determine the "From/To ratio" for each machine.* This is accomplished by summing all of the "From" trips and "To" trips for each machine (or operation). The "From" sum for a machine is determined by adding the entries in the corresponding row, and the "To" sum is determined by adding the entries in the corresponding column. For each machine, the "From/To ratio" is calculated by taking the "From" sum for each machine and dividing by the respective "To" sum.

3. *Arrange machines in order of decreasing From/To ratio.* Machines with a high From/To ratio distribute more work to other machines in the cell but receive less work from other machines. Conversely, machines with a low From/To ratio receive more work than they distribute. Therefore, machines are arranged in order of descending From/To ratio; that is, machines with high ratios are placed at the beginning of the work flow, and machines with low ratios are placed at the end of the work flow. In case of a tie, the machine with the higher "From" value is placed ahead of the machine with a lower value.

EXAMPLE 18.3 Group Technology Machine Sequence Using Hollier Method 2

Suppose that four machines, 1, 2, 3, and 4, belong to a GT machine cell. An analysis of 50 parts processed on these machines has been summarized in the From/To chart shown in Table 18.7. Additional information is that 50 parts enter the machine grouping at machine 3, 20 parts leave after processing at machine 1, and 30 parts leave after machine 4. Determine the most logical machine sequence using the Hollier method.

TABLE 18.7 From/To Chart for Example 18.3

From	To			
	1	2	3	4
1	0	5	0	25
2	30	0	0	15
3	10	40	0	0
4	10	0	0	0

[2]Hollier [19] introduces six heuristic approaches to solving the machine arrangement problem, of which we describe only one. He presents a comparison of the six methods in his paper.

TABLE 18.8 From/To Sums and From/To Ratios for Example 18.3

From	To				"From" sums	From/To ratio
	1	2	3	4		
1	0	5	0	25	30	0.60
2	30	0	0	15	45	1.0
3	10	40	0	0	50	∞
4	10	0	0	0	10	0.25
"To" sums	50	45	0	40	135	

Solution: Summing the From trips and To trips for each machine yields the "From" and "To" sums in Table 18.8. The From/To ratios are listed in the last column on the right. Arranging the machines in order of descending From/To ratio, the machines in the cell should be sequenced as follows:

$$3 \rightarrow 2 \rightarrow 1 \rightarrow 4$$

It is helpful to use a graphical technique, such as the network diagram (Section 10.3.1), to conceptualize the work flow in the cell. The network diagram for the machine arrangement in Example 18.3 is presented in Figure 18.12. The flow is mostly in-line; however, there is some bypassing and backtracking of parts that must be considered in the design of any material handling system that might be used in the cell. A powered conveyor would be appropriate for the forward flow between machines, with manual handling for the back flow.

Three performance measures can be defined to rate solutions to the machine sequencing problem: (1) percentage of in-sequence moves, (2) percentage of bypassing moves, and (3) percentage of backtracking moves. Each measure is computed by adding all of the values representing that type of move and dividing by the total number of moves. It is desirable for the percentage of in-sequence moves to be high, and for the percentage of backtracking moves to be low. The Hollier method is designed to achieve these goals. Bypassing moves are less desirable than in-sequence moves, but certainly better than backtracking.

EXAMPLE 18.4 Performance Measures for Machine Sequences in a GT Cell

Compute (a) the percentage of in-sequence moves, (b) the percentage of bypassing moves, and (c) the percentage of backtracking moves for the solution in Example 18.3.

Solution: From Figure 18.12, the number of in-sequence moves = 40 + 30 + 25 = 95, the number of bypassing moves = 10 + 15 = 25, and the number of backtracking

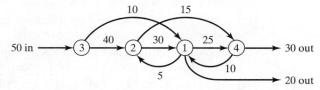

Figure 18.12 Network diagram for machine cell in Example 18.3. Flow of parts into and out of the cells is included.

moves = 5 + 10 = 15. The total number of moves = 135 (totaling either the "From" sums or the "To" sums). Thus,

(a) Percentage of in-sequence moves = 95/135 = 0.704 = 70.4%

(b) Percentage of bypassing moves = 25/135 = 0.185 = 18.5%

(c) Percentage of backtracking moves = 15/135 = 0.111 = 11.1%

REFERENCES

[1] ANEKE, N. A. G., and A. S. CARRIE, "A Design Technique for the Layout of Multi-Product Flowlines," *International Journal of Production Research*, Volume 24, 1986, pp. 471–481.

[2] ASKIN, R. G., H. M. SELIM, and A. J. VAKHARIA, "A Methodology for Designing Flexible Cellular Manufacturing Systems," *IIE Transactions*, Vol. 29, 1997, pp. 599–610.

[3] BEAULIEU, A., A. GHARBI, and A. AIT-KADI, "An Algorithm for the Cell Formation and the Machine Selection Problems in the Design of a Cellular Manufacturing System," *International Journal of Production Research*, Volume 35, 1997, pp. 1857–1874.

[4] BLACK, J. T., "An Overview of Cellular Manufacturing Systems and Comparison to Conventional Systems," *Industrial Engineering*, November 1983, pp. 36–48.

[5] BLACK, J. T., *The Design of the Factory with a Future*, McGraw-Hill Book Company, NY, 1990.

[6] BLACK, J T., and S. L. HUNTER, *Lean Manufacturing Systems and Cell Design*, Society of Manufacturing Engineers, Dearborn, MI, 2003.

[7] BURBIDGE, J. L., "Production Flow Analysis," *Production Engineer*, Volume 41, 1963, p. 742.

[8] BURBIDGE, J. L., *The Introduction of Group Technology*, John Wiley & Sons, NY, 1975.

[9] BURBIDGE, J. L., "A Manual Method of Production Flow Analysis," *Production Engineer*, Volume 56, 1977, p. 34.

[10] BURBIDGE, J. L., *Group Technology in the Engineering Industry*, Mechanical Engineering Publications Ltd., London, UK, 1979.

[11] BURBIDGE, J. L., "Change to Group Technology: Process Organization is Obsolete," *International Journal of Production Research*, Volume 30, 1992, pp. 1209–1219.

[12] CANTAMESSA, M., and A. TURRONI, "A Pragmatic Approach to Machine and Part Grouping in Cellular Manufacturing Systems Design," *International Journal of Production Research*, Volume 35, 1997, pp. 1031–1050.

[13] CHANDRASEKHARAN, M. P., and R. RAJAGOPALAN, "ZODIAC: An Algorithm for Concurrent Formation of Part Families and Machine Cells," *International Journal of Production Research*, Volume 25, 1987, pp. 835–850.

[14] ESPINOSA, A., "The New Shape of Manufacturing," *Assembly*, October 2003, pp. 52–54.

[15] GALLAGHER, C. C., and W. A. KNIGHT, *Group Technology*, Butterworth & Co. Ltd., London, UK, 1973.

[16] GROOVER, M. P., *Fundamentals of Modern Manufacturing: Materials, Processes, and Systems*, 3d ed., John Wiley & Sons, Inc., Hoboken, NJ, 2007.

[17] HAM, I., "Introduction to Group Technology," *Technical Report MMR76-03*, Society of Manufacturing Engineers, Dearborn, MI, 1976.

[18] HAM, I., K. HITOMI, and T. YOSHIDA, *Group Technology: Applications to Production Management*, Kluwer-Nijhoff Publishing, Boston, MA, 1985.

[19] HOLLIER, R. H., "The Layout of Multi-Product Lines," *International Journal of Production Research*, Volume 2, 1963, pp. 47–57.

[20] HOLTZ, R. D., "GT and CAPP Cut Work-in-Process Time 80%," *Assembly Engineering*, Part 1: June 1978, pp. 24–27; Part 2: July 1978, pp. 16–19.

[21] HYER, N. L., and U. WEMMERLOV, "Group Technology in the U.S. Manufacturing Industry: A Survey of Current Practices," *International Journal of Production Research*, Volume 27, 1989, pp. 1287–1304.

[22] HYER, N. L., and U. WEMMERLOV, *Reorganizing the Factory: Competing through Cellular Manufacturing*, Productivity Press, Portland OR, 2002.

[23] IRANI, S. A., T. M. CAVALIER, and P. H. COHEN, "Virtual Manufacturing Cells: Exploiting Layout Design and Intercell Flows for the Machine Sharing Problem," *International Journal of Production Research*, Volume 31, 1993, pp. 791–810.

[24] KING, J. R., "Machine-Component Grouping in Production Flow Analysis: An Approach Using a Rank Order Clustering Algorithm," *International Journal of Production Research*, Volume 18, 1980, pp. 213–222.

[25] KUSIAK, A., "EXGT-S: A Knowledge Based System for Group Technology," *International Journal of Production Research*, Volume 26, 1988, pp. 1353–1367.

[26] LYNCH, M., *Computer Numerical Control for Machining*, McGraw-Hill, Inc., NY, 1992.

[27] MONDEN, Y., *Toyota Production System*, Industrial Engineering and Management Press, Institute of Industrial Engineers, Norcross, GA, 1983.

[28] MOODIE, C., R. UZSOY, and Y. YIH, *Manufacturing Cells: A Systems Engineering View*, Taylor & Francis Ltd., London, UK, 1995.

[29] OPITZ, H., *A Classification System to Describe Workpieces*, Pergamon Press, Oxford, UK, 1970.

[30] OPITZ, H., and H. P. WIENDAHL, "Group Technology and Manufacturing Systems for Medium Quantity Production," *International Journal of Production Research*, Volume 9, No. 1, 1971, pp. 181–203.

[31] SINGH, N., and D. RAJAMANI, *Cellular Manufacturing Systems: Design, Planning, and Control*, Chapman & Hall, London, UK, 1996.

[32] WEMMERLOV, U., and N. L. HYER, "Cellular Manufacturing in U.S. Industry: A Survey of Users," *International Journal of Production Research*, Volume 27, 1989, pp. 1511–1530.

[33] WEMMERLOV, U., and N. L. HYER, "Group Technology," *Handbook of Industrial Engineering*, G. Salvendy (ed.), John Wiley & Sons, Inc., NY, 1992, pp. 464–488.

[34] WILD, R., *Mass Production Management*, John Wiley & Sons Ltd., London, UK, 1972.

REVIEW QUESTIONS

18.1 What is group technology?

18.2 What is cellular manufacturing?

18.3 What are the production conditions under which group technology and cellular manufacturing are most applicable?

18.4 What are the two major tasks that a company must undertake when it implements group technology?

18.5 What is a part family?

18.6 What are the three methods for solving the problem of grouping parts into part families?

18.7 What is the difference between a hierarchical structure and a chain-type structure in a classification and coding scheme?

18.8 What is production flow analysis?

18.9 What are the typical objectives when implementing cellular manufacturing?

18.10 What is the composite part concept, as the term is applied in group technology?

18.11 What are the four common GT cell configurations, as identified in the text?

18.12 What is the key machine concept in cellular manufacturing?

18.13 What is the difference between a virtual machine cell and a formal machine cell?

18.14 What is the principal application of group technology in product design?

18.15 What is the application of rank order clustering?

PROBLEMS

Parts Classification and Coding

18.1 Develop the form code (first five digits) in the Opitz System for the part illustrated in Figure P18.1.

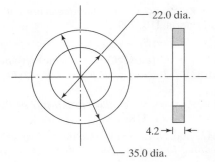

4.2

35.0 dia.

22.0 dia.

Figure P18.1 Part for Problem 18.1. Dimensions are in millimeters.

18.2 Develop the form code (first five digits) in the Opitz System for the part illustrated in Figure P18.2.

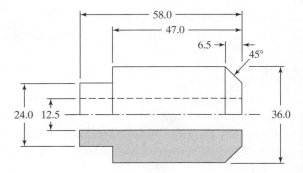

58.0

47.0

6.5

45°

24.0 12.5

36.0

Figure P18.2 Part for Problem 18.2. Dimensions are in millimeters.

18.3 Develop the form code (first five digits) in the Opitz System for the part illustrated in Figure P18.3.

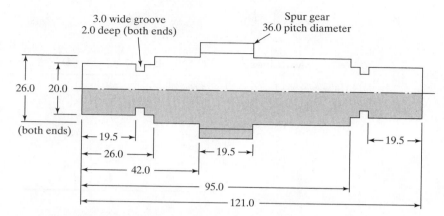

Figure P18.3 Part for Problem 18.3. Dimensions are in millimeters.

Rank Order Clustering

18.4 Apply the rank order clustering technique to the part-machine incidence matrix in the following table to identify logical part families and machine groups. Parts are identified by letters, and machines are identified numerically.

Machines	Parts				
	A	B	C	D	E
1	1				
2		1			1
3	1			1	
4		1	1		
5				1	

18.5 Apply the rank order clustering technique to the part-machine incidence matrix in the following table to identify logical part families and machine groups. Parts are identified by letters, and machines are identified numerically.

Machines	Parts					
	A	B	C	D	E	F
1	1				1	
2				1		1
3	1	1				
4			1	1		
5		1			1	
6			1	1		1

545

18.6 Apply the rank order clustering technique to the part-machine incidence matrix in the table that follows to identify logical part families and machine groups. Parts are identified by letters, and machines are identified numerically.

Machines	A	B	C	D	E	F	G	H	I
1	1								1
2		1					1		
3			1		1			1	
4		1				1	1		
5			1					1	
6						1	1		
7	1			1					
8			1		1				

Parts

18.7 Apply the rank order clustering technique to the part-machine incidence matrix in the table that follows to identify logical part families and machine groups. Parts are identified by letters, and machines are identified numerically.

Machines	A	B	C	D	E	F	G	H	I
1			1	1	1				
2	1	1					1	1	1
3						1	1	1	
4	1	1		1					
5			1		1				
6		1						1	1
7	1		1	1					
8		1				1		1	1

Parts

18.8 The following table lists the weekly quantities and routings of ten parts that are being considered for cellular manufacturing in a machine shop. Parts are identified by letters and machines are identified numerically. For the data given, (a) develop the part-machine incidence matrix, and (b) apply the rank order clustering technique to the part-machine incidence matrix to identify logical part families and machine groups.

Part	Weekly Quantity	Machine Routing	Part	Weekly Quantity	Machine Routing
A	50	$3 \rightarrow 2 \rightarrow 7$	F	60	$5 \rightarrow 1$
B	20	$6 \rightarrow 1$	G	5	$3 \rightarrow 2 \rightarrow 4$
C	75	$6 \rightarrow 5$	H	100	$3 \rightarrow 2 \rightarrow 4 \rightarrow 7$
D	10	$6 \rightarrow 5 \rightarrow 1$	I	40	$2 \rightarrow 4 \rightarrow 7$
E	12	$3 \rightarrow 2 \rightarrow 7 \rightarrow 4$	J	15	$5 \rightarrow 6 \rightarrow 1$

Machine Cell Organization and Design

18.9 Four machines used to produce a family of parts are to be arranged into a GT cell. The From/To data for the parts processed by the machines are shown in the table below. (a) Determine the most logical sequence of machines for this data. (b) Construct the network diagram for the data, showing where and how many parts enter and exit the system. (c) Compute the percentages of in-sequence moves, bypassing moves, and backtracking moves in the solution. (d) Develop a feasible layout plan for the cell.

		To		
From	1	2	3	4
1	0	10	0	40
2	0	0	0	0
3	50	0	0	20
4	0	50	0	0

18.10 In Problem 18.8, two logical machine groups are identified by rank order clustering. For each machine group, (a) determine the most logical sequence of machines for this data, (b) construct the network diagram for the data, and (c) compute the percentages of in-sequence moves, bypassing moves, and backtracking moves in the solution.

18.11 Five machines constitute a GT cell. The From/To data for the machines are shown in the table below. (a) Determine the most logical sequence of machines for this data, and construct the network diagram, showing where and how many parts enter and exit the system. (b) Compute the percentages of in-sequence moves, bypassing moves, and backtracking moves in the solution. (c) Develop a feasible layout plan for the cell based on the solution.

			To		
From	1	2	3	4	5
1	0	10	80	0	0
2	0	0	0	85	0
3	0	0	0	0	0
4	70	0	20	0	0
5	0	75	0	20	0

18.12 A GT machine cell contains three machines. Machine 1 feeds machine 2, the key machine in the cell. Machine 2 feeds machine 3. The cell is set up to produce a family of five parts (A, B, C, D, and E). The operation times for each part at each machine are given in the table below. The products are to be produced in the ratios 4:3:2:2:1, respectively. If the machine cell runs for 35 hours per week, (a) how many of each product will be made by the cell, and (b) what is the utilization of each machine in the cell?

	Operation Time (min)		
Part	Machine 1	Machine 2	Machine 3
A	4.0	15.0	10.0
B	15.0	18.0	7.0
C	26.0	20.0	15.0
D	15.0	20.0	10.0
E	8.0	16.0	10.0

18.13 A GT cell will machine the components for a family of parts. The parts come in several different sizes and the cell will be designed to quickly change over from one size to the next. This will be accomplished using fast-change fixtures and downloading the part programs from the plant computer to the CNC machines in the cell. The parts are rotational type, so the cell must be able to perform turning, boring, facing, drilling, and cylindrical grinding operations. Accordingly, there will be several machine tools in the cell, of types and numbers specified by the designer. To transfer parts between machines in the cell, the designer may use a belt or similar conveyor system. Any conveyor equipment of this type will be 0.4 meters wide. The arrangement of the various pieces of equipment in the cell is the principal problem. The raw workparts will be delivered into the machine cell on a belt conveyor. The finished parts must be deposited onto a conveyor that delivers them to the assembly department. The input and output conveyors are 0.4 meters wide, and the designer must specify where they enter and exit the cell. The parts are currently machined by conventional methods in a process-type layout. In the current production method, there are seven machines involved but two of the machines are duplicates. From/To data have been collected for the jobs that are relevant to this problem.

	To							
From	1	2	3	4	5	6	7	Parts out
1	0	112	0	61	59	53	0	0
2	12	0	0	0	0	226	0	45
3	74	0	0	35	31	0	180	0
4	0	82	0	0	0	23	5	16
5	0	73	0	0	0	23	0	14
6	0	0	0	0	0	0	0	325
7	174	16	20	30	20	0	0	0
Parts in	25	0	300	0	0	0	75	

The From/To data indicate the number of workparts moved between machines during a typical 40-hour week. The data refer to the parts considered in the case. The two categories "parts in" and "parts out" indicate parts entering and exiting the machine group. Each week, an average 400 parts are processed through the seven machines. However, as indicated by the data, not all 400 parts are processed by every machine. Machines 4 and 5 are identical and assignment of parts to these machines is arbitrary. Average production rate capacity on each of the machines for the particular distribution of this parts family is given in the table below. Also given are the floor space dimensions of each machine in meters. Assume that all loading and unloading operations take place in the center of the machine.

Machine	Operation	Production Rate (pc/hr)	Machine Dimensions
1	Turn outside diameter	9	3.5 m × 1.5 m
2	Bore inside diameter	15	3.0 m × 1.6 m
3	Face ends	10	2.5 m × 1.5 m
4	Grind outside diameter	12	2.5 m × 1.5 m
5	Grind outside diameter	12	3.0 m × 1.5 m
6	Inspect	5	Bench 1.5 m × 1.5 m
7	Drill	9	1.5 m × 2.5 m

Operation 6 is currently a manual inspection operation. It is anticipated that this manual station will be replaced by a coordinate measuring machine (CMM). This automated inspection machine will triple throughput rate to 15 parts per hour. The floor space dimensions of the CMM are 2.0 m × 1.6 m. All other machines are candidates for inclusion in the new machine cell. (a) Analyze the problem and determine the most appropriate sequence of machines in the cell using the data contained in the From/To chart. (b) Construct the network diagram for the cell, showing where and how many parts enter and exit the cell. (c) Determine the utilization and production capacity of the machines in the cell as you have designed it. (d) Prepare a layout (top view) drawing of the GT cell, showing the machines, the robot(s), and any other pieces of equipment in the cell. (e) Write a one-page (or less) description of the cell, explaining the basis of your design and why the cell is arranged as it is.

Chapter 19

Flexible Manufacturing Systems

CHAPTER CONTENTS

The flexible manufacturing system (FMS) was identified in the previous chapter as one of the machine cell types used to implement cellular manufacturing. It is the most automated and technologically sophisticated of the group technology cells. An FMS typically

From Chapter 19 of *Automation, Production Systems, and Computer-Integrated Manufacturing*, Third Edition.
Mikell P. Groover. Copyright © 2008 by Pearson Education, Inc. Publishing as Prentice Hall. All rights reserved.

possesses multiple automated stations and is capable of variable routings among stations. Its flexibility allows it to operate as a mixed model system. An FMS integrates into one highly automated manufacturing system many of the concepts and technologies discussed in previous chapters, including flexible automation (Section 1.2.1), CNC machines (Chapter 7), distributed computer control (Section 5.3.3), automated material handling and storage (Chapters 10 and 11), and group technology (Chapter 18). The concept for flexible manufacturing systems originated in Britain in the early 1960s (Historical Note 19.1). The first FMS installations in the United States occurred around 1967. These initial systems performed machining operations on families of parts using NC machine tools.

FMS technology can be applied in production situations similar to those identified for cellular manufacturing:

- Presently, the plant either produces parts in batches or uses manned GT cells, and management wants to automate.
- It is possible to group a portion of the parts made in the plant into part families, whose similarities permit them to be processed on the machines in the flexible manufacturing system. Part similarities can be interpreted to mean that (1) the parts belong to a common product and/or (2) the parts possess similar geometries. In either

---

Historical Note 19.1 Flexible manufacturing systems [19], [20], [21]

The flexible manufacturing system was first conceptualized for machining, and it required the prior development of numerical control. The concept is credited to David Williamson, a British engineer employed by Molins during the mid 1960s. Molins applied for a patent for the invention that was granted in 1965. The concept was called *System 24* because it was believed that the group of machine tools comprising the system could operate 24 hours a day, 16 hours of which would be unattended by human workers. The original concept included computer control of the NC machines, production of a variety of parts, and use of tool magazines that could hold various tools for different machining operations.

One of the first flexible manufacturing systems to be installed in the United States was a machining system at Ingersoll-Rand Company in Roanoke, Virginia, in the late 1960s by Sundstrand, a machine tool builder. Other systems introduced soon after included a Kearney & Trecker FMS at Caterpillar Tractor and Cincinnati Milacron's "Variable Mission System." Most of the early FMS installations in the United States were in large companies, such as Ingersoll-Rand, Caterpillar, John Deere, and General Electric Company. These large companies had the financial resources to make the major investments necessary, and they also possessed the prerequisite experience in NC machine tools, computer systems, and manufacturing systems to pioneer the new FMS technology. Flexible manufacturing systems were also installed in other countries around the world. In the Federal Republic of Germany (West Germany, now Germany), a manufacturing system was developed in 1969 by Heidleberger Druckmaschinen in cooperation with the University of Stuttgart. In the USSR (now Russia) a flexible manufacturing system was demonstrated at the 1972 Stanki Exhibition in Moscow. The first Japanese FMS was installed around the same time by Fuji Xerox. By around 1985, the number of FMS installations throughout the world had increased to about 300. About 20 to 25% of these were located in the United States. In recent years, there has been an emphasis on smaller, less expensive flexible manufacturing cells.

case, the processing requirements of the parts must be sufficiently similar to allow them to be made on the FMS.

- The parts or products made by the facility are in the mid-volume, mid-variety production range. The appropriate production volume range is 5,000 to 75,000 parts per year [14]. If annual production is below this range, an FMS is likely to be an expensive alternative. If production volume is above this range, then a more specialized production system should probably be considered.

The differences between installing a flexible manufacturing system and implementing a manually operated machine cell are the following: (1) the FMS requires a significantly greater capital investment because new equipment is being installed, whereas the manually operated machine cell might only require existing equipment to be rearranged, and (2) the FMS is technologically more sophisticated for the human resources who must make it work. However, the potential benefits are substantial. They include increased machine utilization, reduced factory floor space, greater responsiveness to change, lower inventory and manufacturing lead times, and higher labor productivity. We elaborate on these benefits in Section 19.3.2.

In this chapter, we define and discuss flexible manufacturing systems: what makes them flexible, what their components and applications are, and how to implement the technology. In the final section we present a mathematical model for assessing the performance of flexible manufacturing systems.

19.1 WHAT IS A FLEXIBLE MANUFACTURING SYSTEM?

A flexible manufacturing system (FMS) is a highly automated GT machine cell, consisting of a group of processing workstations (usually CNC machine tools), interconnected by an automated material handling and storage system, and controlled by a distributed computer system. The reason the FMS is called flexible is that it is capable of processing a variety of different part styles simultaneously at the various workstations, and the mix of part styles and quantities of production can be adjusted in response to changing demand patterns. The FMS is most suited for the mid-variety, mid-volume production range (refer to Figure 1.5).

The initials FMS are sometimes used to denote the term *flexible machining system*. The machining process is presently the largest application area for FMS technology. However, it seems appropriate to interpret FMS in its broader meaning, allowing for a wide range of possible applications beyond machining.

An FMS relies on the principles of group technology. No manufacturing system can be completely flexible. There are limits to the range of parts or products that can be made in an FMS. Accordingly, a flexible manufacturing system is designed to produce parts (or products) within a defined range of styles, sizes, and processes. In other words, an FMS is capable of producing a single part family or a limited range of part families.

A more appropriate term for FMS would be *flexible automated manufacturing system*. The use of the word "automated" would distinguish this type of production technology from other manufacturing systems that are flexible but not automated, such as a manned GT machine cell. The word "flexible" would distinguish it from other manufacturing systems that are highly automated but not flexible, such as a conventional transfer line. However, the existing terminology is well established.

19.1.1 Flexibility

The issue of manufacturing system flexibility was discussed previously in Section 13.2.5. In that discussion, we identified three capabilities that a manufacturing system must possess in order to be flexible: (1) the ability to identify and distinguish among the different incoming part or product styles processed by the system, (2) quick changeover of operating instructions, and (3) quick changeover of physical setup. Flexibility is an attribute that applies to both manual and automated systems. In manual systems, the human workers are often the enablers of the system's flexibility.

To develop the concept of flexibility in an automated manufacturing system, consider a machine cell consisting of two CNC machine tools that are loaded and unloaded by an industrial robot from a parts carousel, perhaps in the arrangement depicted in Figure 19.1. The cell operates unattended for extended periods of time. Periodically, a worker must unload completed parts from the carousel and replace them with new workparts. By any definition, this is an automated manufacturing cell, but is it a flexible manufacturing cell? One might argue yes, it is flexible since the cell consists of CNC machine tools, and CNC machines are flexible because they can be programmed to machine different part configurations. However, if the cell only operates in a batch mode, in which the same part style is produced by both machines in lots of several hundred units, then this does not qualify as flexible manufacturing.

To qualify as being flexible, a manufacturing system should satisfy several criteria. The following are four reasonable tests of flexibility in an automated manufacturing system:

1. *Part variety test.* Can the system process different part styles in a non-batch mode?
2. *Schedule change test.* Can the system readily accept changes in production schedule, that is, changes in part mix and/or production quantities?
3. *Error recovery test.* Can the system recover gracefully from equipment malfunctions and breakdowns, so that production is not completely disrupted?

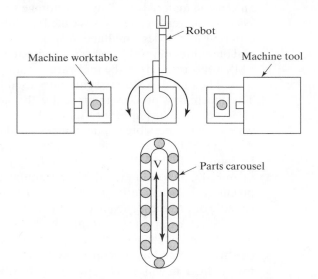

Figure 19.1 Automated manufacturing cell with two machine tools and robot. Is it a flexible cell?

4. *New part test.* Can new part designs be introduced into the existing product mix with relative ease?

If the answer to all of these questions is "yes" for a given manufacturing system, then the system can be considered flexible. The most important criteria are (1) and (2). Criteria (3) and (4) are softer and can be implemented at various levels. In fact, introduction of new part designs is not a consideration in some flexible manufacturing systems; such systems are designed to produce a part family whose members are all known in advance.

Getting back to our illustration, the robotic work cell satisfies the criteria if it (1) can machine different part configurations in a mix rather than in batches; (2) permits changes in production schedule; (3) is capable of continuing to operate even though one machine experiences a breakdown (for example, while repairs are being made on the broken machine, its work is temporarily reassigned to the other machine), and (4) can accommodate new part designs if the NC part programs are written off-line and then downloaded to the system for execution. This fourth capability requires the new part to be within the part family intended for the FMS, so that the tooling used by the CNC machines as well as the end effector of the robot are suited to the new part design.

19.1.2 Types of FMS

Having considered the issue of flexibility, let us now consider the various types of flexible manufacturing systems. Each FMS is designed for a specific application, that is, a specific family of parts and processes. Therefore, each FMS is custom-engineered and unique. Given these circumstances, one would expect to find a great variety of system designs to satisfy a wide variety of application requirements.

Flexible manufacturing systems can be distinguished according to the kinds of operations they perform: processing operations or assembly operations (Section 2.2.1). An FMS is usually designed to perform one or the other but rarely both. A difference that is applicable to machining systems is whether the system will process rotational parts or nonrotational parts (Section 13.2.1). Flexible machining systems with multiple stations that process rotational parts are much less common than systems that process nonrotational parts. Two other ways to classify flexible manufacturing systems are by number of machines and level of flexibility.

Number of Machines. Flexible manufacturing systems can be distinguished according to the number of machines in the system. The following are typical categories: (1) single machine cell, (2) flexible manufacturing cell, and (3) flexible manufacturing system.

A *single machine cell* consists of one CNC machining center combined with a parts storage system for unattended operation (Section 14.2.2), as in Figure 19.2. Completed parts are periodically unloaded from the parts storage unit, and raw workparts are loaded into it. The cell can be designed to operate in a batch mode, a flexible mode, or a combination of the two. When operated in a batch mode, the machine processes parts of a single style in specified lot sizes and is then changed over to process a batch of the next part style. When operated in a flexible mode, the system satisfies three of the four flexibility tests. It is capable of (1) processing different part styles, (2) responding to changes in production schedule, and (4) accepting new part introductions. Criterion (3), error recovery, cannot be satisfied because if the single machine breaks down, production stops.

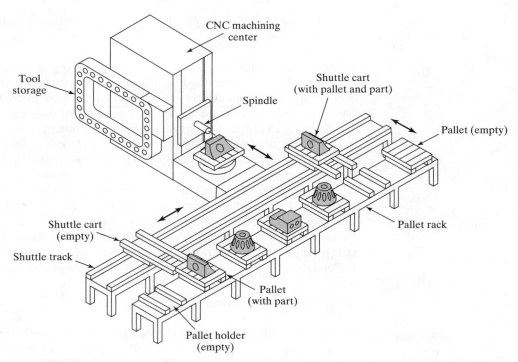

Figure 19.2 Single machine cell consisting of one CNC machining center and parts storage unit.

A *flexible manufacturing cell* (FMC) consists of two or three processing workstations (typically CNC machining centers or turning centers) plus a parts handling system. The parts handling system is connected to a load/unload station. The handling system usually includes a limited parts storage capacity. One possible FMC is illustrated in Figure 19.3. A flexible manufacturing cell satisfies the four flexibility tests discussed previously.

A *flexible manufacturing system* (FMS) has four or more processing stations connected mechanically by a common parts handling system and electronically by a distributed computer system. Thus, an important distinction between a FMS and a FMC is in the number of machines: a FMC has two or three machines, while a FMS has four or more.[1] There are usually other differences as well. One is that the FMS generally includes nonprocessing workstations that support production but do not directly participate in it. These other stations include part/pallet washing stations, coordinate measuring machines, and so on. Another difference is that the computer control system of a FMS is generally larger and more sophisticated, often including functions not always found in a cell, such as diagnostics and tool monitoring. These additional functions are needed more in a FMS than in a FMC because the FMS is more complex.

---

[1]We have defined the dividing line that separates a FMS from a FMC to be four machines. It should be noted that not all practitioners would agree with that dividing line; some might prefer a higher value while a few would prefer a lower number. Also, the distinction between cell and system seems to apply only to flexible manufacturing systems that are automated. The manned counterparts of these systems discussed in the previous chapter are always referred to as cells, no matter how many workstations are included.

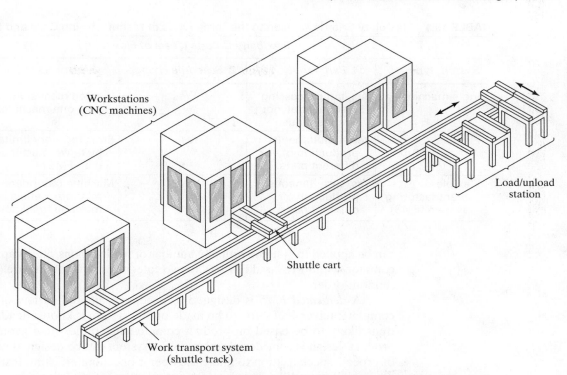

Figure 19.3 A flexible manufacturing cell consisting of three identical processing stations (CNC machining centers), a load/unload station, and a parts handling system.

Some of the distinguishing characteristics of the three categories of flexible manufacturing cells and systems are summarized in Figure 19.4. Table 19.1 compares the three systems in terms of the four flexibility tests.

Level of Flexibility. Another way to classify flexible manufacturing systems is according to the level of flexibility designed into the system. This method of classification

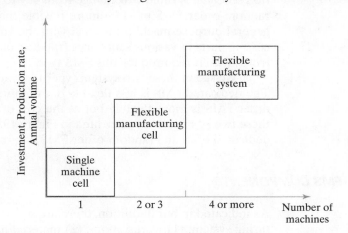

Figure 19.4 Features of the three categories of flexible cells and systems.

TABLE 19.1 Flexibility Criteria Applied to the Three Types of Manufacturing Cells and Systems

| | Flexibility Criteria (Tests of Flexibility) | | | |
System type	1. Part variety	2. Schedule change	3. Error recovery	4. New part
Single machine cell	Yes, but processing is sequential, not simultaneous.	Yes	Limited recovery due to only one machine.	Yes
Flexible manufacturing cell (FMC)	Yes, simultaneous production of different parts.	Yes	Error recovery limited by fewer machines than FMS.	Yes
Flexible manufacturing system (FMS)	Yes, simultaneous production of different parts.	Yes	Machine redundancy minimizes effect of machine breakdowns.	Yes

can be applied to systems with any number of workstations, but its application seems most common with FMCs and FMSs. The two categories of flexibility are (1) dedicated and (2) random-order.

A *dedicated FMS* is designed to produce a limited variety of part styles, and the complete universe of parts to be made on the system is known in advance. The part family is likely to be based on product commonality rather than geometric similarity. The product design is considered stable, so the system can be designed with a certain amount of process specialization to make the operations more efficient. Instead of being general purpose, the machines can be designed for the specific processes required to make the limited part family, thus increasing the production rate of the system. In some instances, the machine sequence may be identical or nearly identical for all parts processed, so a transfer line may be appropriate, in which the workstations possess the necessary flexibility to process the different parts in the mix. Indeed, the term *flexible transfer line* is sometimes used for this case [15].

A *random-order FMS* is more appropriate when the part family is large, there are substantial variations in part configurations, new part designs will be introduced into the system and engineering changes will occur in parts currently produced, and the production schedule is subject to change from day to day. To accommodate these variations, the random-order FMS must be more flexible than the dedicated FMS. It is equipped with general purpose machines to deal with the variations in product and is capable of processing parts in various sequences (random order). A more sophisticated computer control system is required for this FMS type.

We see in these two system types the trade-off between flexibility and productivity. The dedicated FMS is less flexible but capable of higher production rates. The random-order FMS is more flexible but at the cost of lower production rates. A comparison of these two FMS types is presented in Figure 19.5. Table 19.2 presents a comparison of the dedicated FMS and random-order FMS in terms of the four flexibility tests.

19.2 FMS COMPONENTS

As indicated in our definition, there are several basic components of a flexible manufacturing system: (1) workstations, (2) material handling and storage system, and (3) computer control system. In addition, even though a FMS is highly automated, (4) people are required to manage and operate the system.

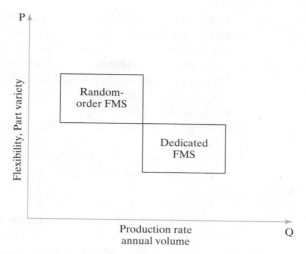

Figure 19.5 Comparison of dedicated and random-order FMS types.

TABLE 19.2 Flexibility Criteria Applied to Dedicated FMS and Random-Order FMS.

| System type | Flexibility Criteria (Tests of Flexibility) | | | |
	1. Part variety	2. Schedule change	3. Error recovery	4. New part
Dedicated FMS	Limited. All parts known in advance.	Limited changes can be tolerated.	Limited by sequential processes.	No. New part introductions difficult.
Random-order FMS	Yes. Substantial part variations possible.	Frequent and significant changes possible.	Machine redundancy minimizes effect of machine breakdowns.	Yes. System designed for new part introductions.

19.2.1 Workstations

The processing or assembly equipment used in a flexible manufacturing system depends on the type of work accomplished by the system. In a system designed for machining operations, the principal types of processing station are CNC machine tools. However, the FMS concept is applicable to various other processes as well. Following are the types of workstations typically found in a FMS.

Load/Unload Stations. The load/unload station is the physical interface between the FMS and the rest of the factory. It is where raw workparts enter the system and finished parts exit the system. Loading and unloading can be accomplished either manually (the most common method) or by automated handling systems. The load/unload station should be ergonomically designed to permit convenient and safe movement of workparts. Mechanized cranes and other handling devices are installed to assist the operator with parts that are too heavy to lift by hand. A certain level of cleanliness must be maintained at the workplace, and air hoses or other washing facilities are often used to flush away chips and ensure clean mounting and locating points. The station is often raised slightly above floor level using an open-grid platform to permit chips and cutting fluid to drop through the openings for subsequent recycling or disposal.

The load/unload station includes a data entry unit and monitor for communication between the operator and the computer system. Through this system the operator receives instructions regarding which part to load onto the next pallet in order to adhere to the production schedule. When different pallets are required for different parts, the correct pallet must be supplied to the station. When modular fixturing is used, the correct fixture must be specified and the required components and tools must be available at the workstation to build it. When the part loading procedure has been completed, the handling system must launch the pallet into the system, but not until then; the handling system must be prevented from moving the pallet while the operator is still working. All of these conditions require communication between the computer system and the operator at the load/unload station.

Machining Stations. The most common applications of flexible manufacturing systems are machining operations. The workstations used in these systems are therefore predominantly CNC machine tools. Most common is the CNC machining center (Section 14.3.3), in particular, the horizontal machining center. CNC machining centers possess features that make them compatible with the FMS, including automatic tool changing and tool storage, use of palletized workparts, CNC, and capacity for DNC (Section 7.3). Machining centers can be ordered with automatic pallet changers that can be readily interfaced with the FMS part handling system. Machining centers are generally used for nonrotational parts. For rotational parts, turning centers are used; and for parts that are mostly rotational but require multi-tooth rotational cutters (milling and drilling), mill-turn centers can be used. These types of equipment are described in Section 14.3.3.

In some machining systems, the types of operations performed are concentrated in a certain category, such as milling or turning. For milling, special *milling machine modules* can be used to achieve higher production levels than a machining center can manage. The milling module can be vertical spindle, horizontal spindle, or multiple spindle. For turning operations, special *turning modules* can be designed for the FMS. In conventional turning, the workpiece is rotated against a tool that is held in the machine and fed in a direction parallel to the axis of work rotation. Parts made on most FMSs are usually nonrotational, but they may require some turning in their process sequence. For these cases, the parts are held in a pallet fixture throughout processing on the FMS, and a turning module is designed to rotate the single point tool around the work.

Other Processing Stations. The flexible manufacturing system concept has been applied to other processing operations in addition to machining. One such application is sheet-metal fabrication, reported in Winship [34]. The processing workstations consist of pressworking operations such as punching, shearing, and certain bending and forming processes. Also, flexible systems have been developed to automate the forging process [32]. Forging is traditionally a very labor intensive operation. The workstations in the system consist principally of a heating furnace, a forging press, and a trimming station.

Assembly. Some flexible manufacturing systems are designed to perform assembly operations. Flexible automated assembly systems are gradually replacing manual labor in the assembly of products typically made in batches. Industrial robots are often used as the automated workstations in these flexible assembly systems. They can be programmed to perform tasks with variations in sequence and motion pattern to accommodate the different product styles assembled in the system. Other examples of flexible assembly workstations are the programmable component placement machines widely used in electronics assembly.

Other Stations and Equipment. Inspection can be incorporated into a flexible manufacturing system, either by including an inspection operation at a processing workstation, or by including a station specifically designed for inspection. Coordinate measuring machines (Section 22.4), special inspection probes that can be used in a machine tool spindle (Section 22.4.5), and machine vision (Section 22.6) are three possible technologies for performing inspection on an FMS. Inspection is particularly important in flexible assembly systems to ensure that components have been properly added at the workstations. We examine the topic of automated inspection in more detail in Chapter 21.

In addition to the above, other operations and functions are often accomplished on a flexible manufacturing system. These include cleaning parts and/or pallet fixtures, central coolant delivery systems for the entire FMS, and centralized chip removal systems often installed below floor level.

19.2.2 Material Handling and Storage System

The second major component of a FMS is its material handling and storage system. In this section we discuss the functions of the handling system, material handling equipment typically used in an FMS, and types of FMS layout.

Functions of the Handling System. The material handling and storage system in a flexible manufacturing system performs the following functions:

- *Allows random, independent movement of workparts between stations.* Parts must be capable of moving from any machine in the system to any other machine, in order to provide various routing alternatives for the different parts and to make machine substitutions when certain stations are busy.
- *Enables handling of a variety of workpart configurations.* For *prismatic* parts (nonrotational parts), this is usually accomplished by using modular pallet fixtures in the handling system. The fixture is located on the top face of the pallet and is designed to accommodate different part configurations by means of common components, quick-change features, and other devices that permit a rapid build-up of the fixture for a given part. The base of the pallet is designed for the material handling system. For rotational parts, industrial robots are often used to load and unload the turning machines and to move parts between stations.
- *Provides temporary storage.* The number of parts in the FMS will typically exceed the number of parts actually being processed at any moment. Thus, each station has a small queue of parts, perhaps only one part, waiting to be processed, which helps to maintain high machine utilization.
- *Provides convenient access for loading and unloading workparts.* The handling system must include locations for load/unload stations.
- *Creates compatibility with computer control.* The handling system must be under the direct control of the computer system which directs it to the various workstations, load/unload stations, and storage areas.

Material Handling Equipment. The types of material handling systems used to transfer parts between stations in a FMS include a variety of conventional material transport equipment (Chapter 10), in-line transfer mechanisms (Section 16.1.2), and industrial robots (Chapter 8). The material handling function in a FMS is often shared between two systems: (1) a primary handling system and (2) a secondary handling system. The *primary handling system* establishes the basic layout of the FMS and is responsible for moving parts between stations in the system. FMS layouts are discussed later in this section.

The *secondary handling system* consists of transfer devices, automatic pallet changers, and similar mechanisms located at the workstations in the FMS. The function of the secondary handling system is to transfer work from the primary system to the machine tool or other processing station and to position the parts with sufficient accuracy and repeatability to perform the processing or assembly operation. Other purposes served by the secondary handling system include (1) reorientation of the workpart if necessary to present the surface that is to be processed, and (2) buffer storage of parts to minimize work change time and maximize station utilization. In some FMS installations, the positioning and registration requirements at the individual workstations are satisfied by the primary work handling system. In these cases, there is no secondary handling system.

FMS Layout Configurations. The material handling system establishes the FMS layout. Most layout configurations found in today's flexible manufacturing systems can be classified into five categories: (1) in-line layout, (2) loop layout, (3) ladder layout, (4) open field layout, and (5) robot-centered cell. The types of material handling equipment utilized in these five layouts are summarized in Table 19.3.

In the *in-line layout*, the machines and handling system are arranged in a straight line. In its simplest form, the parts progress from one workstation to the next in a well-defined sequence with work always moving in one direction and no back-flow, as in Figure 19.6(a). The operation of this type of system is similar to a transfer line (Chapter 16), except that a variety of workparts are processed in the system. For in-line systems requiring greater routing flexibility, a linear transfer system that permits movement in two directions can be installed. One possible arrangement for doing this is shown in Figure 19.6(b), in which a secondary work handling system is provided at each workstation to separate most of the parts from the primary line.

TABLE 19.3 Material Handling Equipment Typically Used as the Primary Handling System for the Five FMS Layouts

Layout Configuration	Typical Material Handling System
In-line layout	In-line transfer system (Section 16.1.2)
	Conveyor system (Section 10.2.4)
	Rail-guided vehicle system (Section 10.2.3)
Loop layout	Conveyor system (Section 10.2.4)
	In-floor towline carts (Section 10.2.4)
Ladder layout	Conveyor system (Section 10.2.4)
	Automated guided vehicle system (Section 10.2.2)
	Rail-guided vehicle system (Section 10.2.3)
Open field layout	Automated guided vehicle system (Section 10.2.2)
	In-floor towline carts (Section 10.2.4)
Robot-centered layout	Industrial robot (Chapter 8)

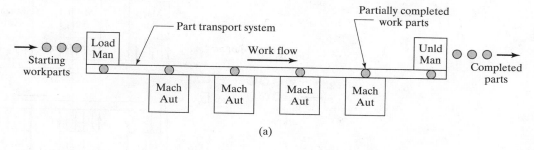

(a)

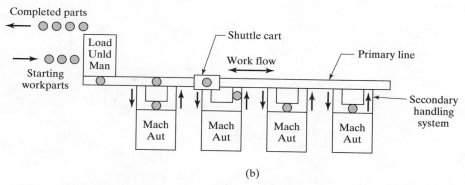

(b)

Figure 19.6 FMS in-line layouts: (a) one direction flow similar to a transfer line, (b) linear transfer system with secondary parts handling system at each station to facilitate flow in two directions. Key: Load = parts loading station, Unld = parts unloading station, Mach = machining station, Man = manual station, Aut = automated station.

In the *loop layout*, the workstations are organized in a loop that is served by a parts handling system in the same shape, as shown in Figure 19.7(a). Parts usually flow in one direction around the loop with the capability to stop and be transferred to any station. A secondary handling system is shown at each workstation to permit parts to move without obstruction around the loop. The load/unload station(s) are typically located at one end of the loop. An alternative form of loop layout is the *rectangular layout*. As shown in Figure 19.7(b), this arrangement might be used to return pallets to the starting position in a straight line machine arrangement.

The *ladder layout* consists of a loop with rungs between the straight sections of the loop, on which workstations are located, as shown in Figure 19.8. The rungs increase the number of possible ways of getting from one machine to the next, and obviate the need for a secondary handling system. This reduces average travel distance and minimizes congestion in the handling system, thereby reducing transport time between stations.

The *open field layout* consists of multiple loops and ladders, and may include sidings as well, as illustrated in Figure 19.9. This layout type is generally appropriate for processing a large family of parts. The number of different machine types may be limited, and parts are routed to different workstations depending on which one becomes available first.

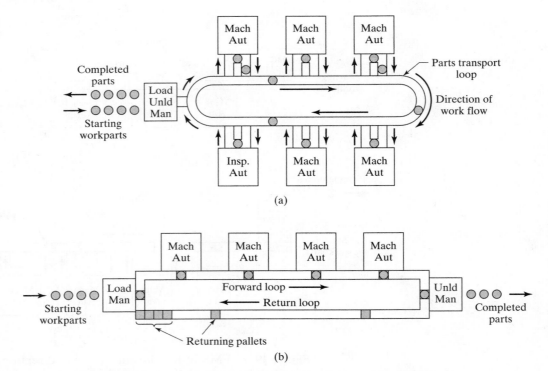

Figure 19.7 (a) FMS loop layout with secondary parts handling system at each station to allow unobstructed flow on the loop, and (b) rectangular layout for recirculation of pallets to the first workstation in the sequence. Key: Load = parts loading station, UnLd = parts unloading station, Mach = machining station, Man = manual station, Aut = automated station.

The *robot-centered layout* (Figure 19.1) uses one or more robots as the material handling system. Industrial robots can be equipped with grippers that make them well suited for the handling of rotational parts, and robot-centered FMS layouts are often used to process cylindrical or disk-shaped parts.

19.2.3 Computer Control System

The FMS includes a distributed computer system that is interfaced to the workstations, material handling system, and other hardware components. A typical FMS computer system consists of a central computer and microcomputers controlling the individual machines and other components. The central computer coordinates the activities of the components to achieve smooth overall operation of the system. Functions performed by the FMS computer control system can be grouped into the following categories:

1. *Workstation control.* In a fully automated FMS, the individual processing or assembly stations generally operate under some form of computer control. CNC controls the individual machine tools in a machining system.

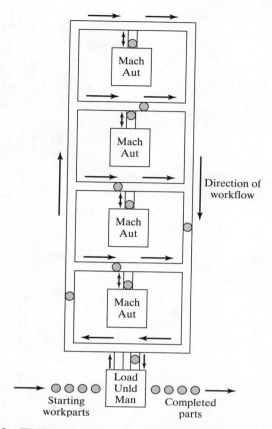

Figure 19.8 FMS ladder layout. Key: Load = parts loading station, Unld = parts unloading station, Mach = machining station, Man = manual station, Aut = automated station.

2. *Distribution of control instructions to workstations.* Some form of central intelligence is required to coordinate the processing at individual stations. In a machining FMS, part programs must be downloaded to machines, and DNC is used for this purpose. The DNC system stores the programs, allows submission of new programs and editing of existing programs as needed, and performs other DNC functions (Section 7.3).

3. *Production control.* The mix and rate at which the various parts are launched into the system must be managed. Input data required for production control includes desired daily production rates per part, numbers of raw workparts available, and number of applicable pallets.[2] The production control function is accomplished by routing an applicable pallet to the load/unload area and providing instructions to the operator to load the desired workpart.

4. *Traffic control.* This refers to the management of the primary material handling system that moves parts between stations. Traffic control is accomplished by actuating

[2]The term *applicable pallet* refers to a pallet that is fixtured to accept a workpart of a given type.

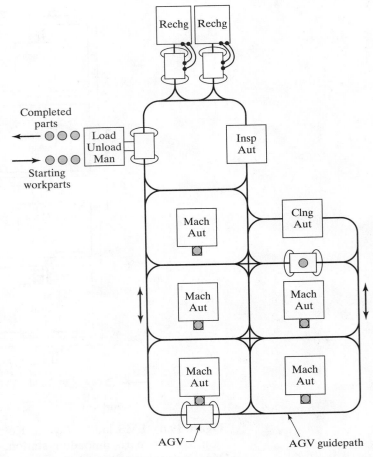

Figure 19.9 FMS open field layout. Key: Load = parts loading station, UnLd = parts unloading station, Mach = machining station, Man = manual station, Aut = automated station, AGV = automated guided vehicle, Rechg = battery recharging station for AGVs.

switches at branches and merging points, stopping parts at machine tool transfer locations, and moving pallets to load/unload stations.

5. *Shuttle control.* This control function is concerned with the operation and control of the secondary handling system at each workstation. Each shuttle must be coordinated with the primary handling system and synchronized with the operation of the machine tool it serves.

6. *Workpiece monitoring.* The computer must monitor the status of each cart and/or pallet in the primary and secondary handling systems, as well as the status of each of the various workpiece types.

7. *Tool control.* Tool control is concerned with managing two aspects of the cutting tools:
   - *Tool location.* This involves keeping track of the cutting tools at each workstation. If one or more tools required to process a particular workpiece is not present at

the station that is specified in the part's routing, the tool control subsystem takes one or both of the following actions: (a) determines whether an alternative workstation that has the required tool is available, and/or (b) notifies the operator responsible for tooling in the system that the tool storage unit at the station must be loaded with the required cutter(s).

- *Tool life monitoring.* In this aspect of tool control, a tool life is specified to the computer for each cutting tool in the FMS. A record of the machining time usage is maintained for each of the tools, and when the cumulative machining time reaches the specified life of the tool, the computer notifies the operator that a tool replacement is needed.

8. *Performance monitoring and reporting.* The computer control system is programmed to collect data on the operation and performance of the flexible manufacturing system. The data are periodically summarized, and reports on system performance are prepared for management. Some of the reports indicating FMS performance are listed in Table 19.4.

9. *Diagnostics.* This function is available to a greater or lesser degree on many manufacturing systems to indicate the probable source of the problem when a malfunction occurs. It can also be used to plan preventive maintenance in the system and to identify impending failures. The purpose of the diagnostics function is to reduce breakdowns and downtime, and to increase availability of the system.

An FMS possesses the characteristic architecture of a distributed numerical control (DNC) system. As in other DNC systems, two-way communication is used. Data and commands are sent from the central computer to the individual machines and other hardware components, and data on execution and performance are transmitted from the components back to the central computer. In addition, an uplink from the FMS to the corporate host computer is provided.

19.2.4 Human Resources

One additional component in the FMS is human labor. Humans are needed to manage the operations of the system. Functions typically performed by humans include (1) loading raw workparts into the system, (2) unloading finished parts (or assemblies) from the

TABLE 19.4 Typical FMS Performance Reports

Type of Report	Description
Availability	Summary of the uptime proportion (reliability) of the workstations. Details such as reasons for downtime are included to identify recurring problem areas.
Utilization	Summary of utilization of each workstation as well as the average utilization of the FMS for specified periods (days, weeks, months).
Production	Daily and weekly quantities of different parts produced by the FMS. Comparison of actual quantities against the production schedule.
Tooling	Information on various aspects of tool control, such as a listing of tools at each workstation and tool life status.
Status	Instantaneous "snapshot" of the present condition of the FMS. Line supervision can request this report at any time to learn the current status of system operating parameters (e.g., trouble spots, utilization, availability, cumulative piece counts, and tooling).

system, (3) changing and setting tools, (4) performing equipment maintenance and repair, (5) performing NC part programming, (6) programming and operating the computer system, and (7) managing the system.

19.3 FMS APPLICATIONS AND BENEFITS

In this section we explore the applications of flexible manufacturing systems and the benefits that result from these applications.

19.3.1 FMS Applications

Flexible automation is applicable to a variety of manufacturing operations. FMS technology is most widely applied in machining operations. Other applications include sheet-metal pressworking, forging, and assembly. In this section, we examine some of the applications using case study examples as illustrations.

Flexible Machining Systems. Historically, most of the applications of flexible machining systems have been in milling and drilling type operations (nonrotational parts), using CNC machining centers. FMS applications for turning (rotational parts) were much less common until recently, and the systems that are installed tend to consist of fewer machines. For example, single machine cells consisting of parts storage units, parts loading robots, and CNC turning centers are widely used today, although not always in a flexible mode. Let us explore some of the issues behind this anomaly in the development of flexible machining systems.

Unlike rotational parts, nonrotational parts are often too heavy for a human operator to easily and quickly load into the machine tool. Accordingly, pallet fixtures were developed so that these parts could be loaded onto the pallet off-line using hoists, and then the part-on-pallet could be moved into position in front of the machine tool spindle. Nonrotational parts also tend to be more expensive than rotational parts, and the manufacturing lead times tend to be longer. These factors provide a strong incentive to produce them as efficiently as possible, using advanced technologies such as FMSs. For these reasons, the technology for FMS milling and drilling applications is more mature today than for FMS turning applications.

EXAMPLE 19.1 Vought Aerospace FMS

A flexible manufacturing system installed at Vought Aerospace in Dallas, Texas, by Cincinnati Milacron is shown in Figure 19.10. The system is used to machine approximately 600 different aircraft components. The FMS consists of eight CNC horizontal machining centers plus inspection modules. Part handling is accomplished by an automated guided vehicle system using four vehicles. Loading and unloading of the system is done at two stations. These load/unload stations consist of storage carousels that permit parts to be stored on pallets for subsequent transfer to the machining stations by the AGVS. The system is capable of processing a sequence of single, one-of-a-kind parts in a continuous mode, so a complete set of components for one aircraft may be made efficiently without batching.

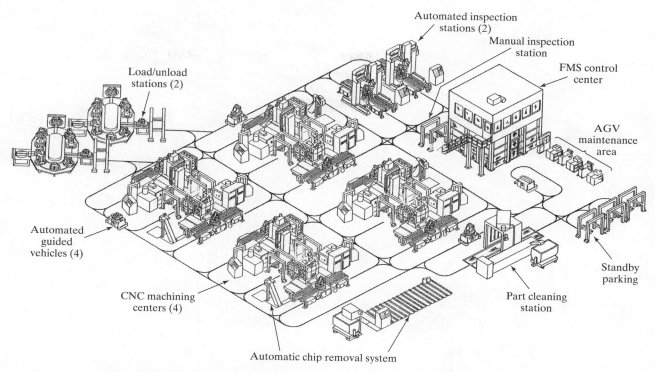

Figure 19.10 FMS at Vought Aircraft (line drawing courtesy of Cincinnati Milacron).

Other FMS Applications. Pressworking and forging are two other manufacturing processes in which efforts are being made to develop flexible automated systems. The FMS technologies involved are described in Vaccari [32] and Winship [34]. The following example illustrates the development efforts in the pressworking area.

EXAMPLE 19.2 Flexible Fabricating System

The term flexible fabricating system (FFS) is sometimes used in connection with systems that perform sheet-metal pressworking operations. One FFS concept by Wiedemann is illustrated in Figure 19.11. The system is designed to unload sheet-metal stock from the automated storage/retrieval system (AS/RS), move the stock by rail-guided cart to the CNC punch press operations, and then move the finished parts back to the AS/RS, all under computer control.

Flexible automation concepts can be applied to assembly operations. Although some examples have included industrial robots to perform the assembly tasks, the following example illustrates a flexible assembly system that makes minimal use of industrial robots.

EXAMPLE 19.3 Assembly FMS at Allen-Bradley

An FMS for assembly installed by Allen-Bradley Company is reported in Waterbury [33]. The "flexible automated assembly line" produces motor

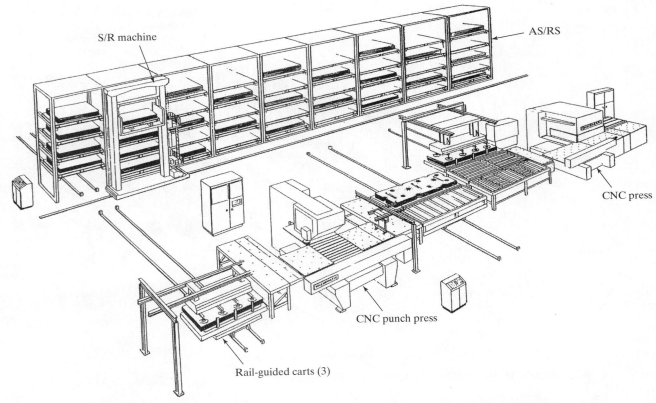

S/R machine

AS/RS

CNC press

CNC punch press

Rail-guided carts (3)

Figure 19.11 Flexible fabricating system for automated sheet-metal processing (based on line drawing provided courtesy of Wiedemann Division, Cross & Trecker Company.)

starters in 125 model styles. The line boasts a one-day manufacturing lead time on lot sizes as low as one and production rates of 600 units per hour. The system consists of 26 workstations that perform all assembly, subassembly, testing, and packaging. The stations are linear and rotary indexing assembly machines with pick-and-place robots performing certain handling functions between the machines. Each step in the process uses 100% automated testing to ensure very high quality levels. The flexible assembly line is controlled by a system of Allen-Bradley programmable logic controllers.

19.3.2 FMS Benefits

A number of benefits can be expected in successful FMS applications. The principal benefits are

- *Increased machine utilization.* Flexible manufacturing systems achieve a higher average utilization than machines in a conventional batch production machine shop. Reasons for this include (1) 24 hour per day operation, (2) automatic tool changing of machine tools, (3) automatic pallet changing at workstations, (4) queues of parts at stations, and (5) dynamic scheduling of production that compensates for irregularities. It should be possible to approach 80% to 90% asset utilization by implementing FMS technology [19].

- *Fewer machines required.* Because of higher machine utilization, fewer machines are required.

- *Reduction in the amount of factory floor space required.* Compared to a job shop of equivalent capacity, a FMS generally requires less floor area. Reductions in floor space requirements are estimated to be 40% to 50% [19].

- *Greater responsiveness to change.* A flexible manufacturing system improves response capability to part design changes, introduction of new parts, changes in production schedule and product mix, machine breakdowns, and cutting tool failures. Adjustments can be made in the production schedule from one day to the next to respond to rush orders and special customer requests.

- *Reduced inventory requirements.* Because different parts are processed together rather than separately in batches, work-in-process is less than in a batch production mode. The inventory of starting and finished parts can be reduced also, by a typical 60% to 80% [19].

- *Lower manufacturing lead times.* Closely correlated with reduced work-in-process is the time spent in process by the parts. This means faster customer deliveries.

- *Reduced direct labor requirements* and *higher labor productivity*. Higher production rates and lower reliance on direct labor mean greater productivity per labor hour with an FMS than with conventional production methods. An FMS can result in labor savings of 30% to 50% [19].

- *Opportunity for unattended production.* The high level of automation in a flexible manufacturing system allows it to operate for extended periods of time without human attention. In the most optimistic scenario, parts and tools are loaded into the system at the end of the day shift, the FMS continues to operate throughout the night, and the finished parts are unloaded the next morning.

19.4 FMS PLANNING AND IMPLEMENTATION ISSUES

Implementation of a flexible manufacturing system represents a major investment and commitment by the user company. It is important that the installation of the system be preceded by thorough planning and design, and that its operation be characterized by good management of all resources: machines, tools, pallets, parts, and people. Our discussion of these issues is organized along these lines: (1) FMS planning and design issues, and (2) FMS operational issues.

19.4.1 FMS Planning and Design Issues

The initial phase of FMS planning must consider the parts that will be produced by the system. The issues are similar to those in cellular manufacturing. They include

- *Part family considerations.* Any flexible manufacturing system must be designed to process a limited range of part or product styles. The boundaries of the range must be decided. In effect, the part family that will be processed on the FMS must be defined. The definition of part families to be processed on the FMS can be based on product commonality as well as part similarity. The term *product commonality* refers to different components used on the same product. Many successful FMS installations are designed to accommodate part families defined by this criterion. This allows all of the components required to assemble a given product unit to be completed just prior to beginning assembly.

- *Processing requirements.* The types of parts and their processing requirements determine the types of processing equipment that will be used in the system. In machining applications, nonrotational parts are produced by machining centers, milling machines, and similar machine tools; rotational parts are machined by turning centers and similar equipment.

- *Physical characteristics of the workparts.* The size and weight of the parts determine the size of the machines at the workstations and the size of the material handling system that must be used.

- *Production volume.* Quantities to be produced by the system determine how many machines of each type will be required. Production volume is also a factor in selecting the most appropriate type of material handling equipment for the system.

After the part family, production volumes, and similar part issues have been decided, design of the system can proceed. Important factors that must be specified in FMS design include

- *Types of workstations.* The types of machines are determined by part processing requirements. Consideration of workstations must also include the load/unload station(s).

- *Variations in process routings* and *FMS layout.* If variations in process sequence are minimal, then an in-line flow is most appropriate. For a system with higher product variety, a loop is more suitable. If there is significant variation in the processing, a ladder layout or open field layout are most appropriate.

- *Material handling system.* Selection of the material handling equipment and layout are closely related, since the type of handling system limits the layout selection. The material handling system includes both primary and secondary handling systems (Section 19.2.2).

- *Work-in-process* and *storage capacity.* The level of work-in-process (WIP) allowed in the FMS is an important variable in determining utilization and efficiency of the FMS. If the WIP level is too low, then stations may become starved for work, causing reduced utilization. If the WIP level is too high, then congestion may result. The WIP level should be planned, not just allowed to happen. Storage capacity in the FMS must be compatible with WIP level.

- *Tooling.* Tooling decisions include types and numbers of tools at each station, and the degree of duplication of tooling at the different stations. Tool duplication at stations tends to increase the flexibility with which parts can be routed through the system.

- *Pallet fixtures.* In machining systems for nonrotational parts, it is necessary to select the number of pallet fixtures used in the system. Factors influencing the decision include levels of WIP allowed in the system and differences in part style and size. Parts that differ too much in configuration and size require different fixturing.

19.4.2 FMS Operational Issues

Once the FMS is installed, then its resources must be optimized to meet production requirements and achieve operational objectives related to profit, quality, and customer satisfaction. The operational problems that must be solved include [20], [22], [29], [30]:

- *Scheduling and dispatching.* Scheduling of production in the FMS is dictated by the master production schedule (Section 25.1). Dispatching is concerned with launching of parts into the system at the appropriate times. Several of the following problem areas are related to scheduling.

- *Machine loading.* This problem is concerned with allocating the operations and tooling resources among the machines in the system to accomplish the required production schedule.

- *Part routing.* Routing decisions involve selecting the routes that should be followed by each part in the production mix in order to maximize use of workstation resources.

- *Part grouping.* Part types must be grouped for simultaneous production, given limitations on available tooling and other resources at workstations.

- *Tool management.* Managing the available tools involves making decisions on when to change tools and how to allocate tooling to workstations in the system.

- *Pallet and fixture allocation.* This problem is concerned with the allocation of pallets and fixtures to the parts being produced in the system. Different parts require different fixtures, and before a given part style can be launched into the system, a fixture for that part must be made available.

19.5 QUANTITATIVE ANALYSIS OF FLEXIBLE MANUFACTURING SYSTEMS

Most of the design and operational problems identified in Section 19.4 can be addressed using quantitative analysis techniques. Flexible manufacturing systems have constituted an active area of interest in operations research, and many of the important contributions are included in our list of references. FMS analysis techniques can be classified into (1) deterministic models, (2) queueing models, (3) discrete event simulation, and (4) other approaches, including heuristics.

Deterministic models are useful in obtaining starting estimates of system performance. Later in this section, we present a deterministic model that is useful in the beginning stages of FMS design to provide rough estimates of system parameters such as production rate, capacity, and utilization. Deterministic models do not permit evaluation of operating characteristics such as the build-up of queues and other dynamics that can impair system performance. Consequently, deterministic models tend to overestimate FMS performance. On the other hand, if actual system performance is much lower than the estimates provided by these models, it may be a sign of either poor system design or poor management of FMS operations.

Queueing models can be used to describe some of the dynamics not accounted for in deterministic approaches. These models are based on the mathematical theory of queues. They permit the inclusion of queues, but only in a general way and for relatively simple system configurations. The performance measures that are calculated are usually average values for steady-state operation of the system. Examples of queueing models to study flexible manufacturing systems are described in several of our references [4], [27], and [30]. Probably the most well known of the FMS queueing models is CAN-Q [25], [26].

In the later stages of design, discrete event simulation probably offers the most accurate method for modeling the specific aspects of a given flexible manufacturing system [24], [35]. The computer model can be constructed to closely resemble the details of a complex FMS operation. Characteristics such as layout configuration, number of pallets in the system, and production scheduling rules can be incorporated into the FMS simulation model. Indeed, the simulation can be helpful in determining optimum values for these parameters.

Other techniques that have been applied to analyze FMS design and operational problems include mathematical programming [28] and various heuristic approaches [1], [13]. Several literature reviews on operations research techniques directed at FMS problems are included among our references [2], [6], [16], and [31].

19.5.1 Bottleneck Model

Important aspects of FMS performance can be mathematically described by a deterministic model called the bottleneck model, developed by Solberg [27].[3] Although it has the limitations of a deterministic approach, the bottleneck model is simple and intuitive. It can be used to provide starting estimates of FMS design parameters such as production rate, number of workstations, and similar measures. The term *bottleneck* refers to the fact that the output of the production system has an upper limit, given that the product mix flowing through the system is fixed. The model can be applied to any production system that possesses this bottleneck feature, for example, a manually operated machine cell or a production job shop. It is not limited to flexible manufacturing systems.

Terminology and Symbols. Let us define the features, terms, and symbols for the bottleneck model, as they might be applied to a flexible manufacturing system:

- *Part mix.* The mix of the various part or product styles produced by the system is defined by p_j, where p_j = the fraction of the total system output that is of style j. The subscript $j = 1, 2, \ldots P$, where P = the total number of different part styles made in the FMS during the time period of interest. The values of p_j must sum to unity, that is,

$$\sum_{j=1}^{P} p_j = 1.0 \qquad (19.1)$$

- *Workstations and servers.* The flexible production system has a number of distinctly different workstations n. In the terminology of the bottleneck model, each workstation may have more than one server, which simply means that it is possible to have two or more machines capable of performing the same operations. Using the terms *stations* and *servers* in the bottleneck model is a precise way of distinguishing between machines that accomplish identical operations and those that accomplish different operations. Let s_i = the number of servers at workstation i, where $i = 1, 2, \ldots, n$. We include the load/unload station as one of the stations in the FMS.

---

[3]We have simplified Solberg's model somewhat and adapted the notation and performance measures to be consistent with our discussion in this chapter.

- *Process routing.* For each part or product, the process routing defines the sequence of operations, the workstations where operations are performed, and the associated processing times. The sequence includes the loading operation at the beginning of processing on the FMS and the unloading operation at the end of processing. Let t_{ijk} = the processing time, which is the total time that a production unit occupies a given workstation server, not counting any waiting time at the station. In the notation for t_{ijk}, the subscript i refers to the station, j refers to the part or product, and k refers to the sequence of operations in the process routing. For example, the fourth operation in the process plan for part A is performed on machine 2 and takes 8.5 minutes; thus, $t_{2A4} = 8.5$ min. Note that process plan j is unique to part j. The bottleneck model does not conveniently allow for alternative process plans for the same part.

- *Work handling system.* The material handling system used to transport parts or products within the FMS can be considered to be a special case of a workstation. Let us designate it as station $n + 1$, and the number of carriers in the system (e.g., conveyor carts, AGVs, monorail vehicles, etc.) is analogous to the number of servers in a regular workstation. Let s_{n+1} = the number of carriers in the FMS handling system.

- *Transport time.* Let t_{n+1} = the mean transport time required to move a part from one workstation to the next station in the process routing. This value could be computed for each individual transport based on transport velocity and distances between stations in the FMS, but it is more convenient to simply use an average transport time for all moves in the FMS.

- *Operation frequency.* The operation frequency is defined as the expected number of times a given operation in the process routing is performed for each work unit. For example, an inspection might be performed on a sampling basis, once every four units; hence, the frequency for this operation would be 0.25. In other cases, the part may have an operation frequency greater than 1.0, for example, for a calibration procedure that may have to be performed more than once on average to be completely effective. Let f_{ijk} = operation frequency for operation k in process plan j at station i.

FMS Operational Parameters. Using the above terms, we can next define certain average operational parameters of the production system. The *average workload* for a given station is defined as the mean total time spent at the station per part. It is calculated as

$$WL_i = \sum_j \sum_k t_{ijk} f_{ijk} p_j \qquad (19.2)$$

where WL_i = average workload for station i, min; t_{ijk} = processing time for operation k in process plan j at station i, min; and f_{ijk} = operation frequency for operation k in part j at station i; and p_j = part mix fraction for part j.

The work handling system (station $n + 1$) is a special case, as noted in our terminology above. The workload of the handling system is the mean transport time multiplied by the average number of transports required to complete the processing of a workpart. The average number of transports is equal to the mean number of operations in the process routing minus one. That is,

$$n_t = \sum_i \sum_j \sum_k f_{ijk} p_j - 1 \qquad (19.3)$$

where n_t = mean number of transports and the other terms are defined above. Let us illustrate this with a simple example.

EXAMPLE 19.4   Determining n_t

Consider a manufacturing system with two stations: (1) a load/unload station and (2) a machining station. The system processes just one part, part A, so the part mix fraction $p_A = 1.0$. The frequency of all operations is $f_{iAk} = 1.0$. The parts are loaded at station 1, routed to station 2 for machining, and then sent back to station 1 for unloading (three operations in the routing). Using Eq. (19.3),

$$n_t = 1(1.0) + 1(1.0) + 1(1.0) - 1 = 3 - 1 = 2$$

Looking at it another way, the process routing is (1) $\rightarrow$ (2) $\rightarrow$ (1). Counting the number of arrows gives us the number of transports: $n_t = 2$.

We are now in a position to compute the workload of the handling system,

$$WL_{n+1} = n_t t_{n+1} \tag{19.4}$$

where WL_{n+1} = workload of the handling system, min; n_t = mean number of transports by Eq. (19.3); and t_{n+1} = mean transport time per move, min.

System Performance Measures. Important measures for assessing the performance of a flexible manufacturing system include production rate of all parts, production rate of each part style, utilization of the different workstations, and number of busy servers at each workstation. These measures can be calculated under the assumption that the FMS is producing at its maximum possible rate. This rate is constrained by the bottleneck station in the system, which is the station with the highest workload per server. The workload per server is simply the ratio WL_i / s_i for each station. Thus, the bottleneck is identified by finding the maximum value of the ratio amongst all stations. The comparison must include the handling system, since it might be the bottleneck in the system.

Let WL^* and s^* equal the workload and number of servers, respectively, for the bottleneck station. The maximum production rate of all parts of the FMS can be determined as the ratio of s^* to WL^*. Let us refer to it as the maximum production rate, because it is limited by the capacity of the bottleneck station,

$$R_p^* = \frac{s^*}{WL^*} \tag{19.5}$$

where R_p^* = maximum production rate of all part styles produced by the system, which is determined by the capacity of the bottleneck station, pc/min; s^* = number of servers at the bottleneck station, and WL^* = workload at the bottleneck station, min/pc. It is not difficult to grasp the validity of this formula as long as all parts are processed through the bottleneck station. A little more thought is required to appreciate that Eq. (19.5) is also valid even when not all the parts pass through the bottleneck station, as long as the product mix (p_j values) remains constant. In other words, if we disallow those parts not passing through the bottleneck from increasing their production rates to reach their respective bottleneck limits, these parts will be limited by the part mix ratios.

The value of R_p^* includes parts of all styles produced in the system. Individual part production rates can be obtained by multiplying R_p^* by the respective part mix ratios. That is,

$$R_{pj}^* = p_j(R_p^*) = p_j \frac{s^*}{WL^*} \tag{19.6}$$

where R_{pj}^* = maximum production rate of part style j, pc/min; and p_j = part mix fraction for part style j.

The mean *utilization* of each workstation is the proportion of time that the servers at the station are working and not idle. This can be computed as:

$$U_i = \frac{WL_i}{s_i}(R_p^*) = \frac{WL_i}{s_i}\frac{s^*}{WL^*} \tag{19.7}$$

where U_i = utilization of station i; WL_i = workload of station i, min/pc; s_i = number of servers at station i; and R_p^* = overall production rate, pc/min. The utilization of the bottleneck station is 100% at R_p^*.

To obtain the average station utilization, one simply computes the average value for all stations, including the transport system. This can be calculated as

$$\overline{U} = \frac{\sum_{i=1}^{n+1} U_i}{n+1} \tag{19.8}$$

where $\overline{U}$ is an unweighted average of the workstation utilizations.

A more useful measure of overall FMS utilization can be obtained using a weighted average, where the weighting is based on the number of servers at each station for the n regular stations in the system, and the transport system is omitted from the average. The argument for omitting the transport system is that the utilization of the processing stations is the important measure of FMS utilization. The purpose of the transport system is to serve the processing stations, and therefore its utilization should not be included in the average. The overall FMS utilization is calculated as

$$\overline{U}_s = \frac{\sum_{i=1}^{n} s_i U_i}{\sum_{i=1}^{n} s_i} \tag{19.9}$$

where $\overline{U}_s$ = overall FMS utilization, s_i = number of servers at station i, and U_i = utilization of station i.

Finally, we are interested in the number of busy servers at each station. All of the servers at the bottleneck station are busy at the maximum production rate, but the servers at the other stations are idle some of the time. The values can be calculated as

$$BS_i = WL_i(R_p^*) = WL_i\frac{s^*}{WL^*} \tag{19.10}$$

where BS_i = number of busy servers on average at station i and WL_i = workload at station i.

Let us present two example problems to illustrate the bottleneck model, the first a simple example whose answers can be verified intuitively, and the second a more complicated problem.

EXAMPLE 19.5 Bottleneck Model on a Simple Problem

A flexible machining system consists of a load/unload station and two machining workstations. Station 1 is the load/unload station. Station 2 performs milling operations and consists of two servers (two identical CNC milling machines). Station 3 has one server that performs drilling (one CNC drill press). The stations are connected by a part handling system that has four work carriers. The mean transport time is 3.0 min. The FMS produces two parts, A and B.

The part mix fractions and process routings for the two parts are presented in the table below. The operation frequency $f_{ijk} = 1.0$ for all operations. Determine (a) maximum production rate of the FMS, (b) corresponding production rates of each product, (c) utilization of each station, and (d) number of busy servers at each station.

Part j	Part mix p_j	Operation k	Description	Station i	Process time t_{ijk} (min)
A	0.4	1	Load	1	4
		2	Mill	2	30
		3	Drill	3	10
		4	Unload	1	2
B	0.6	1	Load	1	4
		2	Mill	2	40
		3	Drill	3	15
		4	Unload	1	2

Solution: (a) To compute the FMS production rate, we first need to compute workloads at each station, so that we can identify the bottleneck station.

$$WL_1 = (4 + 2)(0.4)(1.0) + (4 + 2)(0.6)(1.0) = 6.0 \text{ min}$$

$$WL_2 = 30(0.4)(1.0) + 40(0.6)(1.0) = 36.0 \text{ min}$$

$$WL_3 = 10(0.4)(1.0) + 15(0.6)(1.0) = 13.0 \text{ min}$$

The station routing for both parts is the same: $1 \rightarrow 2 \rightarrow 3 \rightarrow 1$. There are three moves, $n_t = 3$.

$$WL_4 = 3(3.0)(0.4)(1.0) + 3(3.0)(0.6)(1.0) = 9.0 \text{ min}$$

The bottleneck station is identified by finding the largest WL_i/s_i ratio.

For station 1, $WL_1/s_1 = 6.0/1 = 6.0 \text{ min}$

For station 2, $WL_2/s_2 = 36.0/2 = 18.0 \text{ min}$

For station 3, $WL_3/s_3 = 13.0/1 = 13.0 \text{ min}$

For station 4, the part handling system, $WL_4/s_4 = 9.0/4 = 2.25 \text{ min}$

The maximum ratio occurs at station 2, so it is the bottleneck station that determines the maximum production rate of all parts made by the system.

$$R_p^* = 2/36.0 = 0.05555 \text{ pc/min} = 3.333 \text{ pc/hr}$$

(b) To determine production rate of each product, multiply R_p^* by its respective part mix fraction.

$$R_{pA}^* = 3.333(0.4) = 1.333 \text{ pc/hr}$$

$$R_{pB}^* = 3.333(0.6) = 2.00 \text{ pc/hr}$$

(c) The utilization of each station can be computed using Eq. (19.7):

$$U_1 = (6.0/1)(0.05555) = 0.333 \ (33.3\%)$$

$$U_2 = (36.0/2)(0.05555) = 1.0 \ (100\%)$$

$$U_3 = (13.0/1)(0.05555) = 0.722 \ (72.2\%)$$

$$U_4 = (9.0/4)(0.05555) = 0.125 \ (12.5\%)$$

(d) Mean number of busy servers at each station is determined using Eq. (19.10):

$$BS_1 = 6.0(0.05555) = 0.333$$

$$BS_2 = 36.0(0.05555) = 2.0$$

$$BS_3 = 13.0(0.05555) = 0.722$$

$$BS_4 = 9.0(0.05555) = 0.50$$

The preceding example was designed so that most of the results could be verified without resorting to the bottleneck model. For example, it is fairly obvious that station 2 is the limiting station, even with two servers. The processing times at this station are more than twice those at station 3. Given that station 2 is the bottleneck, let us try to verify the maximum production rate of the FMS. To do this, the reader should note that the processing times at station 2 are $t_{2A2} = 30$ min and $t_{2B2} = 40$ min. Note also that the part mix fractions are $p_A = 0.4$ and $p_B = 0.6$. This means that for every unit of B produced, there are $0.4/0.6 = \frac{2}{3}$ units of part A. The corresponding time to process 1 unit of B and $\frac{2}{3}$ unit of A at station 1 is

$$\frac{2}{3}(30) + 1(40) = 20 + 40 = 60 \text{ min}$$

Sixty minutes is exactly the amount of processing time each machine has available in an hour (this is no coincidence; we designed the problem so this would happen). With two servers (two CNC mills), the FMS can produce parts at the following maximum rate:

$$R_p^* = 2\left(\frac{2}{3} + 1\right) = 2(1.6666) = 3.333 \text{ pc/hr}$$

This is the same result obtained by the bottleneck model. Given that the bottleneck station is working at 100% utilization ($U_2 = 1.0$), it is easy to determine the utilizations of the other stations. At station 1, the time needed to load and unload the output of the two servers at station 2 is

$$3.333(4 + 2) = 20.0 \text{ min}$$

As a fraction of 60 min in an hour, this gives a utilization of $U_1 = 0.333$. At station 3, the processing time required to process the output of the two servers at station 2 is

$$\frac{4}{3}(10) + 2(15) = 43.333 \text{ min}$$

As a fraction of the 60 min, we have $U_3 = 43.333/60 = 0.722$. Using the same approach on the part handling system, we have

$$\frac{4}{3}(9.0) + 2(9.0) = 30.0 \text{ min}$$

As a fraction of 60 min, this is 0.50. However, since there are four servers (four work carriers), this fraction is divided by 4 to obtain $U_4 = 0.125$. These are the same utilization values as in our example using the bottleneck model.

EXAMPLE 19.6   Bottleneck Model on a More Complicated Problem

An FMS consists of four stations. Station 1 is a load/unload station with one server. Station 2 performs milling operations with three servers (three identical CNC milling machines). Station 3 performs drilling operations with two servers (two identical CNC drill presses). Station 4 is an inspection station with one server that performs inspections on a sample of the parts. The stations are connected by a part handling system that has two work carriers and a mean transport time = 3.5 min. The FMS produces four parts, A, B, C, and D. The part mix fractions and process routings for the four parts are presented in the table below. Note that the operation frequency at the inspection station (f_{4jk}) is less than 1.0 to account for the fact that only a fraction of the parts are inspected. Determine: (a) maximum production rate of the FMS, (b) corresponding production rate of each part, (c) utilization of each station in the system, and (d) the overall FMS utilization.

Part j	Part mix p_j	Operation k	Description	Station i	Process time t_{ijk} (min)	Frequency f_{ijk}
A	0.1	1	Load	1	4	1.0
		2	Mill	2	20	1.0
		3	Drill	3	15	1.0
		4	Inspect	4	12	0.5
		5	Unload	1	2	1.0
B	0.2	1	Load	1	4	1.0
		2	Drill	3	16	1.0
		3	Mill	2	25	1.0
		4	Drill	3	14	1.0
		5	Inspect	4	15	0.2
		6	Unload	1	2	1.0
C	0.3	1	Load	1	4	1.0
		2	Drill	3	23	1.0
		3	Inspect	4	8	0.5
		4	Unload	1	2	1.0
D	0.4	1	Load	1	4	1.0
		2	Mill	2	30	1.0
		3	Inspect	4	12	0.333
		4	Unload	1	2	1.0

Solution: (a) We first calculate the workloads at the workstations in order to identify the bottleneck station.

$$WL_1 = (4 + 2)(1.0)(0.1 + 0.2 + 0.3 + 0.4) = 6.0 \text{ min}$$

$$WL_2 = 20(1.0)(0.1) + 25(1.0)(0.2) + 30(1.0)(0.4) = 19.0 \text{ min}$$

$$WL_3 = 15(1.0)(0.1) + 16(1.0)(0.2) + 14(1.0)(0.2) + 23(1.0)(0.3)$$
$$= 14.4 \text{ min}$$

$$WL_4 = 12(0.5)(0.1) + 15(0.2)(0.2) + 8(0.5)(0.3) + 12(0.333)(0.4)$$
$$= 4.0 \text{ min}$$

$$n_t = (3.5)(0.1) + (4.2)(0.2) + (2.5)(0.3) + (2.333)(0.4) = 2.873$$

$$WL_5 = 2.873(3.5) = 10.06 \text{ min}$$

The bottleneck station is identified by the largest WL/s ratio:

For station 1, $WL_1/s_1 = 6.0/1 = 6.0$

For station 2, $WL_2/s_2 = 19.0/3 = 6.333$

For station 3, $WL_3/s_3 = 14.4/2 = 7.2$

For station 4, $WL_4/s_4 = 4.0/1 = 4.0$

For the part handling system, $WL_5/s_5 = 10.06/2 = 5.03$

The maximum ratio occurs at station 3, so it is the bottleneck station that determines the maximum rate of production of the system.

$$R_p^* = 2/14.4 = 0.1389 \text{ pc/min} = 8.333 \text{ pc/hr}$$

(b) To determine the production rate of each product, multiply R_p^* by its respective part mix fraction.

$$R_{pA}^* = 8.333(0.1) = 0.8333 \text{ pc/hr}$$

$$R_{pB}^* = 8.333(0.2) = 1.667 \text{ pc/hr}$$

$$R_{pC}^* = 8.333(0.3) = 2.500 \text{ pc/hr}$$

$$R_{pD}^* = 8.333(0.4) = 3.333 \text{ pc/hr}$$

(c) Utilization of each station can be computed using Eq. (19.7):

$$U_1 = (6.0/1)(0.1389) = 0.833 \qquad (83.3\%)$$

$$U_2 = (19.0/3)(0.1389) = 0.879 \qquad (87.9\%)$$

$$U_3 = (14.4/2)(0.1389) = 1.000 \qquad (100\%)$$

$$U_4 = (4.0/1)(0.1389) = 0.555 \qquad (55.5\%)$$

$$U_5 = (10.06/2)(0.1389) = 0.699 \qquad (69.9\%)$$

(d) Overall FMS utilization can be determined using a weighted average of the above values, where the weighting is based on the number of servers per station and the part handling system is excluded from the average, as in Eq. (19.9):

$$\bar{U}_s = \frac{1(0.833) + 3(0.879) + 2(1.0) + 1(0.555)}{7} = 0.861 \ (86.1\%)$$

In the preceding example, it should be noted that the production rate of part D is constrained by the part mix fractions rather than by the bottleneck station (station 3). Part D is not even processed on the bottleneck station. Instead, it is processed through station 2, which has unutilized capacity. It should therefore be possible to increase the output rate of part D by increasing its part mix fraction and at the same time increasing the utilization of station 2 to 100%. The following example illustrates the method for doing this.

EXAMPLE 19.7   Increasing Unutilized Station Capacity

From Example 19.6, $U_2 = 87.9\%$. Determine the production rate of part D that will increase the utilization of station 2 to 100%.

Solution: Utilization of a workstation is calculated using Eq. 19.7. For station 2,

$$U_2 = \frac{WL_2}{3}(0.1389)$$

Setting the utilization of station 2 to 1.0 (100%), we can solve for the corresponding

$$WL_2 = \frac{1.0(3)}{0.1389} = 21.6 \ \text{min}$$

This compares with the previous workload value of 19.0 min computed in Example 19.6. A portion of the workload for both values is accounted for by parts A and B. This portion is

$$WL_2(A + B) = 20(0.1)(1.0) + 25(0.2)(1.0) = 7.0 \ \text{min}$$

The remaining portions of the workloads are due to part D.

For the workload at 100% utilization, $WL_2(D) = 21.6{-}7.0 = 14.6 \ \text{min}$

For the workload at 87.9% utilization, $WL_2(D) = 19.0{-}7.0 = 12.0 \ \text{min}$

We can now use the ratio of these values to calculate the new (increased) production rate for part D:

$$R_{pD} = \frac{14.6}{12.0}(3.333) = 1.2167(3.333) = 4.055 \ \text{pc/hr.}$$

Production rates of the other three products remains the same as before. Accordingly, the production rate of all parts increases to the following:

$$R_p{}^* = 0.833 + 1.667 + 2.500 + 4.055 = 9.055 \ \text{pc/hr}$$

Although the production rates of the other three products are unchanged, the increase in production rate for part D alters the relative part mix fractions.

The new values are

$$p_A = \frac{0.833}{9.055} = 0.092$$

$$p_B = \frac{1.667}{9.055} = 0.184$$

$$p_C = \frac{2.500}{9.055} = 0.276$$

$$p_D = \frac{4.055}{9.055} = 0.448$$

19.5.2 Extended Bottleneck Model

The bottleneck model assumes that the bottleneck station is utilized 100% and that there are no delays due to queues in the system. This implies on the one hand that there are a sufficient number of parts in the system to avoid starving of workstations and on the other hand that there will be no delays due to queueing. Solberg [27] argued that the assumption of 100% utilization makes the bottleneck model overly optimistic and that a queueing model which accounts for process time variations and delays would more realistically and completely describe the performance of a flexible manufacturing system.

An alternative approach, developed by Mejabi [21], addresses some of the weaknesses of the bottleneck model without resorting to queueing computations (which can be difficult). He called his approach the *extended bottleneck model*. This extended model assumes a closed queueing network in which there are always a certain number of workparts in the FMS. Let N = this number of parts in the system. When one part is completed and exits the FMS, a new raw workpart immediately enters the system, so that N remains constant. The new part may or may not have the same process routing as the one that just departed. The process routing of the entering part is determined according to probability p_j.

N plays a critical role in the operation of the production system. If N is small (say, much smaller than the number of workstations), then some of the stations will be idle due to starving, sometimes even the bottleneck station. In this case, the production rate of the FMS will be less than R_p^* calculated in Eq. (19.5). If N is large (say, much larger than the number of workstations), then the system will be fully loaded, with queues of parts waiting in front of the stations. In this case, R_p^* will provide a good estimate of the production capacity of the system. However, work-in-process (WIP) will be high, and manufacturing lead time (MLT) will be long.

In effect, WIP corresponds to N, and MLT is the sum of processing times at the workstations, transport times between stations, and any waiting time experienced by the parts in the system. We can express MLT as

$$MLT = \sum_{i=1}^{n} WL_i + WL_{n+1} + T_w \tag{19.11}$$

where $\sum_{i=1}^{n} WL_i$ = summation of average workloads over all stations in the FMS, min; WL_{n+1} = workload of the part handling system, min; and T_w = mean waiting time experienced by a part due to queues at the stations, min.

WIP (that is, N) and MLT are correlated. If N is small, then MLT will take on its smallest possible value because waiting time will be short or even zero. If N is large, then

MLT will be long and there will be waiting time in the system. Thus we have two alternative cases, and adjustments must be made in the bottleneck model to account for them. To do this, Mejabi found the well-known Little's formula[4] from queueing theory to be useful. Little's formula establishes the relationship between the mean expected time a unit spends in the system, the mean processing rate of items in the system, and the mean number of units in the system. It can be mathematically proved for a single-station queueing system, and its general validity is accepted for multi-station queueing systems. Using our own symbols, Little's formula can be expressed as

$$N = R_p(MLT) \qquad (19.12)$$

where N = number of parts in the system, pc; R_p = production rate of the system, pc/min; and MLT = manufacturing lead time (time spent in the system by a part), min. Now, let us examine the two cases:

Case 1: When N is small, production rate is less than in the bottleneck case because the bottleneck station is not fully utilized. In this case, the waiting time T_w of a unit is theoretically zero, and Eq. (19.11) reduces to

$$MLT_1 = \sum_{i=1}^{n} WL_i + WL_{n+1} \qquad (19.13)$$

where the subscript in MLT_1 is used to identify case 1. Production rate can be estimated using Little's formula:

$$R_p = \frac{N}{MLT_1} \qquad (19.14)$$

and production rates of the individual parts are given by

$$R_{pj} = p_j R_p \qquad (19.15)$$

As indicated waiting time is assumed to be zero:

$$T_w = 0 \qquad (19.16)$$

Case 2: When N is large, the estimate of maximum production rate provided by Eq. (19.5) should be valid: $R_p^* = s^*/WL^*$, where the asterisk (*) denotes that production rate is constrained by the bottleneck station in the system. The production rates of the individual products are given by

$$R_{pj}^* = p_j R_p^* \qquad (19.17)$$

In this case, average manufacturing lead time is evaluated using Little's formula:

$$MLT_2 = \frac{N}{R_p^*} \qquad (19.18)$$

---

[4]Little's formula is usually given as $L = \lambda W$ where L = expected number of units in the system, λ = processing rate of units in the system, and W = expected time spent by a unit in the system. We are substituting our own symbols: L becomes N, the number of parts in the FMS; λ becomes R_p, the production rate of the FMS; and W becomes MLT, the total time in the FMS, which is the sum of processing and transport times plus any waiting time.

TABLE 19.5 Equations and Guidelines for the Extended Bottleneck Model

Case 1: $N < N^* = R_p^* \left(\sum_{i=1}^{n} WL_i + WL_{n+1} \right)$	Case 2: $N \geq N^* = R_p^* \left(\sum_{i=1}^{n} WL_i + WL_{n+1} \right)$
$MLT_1 = \sum_{i=1}^{n} WL_i + WL_{n+1}$	$R_p^* = \dfrac{s^*}{WL^*}$
$R_p = \dfrac{N}{MLT_1}$	$R_{pj}^* = p_j R_p^*$
$R_{pj} = p_j R_p$	$MLT_2 = \dfrac{N}{R_p^*}$
$T_w = 0$	$T_w = MLT_2 - \left(\sum_{i=1}^{n} WL_i + WL_{n+1} \right)$

The mean waiting time a part spends in the system can be estimated by rearranging Eq. (19.11) to solve for T_w:

$$T_w = MLT_2 - \left(\sum_{i=1}^{n} WL_i + WL_{n+1} \right) \tag{19.19}$$

The decision on whether to use case 1 or case 2 depends on the value of N. The dividing line between cases 1 and 2 is determined by whether N is greater than or less than a critical value given by

$$N^* = R_p^* \left(\sum_{i=1}^{n} WL_i + WL_{n+1} \right) = R_p^*(MLT_1) \tag{19.20}$$

where N^* = critical value of N, the dividing line between the bottleneck and nonbottleneck cases. If $N < N^*$, then case 1 applies. If $N \geq N^*$, then case 2 applies. The applicable equations for the two cases are summarized in Table 19.5.

EXAMPLE 19.8   Extended Bottleneck Model

Let us use the extended bottleneck model on the data given in previous Example 19.5 to compute production rate, manufacturing lead time, and waiting time for three values of N: (a) $N = 2$, (b) $N = 3$, and (c) $N = 4$.

Solution: Let us first compute the critical value of N. We have R_p^* from Example 19.5: $R_p^* = 0.05555$ pc/min. We also need the value of MLT_1. Again using previously calculated values from Example 19.5,

$$MLT_1 = 6.0 + 36.0 + 13.0 + 9.0 = 64.0 \text{ min}$$

The critical value of N is given by Eq. (19.20):

$$N^* = 0.05555(64.0) = 3.555$$

(a) $N = 2$ is less than the critical value, so we apply the equations for case 1.

$$MLT_1 = 64.0 \text{ min (calculated several lines earlier)}$$

$$R_p = \frac{N}{MLT_1} = \frac{2}{64} = 0.03125 \text{ pc/min} = 1.875 \text{ pc/hr}$$

$$T_w = 0 \text{ min}$$

(b) $N = 3$ is again less than the critical value, so case 1 applies.

$$MLT_1 = 64.0 \text{ min}$$

$$R_p = \frac{3}{64} = 0.0469 \text{ pc/min} = 2.813 \text{ pc/hr}$$

$$T_w = 0 \text{ minutes}$$

(c) For $N = 4$, case 2 applies, since $N > N^*$.

$$R_p^* = \frac{s^*}{WL^*} = 0.05555 \text{ pc/min} = 3.333 \text{ pc/hr from Example 19.5.}$$

$$MLT_2 = \frac{4}{0.05555} = 72.0 \text{ min}$$

$$T_w = 72.0 - 64.0 = 8.0 \text{ min}$$

The results of this example typify the behavior of the extended bottleneck model, shown in Figure 19.12. Below N^* (Case 1), MLT has a constant value, and R_p decreases proportionally as N decreases. Manufacturing lead time cannot be less than the sum of the processing and transport times, and production rate is adversely affected by low values of N because stations become starved for work. Above N^* (Case 2), R_p has a constant value equal to R_p^* and MLT increases. No matter how large N is, the roduction rate cannot be greater than the output capacity of the bottleneck station. Manufacturing leadtime increases because backlogs build up at the stations.

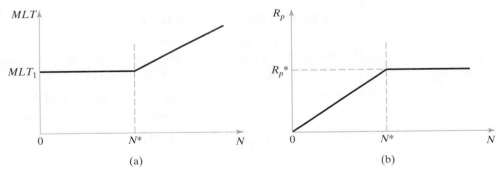

Figure 19.12 General behavior of the extended bottleneck model: (a) manufacturing lead time MLT as a function of N; and (b) production rate R_p as a function of N.

These observations might tempt us to conclude that the optimum N value occurs at N^*, since MLT is at its minimum possible value and R_p is at its maximum possible value. However, caution must be exercised in the use of the extended bottleneck model (and the same caution applies even more so to the conventional bottleneck model, which disregards the effect of N). It is intended to be a rough-cut method to estimate FMS performance in the early phases of FMS design. More reliable estimates of performance can be obtained using computer simulations of detailed models of the FMS—models that include considerations of layout, material handling and storage system, and other system design factors.

19.5.3 Sizing the FMS

The bottleneck model can be used to calculate the number of servers required at each workstation to achieve a specified production rate. Such calculations would be useful during the initial stages of FMS design in determining the "size" (number of workstations and servers) of the system. To make the computation, we need to know the part mix, process routings, and processing times so that workloads can be calculated for each of the stations to be included in the FMS. Given the workloads, the number of servers at each station i is determined as

$$s_i = \text{minimum integer} \geq R_p(WL_i) \qquad (19.22)$$

where s_i = number of servers at station i; R_p = specified production rate of all parts to be produced by the system, pc/min; and WL_i = workload at station i, min. The following example illustrates the procedure.

EXAMPLE 19.9 Sizing the FMS

Suppose the part mix, process routings, and processing times for the family of parts to be machined on a proposed FMS are those given in previous Example 19.6. The FMS will operate 24 hours a day, five days a week, 50 weeks a year. Determine (a) the number of servers that will be required at each station i to achieve an annual production rate of 60,000 parts per year and (b) the utilization of each workstation.

Solution: (a) The number of hours of FMS operation per year will be $24 \times 5 \times 50 = 6000$ hr/yr. The hourly production rate is given by:

$$R_p = \frac{60,000}{6000} = 10.0 \text{ pc/hr} = 0.1667 \text{ pc/min}$$

The workloads at each station were previously calculated in Example 19.6: $WL_1 = 6.0$ min, $WL_2 = 19.0$ min, $WL_3 = 14.4$ min, $WL_4 = 4.0$ min, and $WL_5 = 10.06$ min. Using Eq. (19.22), we have the following number of servers required at each station:

$s_1 = \text{minimum integer} \geq 0.1667(6.0) = 1.000 = 1$ server

$s_2 = \text{minimum integer} \geq 0.1667(19.0) = 3.167$ rounded up to 4 servers

$s_3 = \text{minimum integer} \geq 0.1667(14.4) = 2.40$ rounded up to 3 servers

$s_4 = \text{minimum integer} \geq 0.1667(4.0) = 0.667$ rounded up to 1 server

$s_5 = \text{minimum integer} \geq 0.1754(10.06) = 1.677$ rounded up to 2 servers

(b) The utilization at each workstation is determined as the calculated value of s_i divided by the resulting minimum integer value $\geq s_i$.

$$U_1 = 1.0/1 = 1.0 \qquad (100\%)$$

$$U_2 = 3.167/4 = 0.79233 \quad (79.2\%)$$

$$U_3 = 2.40/3 = 0.800 \qquad (80\%)$$

$$U_4 = 0.667/1 = 0.667 \qquad (66.7\%)$$

$$U_5 = 1.677/2 = 0.839 \qquad (83.9\%)$$

The maximum value is at station 1, the load/unload station. This is the bottleneck station.

Because the number of servers at each workstation must be an integer, station utilization may be less than 100% for most if not all of the stations. In Example 19.9, the load/unload station has a utilization of 100%, but all of the other stations have utilizations less than 100%. The next highest utilization is at station 5, the work transport system. It's a shame that the load/unload station or the work transport system would be the limiting factors on the overall production rate of the FMS. It would be much more desirable for one of the production stations or the inspection station to be the bottleneck. Let us deal with this issue in the following example.

EXAMPLE 19.10   The FMS Sizing Problem Revisited

For previous Example 19.9, (a) make the necessary design changes in the FMS so that the production rate of the system is limited by one of the production or inspection stations rather than by the load/unload station or the transport system. (b) Then, determine the maximum possible annual production rate of the FMS for the 6000 operating hours per year.

Solution: (a) Given that the two highest utilizations in Example 19.9 are $U_1 = 100\%$ and $U_5 = 83.9\%$, it makes sense to increase the number of servers at these stations. This will reduce the utilization for each of these stations, thereby making one of the other stations into the bottleneck. For station 1, currently with 1 server, change s_1 to $s_1 = 2$ servers.

$$U_1 = 1/2 = 0.50 \qquad (50\%)$$

And for station 5, currently with 2 servers, change s_5 to $s_5 = 3$ servers.

$$U_5 = 1.677/3 = 0.559 \quad (55.9\%)$$

With these changes, station 3 ($U_3 = 80\%$) becomes the new bottleneck.

(b) The maximum possible production rate of the FMS can be increased so that the bottleneck station operates at 100% utilization. This is accomplished by dividing the current production rate at 80% utilization ($R_p = 10$ pc/hr) by the current utilization factor ($U_3 = 80\%$).

$$R_p^* = \frac{10.0}{0.80} = 12.5 \text{ pc/hr} = 0.20833 \text{ pc/min}$$

At 6000 operating hours per year, $R_p^* = 12.5(6000) = 75{,}000$ pc/yr

The utilizations at all other stations are affected by this higher production rate. The utilization at each station is calculated using Eq. (19.7), which states $U_i = R_p^*(WL_i/s_i)$:

$$U_1 = 0.20833(6.0/2) = 0.625 \qquad (62.5\%)$$

$$U_2 = 0.20833(19.0/4) = 0.98 \qquad (98.9\%)$$

$$U_3 = 0.20833(14.4/3) = 1.00 \qquad (100\%)$$

$$U_4 = 0.20833(4.0/1) = 0.833 \qquad (83.3\%)$$

$$U_5 = 0.20833(10.06/3) = 0.699 \qquad (69.9\%)$$

19.5.4 What the Equations Tell Us

Despite their limitations, the bottleneck model and extended bottleneck model provide the following practical guidelines on the design and operation of flexible manufacturing systems:

- For a given product or part mix, the total production rate of the FMS is ultimately limited by the productive capacity of the bottleneck station, which is the station with the maximum workload per server.
- If the product or part mix ratios can be relaxed, it may be possible to increase total FMS production rate by increasing the utilization of nonbottleneck workstations.
- The number of parts in the FMS at any time should be greater than the number of servers (processing machines) in the system. A ratio of around two parts per server is probably optimum, assuming that the parts are distributed throughout the FMS to ensure that one part is waiting at every station. This is especially critical at the bottleneck station.
- If work-in-process (number of parts in the system) is kept too low, production rate of the system is impaired.
- If work-in-process is allowed to be too high, then manufacturing lead time will be long with no improvement in production rate.
- As a first approximation, the bottleneck model can be used to estimate the number of servers at each station (number of machines of each type) to achieve a specified overall production rate of the system.

REFERENCES

[1] ASKIN, R. G., H. M. SELIM, and A. J. VAKHARIA, "A Methodology for Designing Flexible Cellular Manufacturing Systems," *IIE Transactions*, Volume 29, 1997, pp. 599–610.

[2] BASNET, C., and J. H. MIZE, "Scheduling and Control of Flexible Manufacturing Systems: A Critical Review," *International Journal of Computer Integrated Manufacturing*, Volume 7, 1994, pp. 340–355.

[3] BROWNE, J., D. DUBOIS, K. RATHMILL, S. P. SETHI, and K. E. STECKE, "Classification of Flexible Manufacturing Systems," *FMS Magazine*, Volume 2, April 1984, pp. 114–117.

[4] Buzacott, J. A. "Modeling Automated Manufacturing Systems," *Proceedings*, 1983 Annual Industrial Engineering Conference, Louisville, Kentucky, May 1983, pp. 341–347.

[5] Buzacott, J. A., and J. G. Shanthikumar, "Models for Understanding Flexible Manufacturing Systems," *AIIE Transactions*, December 1980, pp. 339–349.

[6] Buzacott, J. A., and D. D. Yao, "Flexible Manufacturing Systems: A Review of Analytical Models," *Management Science*, Volume 32, 1986, pp. 890–895.

[7] Falkner, C. H., "Flexibility in Manufacturing Systems," *Proceedings*, Second ORSA/TIMS Conference on Flexible Manufacturing Systems: Operations Research Models and Applications, Edited by K. E. Stecke and R. Suri, Elsevier Science Publishers, NY, 1986, pp. 95–106.

[8] *Flexible Manufacturing Systems Handbook*, prepared by the staff of The Charles Stark Draper Laboratory, Inc., Cambridge, MA, published by Noyes Publications, Park Ridge, NJ, 1984.

[9] Groover, M. P., and O. Mejabi, "Trends in Manufacturing System Design," *Proceedings*, IIE Fall Conference, Nashville, TN, November 1987.

[10] Hartley, J., *FMS at Work*, IFS (Publications) Ltd., Bedford, UK, and North-Holland, Amsterdam, The Netherlands, 1984.

[11] Jablonski, J., "Reexamining FMSs," Special Report 774, *American Machinist*, March 1985, pp. 125–140.

[12] Joshi, S. B., and J. S. Smith, editors, *Computer Control of Flexible Manufacturing Systems*, Chapman & Hall, London, UK, 1994.

[13] Kattan, I. A., "Design and Scheduling of Hybrid Multi-Cell Flexible Manufacturing Systems, *International Journal of Production Research*, Volume 35, 1997, pp. 1239–1257.

[14] Klahorst, H. T., "How To Plan Your FMS," *Manufacturing Engineering*, September 1983, pp. 52–54.

[15] Koelsch, J. R., "A New Look to Transfer Lines," *Manufacturing Engineering*, April 1994, pp. 73–78.

[16] Kouvelis, P., "Design and Planning Problems in Flexible Manufacturing Systems: A Critical Review," *Journal of Intelligent Manufacturing*, Volume 3, 1992, pp. 75–99.

[17] Kusiak, A., and C.-X. Feng, "Flexible Manufacturing," *The Engineering Handbook*, Richard C. Dorf, editor, CRC Press, 1996, pp. 1718–1723.

[18] Lenz, J. E., *Flexible Manufacturing*, Marcel Dekker, Inc., NY, 1989.

[19] Luggen, W. W., *Flexible Manufacturing Cells and Systems*, Prentice Hall, Inc., Englewood Cliffs, NJ, 1991.

[20] Maleki, R. A., *Flexible Manufacturing Systems: The Technology and Management*, Prentice Hall, Inc., Englewood Cliffs, NJ, 1991.

[21] Mejabi, O., "Modeling in Flexible Manufacturing Systems Design," *PhD Dissertation*, Lehigh University, Bethlehem, PA, 1988.

[22] Mohamed, Z. M., *Flexible Manufacturing Systems - Planning Issues and Solutions*, Garland Publishing, Inc., NY, 1994.

[23] Moodie, C., R. Uzsoy, and Y. Yih, editors, *Manufacturing Cells: A Systems Engineering View*, Taylor & Francis, London, UK, 1996.

[24] Rahimifard, S., and S. T. Newman, "Simultaneous Scheduling of Workpieces, Fixtures and Cutting Tools within Flexible Machining Cells," *International Journal of Production Systems*, Volume 15, 1997, pp. 2379–2396.

[25] Solberg, J. J., "A Mathematical Model of Computerized Manufacturing Systems," *Proceedings of the 4th International Conference on Production Research*, Tokyo, Japan, 1977.

[26] Solberg, J. J., "CAN-Q User's Guide," Report No. 9 (Revised), NSF Grant No. APR74-15256, Purdue University, School of Industrial Engineering, West Lafayette, IN, 1980.

[27] SOLBERG, J. J., "Capacity Planning with a Stochastic Workflow Model," *AIIE Transactions*, Volume 13, No. 2, 1981, pp. 116–122.

[28] STECKE, K. E., "Formulation and Solution of Nonlinear Integer Production Planning Problems for Flexible Manufacturing Systems," *Management Science*, Volume 29, 1983, pp. 273–288.

[29] STECKE, K. E., "Design, Planning, Scheduling and Control Problems of FMS," *Proceedings*, First ORSA/TIMS Special Interest Conference on Flexible Manufacturing Systems, Ann Arbor, MI, 1984.

[30] STECKE, K. E., and J. J. Solberg, "The Optimality of Unbalancing Both Workloads and Machine Group Sizes in Closed Queueing Networks of Multiserver Queues," *Operational Research*, Volume 33, 1985, pp. 822–910.

[31] SURI, R., "An Overview of Evaluative Models for Flexible Manufacturing Systems," *Proceedings*, First ORSA/TIMS Special Interest Conference on Flexible Manufacturing Systems, University of Michigan, Ann Arbor, MI, August 1984, pp. 8–15.

[32] VACCARI, J. A., "Forging in the Age of the FMS," Special Report 782, *American Machinist*, January 1986, pp. 101–108.

[33] WATERBURY, R., "FMS Expands into Assembly," *Assembly Engineering*, October 1985, pp. 34–37.

[34] WINSHIP, J. T., "Flexible Sheetmetal Fabrication," Special Report 779, *American Machinist*, August 1985, pp. 95–106.

[35] WU, S. D., and R. A. WYSK, "An Application of Discrete-Event Simulation to On-line Control and Scheduling in Flexible Manufacturing," *International Journal of Production Research*, Volume 27, 1989, pp. 247–262.

REVIEW QUESTIONS

19.1 Name three production situations in which FMS technology can be applied.

19.2 What is a flexible manufacturing system?

19.3 What are the three capabilities that a manufacturing system must possess in order to be flexible?

19.4 Name the four tests of flexibility that a manufacturing system must satisfy in order to be classified as flexible.

19.5 What is the dividing line between a flexible manufacturing cell and a flexible manufacturing system, in terms of the number of workstations in the system?

19.6 What is the difference between a dedicated FMS and a random-order FMS?

19.7 What are the four basic components of a flexible manufacturing system?

19.8 What are three of the five functions of the material handling and storage system in a flexible manufacturing system?

19.9 What is the difference between the primary and secondary handling systems in flexible manufacturing systems?

19.10 Name four of the five categories of layout configurations that are found in a flexible manufacturing system.

19.11 Name four of the seven functions performed by human resources in an FMS.

19.12 What are four benefits that can be expected from a successful FMS installation?

PROBLEMS

Bottleneck Model

19.1 A flexible manufacturing cell consists of a load/unload station and two machining workstations. The load/unload station is station 1. Station 2 performs milling operations and consists of one server (one CNC milling machine). Station 3 has one server that performs drilling (one CNC drill press). The three stations are connected by a part handling system that has one work carrier. The mean transport time is 2.5 min. The FMC produces three parts, A, B, and C. The part mix fractions and process routings for the three parts are presented in the table below. The operation frequency $f_{ijk} = 1.0$ for all operations. Determine (a) maximum production rate of the FMC, (b) corresponding production rates of each product, (c) utilization of each machine in the system, and (d) number of busy servers at each station.

Part j	Part mix p_j	Operation k	Description	Station i	Process time t_{ijk} (min)
A	0.2	1	Load	1	3
		2	Mill	2	20
		3	Drill	3	12
		4	Unload	1	2
B	0.3	1	Load	1	3
		2	Mill	2	15
		3	Drill	3	30
		4	Unload	1	2
C	0.5	1	Load	1	3
		2	Drill	3	14
		3	Mill	2	22
		4	Unload	1	2

19.2 Solve Problem 19.1 assuming the number of servers at station 2 (CNC milling machines) = 3 and the number of servers at station 3 (CNC drill presses) = 2. Note that with the increase in the number of machines from two to five, the FMC is now a FMS according to our definitions in Section 19.1.2.

19.3 A FMS consists of a load/unload station and three workstations. Station 1 loads and unloads parts using two servers (material handling workers). Station 2 performs horizontal milling operations with two servers (identical CNC horizontal milling machines). Station 3 performs vertical milling operations with three servers (identical CNC vertical milling machines). Station 4 performs drilling operations with two servers (identical drill presses). The machines are connected by a part handling system that has two work carriers and a mean transport time = 3.5 min. The FMS produces four parts, A, B, C, and D, whose part mix fractions and process routings are presented in the table below. The operation frequency $f_{ijk} = 1.0$ for all operations. Determine (a) maximum production rate of the FMS, (b) utilization of each machine in the system, and (c) average utilization of the system 19.4 $\overline{U}_s$.

Part j	Part mix p_j	Operation k	Description	Station i	Process time t_{ijk} (min)
A	0.2	1	Load	1	4
		2	H. Mill	2	15
		3	V. Mill	3	14
		4	Drill	4	13
		5	Unload	1	3
B	0.2	1	Load	1	4
		2	Drill	4	12
		3	H. Mill	2	16
		4	V. Mill	3	11
		5	Drill	4	17
		6	Unload	1	3
C	0.25	1	Load	1	4
		2	H. Mill	2	10
		3	Drill	4	9
		4	Unload	1	3
D	0.35	1	Load	1	4
		2	V. Mill	3	18
		3	Drill	4	8
		4	Unload	1	3

19.4 Solve Problem 19.3 using the number of carriers in the part handling system = 3.

19.5 Suppose it is decided to increase the utilization of the two non-bottleneck machining stations in the FMS of Problem 19.4 by introducing a new part, part E, into the mix. If the new product will be produced at a rate of two units per hour, what would be the ideal process routing (sequence and processing times) for part E that would increase the utilization of the two non-bottleneck machining stations to 100% each? The respective production rates of parts A, B, C, and D will remain the same as in Problem 19.4. Disregard the utilizations of the load/unload station and the part handling system.

19.6 A semi-automated flexible manufacturing cell is used to produce three products. The products are made by two automated processing stations followed by an assembly station. There is also a load/unload station. Material handling between stations in the FMC is accomplished by mechanized carts that move tote bins containing the particular components to be processed and then assembled into a given product. The carts transfer tote bins between stations. In this way the carts are kept busy while the tote bins are queued in front of the workstations. Each tote bin remains with the product throughout processing and assembly. The details of the FMC can be summarized as follows:

Station	Description	Number of Servers
1	Load and unload	2 human workers
2	Process X	1 automated server
3	Process Y	1 automated server
4	Assembly	2 human workers
5	Transport	Number of carriers to be determined

The product mix fractions and station processing times for the parts are presented in the table that follows. The same station sequence is followed by all products: $1 \rightarrow 2 \rightarrow 3 \rightarrow 4 \rightarrow 1$.

Product j	Product mix p_j	Station 1 (min)	Station 2 (min)	Station 3 (min)	Station 4 (min)	Station 1 (min)
A	0.35	3	9	7	5	2
B	0.25	3	5	8	5	2
C	0.40	3	4	6	8	2

The average cart transfer time between stations is 4 minutes. (a) What is the bottleneck station in the FMC, assuming that the material handling system is not the bottleneck? (b) At full capacity, what is the overall production rate of the system and the rate for each product? (c) What is the minimum number of carts in the material handling system required to keep up with the production workstations? (d) Compute the overall utilization of the FMC. (e) What recommendations would you make to improve the efficiency and/or reduce the cost of operating the FMC?

Extended Bottleneck Model

19.7 Use the extended bottleneck model to solve problem 19.1 with the following number of parts in the system: (a) $N = 2$ parts and (b) $N = 4$ parts. Also determine the manufacturing lead time for the two cases of N in (a) and (b).

19.8 Use the extended bottleneck model to solve problem 19.2 with the following number of parts in the system: (a) $N = 3$ parts and (b) $N = 6$ parts. Also determine the manufacturing lead time for the two cases of N in (a) and (b).

19.9 Use the extended bottleneck model to solve problem 19.3 with the following number of parts in the system: (a) $N = 5$ parts, (b) $N = 8$ parts, and (c) $N = 12$ parts. Also determine the manufacturing lead time for the three cases of N in (a), (b), and (c).

19.10 Use the extended bottleneck model to solve Problem 19.4 with the following number of parts in the system: (a) $N = 5$ parts, (b) $N = 8$ parts, and (c) $N = 12$ parts. Also determine the manufacturing lead time for the three cases of N in (a), (b), and (c).

19.11 For the data given in Problem 19.6, use the extended bottleneck model to develop the relationships for production rate R_p and manufacturing lead time MLT each as a function of the number of parts in the system N. Plot the relationships in the format of Figure 19.12.

19.12 A flexible manufacturing system is used to produce three products. The FMS consists of a load/unload station, two automated processing stations, an inspection station, and an automated conveyor system with an individual cart for each product. The conveyor carts remain with the parts during their time in the system, so the mean transport time includes not only the move time, but also the average total processing time per part. The number of servers at each station is given in the following table:

Station 1	Load and unload	2 workers
Station 2	Process X	3 servers
Station 3	Process Y	4 servers
Station 4	Inspection	1 server
Transport system	Conveyor	8 carriers

All parts follow either of two routings, which are $1 \rightarrow 2 \rightarrow 3 \rightarrow 4 \rightarrow 1$ or $1 \rightarrow 2 \rightarrow 3 \rightarrow 1$, the difference being that inspections at station 4 are performed on only one part in four for each product ($f_{4jk} = 0.25$). The product mix and process times for the parts are presented in the table below:

Product j	Part mix p_j	Station 1 (min)	Station 2 (min)	Station 3 (min)	Station 4 (min)	Station 1 (min)
A	0.2	5	15	25	20	4
B	0.3	5	10	30	20	4
C	0.5	5	20	10	20	4

The move time between stations is four minutes. (a) Using the bottleneck model, show that the conveyor system is the bottleneck in the present FMS configuration, and determine the overall production rate of the system. (b) Determine how many carts are required to eliminate the conveyor system as the bottleneck. (c) With the number of carts determined in (b), use the extended bottleneck model to determine the production rate for the case when $N = 8$; that is, when only eight parts are allowed in the system even though the conveyor system has enough carriers to handle more than eight. (d) How close are your answers in (a) and (c)? Why?

19.13 A group technology cell is organized to produce a particular family of products. The cell consists of three processing stations, each with one server; an assembly station with three servers; and a load/unload station with two servers. A mechanized transfer system moves the products between stations. The transfer system has a total of six transfer carts. Each cart includes a workholder that holds the products during their processing and assembly, and therefore, each cart must remain with the product throughout processing and assembly. The cell resources can be summarized as follows:

Station	Description	Number of Servers
1	Load and unload	2 workers
2	Process X	1 server
3	Process Y	1 server
4	Process Z	1 server
5	Assembly	3 workers
6	Transport system	6 carriers

The GT cell is currently used to produce four products. All products follow the same routing, which is $1 \rightarrow 2 \rightarrow 3 \rightarrow 4 \rightarrow 5 \rightarrow 1$. The product mix and station times for the parts are presented in the table below:

Product j	Product mix p_j	Station 1 (min)	Station 2 (min)	Station 3 (min)	Station 4 (min)	Station 5 (min)	Station 1 (min)
A	0.35	4	8	5	7	18	2.5
B	0.25	4	4	8	6	14	2.5
C	0.10	4	2	6	5	11	2.5
D	0.30	4	6	7	10	12	2.5

The average transfer between stations takes 2 minutes in addition to the time spent at the workstation. (a) Determine the bottleneck station in the GT cell and the critical valwe of N. Compute the overall production rate and manufacturing lead time of the cell, given tiat the number of parts in the system $= N^*$. If N^* is not an integer, use the integer that is closest to N^*. (b) Compute the overall production rate and manufacturing lead time of the cell, given that the number of parts in the system $= N^* + 10$. If N^* is not an integer, use the integer that is closest to $N^* + 10$. (c) Compute the utilizations of the six stations.

19.14 In Problem 19.13, compute the average manufacturing lead times for each product for the two cases: (a) $N = N^*$, and (b) $N = N^* + 10$. If N^* is not an integer, use the integers that are closest to N^* and $N^* + 10$, respectively.

19.15 In Problem 19.13, what could be done to (a) increase the production rate and/or (b) reduce the operating costs of the cell in light of your analysis? Support your answers with calculations.

19.16 A flexible manufacturing cell consists of a manual load/unload station, three CNC machines, and an automated guided vehicle system (AGVS) with two vehicles. The vehicles deliver parts to the individual machines and then go perform other work. The workstations are listed in the table below, where the AGVS is listed as station 5.

Station	Description	Servers
1	Load and unload	1 worker
2	Milling	1 CNC milling machine
3	Drilling	1 CNC drill press
4	Grinding	1 CNC grinding machine
5	AGVS	2 vehicles

The FMC is used to machine four workparts. The product mix, routings, and processing times for the parts are presented in this table:

Part j	Part Mix p_j	Station Routing	Station 1 (min)	Station 2 (min)	Station 3 (min)	Station 4 (min)	Station 1 (min)
A	0.25	$1 \to 2 \to 3 \to 4 \to 1$	4	8	7	18	2
B	0.33	$1 \to 3 \to 2 \to 1$	4	9	10	0	2
C	0.12	$1 \to 2 \to 4 \to 1$	4	10	0	14	2
D	0.30	$1 \to 2 \to 4 \to 3 \to 1$	4	6	12	16	2

The mean travel time of the AGVS between any two stations in the FMC is 3 min which includes the time required to transfer loads to and from the stations. Given that the loading on the system is maintained at 10 parts (10 workparts in the system at all times), use the extended bottleneck model to determine (a) the bottleneck station, (b) the production rate of the system and the average time to complete a unit of production, and (c) the overall utilization of the system, not including the AGVS.

Sizing the FMS

19.17 A flexible manufacturing system is used to produce four parts. The FMS consists of one load/unload station and two automated processing stations (processes X and Y). The number of servers for each station type is to be determined. The FMS also includes an automated conveyor system with individual carts to transport parts between servers. The carts move the parts from one server to the next, drop them off, and proceed to the next delivery task. Average time required per transfer is 3.5 minutes. The following table summarizes the FMS:

Station 1	Load and unload	Number of human servers (workers) to be determined
Station 2	Process X	Number of automated servers to be determined
Station 3	Process Y	Number of automated servers to be determined
Station 4	Transport system	Number of carts to be determined

All parts follow the same routing, which is $1 \rightarrow 2 \rightarrow 3 \rightarrow 1$. The product mix and processing times at each station are presented in this table:

Product j	Product mix p_j	Station 1 (min)	Station 2 (min)	Station 3 (min)	Station 1 (min)
A	0.1	3	15	25	2
B	0.3	3	40	20	2
C	0.4	3	20	10	2
D	0.2	3	30	5	2

Required production is 10 parts per hour, distributed according to the product mix indicated. Use the bottleneck model to determine (a) the minimum number of servers at each station and the minimum number of carts in the transport system that are required to satisfy production demand and (b) the utilization of each station for the answers above.

19.18 A flexible machining system is being planned that will consist of four workstations plus a part handling system. Station 1 will be a load/unload station. Station 2 will consist of horizontal machining centers. Station 3 will consist of vertical machining centers. Station 4 will be an inspection station. For the part mix that will be processed by the FMS, the workloads at the four stations are as follows: $WL_1 = 7.5$ min, $WL_2 = 22.0$ min, $WL_3 = 18.0$ min, and $WL_4 = 10.2$ min. The workload of the part handling system $WL_5 = 8.0$ min. The FMS will be operated 16 hours per day, 250 days per year. Maintenance will be performed during nonproduction hours, so uptime proportion (availability) is expected to be 97%. Annual production of the system will be 50,000 parts. Determine the number of machines (servers) of each type (station) required to satisfy production requirements.

19.19 In Problem 19.18, determine (a) the utilizations of each station in the system for the specified production requirements, and (b) the maximum possible production rate of the system if the bottleneck station were to operate at 100% utilization.

19.20 Given the part mix, process routings, and processing times for the three parts in Problem 19.1: The FMS planned for this part family will operate 250 days per year and the anticipated availability of the system is 90%. Determine how many servers at each station will be required to achieve an annual production rate of 40,000 parts per year if (a) the FMS operates 8 hours per day, (b) 16 hours per day, and (c) 24 hours per day. (d) Which system configuration is preferred, and why?

19.21 Given the part mix, process routings, and processing times for the four parts in Problem 19.3: The FMS proposed to machine these parts will operate 20 hours per day, 250 days per year. Assume system availability = 95%. Determine (a) the number of servers at each station required to achieve an annual production rate of 75,000 parts per year, and (b) the utilization of each workstation. (c) What is the maximum possible annual production rate the system if the bottleneck station were to operate at 100% utilization?

Quality Programs
for Manufacturing

CHAPTER CONTENTS

From Chapter 20 of *Automation, Production Systems, and Computer-Integrated Manufacturing*, Third Edition.
Mikell P. Groover. Copyright © 2008 by Pearson Education, Inc. Publishing as Prentice Hall. All rights reserved.

In the United States, quality control (QC) has traditionally been concerned with detecting poor quality in manufactured products and taking corrective action to eliminate it. Operationally, QC has often been limited to inspecting the product and its components, and deciding whether the dimensions and other features conformed to design specifications. If they did, the product was shipped. The modern view of quality control encompasses a broader scope of activities that are accomplished throughout the enterprise, not just by the inspection department. The quality programs described in this chapter reflect this modern view. The common objective of these programs is to assure that a product will satisfy or even surpass the needs and requirements of the customer.

This part of the book contains three chapters dealing with quality control systems. The position of the quality control systems in the larger production system is shown in Figure 20.1. Our block diagram shows QC as a manufacturing support system, but it also includes inspection procedures and equipment used in the factory. Inspection is the subject of Chapters 21 and 22. Chapter 21 examines inspection principles and practices used in manufacturing systems, while Chapter 22 describes the various technologies used to accomplish inspection and measurement. The present chapter discusses several quality-related programs that are widely used throughout industry. The list of these programs is indicated in our chapter contents above. We begin our coverage with some general issues on quality and QC.

20.1 QUALITY IN DESIGN AND MANUFACTURING

Two aspects of quality in a manufactured product must be distinguished [12]: (1) product features, and (2) freedom from deficiencies. *Product features* are the characteristics of a product that result from design; they are the functional and aesthetic features of the product intended to appeal to and provide satisfaction to the customer. In an automobile, these

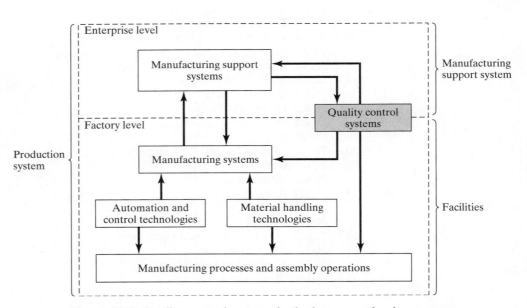

Figure 20.1 Quality control systems in the larger production system.

features include the size of the car, the arrangement of the dashboard, the fit and finish of the body, and similar aspects. They also include the available options for the customer to choose. Table 20.1 lists some of the important general product features.

The sum of the features of a product usually defines its *grade*, which relates to the level in the market at which the product is aimed. Cars and most other products come in different grades. Certain cars provide basic transportation because that is what some customers want, while others are upscale for consumers willing to spend more to own a "better product." The features are decided in design, and they generally determine the inherent cost of the product. Superior features and more of them translates to higher cost.

Freedom from deficiencies means that the product does what it is supposed to do (within the limitations of its design features) and that it is absent of defects and out-of-tolerance conditions (see Table 20.1). This aspect of quality applies to the individual components of the product as well as the product itself. Achieving freedom from deficiencies means producing the product in conformance with design specifications, which is the responsibility of the manufacturing departments. Although the inherent cost to make a product is a function of its design, minimizing the product's cost to the lowest possible level within the limits set by its design is largely a matter of avoiding defects, tolerance deviations, and other errors during production. Costs of these deficiencies include scrapped parts, larger lot sizes for scrap allowances, rework, reinspection, sortation, customer complaints and returns, warranty costs and customer allowances, lost sales, and lost goodwill in the marketplace.

To summarize, product features are the aspect of quality for which the design department is responsible. Product features determine to a large degree the price that a company can charge for its products. Freedom from deficiencies is the quality aspect for which the manufacturing departments are responsible. The ability to minimize these deficiencies strongly influences the cost of the product. These are generalities that oversimplify the way things work, because the responsibility for high quality extends well beyond the design and manufacturing functions in an organization.

20.2 TRADITIONAL AND MODERN QUALITY CONTROL

The principles and approaches to quality control have evolved during the 20th century. Early applications of QC were associated with the developing field of statistics. Since the 1980s, global competition and the demand of the consuming public for high quality

TABLE 20.1 Two Aspects of Quality
(Compiled from Juran and Gryna [12] and Other Sources)

Product Features	Freedom From Deficiencies
Design configuration, size, weight	Absence of defects
Function and performance	Conformance to specifications
Distinguishing features of the model	Components within tolerance
Aesthetic appeal	No missing parts
Ease of use	No early failures
Availability of options	
Reliability and dependability	
Durability and long service life	
Serviceability	
Reputation of product and producer	

products have resulted in a modern view of quality control, which includes programs such as statistical process control, Six Sigma, and ISO 9000.

20.2.1 Traditional Quality Control

Traditional QC focused on inspection. In many factories, the only department responsible for quality control was the inspection department. Much attention was given to sampling and statistical methods, which were termed statistical quality control. In *statistical quality control* (SQC), inferences are made about the quality of a population of manufactured items (e.g., components, subassemblies, products) based on a sample taken from the population. The sample consists of one or more of the items drawn at random from the population. Each item in the sample is inspected for certain quality characteristics of interest. In the case of a manufactured part, these characteristics relate to the process or processes just completed. For example, a cylindrical part may be inspected for diameter following the turning operation that generated it.

Two statistical sampling methods dominate the field of statistical quality control: (1) control charts and (2) acceptance sampling. A *control chart* is a graphical technique in which statistics on one or more process parameters of interest are plotted over time to determine if the process is behaving normally or abnormally. The chart has a central line that indicates the value of the process mean under normal operation. Abnormal process behavior is identified when the process parameter strays significantly from the process mean. Control charts are widely used in statistical process control, the topic of Section 20.4.

Acceptance sampling is a statistical technique in which a sample drawn from a batch of parts is inspected, and a decision is made whether to accept or reject the batch on the basis of the quality of the sample. Acceptance sampling is traditionally used for various purposes: (1) verifying quality of raw materials received from a vendor, (2) deciding whether or not to ship a batch of parts or products to a customer, and (3) inspecting parts between steps in a manufacturing sequence.

In statistical sampling, which includes both control charts and acceptance sampling, there are risks that defects will slip through the inspection process and that defective products will be delivered to the customer. With the growing demand for 100% good quality rather than even a small fraction of defective product, the use of sampling procedures has declined over the past several decades in favor of 100% automated inspection. We discuss these inspection principles in Chapter 21 and the associated technologies in Chapter 22.

The management principles and practices that characterized traditional quality control included the following [7]:

- Customers are external to the organization. The sales and marketing department is responsible for relations with customers.
- The company is organized by functional departments. There is little appreciation of the interdependence of the departments in the larger enterprise. The loyalty and viewpoint of each department tends to be centered on itself rather than on the corporation.
- Quality is the responsibility of the inspection department. The quality function in the organization emphasizes inspection and conformance to specifications. Its objective is simple: eliminate defects.
- Inspection follows production. The objectives of production (to ship product) often clash with the objectives of quality control (to ship only good product).

- Knowledge of statistical quality control techniques reside only in the minds of the QC experts in the organization. Workers' responsibilities are limited to following instructions. Managers and technical staff do all the planning.

20.2.2 The Modern View of Quality Control

High quality is achieved by a combination of good management and good technology. The two factors must be integrated to achieve an effective quality system in an organization. The management factor is captured in the frequently used term "total quality management." The technology factor includes traditional statistical tools combined with modern measurement and inspection technologies.

Total Quality Management. *Total quality management* (TQM) is a management approach that pursues three main objectives: (1) achieving customer satisfaction, (2) continuously improving, and (3) involving the entire work force. These objectives contrast sharply with the practices of traditional management regarding the quality control function. Compare the following factors, which reflect the modern view of quality management, with the preceding list that characterizes the traditional approach to quality management:

- Quality is focused on customer satisfaction. Products are designed and manufactured with this quality focus. "Quality is customer satisfaction" defines the requirement for any product.[1] The product features must be established to achieve customer satisfaction. The product must be manufactured free of deficiencies.
- Included in the focus on customers is the notion that there are internal customers as well as external customers. External customers are those who buy the company's products. Internal customers are departments or individuals inside the company who are served by other departments and individuals in the organization. The final assembly department is the customer of the parts production departments, the engineer is the customer of the technical staff support group, and so forth.
- The quality goals of the organization are driven by top management, which determines the overall attitude toward quality in a company. The quality goals of a company are not established in manufacturing; they are defined at the highest levels of the organization. Does the company want to simply meet specifications set by the customer, or does it want to make products that go beyond the technical specifications? Does it want to be known as the lowest price supplier, or as the highest quality producer in its industry? Answers to these kinds of questions define the quality goals of the company. These must be set by top management. Through the goals they define, the actions they take, and the examples they set, top management determines the overall attitude toward quality in the company.
- Quality control is not just the job of the inspection department, it is pervasive in the organization. It extends from the top of the organization through all levels. It is understood that product design has an important influence on product quality. Decisions made in product design directly impact the quality that can be achieved in manufacturing.
- In manufacturing, the view is that inspecting the product after it is made is not good enough. Quality must be built into the product. Production workers must

[1]The statement is attributed to J. M. Juran [Juran 1993].

inspect their own work and not rely on the inspection department to find their mistakes.

- The pursuit of high quality extends outside the immediate organization to the suppliers. One of the tenets of a modern QC system is to develop close relationships with suppliers.
- High product quality is a process of continuous improvement. It is a never-ending chase to design better products and then to manufacture them better.

Quality Control Technologies. Good technology also plays an important role in achieving high quality. Modern technologies in quality control include (1) quality engineering and (2) quality function deployment. Quality engineering is discussed in Section 20.7. The topic of quality function deployment is related to product design and we discuss it in Chapter 23 (Section 23.4). Other technologies in modern quality control include (3) 100% automated inspection, (4) on-line inspection, (5) coordinate measurement machines for dimensional measurement, and (6) non contact sensors such as machine vision for inspection. These topics are discussed in the following two chapters.

20.3 PROCESS VARIABILITY AND PROCESS CAPABILITY

Before describing the various quality programs, it is appropriate to discuss process variability, the reason for needing these programs. In any manufacturing operation, variability exists in the process output. In a machining operation, which is one of the most accurate manufacturing processes, the machined parts may appear to be identical, but close inspection reveals dimensional differences from one part to the next.

20.3.1 Process Variations

Manufacturing process variations can be divided into two types: (1) random and (2) assignable. *Random variations* result from intrinsic variability in the process, no matter how well designed or well controlled it is. All processes are characterized by these kinds of variations, if one looks closely enough. Random variations cannot be avoided; they are caused by factors such as inherent human variability from one operation cycle to the next, minor variations in raw materials, and machine vibration. Individually, these factors may not amount to much, but collectively the errors can be significant enough to cause trouble unless they are within the tolerances specified for the part. Random variations typically form a normal statistical distribution. The output of the process tends to cluster about the mean value, in terms of the product's quality characteristic of interest, such as part length or diameter. A large proportion of the population is centered around the mean, with fewer parts away from the mean. When the only variations in the process are of this type, the process is said to be *in statistical control*. This kind of variability will continue so long as the process is operating normally. It is when the process deviates from this normal operating condition that variations of the second type appear.

Assignable variations indicate an exception from normal operating conditions. Something has occurred in the process that is not accounted for by random variations. Reasons for assignable variations include operator mistakes, defective raw materials, tool failures, and equipment malfunctions. Assignable variations in manufacturing usually betray themselves by causing the output to deviate from the normal distribution. The process is suddenly *out of statistical control*.

Let us expand on our descriptions of random and assignable variations with reference to Figure 20.2. The variation of some part characteristic of interest is shown at four points in time, t_0, t_1, t_2, and t_3. These are the times during operation of the process when samples are taken to assess the distribution of values of the part characteristic. At sampling time t_0 the process is operating in statistical control, and the variation in the part characteristic follows a normal distribution whose mean $= \mu_0$ and standard deviation $= \sigma_0$. This represents the inherent variability of the process during normal operation. The process is in statistical control. At sampling time t_1 an assignable variation has been introduced into the process, which is manifested by an increase in the process mean ($\mu_1 > \mu_0$). The process standard deviation seems unchanged ($\sigma_1 = \sigma_0$). At time t_2 the process mean seems to have assumed its normal value ($\mu_2 = \mu_0$), but the variation about the process mean has increased ($\sigma_2 > \sigma_0$). Finally, at sampling time t_3, both the mean and standard deviation of the process are observed to have increased ($\mu_3 > \mu_0$ and $\sigma_3 > \sigma_0$).

Using statistical methods based on the preceding distinction between random and assignable variations, it should be possible to periodically observe the process by collecting measurements of the part characteristic of interest and thereby detect when the process has gone out of statistical control. The most applicable statistical method for doing this is the control chart (Section 20.4.1).

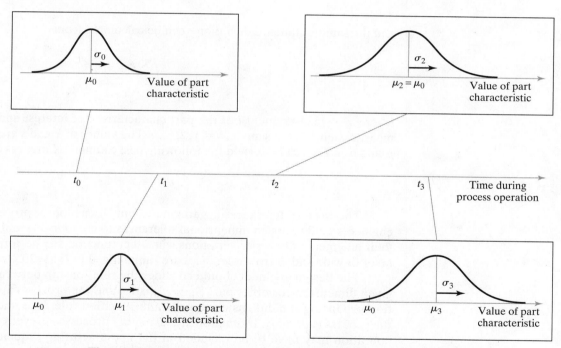

Figure 20.2 Distribution of values of a part characteristic of interest at four times during process operation: at t_0 process is in statistical control; at t_1 process mean has increased; at t_2 process standard deviation has increased; and at t_3 both process mean and standard deviation have increased.

20.3.2 Process Capability and Tolerances

Process capability relates to the normal variations inherent in the output when the process is in statistical control. By definition, *process capability* equals ±3 standard deviations about the mean output value (a total range of 6 standard deviations):

$$PC = \mu \pm 3\sigma \tag{20.1}$$

where PC = process capability; μ = process mean, which is set at the nominal value of the product characteristic; and σ = standard deviation of the process. Assumptions underlying this definition are that (1) the output is normally distributed, and (2) steady state operation has been achieved and the process is in statistical control. Under these assumptions, 99.73% of the parts produced will have output values that fall within ±3.0σ of the mean.

The process capability of a given manufacturing operation is not always known (in fact, it is rarely known), and the characteristic of interest must be measured to assess it. These measurements form a sample, and so the parameters μ and σ in Eq. (20.1) must be estimated from the sample average and the sample standard deviation, respectively. The sample average $\overline{x}$ is given by

$$\overline{x} = \frac{\sum_{i=1}^{n} x_i}{n} \tag{20.2}$$

and the sample standard deviation s can be calculated from

$$s = \sqrt{\frac{\sum_{i=1}^{n} (x_i - \overline{x})^2}{n-1}} \tag{20.3}$$

where x_i = measurement i of the part characteristic of interest and n = the number of measurements in the sample, $i = 1, 2, \ldots n$. The values of $\overline{x}$ and s are then substituted for μ and σ in Eq. (20.1) to yield the following best estimate of process capability:

$$PC = \overline{x} \pm 3s \tag{20.4}$$

The issue of tolerances is germane to our discussion of process capability. Design engineers tend to assign dimensional tolerances to components and assemblies based on their judgment of how size variations will affect function and performance. The factors in favor of wide and narrow tolerances are summarized in Table 20.2.

The design engineer should consider the relationship between the tolerance on a given dimension (or other part characteristic) and the process capability of the operation producing the dimension. Ideally, the specified tolerance should be greater than the process capability. If function and available processes prevent this, then a sortation operation may have to be included in the manufacturing sequence to separate parts that are within the tolerance from those that are beyond. This sortation step increases part cost.

When design tolerances are specified as being equal to process capability, then the upper and lower boundaries of this range define the *natural tolerance limits*. It is useful to

TABLE 20.2 Factors in Favor of Wide and Narrow Tolerances

Factors in favor of wide (loose) tolerances	Factors in favor of narrow (tight) tolerances
Yield in manufacturing is increased. Fewer defects are produced.	Parts interchangeability is increased in assembly. Fit and finish of the assembled product is better, for greater aesthetic appeal.
Fabrication of special tooling (dies, jigs, molds, etc.) is easier. Tools are therefore less costly.	Product functionality and performance are likely to be improved.
Setup and tooling adjustment is easier.	
Fewer production operations may be needed.	Durability and reliability of the product may be increased.
Less skilled, lower cost labor can be used.	
Machine maintenance may be reduced.	Serviceability of the product in the field is likely to be improved due to increased parts interchangeability.
The need for inspection may be reduced.	
Overall manufacturing cost is reduced.	Product may be safer in use.

know the ratio of the specified tolerance range relative to the process capability. This ratio, called the *process capability index*, is defined as

$$PCI = \frac{UTL - LTL}{6\sigma} \qquad (20.5)$$

where PCI = process capability index, UTL = upper tolerance limit of the tolerance range, LTL = lower tolerance limit, and 6σ = range of the natural tolerance limits. The underlying assumptions in this definition are that (1) bilateral tolerances are used and (2) the process mean is set equal to the nominal design specification, so that the numerator and denominator in Eq. (20.5) are centered about the same value.

Table 20.3 shows how defect rate (fraction of parts that are out of tolerance) varies with process capability index. It is clear that any increase in the tolerance range will reduce the percentage of nonconforming parts. The desire to achieve very low fraction defect rates has led to the popular notion of "six sigma" limits in quality control (bottom row in Table 20.1). Achieving six sigma limits virtually eliminates defects in manufactured product. We discuss the Six Sigma quality program in Sections 20.5 and 20.6.

TABLE 20.3 Defect Rate as a Function of Process Capability Index for a Process Operating in Statistical Control

Process Capability Index (PCI)	Tolerance = Number of Standard Deviations	Defect Rate(%)	Defective Parts per Million	Comments
0.333	± 1.0	31.74	317,400	Sortation required
0.667	± 2.0	4.56	45,600	Sortation required
1.000	± 3.0	0.27	2,700	Tolerance = process capability
1.333	± 4.0	0.0063	63	Low occurrence of defects
1.667	± 5.0	0.000057	0.57	Rare occurrence of defects
2.000	± 6.0	0.0000002	0.002	Defects almost never occur

20.4 STATISTICAL PROCESS CONTROL

Statistical process control (SPC) involves the use of various methods to measure and analyze a process. SPC methods are applicable in both manufacturing and nonmanufacturing situations, but most of the applications are in manufacturing. The overall objectives of SPC are to (1) improve the quality of the process output, (2) reduce process variability and achieve process stability, and (3) solve processing problems. There are seven principal methods or tools used in statistical process control, sometimes referred to as the "magnificent seven" [15]: (1) control charts, (2) histograms, (3) Pareto charts, (4) check sheets, (5) defect concentration diagrams, (6) scatter diagrams, and (7) cause and effect diagrams. Most of these tools are statistical and/or technical in nature. However, it should be mentioned that statistical process control includes not only the magnificent seven tools. There are also nontechnical aspects in the implementation of SPC. To be successful, statistical process control must include a commitment to quality that pervades the organization from senior management to the starting worker on the production line.

Our discussion in this section will emphasize the seven SPC tools. A more detailed treatment of statistical process control is presented in several of our references. For coverage of control charts we recommend [12], [15], [18], and [21], and for the other six SPC tools [7] and [8].

20.4.1 Control Charts

Control charts are the most widely used method in statistical process control. The underlying principle of control charts is that the variations in any process divide into two types, as described in Section 20.3.1: (1) random variations, which are the only variations present if the process is in statistical control; and (2) assignable variations, which indicate a departure from statistical control. The purpose of a control chart is to identify when the process has gone out of statistical control, thus signaling the need for corrective action.

A *control chart* is a graphical technique in which statistics computed from measured values of a certain process characteristic are plotted over time to determine if the process remains in statistical control. The general form of the control chart is illustrated in Figure 20.3.

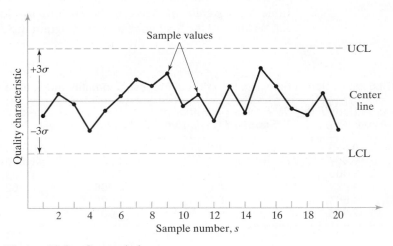

Figure 20.3   Control chart.

The chart consists of three horizontal lines that remain constant over time: a center, a lower control limit (LCL), and an upper control limit (UCL). The center is usually set at the nominal design value, and the upper and lower control limits are generally set at ± 3 standard deviations of the sample mean.

It is highly unlikely that a sample drawn from the process lies outside the upper or lower control limits while the process is in statistical control. Therefore, if it happens that a sample value does fall outside these limits, it is interpreted to mean that the process is out of control. After an investigation to determine the reason for the out-of-control condition, appropriate corrective action is taken to eliminate the condition. Alternatively, if the process is operating in statistical control, and there is no evidence of undesirable trends in the data, then no adjustments should be made since they would introduce an assignable variation to the process. The philosophy "if it ain't broke, don't fix it" is applicable in control charts.

There are two basic types of control charts: (1) control charts for variables, and (2) control charts for attributes. Control charts for variables require a measurement of the quality characteristic of interest. Control charts for attributes simply require a determination of either the fraction of defects in the sample or the number of defects in the sample.

Control Charts for Variables. A process that is out of statistical control manifests this condition in the form of significant changes in (1) process mean, and/or (2) process variability. Corresponding to these possibilities, there are two principal types of control charts for variables: (1) $\bar{x}$ chart, and (2) R chart. The $\bar{x}$ *chart* (call it "x-bar chart") is used to plot the average measured value of a certain quality characteristic for each of a series of samples taken from the production process. It indicates how the process mean changes over time. The R *chart* plots the range of each sample, thus monitoring the variability of the process and indicating whether it changes over time.

A suitable quality characteristic of the process must be selected as the variable to be monitored in the $\bar{x}$ and R charts. In a mechanical process, this might be a shaft diameter or other critical dimension. Measurements of the process itself must be used to construct the two control charts.

With the process operating smoothly and absent of assignable variations, a series of samples ($m = 20$ or more is generally recommended) of small size (for example, $n = 5$ parts per sample) are collected and the characteristic of interest is measured for each part. The following procedure is used to construct the center, LCL, and UCL for each chart:

1. Compute the mean $\bar{x}$ and range R for each of the m samples.
2. Compute the grand mean $\bar{\bar{x}}$, which is the mean of the $\bar{x}$ values for the m samples; this will be the center for the $\bar{x}$ chart.
3. Compute $\bar{R}$ which is the mean of the R values for the m samples; this will be the center for the R chart.
4. Determine the upper and lower control limits, LCL and UCL, for the $\bar{x}$ and R charts. Values of standard deviation can be estimated from the sample data using Eq. (20.3) to compute these control limits. However, an easier approach is based on statistical factors tabulated in Table 20.4 that have been derived specifically for these control charts. Values of the factors depend on sample size n. For the $\bar{x}$ chart,

$$LCL = \bar{\bar{x}} - A_2\bar{R} \tag{20.6a}$$

$$UCL = \bar{\bar{x}} + A_2\bar{R} \tag{20.6b}$$

TABLE 20.4 Constants for the $\bar{x}$ and R Charts

Sample size	$\bar{x}$ chart	R chart	
n	A_2	D_3	D_4
3	1.023	0	2.574
4	0.729	0	2.282
5	0.577	0	2.114
6	0.483	0	2.004
7	0.419	0.076	1.924
8	0.373	0.136	1.864
9	0.337	0.184	1.816
10	0.308	0.223	1.777

and for the R chart,

$$LCL = D_3\overline{R} \tag{20.7a}$$

$$UCL = D_4\overline{R} \tag{20.7b}$$

EXAMPLE 20.1 $\bar{x}$ and R Charts

Although 20 or more samples are recommended, let us use a smaller number here to illustrate the calculations. Suppose eight samples ($m = 8$) of size 5 ($n = 5$) have been collected from a manufacturing process that is in statistical control, and the dimension of interest has been measured for each part. It is desired to determine the values of the center, LCL, and UCL to construct the $\bar{x}$ and R charts. The calculated values of $\bar{x}$ and R for each sample are given here (measured values are in centimeters), which is step (1) in our procedure:

s	1	2	3	4	5	6	7	8
$\bar{x}$	2.008	1.998	1.993	2.002	2.001	1.995	2.004	1.999
R	0.027	0.011	0.017	0.009	0.014	0.020	0.024	0.018

Solution: In step (2), we compute the grand mean of the sample averages.

$$\bar{\bar{x}} = \frac{2.008 + 1.998 + 1.993 + 2.002 + 2.001 + 1.995 + 2.004 + 1.999}{8}$$

$$= 2.000 \text{ centimeters}$$

In step (3), the mean value of R is computed.

$$R = \frac{0.027 + 0.011 + 0.017 + 0.009 + 0.014 + 0.020 + 0.024 + 0.018}{8}$$

$$= 0.0175 \text{ centimeters}$$

In step (4), the values of LCL and UCL are determined based on factors in Table 20.4. First, using Eq. (20.6) for the $\bar{x}$ chart,

$$LCL = 2.000 - 0.577(0.0175) = 1.9899$$

$$UCL = 2.000 + 0.577(0.0175) = 2.0101$$

And for the R chart using Eq. (20.7),

$$LCL = 0(0.0175) = 0$$

$$UCL = 2.114(0.0175) = 0.0370$$

The two control charts are constructed in Figure 20.4 with the sample data plotted in the charts.

Control Charts for Attributes. Control charts for attributes monitor the fraction defect rate or the number of defects in the sample as the plotted statistic. Examples of these kinds of attributes include fraction of nonconforming parts in a sample, proportion of plastic molded parts that have flash, number of defects per automobile, and number of

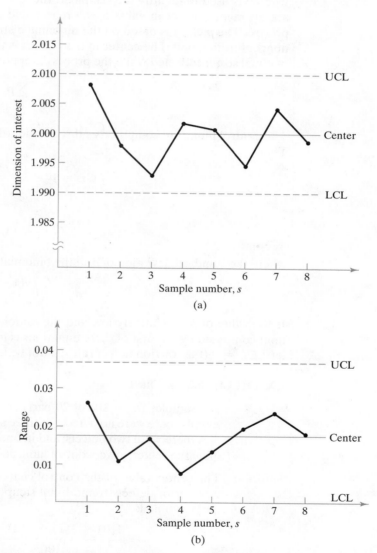

Figure 20.4 Control charts for Example 20.1: (a) $\bar{x}$ chart, and (b) R chart.

flaws in a roll of sheet steel. Inspection procedures that involve GO/NO-GO gaging are included in this group since they determine whether a part is good or bad.

The two principal types of control charts for attributes are (1) the *p chart*, which plots the fraction defect rate in successive samples, and (2) the *c chart*, which plots the number of defects, flaws, or other nonconformities per sample.

In the *p* chart, the quality characteristic of interest is the proportion (*p* for proportion) of nonconforming or defective units. For each sample, this proportion p_i is the ratio of the number of nonconforming or defective items d_i over the number of units in the sample *n* (we assume samples are of equal size in constructing and using the control chart),

$$p_i = \frac{d_i}{n} \tag{20.8}$$

where *i* is used to identify the sample. If the p_i values for a sufficient number of samples are averaged, the mean value $\bar{p}$ is a reasonable estimate of the true value of *p* for the process. The *p* chart is based on the binomial distribution, where *p* is the probability of a nonconforming unit. The center in the *p* chart is the computed value of $\bar{p}$ for *m* samples of equal size *n* collected while the process is operating in statistical control:

$$\bar{p} = \frac{\sum_{i=1}^{m} p_i}{m} \tag{20.9}$$

The control limits are computed as three standard deviations on either side of the center. Thus,

$$LCL = \bar{p} - 3\sqrt{\frac{\bar{p}(1-\bar{p})n}{}} \tag{20.10a}$$

$$UCL = \bar{p} + 3\sqrt{\frac{\bar{p}(1-\bar{p})}{n}} \tag{20.10b}$$

where the standard deviation of $\bar{p}$ in the binomial distribution is given by

$$\sigma_p = \sqrt{\frac{\bar{p}(1-\bar{p})}{n}} \tag{20.11}$$

If the value of $\bar{p}$ is relatively low and the sample size *n* is small, then the lower control limit computed by the first of these equations is likely to be a negative value. In this case, let $LCL = 0$ (the fraction defect rate cannot be less than zero).

EXAMPLE 20.2   *p* Chart

Ten samples ($m = 10$) of 20 parts each ($n = 20$) have been collected. In one sample there were no defects; in three samples there was one defect; in five samples there were two defects; and in one sample there were three defects. Determine the center, lower control limit, and upper control limit for the *p* chart.

Solution: The center value of the control chart can be calculated by summing the total number of defects found in all samples and dividing by the total number of parts sampled:

$$\bar{p} = \frac{1(0) + 3(1) + 5(2) + 1(3)}{10(20)} = \frac{16}{200} = 0.08 = 8\%$$

The lower control limit is given by Eq. (20.10a):

$$LCL = 0.08 - 3\sqrt{\frac{0.08(1 - 0.08)}{20}} = 0.08 - 3(0.06066) = 0.08 - 0.182 \rightarrow 0$$

The upper control limit, by Eq. (20.10b):

$$UCL = 0.08 + 3\sqrt{\frac{0.08(1 - 0.08)}{20}} = 0.08 + 3(0.06066)$$

$$= 0.08 + 0.182 = 0.262$$

In the c chart (c for count), the number of defects in the sample is plotted over time. The sample may be a single product such as an automobile, and c = number of quality defects found during final inspection, or the sample may be a length of carpeting at the factory prior to cutting, and c = number of imperfections discovered per one hundred meters. The c chart is based on the Poisson distribution, where c = parameter representing the number of events occurring within a defined sample space (e.g., defects per car, imperfections per unit length of carpet). Our best estimate of the true value of c is the mean value over a large number of samples drawn while the process is in statistical control:

$$\bar{c} = \frac{\sum_{i=1}^{m} c_i}{m} \tag{20.12}$$

This value of $\bar{c}$ is used as the center for the control chart. In the Poisson distribution, the standard deviation is the square root of parameter c. Thus, the control limits are

$$LCL = \bar{c} - 3\sqrt{\bar{c}} \tag{20.13a}$$
$$UCL = \bar{c} + 3\sqrt{\bar{c}} \tag{20.13b}$$

EXAMPLE 20.3 c Chart

A continuous plastic extrusion process is operating in statistical control. Eight hundred meters of the extrudate have been examined and a total of 14 surface defects have been detected in that length. Develop a c chart for the process, using defects per hundred meters as the quality characteristic of interest.

Solution: The average value of the parameter c can be determined by using Eq. (20.12):

$$\bar{c} = \frac{14}{8} = 1.75$$

This will be used as the center for the control chart. The lower and upper control limits are given by Eqs. (13a) and (13b):

$$LCL = 1.75 - 3\sqrt{1.75} = 1.75 - 3(1.323) = 1.75 - 3.969 \rightarrow 0$$
$$UCL = 1.75 + 3\sqrt{1.75} = 1.75 + 3(1.323) = 1.75 + 3.969 = 5.719$$

Interpreting the Control Charts.   When control charts are used to monitor production quality, random samples are drawn from the process of the same size n used to construct the charts. For $\bar{x}$ and R charts, the $\bar{x}$ and R values of the measured characteristic are plotted on the control chart. By convention, the points are usually connected as in our figures. To interpret the data, one looks for signs that indicate the process is not in statistical control. The most obvious sign is when $\bar{x}$ or R (or both) lie outside the LCL or UCL limits. This indicates an assignable cause such as bad starting materials, an experienced operator, wrong equipment setting, broken tooling, or similar factors. An out-of-limit $\bar{x}$ indicates a shift in the process mean. An out-of-limit R shows that the variability of the process has probably changed. The usual effect is that R increases, indicating variability has risen.

Less obvious conditions may be revealed even though the sample points lie within $\pm 3\sigma$ limits. These conditions include (1) trends or cyclical patterns in the data, which may mean wear or other factors that occur as a function of time; (2) sudden changes in the average values of the data; and (3) points consistently near the upper or lower limits. The same kinds of interpretations that apply to the $\bar{x}$ chart and R chart also apply to the p chart and c chart.

Montgomery [15] provides a list of indicators that a process is likely to be out of statistical control and that corrective action should be taken. The indicators are the following: (1) one point that lies outside the UCL or LCL, (2) two out of three consecutive points that lie beyond ± 2 sigma on one side of the center line of the control chart, (3) four out of five consecutive points that lie beyond ± 1 sigma on one side of the center line of the control chart, (4) eight consecutive points that lie on one side of the center line, and (5) six consecutive points in which each point is always higher or always lower than its predecessor.

Control charts serve as the feedback loop in statistical process control, as suggested by Figure 20.5. They represent the measurement step in process control. If the control chart indicates that the process is in statistical control, then no action is taken. However, if the process is identified as being out of statistical control, then the cause of the problem must be identified and corrective action must be taken.

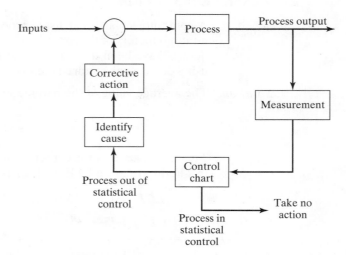

Figure 20.5 Control charts used as the feedback loop in statistical process control.

20.4.2 Other SPC Tools

Although control charts are the most commonly used tool in statistical process control, other tools are also important. Each has its own area of application. In this section we discuss the remaining six of the magnificent seven.

Histograms. The histogram is a basic graphical tool in statistics. After the control chart, it is probably the most important member of the SPC tool kit. A *histogram* is a statistical graph consisting of bars representing different values or ranges of values, in which the length of each bar is proportional to the frequency or relative frequency of the value or range, as shown in Figure 20.6. It is a graphical display of the frequency distribution of the numerical data. What makes the histogram such a useful statistical tool is that it enables an analyst to quickly visualize the features of a complete set of data. These features include (1) the shape of the distribution, (2) any central tendency exhibited by the distribution, (3) approximations of the mean and mode of the distribution, and (4) the amount of scatter or spread in the data. With regard to Figure 20.6, we can see that the distribution is normal (in all likelihood), and that the mean is around 2.00. We can approximate the standard deviation by taking the range of values shown in the histogram ($2.025 - 1.975$) and dividing by 6, based on the fact that nearly the entire distribution (99.73%) is contained within $\pm 3\sigma$ of the mean value. This gives a σ value of around 0.008.

Pareto Charts. A *Pareto chart* is a special form of histogram, illustrated in Figure 20.7, in which attribute data are arranged according to some criteria such as cost or value. When appropriately used, it provides a graphical display of the tendency for a small proportion of a given population to be more valuable than the much larger majority. This tendency is sometimes referred to as *Pareto's Law*, which can be succinctly stated, "the vital

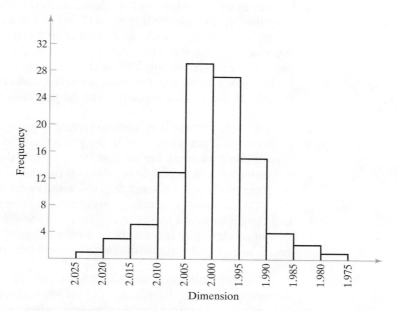

Figure 20.6 Histogram of data collected from the process in Example 20.1.

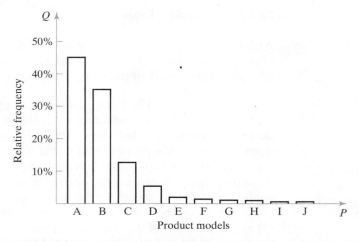

Figure 20.7 Typical (hypothetical) Pareto distribution of a factory's production output. Although there are ten models produced, two of the models account for 80% of the total units. This chart is sometimes referred to as a *P-Q* chart, where *P* = products and *Q* = quantity of production.

few and the trivial many."[2] The "law" was identified by Vilfredo Pareto (1848–1923), an Italian economist and sociologist who studied the distribution of wealth in Italy and found that most of it was held by a small percentage of the population.

Pareto's Law applies not only to the distribution of wealth, but to many other distributions as well. The law is often identified as the 80–20% rule (although exact percentages may differ from 80 and 20): 80% of the wealth of a nation is in the hands of 20% of its people; 80% of inventory value is accounted for by 20% of the items in inventory; 80% of sales revenues are generated by 20% of the customers; and 80% of a factory's production output is concentrated in only 20% of its product models (as in Figure 20.7). What is suggested by Pareto's Law is that the most attention and effort in any study or project should be focused on the smaller proportion of the population that is seen to be the most important.

Check Sheets. The *check sheet* (not to be confused with "check list") is a data-gathering tool generally used in the preliminary stages of the study of a quality problem. The operator running the process (for example the machine operator) is often given the responsibility for recording the data on the check sheet, and the data is often recorded in the form of simple check marks (hence, the check sheet's name).

Check sheets can take many different forms, depending on the problem situation and the ingenuity of the analyst. The form should be designed to allow some interpretation of results directly from the raw data, although subsequent data analysis may be necessary to recognize trends, diagnose the problem, or identify areas for further study.

Defect Concentration Diagrams. The *defect concentration diagram* is a drawing of the product (or part), with all relevant views displayed, onto which have been sketched the various defect types at the locations where each occurred. An analysis of the defect types and corresponding locations can identify the underlying causes of the defects.

[2]The statement is attributed to J. Juran [5].

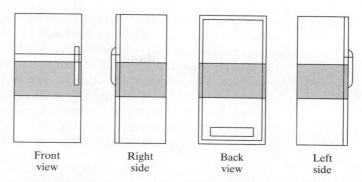

| Front view | Right side | Back view | Left side |

Figure 20.8 Defect concentration diagram showing four views of refrigerator with locations of surface defects indicated in cross-hatched areas.

Montgomery [15] describes a case study involving the final assembly of refrigerators that were plagued by surface defects. A defect concentration diagram (Figure 20.8) was utilized to analyze the problem. The defects were clearly shown to be concentrated around the middle sections of the refrigerators. Upon investigation, it was learned that a belt was wrapped around each unit for material handling purposes. It became evident that the defects were caused by the belt, and corrective action was taken to improve the handling method.

Scatter Diagrams. In many industrial manufacturing operations, it is desired to identify a possible relationship between two process variables. The scatter diagram is useful in this regard. A *scatter diagram* is an *x-y* plot of the data taken of the two variables in question, as illustrated in Figure 20.9. The data are plotted as pairs; for each x_i value, there is a corresponding y_i value. The shape of the data points considered in aggregate often reveals a pattern or relationship between the two variables. For example, the scatter diagram in Figure 20.9 indicates that a negative correlation exists between cobalt content and wear resistance of a cemented carbide cutting tool. As cobalt content increases, wear

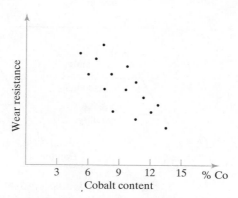

Figure 20.9 Scatter diagram showing the effect of cobalt binder content on wear resistance of a cemented carbide cutting tool insert.

resistance decreases. One must be circumspect in using scatter diagrams and in extrapolating the trends that might be indicated by the data. For instance, it might be inferred from our diagram that a cemented carbide tool with zero cobalt content would possess the highest wear resistance of all. However, cobalt serves as an essential binder in the pressing and sintering process used to fabricate cemented carbide tools, and a minimum level of cobalt is necessary to hold the tungsten carbide particles together in the final product. Another reason why caution is recommended in the use of the scatter diagram is that only two variables are plotted. There may be other variables in the process whose importance in determining the output is far greater than the two variables displayed.

Cause-and-Effect Diagrams. The *cause-and-effect diagram* is a graphical-tabular chart used to list and analyze the potential causes of a given problem. It is not really a statistical tool like the preceding tools. As shown in Figure 20.10, the diagram consists of a central stem leading to the effect (the problem), with multiple branches coming off the stem listing the various groups of possible causes of the problem. Owing to its characteristic appearance, the cause-and-effect diagram is also known as a *fishbone diagram*. In application, the cause-and-effect diagram is developed by a quality team. The team then attempts to determine which causes are most consequential and how to take corrective action against them.

20.4.3 Implementing SPC

There is more to successful implementation of statistical process control than the seven SPC tools. The tools provide the mechanism by which SPC can be implemented, but the mechanism requires a driving force. The driving force in implementing SPC is

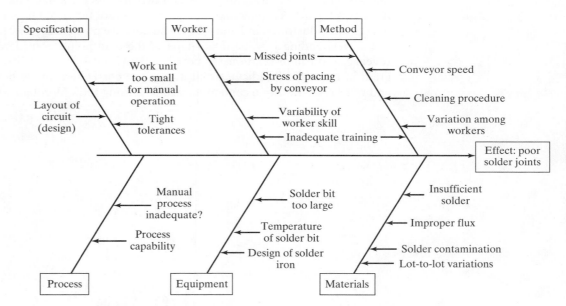

Figure 20.10 Cause-and-effect diagram for a manual soldering operation. The diagram indicates the effect (the problem is poor solder joints) at the end of the arrow and the possible causes are listed on the branches leading toward the effect.

management's commitment to quality and the process of continuous improvement. Through its involvement and example, management drives the successful implementation of SPC. Although management is the most important ingredient, there are other factors that play a role. Five elements usually present in a successful SPC program can be identified as follows in their order of importance, based on Montgomery [15]:

1. *Management commitment and leadership.* This is the most important element. Management sets the example for others in the organization to follow. Continuous quality improvement is a management-driven process.

2. *Team approach to problem solving.* Solving quality problems in production usually requires the attention and expertise of more than one person. It is difficult for one individual, acting alone, to make the necessary changes to solve a quality problem. Teams whose members contribute a broad pool of knowledge and skills are the most effective approach to problem solving.

3. *SPC training for all employees.* Employees at all levels in the organization from the chief executive officer to the starting production worker must be knowledgeable about the tools of SPC so that they can apply the tools in all functions of the enterprise.

4. *Emphasis on continuous improvement.* Due to the commitment and example of management, the process of continuous improvement is pervasive throughout the organization.

5. *A recognition and communication system.* Finally, there should be a mechanism for recognizing successful SPC efforts and communicating them throughout the organization.

20.5 SIX SIGMA[3]

Six Sigma is the name of a quality-focused program that utilizes worker teams to accomplish projects aimed at improving an organization's operational performance. The first Six Sigma program was developed and implemented by Motorola Corporation around 1980. It has been widely adopted by many companies in the United States. In the Normal distribution, six sigma implies near perfection in a process, and that is the goal of a Six Sigma program. To operate at the six sigma level over the long term, a process must be capable of producing no more than 3.4 defects per million, where a defect refers to anything that is outside of customer specifications. Six Sigma projects can be applied to any manufacturing, service, or business processes that affect customer satisfaction. Customers are both internal and external. There is a strong emphasis on customer satisfaction in Six Sigma.

The general goals of Six Sigma and the projects that are performed under its banner are (1) better customer satisfaction, (2) high quality products and services, (3) reduced defects, (4) improved process capability through reduction in process variations, (5) continuous improvement, and (6) cost reduction through more effective and efficient processes.

Worker teams who participate in a Six Sigma project are trained in the use of statistical and problem-solving tools as well as project management techniques to define,

[3]This section and Section 20.6 are based largely on Chapter 21 in Groover [9].

measure, analyze, and make improvements in the operations of the organization by eliminating defects and variability in its processes. The teams are empowered by management, whose responsibility is to identify the important problems in the processes of the organization and to sponsor the teams to address those problems.

A central concept of Six Sigma is that defects in a given process can be measured and quantified. Once they are quantified, the underlying causes of the defects can be identified, and corrective action can be taken to fix the causes and eliminate the defects. The results of the improvement effort can be seen using the same measurement procedures to make a before-and-after comparison. The comparison is often expressed in terms of sigma level. For example, the process was originally operating at the 3-sigma level, but now it is operating at the 5-sigma level. In terms of defect levels, this means that the process was previously producing 66,807 defects per 1,000,000, and now it is producing only 233 defects per 1,000,000. Various other sigma levels and corresponding defects per million (DPM) and other measures are listed in Table 20.5.

TABLE 20.5 Sigma Levels and Corresponding Defects per Million, Fraction Defect Rate, and Yield in a Six Sigma Program

Sigma Level	Defects per Million[a]	Fraction Defect Rate q	Yield Y
6.0 σ	3.4	0.0000034	99.99966%
5.8 σ	8.5	0.0000085	99.99915%
5.6 σ	21	0.000021	99.9979%
5.4 σ	48	0.000048	99.9952%
5.2 σ	108	0.000108	99.9892%
5.0 σ	233	0.000233	99.9770%
4.8 σ	483	0.000483	99.9517%
4.6 σ	968	0.000968	99.9032%
4.4 σ	1,866	0.001866	99.813%
4.2 σ	3,467	0.003467	99.653%
4.0 σ	6,210	0.006210	99.379%
3.8 σ	10,724	0.01072	98.93%
3.6 σ	17,864	0.01768	98.23%
3.4 σ	28,716	0.02872	97.13%
3.2 σ	44,565	0.04457	95.54%
3.0 σ	66,807	0.06681	93.32%
2.8 σ	96,801	0.09680	90.32%
2.6 σ	135,666	0.13567	86.43%
2.4 σ	184,060	0.18406	81.59%
2.2 σ	241,964	0.2420	75.80%
2.0 σ	308,538	0.3085	69.15%
1.8 σ	382,089	0.3821	61.79%
1.6 σ	460,172	0.4602	53.98%
1.4 σ	539,828	0.5398	46.02%
1.2 σ	617,911	0.6179	38.21%
1.0 σ	691,462	0.6915	30.85%

Source: Compiled from Eckes [6], Appendix.
[a]Can be used for defective units per million *DUPM* or defects per million opportunities *DPMO*.

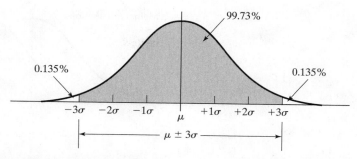

Figure 20.11 Normal distribution of process output variable, showing the ±3σ limits.

The traditional metric for good process quality is ±3σ (three sigma level). As discussed in Section 20.3, if a process is stable and in statistical control for a given output variable of interest, and this variable is normally distributed, then 99.73% of the process output will be within the range defined by ±3σ. This situation is illustrated in Figure 20.11. It means that there will be 0.27% (0.135% in each tail) of the output that lies beyond these limits, or 2,700 parts per million produced.

Let us compare this with a process that operates at the six sigma level. Under the same assumptions as before (normally distributed stable process in statistical control), the proportion of the output that lies within the range ±6σ is 99.9999998%, which corresponds to a defect rate of only 0.002 defects per million. This situation is illustrated in Figure 20.12.

The reader has probably noticed that this defect rate does not match the rate associated with six sigma in Table 20.5. The rate shown in the table is 3.4 defects per million, which corresponds to a yield of 99.99966%. Why is there a difference? Which is correct? If one looks up the proportion of the population that lies within ±6σ in a standard normal probability table (if one could find a table that goes that high), one would find that 99.9999998% is the correct value, not 99.99966%. Admittedly, the difference between the two yields does not seem like much. But the difference between 0.002 defects per million and 3.4 defects per million is significant.

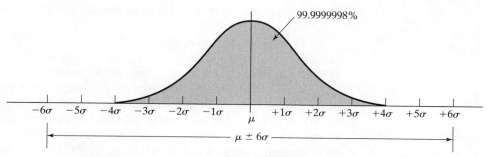

Figure 20.12 Normal distribution of process output variable, showing the ±6σ limits.

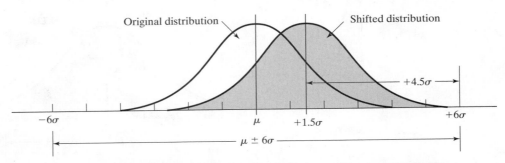

Figure 20.13 Normal distribution shift by a distance of 1.5σ from the original mean.

The explanation for this anomaly is that when the engineers at Motorola devised the Six Sigma standard, they considered processes that operate over the long run, and processes over the long run tend to deviate from the original process mean. While data are collected from a process over a relatively short period of time (e.g., a few weeks or months) to determine the mean and standard deviation, the same process may run for years. During the long-term operation of that process, its mean will likely shift to the right or left. To compensate for these likely shifts, Motorola selected to use 1.5σ as the magnitude of the shift, while leaving the original $\pm 6\sigma$ limits in place for the process. The effect of this shift is shown in Figure 20.13. Accordingly, when 6σ is used in Six Sigma, it really refers to 4.5σ in the normal probability tables.

20.6 THE SIX SIGMA DMAIC PROCEDURE

Six Sigma teams use a problem-solving approach called DMAIC, sometimes pronounced "duh-may-ick." It consists of five steps:

1. *Define* the project goals and customer requirements
2. *Measure* the process to assess its current performance
3. *Analyze* the process and determine root causes of variations and defects
4. *Improve* the process by reducing variations and defects
5. *Control* the future process performance by institutionalizing the improvements.

These are the basic steps in an improvement procedure intended for existing processes that are currently operating at low sigma levels and need improvement. DMAIC provides the worker team with a systematic and data-driven approach to solve an identified problem. It is a roadmap that guides the team towards improvement in the process of interest. Although the approach seems very sequential (step 1, then step 2, and so on), an iterative implementation of DMAIC is sometimes required. For example, in the analyze step (step 3), the team may discover that it did not collect the right data in the measure step (step 2). Therefore, it must repeat the previous step to correct the deficiency.

The following paragraphs describe the five steps of the DMAIC approach and some of the typical tools that might be applied in each step.

20.6.1 Define

The first step in DMAIC consists of (1) organizing the project team, (2) providing it with a charter (the problem to solve), (3) identifying the customers served by the process, and (4) developing a high-level process map.

Organizing the Project Team. Members of the project team are selected on the basis of their knowledge of the problem area and other skills. The team members, at least some of them, have had Six Sigma training. Some of them are the workers who operate the process of interest. Team leaders in a Six Sigma project are called *black belts*; they are the project managers. They have had detailed training in the entire range of Six Sigma problem-solving techniques. Assisting them are *green belts*, other team members who have been trained in some Six Sigma techniques. Providing technical resources and serving as consultants and mentors for the black belts are *master black belts*. Master black belts are generally full-time positions, and they are selected for their teaching aptitudes, quantitative skills, and experience in Six Sigma.

Participating in the formation of a Six Sigma project team is an individual known in Six Sigma terminology as the *champion*, who is typically a member of management. The champion is often the owner of the process, and the process needs improvement.

The Charter. The charter is the documentation that justifies the project. Much of the substance of the charter is provided by the champion, the one with the problem. The charter documentation usually includes the following:

- Problem statement and background. What is wrong with the process? How long has the problem existed? How does the process currently operate, and how should it be operating? This section attempts to define the problem in quantitative terms.
- Objectives of the project. What should the project team be able to accomplish within a certain time frame, say six months? What will be the benefits of the project?
- Scope of the project. What areas should the project team focus on, and what areas should it avoid?
- Business case for solving the problem. How is the project justified in economic terms? What is the potential return by accomplishing the project? Why is this problem more important than other problems?
- Project schedule. What are the logical milestones in the project? These are often defined in terms of the five steps in the DMAIC procedure. When should the *define* step be completed? How many weeks should the *measure* step take?

Identifying the Customer(s). Every process serves customers. The output of the process (e.g., the product or part produced or the service delivered) has one or more customers. Otherwise, there would be no need for the process. Customers are the recipients of the process output and are directly affected by its quality, either positively or negatively. Customers have needs and requirements that must be satisfied or exceeded. An important function in the define step is to identify exactly who the customers of the process are and what their requirements are.

When the team is identifying the customers, it is particularly useful for it to determine those characteristics of the process output that are critical to quality (CTQ) from a customer's viewpoint. The CTQ characteristics are the features or elements of the process and its output that directly impact the customer's perception of quality. Typical CTQ characteristics include the reliability of a product (e.g., automobile, appliance, lawn mower) or the timeliness of a service (e.g., fast food delivery, plumbing repairs). Identifying the CTQ characteristics allows the Six Sigma team to focus on what's important and not to dissipate its energy on what's not important.

High-Level Process Map. The final task in the define step is to develop a high-level process map. *Process mapping* is a graphical technique that can be used to depict the sequence of steps that operate on the inputs to the process and produce the output. An example is illustrated in Figure 20.14. Process maps provide a detailed picture of the process or system of interest. They help the team members understand the issue and communicate with one another.

Process mapping is the preferred technique in a Six Sigma project because it can be used to portray a process at various levels of detail. In the define step, when the improvement project is just getting underway, it is appropriate to visualize the process at a high level, absent of the details that will be examined in subsequent steps. The process map developed here should include the suppliers, inputs, process, outputs, and customers.

In addition to the high level viewpoint, the process map should also be an "as is" picture of the current process, unimproved. Viewing the status quo process map may very well provide leads that will result in improvements. The "as is" process map provides a benchmark against which to compare the subsequent improvements.

20.6.2 Measure

The second step in DMAIC consists of (1) collecting data and (2) measuring the current sigma level of the process. Assessing the sigma level of the current process allows the team to make comparisons later, after improvements have been added.

Data Collection. The first step in data collection is deciding what should be measured. This decision should be made with reference to the process map and the critical-to-quality (CTQ) characteristics developed in the define step. The measurements can be classified into three categories:

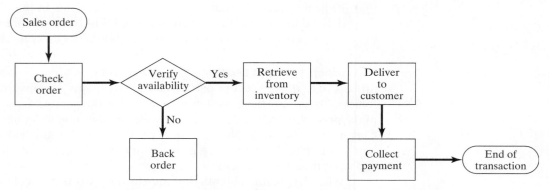

Figure 20.14 An example of a process map showing the sequence of steps and their interrelationships.

- *Input measures.* These are variables related to the process inputs, which are provided by its suppliers. What are the important quality measures to assess the performance of the suppliers?

- *Process measures.* These are the internal variables of the process itself. In general, they deal with efficiency measures such as cycle time and waiting time, and quality measures such as dimensional variables and fraction defect rate.

- *Output measures.* These are the measures seen by the customer. They indicate how well customer requirements and expectations are being satisfied. They are functionally related to the input measures and process measures.

Every Six Sigma project is different, and different types of data must be collected for each one. It is usually necessary to design data collection forms for the project, perhaps using check sheets from statistical process control (Section 20.4.2).

Once the decisions have been made regarding which variables to measure, and data collection forms have been designed, then the actual data collection begins. Adjustments are sometimes required as problems are encountered in the collection procedure. The problems may be related to occurrences or variables in the process that were not anticipated (e.g., identification of an important variable that was previously overlooked) or design of the data collection forms (e.g., no space on the form to note unusual events). An important rule of data collection in a Six Sigma project is that the team itself should be involved in the data collection so that it can recognize these problems and take the appropriate corrective actions [20].

Measuring the Current Sigma Level. After data collection has been completed, the team is in a position to analyze the current sigma level of the "as-is" process. This provides a starting point for making improvements and measuring their effects on the process. It allows a before-and-after comparison. The first step in assessing the current sigma level of the process is to determine the number of defects per million. This value is then converted to the corresponding sigma level using Table 20.5.

There are several alternative measures of defects per million that can be used in a Six Sigma program. The most appropriate measure is probably the *defects per million opportunities* ($DPMO$), which refers to the fact that there may be more than one opportunity for defects to occur in each unit. Thus, the number of opportunities takes into account the complexity of the product or service so that entirely different types of products can be compared on the same sigma scale. Defects per million opportunities is calculated as

$$DPMO = 1,000,000 \frac{N_d}{N_u N_o} \qquad (20.14)$$

where $DPMO$ = defects per million opportunities, N_d = number of defects, N_u = number of units in the population of interest, and N_o = number of opportunities for a defect per unit. The factor 1,000,000 converts the proportion into defects per million.

Other common measures include defects per million (DPM) and defective units per million ($DUPM$). *Defects per million* measures all of the defects encountered in the population, considering that there is more than one opportunity for a defect per defective unit:

$$DPM = 1,000,000 \frac{N_d}{N_u} \qquad (20.15)$$

Defective units per million is the count of defective units in the population of interest, considering that a defective unit may contain more than one defect:

$$DUPM = 1,000,000 \frac{N_{du}}{N_u} \qquad (20.16)$$

where N_{du} = number of defective units. The following example illustrates the procedure for determining $DPMO$, DPM, and $DUPM$, as well as the corresponding sigma levels.

EXAMPLE 20.4   Determining the Sigma Level of a Process

A refrigerator final assembly plant inspects its completed products for 37 features that are considered critical-to-quality (CTQ). During the previous three-month period, 31,487 refrigerators were produced, among which 1,690 had defects of the 37 CTQ features, and 902 refrigerators had one or more defects. Determine (a) the defects per million opportunities and corresponding sigma level, (b) the defects per million and corresponding sigma level, and (c) the defective units per million and corresponding sigma level.

Solution: Summarizing the data: N_o = 37 defect opportunities per product, N_u = 31,487 product units, N_d = 1,690 defects, and N_{du} = 902 defective units.

(a) $DPMO = 1,000,000 \dfrac{1,690}{31,487(37)} = 1,451$ defects per million opportunities

This corresponds to the 4.5 sigma level (interpolating between the 4.4 and 4.6 sigma levels in Table 20.5).

(b) $DPM = 1,000,000 \dfrac{1,690}{31,487} = 53,673$ defects per million

This corresponds to the 3.1 sigma level.

(c) $DUPM = 1,000,000 \dfrac{902}{31,487} = 28,647$ defective units per million

This corresponds to the 3.4 sigma level.

20.6.3 Analyze

The analyze step in DMAIC can be divided into the following phases: (1) basic data analysis, (2) process analysis, and (3) root cause analysis. The analyze step is a bridge between measure (step 2) and improve (step 4). Analyze takes the data collected in step 2 and provides a quantitative basis for developing improvements in step 4. The analyze phase seeks to identify where the improvement opportunities lie.

Basic Data Analysis.   The purpose of the basic data analysis is to present the collected data in a way that lends itself to making inferences. This usually means graphical displays of the data, borrowing tools from SPC such as histograms, Pareto charts, and scatter diagrams. Additional statistical analysis tools often used for data analysis include regression analysis (least squares analysis), analysis of variance, and hypothesis testing.

Process Analysis.   Process analysis is concerned with interpreting the results of the basic data analysis and developing a more detailed picture of the way the process operates and what is wrong with it. The more detailed picture usually includes a series of process maps that focus on the individual steps in the high-level process map created earlier in the define step (step 1 in DMAIC). The low-level process maps are useful in better understanding the inner workings of the process. The Six Sigma team progresses through a process analysis by asking questions like

- What are the value-adding steps in the process?
- What are the non-value-adding but necessary steps?
- What are the steps that add no value and could be eliminated?
- What are the steps that generate variations, deviations, and errors in the process?
- Which steps are efficient and which steps are inefficient (in terms of time, labor, equipment, materials, and other resources)? The inefficient steps merit further scrutiny.
- Why is so much waiting required in this process?
- Why is so much material handling required?

Root Cause Analysis.   Root cause analysis attempts to identify the significant factors that affect process performance. The situation can be depicted using the following general equation:

$$y = f(x_1, x_2, \ldots, x_i, \ldots, x_n) \tag{20.17}$$

where y = some output variable of interest in the project (e.g., some quality feature of importance to the customer) and $x_1, x_2, \ldots, x_i, \ldots x_n$ are the independent variables in the process that may affect the output variable. The value of y is a function of the values of x_i. In root cause analysis, the team attempts to determine which of the x_i variables are most important and how they influence y. In all likelihood, there is more than one y variable of interest. For each y, there is likely to be a different set of x_i variables.

Root cause analysis consists of the following phases: (1) brainstorming of hypotheses, (2) eliminating the unlikely hypotheses, and (3) validating the remaining hypotheses. In general, *brainstorming* is a group problem-solving activity that consists of group members spontaneously contributing ideas on a subject of mutual interest. The cause-and-effect diagram (Section 20.4.2) is a tool that is sometimes used to focus thoughts in brainstorming.

At the end of the brainstorming phase, there is a large list of hypotheses, some of which are less likely to be valid than others. The elimination phase begins. The team must use its collective wisdom and knowledge of the process to identify which hypotheses are highest in priority and which ones should be eliminated from further consideration. The list of hypotheses is reduced from a large number to a much smaller number. The important x_i variables are identified, and the relationships of $y = f(x)$ are conjectured.

The final phase of root cause analysis is concerned with validating the reduced list of hypotheses. It involves testing these hypotheses and determining the mathematical relationships for $y = f(x_1, x_2, \ldots, x_i, \ldots, x_n)$. Scatter diagrams (Section 20.4.2) can be especially useful in determining the shape of the relationship and the form of the mathematical model for the process. In some cases, the team must collect additional data on particular variables that have been identified as significant. It may conduct experiments to ensure that the desired information is extracted from the data collection procedure in the most efficient way.

20.6.4 Improve

The fourth step of DMAIC consists of the following phases: (1) generation of alternative improvements, (2) analysis and prioritization of the alternative improvements, and (3) implementation of the improvements.

Generation of Alternative Improvements. The preceding root cause analysis should indicate the areas in which potential improvements and problem solutions are likely to be found. The Six Sigma team uses brainstorming sessions to generate and refine the alternatives. The team searches for improvements and solutions that will reduce defects, increase customer satisfaction, improve the quality of the product or service, reduce variation, and increase process efficiency.

Analysis and Prioritization. In all likelihood, the team has generated more alternatives for improvement than feasibly can be implemented. At this point the alternatives must be analyzed and prioritized, and those alternatives that are deemed impractical must be discarded. Process mapping is a useful technique that can be used to analyze the alternatives.

The process map developed in this phase is a "should be" description of the process. It incorporates the potential improvements and solutions into the current process to allow visualization and provide graphical documentation of how it would work after the changes have been made. This allows the proposed improvements to be analyzed and refined prior to implementation.

Implementing the Improvements. Having prioritized the proposed improvements, the team moves on to the next phase in the DMAIC improve step, implementation. The priority list of proposed improvements determines where to start. Implementation can proceed one proposal at a time or in groups of proposals, depending on how the proposed changes relate to each other. For example, if the changes in the process required by two different proposals are very similar, it may make sense to implement both at the same time, even though one proposal occupies a much higher priority than the other. Also, if the objectives of the project are to achieve a certain level of overall improvement in the process that is deemed sufficient, then it may not be necessary to implement all of the proposals on the list.

To determine the overall process improvement, the same quality performance measurements should be made as in the original sigma level assessment (Example 20.4). This will provide the project team with a before-and-after comparison to gage the effect of the various changes.

20.6.5 Control

Sometimes after process improvements are made, they are gradually discarded and the improvement benefits are eroded over time. Reasons for this phenomenon include human resistance to change, familiarity and comfort associated with the former method, absence of standard procedures detailing the new method, and lack of attention by supervisory personnel. The purpose of the control step in DMAIC is to avoid this potential erosion and to maintain the improved performance that was achieved through implementation of the proposed changes. The control step consists of the following actions: (1) develop a control plan, (2) transfer responsibility back to original owner, and (3) disband the Six Sigma team.

Development of a Control Plan. The final task of the Six Sigma team is to document the results of the project and develop a control plan that will sustain the improvements that have been made in the process. The control plan documentation establishes the *standard operating procedure* (SOP) for the improved process. It should address issues and questions such as the following:

- Details of the process control relationships. This refers to the various $y = f(x_i)$ relationships that have been developed by the Six Sigma team. These relationships indicate how control of the process is achieved and which variables (x_i) are important to achieve it.
- What input variables must be measured and monitored?
- What process variables must be measured and monitored?
- What output variables must be measured and monitored?
- Who is responsible for these measurements?
- What are the corrective action procedures that should be followed in the event that something goes wrong in the process?
- What institutional procedures must be established to maintain the improvements?
- What are the worker training requirements to sustain the improvements?

Transferring Responsibility and Disbanding the Team. The Six Sigma team has been actively involved in the operation of the process for an extended period of time by now. Its work is nearly complete. One of its final actions is to turn whatever responsibility the team had for operating the process back to its original owner (e.g., the champion). The team must make sure that the owner understands the control plan and that it will be continuously implemented.

Once responsibility reverts back to the original owner, the team is no longer needed. It is therefore disbanded, and the master black belt is assigned to a new team and the next project.

20.7 TAGUCHI METHODS IN QUALITY ENGINEERING

The term *quality engineering* encompasses a broad range of engineering and operational activities whose aim is to ensure that a product's quality characteristics are at their nominal or target values. The field of quality engineering owes much to Genichi Taguchi, who has had an important influence on its development, especially in the design area—both product and process design. In this section we review two of the Taguchi methods: (1) robust design and (2) the Taguchi loss function. More complete treatments of Taguchi's methods can be found among our references [7], [14], [17], [22].

20.7.1 Robust Design

An important Taguchi principle is to set specifications on product and process parameters to create a design that resists failure or reduced performance in the face of variations. Taguchi calls the variations noise factors. A *noise factor* is a source of variation that is impossible or difficult to control and that affects the functional characteristics of the product. Three types of noise factors can be distinguished:

1. *Unit-to-unit noise factors.* These are inherent random variations in the process and product caused by variability in raw materials, machinery, and human participation. They are associated with a production process that is in statistical control.

2. *Internal noise factors.* These sources of variation are internal to the product or process. They include (1) time-dependent factors such as wear of mechanical components, spoilage of raw materials, and fatigue of metal parts; and (2) operational errors, such as improper settings on the product or machine tool.

3. *External noise factors.* An external noise factor is a source of variation that is external to the product or process, such as outside temperature, humidity, raw material supply, and input voltage. Internal and external noise factors constitute what we have previously called assignable variations. Taguchi distinguishes between internal and external noise factors because external noise factors are generally more difficult to control.

A *robust design* is one in which the function and performance of the product or process are relatively insensitive to variations in any of the above noise factors. In product design, robustness means that the product can maintain consistent performance with minimal disturbance due to variations in uncontrollable factors in its operating environment. In process design, robustness means that the process continues to produce good product with minimal effect from uncontrollable variations in its operating environment. Examples of robust designs are presented in Table 20.6.

20.7.2 The Taguchi Loss Function

The Taguchi loss function is a useful concept in tolerance design. Taguchi defines poor quality as "the loss a product costs society from the time the product is released for shipment" [22]. Loss includes costs to operate, failure to function, maintenance and repair costs,

TABLE 20.6 Robust Designs in Products and Processes

Product Design
An airplane that flies as well in stormy weather as in clear weather.
A car that starts as well in Fairbanks, Alaska in January as in Phoenix, Arizona in July.
A tennis racket that returns the ball just as well when hit near the rim as when hit in dead center.
A hospital operating room that maintains lighting and life support systems when the electric power to the hospital is interrupted.

Process Design
A turning operation that produces a good surface finish throughout a wide range of cutting speeds.
A plastic injection molding operation that molds a good part despite variations in ambient temperature and humidity in the factory.
A metal forging operation that presses good parts in spite of variations in starting temperature of the raw billet.

Other
A biological species that survives unchanged for millions of years despite significant climatic changes in the world in which it lives.

customer dissatisfaction, injuries caused by poor design, and similar costs. Some of these losses are difficult to quantify in monetary terms, but they are nevertheless real. Defective products (or their components) that are detected, repaired, reworked, or scrapped before shipment are not considered part of this loss. Instead, any expense to the company resulting from scrap or rework of defective product is a manufacturing cost rather than a quality loss.

Loss occurs when a product's functional characteristic differs from its nominal or target value. Although functional characteristics do not translate directly into dimensional features, the loss relationship is most readily understood in terms of dimensions. When the dimension of a component deviates from its nominal value, the component's function is adversely affected. No matter how small the deviation, there is some loss in function. The loss increases at an accelerating rate as the deviation grows, according to Taguchi. If we let x = the quality characteristic of interest and N = its nominal value, then the loss function will be a U-shaped curve as in Figure 20.15. Taguchi uses a quadratic equation to describe the curve

$$L(x) = k(x - N)^2 \qquad (20.18)$$

where $L(x)$ = loss function, k = constant of proportionality, and x and N are defined above. At some level of deviation $(x_2 - N) = -(x_1 - N)$, the loss will be prohibitive, and it will be necessary to scrap or rework the product. This level identifies one possible way of specifying the tolerance limit for the dimension. But even within these limits, there is also a loss, as suggested by our cross-hatching.

In the traditional approach to quality control, tolerance limits are defined and any product within those limits is acceptable. Whether the quality characteristic (for example, the dimension) is close to the nominal value or close to one of the tolerance limits, it is acceptable. Trying to visualize this approach in terms analogous to the preceding relation, we obtain the discontinuous loss function in Figure 20.16. In this approach any value within the upper tolerance limit (UTL) and lower tolerance limit (LTL) is acceptable. The reality is that products closer to the nominal specification are better quality and will work better, look better, last longer, and have components that fit better. In short, products made closer to nominal specifications will provide greater customer satisfaction. In order to improve quality and customer satisfaction, one must attempt to reduce the loss by designing the product and process to be as close as possible to the target value.

It is possible to make calculations based on the Taguchi loss function, if one accepts the assumption of the quadratic loss equation, Eq. (20.18). In the following examples, we

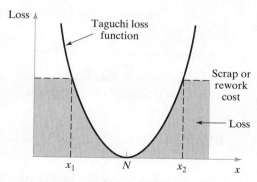

Figure 20.15 The quadratic quality loss function.

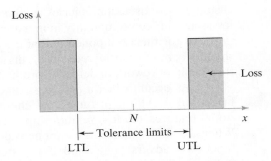

Figure 20.16 Loss function implicit in traditional tolerance specification.

illustrate several aspects of its application: (1) estimating the constant k in the loss function, Eq. (20.18), based on known cost data, (2) using the Taguchi loss function to estimate the cost of alternative tolerances, and (3) comparing the expected loss for alternative manufacturing processes that have different process distributions.

EXAMPLE 20.5 Estimating the Constant k in the Taguchi Loss Function

Suppose that a certain part dimension is specified as 100.0 ± 0.20 mm. To investigate the impact of this tolerance on product performance, the company has studied its repair records to discover that if the ± 0.20 mm tolerance is exceeded, there is a 60% chance that the product will be returned for repairs, at a cost of $100 to the company (during the warranty period) or to the customer (beyond the warranty period). Estimate the Taguchi loss function constant k for this data.

Solution: In Eq. (20.17) for the loss function, the value of $(x - N)$ is the tolerance value 0.20. The loss is the expected cost of the repair, which can be calculated as follows:

$$E\{L(x)\} = 0.60(\$100) + 0.40(0) = \$60$$

Using this cost in Eq. (20.18), we have

$$60 = k(0.20)^2 = k(0.04)$$

$$k = \frac{60}{0.04} = \$1,500$$

Therefore, the Taguchi loss function for this case is the following:

$$L(x) = 1500(x - N)^2 \qquad (20.19)$$

The Taguchi loss function can be used to evaluate the relative costs of alternative tolerances that might be applied to the component in question, as illustrated in the following example.

EXAMPLE 20.6 Using the Taguchi Loss Function to Estimate the Cost of Alternative Tolerances

Let us use the Taguchi quadratic loss function, Eq. (20.19), to evaluate the cost of alternative tolerances for the same data as in previous Example 20.5.

Specifically, given the nominal dimension of 100, as before, determine the cost (value of the loss function) for tolerances of (a) ± 0.10 mm and (b) ± 0.05 mm.

Solution: (a) For a tolerance of ± 0.10 mm, the value of the loss function is

$$L(x) = 1500(0.10)^2 = 1500(0.01) = \$15.00$$

(b) For a tolerance of ± 0.05 mm, the value of the loss function is:

$$L(x) = 1500(0.05)^2 = 1500(0.0025) = \$3.75$$

The loss function can be figured into production piece cost computations, if certain characteristics of the process are known, namely: (1) the applicable Taguchi loss function, (2) the production cost per piece, (3) the probability distribution of the process for the product parameter of interest, and (4) the cost of sortation, rework, and/or scrap for an out of tolerance piece. Combining these terms, we have the total piece cost as

$$C_{pc} = C_p + C_s + qC_r + L(x) \tag{20.20}$$

where C_{pc} = total cost per piece, C_p = production cost per piece, C_s = inspection and sortation cost per piece, q = proportion of parts falling outside of the tolerance limits and needing rework, C_r = rework cost per piece for those parts requiring rework, and $L(x)$ = Taguchi loss function cost per piece. Owing to the probability distribution associated with the production process, the analysis requires the use of expected costs. In the case of the normal distribution, it can be shown that the expected value of $(x - N)^2$ is the variance of the distribution σ^2. Thus, the expected value of the Taguchi loss function for this case is given by

$$E\{L(x)\} = k\sigma^2 \tag{20.21}$$

where σ^2 = the variance of the production process, and its square root is the standard deviation σ of the process.

EXAMPLE 20.7 Comparing the Expected Cost for Alternative Manufacturing Processes

Suppose that the part in Examples 20.5 and 20.6 can be produced by two alternative manufacturing processes. Both processes can produce parts with an average dimension at the desired nominal value of 100 mm. The distribution of the output is normal for each process, but their standard deviations are different. The relevant data for the two processes are given in the following table:

	Process A	Process B
Production cost per piece	$5.00	$10.00
Cost of sortation per piece	$1.00	$1.00
Rework cost per piece if tolerance exceeded	$20.00	$20.00
Taguchi loss function	Eq. (20.19)	Eq. (20.19)
Process standard deviation (mm)	0.08	0.04

Determine the expected cost per piece for the two processes.

Solution: The total cost per piece includes the other costs, namely the production cost per piece, inspection and sortation cost, and rework cost, if there is any rework, in addition to the loss function cost. For process A, the production cost per piece is $5.00, the sortation cost is $1.00 per piece, and the rework cost is $20.00. However, the rework cost is only applicable to those parts that fall outside the specified tolerance of ± 0.20 mm. The proportion of parts that lie beyond this interval can be found by computing the standard normal z statistic and determining the associated probability. The z-value is $0.20/0.08 = 2.5$, and the probability (from standard normal tables) is 0.0124. The Taguchi loss function is given by Eq. (20.19), but we substitute the standard deviation in the loss equation: $E\{L(x)\} = 1,500(0.08)^2 = \$9.60$. The total cost per piece is calculated as follows:

$$C_{pc} = 5.00 + 1.00 + 0.0124(20.00) + 9.60 = \$15.85 \text{ per piece}$$

For process B, although its production piece cost is much higher than for process A, there are virtually no out-of-tolerance units produced (as long as the process is in statistical control, which can be verified by statistical sampling). We should take advantage of this fact by omitting the sortation step. Also, there is no rework. The Taguchi loss function $E\{L(x)\} = 1500(0.04)^2 = \$2.40$. The total cost per piece for process B is calculated as follows:

$$C_{pc} = 10.00 + 0 + 0 + 2.40 = \$12.40 \text{ per piece}$$

Owing to a much smaller Taguchi loss function cost, process B is the lower cost production method.

Eq. (20.21) represents a special case of the more general situation. The special case is when the process mean μ, which is the average of all x_i, is centered about the nominal value N. The more general case is when the process mean μ may or may not be centered about the nominal value. In this more general case, the calculation of the value of the Taguchi loss function becomes

$$E\{L(x)\} = k[(\mu - N)^2 + \sigma^2] \qquad (20.22)$$

If the process mean is centered at the nominal value, so that $\mu = N$, then Eq. (20.22) reduces to Eq. (20.21).

20.8 ISO 9000

This chapter on quality control would not be complete without mention of the principal standard devoted to this subject. ISO 9000 is a set of international standards on quality developed by the International Organization for Standardization (ISO), based in Geneva, Switzerland and representing virtually all industrialized nations. The U.S. representative to ISO is the American National Standards Institute (ANSI). The American Society for Quality (ASQ) is the ANSI member organization that is responsible for quality standards. ASQ publishes and disseminates ANSI/ASQ Q9000, which is the U.S. version of ISO 9000.

ISO 9000 establishes standards for the systems and procedures used by a facility that affect the quality of the products and services produced by the facility. It is not a standard for the products or services themselves. ISO 9000 is not just one standard; it is a family of standards. The family includes a glossary of quality terms, guidelines for selecting and using the various standards, models for quality systems, and guidelines for auditing quality systems.

The ISO standards are generic rather than industry-specific. They are applicable to any facility producing any product and/or providing any service, no matter what the market. As mentioned, the focus of the standards is on the facility's quality system rather than its products or services. In the ISO standards, a *quality system* is defined as "the organizational structure, responsibilities, procedures, processes, and resources needed to implement quality management." ISO 9000 is concerned with the set of activities undertaken by a facility to ensure that its output provides customer satisfaction. It does not specify methods or procedures for achieving customer satisfaction; instead it describes concepts and objectives for achieving it.

ISO 9000 can be applied in a facility in two ways. The first is to implement the standards or selected portions of the standards simply for the sake of improving the firm's quality systems. Improving the procedures and systems for delivering high quality products and/or services is a worthwhile accomplishment, whether or not formal recognition is awarded. Implementation of ISO 9000 requires that all of a facility's activities affecting quality be carried out in a three-phase cycle that continues indefinitely. The three phases are (1) planning the activities and procedures that affect quality, (2) controlling the activities that affect quality to ensure that customer specifications are satisfied and that corrective action is taken on any deviations from specifications, and (3) documenting the activities and procedures affecting quality to ensure that quality objectives are understood by employees, feedback is provided for planning, and evidence of quality system performance is available for managers, customers, and for certification purposes.

The second way to apply ISO 9000 is to become registered. ISO 9000 registration not only improves the facility's quality systems, but it also provides formal certification that the facility meets the requirements of the standard. This benefits the firm in several ways. Two significant benefits are (1) reducing the frequency of quality audits performed by the facility's customers and (2) qualifying the facility for business partnerships with companies that require ISO 9000 registration. This latter benefit is especially important for firms doing business in the European Community, where certain products are classified as regulated and ISO 9000 registration is required for companies making these products as well as their suppliers.

Registration is obtained by having the facility certified by an accredited third-party agency. The certification process consists of on-site inspections and review of the firm's documentation and procedures so that the agency is satisfied that the facility conforms to the ISO 9000 standard. If the outside agency finds the facility nonconforming in certain areas, then it will notify the facility about which areas need upgrading, and schedule a repeat visit. Once the facility is registered, the external agency will periodically audit the facility to verify continuing conformance. The facility must pass these audits in order to retain ISO 9000 registration.

REFERENCES

[1] ARNOLD, K. L., *The Manager's Guide to ISO 9000*, The Free Press, NY, 1994.

[2] BASU, R., "Six Sigma to Fit Sigma," *IIE Solutions*, July 2001, pp. 28–33.

[3] BESTERFIELD, D. H., C. BESTERFIELD-MICHNA, G. H., Besterfield and M. BESTERFIELD-SACRE., *Total Quality Management*, 3d ed., Prentice Hall, Upper Saddle River, New Jersey, 2003.

[4] BOX, G. E. P., and N. R. DRAPER, *Evolutionary Operation*, John Wiley & Sons, Inc., NY, 1969.

[5] CROSBY, P. B., *Quality is Free*, McGraw-Hill Book Company, NY, 1979.

[6] ECKES, G., *Six Sigma for Everyone*, John Wiley & Sons, Inc., Hoboken, NJ, 2003.

[7] EVANS, J. R., and W. M. LINDSAY *The Management and Control of Quality*, 3d ed., West Publishing Company, St. Paul, MN, 1996.

[8] GOETSCH, D. L., and S. B. DAVIS, *Quality Management*, 5th ed., Prentice Hall, Upper Saddle River, NJ, 2006.

[9] GROOVER, M. P., *Work Systems and the Methods, Measurement, and Management of Work*, Pearson Prentice Hall, Upper Saddle River, NJ, 2007.

[10] JING, G. G., and L. NING, "Claiming Six Sigma," *Industrial Engineer*, February 2004, pp. 37–39.

[11] JOHNSON, P. L., *ISO 9000: Meeting the New International Standards*, McGraw-Hill, Inc., NY, 1993.

[12] JURAN, J. M., and F. M. GRYNA, *Quality Planning and Analysis*, 3d ed., McGraw-Hill, Inc., NY, 1993.

[13] KANTNER, R., *The ISO 9000 Answer Book*, Oliver Wight Publications, Inc., Essex Junction, VT, 1994.

[14] LOCHNER, R. H., and J. E. MATAR *Designing for Quality*, ASQC Quality Press, Milwaukee, WI, 1990.

[15] MONTGOMERY, D., *Introduction to Statistical Quality Control*, 5th ed., John Wiley & Sons, Inc., NY, 2005.

[16] OKES, D., "Improve Your Root Cause Analysis," *Manufacturing Engineering*, March 2005, pp. 171–178.

[17] PEACE, G. S., *Taguchi Methods*, Addison-Wesley Publishing Company, Inc., Reading, MA, 1993.

[18] PYZDEK, T., and R. W. BERGER, *Quality Engineering Handbook*, Marcel Dekker, Inc., NY, and ASQC Quality Press, Milwaukee, WI, 1992.

[19] ROBISON, J., "Integrate Quality Cost Concepts into Team's Problem-Solving Efforts," *Quality Progress*, March 1997, pp. 25–30.

[20] STAMATIS, D. H., *Six Sigma Fundamentals—A Complete Guide to the System, Methods, and Tools*, Productivity Press, NY, 2004.

[21] SUMMERS, D. C. S., *Quality*, Prentice Hall, Upper Saddle River, NJ, 1997.

[22] TAGUCHI, G., E. A. ELSAYED, and T. C. HSIANG, *Quality Engineering in Production Systems*, McGraw-Hill Book Company, NY, 1989.

[23] TITUS, R., "Total Quality Six Sigma Overview," Slide presentation, Lehigh University, Bethlehem, PA, May 2003.

[24] VEACH, C., "Real-Time SPC: The Rubber Meets the Road," *Manufacturing Engineering*, July, 1999, pp. 58–64.

[25] Website: *www.isixsigma.com*

[26] Website: *www.ge.com/sixsigma*

[27] Website: *www.motorola.com/sixsigma*

REVIEW QUESTIONS

20.1 What are the two aspects of quality in a manufactured product? List some of the product characteristics in each category.

20.2 Discuss the differences between the traditional view of quality control and the modern view.

20.3 What are the three main objectives of Total Quality Management?

20.4 What do the terms external customer and internal customer mean?

20.5 Manufacturing process variations can be divided into two types: (1) random and (2) assignable. Distinguish between these two types.

20.6 What is meant by the term *process capability*?

20.7 What is a control chart?

20.8 What are the two basic types of control charts?

20.9 What is a histogram?

20.10 What is a Pareto chart?

20.11 What is a defect concentration diagram?

20.12 What is a scatter diagram?

20.13 What is a cause-and-effect diagram?

20.14 What is Six Sigma?

20.15 What are the general goals of Six Sigma?

20.16 Why does 6σ in Six Sigma really mean 4.5σ?

20.17 What does DMAIC stand for?

20.18 What is the define step in DMAIC? What is accomplished during the define step?

20.19 What are master black belts in the Six Sigma hierarchy?

20.20 What is a CTQ characteristic?

20.21 What is the measure step in DMAIC?

20.22 Why is defects per million (DPM) not necessarily the same as defects per million opportunities ($DPMO$)?

20.23 What is the analyze step in DMAIC?

20.24 What is root cause analysis?

20.25 What is the improve step in DMAIC?

20.26 What is the control step in DMAIC?

20.27 What is a robust design in Taguchi's quality engineering?

20.28 What is ISO 9000?

PROBLEMS

Process Capability

(**Note:** Problems 20.2 and 20.5 require the use of standard normal distribution tables not included in this book.)

20.1 A turning process is in statistical control and the output is normally distributed, producing parts with a mean diameter = 30.020 mm and a standard deviation = 0.040 mm. Determine the process capability.

20.2 In previous problem 20.1, the design specification on the part is that the diameter = 30.000 ± 0.150 mm. (a) What proportion of parts fall outside the tolerance limits? (b) If the process is adjusted so that its mean diameter = 30.000 mm and the standard deviation remains the same, what proportion of parts fall outside the tolerance limits?

20.3 An automated tube bending operation produces parts with an included angle = 91.2°. The process is in statistical control and the values of the included angle are normally distributed with a standard deviation = 0.55°. The design specification on the angle = 90.0° ± 2.0°. (a) Determine the process capability. (b) If the process could be adjusted so that its mean = 90.0°, what would the value of the process capability index be?

20.4 A plastic extrusion process is in statistical control and the output is normally distributed. Extrudate is produced with a critical cross-sectional dimension = 28.6 mm and standard deviation = 0.53 mm. Determine the process capability.

20.5 In previous problem 20.4, the design specification on the part is that the critical cross-sectional dimension = 28.0 ± 2.0 mm. (a) What proportion of parts fall outside the tolerance limits? (b) If the process were adjusted so that its mean diameter = 28.0 mm and the standard deviation remained the same, what proportion of parts would fall outside the tolerance limits? (c) With the adjusted mean at 28.0 mm, determine the value of the process capability index.

Control Charts

20.6 Seven samples of five parts each have been collected from an extrusion process which is in statistical control, and the diameter of the extrudate has been measured for each part. (a) Determine the values of the center, LCL, and UCL for $\bar{x}$ and R charts. The calculated values of $\bar{x}$ and R for each sample are given below (measured values are in inches). (b) Construct the control charts and plot the sample data on the charts.

s	1	2	3	4	5	6	7
$\bar{x}$	1.002	0.999	0.995	1.004	0.996	0.998	1.006
R	0.010	0.011	0.014	0.020	0.008	0.013	0.017

20.7 Ten samples of size $n = 8$ have been collected from a process in statistical control, and the dimension of interest has been measured for each part. (a) Determine the values of the center, LCL, and UCL for the $\bar{x}$ and R charts. The calculated values of $\bar{x}$ and R for each sample are given below (measured values are in mm). (b) Construct the control charts and plot the sample data on the charts.

s	1	2	3	4	5	6	7	8	9	10
$\bar{x}$	9.22	9.15	9.20	9.28	9.19	9.12	9.20	9.24	9.17	9.23
R	0.24	0.17	0.30	0.26	0.27	0.19	0.21	0.32	0.21	0.23

20.8 In 12 samples of size $n = 7$, the average value of the sample means is $\bar{x} = 6.860$ inch for the dimension of interest, and the mean of the ranges of the samples is $\bar{R} = 0.027$ inch. Determine: (a) lower and upper control limits for the $\bar{x}$ chart and (b) lower and upper control limits for the R chart.

20.9 In nine samples each of size $n = 10$, the grand mean of the samples is $\bar{x} = 100$ for the characteristic of interest, and the mean of the ranges of the samples is $\bar{R} = 8.5$. Determine: (a) lower and upper control limits for the $\bar{x}$ chart and (b) lower and upper control limits for the R chart.

20.10 A p chart is to be constructed. Six samples of 25 parts each have been collected, and the average number of defects per *sample* = 2.75. Determine the center, LCL and UCL for the p chart.

20.11 Ten samples of equal size are taken to prepare a p chart. The total number of parts in these ten samples was 900 and the total number of defects counted was 117. Determine the center, LCL and UCL for the p chart.

20.12 The yield of good chips during a certain step in silicon processing of integrated circuits averages 91%. The number of chips per wafer is 200. Determine the center, LCL, and UCL for the p chart that might be used for this process.

20.13 The upper and lower control limits for a p chart are: $LCL = 0.10$ and $UCL = 0.24$. Determine the sample size n that is used with this control chart.

20.14 The upper and lower control limits for a p chart are $LCL = 0$ and $UCL = 0.20$. The center line of the p chart is at 0.10. Determine the sample size n that is compatible with this control chart.

20.15 Twelve cars were inspected after final assembly. The number of defects found ranged between 87 and 139 defect per car with an average of 116. Determine the center and upper and lower control limits for the c chart that might be used in this situation.

20.16 For each of the three control charts in Figure P20.16, identify whether or not there is evidence that the process depicted is out of control.

Determining Sigma Level in Six Sigma

20.17 A garment manufacturer produces 22 different coat styles, and every year new coat styles are introduced and old styles are discarded. The final inspection department checks each coat, regardless of its style, before it leaves the factory for nine features that are considered critical-to-quality (CTQ) characteristics for customer satisfaction. The inspection report for last month indicated that a total of 366 deficiencies of the nine features were found among 8,240 coats produced. Determine (a) defects per million opportunities and (b) sigma level for the manufacturer's production performance.

20.18 A producer of cell phones checks each phone prior to packaging, using seven critical-to-quality (CTQ) characteristics that are deemed important to customers. Last year, out of 205,438 phones produced by the company, a total of 578 phones had at least one defect, and the total number of defects among these 578 phones was 1692. Determine (a) the number of defects per million opportunities and corresponding sigma level, (b) the number of defects per million and corresponding sigma level, and (c) the number of defective units per million and corresponding sigma level.

20.19 The inspection department in an automobile final assembly plant checks cars coming off the line against 85 features that are considered critical-to-quality characteristics for customer satisfaction. During a one-month period, a total of 16,578 cars were produced. For those cars, a total of 1,989 defects of various types were found, and the total number of cars that had one or more defects was 512. Determine (a) the number of defects per million opportunities and corresponding sigma level, (b) the number of defects per million and corresponding sigma level, and (c) the number of defective units per million and corresponding sigma level.

20.20 A digital camera maker produces three different models: (1) base model, (2) zoom model, and (3) zoom model with extra memory. Data for the three models are shown in the table below. The three models have been on the market for one year, and the first year's sales are

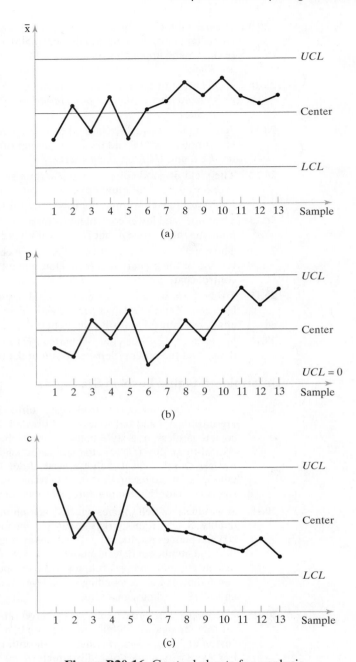

Figure P20.16 Control charts for analysis.

given in the table. Also given are critical-to-quality (CTQ) characteristics and total defects that have been tabulated for the products sold. Higher model numbers have more CTQ characteristics (opportunities for defects) because they are more complex. The category of total defects refers to the total number of defects of all CTQ characteristics for each model. For each of the three models, determine (a) the number of defects per million opportunities and corresponding sigma level, (b) the number of defects per million and corresponding sigma level, and (c) the number of defective units per million and corresponding sigma level. (d) Is any one model produced at a higher quality level than the others? (e) Determine aggregate values for $DPMO, DPM$, and $DUPM$ and their corresponding sigma levels for all models made by the camera maker.

Model	Annual Sales	CTQ Characteristics	Number of Defective Cameras	Total Number of Defects
1	62,347	16	127	282
2	31,593	23	109	429
3	18,662	29	84	551

Taguchi Loss Function

20.21 A certain part dimension on a power garden tool is specified as 25.50 ± 0.30 mm. Company repair records indicate that if the ±0.30 mm tolerance is exceeded, there is a 75% chance that the product will be returned for replacement. The cost associated with replacing the product, which includes not only the product cost itself but also the additional paperwork and handling associated with replacement, is estimated to be $300. Determine the constant k in the Taguchi loss function for this data.

20.22 The design specification on the resistance setting for an electronic component is 0.50 ± 0.02 ohm. If the component is scrapped, the company suffers a $200 cost. (a) What is the implied value of the constant k in the Taguchi quadratic loss function? (b) If the output of the process that sets the resistance is centered on 0.50 ohm, with a standard deviation of 0.01 ohm, what is the expected loss per unit?

20.23 The Taguchi quadratic loss function for a particular component in a piece of earth moving equipment is $L(x) = 3500(x - N)^2$, where x = the actual value of a critical dimension and N is the nominal value. If $N = 150.00$ mm, determine the value of the loss function for tolerances of (a) ±0.20 mm and (b) ±0.10 mm.

20.24 The Taguchi loss function for a certain component is given by $L(x) = 8000(x - N)^2$, where x = the actual value of a dimension of critical importance and N is its nominal value. Company management has decided that the maximum loss that can be accepted is $10. (a) If the nominal dimension is 30.00 mm, at what value should the tolerance on this dimension be set? (b) Does the value of the nominal dimension have any effect on the tolerance that should be specified?

20.25 Two alternative manufacturing processes, A and B, can be used to produce a certain dimension on one of the parts in an assembled product. Both processes can produce parts with an average dimension at the desired nominal value. The tolerance on the dimension is ±0.15 mm. The output of each process follows a normal distribution. However, the standard deviations are different. For process A, $\sigma = 0.12$ mm; and for process B, $\sigma = 0.07$ mm. Production costs per piece for A and B are $7.00 and $12.00, respectively. If inspection and sortation is required, the cost is $0.50 per piece. If a part is found to be defective, it must be scrapped at a cost equal to its production cost. The Taguchi loss function for this component is given by $L(x) = 2500(x - N)^2$, where x = value of the dimension and N is its nominal value. Determine the average cost per piece for the two processes.

20.26 Solve previous problem 20.25, using a tolerance on the dimension of ±0.30 mm rather than ±0.15 mm.

20.27 Solve previous problem 20.25, assuming the average value of the dimension produced by process B is 0.10 mm greater than the nominal value specified. The average value of the dimension produced by process A remains at the nominal value N.

20.28 Two different manufacturing processes, A and B, can be used to produce a certain component. The specification on the dimension of interest is 100.00 mm $\pm$ 0.20 mm. The output of process A follows the normal distribution, with $\mu = 100.00$ mm and $\sigma = 0.10$ mm. The output of process B is a uniform distribution defined by $f(x) = 2.0$ for $99.75 \leq x \leq 100.25$ mm. Production costs per piece for processes A and B are each $5.00. Inspection and sortation cost is $0.50 per piece. If a part is found to be defective, it must be scrapped at a cost equal to twice its production cost. The Taguchi loss function for this component is given by $L(x) = 2500(x - N)^2$, where $x =$ value of the dimension and N is its nominal value. Determine the average cost per piece for the two processes.

Chapter 21

Inspection Principles and Practices

In quality control, inspection is the means by which poor quality is detected and good quality is assured. Inspection is traditionally accomplished using labor-intensive methods that are time-consuming and costly. Consequently, manufacturing lead time and product cost are increased without adding any real value to the products. In addition, manual inspection

From Chapter 21 of *Automation, Production Systems, and Computer-Integrated Manufacturing*, Third Edition.
Mikell P. Groover. Copyright © 2008 by Pearson Education, Inc. Publishing as Prentice Hall. All rights reserved.

is performed after the process is complete, often after a significant time delay. Therefore, if a bad product has been made, it is too late to correct the defect(s) during regular processing. Parts already manufactured that do not meet specified quality standards must either be scrapped or reworked at additional cost.

New approaches to quality control are addressing these problems and drastically altering the way inspection is accomplished. The new approaches include

- 100% automated inspection rather than sampling inspection using manual methods
- Use of on-line sensor systems to accomplish inspection during or immediately after the manufacturing process, instead of off-line inspection performed later
- Feedback control of the manufacturing operation, monitoring process variables that determine product quality rather than the product itself
- Use of software tools to track and analyze the sensor measurements over time for statistical process control
- Advanced inspection and sensor technologies, combined with computer-based systems to automate the operation of the sensor systems.

In this chapter, we examine some of these modern approaches to inspection with an emphasis on automating the inspection function. In the following chapter, we discuss the relevant inspection technologies such as coordinate measuring machines and machine vision.

21.1 INSPECTION FUNDAMENTALS

The term *inspection* refers to the activity of examining the product, its components, subassemblies, or the raw materials to determine whether they conform to design specifications. The design specifications are defined by the product designer.

21.1.1 Types of Inspection

Inspections can be classified into two types, according to the amount of information derived from the inspection procedure about the item's conformance to specification:

1. *Inspection for variables*, in which one or more quality characteristics of interest are measured using an appropriate measuring instrument or sensor. We discuss measurement principles in Section 22.1 of the following chapter.
2. *Inspection for attributes*, in which the part or product is inspected to determine whether it conforms to the accepted quality standard. The determination is sometimes based simply on the judgment of the inspector. In other cases, the inspector uses a gage to aid in the decision. Inspection by attributes can also involve counting the number of defects in a product.

Examples of the two types of inspection are listed in Table 21.1. To relate these differences to our discussion of control charts in the previous chapter, inspection for variables uses the $\bar{x}$ chart and R chart, whereas inspection for attributes uses the p chart or c chart.

The advantage of inspection for variables is that more information is obtained from the inspection procedure about the item's conformance to design specification. The inspection yields a quantitative value. Data can be collected and recorded over time to

TABLE 21.1 Examples of Inspection for Variables and Inspection for Attributes

Examples of Inspection for Variables	Examples of Inspection for Attributes
Measuring the diameter of a cylindrical part	Gaging a cylindrical part with a GO/NO-GO gage to determine if it is within tolerance
Measuring the temperature of a toaster oven to see if it is within the range specified by design engineering	Determining the fraction defect rate of a sample of production parts
Measuring the electrical resistance of an electronic component	Counting the number of defects in an automobile as it leaves the final assembly plant
Measuring the specific gravity of a fluid chemical product	Counting the number of imperfections in a production run of carpeting

observe trends in the process that makes the part. The data can be used to fine-tune the process so that future parts are produced with dimensions closer to the nominal design value. In attributes inspection (e.g., when a dimension is simply checked with a gage), all that is known is whether the part is acceptable and perhaps whether it is too big or too small. On the other hand, inspection for attributes does have the advantage that it can be done quickly and therefore at lower cost. Measuring the quality characteristic is a more involved procedure that takes more time.

21.1.2 Inspection Procedure

A typical inspection procedure performed on an individual item, such as a part, subassembly, or final product, consists of the following steps [2]:

1. *Presentation*—The item is presented for examination.
2. *Examination*—The item is examined for nonconforming feature(s). In inspection for variables, examination consists of measuring a dimension or other attribute of the part or product. In inspection for attributes, this involves gaging one or more dimensions or searching the item for flaws.
3. *Decision*—Based on the examination, a decision is made whether the item satisfies the defined quality standards. The simplest case involves a binary decision, in which the item is deemed either acceptable or unacceptable. In more complicated cases, the decision may involve grading the item into one of more than two possible quality categories, such as grade A, grade B, and unacceptable.
4. *Action*—The decision should result in some action, such as accepting or rejecting the item, or sorting the item into the most appropriate quality grade. It may also be desirable to take action to correct the manufacturing process to minimize the occurrence of future defects.

The inspection procedure is traditionally performed by a human worker (referred to as *manual inspection*), but automated inspection systems are increasingly being used as sensor and computer technologies are developed and refined for the purpose. In some production situations only one item is produced (e.g., a one-of-a-kind machine or a prototype), and the inspection procedure is applied only to the one item. In other situations, such as batch production and mass production, the inspection procedure is repeated either on all of the items in the production run (*100% inspection*, sometimes called *screening*) or on only a sample taken from the population of items (*sampling inspection*). Manual inspection is more likely to be used when only one item or a sample of parts from a larger batch is inspected, whereas automated systems are more common for 100% inspection in mass production.

In the ideal inspection procedure, all of the specified dimensions and attributes of the part or product would be inspected. However, inspecting every dimension is time consuming and expensive. In general, it is unnecessary. As a practical matter, certain dimensions and specifications are more important than others in terms of assembly or function of the product. These important specifications are called *key characteristics* (KCs). They are the specifications that should be recognized as important in design; they are identified as KCs on the part drawings and in the engineering specifications, given the most attention in manufacturing, and inspected in quality control. Examples of KCs include matching dimensions of assembled components, surface roughness on bearing surfaces, straightness and concentricity of high speed rotating shafts, and finishes of exterior surfaces of consumer products. The inspection procedure should be designed to focus on these KCs. It usually turns out that if the processes responsible for the KCs are maintained in statistical control (Section 20.3.1), then the other dimensions of the part will also be in statistical control. If these less important part features deviate from their nominal values, the consequences, if any, are less severe than if a KC deviates.

21.1.3 Inspection Accuracy

Errors sometimes occur in the inspection procedure during the examination and decision steps. Items of good quality are incorrectly classified as not conforming to specifications, and nonconforming items are mistakenly classified as conforming. These two kinds of mistakes are called Type I and Type II errors. A *Type I error* occurs when an item of good quality is incorrectly classified as being defective. It is a "false alarm." A *Type II error* is when an item of poor quality is erroneously classified as being good. It is a "miss." These error types are portrayed graphically in Table 21.2.

Inspection errors do not always neatly follow the above classification. For example, in inspection by variables, a common inspection error consists of incorrectly measuring a part dimension. As another example, a form of inspection by attributes involves counting the number of nonconforming features on a given product, such as the number of defects on a new automobile coming off the final assembly line. An error is made if the inspector misses one of the defects. In both of these examples, an error may result in either a conforming feature being classified as nonconforming (Type I error) or a nonconforming feature being classified as conforming (Type II error).

TABLE 21.2   Type I and Type II Inspection Errors

Decision	Conforming Item	Nonconforming Item
Accept item	Good decision	Type II error "Miss"
Reject item	Type I error "False alarm"	Good decision

In manual inspection, these errors result from factors such as (1) complexity and difficulty of the inspection task, (2) inherent variations in the inspection procedure, (3) judgment required by the human inspector, (4) mental fatigue in the human inspector, and (5) inaccuracies or problems with the gages or measuring instruments used in the inspection procedure. When the procedure is accomplished by an automated system, inspection errors occur due to factors such as (1) complexity and difficulty of the inspection task, (2) resolution of the inspection sensor, which is affected by "gain" and similar control parameter settings, (3) equipment malfunctions, and (4) faults or "bugs" in the computer program controlling the inspection procedure.

The term *inspection accuracy* refers to the capability of the inspection process to avoid these types of errors. Inspection accuracy is high when few or no errors are made. Measures of inspection accuracy are suggested by Drury [2] for the case in which parts are classified by an inspector (or automatic inspection system) into either of two categories, conforming or nonconforming. Considering this binary case, let p_1 = proportion of times (or probability) that a conforming item is classified as conforming, and let p_2 = proportion of times (or probability) that a nonconforming item is classified as nonconforming. In other words, both of these proportions (or probabilities) correspond to correct decisions. Thus, $(1 - p_1)$ = probability that a conforming item is classified as nonconforming (Type I error), and $(1 - p_2)$ = probability that a nonconforming item is classified as conforming (Type II error).

If we let q = actual fraction defect rate in the batch of items, a table of possible outcomes can be constructed as in Table 21.3 to show the fraction of parts correctly and incorrectly classified and for those incorrectly classified, whether the error is Type I or Type II.

These proportions (probabilities) would have to be assessed empirically for individual inspectors by determining the proportion of correct decisions made in each of the two cases of conforming and nonconforming items in a parts batch of interest. Unfortunately, the proportions vary for different inspection tasks. The error rates are generally higher (lower p_1 and p_2 values) for more difficult inspection tasks. Also, different inspectors tend to have different p_1 and p_2 rates. Typical values of p_1 range between 0.90 and 0.99, and typical p_2 values range between 0.80 and 0.90, but values as low as 0.50 for both p_1 and p_2 have been reported [2]. For human inspectors, p_1 is inclined to be higher than p_2 because inspectors are usually examining items that are mostly good quality and tend to be on the lookout for defects.

Values of p_1 and p_2 are workable measures of inspection accuracy for a human inspector or an automated inspection system. Each measure taken separately provides useful information because the p_1 and p_2 values would be expected to vary independently to some degree depending on the inspection task and the person or system performing the

TABLE 21.3 Table of Possible Outcomes in Inspection Procedure, Given q, p_1, and p_2

Decision	True State of Item		
	Conforming	Nonconforming	Total
Accept item	$p_1(1 - q)$	$(1 - p_2)q$ Type II error	$p_1 + q(1 - p_1 - p_2)$
Reject item	$(1 - p_1)(1 - q)$ Type I error	$p_2 q$	$1 - p_1 - q(1 - p_1 - p_2)$
Total	$(1 - q)$	q	1.0

inspection. A practical difficulty in applying the measures is determining the true values of p_1 and p_2. These values would have to be determined by an alternative inspection process, which would itself be prone to the same errors as the first process whose accuracy is being assessed.

EXAMPLE 21.1 Inspection Accuracy

A human worker has inspected a batch of 100 parts and reported a total of 12 defects in the batch. On careful reexamination, it was found that four of these reported defects were in fact good pieces (four false alarms), whereas six defective units in the batch were undetected by the inspector (six misses). What is the inspector's accuracy in this instance? Specifically, what are the values of p_1 and p_2?

Solution: Of the 12 reported defects, four are good, leaving eight defects among those reported. In addition, six other defects were found among the reportedly good units. Thus, the total number of defects in the batch of 100 is $8 + 6 = 14$. This means there were $100 - 14 = 86$ good units in the batch. We can assess the values of p_1 and p_2 on the basis of these numbers.

To assess p_1, we note that the inspector reported 12 defects, leaving 88 that were reported as acceptable. Of these 88, six were actually defects, thus leaving $88 - 6 = 82$ actual good units reported by the inspector. Thus, the proportion of good parts reported as conforming is

$$p_1 = \frac{82}{86} = 0.9535$$

There are 14 defects in the batch, of which the inspector correctly identified eight. Thus, the proportion of defects reported as nonconforming is

$$p_2 = \frac{8}{14} = 0.5714$$

21.1.4 Inspection vs. Testing

Quality control utilizes both inspection and testing procedures to detect whether a part or product is within design specifications. Both activities are important in a company's quality control program. Whereas inspection is used to assess the quality of the product relative to design specifications, testing is a term in quality control that refers to assessment of the functional aspects of the product: Does the product operate the way it is supposed to operate? Will it continue to operate for a reasonable period of time? Will it operate in environments of extreme temperature and humidity? Accordingly, *QC testing* is a procedure in which the item being tested (product, subassembly, part, or material) is observed during actual operation or under conditions that might be present during operation. For example, a product might be tested by running it for a certain period of time to determine whether it functions properly. If the product passes the test, it is approved for shipment to the customer. As another example, a part, or the material out of which the part is to be made, might be tested by subjecting it to a stress load that is equivalent to or greater than the load anticipated during normal service.

Sometimes the testing procedure is damaging or destructive to the item. To ensure that the majority of the items (e.g., raw materials or finished products) are of satisfactory quality, a limited number of the items are sacrificed. However, the expense of destructive testing is significant enough that great efforts are made to devise methods that do not result in the destruction of the item. These methods are referred to as *nondestructive testing* (NDT) and *nondestructive evaluation* (NDE).

Another type of testing procedure involves not only the testing of the product to see that it functions properly; it also requires an adjustment or *calibration* of the product that depends on the outcome of the test. During the testing procedure, one or more operating variables of the product are measured, and adjustments are made in certain inputs that influence the performance of the operating variables. For example, in the testing of certain appliances with heating elements, if the measured temperature is too high or too low after a specified time, adjustments can be made in the control circuitry (e.g., changes in potentiometer settings) to bring the temperature within the acceptable operating range.

21.2 SAMPLING VS. 100% INSPECTION

The primary focus of this chapter is inspection rather than testing. As suggested by the preceding descriptions of the two functions, inspection is more closely associated with manufacturing operations. Inspection can be performed using statistical sampling or 100%.

21.2.1 Sampling Inspection

Inspection is traditionally accomplished using manual labor. The work is often boring and monotonous, yet the need for precision and accuracy is great. Sometimes it takes hours to measure the important dimensions of only one workpart. Because of the time and expense involved in inspection work, sampling procedures are often used to reduce the need to inspect every part. The statistical sampling procedures are known by the terms *acceptance sampling* or *lot sampling*.

Types of Sampling Plans. There are two basic types of acceptance sampling: (1) variables sampling and (2) attributes sampling, corresponding to inspection by variables and inspection by attributes previously described (Section 21.1.1). In a *variables sampling plan*, a random sample is taken from the population, and the quality characteristic of interest (e.g., a part dimension) is measured for each unit in the sample. These measurements are then averaged, and the mean value is compared with an allowed value for the plan. The batch is then accepted or rejected depending on the results of this comparison. The allowed value used in the comparison is chosen so that the probability that the batch will be rejected is small unless the actual quality level in the population is indeed poor.

In an *attributes sampling plan*, a random sample is drawn from the batch, and the units in the sample are classified as acceptable or defective, depending on the quality criterion being used. The batch is accepted if the number of defects does not exceed a certain value, called the *acceptance number*. If the number of defects found in the sample is greater than the acceptance number, the batch is rejected. As in variables sampling, the value of the acceptance number is selected so that the probability that the batch will be rejected is small unless the overall quality of the parts in the batch is poor.

In sampling, there is always a probability that the batch will be rejected even if the overall quality is acceptable. Similarly, there is a probability that the batch will be accepted even if the overall quality level in the batch is not acceptable. Statistical errors are a fact of life in statistical sampling. Let us explore what is meant by the word "acceptable" in the context of acceptance sampling and at the same time examine the risks associated with committing a statistical error. Our focus will be on attributes sampling, but the same basic notions apply to variables sampling. Ideally, a batch of parts would be absolutely free of defects. However, such perfection is difficult if not impossible to attain in practice. Accordingly, the customer and the supplier agree that a certain level of quality is acceptable, even though that quality is less than perfect. This *acceptable quality level* (*AQL*) is defined in terms of proportion of defects, or fraction defect rate q_o. Alternatively, there is another level of quality, again defined in terms of fraction defect rate q_1, where $q_1 > q_o$, which the customer and supplier agree is unacceptable. This q_1 level is called the *lot tolerance percent defective* (*LTPD*).

Statistical Errors in Sampling. Two possible statistical errors can occur in acceptance sampling. The first is rejecting a batch of product that is equal to or better than the AQL (meaning that the actual $q \leq q_o$). This is a *Type I error*, and the probability of committing this type of error is called the *producer's risk* α. The second error is accepting a batch of product whose quality is worse than the *LTPD* ($q \geq q_1$). This is a *Type II error*, and the probability of this error is called the *consumer's risk* β. These errors are depicted in Table 21.4. Sampling errors should not be confused with the inspection errors previously described in our discussion of inspection accuracy (Section 21.1.3). Sampling errors occur because only a fraction of the total population has been inspected. We are at the mercy of the laws of probability as to whether the sample is an accurate reflection of the population. Inspection errors, on the other hand, occur when an individual item is wrongly classified as being defective when it is good (Type I error) or good when it is defective (Type II error).

Design of an acceptance sampling plan involves determining values of the sample size Q_s and the acceptance number N_a that provide the agreed-on *AQL* and *LTPD*, together with the associated probabilities α and β (producer's and consumer's risks). Procedures for determining Q_s and N_a based on *AQL*, *LTPD*, α, and β are described in texts on quality control, such as [3], and [4]. Also, standard sampling plans have been developed, such as MIL-STD-105D (also known as ANSI/ASQC Z1.4, the U.S. standard, and ISO/DIS-2859, the international standard).

Operating Characteristic Curve. Much information about a sampling plan can be obtained from its operating characteristic curve (OC curve). The operating characteristic curve for a given sampling plan gives the probability that a batch will be accepted as a function of the possible fraction defect rates that might exist in the batch. The general shape of the OC curve is shown in Figure 21.1. In effect, the OC curve indicates the degree of protection provided by the sampling plan for various quality levels of incoming lots. If the incoming batch has a high quality level (low q), then the probability of acceptance is high. If the quality level of the incoming batch is poor (high q), then the probability of acceptance is low.

TABLE 21.4 Type I and Type II Sampling Errors

Decision	Acceptable Batch	Unacceptable Batch
Accept batch	Good decision	Type II error (β) Consumer's risk
Reject batch	Type I error Producer's risk (α)	Good decision

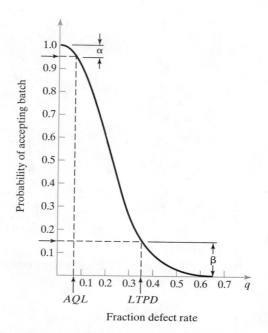

Figure 21.1 The operating characteristic (OC) curve for a given sampling plan shows the probability of accepting the lot for different fraction defect rates of incoming batches.

When a batch is rejected as a result of a sampling procedure, several possible actions might be taken. One possibility is to send the parts back to the supplier. If there is an immediate need for the parts in production, this action may be impractical. A more appropriate action may be to inspect the batch 100% and sort out the defects, which are sent back to the supplier for replacement or credit. A third possible action is to sort out the defects and rework or replace them at the supplier's expense. Whatever the action, rejecting a batch leads to corrective action that has the effect of improving the overall quality of the batch exiting the inspection operation. A given sampling plan can be described by its *average outgoing quality* curve (*AOQ* curve), the typical shape of which is illustrated in Figure 21.2. The *AOQ* curve shows the average quality of batches passing through the sampling inspection plan as a function of incoming lot quality (before inspection). As one would expect, when the incoming quality is good (low q), the average outgoing quality is good (low *AOQ*). When the incoming quality is poor (high q), the *AOQ* is also low because there is a strong probability of rejecting the batch, with the resulting action that defectives in the batch are sorted out and replaced with good parts. It is in the intermediate range, between the *AQL* and *LTPD*, that the outgoing batch quality of the sampling plan is the poorest. As shown in our plot, the highest *AOQ* level will be found at some intermediate value of q, and this *AOQ* is called the *average outgoing quality limit* (*AOQL*) of the plan.

21.2.2 100% Manual Inspection

When sampling inspection is conducted, the sample size is often small compared with the size of the population. The sample size may represent only 1% or fewer of the number of parts in the batch. Because only a portion of the items in the population is inspected in a statistical sampling procedure, there is a risk that some defective parts will slip through

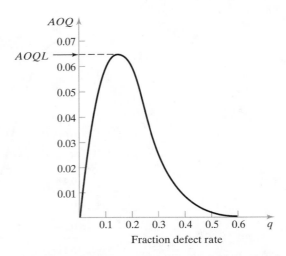

Figure 21.2   Average outgoing quality (AOQ) curve for a sampling plan.

the inspection screen. As indicated in our preceding discussion of average outgoing quality, one of the objectives in statistical sampling is to define the expected risk, that is, to determine the average fraction defect rate that will pass through the sampling inspection procedure over the long run, under the assumption that the manufacturing process remains in statistical control. The frequency with which samples are taken, the sample size, and the permissible quality level (AQL) are three important factors that affect the level of risk involved. But the fact remains that something less than 100% good quality must be tolerated as the price to be paid for using statistical sampling procedures.

In principle, the only way to achieve 100% acceptable quality is to use 100% inspection. It is instructive to compare the OC curve of a 100% inspection plan, shown in Figure 21.3, with the OC curve of a sampling plan as in Figure 21.1. The advantage of 100% inspection is that the probability the batch will be accepted is 1.0 if its quality is equal to or better than the AQL and zero if the quality is lower than the AQL. One might logically argue that the term "acceptable quality level" has less meaning in 100% inspection, since a target of zero defects should be attainable if every part in the batch is inspected; in other words, the AQL should be set at $q = 0$. However, one must distinguish between the output of the manufacturing process that makes the parts and the output of the inspection procedure that sorts the parts. It may be possible to separate out all of the defects in the batch so that only good parts remain after inspection (AOQ = zero defects), whereas the manufacturing process still produces defects at a certain fraction defect rate q ($q > 0$).

Theoretically, 100% inspection allows only good quality parts to pass through the inspection procedure. However, when 100% inspection is done manually, two problems arise: First, the obvious problem is the expense involved. Instead of dividing the time of inspecting the sample over the number of parts in the production run, the inspection time per piece is applied to every part. The inspection cost sometimes exceeds the cost of making the part. Second, with 100% manual inspection, there is the problem of inspection accuracy (Section 21.1.3). There are almost always errors associated with 100% inspection (Type I and II errors), especially when the inspection procedure is performed by human inspectors.

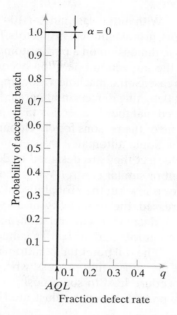

Figure 21.3 Operating characteristic curve of a 100% inspection plan.

Because of these human errors, 100% inspection using manual methods is no guarantee of 100% good quality product.

21.3 AUTOMATED INSPECTION

An alternative to manual inspection is automated inspection. Automation of the inspection procedure will almost always reduce inspection time per piece, and automated machines do not experience the fatigue and mental errors suffered by human inspectors. Economic justification of an automated inspection system depends on whether the savings in labor cost and improvement in accuracy will more than offset the investment and/or development costs of the system.

Automated inspection can be defined as the automation of one or more of the steps involved in the inspection procedure. There are a number of alternative ways in which automated or semiautomated inspection can be implemented:

1. Automated *presentation* of parts by an automatic handling system with a human operator still performing the *examination* and *decision* steps.
2. Automated *examination* and *decision* by an automatic inspection machine, with manual loading (*presentation*) of parts into the machine.
3. Completely automated inspection system in which parts *presentation*, *examination*, and *decision* are all performed automatically.

In the first case, the inspection procedure is performed by a human worker, with all of the possible errors in this form of inspection. In cases (2) and (3), the actual inspection operation is accomplished by an automated system. These latter cases are our primary interest here.

As in manual inspection, automated inspection can be performed using statistical sampling or 100%. When statistical sampling is used, sampling errors are possible.

With either sampling or 100% inspection, automated systems can commit inspection errors, just like human inspectors. For simple inspection tasks, such as automatic gaging of a single dimension on a part, automated systems operate with high accuracy (low error rate). As the inspection operation becomes more complex and difficult, the error rate tends to increase. Some machine vision applications (Section 22.6) fall into this category; for example, detecting defects in integrated circuit chips or printed circuit boards. It should be mentioned that these inspection tasks are also complex and difficult for human workers, and this is one of the reasons for developing automated inspection systems that can do the job.

Some automated inspection systems can be adjusted to increase their sensitivity to the defect they are designed to detect. This is accomplished by means of a "gain" adjustment or similar control. When the sensitivity adjustment is low, the probability of a Type I error is low but the probability of a Type II error is high. When the sensitivity adjustment is increased, the probability of a Type I error increases, while the probability of a Type II error decreases. This relationship is portrayed in Figure 21.4. Because of these errors, 100% automated inspection cannot guarantee 100% good quality product.

The full potential of automated inspection is best achieved when it is integrated into the manufacturing process, when 100% inspection is used, and when the results of the procedure lead to some positive action. The positive actions can take either or both of two possible forms, as illustrated in Figure 21.5:

(a) *Feedback process control*. In this case, data are fed back to the preceding manufacturing process responsible for the quality characteristics being evaluated or gaged in the inspection operation. The purpose of feedback is to allow compensating adjustments to be made in the process to reduce variability and improve quality. If the measurements from the automated inspection indicate that the output of the process is beginning to drift toward the high side of the tolerance (e.g., tool wear

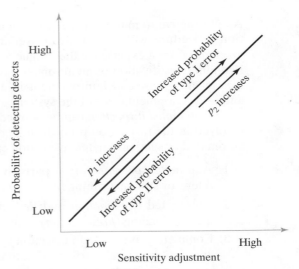

Figure 21.4 Relationship between sensitivity of an automated inspection system and the probability of Type I and Type II errors: p_1 = the probability that a conforming item is correctly classified, and p_2 = the probability that a nonconforming item is correctly classified.

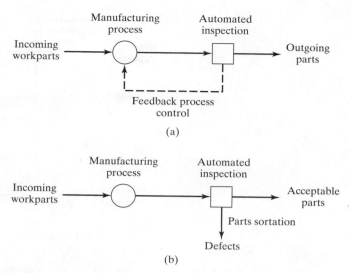

Figure 21.5 Action steps resulting from automated inspection: (a) feedback process control and (b) sortation of parts into two or more quality levels.

might cause a part dimension to increase over time), corrections can be made in the input parameters to bring the output back to the nominal value. In this way, average quality is maintained within a smaller variability range than is possible with sampling inspection methods. In effect, process capability is improved.

(b) *Parts sortation.* In this case, the parts are sorted according to quality level: acceptable versus unacceptable quality. There may be more than two levels of quality appropriate for the process (e.g., acceptable, reworkable, and scrap). Sortation and inspection may be accomplished in several ways. One alternative is to both inspect and sort at the same station. Other installations locate one or more inspections along the processing line, with a single sortation station near the end of the line. Inspection data are analyzed and instructions are forwarded to the sortation station indicating what action is required for each part.

21.4 WHEN AND WHERE TO INSPECT

Inspection can be performed at any of several places in production: (1) receiving, when raw materials and parts are received from suppliers, (2) various stages of manufacture, and (3) before shipment to the customer. In this section our principal focus is on case (2), that is, when and where to inspect during production.

21.4.1 Off-Line and On-Line Inspection

The timing of the inspection procedure in relation to the manufacturing process is an important consideration in quality control. Three alternative situations can be distinguished, shown in Figure 21.6: (a) off-line inspection, (b) on-line/in-process inspection, and (c) on-line/post-process inspection.

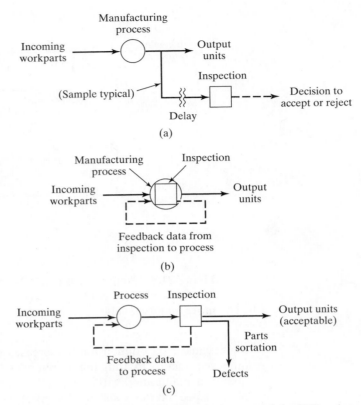

Figure 21.6 Three inspection alternatives are (a) off-line inspection, (b) on-line/in-process inspection, and (c) on-line/post-process inspection.

Off-Line Inspection. Off-line inspection is performed away from the manufacturing process, and there is generally a time delay between processing and inspection. Off-line inspection is often accomplished using statistical sampling methods. Manual inspection is common. Factors that tend to promote the use of off-line inspection are that: (1) variability of the process is well within design tolerance, (2) processing conditions are stable and the risk of significant deviations in the process is small, and (3) cost of inspection is high relative to the cost of a few defective parts. The disadvantage of off-line inspection is that the parts have already been made by the time poor quality is detected. When sampling is used, an additional disadvantage is that defective parts can pass through the sampling procedure.

On-Line Inspection. The alternative to off-line inspection is on-line inspection performed when the parts are made, either as an integral step in the processing or assembly operation, or immediately afterward. Two on-line inspection procedures can be distinguished: on-line/in-process and on-line/post-process, illustrated in Figure 21.6(b) and (c).

On-line/in-process inspection is achieved by performing the inspection procedure during the manufacturing operation. As the parts are being made, the inspection procedure simultaneously measures or gages their dimensions. The benefit of in-process inspection is that it may be possible to influence the operation that is making the current part, thereby correcting a potential quality problem before the part is completed. When on-line/in-process inspection is performed manually, it means that the worker who is performing the manufacturing process is also performing the inspection procedure. For automated manufacturing systems, this on-line inspection method is typically done on a 100% basis using automated sensor methods. Technologically, automated on-line/in-process inspection of the product is usually difficult and expensive to implement.

With *on-line/post-process inspection*, the measurement or gaging procedure is accomplished immediately following the production process. Even though it follows the process, it is still considered an on-line method because it is integrated with the manufacturing workstation, and the results of the inspection can immediately influence the production operation. The limitation of on-line/post-process inspection is that the part has already been made, and it is therefore impossible to make corrections that will influence its processing. The best that can be done is to influence the production of the next part.

On-line/post-process inspection can be performed as either a manual or an automated procedure. When done manually, it can be accomplished using either sampling or 100% inspection (with all of the risks associated with each). Gaging of part dimensions at the production machine with go/no-go gages is a common example of on-line/post-process inspection. When on-line/post-process inspection is automated, it is typically performed on a 100% basis. Whether manual or automated, the inspection procedure generates data that can be analyzed by using statistical process control techniques (Section 20.4).

Either form of on-line inspection should drive some action in the manufacturing operation, either feedback process control or parts sortation. If on-line inspection results in no action, then off-line inspection might as well be utilized instead of on-line technologies.

21.4.2 Product Inspection vs. Process Monitoring

In the preceding discussion of inspection issues, we have implicitly assumed that it was the product itself that was being measured or gaged, either during or after the manufacturing process. An alternative approach is to measure the process rather than the product, that is, to monitor the key parameters of the manufacturing process that determine product quality. The advantage of this approach is that an on-line/in-process measurement system is much more likely to be practicable for process variables than for product variables. Such a measurement procedure could be readily incorporated into an on-line feedback control system, permitting any required corrective action to be taken while the product is still being processed and theoretically preventing defective units from being made. If this arrangement was entirely reliable, it would avoid, or at least reduce, the need for subsequent off-line inspection of the actual product.

Use of process monitoring as an alternative to product inspection relies on the assumption of a deterministic cause-and-effect relationship between the process parameters that can be measured and the quality characteristics that must be maintained within tolerance. Accordingly, by controlling the process parameters, indirect control of product quality is achieved. This assumption is most applicable under the following circumstances: (1) the process is well behaved, meaning that it is ordinarily in statistical control and that deviations from this normal condition are rare; (2) process capability is good, meaning that the standard deviation of each process variable of interest under normal operating conditions is small; and (3) the process has been studied to establish the cause-and-effect relationships between process variables and product quality characteristics, and mathematical models of these relationships have been derived.

Although the approach of controlling product quality indirectly through the use of process monitoring is uncommon in piece parts production, it is quite prevalent in the continuous process industries such as chemicals and petroleum. In these continuous processes, it is usually difficult to directly measure the product quality characteristics of interest, except by periodic sampling. To maintain uninterrupted control over product quality, the related process parameters are monitored and regulated continuously. Typical production variables in the process industries include temperature, pressure, flow rates, and similar variables that can easily be measured (chemical engineers might dispute how easily these variables can be measured) and can readily be combined into mathematical equations to predict product parameters of interest. Variables in discrete product manufacturing are generally more difficult to measure, and mathematical models that relate them to product quality are not as easy to derive. Examples of process variables in the parts production industries include [1] tool wear, deflection of production machinery components, part deflection during processing, vibration frequencies and amplitudes of machinery, and temperature profiles of production machinery and piece parts during processing.

21.4.3 Distributed Inspection vs. Final Inspection

When inspection stations are located along the line of work flow in a factory, this is referred to as *distributed inspection*. In the most extreme case, inspection and sortation operations are located after every processing step. However, a more common and cost-effective approach is for inspections to be strategically placed at critical points in the manufacturing sequence, with several manufacturing operations between each inspection. The function of a distributed inspection system is to identify defective parts or products soon after they have been made so that the defects can be excluded from further processing. The goal of this inspection strategy is to prevent unnecessary cost from being invested in defective units. This is especially relevant in assembled products where many components are combined into a single entity that cannot easily be taken apart. If one defective component would render the assembly defective, then it is obviously better to catch the defect before it is assembled. These situations are found in electronics manufacturing operations. Printed circuit board (PCB) assembly is a good example. An assembled PCB may consist of 100 or more electronic components that have been soldered to the base board. If only one of the components is defective, the entire board may be useless unless repaired at substantial additional cost. In these kinds of cases, it is important to discover and remove the defects from the production line before further processing or assembly is accomplished. 100% on-line automated inspection is most appropriate in these situations.

Another approach, sometimes considered an alternative to distributed inspection, is *final inspection*, which involves one comprehensive inspection procedure on the product immediately before shipment to the customer. The motivation behind this approach is that it is more efficient, from an inspection viewpoint, to perform all of the inspection tasks in one step, rather than distribute them throughout the plant. Final inspection is more appealing to the customer because, in principle, if done effectively, it offers the greatest protection against poor quality.

However, exclusive implementation of the final inspection approach (without some intermediate inspection of the product as it is being made) is potentially very expensive to the producer for two reasons: (1) the wasted cost of defective units made in early processing steps being processed in subsequent operations and (2) the cost of final inspection itself. The first issue, cost of processing defective units, has been discussed. The second issue, inspection cost, will benefit from elaboration. Final inspection, when performed on a 100% basis, can be very costly since every unit of product is subjected to an inspection procedure that must be designed to detect all possible defects. The procedure often requires functional testing as well as inspection. If the inspection is performed manually on a 100% basis, as at least a portion of it is likely to be done, it is subject to the risks of 100% manual inspection (Section 21.2.2). Because of these costs and risks, the producer often resorts to sampling inspection, with the associated statistical risks of defective product slipping around the sample to the customer (Section 21.2.1). Thus, final product inspection is potentially costly, potentially ineffective, or both.

Quality conscious manufacturers combine the two strategies. Distributed inspection is used for operations in the plant with high defect rates to prevent processing of bad parts in later operations and to ensure that only good components are assembled in the product, and some form of final inspection is used on the finished units to ensure that only the highest quality product is delivered to the customer.

21.5 QUANTITATIVE ANALYSIS OF INSPECTION

Mathematical models can be developed to analyze certain performance aspects of production and inspection. In this section, we examine three areas: (1) effect of defect rate on production quantities in a sequence of production operations, (2) final inspection versus distributed inspection, and (3) when to inspect and when not to inspect.

21.5.1 Effect of Defect Rate in Serial Production

Let us define the basic element in the analysis as the unit operation for a manufacturing process, illustrated in Figure 21.7. In the figure, the process is depicted by a node, the input to which is a starting quantity of raw material. Let Q_o = the starting quantity or batch size to be processed. The process has a certain fraction defect rate q (stated another way, q = probability of producing a defective piece each cycle of operation), so the quantity of good pieces produced is diminished in size as

$$Q = Q_o(1 - q) \tag{21.1}$$

where Q = quantity of good products made in the process, Q_o = original or starting quantity, and q = fraction defect rate. The number of defects is given by

$$D = Q_o q \tag{21.2}$$

where D = number of defects made in the process.

Most manufactured parts require more than one processing operation. The operations are performed in sequence on the parts, as depicted in Figure 21.8. Each process has a fraction defect rate q_i, so the final quantity of defect-free parts made by a sequence of n unit operations is given by

$$Q_f = Q_o \prod_{i=1}^{n} (1 - q_i) \tag{21.3}$$

where Q_f = final quantity of defect-free units produced by the sequence of n processing operations, and Q_o is the starting quantity. If all q_i are equal, which is unlikely but nevertheless convenient for conceptualization and computation, then the preceding equation becomes

$$Q_f = Q_o(1 - q)^n \tag{21.4}$$

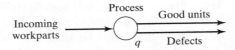

Figure 21.7 The unit operation for a manufacturing process, represented as an input-output model in which the process has a certain fraction defect rate.

where q = fraction defect rate for all n processing operations. The total number of defects produced by the sequence is most easily computed as

$$D_f = Q_o - Q_f \qquad (21.5)$$

where D_f = total number of defects produced.

EXAMPLE 21.2 Compounding Effect of Defect Rate in a Sequence of Operations

A batch of 1,000 raw work units is processed through ten operations, each of which has a fraction defect rate of 0.05. How many defect-free units and how many defects are in the final batch?

Solution: Eq. (21.4) can be used to determine the quantity of defect-free units in the final batch:

$$Q_f = 1,000 \, (1 - .05)^{10} = 1,000 \, (0.95)^{10} = 1,000 \, (0.59874) = 599 \text{ good units}$$

The number of defects is given by Eq. (21.5):

$$D_f = 1,000 - 599 = 401 \text{ defective units.}$$

The binomial expansion can be used to determine the allocation of defects associated with each processing operation i. Given that q_i = probability of a defect being produced in operation i, let p_i = probability of a good unit being produced in the sequence; thus, $p_i + q_i = 1$. Expanding this for n operations, we have

$$\prod_{i=1}^{n} (p_i + q_i) = 1 \qquad (21.6)$$

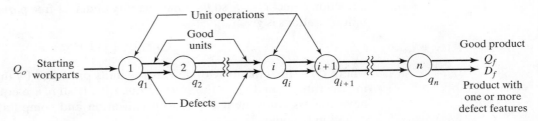

Figure 21.8 A sequence of n unit operations used to produce a part. Each process has a certain fraction defect rate.

To illustrate, consider the case of two operations in sequence ($n = 2$). The binomial expansion yields the expression

$$(p_1 + q_1)(p_2 + q_2) = p_1p_2 + p_1q_2 + p_2q_1 + q_1q_2$$

where p_1p_2 = proportion of defect-free parts, p_1q_2 = proportion of parts that have no defects from operation 1 but a defect from operation 2, p_2q_1 = proportion of parts that have no defects from operation 2 but a defect from operation 1, and q_1q_2 = proportion of parts that have both types of defect.

21.5.2 Final Inspection vs. Distributed Inspection

The preceding model portrays a sequence of operations, each with its own fraction defect rate, whose output forms a distribution of parts possessing either (1) no defects or (2) one or more defects, depending on how the defect rates from the different unit operations combine. The model makes no provision for separating the good units from the defects; thus, the final output is a mixture of the two categories. This is a problem. To deal with the problem, let us expand our model to include inspection operations, either one final inspection at the end of the sequence or distributed inspection, in which each production step is followed by an inspection.

Final Inspection. In the first case, one final inspection and sortation operation is located at the end of the production sequence, as represented by the square in Figure 21.9. In this case, the output of the process is 100% inspected to identify and separate defective units. The inspection screen is assumed to be 100% accurate, meaning that there are no Type I or Type II inspection errors.

The probabilities in this new arrangement are pretty much the same as before. Defects are still produced. The difference is that the defective units D_f have been completely and accurately isolated from the good units Q_f by the final inspection procedure. Obviously, there is a cost associated with the inspection and sortation operation that is added to the regular cost of processing. The costs of processing and then sorting a batch of Q_o parts as indicated in Figure 21.9 can be expressed as

$$C_b = Q_o \sum_{i=1}^{n} C_{pri} + Q_o C_{sf} = Q_o \left(\sum_{i=1}^{n} C_{pri} + C_{sf} \right) \tag{21.7}$$

where C_b = cost of processing and sorting the batch, Q_o = number of parts in the starting batch, C_{pri} = cost of processing a part at operation i, and C_{sf} = cost of the final

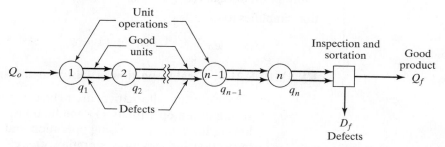

Figure 21.9 A sequence of n unit operations with one final inspection and sortation operation to separate the defects.

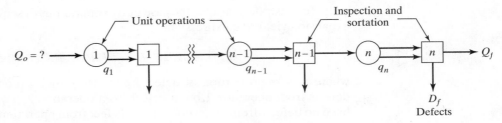

Figure 21.10 Distributed inspection, consisting of a sequence of unit operations with an inspection and sortation after each operation.

inspection and sortation per part. The processing cost C_{pr} is applicable to every unit for each of the n operations; hence the summation from 1 to n. The final inspection is done once for each unit. We have neglected consideration of material cost. For the special case in which every processing cost is equal ($C_{pri} = C_{pr}$ for all i), we have

$$C_b = Q_o(nC_{pr} + C_{sf}) \tag{21.8}$$

Note that the fraction defect rate does not figure into total cost in either of these equations, since no defective units are sorted from the batch until after the final processing operation. Therefore, every unit in Q_o is processed through all operations, whether it is good or defective, and every unit is inspected and sorted.

Distributed Inspection. Next, let us consider a distributed inspection strategy, in which every operation in the sequence is followed by an inspection and sortation step, as seen in Figure 21.10. In this arrangement, the defects produced in each processing step are sorted from the batch immediately after they are made, so that only good parts are permitted to advance to the next operation. In this way, no defective units are processed in subsequent operations, thereby saving the processing cost of those units. Our model of distributed inspection must take the defect rate at each operation into account as

$$C_b = Q_o(C_{pr1} + C_{s1}) + Q_o(1 - q_1)(C_{pr2} + C_{s2})$$
$$+ Q_o(1 - q_1)(1 - q2)(C_{pr3} + C_{s3}) + \cdots + Q_o\prod_{i=1}^{n-1}(1 - q_i)(C_{prn} + C_{sn}) \tag{21.9}$$

where $C_{s1}, C_{s2}, \ldots, C_{si}, \ldots, C_{sn}$ = costs of inspection and sortation at each station, respectively. In the special case where $q_i = q$, $C_{pri} = C_{pr}$, and $C_{si} = C_s$ for all i, the above equation simplifies to

$$C_b = Q_o(1 + (1 - q) + (1 - q)^2 + \cdots + (1 - q)^{n-1})(C_{pr} + C_s) \tag{21.10}$$

EXAMPLE 21.3   Final Inspection vs. Distributed Inspection

Two inspection alternatives are to be compared for a processing sequence consisting of ten operations: (1) one final inspection and sortation operation following the tenth processing operation and (2) distributed inspection with an inspection and sortation operation after each of the ten processing operations. The batch size $Q_o = 1,000$ pieces. The cost of each processing operation $C_{pr} = \$1.00$. The fraction defect rate at each operation $q = 0.05$. The cost of the single final inspection and sortation operation in alternative (1) is $C_{sf} = \$2.50$. The cost of each inspection and sortation operation in alternative (2) is $C_s = \$0.25$. Compare total processing and inspection costs for the two cases.

Solution: For the final inspection alternative, we can use Eq. (21.8) to determine the batch cost:

$$C_b = 1,000(10 \times 1.00 + 2.50) = 1,000(12.50) = \$12,500$$

For the distributed inspection alternative, we can use Eq. (21.10) to solve for the batch cost:

$$C_b = 1,000(1 + (.95) + (.95)^2 + \cdots + (.95)^9)(1.00 + 0.25)$$
$$= 1,000(8.0252)(1.25) = \$10,032$$

We see that the cost of distributed inspection is less for the cost data given in Example 21.3. A savings of $2,468 or nearly 20% is achieved by using distributed inspection. The reader might question why the cost of one final inspection ($2.50) is so much more than the cost of an inspection in distributed inspection ($0.25). We offer both a logical answer and a practical answer to the question. The logical answer goes like this: Each processing step produces its own unique defect feature (at fraction defect rate q_i), and the inspection procedure must be designed to inspect for that feature. For ten processing operations with ten different defect features, the cost to inspect for these features is the same whether the inspection is accomplished after each processing step or all at once after the final processing step. If the cost of inspecting for each defect feature is $0.25, it follows that the cost of inspecting for all ten defect features is simply $10(\$0.25) = \$2.50$. In general, this relationship can be expressed as

$$C_{sf} = \sum_{i=1}^{n} C_{si} \qquad (21.11)$$

For the special case where all C_{si} are equal ($C_{si} = C_s$ for all i), as in Example 21.3,

$$C_{sf} = n\, C_s \qquad (21.12)$$

Given this multiplicative relationship between the single final inspection cost and the unit inspection cost in distributed inspection, it is readily seen that the total cost advantage of distributed inspection in Example 21.3 derives entirely from the fact that the number of parts that are processed and inspected is reduced after each processing step due to the sortation of defective parts from the batch during production rather than afterward.

Notwithstanding the logic of Eqs. (21.11) and (21.12), we are sure that in practice there is some economy in performing one inspection procedure at a single location, even if the procedure includes scrutinizing the product for ten different defect features. Thus, the actual final inspection cost per unit C_{sf} is likely to be less than the sum of the unit costs in distributed inspection. Nevertheless, the fact remains that distributed inspection and sortation reduces the number of units processed, thus avoiding the waste of valuable production resources on the processing of defective units.

Partially Distributed Inspection. A distributed inspection strategy can be followed in which inspections are located between groups of processes rather than after every processing step as in Example 21.3. If there is any economy in performing multiple inspections at a single location, as argued in the preceding paragraph, then this might be a worthwhile way to exploit this economy while preserving at least some of the advantages of distributed inspection. Let us use Example 21.4 to illustrate the grouping of unit operations for inspection purposes. As expected, the total batch cost lies between the two cases of fully distributed inspection and final inspection for the data in our example.

EXAMPLE 21.4 Partially Distributed Inspection

For comparison, let us use the same sequence of ten processing operations as before, where the fraction defect rate of each operation is $q = 0.05$. Instead of inspecting and sorting after every operation, the ten operations will be divided into groups of five, with inspections after operations 5 and 10. Following the logic of Eq. (21.12), the cost of each inspection will be five times the cost of inspecting for one defect feature; that is, $C_{s5} = C_{s10} = 5(\$0.25) = \$1.25$ per unit inspected. Processing cost per unit for each process remains the same as before at $C_{pr} = \$1.00$, and $Q_o = 1,000$ units.

Solution: The batch cost is the processing cost for all 1,000 pieces for the first five operations, after which the inspection and sortation procedure separates the defects produced in those first five operations from the rest of the batch. This reduced batch quantity then proceeds through operations 6–10, followed by the second inspection and sortation procedure. The equation for this is the following:

$$C_b = Q_o\left(\sum_{i=1}^{5} C_{pri} + C_{s5}\right) + Q_o\prod_{i=1}^{5}(1 - q_i)\left(\sum_{i=6}^{10} C_{pri} + C_{s10}\right) \tag{21.13}$$

Since all C_{pri} are equal ($C_{pri} = C_{pr}$ for all i), and all q are equal ($q_i = q$ for all i), this equation can be simplified to

$$C_b = Q_o(5C_{pr} + C_{s5}) + Q_o(1 - q)^5(5C_{pr} + C_{s10}) \tag{21.14}$$

Using our values for this example, we have

$$C_b = 1,000(5 \times 1.00 + 1.25) + 1,000(.95)^5(5 \times 1.00 + 1.25)$$

$$= 1,000(6.25) + 1,000(0.7738)(6.25) = \$11,086$$

This is a savings of $1,414 or 11.3% compared with the $12,500 cost of one final inspection. Note that we have been able to achieve a significant portion of the total savings from fully distributed inspection by using only two inspection stations rather than ten. Our savings here of $1,414 is about 57% of the $2468 savings from the previous example, with only 20% of the inspection stations. This suggests that it may not be advantageous to locate an inspection operation after every production step, but instead to place them after groups of operations. The "law of diminishing returns" is applicable in distributed inspection.

21.5.3 Inspection or No Inspection

A relatively simple model for deciding whether to inspect at a certain point in the production sequence is proposed in Juran and Gryna [3]. The model uses the fraction defect rate in the production batch, the inspection cost per unit inspected, and the cost of damage that one defective unit would cause if it were not inspected. The total cost per batch of 100% inspection can be formulated as

$$C_b(100\% \text{ inspection}) = QC_s \qquad (21.15)$$

where C_b = total cost for the batch under consideration, Q = quantity of parts in the batch, and C_s = inspection and sortation cost per part. The total cost of no inspection, which leads to a damage cost for each defective unit in the batch, would be

$$C_b(\text{no inspection}) = QqC_d \qquad (21.16)$$

where C_b = batch cost, as before; Q = number of parts in the batch; q = fraction defect rate; and C_d = damage cost for each defective part that proceeds to subsequent processing or assembly. This damage cost may be high, for example, in the case of an electronics assembly where one defective component might render the entire assembly defective and rework would be expensive.

Finally, if sampling inspection is used on the batch, we must include the sample size and the probability that the batch will be accepted by the inspection sampling plan that is used. This probability can be obtained from the OC curve (Figure 21.1) for a given fraction defect rate q. The resulting expected cost of the batch is the sum of three terms: (1) cost of inspecting the sample of size Q_s, (2) expected damage cost of those parts that are defective if the sample passes inspection, and (3) expected cost of inspecting the remaining parts in the batch if the sample does not pass inspection. In equation form,

$$C_b(\text{sampling}) = C_s Q_s + (Q - Q_s) q C_d P_a + (Q - Q_s) C_s (1 - P_a) \qquad (21.17)$$

where C_b = batch cost, C_s = cost of inspecting and sorting one part, Q_s = number of parts in the sample, Q = batch quantity, q = fraction defect rate, C_d = damage cost per defective part, and P_a = probability of accepting the batch based on the sample.

A simple decision rule can be established to decide whether to inspect the batch. The decision is based on whether the expected fraction defect rate in the batch is greater than or less than a critical defect level q_c, which is the ratio of the inspection cost to the damage cost. This critical value represents the break-even point between inspection or no inspection. In equation form, q_c is defined as

$$q_c = \frac{C_s}{C_d} \qquad (21.18)$$

where C_s = cost of inspecting and sorting one part, and C_d = damage cost per defective part. If, based on past history with the component, the batch fraction defect rate q is less than this critical level, then no inspection is indicated. On the other hand, if it is expected that the fraction defect rate will be greater than q_c, then the total cost of production and inspection will be less if 100% inspection and sortation is performed prior to subsequent processing.

EXAMPLE 21.5 Inspection or No Inspection

A facility has completed a production run of 10,000 parts and management must decide whether to 100% inspect the batch. Past history with this part suggests that the fraction defect rate is around 0.03. Inspection cost per part is $0.25. If the batch is passed on for subsequent processing, the damage cost for each defective unit in the batch will be $10.00. Determine (a) batch cost for 100% inspection and (b) batch cost if no inspection is performed. (c) What is the critical fraction defect value for deciding whether to inspect?

Solution: (a) Batch cost for 100% inspection is given by Eq. (21.15):

$$C_b(100\% \text{ inspection}) = QC_s = 10,000(\$0.25) = \$2,500$$

(b) Batch cost for no inspection can be calculated by Eq. (21.16):

$$C_b(\text{no inspection}) = QqC_d = 10,000(0.03)(\$10.00) = \$3,000$$

(c) The critical fraction defect value for deciding whether to inspect is determined from Eq. (21.18):

$$q_c = \frac{C_s}{C_d} = \frac{0.25}{10.00} = 0.025$$

Since the anticipated defect rate in the batch is $q = 0.03$, the decision should be to inspect. Note that this decision is consistent with the two batch costs calculated for no inspection and 100% inspection. The lowest cost is attained when 100% inspection is used.

EXAMPLE 21.6   Cost of Sampling Inspection

Given the data from the preceding example, suppose that sampling inspection is being considered as an alternative to 100% inspection. The sampling plan calls for a sample of 100 parts to be drawn at random from the batch. Based on the OC curve for this sampling plan, the probability of accepting the batch is 92% at the given defect rate of $q = 0.03$. Determine the batch cost for sampling inspection.

Solution: The batch cost for sampling inspection is given by Eq. (21.17):

$$C_b(\text{sampling}) = C_s Q_s + (Q - Q_s)\,qC_d P_a + (Q - Q_s)C_s(1 - P_a)$$

$$= \$0.25(100) + (10,000 - 100)(0.03)(\$10.00)(0.92)$$

$$+ (10,000 - 100)(\$0.25)(1 - 0.92)$$

$$= \$25.00 + 2732.40 + 198.00 = \$2955.40$$

The significance of Example 21.6 must not be overlooked. The total cost of sampling inspection for our data is greater than the cost of 100% inspection and sortation. If only the cost of the inspection procedure is considered, then sampling inspection is much less expensive ($25 versus $2500). But if total costs, which include the damage that results from defects passing through sampling inspection, are considered, then sampling inspection is not the least expensive inspection alternative. We might consider the question: what if the ratio C_s/C_d in Eq. (21.18) had been greater than the fraction defect rate of the batch, in other words, the opposite of the case in Examples 21.5 and 21.6? The answer is that if q_c were greater than the batch defect rate q, then the cost of no inspection would be less than the cost of 100% inspection, and the cost of sampling inspection would again lie between the two cost values. The cost of sampling inspection will always lie between the cost of 100% inspection and no inspection, whichever of these two alternatives is greater. If this argument is followed to its logical end, then the conclusion is that either no inspection or 100% inspection is preferred over sampling inspection, and it is just a matter of deciding whether none or all is the better alternative.

21.5.4 What the Equations Tell Us

Several lessons can be inferred from the above mathematical models and examples. These lessons should be useful in designing inspection systems for production.

- Distributed inspection/sortation reduces the total number of parts processed in a sequence of production operations compared with one final inspection at the end of the sequence. This reduces waste of processing resources.
- Partially distributed inspection is less effective than fully distributed inspection at reducing the waste of processing resources. However, if there is an economic advantage in combining several inspection steps at one location, then partially distributed inspection may reduce total batch costs compared with fully distributed inspection.
- The "law of diminishing returns" operates in distributed inspection systems, meaning that each additional inspection station added in distributed inspection yields less savings than the previous station added, other factors being equal.
- As the ratio of unit processing cost to unit inspection cost increases, the advantage of distributed inspection over final inspection increases.
- Inspections should be performed immediately following processes that have a high fraction defect rate.
- Inspections should be performed prior to high cost processes.
- When expected damage cost (of those defects that pass around the inspection plan when the batch is accepted) and expected cost of inspecting the entire batch (when the batch is rejected) are considered, sampling inspection is not the lowest cost inspection alternative. Either no inspection or 100% inspection is a more appropriate alternative, depending on the relative values of inspection/sortation cost and damage cost for a defective unit that proceeds to the next stage of processing.

REFERENCES

[1] BARKMAN, W. E., *In-Process Quality Control for Manufacturing*, Marcel Dekker, Inc., NY, 1989.

[2] DRURY, C. G., "Inspection Performance," *Handbook of Industrial Engineering*, 2d ed., G. Salvendy, editor, John Wiley & Sons, Inc., NY, 1992, pp. 2282–2314.

[3] JURAN, J. M., and GRYNA, F. M., *Quality Planning and Analysis*, 3d ed., McGraw-Hill, Inc., NY, 1993.

[4] MONTGOMERY, D. C., *Introduction to Statistical Quality Control*, 5th ed., John Wiley & Sons, Inc., NY, 2005.

[5] MURPHY, S. D., *In-Process Measurement and Control*, Marcel Dekker, Inc., NY, 1990.

[6] STOUT, *Quality Control in Automation*, Prentice Hall, Inc., Englewood Cliffs, NJ, 1985.

[7] TANNOCK, J. D. T., *Automating Quality Systems*, Chapman & Hall, London, UK, 1992.

[8] WICK, C., and VEILLEUX, R. F., *Tool and Manufacturing Engineers Handbook*, 4th ed., Volume IV, *Quality Control and Assembly*, Society of Manufacturing Engineers, Dearborn, MI, 1987, Section 1.

[9] WINCHELL, W., *Inspection and Measurement*, Society of Manufacturing Engineers, Dearborn, MI, 1996.

[10] YURKO, J., "The Optimal Placement of Inspections Along Production Lines," *Masters Thesis*, Industrial Engineering Department, Lehigh University, 1986.

REVIEW QUESTIONS

21.1 What is inspection?

21.2 Briefly define the two basic types of inspection.

21.3 What are the four steps in a typical inspection procedure?

21.4 What are the Type I and Type II errors that can occur in inspection?

21.5 What is quality control testing as distinguished from inspection?

21.6 What are the Type I and Type II statistical errors that can occur in acceptance sampling?

21.7 Describe what an operating characteristic curve is in acceptance sampling.

21.8 What are the two problems associated with 100% manual inspection?

21.9 What are the three ways in which an inspection procedure can be automated?

21.10 What is the difference between off-line inspection and on-line inspection?

21.11 Under what circumstances is process monitoring a suitable alternative to actual inspection of the quality characteristic of the part or product?

21.12 What is the difference between distributed inspection and final inspection in quality control?

PROBLEMS

Inspection Accuracy

21.1 An inspector reported a total of 18 defects out of a total batch size of 250 parts. On closer examination, it was determined that five of these reported defects were in fact good pieces, while a total of nine defective units were undetected by the inspector. What is the inspector's accuracy in this instance? Specifically, what are the values of p_1 and p_2? What was the true fraction defect rate q?

21.2 For the preceding problem, develop a table of outcomes similar in format to Table 21.3 in the text. The entries in the table should represent the probabilities of the various possible outcomes in the inspection operation.

21.3 For Example 21.1 in the text, develop a table of outcomes similar in format to Table 21.3 in the text. The entries will be the probabilities of the various possible outcomes in the inspection operation.

21.4 An inspector's accuracy is as follows: $p_1 = 0.94$ and $p_2 = 0.80$. The inspector is given the task of inspecting a batch of 200 parts and sorting out the defects from good units. If the actual defect rate in the batch is $q = 0.04$, determine (a) the expected number of Type I and (b) Type II errors the inspector will make. (c) Also, what is the expected fraction defect rate that the inspector will report at the end of the inspection task?

21.5 An inspector must 100% inspect a production batch of 500 parts using a gaging method. If the actual fraction defect rate in the batch is $q = 0.02$, and the inspector's accuracy is given by

$p_1 = 0.96$ and $p_2 = 0.84$, determine (a) the number of defects the inspector can be expected to report and (b) the expected number of Type I and Type II errors the inspector will make.

Effect of Fraction Defect Rate

21.6 A batch of 10,000 raw work units is processed through 15 operations, each of which has a fraction defect rate of 0.03. How many defect-free units and how many defects are in the final batch?

21.7 A silicon wafer has a total of 400 integrated circuits (ICs) at the beginning of its fabrication sequence. A total of 80 operations are used to complete the integrated circuits, each of which inflicts damages on 1.5% of the ICs. The damages compound, meaning that an IC that is already damaged has the same probability of being damaged by a subsequent process as a previously undamaged IC. How many defect-free ICs remain at the end of the fabrication sequence?

21.8 A batch of workparts is processed through a sequence of nine processing operations, which have fraction defect rates of 0.03, 0.05, 0.02, 0.04, 0.06, 0.01, 0.03, 0.04, and 0.07, respectively. A total of 5000 completed parts are produced by the sequence. What was the starting batch quantity?

21.9 A production line consists of six workstations, as shown in the accompanying figure. The six stations are as follows: (1) first manufacturing process, scrap rate is $q_1 = 0.10$; (2) inspection for first process, separates all defects from first process; (3) second manufacturing process, scrap rate is $q_3 = 0.20$; (4) inspection for second process, separates all defects from second process; (5) rework, repairs defects from second process, recovering 70% of the defects from the preceding operation and leaving 30% of the defects as still defective; (6) third manufacturing process, scrap rate $q_6 =$ zero. If the output from the production line is to be 100,000 defect-free units, what quantity of raw material units must be launched onto the front of the line?

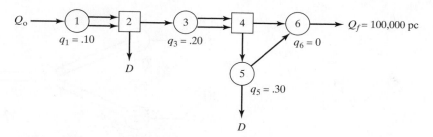

Figure P21.9 Production line for Problem 21.9.

21.10 A certain industrial process can be depicted by the diagram below. Operation 1 is a disassembly process in which each unit of raw material is separated into one unit each of parts A and B. These parts are then processed separately in operations 2 and 3, respectively, which have scrap rates of $q_2 = 0.05$ and $q_3 = 0.10$. Inspection stations 4 and 5 sort good units from bad for the two parts. Then the parts are assembled back together in operation 6, which has a fraction defect rate $q_6 = 0.15$. Final inspection station 7 sorts good units from bad. The desired final output quantity is 100,000 units. (a) What is the required starting quantity (into operation 1) to achieve this output? (b) Will there be any leftover units of parts A or B, and if so, how many?

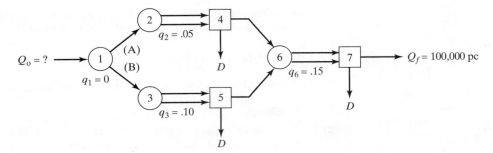

Figure P21.10 Production line for Problem 21.10.

21.11 A certain component is produced in three sequential operations. Operation 1 produces defects at a rate $q_1 = 5\%$. Operation 2 produces defects at a rate $q_2 = 8\%$. Operation 3 produces defects at a rate $q_3 = 10\%$. Operations 2 and 3 can be performed on units that are already defective. If 10,000 starting parts are processed through the sequence, (a) how many units are expected to be defect-free, (b) how many units are expected to have exactly one defect, and (c) how many units are expected to have all three defects?

21.12 An industrial process can be depicted as in the diagram below. Two components are made, respectively, by operations 1 and 2, and then assembled together in operation 3. Scrap rates are as follows: $q_1 = 0.20$, $q_2 = 0.10$, and $q_3 = 0$. Input quantities of raw components at operations 1 and 2 are 25,000 and 20,000, respectively. One of each component is required in the assembly operation. The trouble is that defective components can be assembled just as easily as good components, so inspection and sortation is required in operation 4. Determine (a) how many defect-free assemblies will be produced and (b) how many assemblies will be made with one or more defective components. (c) Will there be any leftover units of either component, and if so, how many?

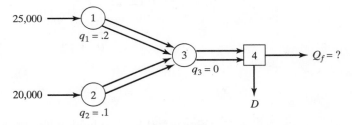

Figure P21.12 Production line for Problem 21.12.

Inspection Costs

21.13 Two inspection alternatives are to be compared for a processing sequence consisting of 20 operations performed on a batch of 100 starting parts: (1) one final inspection and sortation operation following the last processing operation, and (2) distributed inspection with an inspection and sortation operation after each processing operation. The cost of each processing operation $C_{pr} = \$1.00$ per unit processed. The fraction defect rate at each operation $q = 0.03$. The cost of the single final inspection and sortation operation in alternative (1) is $C_{sf} = \$2.00$ per unit. The cost of each inspection and sortation operation in alternative (2) is $C_s = \$0.10$ per unit. Compare total processing and inspection costs per batch for the two cases.

21.14 In the preceding problem, instead of inspecting and sorting after every operation, the 20 operations will be divided into groups of five, with inspections after operations 5, 10, 15, and 20. Following the logic of Eq. (21.11), the cost of each inspection will be five times the cost of inspecting for one defect feature; that is, $C_{s5} = C_{s10} = C_{s15} = C_{s20} = 5(\$0.10) = \$0.50$ per unit inspected. Processing cost per unit for each operation remains the same as before at $C_{pr} = \$1.00$, and $Q_o = 100$ parts. What is the total processing and inspection cost per batch for this partially distributed inspection system?

21.15 A processing sequence consists of 10 operations, each of which is followed by an inspection and sortation operation to detect and remove defects generated in the processing operation. Defects in each process occur at a rate of $q = 0.04$. Each processing operation costs $\$1.00$ per unit processed, and the inspection/sortation operation costs $\$0.30$ per unit. (a) Determine the total processing and inspection costs for this distributed inspection system. (b) A proposal is being considered to combine all of the inspections into one final inspection and sortation station following the last processing operation. Determine the cost per unit of this final inspection and sortation station that would make the total cost of this system equal to that of the distributed inspection system.

21.16 This problem is intended to show the merits of a partially distributed inspection systems in which inspections are placed after processing steps that generate a high fraction defect rate. The processing sequence consists of eight operations with fraction defect rates for each operation as follows:

Operation	1	2	3	4	5	6	7	8
Defect rate q	0.01	0.01	0.01	0.11	0.01	0.01	0.01	0.11

The company is considering three alternatives: (1) fully distributed inspection, with an inspection after every operation; (2) partially distributed inspection, with inspections following operations 4 and 8 only; and (3) one final inspection station after operation 8. All inspections include sortations. In alternative (2), each inspection procedures is designed to detect all of the defects for the preceding four operations. The cost of processing is $C_{pr} = \$1.00$ for each of operations 1 through 8. Inspection/sortation costs for each alternative are given in the next table. Compare total processing and inspection costs for the three cases.

Alternative	Inspection and Sortation Cost
(1)	$C_s = \$0.10$ per unit for each of the eight inspection stations
(2)	$C_s = \$0.40$ per unit for each of the two inspection stations
(3)	$C_s = \$0.80$ per unit for the one final inspection station

Inspection or No Inspection

21.17 A batch of 1000 parts has been produced and management must decide whether to 100% inspect the batch or not. Past history with this part suggests that the fraction defect rate is around 0.02. Inspection cost per part is $\$0.20$. If the batch is passed on for subsequent processing, the damage cost for each defective unit in the batch is $\$8.00$. Determine (a) batch cost for 100% inspection and (b) batch cost if no inspection is performed. (c) What is the critical fraction defect value for deciding whether to inspect?

21.18 Given the data from the preceding problem, sampling inspection is being considered as an alternative to 100% inspection. The sampling plan calls for a sample of 50 parts to be drawn at random from the batch. Based on the operating characteristic curve for this sampling plan, the probability of accepting the batch is 95% at the given defect rate of $q = 0.02$. Determine the batch cost for sampling inspection.

Chapter 22

Inspection Technologies

CHAPTER CONTENTS

From Chapter 22 of *Automation, Production Systems, and Computer-Integrated Manufacturing*, Third Edition.
Mikell P. Groover. Copyright © 2008 by Pearson Education, Inc. Publishing as Prentice Hall. All rights reserved.

The inspection procedures described in the previous chapter are enabled by various sensors, instruments, and gages. Some of these inspection techniques involve manually operated devices that have been used for more than a century, for example, micrometers, calipers, protractors, and go/no-go gages. Other techniques are based on modern technologies such as coordinate measuring machines and machine vision. These newer techniques require computer systems to control their operation and analyze the data collected. The computer-based technologies allow the inspection procedures to be automated. In some cases they permit 100% inspection to be accomplished economically. Our coverage in this chapter will emphasize these modern technologies. Let us begin by discussing a prerequisite topic in inspection technology: metrology.

22.1 INSPECTION METROLOGY[1]

Measurement is a procedure in which an unknown quantity is compared to a known standard, using an accepted and consistent system of units. The measurement may involve a simple linear rule to scale the length of a part, or it may require measurement of force versus deflection during a tension test. Measurement provides a numerical value of the quantity of interest, within certain limits of accuracy and precision. It is the means by which *inspection for variables* is accomplished (Section 21.1.1).

Metrology is the science of measurement. It is concerned with seven basic quantities: length, mass, time, electric current, temperature, luminous intensity, and matter. From these basic quantities, other physical quantities are derived, such as area, volume, velocity, acceleration, force, electric voltage, energy, and so on. In manufacturing metrology, our main concern is usually with measuring the length quantity in the many ways in which it manifests itself in a part or product. These include length, width, depth, diameter, straightness, flatness, and roundness. Even surface roughness (Section 22.5) is defined in terms of length quantities.

22.1.1 Characteristics of Measuring Instruments

All measuring instruments possess certain characteristics that make them useful in the particular applications they serve. Primary among these are accuracy and precision, but other features include speed of response, operating range, and cost. These attributes can be used as criteria in selecting a measuring device. No measuring instrument scores perfect marks in all of the criteria. The choice of a device for a given application should emphasize those criteria that are most important.

Accuracy and Precision. Measurement *accuracy* is the degree to which the measured value agrees with the true value of the quantity of interest. A measurement procedure is accurate when it is absent of *systematic errors*, which are positive or negative deviations from the true value that are consistent from one measurement to the next.

Precision is a measure of repeatability in a measurement process. Good precision means that random errors in the measurement procedure are minimized. Random errors are often due to human participation in the measurement process. Examples include variations in the setup, imprecise reading of the scale, round-off approximations, and so on. Nonhuman contributors to random error include changes in temperature, gradual wear

---

[1]This section is based on Groover [10], Section 45.1.

and/or misalignment in the working elements of the device, and other variations. It is generally assumed that random errors obey a normal statistical distribution with a mean of zero and a standard deviation (σ) that indicates the amount of dispersion that exists in the measurement. The normal distribution has certain well-defined properties, including the fact that 99.73% of the population is included within $\pm 3\sigma$ of the population mean. A measuring instrument's precision is often defined as $\pm 3\sigma$.

The distinction between accuracy and precision is depicted in Figure 22.1. In (a), the random error in the measurement is large, indicating low precision, but the mean value of the measurement coincides with the true value, indicating high accuracy. In (b), the measurement error is small (good precision), but the measured value differs substantially from the true value (low accuracy). In (c), both accuracy and precision are good.

No measuring instrument has perfect accuracy (no systematic error) and perfect precision (no random error). Perfection in measurement, as in anything else, is impossible. Accuracy of the instrument is maintained by proper and regular calibration (explained below). Precision is achieved by selecting the proper instrument technology for the application. A guideline often applied to determine the right level of precision is the *rule of 10*, which means that the measuring device must be ten times more precise than the specified tolerance. Thus, if the tolerance to be measured is ± 0.25 mm (± 0.010 in), then the measuring device must have a precision of ± 0.025 mm (± 0.001 in).

Other Features of Measuring Instruments. Another aspect of a measuring instrument is its capacity to distinguish very small differences in the quantity of interest. The indication of this characteristic is the smallest variation of the quantity that can be detected by the instrument. The terms *resolution* and *sensitivity* describe this attribute of a measuring device. Other desirable features include stability, speed of response, wide operating range, high reliability, and low cost.

Some measurements, especially in a manufacturing environment, must be made quickly. The ability of a measuring instrument to indicate the quantity with a minimum time lag is called its *speed of response*. Ideally, the time lag should be zero, but this is an impossible ideal. For an automatic measuring device, speed of response is usually taken to be the time lapse between the moment when the quantity of interest changes and the moment when the device is able to indicate the change within a certain small percentage of the true value.

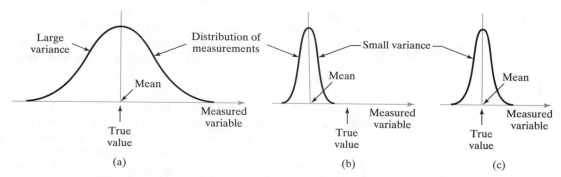

Figure 22.1 Accuracy versus precision in measurement: (a) high accuracy but low precision, (b) low accuracy but high precision, and (c) high accuracy and high precision.

The measuring instrument should possess a *wide operating range*, or capability to measure the physical variable throughout the entire span of practical interest to the user. *High reliability*, which can be defined as the absence of frequent malfunctions and failures of the device, and low cost are of course desirable attributes of any engineering equipment.

Analog Versus Digital Instruments. An analog measuring instrument provides an output that is analog; that is, the output signal of the instrument varies continuously with the variable being measured. Because the output varies continuously, it can take on any of an infinite number of possible values over the range in which it is designed to operate. Of course, when the output is read by the human eye, there are limits on the resolution that can be discriminated. When analog measuring devices are used for process control, the common output signal is voltage. Since most modern process controllers are based on the digital computer, the voltage signal must be converted to digital form by means of an analog-to-digital converter (ADC, Section 6.3).

A digital measuring instrument provides an output that is digital; that is, it can assume any of a discrete number of incremental values corresponding to the value of the quantity being measured. The number of possible output values is finite. The digital signal may consist of a set of parallel bits in a storage register or a series of pulses that can be counted. When parallel bits are used, the number of possible output values is determined by the number of bits as:

$$n_o = 2^B \qquad (22.1)$$

where n_o = number of possible output values of the digital measuring device, and B = number of bits in the storage register. The resolution of the measuring instrument is given by

$$MR = \frac{L}{n_o - 1} = \frac{L}{2^B - 1} \qquad (22.2)$$

where MR = measurement resolution, the smallest increment that can be distinguished by the device; L = its measuring range; and B = number of bits used by the device to store the reading, as before. Although a digital measuring instrument can provide only a finite number of possible output values, this is hardly a limitation in practice, since the storage register can be designed with a sufficient number of bits to achieve the required resolution for nearly any application.

Digital measuring devices are finding increased utilization in industrial practice for two good reasons: (1) they can be read easily as stand-alone instruments; and (2) most digital devices can be directly interfaced with a digital computer, avoiding the need for analog-to-digital conversion.

Calibration. Measuring devices must be calibrated periodically. Calibration is a procedure in which the measuring instrument is checked against a known standard. For example, calibrating a thermometer might involve checking its reading in boiling (pure) water at standard atmospheric pressure, under which conditions the temperature is known to be 100°C (212°F). The calibration procedure should include checking the instrument over its entire operating range. The known standard, if it is a physical instrument, should be used only for calibration purposes; it should not serve as a spare shop floor instrument when an extra is needed.

For convenience, the calibration procedure should be as quick and uncomplicated as possible. Once calibrated, the instrument should be capable of retaining its calibration — continuing to measure the quantity without deviating from the standard for an extended period of time. This capability to retain calibration is called *stability*, and the tendency of the device to gradually lose its accuracy relative to the standard is called *drift*. Reasons for drift include factors such as (1) mechanical wear, (2) dirt and dust, (3) fumes and chemicals in the environment, and (4) aging of the materials out of which the instrument is made. Good coverage of the measurement calibration issue is provided in Morris [14].

22.1.2 Measurement Standards and Systems

A common feature of any measurement procedure is comparison of the unknown value with a known standard. Two aspects of a standard are critical: (1) it must be constant, not changing over time; and (2) it must be based on a system of units that is consistent and accepted by users. In modern times, standards for length, mass, time, electric current, temperature, light, and matter are defined in terms of physical phenomena that can be relied upon to remain unchanged. These standards are defined by international agreement. For the edification and amusement of our readers, we present these standards in Table 22.1.

Two systems of units have evolved into predominance in the world: (1) the U.S. customary system (U.S.C.S.); and (2) the International System of Units (or SI, for Le Système International d'Unités), more popularly known as the metric system. Both of these systems are well known. We use both in parallel throughout this book. The metric system (Table 22.1) is widely accepted in nearly every part of the industrialized world except the United States, which has stubbornly clung to its U.S.C.S. Gradually, the United States is going metric and adopting the SI.

TABLE 22.1 Standard Units for Basic Physical Quantities (System International)

Quantity	Standard Unit	Symbol	Standard Unit Defined
Length	Meter	m	The distance traveled by light in a vacuum in 1/299,792,458th of a second.
Mass	Kilogram	kg	A cylinder of platinum-iridium alloy that is kept by the International Bureau of Weights and Measures in Paris. A "duplicate" is retained by the National Institute of Standards and Technology (NIST) near Washington, DC.
Time	Second	s	Duration of 9,192,631,770 cycles of the radiation associated with a change in energy level of the cesium atom.
Electric current	Ampere	A	Magnitude of current which, when flowing through each of two long parallel wires a distance of 1 m apart in free space, results in a magnetic force between the wires of 2×10^{-7} N for each meter of length.
Thermo-dynamic temperature	Kelvin	K	The kelvin temperature scale has its zero point at absolute zero and has a fixed point of 273.15 K at the triple point of water, which is the temperature and pressure at which ice, liquid water, and water vapor are in equilibrium. The Celsius temperature scale is derived from the kelvin as $C = K - 273.15$.
Light intensity	Candela	cd	Defined as the luminous intensity of 1/600,000 of a square meter of a radiating cavity at the melting temperature of platinum (1,769°C).
Matter	Mole	mol	Defined as the number of atoms in 0.012 kg mass of carbon 12.

22.2 CONTACT VS. NONCONTACT INSPECTION TECHNIQUES

Inspection techniques can be divided into two broad categories: (1) contact and (2) noncontact. In contact inspection, physical contact is made between the object and the measuring or gaging instrument, whereas in noncontact inspection no physical contact is made.

22.2.1 Contact Inspection Techniques

Contact inspection involves the use of a mechanical probe or other device that makes contact with the object being inspected. The purpose of the probe is to measure or gage the object in some way. By its nature, contact inspection is often concerned with some physical dimension of the part. Accordingly, these techniques are widely used in the manufacturing industries, in particular in the production of metal parts (machining, stamping, and other metalworking processes). Contact inspection is also used in electrical circuit testing. The principal contact inspection technologies are

- Conventional measuring and gaging instruments
- Coordinate measuring machines (CMMs) and related techniques to measure mechanical dimensions
- Stylus type surface texture measuring machines to measure surface characteristics such as roughness and waviness
- Electrical contact probes for testing integrated circuits and printed circuit boards.

Conventional techniques and CMMs compete with each other in the measurement and inspection of part dimensions. The general application ranges for the different types of inspection and measurement equipment are presented in the *PQ* chart of Figure 22.2, where *P* and *Q* refer to the variety and quantity of parts inspected.

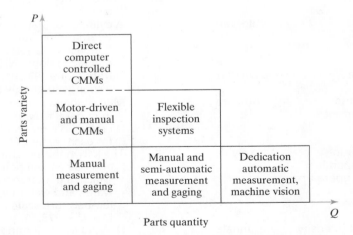

Figure 22.2 *PQ* chart indicating most appropriate measurement equipment as a function of parts variety and quantity (adapted from Bosch [3]).

Reasons these contact inspection methods are technologically and commercially important include the following:

- They are the most widely used inspection technologies today.
- They are accurate and reliable.
- In many cases, they represent the only methods available to accomplish the inspection.

22.2.2 Noncontact Inspection Technologies

Noncontact inspection methods utilize a sensor located at a certain distance from the object to measure or gage the desired features. The noncontact inspection technologies can be classified into two categories: (1) optical and (2) nonoptical. *Optical inspection technologies* use light to accomplish the measurement or gaging cycle. The most important optical technology is machine vision; however, other optical techniques are important in certain industries. *Nonoptical inspection technologies* utilize energy forms other than light to perform the inspection; these other energies include various electrical fields, radiation (other than light), and ultrasonics.

Noncontact inspection offers certain advantages over contact inspection techniques. The advantages include the following:

- They avoid damage to the surface that might result from contact inspection.
- Inspection cycle times are inherently faster. Contact inspection procedures require the contacting probe to be positioned against the part, which takes time. Most of the noncontact methods use a stationary probe that does not need repositioning for each part.
- Noncontact methods can often be accomplished on the production line without the need for any additional handling of the parts, whereas contact inspection usually requires special handling and positioning of the parts.
- It is more feasible to conduct 100% automated inspection, since noncontact methods have faster inspection cycle times and reduced need for special handling.

A comparison of some of the features of the various contact and noncontact inspection technologies is presented in Table 22.2.

22.3 CONVENTIONAL MEASURING AND GAGING TECHNIQUES[2]

Conventional measuring and gaging techniques use manually operated devices for linear dimensions such as length, depth, and diameter, as well as features such as angles, straightness, and roundness. *Measuring devices* provide a quantitative value of the part feature of interest, while *gages* determine whether the part feature (usually a dimension) falls within a certain acceptable range of values. Measuring takes more time but provides more information about the part feature. Gaging can be accomplished more quickly but does not provide as much information. Both techniques are widely used for post-process inspection of piece parts in manufacturing.

Measuring devices tend to be used on a sampling inspection basis. Some devices are portable and can be used at the production process. Others require bench setups that are

---

[2]This section is based on Groover [10], Section 45.3.

TABLE 22.2 Comparison of Resolution and Relative Speed of Several Inspection Technologies

Inspection Technology	Typical Resolution	Relative Speed of Application
Conventional instruments:		
Steel rule	0.25 mm (0.01 in)	Medium speed (medium cycle time)
Vernier caliper	0.025 mm (0.001 in)	Slow speed (high cycle time)
Micrometer	0.0025 mm (0.0001 in)	Slow speed (high cycle time)
Coordinate measuring machine	0.0005 mm (0.00002 in)*	Slow cycle time for single measurement. High speed for multiple measurements on same object.
Machine vision	0.25 mm (0.01 in)**	High speed (very low cycle time per piece)

*See Table 22.5 for other parameters on coordinate measuring machines.
**Precision in machine vision is highly dependent on the camera lens system and magnification used in the applications.

remote from the process, where the measuring instruments can be set up accurately on a flat reference surface called a *surface plate*. Gages are used either for sampling or 100% inspection. They tend to be more portable and lend themselves to application in the production process. Certain measuring and gaging techniques can be incorporated into automated inspection systems to permit feedback control of the process, or for statistical process control purposes.

The ease of use and precision of measuring instruments and gages have been enhanced in recent years by electronics. *Electronic gages* are a family of measuring and gaging instruments based on transducers capable of converting a linear displacement into a proportional electrical signal. The electrical signal is then amplified and transformed into a suitable data format such as a digital readout. For example, modern micrometers and graduated calipers are available with a digital display of the measurement of interest. These instruments are easier to read and eliminate much of the human error associated with reading conventional graduated devices.

Applications of electronic gages have grown rapidly in recent years, driven by advances in microprocessor technology. They are steadily replacing many of the conventional measuring and gaging devices. Advantages of electronic gages include (1) good sensitivity, accuracy, precision, repeatability, and speed of response; (2) ability to sense very small dimensions—down to 1 microinch (0.025 micron); (3) ease of operation; (4) reduced human error; (5) ability to display electrical signal in various formats; and (6) capability to be interfaced with computer systems for data processing.

For reference, we list the common conventional measuring instruments and gages with brief descriptions in Table 22.3. It is not our purpose in this book to provide an exhaustive discussion of these devices. A comprehensive survey can be found in books on metrology, such as [4] or [7], or for a more concise treatment [10]. Our purpose here is to focus on more modern technologies, such as coordinate measuring machines.

22.4 COORDINATE MEASURING MACHINES

Coordinate metrology is concerned with measuring the actual shape and dimensions of an object and comparing these results with the desired shape and dimensions, as might be specified on a part drawing. In this sense, coordinate metrology consists of the evaluation

TABLE 22.3 Common Conventional Measuring Instruments and Gages (Adapted from Groover [10]) (Some of These Devices Can Be Incorporated into Automated Inspection Systems)

	Instrument and Description
Steel rule	Linear graduated measurement scale used to measure linear dimensions. Available in various lengths, typically ranging from 150 to 1000 mm, with graduations of 1 or 0.5 mm. (U.S.C.S. rules available from 6 to 36 in, with graduations of 1/32 in or 0.01 in.)
Calipers	Family of graduated and nongraduated measuring devices consisting of two legs joined by a hinge mechanism. The ends of the legs contact the surfaces of the object to provide a comparative measure. Can be used for internal (e.g., inside diameter) or external (e.g., outside diameter) measurements.
Slide caliper	Steel rule to which two jaws are added, one fixed and the other movable. Jaws are forced to contact part surfaces to be measured, and the location of the movable jaw indicates the dimension of interest. Can be used for internal or external measurements.
Vernier caliper	Refinement of the slide caliper, in which a vernier scale is used to obtain more precise measurements (as close as 0.001 in).
Micrometer	Common device consisting of a spindle and C-shaped anvil (similar to a C-clamp). The spindle is closed relative to the fixed anvil by means of a screw thread to contact the surfaces of the object being measured. A vernier scale is used to obtain precisions of 0.01 mm in S.I. (0.0001 in in U.S.C.S.). Available as *outside micrometers, inside micrometers,* or *depth micrometers.* Also available as electronic gages to obtain a digital readout of the dimension of interest.
Dial indicator	Mechanical gage that converts and amplifies the linear movement of a contact pointer into rotation of a dial needle. The dial is graduated in units of 0.01 mm in S.I. (0.001 in in U.S.C.S.). Can be used to measure straightness, flatness, squareness, and roundness.
Gages	Family of gages, usually of the go/no-go type, that check whether a part dimension lies within acceptable limits defined by tolerance specified in part drawing. Includes: (1) *snap gages* for external dimensions such as a thickness, (2) *ring gages* for cylindrical diameters, (3) *plug gages* for hole diameters, and (4) *thread gages.*
Protractor	Device for measuring angles. *Simple protractor* consists of a straight blade and a semicircular head graduated in angular units (e.g., degrees). *Bevel protractor* consists of two straight blades that pivot one to the other; the pivot mechanism has a protractor scale to measure the angle of the two blades.

of the location, orientation, dimensions, and geometry of the part or object. A *coordinate measuring machine* (CMM) is an electromechanical system designed to perform coordinate metrology. A CMM has a contact probe that can be positioned in three dimensions relative to the surfaces of a workpart. The x, y, and z coordinates of the probe can be accurately and precisely recorded to obtain dimensional data about the part geometry. See Figure 22.3. The technology of CMMs dates from the mid 1950s.

To accomplish measurements in three-dimensional space, the basic CMM consists of the following components:

- Probe head and probe to contact the workpart surfaces
- Mechanical structure that provides motion of the probe in three Cartesian axes and displacement transducers to measure the coordinate values of each axis.

In addition, many CMMs have the following components:

- Drive system and control unit to move each of the three axes
- Digital computer system with application software.

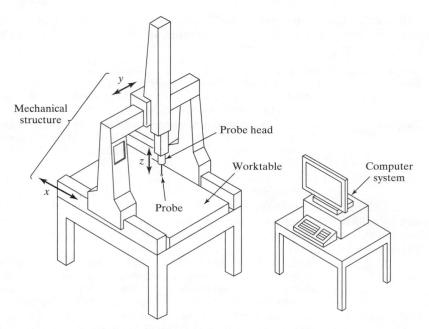

Figure 22.3 Coordinate measuring machine.

In this section, we discuss (1) the construction features of a CMM, (2) operation and programming of the machine, (3) the kinds of application software that enable it to measure more than just *x-y-z* coordinates, (4) applications and benefits of the CMM over manual inspection, (5) use of contact inspection probes on machine tools, and (6) portable CMMs.

22.4.1 CMM Construction

In the construction of a CMM, the probe is fastened to a mechanical structure that allows movement of the probe relative to the part. The part is usually located on a worktable that is connected to the structure. Let us examine the two basic components of the CMM: (1) its probe and (2) its mechanical structure.

Probe. The contact probe is a key component of a CMM. It indicates when contact has been made with the part surface during measurement. The tip of the probe is usually a ruby ball. Ruby is a form of corundum (aluminum oxide), whose desirable properties in this application include high hardness for wear resistance and low density for minimum inertia. Probes can have either a single tip, as in Figure 22.4(a), or multiple tips as in Figure 22.4(b).

Most probes today are *touch-trigger probes*, which actuate when the probe makes contact with the part surface. Commercially available touch-trigger probes utilize any of various triggering mechanisms, including the following: (1) a highly sensitive electrical contact switch that emits a signal when the tip of the probe is deflected from its neutral position, (2) a contact switch that permits actuation only when electrical contact is established between the probe and the (metallic) part surface, or (3) a piezoelectric sensor that generates a signal based on tension or compression loading of the probe.

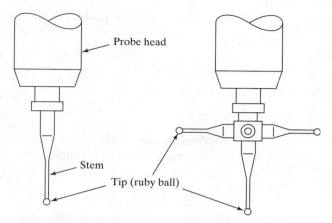

Figure 22.4 Contact probe configurations: (a) single tip and (b) multiple tips.

Immediately after contact is made between the probe and the surface of the object, the coordinate positions of the probe are accurately measured by displacement transducers associated with each of the three linear axes and recorded by the CMM controller. Compensation is made for the radius of the probe tip, as indicated in our Example 22.1, and any limited overtravel of the probe quill due to momentum is neglected. After the probe has been separated from the contact surface, it returns to its neutral position.

EXAMPLE 22.1 Dimensional Measurement with Probe Tip Compensation

The part dimension L in Figure 22.5 is to be measured. The dimension is aligned with the x-axis, so it can be measured using only x-coordinate locations. When the probe is moved toward the part from the left, contact made at $x = 68.93$ is recorded (mm). When the probe is moved toward the opposite side of the part from the right, contact made at $x = 137.44$ is recorded. The probe tip diameter is 3.00 mm. What is the dimension L?

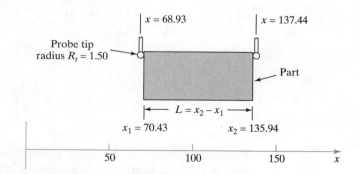

Figure 22.5 Setup for CMM measurement in Example 22.1.

Solution: Given that the probe tip diameter $D_t = 3.00$ mm, the radius $R_t = 1.50$ mm. Each of the recorded x values must be corrected for this radius:

$$x_1 = 68.93 + 1.50 = 70.43 \text{ mm}$$

$$x_2 = 137.44 - 1.50 = 135.94 \text{ mm}$$

$$L = x_1 - x_2 = 135.94 - 70.43 = 65.51 \text{ mm}$$

Mechanical Structure. There are various physical configurations for achieving the motion of the probe, each with advantages and disadvantages. Nearly all CMMs have a mechanical configuration that fits into one of the following six types, illustrated in Figure 22.6:

a. *Cantilever.* In the cantilever configuration, illustrated in Figure 22.6(a), the probe is attached to a vertical quill that moves in the z-axis direction relative to a horizontal arm that overhangs a fixed worktable. The quill can also be moved along the length of the arm to achieve y-axis motion, and the arm can be moved relative to the worktable to

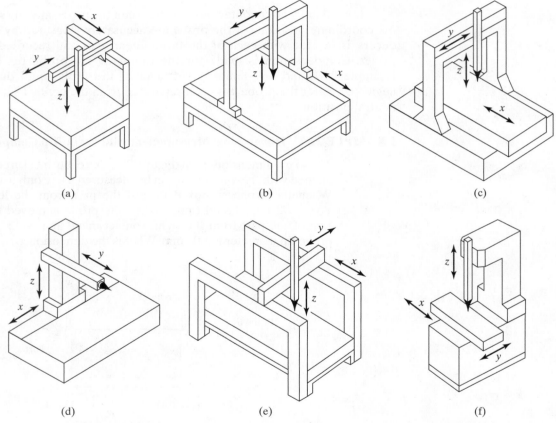

Figure 22.6 Six types of CMM construction: (a) cantilever, (b) moving bridge, (c) fixed bridge, (d) horizontal arm (moving ram type), (e) gantry, and (f) column.

achieve x-axis motion. The advantages of this construction are (1) convenient access to the worktable, (2) high throughput—the rate at which parts can be mounted and measured on the CMM, (3) capacity to measure large workparts (on large CMMs), and (4) relatvely small floor space requirements. The disadvantage is lower rigidity than most other CMM constructions.

b. *Moving bridge.* In the moving bridge design, Figure 22.6(b), the probe is mounted on a bridge structure that is moved relative to a stationary table on which is positioned the part to be measured. This provides a more rigid structure than the cantilever design, and its advocates claim that this makes the moving bridge CMM more accurate. However, one of the problems encountered with the moving bridge design is *yawing* (also known as *walking*), in which the two legs of the bridge move at slightly different speeds, resulting in twisting of the bridge. This phenomenon degrades the accuracy of the measurements. Yawing is reduced on moving bridge CMMs when dual drives and position feedback controls are installed for both legs. The moving bridge design is the most widely used in industry. It is well suited to the size range of parts commonly encountered in production machine shops.

c. *Fixed bridge.* In this configuration, Figure 22.6(c), the bridge is attached to the CMM bed, and the worktable is moved in the x-direction beneath the bridge. This construction eliminates the possibility of yawing, hence increasing rigidity and accuracy. However, throughput is adversely affected because of the additional energy needed to move the heavy worktable with the part mounted on it.

d. *Horizontal arm.* The horizontal arm configuration consists of a cantilevered horizontal arm mounted to a vertical column. The arm moves vertically and in and out to achieve y-axis and z-axis motions. To achieve x-axis motion, either the column is moved horizontally past the worktable (called the *moving ram* design), or the worktable is moved past the column (called the *moving table* design). The moving ram design is illustrated in Figure 22.6(d). The cantilever design of the horizontal arm configuration makes it less rigid and therefore less accurate than other CMM structures. On the positive side, it allows good accessibility to the work area. Large horizontal arm machines are suited to the measurement of automobile bodies, and some CMMs are equipped with dual arms so that independent measurements can be taken on both sides of the car body at the same time.

e. *Gantry.* This construction, illustrated in Figure 22.6(e), is generally intended for inspecting large objects. The probe quill (z-axis) moves relative to the horizontal arm extending between the two rails of the gantry. The workspace in a large gantry type CMM can be as great as 25 m (82 ft) in the x-direction by 8 m (26 ft) in the y-direction by 6 m (20 ft) in the z-direction.

f. *Column.* This configuration, in Figure 22.6(f), is similar to the construction of a machine tool. The x- and y-axis movements are achieved by moving the worktable, while the probe quill is moved vertically along a rigid column to achieve z-axis motion.

In all of these constructions, special design features are used to build high accuracy and precision into the frame. These features include precision rolling-contact bearings and hydrostatic air-bearings, installation mountings to isolate the CMM and reduce vibrations in the factory from being transmitted through the floor, and various schemes to counterbalance the overhanging arm of the cantilever construction [6], [17].

22.4.2 CMM Operation and Programming

Positioning the probe relative to the part can be accomplished in several ways, ranging from manual operation to direct computer control (DCC). Computer-controlled CMMs operate much like CNC machine tools, and these machines must be programmed. In this section, we consider (1) types of CMM controls and (2) programming of computer-controlled CMMs.

CMM Controls. The methods of operating and controlling a CMM can be classified into four main categories: (1) manual drive, (2) manual drive with computer-assisted data processing, (3) motor drive with computer-assisted data processing, and (4) DCC with computer-assisted data processing.

In a manual drive CMM, the human operator physically moves the probe along the machine's axes to make contact with the part and record the measurements. The three orthogonal slides are designed to be nearly frictionless to permit the probe to float freely in the x-, y-, and z-directions. The measurements are provided by a digital readout, which the operator can record either manually or with paper printout. Any calculations on the data (e.g., calculating the center and diameter of a hole) must be made by the operator.

A CMM with manual drive and computer-assisted data processing provides some data processing and computational capability for performing the calculations required to evaluate a given part feature. The types of data processing and computations range from simple conversions between U.S. customary units and metric to more complicated geometry calculations, such as determining the angle between two planes. The probe is still free floating to permit the operator to bring it into contact with the desired part surfaces.

A motor-driven CMM with computer-assisted data processing uses electric motors to drive the probe along the machine axes under operator control. An operator controls the motion using a joystick or similar device. Features such as low-power stepping motors and friction clutches are utilized to reduce the effects of collisions between the probe and the part. The motor drive can be disengaged to permit the operator to physically move the probe as in the manual control method. Motor-driven CMMs are generally equipped with data processing to accomplish the geometric computations required in feature assessment.

A CMM with direct computer control (DCC) operates like a CNC machine tool. It is motorized, and the movements of the coordinate axes are controlled by a dedicated computer under program control. The computer also performs the various data processing and calculation functions and compiles a record of the measurements made during inspection. As with a CNC machine tool, the DCC CMM requires part programming.

DCC Programming. There are two principle methods of programming a DCC measuring machine: (1) manual leadthrough and (2) off-line programming. In the *manual leadthrough* method, the operator leads the CMM probe through the various motions required in the inspection sequence, indicating the points and surfaces that are to be measured and recording these into the control memory. This is similar to the robot programming technique of the same name (Section 8.6.1). During regular operation, the CMM controller plays back the program to execute the inspection procedure.

Off-line programming is accomplished in the manner of computer-assisted NC part programming. The program is prepared off-line based on the part drawing and then downloaded to the CMM controller for execution. This permits the programming to be accomplished on new jobs while the CMM itself is working on jobs that have been previously programmed. Programming statements for a computer-controlled CMM include motion commands, measurement commands, and report formatting commands. The motion commands are used to direct the probe to a desired inspection location, the same way that a cutting tool is directed in a machining operation. The measurement statements are used to control the measuring and inspection functions of the machine, calling the various data processing and calculation routines into play. Finally, the formatting statements permit the specification of the output reports to document the inspection.

Most off-line programming of CMMs today is based on CAD systems [24], in which the measurement cycle is generated from CAD (Computer-Aided Design, Chapter 23) geometric data representing the part rather than from a hard copy part drawing. Off-line programming on a CAD system is facilitated by the *Dimensional Measuring Interface Specification* (DMIS), an ANSI standard. DMIS is a protocol that permits two-way communication between CAD systems and CMMs. Use of DMIS has the following advantages [3]: (1) It allows any CAD system to communicate with any CMM, (2) it reduces software development costs for CMM and CAD companies because only one translator is required to communicate with the DMIS, (3) users have greater choice among CMM suppliers, and (4) user training requirements are reduced.

22.4.3 Other CMM Software

CMM software is the set of programs and procedures (with supporting documentation) used to operate the CMM and its associated equipment. In addition to part programming software used for programming DCC machines, discussed above, other software is also required to achieve full functionality of a CMM. Indeed, it is software that has enabled the CMM to become the workhorse inspection machine that it is. Additional software can be divided into the following categories [3]: (1) core software other than DCC programming, (2) post-inspection software, and (3) reverse engineering and application-specific software.

Core Software Other than DCC Programming. Core software consists of the minimum basic programs required for the CMM to function, other then part programming software, which applies only to DCC machines. This software is generally applied either before or during the inspection procedure. Core programs normally include the following:

- *Probe calibration.* This function is required to define the parameters of the probe (such as tip radius, tip positions for a multi-tip probe, and elastic bending coefficients). Probe calibration allows coordinate measurements to automatically compensate for the probe dimensions when the tip contacts the part surface, avoiding the need for the probe tip calculations in Example 22.1. Calibration is usually accomplished by making the probe contact a cube or sphere of known dimensions.

- *Part coordinate system definition.* This software permits measurements of the part to be made without requiring a time-consuming part alignment procedure on the CMM worktable. Instead of physically aligning the part to the CMM axes, the measurement axes are mathematically aligned relative to the part.

- *Geometric feature construction.* This software addresses the problems associated with geometric features whose evaluation requires more than one point measurement. These features include flatness, squareness, determining the center of a hole or the axis of a cylinder, and so on. The software integrates the multiple measurements so that a given geometric feature can be evaluated. Table 22.4 lists a number of the

TABLE 22.4 Geometric Features Requiring Multiple Point Measurements to Evaluate:
Subroutines for Evaluating These Features Are Commonly Available Among CMM Software

---

Dimensions. A dimension of a part can be determined by taking the difference between the two surfaces defining the dimension. The two surfaces can be defined by a point location on each surface. In two axes $(x-y)$, the distance L between two point locations (x_1, y_1) and (x_2, y_2) is given by

$$L = \pm\sqrt{(x_2 - x_1)^2 + (y_2 - y_1)^2} \tag{22.3}$$

In three axes $(x-y-z)$, the distance L between two point locations (x_1, y_1, z_1) and (x_2, y_2, z_2) is given by

$$L = \pm\sqrt{(x_2 - x_1)^2 + (y_2 - y_1)^2 + (z_2 - z_1)^2} \tag{22.4}$$

See Example 22.1.

Hole location and diameter. By measuring three points around the surface of a circular hole, the "best-fit" center coordinates (a, b) of the hole and its radius R can be computed. The diameter = twice the radius. In the $x-y$ plane, the coordinate values of the three point locations are used in the following equation for a circle to set up three equations with three unknowns:

$$(x - a)^2 + (y - b)^2 = R^2 \tag{22.5}$$

where a = x-coordinate of the hole center, b = y-coordinate of the hole circle, and R = radius of the hole circle. Solving the three equations yields the values of a, b, and R. $D = 2R$. See Example 22.2.

Cylinder axis and diameter. This is similar to the preceding problem except that the calculation deals with an outside surface rather than an internal (hole) surface.

Sphere center and diameter. By measuring four points on the surface of a sphere, the best-fit center coordinates (a, b, c) and the radius R (diameter $D = 2R$) can be calculated. The coordinate values of the four point locations are used in the following equation for a sphere to set up four equations with four unknowns:

$$(x - a)^2 + (y - b)^2 + (z - c)^2 = R^2 \tag{22.6}$$

where a = x-coordinate of the sphere, b = y-coordinate of the sphere, c = z-coordinate of the sphere, and R = radius of the sphere. Solving the four equations yields the values of a, b, c, and R.

Definition of a line in x-y plane. Based on a minimum of two contact points on the line, the best-fit line is determined. For example, the line might be the edge of a straight surface. The coordinate values of the two point locations are used in the following equation for a line to set up two equations with two unknowns:

$$x + Ay + B = 0 \tag{22.7}$$

where A is a parameter indicating the slope of the line in the y-axis direction and B is a constant indicating the x-axis intercept. Solving the two equations yields the values of A and B, which defines the line. This form of equation can be converted into the more familiar conventional equation of a straight line, which is

$$y = mx + b \tag{22.8}$$

where slope $m = -1/A$ and y-intercept $b = -B/A$.

Angle between two lines. Based on the conventional form equations of the two lines, that is, Eq. (22.8), the angle between the two lines relative to the positive x-axis is given by:

$$\text{Angle between line 1 and line 2} = \alpha - \beta \tag{22.9}$$

where $\alpha = \tan^{-1}(m_1)$, where m_1 = slope of line 1; and $\beta = \tan^{-1}(m_2)$, where m_2 = slope of line 2.

Definition of a plane. Based on a minimum of three contact points on a plane surface, the best-fit plane is determined. The coordinate values of the three point locations are used in the following equation for a plane to set up three equations with three unknowns:

$$x + Ay + Bz + C = 0 \tag{22.10}$$

where A and B are parameters indicating the slopes of the plane in the y- and z-axis directions, and C is a constant indicating the x-axis intercept. Solving the three equations yields the values of A, B, and C, which defines the plane.

Flatness. By measuring more than three contact points on a supposedly plane surface, the deviation of the surface from a perfect plane can be determined.

Angle between two planes. The angle between two planes can be found by defining each of two planes using the plane definition method above and calculating the angle between them.

Parallelism between two planes. This is an extension of the previous function. If the angle between two planes is zero, then the planes are parallel. The degree to which the planes deviate from parallelism can be determined.

Angle and point of intersection between two lines. Given two lines known to intersect (e.g., two edges of a part that meet in a corner), the point of intersection and the angle between the lines can be determined based on two points measured for each line (a total of four points).

common geometric features, indicating how the features might be assessed by the CMM software. Examples 22.2 and 22.3 illustrate the application of two of the feature evaluation techniques. For increased statistical reliability, it is common to measure more than the theoretically minimum number of points needed to assess the feature and then to apply curve-fitting algorithms (such as least squares) in calculating the best estimate of the geometric feature's parameters. A review of CMM form-fitting algorithms is presented in Lin et al. [13].

- *Tolerance analysis.* This software compares measurements taken on the part with the dimensions and tolerances specified on the engineering drawing.

EXAMPLE 22.2 Computing a Linear Dimension

The coordinates at the two ends of a certain length dimension of a machined component have been measured by a CMM. The coordinates of the first end are (23.47, 48.11, 0.25), and the coordinates of the opposite end are (73.52, 21.70, 60.38), where the units are millimeters. The given coordinates have been corrected for probe radius. Determine the length dimension that would be computed by the CMM software.

Solution: Using Eq. (22.4) in Table 22.4, we have

$$L = \sqrt{(23.47 - 73.52)^2 + (48.11 - 21.70)^2 + (0.25 - 60.38)^2}$$
$$= \sqrt{(-50.05)^2 + (26.41)^2 + (-60.13)^2}$$
$$= \sqrt{2505.0025 + 697.4881 + 3615.6169} = \sqrt{6818.1075} = 82.57 \text{ mm}$$

EXAMPLE 22.3 Determining the Center and Diameter of a Drilled Hole

Three point locations on the surface of a drilled hole have been measured by a CMM in the *x-y* axes. The three coordinates are: (34.41, 21.07), (55.19, 30.50), and (50.10, 13.18) millimeters. The given coordinates have been corrected for probe radius. Determine: (a) coordinates of the hole center and (b) hole diameter, as they would be computed by the CMM software.

Solution: To determine the coordinates of the hole center, we must establish three equations patterned after Eq. (22.5) in Table 22.4:

$$(34.41 - a)^2 + (21.07 - b)^2 = R^2 \qquad \text{(i)}$$

$$(55.19 - a)^2 + (30.50 - b)^2 = R^2 \qquad \text{(ii)}$$

$$(50.10 - a)^2 + (13.18 - b)^2 = R^2 \qquad \text{(iii)}$$

Expanding each of the equations, we have

$$1184.0481 - 68.82a + a^2 + 443.9449 - 42.14b + b^2 = R^2 \qquad \text{(i)}$$

$$3045.9361 - 110.38a + a^2 + 930.25 - 61b + b^2 = R^2 \qquad \text{(ii)}$$

$$2510.01 - 100.2a + a^2 + 173.7124 - 26.36b + b^2 = R^2 \qquad \text{(iii)}$$

Simultaneous solution of the three equations yields the following values: $a = 45.66$ mm, $b = 23.89$ mm, and $R = 11.60$ mm. Thus, the center of the hole is located at $x = 45.66$ and $y = 23.89$, and the hole diameter is $D = 23.20$ mm.

Post-Inspection Software. Post-inspection software is the set of programs that are applied after the inspection procedure. Such software often adds significant utility and value to the inspection function. Among the programs included in this group are the following:

- *Statistical analysis.* This software is used to carry out any of various statistical analyses on the data collected by the CMM. For example, part dimension data can be used to assess process capability (Section 20.3.2) of the associated manufacturing process or perform statistical process control (Section 20.4). Two alternative approaches have been adopted by CMM makers in this area. The first approach is to provide software that creates a database of the measurements taken and facilitates export of the database to other software packages. What makes this feasible is that the data collected by a CMM are already coded in digital form. This approach permits the user to select among many statistical analysis packages that are commercially available. The second approach is to include a statistical analysis program among the software supplied by the CMM builder. This approach is generally quicker and easier, but the range of analyses available is more limited.
- *Graphical data representation.* The purpose of this software is to display the data collected during the CMM procedure in a graphical or pictorial way, to allow easier visualization of form errors and other data by the user.

Reverse Engineering and Application-Specific Software. Reverse engineering software is designed to take an existing physical part and construct a computer model of the part geometry based on a large number of measurements of its surface by a CMM. The simplest approach is to use the CMM in the manual mode of operation, in which the operator moves the probe by hand and scans the physical part to create a digitized three-dimensional (3-D) surface model. Manual digitization can be quite time-consuming for complex part geometries. More automated methods are being developed, in which the CMM explores the part surfaces with little or no human intervention to construct the 3-D model. The challenge here is to minimize the exploration time of the CMM, yet capture the details of a complex surface contour and avoid collisions that would damage the probe. Significant potential exists for using noncontacting probes (such as lasers) in reverse engineering applications.

Application-specific software refers to programs written for certain types of parts and/or products, whose applications are generally limited to specific industries. Several important examples are [3], [4]

- *Gear checking.* These programs are used on a CMM to measure the geometric features of a gear, such as tooth profile, tooth thickness, pitch, and helix angle.
- *Thread checking.* These are used for inspection of cylindrical and conical threads.
- *Cam checking.* This specialized software is used to evaluate the accuracy of physical cams relative to design specifications.
- *Automobile body checking.* This software is designed for CMMs used to measure sheet metal panels, subassemblies, and complete car bodies in the automotive industry. Unique measurement issues arise in this application that distinguish it from the measurement of machined parts: (1) large sheet metal panels lack rigidity, (2) compound curved surfaces are common, and (3) surface definition cannot be determined without measuring a great number of points.

Also included in the category of application-specific software are programs to operate accessory equipment associated with the CMM. Some types of accessory equipment that require their own application software include probe changers, rotary worktables used on the CMM, and automatic part loading and unloading devices.

22.4.4 CMM Applications and Benefits

Many applications of CMMs have been indicated by our previous discussion of CMM software. The most common applications are off-line inspection and on-line/post-process inspection (Section 21.4.1). Machined components are frequently inspected using CMMs. One common application is to check the first part machined on a numerically controlled machine tool. If the first part passes inspection, then the remaining parts produced in the batch are assumed to be identical to the first. Gears and automobile bodies are two examples previously mentioned in the context of application-specific software (Section 22.4.3).

Inspection of parts and assemblies on a CMM is generally accomplished using sampling techniques. One reason for this is the time required to perform the measurements. It often takes more time to inspect a part than it does to produce it. On the other hand, CMMs are sometimes used for 100% inspection if the inspection cycle is compatible with the production cycle and the CMM can be dedicated to the process. Whether the CMM is used for sampling inspection or 100% inspection, the CMM measurements are frequently used for statistical process control.

Other CMM applications include audit inspection and calibration of gages and fixtures. *Audit inspection* refers to the inspection of incoming parts from a vendor to ensure that the vendor's quality control systems are reliable. This is usually done on a sampling basis. In effect, this application is the same as post-process inspection. *Gage and fixture calibration* involves the measurement of various gages, fixtures, and other inspection and production tooling to validate their continued use.

One of the factors that makes a CMM so useful is its accuracy and repeatability. Typical values of these measures are given in Table 22.5 for a moving bridge CMM. It can be seen that these performance measures degrade as the size of the machine increases.

Coordinate measuring machines are most appropriate for applications possessing the following characteristics:

1. *Many inspectors are currently performing repetitive manual inspection operations.* If the inspection function represents a significant labor cost to the plant, then automating the inspection procedures will reduce labor cost and increase throughput.

TABLE 22.5 Typical Accuracy and Repeatability Measures for Two Different Sizes of CMM; Data Apply to a Moving Bridge CMM

CMM Feature		Small CMM	Large CMM
Measuring range:	x	650 mm (25.6 in)	900 mm (35.4 in)
	y	600 mm (23.6 in)	1200 mm (47.2 in)
	z	500 mm (19.7 in)	850 mm (33.5 in)
Accuracy:	x	0.004 mm (0.00016 in)	0.006 mm (0.00024 in)
	y	0.004 mm (0.00016 in)	0.007 mm (0.00027 in)
	z	0.0035 mm (0.00014 in)	0.0065 mm (0.00026 in)
Repeatability		0.0035 mm (0.00014 in)	0.004 mm (0.00016 in)
Resolution		0.0005 mm (0.00002 in)	0.0005 mm (0.00002 in)

Source: Bosch [3].

2. *The application involves post-process inspection.* CMMs are useful only in inspection operations performed after the manufacturing process.

3. *Measurement of geometric features requires multiple contact points.* These kinds of features are identified in Table 22.4, and available CMM software facilitates evaluation of these features.

4. *Multiple inspection setups would be required if parts were manually inspected.* Manual inspections are generally performed on surface plates using gage blocks, height gages, and similar devices, and a different setup is often required for each measurement. The same group of measurements on the part can usually be accomplished in one setup on a CMM.

5. *The part geometry is complex.* If many measurements are to be made on a complex part, and many contact locations are required, then the cycle time of a DCC CMM will be significantly less than the corresponding time for a manual procedure.

6. *A wide variety of parts must be inspected.* A DCC CMM is a programmable machine, capable of dealing with high parts variety.

7. *Repeat orders are common.* Once the part program has been prepared for the first part, subsequent parts from repeat orders can be inspected using the same program.

When applied in the appropriate parts quantity-parts variety range, the advantages of using CMMs over manual inspection methods are the following [17]:

- *Reduced inspection cycle time.* Because of the automated techniques included in the operation of a CMM, inspection procedures are faster and labor productivity is improved. A DCC CMM is capable of accomplishing many of the measurement tasks listed in Table 22.4 in one-tenth the time or less, compared with manual techniques. Reduced inspection cycle time translates into higher throughput.

- *Flexibility.* A CMM is a general-purpose machine that can be used to inspect a variety of different part configurations with minimal changeover time. In the case of the DCC machine, where programming is performed off-line, changeover time on the CMM involves only the physical setup.

- *Reduced operator errors.* Automating the inspection procedure has the obvious effect of reducing human errors in measurements and setups.

- *Greater inherent accuracy and precision.* A CMM is inherently more accurate and precise than the manual surface plate methods that are traditionally used for inspection.

- *Avoidance of multiple setups.* Traditional inspection techniques often require multiple setups to measure multiple part features and dimensions. In general, all measurements can be made in a single setup on a CMM, thereby increasing throughput and measurement accuracy.

22.4.5 Inspection Probes on Machine Tools

In recent years there has been a significant growth in the use of tactile probes as on-line inspection systems in CNC machining center applications. Called "on-machine inspection," the probes in these systems are mounted in toolholders, inserted into the machine tool spindle, stored in the tool drum, and handled by the automatic tool changer in the same way that cutting tools are handled. When the probe is mounted in the spindle, the machine tool is controlled very much like a CMM. Sensors in the probe determine when contact has been made with the part surface. Signals from the sensor are transmitted to the controller that performs the required data processing to interpret and utilize the signal.

Touch-sensitive probes are sometimes referred to as in-process inspection devices, but by our definitions they are on-line/post-process devices (Section 21.4.1) because they are employed immediately following the machining operation rather than during cutting. However, these probes are sometimes used between machining steps in the same setup, for example, to establish a datum reference either before or after initial machining so that subsequent cuts can be performed with greater accuracy. Some of the other calculation features of machine-mounted inspection probes are similar to the capabilities of CMMs with computer-assisted data processing. These features include determining the centerline of a cylindrical part or a hole and determining the coordinates of an inside or outside corner. Given the appropriate applications, use of the probes permits machining and inspection to be accomplished in one setup rather than two.

One of the controversial aspects of machine-mounted inspection probes is that the same machine tool making the part is also performing the inspection. The argument is that certain errors inherent in the cutting operation will also be manifested in the measuring operation. For example, if there is misalignment between the machine tool axes that is producing out-of-square parts, this condition will not be identified by the machine-mounted probe because the movement of the probe is affected by the same axis misalignment. To generalize, errors that are common to both the production process and the measurement procedure will go undetected by a machine-mounted inspection probe. These errors include machine tool geometry errors (such as the axis misalignment problem identified above), thermal distortions in the machine tool axes, and errors in any thermal correction procedures applied to the machine tool [3]. Errors that are not common to both systems should be detectable by the measurement probe. These measurable errors include tool and/or toolholder deflection, workpart deflection, tool offset errors, and effects of tool wear on the workpart. In practice, the use of machine-mounted inspection probes has proved to be effective in improving quality and saving time as an alternative to expensive off-line inspection operations.

Another objection to the use of machine-mounted inspection probes is that they take time above and beyond the regular machining cycle [5], [19]. Time is required to program the inspection routines, and time is lost during the cutting sequence for the probe to perform its measurement function. Software suppliers have developed advanced packages to streamline the programming task, but the interruptions during the machining cycle remain an impediment to potential users. These time losses must be weighed against the additional time that would be required to perform a separate inspection of the part at the end of the machine cycle and the cost of rework or scrap if the part is machined incorrectly.

22.4.6 Portable CMMs

In the conventional application of a coordinate measurement machine, parts must be removed from the production machine where they are made and taken to a special inspection room where the CMM is located. New coordinate measuring devices allow the inspection procedures to be performed at the site where the parts are made, eliminating the need to move the parts. The leading products in this area at time of writing are the Faro gage and the Faro arm, both products of the European firm, Faro. The Faro gage, nicknamed the Personal CMM, is a six-jointed articulated arm, whose configuration is similar to the human upper arm, forearm, and wrist. Fully extended it has a reach of about 1.2 m (47 in). At the end of the arm is a touch probe to perform the coordinate measurements, similar to a CMM. The difference is that the Faro gage mounts onto the machine tool that makes the parts. Thus, the inspection procedure can be carried out right at the machine. *In situ* inspection has the following advantages:

- It is no longer necessary to move the parts from the machine tool to the CMM and back. Material handling is reduced.
- The results of the inspection procedure are known immediately.
- The machinist who makes the part also performs the inspection procedure (a minimum of training is required to use the Faro gage).
- Because the part is still attached to the machine while it is being inspected, datum reference locations established during the machining operation are not lost. Any further machining uses the same references without the need to refixture the part.

Precision capability of the Faro gage is claimed to be 5 μm (0.0002 in). This accuracy is achieved through the use of highly accurate shaft encoders in the arm joints. A computer uses the encoder values to calculate the position of the probe in x-y-z space. Probes can be interchanged readily for various measurement tasks, just as they can when using a conventional CMM. Various types of mounting are available, including fixed attachment to the machine, and magnetic or vacuum mounts.

Closely related to the Faro gage is the Faro arm, which has a longer reach than the smaller unit, but has a similar six-jointed articulated-arm configuration. Several different sizes are available, with the longest reach being 3.7 m (145 in). Precision and repeatability are reduced as the reach increases. The larger size of the Faro arm enables it to be used on much larger products, such as automobile and truck bodies.

22.5 SURFACE MEASUREMENT [3]

The measurement and inspection technologies discussed in Sections 22.3 and 22.4 are concerned with evaluating dimensions and related characteristics of a part or product. Another measurable attribute of a part or product is its surface. The measurement of surfaces is usually

---

[3]Portions of this section are based on Groover [10], Section 5.2 and 45.4.

accomplished by instruments that use a contacting stylus. Hence, surface metrology is most appropriately included within the scope of contact inspection technologies.

22.5.1 Stylus Instruments

Stylus-type instruments are commercially available to measure surface roughness. These electronic devices have a cone-shaped diamond stylus with point radius of about 0.005 mm (0.0002 in) and a 90° tip angle that is traversed across the test surface at a constant slow speed. The operation is depicted in Figure 22.7. As the stylus head moves horizontally, it also moves vertically to follow the surface deviations. The vertical movements are converted into an electronic signal that represents the topography of the surface along the path taken by the stylus. This can be displayed as either (1) a profile of the surface or (2) an average roughness value.

Profiling devices use a separate flat plane as the nominal reference against which deviations are measured. The output is a plot of the surface contour along the line traversed by the stylus. This type of system can identify roughness, waviness, and other measures of the test surface. By traversing successive lines parallel and closely spaced with each other, the devices can create a "topographical map" of the surface.

Averaging devices reduce the vertical deviations to a single value of surface roughness. As illustrated in Figure 22.8, *surface roughness* is defined as the average of the vertical deviations from the nominal surface over a specified surface length. An arithmetic average (AA) is generally used, based on the absolute values of the deviations. In equation form,

$$R_a = \int_0^L \frac{|y|}{L} dx \qquad (22.11)$$

where R_a = arithmetic mean value of roughness (m, in); y = vertical deviation from the nominal surface converted to absolute value (m, in); and L = sampling distance, called the *cutoff length*, over which the surface deviations are averaged. The distance L_m in Figure 22.8 is the total measurement distance that is traced by the stylus. A stylus-type averaging device performs Eq. (22.11) electronically. To establish the nominal reference plane, the device uses skids riding on the actual surface. The skids act as a mechanical filter to reduce the effect of waviness in the surface.

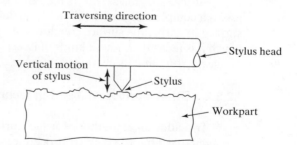

Figure 22.7 Sketch illustrating the operation of stylus-type instrument. Stylus head traverses horizontally across surface, while stylus moves vertically to follow surface profile. Vertical movement is converted into either: (1) a profile of the surface or (2) the average roughness value (source: Groover [10]).

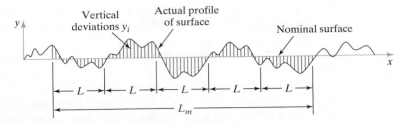

Figure 22.8 Deviations from nominal surface used in the definition of surface roughness (source: Groover [10]).

One of the difficulties in surface roughness measurement is the possibility that waviness can be included in the measurement of R_a. To deal with this problem, the cutoff length is used as a filter that separates waviness from roughness deviations. As defined above, the cutoff length is a sampling distance along the surface. It can be set at any of several values on the measurement device, usually ranging between 0.08 mm (0.0030 in) and 2.5 mm (0.10 in). A cutoff length shorter than the waviness width eliminates the vertical deviations associated with waviness and only includes those associated with roughness. The most common cutoff length used in practice is 0.8 mm (0.030 in). The cutoff length should be set at a value that is at least 2.5 times the distance between successive roughness peaks. The measuring length L_m is normally set at about five times the cutoff length.

An approximation of Eq. (22.11), perhaps easier to visualize, is given by

$$R_a = \frac{\sum_{i=1}^{n} |y_i|}{n} \tag{22.12}$$

where R_a has the same meaning as above; y_i = vertical deviations identified by the subscript i, converted to absolute value (m, in); and n = the number of deviations included in L. We have indicated the units in these equations to be meters (inches). However, the scale of the deviations is very small, so more appropriate units are microns, which equal 10^{-6} m or 10^{-3} mm, or microinches, which equal 10^{-6} in. These are the units commonly used to express surface roughness.

Surface roughness suffers the same kinds of deficiencies of any single measure used to assess a complex physical attribute. One deficiency is that it fails to account for the lay of the surface pattern; thus, surface roughness may vary significantly depending on the direction in which it is measured. These kinds of issues are addressed in books that deal specifically with surface texture and its characterization and measurement, such as Mummery [15].

22.5.2 Other Surface Measuring Techniques

Two additional methods for measuring surface roughness and related characteristics are available. One is a contact procedure of sorts, while the other is a noncontact method. We mention them here for completeness of coverage.

The first technique involves a subjective comparison of the part surface with standard surface finish blocks that are produced to specified roughness values. In the United States, these blocks have surfaces with roughness values of 2, 4, 8, 16, 32, 64, and 128 microinches. To estimate the roughness of a given test specimen, the surface is compared to the standard both visually and by using a "fingernail test." In this test, the user gently scratches the surfaces of the specimen and the standard, judging which standard is closest to the specimen. Standard test surfaces are a convenient way for a machine operator to obtain an estimate of surface roughness. They are also useful for product design engineers in judging what value of surface roughness to specify on the part drawing. The drawback of this method is its subjectivity.

Most other surface measuring instruments employ optical techniques to assess roughness. These techniques are based on light reflectance from the surface, light scatter or diffusion, and laser technology. They are useful in applications where stylus contact with the surface is undesirable. Some of the techniques permit very high speed operation, thus making 100% parts inspection feasible. One system described in Aronson [2] utilizes a laser to scan a 300 mm by 300 mm surface area in one minute and provide a three-dimensional colored hologram of the surface. The image consists of more than four million data points, readily shows surface variations, and permits measurements of the deviations to be made. One drawback of optical techniques is that their measured values do not always correlate well with roughness metrics obtained by stylus-type instruments.

22.6 MACHINE VISION

Machine vision is the acquisition of image data, followed by the processing and interpretation of these data by computer for some useful application. Machine vision (also called *computer vision*, since a digital computer is required to process the image data) is a growing technology, with its principal applications in industrial inspection. In this section, we examine how machine vision works and discuss its applications in QC inspection and other areas.

Vision systems are classified as being either 2-D or 3-D. Two-dimensional systems view the scene as a 2-D image. This is quite adequate for most industrial applications, since many situations involve a 2-D scene. Examples include dimensional measuring and gaging, verifying the presence of components, and checking for features on a flat (or semi-flat) surface. Other applications require 3-D analysis of the scene, and 3-D vision systems are required for this purpose. Our discussion will emphasize the simpler 2-D systems, although many of the techniques used for 2-D are also applicable in 3-D vision work.

The operation of a machine vision system can be divided into the following three functions: (1) image acquisition and digitization, (2) image processing and analysis, and (3) interpretation. These functions and their relationships are illustrated schematically in Figure 22.9.

22.6.1 Image Acquisition and Digitization

Image acquisition and digitization is accomplished using a video camera and a digitizing system to store the image data for subsequent analysis. The camera is focused on the subject of interest, and an image is obtained by dividing the viewing area into a matrix of discrete picture elements (called pixels), in which each element has a value that is proportional to the light intensity of that portion of the scene. The intensity value for each pixel is converted into its equivalent digital value by an ADC (Section 6.3). The operation of viewing a scene consisting of a simple object that contrasts substantially with its background, and dividing the scene into a corresponding matrix of picture elements, is depicted in Figure 22.10.

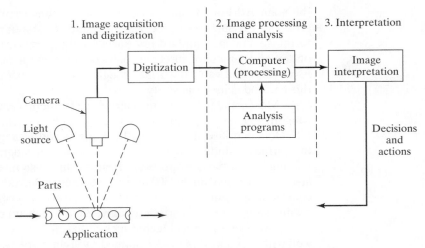

Figure 22.9 Basic functions of a machine vision system.

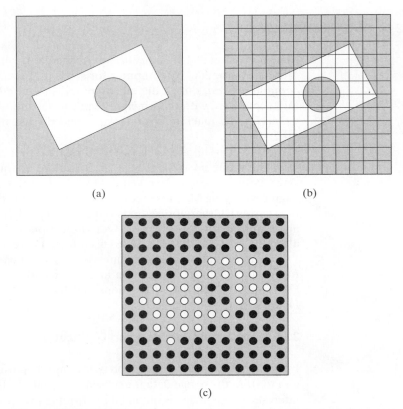

Figure 22.10 Dividing the image into a matrix of picture elements, where each element has a light intensity value corresponding to that portion of the image: (a) the scene; (b) 12 × 12 matrix superimposed on the scene; and (c) pixel intensity values, either black or white, for the scene.

The figure illustrates the likely image obtained from the simplest type of vision system, called a binary vision system. In *binary vision*, the light intensity of each pixel is ultimately reduced to either of two values, white or black, depending on whether the light intensity exceeds a given threshold level. A more sophisticated vision system is capable of distinguishing and storing different shades of gray in the image. This is called a *grayscale system*. This type of system can determine not only an object's outline and area characteristics, but also its surface characteristics such as texture and color. Grayscale vision systems typically use 4, 6, or 8 bits of memory. Eight bits corresponds to $2^8 = 256$ intensity levels, which is generally more levels than the video camera can really distinguish and certainly more than the human eye can discern.

Each set of digitized pixel values is referred to as a frame. Each frame is stored in a computer memory device called a *frame buffer*. The process of reading all the pixel values in a frame is performed with a frequency of 30 times per second. Very high-resolution cameras often operate at slower frequencies (e.g., 15 frames per second).

Cameras. Solid-state cameras are the principal types used in machine vision applications. They have largely replaced the vidicon cameras traditionally used as television cameras. Solid-state cameras operate by focusing the image onto a 2-D array of very small, finely spaced photosensitive elements. The photosensitive elements form the matrix of pixels shown in Figure 22.10. An electrical charge is generated by each element according to the intensity of light striking the element. The charge is accumulated in a storage device consisting of an array of storage elements corresponding one-to-one with the photosensitive picture elements. These charge values are read sequentially in the data processing and analysis function of machine vision.

Comparing the vidicon camera and solid-state camera, the latter is physically smaller and more rugged, and the image produced is more stable. These advantages have resulted in the growing dominance of their use in machine vision systems. Types of solid-state cameras include (1) the charge-coupled-device (CCD), (2) the charge-injected device (CID), and (3) the charge-priming device (CPD). These types are compared in Galbiati [8].

Typical square pixel arrays are 640 (horizontal) × 480 (vertical), 1024 × 768, and 1040 × 1392 picture elements. The *resolution* of the vision system is its ability to sense fine details and features in the image. Resolution depends on the number of picture elements used; the more pixels designed into the vision system, the higher its resolution. However, the cost of the camera increases as the number of pixels is increased. Even more important, the time required to sequentially read the picture elements and process the data increases as the number of pixels grows. The following example illustrates the problem.

EXAMPLE 22.4 Machine Vision

A video camera has a 640 × 480 pixel matrix. Each pixel must be converted from an analog signal to the corresponding digital signal by an ADC. The analog-to-digital conversion takes 0.1 microsecond $(0.1 \times 10^{-6}\,\text{sec})$ to complete, including the time to move between pixels. How long will it take to collect the image data for one frame, and is this time compatible with processing at the rate of 30 frames per second?

Solution: There are 640 × 480 = 307,200 pixels to be scanned and converted. The total time to complete the analog-to-digital conversion process is

$$(307{,}200 \text{ pixels})(0.1 \times 10^{-6}\,\text{sec}) = 0.0307 \text{ sec}$$

At a processing rate of 30 frames per second, the processing time for each frame is 0.0333 sec, which is longer than the 0.0307 sec required to perform the 307,200 analog-to-digital conversions.

Illumination. Another important aspect of machine vision is illumination. The scene viewed by the vision camera must be well illuminated, and the illumination must be constant over time. This almost always requires that special lighting be installed for a machine vision application rather than relying on ambient lighting in the facility.

Five categories of lighting can be distinguished for machine vision applications, as depicted in Figure 22.11: (a) front lighting, (b) back lighting, (c) side lighting, (d) structured lighting, and (e) strobe lighting. These categories represent differences in the positions of the light source relative to the camera as much as they do differences in lighting technologies. The lighting technologies include incandescent lamps, fluorescent lamps, sodium vapor lamps, and lasers.

In front lighting, the light source is located on the same side of the object as the camera. This produces a reflected light from the object that allows inspection of surface features such as printing on a label and surface patterns such as solder lines on a printed circuit board. In back lighting, the light source is placed behind the object being viewed by the camera. This creates a dark silhouette of the object that contrasts sharply with the light background. This type of lighting can be used for binary vision systems to inspect part dimensions and to distinguish between different part outlines. Side lighting causes irregularities in an otherwise plane smooth surface to cast shadows that can be identified by the vision system. This can be used to inspect for defects and flaws in the surface of an object.

Structured lighting involves the projection of a special light pattern onto the object to enhance certain geometric features. Probably the most common structured light pattern is a planar sheet of highly focused light directed against the surface of the object at a certain known angle, as in Figure 22.11(d). The sheet of light forms a bright line where the beam intersects the surface. In our sketch, the vision camera is positioned with its line of sight perpendicular to the surface of the object, so that any variations from the general plane of the part appear as deviations from a straight line. The distance of the deviation can be determined by optical measurement, and the corresponding elevation differences can be calculated using trigonometry.

In strobe lighting, the scene is illuminated by a short pulse of high-intensity light, which causes a moving object to appear stationary. The moving object might be a part moving past the vision camera on a conveyor. The pulse of light can last 5–500 microseconds [8]. This is sufficient time for the camera to capture the scene, although the camera actuation must be synchronized with that of the strobe light.

22.6.2 Image Processing and Analysis

The second function in the operation of a machine vision system is image processing and analysis. As indicated by Example 22.4, the amount of data that must be processed is significant. The data for each frame must be analyzed within the time required to complete one scan (typically 1/30 sec). A number of techniques have been developed for analyzing the image data in a machine vision system. One category of techniques in image processing and analysis is called segmentation. *Segmentation* techniques are intended to define and separate regions of interest within the image. Two of the common segmentation techniques are thresholding and edge detection. *Thresholding* involves the conversion of each pixel intensity level

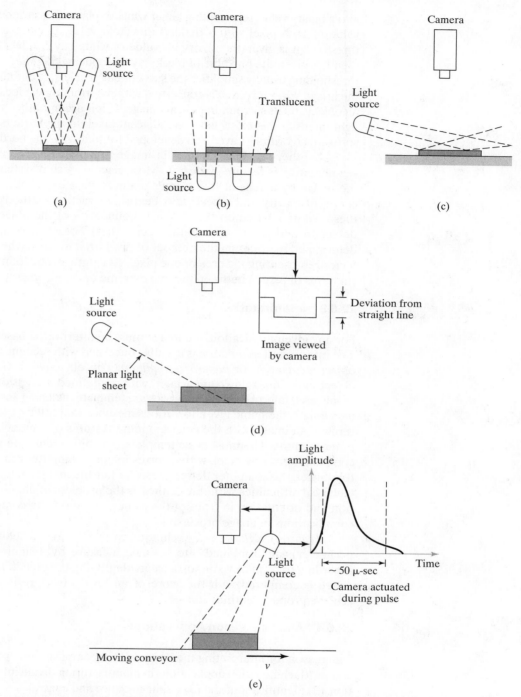

Figure 22.11 Types of illumination in machine vision: (a) front lighting, (b) back lighting, (c) side lighting, (d) structured lighting using a planar sheet of light, and (e) strobe lighting.

into a binary value, representing either white or black. This is done by comparing the intensity value of each pixel with a defined threshold value. If the pixel value is greater than the threshold, it is given the binary bit value of white, say 1; if less than the defined threshold, then it is given the bit value of black, say 0. Reducing the image to binary form by means of thresholding usually simplifies the subsequent problem of defining and identifying objects in the image. *Edge detection* is concerned with determining the location of boundaries between an object and its surroundings in an image. This is accomplished by identifying the contrast in light intensity that exists between adjacent pixels at the borders of the object. A number of software algorithms have been developed for following the border around the object.

Another set of techniques in image processing and analysis that normally follows segmentation is *feature extraction*. Most machine vision systems characterize an object in the image by means of the object's features: its area, length, width, diameter, perimeter, center of gravity, and aspect ratio. Feature extraction methods are designed to determine these features based on the area and boundaries of the object (using thresholding, edge detection, and other segmentation techniques). For example, the area of the object can be determined by counting the number of pixels that make up the object and multiplying by a factor representing the area of one pixel. Its length can be found by measuring the distance (in terms of pixels) between the two extreme opposite edges of the part.

22.6.3 Interpretation

For any given application, the image must be interpreted based on the extracted features. The interpretation function is usually concerned with recognizing the object, a task termed *object recognition* or *pattern recognition*. The objective in these tasks is to identify the object in the image by comparing it with predefined models or standard values. Two commonly used interpretation techniques are template matching and feature weighting. *Template matching* is the name given to various methods that attempt to compare one or more features of an image with the corresponding features of a model or template stored in computer memory. The most basic template matching technique is one in which the image is compared, pixel by pixel, with a corresponding computer model. Within certain statistical tolerances, the computer determines whether the image matches the template. One of the technical difficulties with this method is the problem of aligning the part in the same position and orientation in front of the camera, to allow the comparison to be made without complications in image processing.

Feature weighting is a technique in which several features (e.g., area, length, and perimeter) are combined into a single measure by assigning a weight to each feature according to its relative importance in identifying the object. The score of the object in the image is compared with the score of an ideal object residing in computer memory to achieve proper identification.

22.6.4 Machine Vision Applications

The reason for interpreting the image is to accomplish some practical objective in an application. Machine vision applications in manufacturing divide into three categories: (1) inspection, (2) identification, and (3) visual guidance and control.

Inspection. By far, quality control inspection is the biggest category. Machine vision installations in industry perform a variety of automated inspection tasks, most of which are either on-line/in-process or on-line/post-process. The applications are almost always in mass production where the time required to program and set up the vision system can be spread over many thousands of units. Typical industrial inspection tasks include the following:

- *Dimensional measurement.* These applications involve determining the size of certain dimensional features of parts or products usually moving at relatively high speeds on a moving conveyor. The machine vision system must compare the features (dimensions) with the corresponding features of a computer-stored model and determine the size value.
- *Dimensional gaging.* This is similar to the preceding except that a gaging function rather than a measurement is performed.
- *Verification of the presence of components.* This is done in an assembled product.
- *Verification of hole location and number of holes.* Operationally, this task is similar to dimensional measurement and verification of components.
- *Detection of surface flaws and defects.* Flaws and defects on the surface of a part or material often reveal themselves as a change in reflected light. The vision system can identify the deviation from an ideal model of the surface.
- *Detection of flaws in a printed label.* The defect can be in the form of a poorly located label or poorly printed text, numbering, or graphics on the label.

All of the preceding inspection applications can be accomplished using 2-D vision systems. Certain applications require 3-D vision, such as scanning the contour of a surface, inspecting cutting tools to check for breakage and wear, and checking solder paste deposits on surface mount circuit boards. Three-dimensional systems are being used increasingly in the automotive industry to inspect surface contours of parts such as body panels and dashboards. Vision inspection can be accomplished at much higher speeds than can traditional inspection with CMMs.

Other Machine Vision Applications. Part identification applications use a vision system to recognize and perhaps distinguish parts or other objects so that some action can be taken. The applications include part sorting, counting different types of parts flowing past along a conveyor, and inventory monitoring. Part identification can usually be accomplished by 2-D vision systems. Reading of two-dimensional bar codes and character recognition (Chapter 12) represent additional identification applications performed by 2-D vision systems.

Visual guidance and control involves applications in which a vision system is teamed with a robot or similar machine to control the movement of the machine. The term vision-guided robotic (VGR) system is used in connection with this technology [28]. Examples of VGR applications include seam tracking in continuous arc welding, part positioning and/or reorientation, picking parts from moving conveyors or stationary bins, collision avoidance, machining operations, and assembly tasks. These applications have been encouraged by recent improvements in the software that coordinates the operations of the vision system and robot.

22.7 OTHER OPTICAL INSPECTION METHODS

Machine vision is a well-publicized technology, perhaps because it is similar to one of the important human senses. It has potential for many applications in industry. However, there are also other optical sensing techniques that are used for inspection. This section surveys these technologies. The dividing line between machine vision and these techniques is sometimes blurred (excuse the pun). The distinction is that machine vision tends to imitate the capabilities of the human optical sensory system, which includes not only the eyes but also the complex interpretive powers of the brain. The techniques described below have a much simpler mode of operation.

Conventional Optical Instruments.   These conventional instruments include optical comparators and microscopes [24]. An *optical comparator* projects the shadow of an object (e.g., a workpart) against a large screen in front of an operator. The object can be moved in the *x-y* directions, permitting the operator to obtain dimensional data using cross hairs on the screen. Modern comparators feature edge-detection capabilities and advanced software that enable measurements to be taken accurately and quickly. Also known as contour projectors and shadowgraphs, they are easier to use than coordinate measuring machines and can be attractive alternatives for the more sophisticated technology in many applications requiring measurements in only two dimensions. The price of an optical comparator is about half the price of the least expensive CMM.

An alternative to the optical comparator is the conventional microscope. While the comparator is generally a unit that stands on the floor, a microscope is usually a bench-top unit, thus requiring less space in the shop floor. Microscopes can be equipped with an optical projection system instead of an eyepiece, providing ergonomic benefits for the operator. A significant advantage over the optical comparator is that the projection system shows the actual surface of the object rather than its shadow. The user can see its color, texture, and other features rather than just an outline.

Laser Systems.   The unique feature of a laser (*laser* stands for *l*ight *a*mplification by *s*timulated *e*mission of *r*adiation) is that it uses a coherent beam of light that can be projected with minimum diffusion. Because of this feature, lasers have been used in a number of industrial processing and measuring applications. High-energy laser beams are used for welding and cutting of materials, and low-energy lasers are utilized in various measuring and gaging situations.

The scanning laser device falls into the latter category. As shown in Figure 22.12, the scanning laser uses a laser beam that is deflected by a rotating mirror to produce a beam of light that can be focused to sweep past an object. A photodetector on the far side of the object

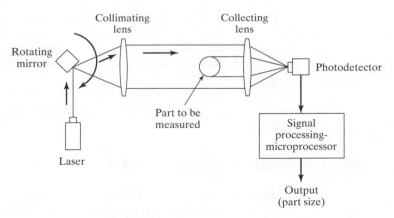

Figure 22.12 Diagram of scanning laser device.

senses the light beam except for the time period during the sweep when it is interrupted by the object. This time period can be measured with great accuracy and related to the size of the object in the path of the laser beam. The scanning laser beam device can complete its measurement in a very short time. Hence, the scheme can be applied in high-production on-line/post-process inspection or gaging. A microprocessor counts the time interruption of the scanning laser beam as it sweeps past the object, makes the conversion from time to a linear dimension, and signals other equipment to make adjustments in the manufacturing process and/or activate a sortation device on the production line. Applications of the scanning laser technique include rolling mill operations, wire extrusion, and machining and grinding processes.

More sophisticated applications of laser inspection systems are found in the automotive industry for measuring the contour and fit of car bodies and their component sheet metal parts. These applications require very large numbers of measurements to be taken in order to capture the shapes of complex geometric contours. Tolinski [21] describes three components in the inspection systems that perform these measurements. The first is a laser scanner capable of collecting more than 15,000 geometric data points per second. The second component is a mobile coordinate measuring machine to which the laser device is attached. The function of the CMM is to accurately locate the scanned points in three-dimensional space. The third component is a computer system that is programmed to compare the data points to a geometric model of the desired shape.

Linear Array Devices. The operation of a linear array for automated inspection is similar in some respects to machine vision, except that the pixels are arranged in only one dimension rather than two. A schematic diagram showing one possible arrangement of a linear array device is presented in Figure 22.13. The device consists of a light source that emits a planar sheet of light directed at an object. On the opposite side of the object is a linear array of closely spaced photo diodes. Typical numbers of diodes in the array are 256, 1024, and 2048 [23]. The sheet of light is blocked by the object, and this blocked light is measured by the photo diode array to indicate the object's dimension of interest.

The linear array measuring scheme has the advantages of simplicity, accuracy, and speed. It has no moving parts and is claimed to possess a resolution as small as 50 millionths of an inch [20]. It can complete a measurement in a much smaller time cycle than either machine vision or the scanning laser beam technique.

Optical Triangulation Techniques. Triangulation techniques are based on the trigonometric relationships of a right triangle. Triangulation is used for range-finding, that is, determining the distance or range of an object from two known points. Use of the principle in an

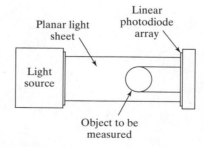

Figure 22.13 Operation of a linear array measuring device.

optical measuring system is explained with reference to Figure 22.14. A light source (typically a laser) is used to focus a narrow beam at an object to form a spot of light on the object. A linear array of photo diodes or other position-sensitive optical detector is used to determine the location of the spot. The angle A of the beam directed at the object is fixed and known, and so is the distance L between the light source and the photosensitive detector. Accordingly, the range R of the object from the base line defined by the light source and the photosensitive detector in Figure 22.14 can be determined as a function of the angle from trigonometric relationships as follows:

$$R = L \cot A \qquad (22.13)$$

22.8 NONCONTACT NONOPTICAL INSPECTION TECHNIQUES

In addition to noncontact optical inspection methods, there is also a variety of nonoptical techniques used for inspection tasks in manufacturing. Examples include sensor techniques based on electrical fields, radiation, and ultrasonics. This section briefly reviews these technologies as they might be used for inspection. They are important because they are *nondestructive evaluation methods*.

Electrical Field Techniques. Under certain conditions, an electrically active probe can create an electrical field. The field is affected by an object in the vicinity of the probe. Examples of electrical fields include reluctance, capacitance, and inductance. In the typical application, the object (workpart) is positioned in a defined relation with respect to the probe. A measurement of the object's effect on the electrical field allows an indirect measurement or gaging of certain part characteristics to be made, such as dimensional features, thickness of sheet material, and in some cases, flaws (cracks and voids below the surface) in the material.

Radiation Techniques. Radiation techniques utilize X-ray radiation to accomplish noncontact inspection procedures on metals and weld-fabricated products. The amount of radiation absorbed by the metal object can be used to indicate thickness and presence of

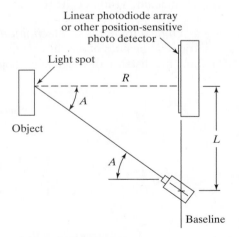

Figure 22.14 Principle of optical triangulation sensing.

flaws in the metal part or welded section. An example is the use of X-ray inspection techniques to measure thickness of sheet metal made in a rolling mill. The inspection is performed as an on-line/post-process procedure, with information from the inspection used to make adjustments in the opening between rolls in the rolling mill.

Ultrasonic Inspection Methods. Ultrasonic techniques make use of very high frequency sound (greater than 20,000 Hz) for various inspection tasks. Some of the techniques are performed manually, whereas others are automated. One of the automated methods involves emitting ultrasonic waves from a probe and reflecting them off the object to be inspected. In the setup of the inspection procedure, an ideal test part is placed in front of the probe to obtain a reflected sound pattern. This sound pattern becomes the standard against which production parts are later compared. If the reflected pattern from a given production part matches the standard (within an allowable statistical variation), the part is considered acceptable; otherwise, it is rejected. One technical problem with this technique involves the presentation of production parts in front of the probe. To avoid extraneous variations in the reflected sound patterns, the parts must always be placed in the same position and orientation relative to the probe.

REFERENCES

[1] ARONSON, R. B., "Shop-Hardened CMMs," *Manufacturing Engineering*, April 1998, pp. 62–68.

[2] ARONSON, R. B., "Finding the Flaws," *Manufacturing Engineering*, November 2006, pp. 81–88.

[3] BOSCH, A., editor, *Coordinate Measuring Machines and Systems*, Marcel Dekker, Inc., NY, 1995.

[4] BROWN & SHARPE, *Handbook of Metrology*, North Kingston, RI, 1992.

[5] DESTAFANI, J., "On-Machine Probing," *Manufacturing Engineering*, November 2004, pp. 51–57.

[6] DOEBLIN, E. O., *Measurement Systems: Applications and Design*, 4th ed., McGraw-Hill, Inc., NY, 1990.

[7] FARAGO, F. T., *Handbook of Dimensional Measurement*, 2d ed., Industrial Press Inc., NY, 1982.

[8] GALBIATI, L. J., Jr., *Machine Vision and Digital Image Processing Fundamentals*, Prentice Hall, Englewood Cliffs, NJ, 1990.

[9] GROOVER, M. P., M. WEISS, R. N. NAGEL, and N. G. ODREY, *Industrial Robotics: Technology, Programming, and Applications*, McGraw-Hill Book Co., NY, 1986, Chapter 7.

[10] GROOVER, M. P., *Fundamentals of Modern Manufacturing—Materials, Processes, and Systems*, 3d ed., John Wiley & Sons, Inc., Hoboken, NJ, 2007, Chapter 45.

[11] HOGARTH, S., "Machines with Vision," *Manufacturing Engineering*, April, 1999, pp. 100–107.

[12] KUBEL, E., "Machine Vision: Eyes for Industry," *Manufacturing Engineering*, April, 1998, pp. 42–51.

[13] LIN, S-S., P. VARGHESE, C. ZHANG, and H-P. B. WANG, "A Comparative Analysis of CMM Form-Fitting Algorithms," *Manufacturing Review*, Volume 8, No. 1, March 1995, pp. 47–58.

[14] MORRIS, A. S., *Measurement and Calibration for Quality Assurance*, Prentice Hall, Englewood Cliffs, NJ, 1991.

[15] MUMMERY, L., *Surface Texture Analysis—The Handbook;* Hommelwerke Gmbh, Germany, 1990.

[16] SAUNDERS, M., "Keeping in Touch with Probing," *Manufacturing Engineering*, October, 1998, pp. 52–58.

[17] SCHAFFER, G. H., "Taking the Measure of CMMs," Special Report 749, *American Machinist*, October 1982, pp. 145–160.

[18] SCHAFFER, G. H., "Machine Vision: A Sense for CIM," Special Report 767, *American Machinist*, June 1984, pp. 101–120.

[19] SHARKE, P., "On-Machine Inspecting," *Mechanical Engineering*, April 2005, pp. 30–33.

[20] Sheffield Measurement Division, *66 Centuries of Measurement*, Cross & Trecker Corporation, Dayton, OH, 1984.

[21] S. Starrett Company, *Tools and Rules*, Athol, MA, 1992.

[22] TOLINSKI, M., "Hands-Off Inspection," *Manufacturing Engineering*, September 2005, pp. 117–130.

[23] VERNON, D., *Machine Vision—Automated Visual Inspection and Robot Vision*, Prentice Hall International (UK) Ltd., London, 1991.

[24] WAURZYNIAK, P., "Programming CMMs," *Manufacturing Engineering*, May 2004, pp. 117–126.

[25] WAURZYNIAK, P., "Optical Inspection," *Manufacturing Engineering*, July 2004, pp. 107–114.

[26] Website: *www. Faro. com*

[27] WICK, C., and R. F. VEILLEUX, Editors, *Tool and Manufacturing Engineers Handbook*, 4th ed., Volume IV, *Quality Control and Assembly*, Society of Manufacturing Engineers, Dearborn, MI, 1987.

[28] ZENS, Jr., R. G., "Guided by Vision," *Assembly*, September 2005, pp. 52–58.

REVIEW QUESTIONS

22.1 Define the term measurement.

22.2 What is metrology?

22.3 What are the seven basic quantities used in metrology from which all other variables are derived?

22.4 What is the difference between accuracy and precision in measurement? Define these two terms.

22.5 With respect to measuring instruments, what is calibration?

22.6 What is meant by the term contact inspection?

22.7 What are some of the advantages of noncontact inspection?

22.8 What is meant by the term coordinate metrology?

22.9 What are the two basic components of a coordinate measuring machine?

22.10 Name the four categories into which the methods of operating and controlling a CMM can be classified.

22.11 What does the term reverse engineering mean in the context of coordinate measuring machines?

22.12 Name four of the seven characteristics of potential applications for which CMMs are most appropriate.

22.13 What are some of the arguments and objections to the use of inspection probes mounted in toolholders on machine tools?

22.14 What is the most common method used to measure surfaces of a part?

22.15 What is machine vision?

22.16 The operation of a machine vision system can be divided into three functions. Name and briefly describe them.

22.17 What is the main application of machine vision in industry?

22.18 What is an optical comparator?

22.19 The word *laser* is an acronym for what?

PROBLEMS

Coordinate Measuring Machines

(For ease of computation, numerical values in the following problems are given at a lower level of precision than the level of which most CMMs would be capable.)

22.1 Two point locations corresponding to a certain length dimension have been measured by a coordinate measuring machine in the x-y plane. The coordinates of the first end are (12.511, 2.273), and the coordinates of the opposite end are (4.172, 1.985), where the units are in inches. The coordinates have been corrected for probe radius. Determine the length dimension that would be computed by the CMM software.

22.2 The coordinates at the two ends of a certain length dimension have been measured by a CMM. The coordinates of the first end are (120.5, 50.2, 20.2), and the coordinates of the opposite end are (23.1, 11.9, 20.3), where the units are in millimeters. The given coordinates have been corrected for probe radius. Determine the length dimension that would be computed by the CMM software.

22.3 Three point locations on the surface of a drilled hole have been measured by a CMM in the x-y axes. The three coordinates are (16.42, 17.17), (20.20, 11.85), and (24.08, 16.54), where the units are millimeters. These coordinates have been corrected for probe radius. Determine (a) the coordinates of the hole center and (b) the hole diameter, as they would be computed by the CMM software.

22.4 Three point locations on the surface of a cylinder have been measured by a coordinate measuring machine. The cylinder is positioned so that its axis is perpendicular to the x-y plane. The three coordinates in the x-y axes are (5.242, 0.124), (0.325, 4.811), and (−4.073, −0.544), where the units are inches. The coordinates have been corrected for probe radius. Determine (a) the coordinates of the cylinder axis and (b) the cylinder diameter, as they would be computed by the CMM software.

22.5 Two points on a line have been measured by a CMM in the x-y plane. The point locations have the coordinates (12.257, 2.550) and (3.341, −10.294), where the units are inches and the coordinates have been corrected for probe radius. Find the equation for the line in the form of Eq. (23.7).

22.6 Two points on a line are measured by a CMM in the x-y plane. The points have the coordinates (100.24, 20.57) and (50.44, 60.46), where the units are millimeters. The given coordinates have been corrected for probe radius. Determine the equation for the line in the form of Eq. (23.7).

22.7 The coordinates of the intersection of two lines are to be determined using a CMM to define the equations for the two lines. The two lines are the edges of a machined part, and the intersection represents the corner where the two edges meet. Both lines lie in the x-y plane. Measurements are in inches. Two points are measured on the first line to have coordinates of (5.254, 10.430) and (10.223, 6.052). Two points are measured on the second line to have coordinates of (6.101, 0.657) and (8.970, 3.824). The coordinate values have been corrected for probe radius. (a) Determine the equations for the two lines in the form of Eq. (23.7). (b)

What are the coordinates of the intersection of the two lines? (c) The edges represented by the two lines are specified to be perpendicular to each other. Find the angle between the two lines to determine if the edges are perpendicular.

22.8   Two of the edges of a rectangular part are represented by two lines in the x-y plane on a CMM worktable, as illustrated in Figure P22.8. We wish to mathematically redefine the coordinate system so that the two edges are used as the x- and y-axes, rather than the regular x-y axes of the CMM. To define the new coordinate system, two parameters must be determined: (a) the origin of the new coordinate system must be located in the existing CMM axis system; and (b) the angle of the x-axis of the new coordinate system must be determined relative to the CMM x-axis. Two points on the first edge (line 1) have been measured by the CMM and the coordinates are (46.21, 22.98) and (90.25, 32.50), where the units are millimeters. Also, two points on the second edge (line 2) have been measured by the CMM and the coordinates are (26.53, 40.75) and (15.64, 91.12). The coordinates have been corrected for the radius of the probe. Find (a) the coordinates of the new origin relative to the CMM origin and (b) degrees of rotation of the new x-axis relative to the CMM x-axis. (c) Are the two lines (part edges) perpendicular?

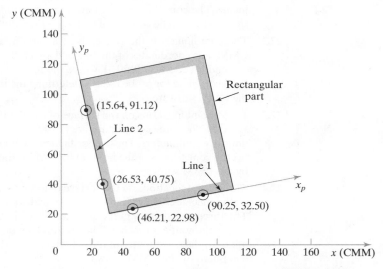

Figure P22.8 Overhead view of part relative to CMM axes.

22.9   Three point locations on the flat surface of a part have been measured by a CMM. The three point locations are (225.21, 150.23, 40.17), (14.24, 140.92, 38.29), and (12.56, 22.75, 38.02), where the units are millimeters. The coordinates have been corrected for probe radius. (a) Determine the equation for the plane in the form of Eq. (23.10). (b) To assess flatness of the surface, the CMM measures a fourth point. If its coordinates are (120.22, 75.34, 39.26), what is the vertical deviation of this point from the perfectly flat plane determined in (a)?

Optical Inspection Methods

22.10   A solid-state camera has a 256×256 pixel matrix. The analog-to-digital converter takes 0.20 microseconds (0.20×10^{-6} sec) to convert the analog charge signal for each pixel into the corresponding digital signal. If there is no time loss in switching between pixels, determine the following: (a) the amount of time required to collect the image data for one frame, and (b) whether the time determined in part (a) is compatible with the processing rate of 30 frames per second.

22.11 The pixel count of a solid-state camera is 500×582. Each pixel is converted from an analog voltage signal to the corresponding digital signal by an analog-to-digital converter. The conversion process takes 0.08 microseconds (0.08×10^{-6} seconds) to complete. Given this time, how long will it take to collect and convert the image data for one frame? Can this be done 30 times per second?

22.12 A high-resolution solid state camera is to have a 1035×1320 pixel matrix. An image processing rate of 30 times per second must be achieved, or 0.0333 sec per frame. To allow for time lost in other data processing per frame, the total ADC time per frame must be 80% of the 0.0333 sec, or 0.0267 sec. In order to be compatible with this speed, in what time period must the analog-to-digital conversion be accomplished per pixel?

22.13 A solid-state camera system has 512×512 picture elements. All pixels are converted sequentially by an ADC and read into the frame buffer for processing. The machine vision system will operate at the rate of 30 frames per second. However, in order to allow time for data processing of the contents of the frame buffer, the analog-to-digital conversion of all pixels by the ADC must be completed in 1/80 second. Assuming that 10 nanoseconds (10×10^{-9} sec.) are lost in switching from one pixel to the next, determine the time required to carry out the analog-to-digital conversion process for each pixel, in nanoseconds.

22.14 A scanning laser device similar to the one shown in Figure 22.12 is to be used to measure the diameter of shafts that are ground in a centerless grinding operation. The part has a diameter of 0.475 inch with a tolerance of ± 0.002 inch. The four-sided mirror of the scanning laser beam device rotates at 250 rev/min. The collimating lens focuses 30° of the sweep of the mirror into a swath that is 1.000 inch wide. It is assumed that the light beam moves at a constant speed across this swath. The photodetector and timing circuitry is capable of resolving time units as fine as 100 nanoseconds (100×10^{-9} sec.). This resolution should be equivalent to no more than 10% of the tolerance band (0.004 inch). (a) Determine the interruption time of the scanning laser beam for a part whose diameter is equal to the nominal size. (b) How much of a difference in interruption time is associated with the tolerance of ± 0.002 inch? (c) Is the resolution of the photodetector and timing circuitry sufficient to achieve the 10% rule on the tolerance band?

22.15 Triangulation computations are to be used to determine the distance of parts moving on a conveyor. The setup of the optical measuring apparatus is as illustrated in the text in Figure 22.14. The angle between the beam and the surface of the part is 25°. Suppose for one given part passing on the conveyor, the baseline distance is 6.55 inches, as measured by the linear photosensitive detection system. What is the distance of this part from the baseline?

Product Design and CAD/CAM in the Production System

This final part of the book is concerned with manufacturing support systems that operate at the enterprise level, as indicated in Figure 23.1. The *manufacturing support systems* are the procedures and systems used by the firm to manage production and solve the technical and logistics problems associated with designing the products, planning the processes, ordering materials, controlling work-in-process as it moves through the plant, and delivering products to customers. Many of these functions can be automated using computer systems, as suggested by terms like *computer-aided design* and *computer-integrated manufacturing*. Whereas most of our previous discussion on automation has emphasized the flow of the physical product through the factory, the enterprise level is more concerned with the flow of information in the factory and throughout the firm. Most of the topics in Part VI deal with computerized systems, but we also describe some systems and procedures that are labor intensive in their operation. Even the computer-automated systems include people. People make the production systems work.

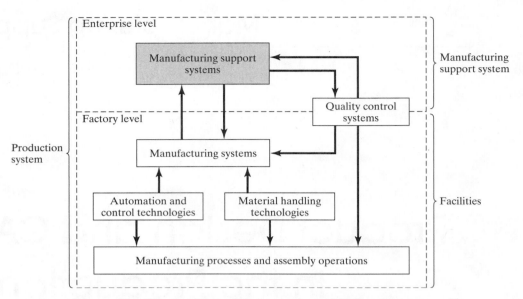

Figure 23.1 The position of the manufacturing support systems in the larger production system.

The present chapter is concerned with product design and the various technologies that are used to augment and automate the design function. CAD/CAM (computer-aided design and computer-aided manufacturing) is one of those technologies. It uses a digital computer to accomplish certain functions in product design and production. CAD uses the computer to support the design engineering function, and CAM uses the computer to support manufacturing engineering activities. The combination CAD/CAM is symbolic of efforts to integrate the design and manufacturing functions of a firm into a continuum of activities rather than to treat them as two separate and disparate activities, as they had been considered in the past. CIM (computer-integrated manufacturing) includes all of CAD/CAM but also embraces the business functions of a manufacturing firm. CIM implements computer technology in all of the operational and information processing activities related to manufacturing. In the final section of the chapter, we discuss a systematic method for approaching a product design project, called quality function deployment.

Chapters 24 through 26 are concerned with topics in production systems and CIM other than product design. Chapter 24 deals with process planning and how it can be automated using computer systems. We also discuss ways in which product design and manufacturing and other functions can be integrated using an approach called concurrent engineering. An important issue in concurrent engineering is design for manufacturing; that is, how can a product be designed to make it easier (and cheaper) to produce? Chapter 25 discusses the various methods used to implement production planning and control, through material requirements planning, shop floor control, and enterprise resource planning (ERP). Finally, Chapter 26 is concerned with just-in-time production and lean production, the techniques that were developed and perfected by the Toyota Motor Company in Japan.

23.1 PRODUCT DESIGN AND CAD

Product design is a critical function in the production system. The quality of the product design (i.e., how well the design department does its job) is probably the single most important factor in determining the commercial success and societal value of a product. If the product design is poor, no matter how well it is manufactured, the product is very likely doomed to contribute little to the wealth and well-being of the firm that produced it. If the product design is good, there is still the question of whether the product can be produced at sufficiently low cost to contribute to the company's profits and success. One of the facts of life about product design is that a very significant portion of the cost of the product is determined by its design. Design and manufacturing cannot be separated in the production system. They are bound together functionally, technologically, and economically.

Let us begin our discussion of product design by describing the general process of design. We then examine how computers are used in the design process.

23.1.1 The Design Process

The general process of design is characterized as an iterative process consisting of six phases [14]: (1) recognition of need, (2) problem definition, (3) synthesis, (4) analysis and optimization, (5) evaluation, and (6) presentation. These six steps, and the iterative nature of the sequence in which they are performed, are depicted in Figure 23.2(a).

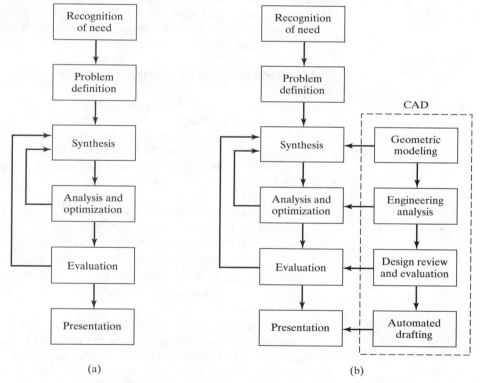

Figure 23.2 (a) Design process as defined by Shigley [14]. (b) The design process using computer-aided design (CAD).

Recognition of need (1) involves the realization by someone that a problem exists that a thoughtful design could solve. This recognition might mean identifying some deficiency in a current machine design by an engineer or perceiving some new product opportunity by a salesperson. Problem definition (2) involves a thorough specification of the item to be designed. This specification includes the physical characteristics, function, cost, quality, and operating performance.

Synthesis (3) and analysis (4) are closely related and highly interactive. Consider the development of a certain product design: Each of the subsystems of the product must be conceptualized by the designer, analyzed, improved through this analysis procedure, redesigned, analyzed again, and so on. The process is repeated until the design has been optimized within the constraints imposed on the designer. The individual components are then synthesized and analyzed into the final product in a similar manner.

Evaluation (5) is concerned with measuring the design against the specifications established in the problem definition phase. This evaluation often requires the fabrication and testing of a prototype model to assess operating performance, quality, reliability, and other criteria. The final phase in the design procedure is the presentation of the design. Presentation (6) is concerned with documenting the design by means of drawings, material specifications, assembly lists, and so on. In essence, documentation means that the design database is created.

The principles of lean production (Chapter 26) can be applied in the product design process. Lean product development seeks to achieve objectives of efficiency, simplicity, and effectiveness similar to those obtained in lean production by integrating departmental functions, reducing redundancy in design procedures (without eliminating the natural iterative processes of design), encouraging collaboration, and minimizing waiting [13]. The use of computers in design can foster lean product development.

23.1.2 Application of Computers in Design

Computer-aided design (CAD) is defined as any design activity that involves the effective use of a computer to create, modify, analyze, or document an engineering design. CAD is most commonly associated with the use of an interactive computer graphics system, referred to as a CAD system. The term CAD/CAM system is also used if it includes manufacturing as well as design applications. Companies using CAD systems have reaped the following benefits (based on [10] and [17]):

- *Increased design productivity.* The use of CAD helps the designer conceptualize the product and its components, which in turn helps reduce the time required by the designer to synthesize, analyze, and document the design.
- *Increased available geometric forms in the design.* CAD permits the designer to select among a wider range of shapes, such as mathematically defined contours, blended angles, and similar forms that would be difficult to create by manual drafting techniques.
- *Improved quality of the design.* The use of a CAD system with appropriate hardware and software capabilities permits the designer to do a more complete engineering analysis and to consider a larger number and variety of design alternatives. The quality of the resulting design is thereby improved.
- *Improved design documentation.* The graphical output of a CAD system results in better documentation of the design than what is practical with manual drafting.

The engineering drawings are superior, with more standardization among the drawings, fewer drafting errors, and greater legibility. In addition, most CAD packages provide automatic documentation of design changes, which includes who made the changes, as well as when and why the changes were made.

- *Creation of a manufacturing data base.* In the process of creating the documentation for the product design (geometric specification of the product, dimensions of the components, materials specifications, bill of materials, etc.), much of the required data base to manufacture the product is also created.

- *Design standardization.* Design rules can be included in CAD software to encourage the designer to utilize company-specified models for certain design features; for example, to limit the number of different hole sizes that can be used in the design. This simplifies the hole specification procedure for the designer and reduces the number of drill bit sizes that must be inventoried in manufacturing.

The output of the creative design process includes huge amounts of data that must be stored and managed. These functions are often accomplished in a modern CAD system by a product data management module. A *product data management* (PDM) system consists of computer software that provides links between users (e.g., designers) and a central data base. The data base stores engineering design data such as geometric models, product structures (e.g., bills of material), and related documentation. The software also manages the data base by tracking the identity of users, facilitating and documenting engineering changes, recording a history of the engineering changes on each part and product, and providing similar documentation functions.

With reference to the six phases of design defined previously, a CAD system can facilitate four of the design phases, as indicated in Table 23.1 and illustrated in Figure 23.2(b) as an overlay on the design process.

Geometric Modeling. Geometric modeling involves the use of a CAD system to develop a mathematical description of the geometry of an object. The mathematical description, called a geometric model, is contained in computer memory. This permits the user of the CAD system to display an image of the model on a graphics terminal and to perform certain operations on the model. These operations include creating new geometric models from basic building blocks available in the system, moving the images around on the screen, zooming in on certain features of the image, and so forth. These capabilities permit the designer to construct a model of a new product (or its components) or to modify an existing model.

There are various types of geometric models used in CAD. One classification distinguishes between two-dimensional (2-D) and three-dimensional (3-D) models. Two-dimensional models are best utilized for design problems, such as flat objects and layouts of buildings. In the first CAD systems developed in the early 1970s, 2-D systems were

TABLE 23.1 Computer-Aided Design Applied to Four of the Shigley Design Phases

Design Phase	CAD Function
3. Synthesis	Geometric modeling
4. Analysis and optimization	Engineering analysis
5. Evaluation	Design review and evaluation
6. Presentation	Automated drafting

used principally as automated drafting systems. They were often used for 3-D objects, and it was left to the designer or draftsman to properly construct the various views of the object. Three-dimensional CAD systems are capable of modeling an object in three dimensions according to user instructions. This is helpful in conceptualizing the object since the true 3-D model can be displayed in various views and from different angles.

Geometric models in CAD can also be classified as being either wire-frame models or solid models. A wire-frame model uses interconnecting lines (straight line segments) to depict the object as illustrated in Figure 23.3(a). Wire-frame models of complicated geometries can become somewhat confusing because all of the lines depicting the shape of the object are usually shown, even the lines representing the other side of the object. Techniques are available for removing these so-called hidden lines, but even with this improvement, wire-frame representation is still often confusing. In solid modeling, Figure 23.3(b), an object is modeled in solid three dimensions, providing the user with a vision of the object that is similar to the way it would be seen in real life. More important for engineering purposes, the geometric model is stored in the CAD system as a 3-D solid model, providing a more accurate representation of the object. This is useful for calculating mass properties, in assembly to perform interference checking between mating components, and in other engineering calculations.

Two other features in CAD system models are color and animation. The value of color is largely to enhance the ability of the user to visualize the object on the graphics screen. For example, the various components of an assembly can be displayed in different colors, permitting the parts to be more readily distinguished. An1imation capability permits the operation of mechanisms and other moving objects to be displayed on the graphics monitor.

Engineering Analysis.   After a particular design alternative has been developed, some form of engineering analysis often must be performed as part of the design process. The analysis may take the form of stress-strain calculations, heat transfer analysis, or dynamic simulation. The computations are often complex and time consuming, and before the advent of the digital computer, these analyses were usually greatly simplified or even omitted in the design procedure. The availability of software for engineering analysis on a CAD system greatly increases the designer's ability and willingness to perform a more thorough analysis of a proposed design. The term *computer-aided engineering* (CAE) is often used for engineering analyses performed by computer. Examples of engineering analysis software in common use on CAD systems include

- *Mass properties analysis.* This involves the computation of such features of a solid object as its volume, surface area, weight, and center of gravity. It is especially applicable in mechanical design. Prior to CAD, determination of these properties often required painstaking and time-consuming calculations by the designer.

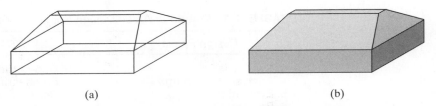

(a)   (b)

Figure 23.3 (a) Wire-frame model. (b) Solid model of the same object.

- *Interference checking.* This CAD software examines 3-D geometric models consisting of multiple components to identify interferences between the components. It is useful in analyzing mechanical assemblies, chemical plants, and similar multicomponent designs.

- *Tolerance analysis.* Software for analyzing the specified tolerances of a product's components is used (1) to assess how the tolerances may affect the product's function and performance, (2) to determine how tolerances may influence the ease or difficulty of assembling the product, and (3) to assess how variations in component dimensions may affect the overall size of the assembly.

- *Finite element analysis.* Software for finite element analysis (FEA), also known as finite element modeling (FEM), is available for use on CAD systems to aid in stress-strain, heat transfer, fluid flow, and other engineering computations. Finite element analysis is a numerical analysis technique for determining approximate solutions to physical problems described by differential equations that are very difficult or impossible to solve. In FEA, the physical object is modeled by an assemblage of discrete interconnected nodes (finite elements), and the variable of interest (e.g., stress, strain, temperature) in each node can be described by relatively simple mathematical equations. Solving the equations for each node provides the distribution of values of the variable throughout the physical object.

- *Kinematic and dynamic analysis.* Kinematic analysis studies the operation of mechanical linkages and analyzes their motions. A typical kinematic analysis specifies the motion of one or more driving members of the subject linkage, and the resulting motions of the other links are determined by the analysis package. Dynamic analysis extends kinematic analysis by including the effects of the mass of each linkage member and the resulting acceleration forces as well as any externally applied forces.

- *Discrete-event simulation.* This type of simulation is used to model complex operational systems, such as a manufacturing cell or a material handling system, as events occur at discrete moments in time and affect the status and performance of the system. For example, discrete events in the operation of a manufacturing cell include parts arriving for processing or a machine breakdown in the cell. Measures of the status and performance include the status of a given machine in the cell (idle or busy) and the overall production rate of the cell. Current discrete-event simulation software usually includes an animated graphics capability that enhances visualization of the system's operation.

Design Evaluation and Review.   Design evaluation and review procedures can be augmented by CAD. Some of the CAD features that are helpful in evaluating and reviewing a proposed design include

- *Automatic dimensioning.* These routines determine precise distance measures between surfaces on the geometric model identified by the user.

- *Error checking.* This term refers to CAD algorithms that are used to review the accuracy and consistency of dimensions and tolerances and to assess whether the proper design documentation format has been followed.

- *Animation of discrete-event simulation solutions.* Discrete-event simulation was described above in the context of engineering analysis. Displaying the solution of the discrete-event simulation in animated graphics is a helpful means of presenting and evaluating the solution. Input parameters, probability distributions, and other factors can be changed to assess their effect on the performance of the system being modeled.

- *Plant layout design scores.* A number of software packages are available for facilities design, that is, designing the floor layout and physical arrangement of equipment in a facility. Some of these packages provide one or more numerical scores for each plant layout design, which allow the user to assess the merits of the alternative with respect to material flow, closeness ratings, and similar factors.

The traditional procedure in designing a new product includes fabrication of a prototype before approval and release of the product for production. The prototype serves as the "acid test" of the design, permitting the designer and others to see, feel, operate, and test the product for any last-minute changes or enhancements of the design. The problem with building a prototype is that it is traditionally very time consuming; in some cases, months are required to make and assemble all of the parts. Motivated by the need to reduce this lead time for building the prototype, engineers have developed several new approaches that rely on the use of the geometric model of the product residing in the CAD data file. We mention two of these approaches here: (1) rapid prototyping and (2) virtual prototyping.

Rapid prototyping is a general term applied to a family of fabrication technologies that allow engineering prototypes of solid parts to be made in minimum lead time [8]. The common feature of the rapid prototyping processes is that they fabricate the part directly from the CAD geometric model. This is usually done by dividing the solid object into a series of layers of small thickness and then defining the area shape of each layer. For example, a vertical cone would be divided into a series of circular layers, the circles becoming smaller and smaller toward the vertex of the cone. The rapid prototyping processes then fabricate the object by starting at the base and building each layer on top of the preceding layer to approximate the solid shape. The fidelity of the approximation depends on the thickness of each layer. As layer thickness decreases, accuracy increases. There are a variety of layer-building processes used in rapid prototyping. The most common process, called *stereolithography*, uses a photosensitive liquid polymer that cures (solidifies) when subjected to intense light. Curing of the polymer is accomplished using a moving laser beam whose path for each layer is controlled by means of the CAD model. A solid polymer prototype of the part is built up of hardened layers, one on top of another.

Virtual prototyping, based on virtual reality technology, involves the use of the CAD geometric model to construct a digital mock-up of the product, enabling the designer and others to obtain the sensation of the real physical product without actually building the physical prototype. Virtual prototyping has been used in the automotive industry to evaluate new car style designs. The observer of the virtual prototype is able to assess the appearance of the new design even though no physical model is on display. Other applications of virtual prototyping include checking the feasibility of assembly operations, for example, parts mating, access and clearance of parts during assembly, and assembly sequence.

Automated Drafting. The fourth area where CAD is useful (step 6 in the design process) is presentation and documentation. CAD systems can be used as automated drafting machines to prepare highly accurate engineering drawings quickly. It is estimated that a CAD system increases productivity in the drafting function by about fivefold over manual preparation of drawings.

23.2 CAD SYSTEM HARDWARE

The hardware for a typical CAD system consists of the following components: (1) one or more design workstations, (2) a digital computer, (3) plotters, printers, and other output devices, and (4) storage devices. The relationship among the components is illustrated in Figure 23.4. In addition, the CAD system would have a communication interface to permit transmission of data to and from other computer systems, thus enabling the benefits of computer integration.

Design Workstations. The workstation is the interface between computer and user in the CAD system. Its functions are the following: (1) communicate with the CPU, (2) continuously generate a graphic image, (3) provide digital descriptions of the image, (4) translate user commands into operating functions, and (5) facilitate interaction between the user and the system.

The design of the CAD workstation and its available features have an important influence on the convenience, productivity, and quality of the user's output. The workstation must include a graphics display terminal and a set of user input devices. The display terminal must be capable of showing both graphics and alphanumeric text. It is the principal means by which the system communicates with the user. For optimum graphics display, the monitor should have a large color screen with high resolution. Today's CAD monitors show three-dimensional objects as two-dimensional images, using shading and brightness to provide an illusion of 3-D. They do not show true 3-D. New monitors being developed will allow designers to see the depth dimension more realistically, rather than using their imaginations to visualize it on the two-dimension screen. One system reported in Thilmany [16] uses 20 2-D slices of the scene at increasingly more distant positions inside the monitor display to give a perception of depth. The viewer sees a more faithful rendition of the three dimensions.

The user input devices permit the operator to communicate with the system. To operate the CAD system, the user must be able to (1) enter alphanumeric data, (2) enter commands to the system to perform various graphics operations, and (3) control the cursor position on the display screen. To enter alphanumeric data, the user has an alphanumeric keyboard, which can also be used to enter commands and instructions to the system. However, other input devices accomplish this function more conveniently. Special function keypads have been developed to allow entry of a command in only one or two keystrokes. These special keypads have from 10 to 50 function keys, depending on the system. However, each key provides more than one function, depending on the combination of keys pressed or the software being used. Another input device for entering commands to

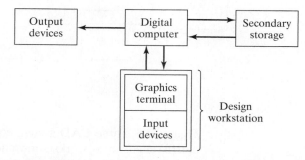

Figure 23.4 Configuration of a typical CAD system.

a CAD system is the electronic tablet, an electronically sensitive board on which an instruction set is displayed, and commands are entered using a mouse or electronic pen.

Cursor control permits the operator to position the cursor on the screen to identify a location where some function is to be executed. For example, to draw a straight line on the screen, the user can locate the cursor in sequence at the two endpoints and give the command to construct the line. Various cursor control devices are used in CAD, including mouses, joysticks, trackballs, thumbwheels, light pens, and electronic tablets.

Digital Computer. CAD applications require a digital computer with a high-speed central processing unit (CPU), math coprocessor to perform computation-intensive operations, and large internal memory. Today's commercial systems have 32-bit or 64-bit processors, which permit high-speed execution of CAD graphics and engineering analysis applications.

Several CAD system configurations are available within the general arrangement shown in Figure 23.5. Let us identify three principal configurations, illustrated in this figure: (a) host and terminal, (b) engineering workstation, and (c) CAD system based on a personal computer (PC).

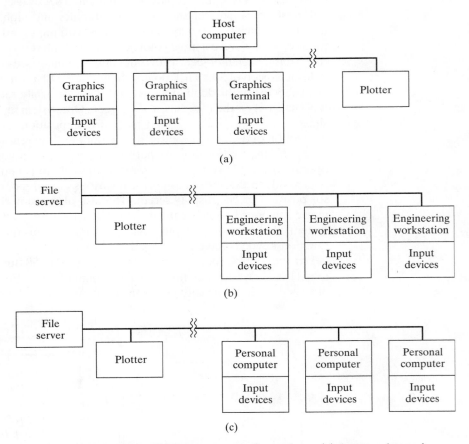

Figure 23.5 Three CAD system configurations: (a) host and terminal, (b) engineering workstation, and (c) CAD system based on a PC.

The host and terminal was the original CAD configuration in the 1970s and early 1980s when the technology was first developing. For many years, it was the only configuration available. In this arrangement, a large mainframe computer or a minicomputer serves as the host for one or more graphics terminals. These systems were expensive, each installation typically representing an investment of a million dollars or more. The powerful microprocessors and high-density memory devices that are so common today were not available at that time. The only way to meet the computational requirements for graphics processing and related CAD applications was to use a mainframe connected to multiple terminals operating on a time-sharing basis. Host and terminal CAD systems are still used today in the automotive industry and other industries in which it is deemed necessary to operate a large central data base.

An *engineering workstation* is a stand-alone computer system that is dedicated to one user and capable of executing graphics software and other programs requiring high-speed computational power. The graphics display is a high-resolution monitor with a large screen. As shown in our figure, engineering workstations are often networked to permit exchange of data files and programs between users and to share plotters and data storage devices.

A *PC-based CAD system* is a PC with a high-performance CPU and high resolution graphics display screen. The computer is equipped with a large random access memory (RAM), math coprocessor, and large-capacity hard disk for storage of the large applications software packages used for CAD. PC-based CAD systems can be networked to share files, output devices, and for other purposes. Starting around 1996, CAD software developers began offering products that utilize the excellent graphics environment of Microsoft Windows [11], thus enhancing the popularity and familiarity of PC-based CAD. Although desktop computers are most widely used, some designers use laptop PCs to accomplish their creative and analytical tasks.

Plotters and Printers. The CRT display is often the only output device physically located at the CAD workstation. There is a need to document the design on paper. The peripherals of the CAD system include one or more output devices for this purpose. Among these output devices are the following:

- *Pen plotters.* These are *x-y* plotters of various types used to produce high accuracy line drawings.

- *Electrostatic plotters.* These are faster devices based on the same technology as photocopying. The resolution of the drawings from electrostatic plotters is generally lower than those made by a pen plotter.

- *Ink-jet printers.* In these printers, images are formed by high-speed jets of ink impacting the paper. Color can be more readily included in the drawings with ink jet technology than with electrostatic printing.

Storage Devices. Storage peripherals are used in CAD systems to store programs and data files. The storage medium is usually a magnetic disk or magnetic tape. Files can be retrieved more quickly from magnetic disks, which facilitates loading and exchange of files between CPU and disk. Magnetic tape is less expensive, but more time is required to access a given file due to the sequential file storage on the tape. It is suited to disk backup, archival files, and data transfer to output devices.

23.3 CAM, CAD/CAM, AND CIM

We have briefly defined CAM, CAD/CAM, and CIM in our introduction. Let us explain and differentiate these terms more thoroughly here. Computer-integrated manufacturing (CIM) is sometimes used interchangeably with CAM and CAD/CAM. Although the terms are closely related, our assertion is that CIM possesses a broader meaning than does either CAM or CAD/CAM.

23.3.1 Computer-Aided Manufacturing

Computer-aided manufacturing (CAM) is the effective use of computer technology in manufacturing planning and control. CAM is most closely associated with functions in manufacturing engineering, such as process planning and numerical control (NC) part programming. With reference to our model of production in Section 1.1.2, the applications of CAM can be divided into two broad categories: (1) manufacturing planning and (2) manufacturing control. We cover these two categories in Chapters 24 and 25, but let us provide a brief discussion of them here to complete our definition of CAM.

Manufacturing Planning. CAM applications for manufacturing planning are those in which the computer is used indirectly to support the production function, but there is no direct connection between the computer and the process. The computer is used "off-line" to provide information for the effective planning and management of production activities. The following list surveys the important applications of CAM in this category:

- *Computer-aided process planning* (CAPP). Process planning is concerned with the preparation of route sheets that list the sequence of operations and work centers required to produce the product and its components. CAPP systems are available today to prepare these route sheets. We discuss CAPP in the following chapter.
- *Computer-assisted NC part programming.* Numerical control part programming was discussed in Chapter 7. For complex part geometries, computer-assisted part programming represents a much more efficient method of generating the control instructions for the machine tool than manual part programming.
- *Computerized machinability data systems.* One of the problems with operating a metal cutting machine tool is determining the speeds and feeds that should be used to machine a given workpart. Computer programs have been written to recommend the appropriate cutting conditions to use for different materials. The calculations are based on data that have been obtained either in the factory or laboratory that relate tool life to cutting conditions. These machinability data systems are described in Groover and Zimmers [10].
- *Computerized work standards.* The time study department has the responsibility for setting time standards on direct labor jobs performed in the factory. Establishing standards by direct time study can be a tedious and time-consuming task. There are several commercially available computer packages for setting work standards. These computer programs use standard time data that have been developed for basic work elements that comprise any manual task. The program sums the times for the individual elements required to perform a new job in order to calculate the standard time for the job. These packages are discussed in Groover [9].

- *Cost estimating.* The task of estimating the cost of a new product has been simplified in most industries by computerizing several of the key steps required to prepare the estimate. The computer is programmed to apply the appropriate labor and overhead rates to the sequence of planned operations for the components of new products. The program then adds up the individual component costs from the engineering bill of materials to determine the overall product cost.

- *Production and inventory planning.* The computer is widely used in many of the functions in production and inventory planning. These functions include maintenance of inventory records, automatic reordering of stock items when inventory is depleted, production scheduling, maintaining current priorities for the different production orders, material requirements planning, and capacity planning. We discuss these functions in Chapter 25.

- *Computer-aided line balancing.* Finding the best allocation of work elements among stations on an assembly line is a large and difficult problem if the line is of significant size. Computer programs have been developed to assist in the solution of this problem (Section 15.3).

Manufacturing Control.   The second category of CAM applications is concerned with developing computer systems to implement the manufacturing control function. Manufacturing control is concerned with managing and controlling the physical operations in the factory. These management and control areas include the following:

- *Process monitoring and control.* Process monitoring and control is concerned with observing and regulating the production equipment and manufacturing processes in the plant. We have previously discussed process control in Chapter 5. The applications of computer process control are pervasive today in automated production systems. They include transfer lines, assembly systems, NC, robotics, material handling, and flexible manufacturing systems. All of these topics have been covered in earlier chapters.

- *Quality control.* Quality control includes a variety of approaches to ensure the highest possible quality levels in the manufactured product. Quality control systems are covered in Part V.

- *Shop floor control.* Shop floor control refers to production management techniques for collecting data from factory operations and using the data to help control production and inventory in the factory. We discuss shop floor control and computerized factory data collection systems in Chapter 25.

- *Inventory control.* Inventory control is concerned with maintaining the most appropriate levels of inventory in the face of two opposing objectives: minimizing the investment and storage costs of holding inventory, and maximizing service to customers. Inventory control is discussed in Chapter 25.

- *Just-in-time production systems.* The term just-in-time (JIT) refers to a production system that is organized to deliver exactly the right number of each component to downstream workstations in the manufacturing sequence just at the time when that component is needed. JIT is one of the pillars of lean production. The term applies not only to production operations but to supplier delivery operations as well. Just-in-time systems and lean production are discussed in Chapter 26.

23.3.2 CAD/CAM

CAD/CAM is concerned with the engineering functions in both design and manufacturing. Product design, engineering analysis, and documentation of the design (e.g., drafting) are design engineering activities. Process planning, NC part programming, and other activities associated with CAM are manufacturing engineering activities. The CAD/CAM systems developed during the 1970s and early 1980s were designed primarily to address these types of engineering problems. Since then, CAM has evolved to include many other functions in manufacturing, such as material requirements planning, production scheduling, computer production monitoring, and computer process control.

It should also be noted that CAD/CAM denotes an integration of design and manufacturing activities by means of computer systems. The method of manufacturing a product is a direct function of its design. With conventional procedures practiced for so many years in industry, engineering drawings were prepared by design draftsmen and later used by manufacturing engineers to develop the process plan. The activities involved in designing the product were separated from the activities associated with process planning. Essentially a two-step procedure was employed, which was time-consuming and duplicated the efforts of design and manufacturing personnel. Using CAD/CAM technology, it is possible to establish a direct link between product design and manufacturing engineering. In effect, CAD/CAM is one of the enabling technologies for concurrent engineering (Section 24.3). It is the goal of CAD/CAM not only to automate certain phases of design and certain phases of manufacturing, but also to automate the transition from design to manufacturing. In the ideal CAD/CAM system, it is possible to take the design specification of the product as it resides in the CAD data base and convert it automatically into a process plan for making the product. A large portion of the processing might be accomplished on a numerically controlled machine tool. As part of the process plan, the NC part program is generated automatically by CAD/CAM. The CAD/CAM system downloads the NC program directly to the machine tool by means of a telecommunications network. Hence, under this arrangement, product design, NC programming, and physical production are all implemented by computer.

23.3.3 Computer-Integrated Manufacturing

Computer-integrated manufacturing includes all of the engineering functions of CAD/CAM, but it also includes the firm's business functions that are related to manufacturing. The ideal CIM system applies computer and communications technology to all the operational functions and information processing functions in manufacturing from order receipt through design and production to product shipment. The scope of CIM, compared with the more limited scope of CAD/CAM, is depicted in Figure 23.6.

The CIM concept is that all of the firm's operations related to production are incorporated in an integrated computer system to assist, augment, and automate the operations. The computer system is pervasive throughout the firm, touching all activities that support manufacturing. In this integrated computer system, the output of one activity serves as the input to the next activity, through the chain of events that starts with the sales order and culminates with shipment of the product. The components of the integrated computer system are illustrated in Figure 23.7. Customer orders are initially entered by the company's sales force or directly by the customer into a computerized order entry system. The orders contain the specifications describing the product. The specifications serve

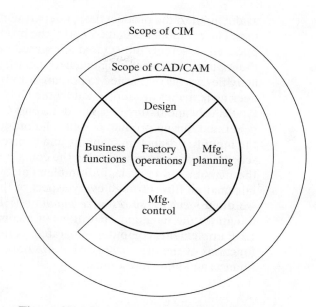

Figure 23.6 The scope of CAD/CAM and CIM.

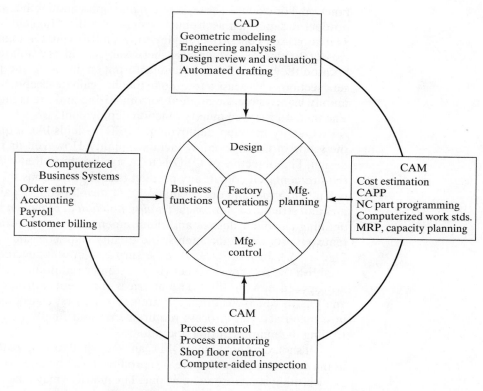

Figure 23.7 Computerized elements of a CIM system.

as the input to the product design department. New products are designed on a CAD system. The components that comprise the product are designed, the bill of materials is compiled, and assembly drawings are prepared. The output of the design department serves as the input to manufacturing engineering, where process planning, tool design, and similar activities are accomplished to prepare for production. Many of these manufacturing engineering activities are supported by the CIM system. Process planning is performed using CAPP. Tool and fixture design is done on a CAD system, making use of the product model generated during product design. The output from manufacturing engineering provides the input to production planning and control, where material requirements planning and scheduling are performed using the computer system, and so it goes, through each step in the manufacturing cycle. Full implementation of CIM results in the automation of the information flow through every aspect of the company's organization. In Section 25.6.2, we discuss *enterprise resource planning* (ERP), which refers to a software system that integrates the data and operations of a company through a central data base. In effect, ERP implements computer-integrated manufacturing. It also includes all of the business functions of the organization that are not related to manufacturing, such as accounting, finance, and human resources.

23.4 QUALITY FUNCTION DEPLOYMENT

A number of concepts and techniques have been developed to aid in the product design function. For example, several of the principles and methods of Taguchi can be applied to product design, such as "robust design" and the "Taguchi loss function" (Section 20.7). The topics of concurrent engineering and design for manufacturing are also related closely to design. We discuss these subjects in the following chapter (Section 24.3) because they also relate to manufacturing engineering and process planning. In the present section, we discuss a technique that has gained acceptance in the product design community as a systematic method for organizing and managing any given design problem. The method is called quality function deployment.

Quality function deployment (QFD) sounds like a quality-related technique, and the scope of QFD certainly includes quality. However, its principal focus is on product design. The objective of QFD is to design products that will satisfy or exceed customer requirements. Of course, any product design project has this objective, but the approach is often informal and unsystematic. QFD, developed in Japan in the mid 1960s, uses a formal and structured approach. *Quality function deployment* is a systematic procedure for defining customer desires and requirements and interpreting them in terms of product features, process requirements, and quality characteristics. The technique is outlined in Figure 23.8. In a QFD analysis, a series of interconnected matrices are developed to establish the relationships between customer requirements and the technical features of a proposed new product. The matrices represent a progression of phases in the QFD analysis, in which customer requirements are first translated into product features, then into manufacturing process requirements, and finally into quality procedures for controlling the manufacturing operations.

It should be noted that QFD can be applied to analyze the delivery of a service as well as the design and manufacture of a product. It can be used to analyze an existing product or service, not just a proposed new one. The matrices may take on different meanings depending on the product or service being analyzed. And the number of matrices used in the analysis

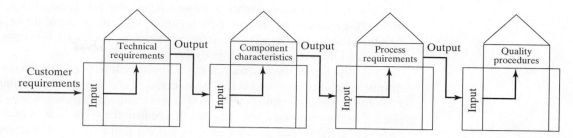

Figure 23.8 Quality function deployment, shown here as a series of matrices that relate customer requirements to successive technical requirements in a typical progression: (1) customer requirements to technical requirements of the product, (2) technical requirements of the product to component characteristics, (3) component characteristics to process requirements, and (4) process requirements to quality procedures.

may also vary, from as few as one (although a single matrix does not fully exploit the potential of QFD) to as many as 30 [4]. QFD is a general framework for analyzing product and process design problems, and it must be adapted to the given problem context.

Each matrix in QFD is similar in format and consists of six sections, as shown in Figure 23.9. On the left-hand side is section 1, a list of input requirements that serve as drivers for the current matrix of the QFD analysis. In the first matrix, these inputs are the needs and desires of the customer. The input requirements are translated into output technical requirements, listed in section 2 of the matrix. These technical requirements indicate

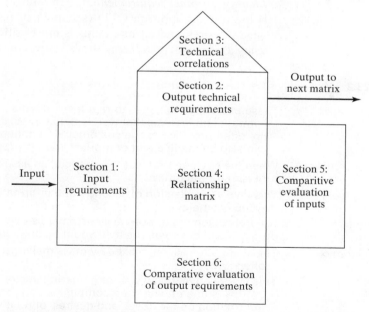

Figure 23.9 General form of each matrix in QFD, known as the *house of quality* in the starting matrix because of its shape.

how the input requirements are to be satisfied in the new product or service. In the starting matrix, they represent the product's technical features or capabilities. The output requirements in the present matrix serve as the input requirements for the next matrix, through to the final matrix in the QFD analysis.

At the top of the matrix is section 3, which depicts technical correlations among the output technical requirements. This section of the matrix uses a diagonal grid to allow each of the output requirements to be compared with all others. The shape of the grid is similar to the roof of a house, and for this reason the term "house of quality" is often used to describe the overall matrix. This term is applied only to the starting matrix in QFD by some authors [4], and the technical correlation section (the roof of the house) may be omitted in subsequent matrices in the analysis. Section 4 is called the relationship matrix; it indicates the relationships between inputs and outputs. Various symbols [1], [4], [11] have been used to define the relationships among pairs of factors in sections 3 and 4. These symbols are subsequently reduced to numerical values.

On the right-hand side of the matrix is section 5, which is used for comparative evaluation of inputs. For example, in the starting matrix, this might be used to compare the proposed new product with competing products already on the market. Finally, at the bottom of the matrix is section 6, used for comparative evaluation of output requirements. The six sections may take on slightly different interpretations for the different matrices of QFD and for different products or services, but our descriptions are adequate as generalities.

Let us illustrate the construction of the house of quality, that is, the matrix used for the first phase of QFD. This is the beginning of the analysis, in which customer requirements and needs are translated into product technical requirements. The procedure can be outlined in the following steps:

1. *Identify customer requirements.* Often referred to as the "voice of the customer," this is the primary input in QFD (section 1 in Figure 23.9). Capturing the customer's needs, desires, and requirements is most critical in the analysis. It is accomplished using a variety of possible methods, several of which are listed in Table 23.2. Selecting

TABLE 23.2 Methods of Capturing Customer Requirements

Comment cards	These allow the customer to rate level of satisfaction of the product or service and to comment on features that were either appreciated or not appreciated. Comment cards are often provided to the customer simultaneously with the product or service. They can also be made a part of product warranty registration.
Customer returns	When the customer returns the product, an associate gathers information about the reason for the return.
Field intelligence	This involves collection of second-hand information from employees who deal directly withcustomers.
Focus groups	Several customers or potential customers serve on a panel. Group dynamics may elicit opinions and observations that would be omitted in one-on-one interviews.
Formal surveys	These are often accomplished by mass mailings. Unfortunately, the response rate is often low.
Internet	This is a relatively new way of gathering customer opinions. Subject-oriented interest groups, some of which are companies and products, can be queried to obtain useful information on the needs and desires of potential customers.
Interviews	One-on-one interviews, either in person or by telephone.
Study of complaints	This allows a statistical review of data on customer complaints.

Source: Compiled from Evans and Lindsay [5], Finch [6], Goetsch and Davis [7], and other sources.

the most appropriate data collection method depends on the product or service situation. In many cases, more than one approach is necessary to identify the full scope of the customer's needs.

2. *Identify product features needed to meet customer requirements.* These are the technical requirements of the product (section 2 in Figure 23.9) corresponding to the requirements and desires expressed by the customer. In effect, these product features are the means by which the voice of the customer is satisfied. Mapping customer requirements into product features often requires ingenuity, sometimes demanding the creation of new features not previously available on competing products.

3. *Determine technical correlations among product features.* This is section 3 in Figure 23.9. The various product features will likely be related to each other. The purpose of this chart is to establish the strength of each of the relationships between pairs of product features. Instead of using symbols, as previously indicated, let us adopt the numerical ratings shown in Table 23.3 for our illustrations. These numerical scores indicate how significant (how strong) the relationship between respective pairs of requirements is.

4. *Develop relationship matrix between customer requirements and product features.* The function of the relationship matrix in the QFD analysis is to show how well the collection of product features is fulfilling individual customer requirements. Identified as section 4 in Figure 23.9, the matrix indicates the relationship between individual factors in the two lists. The numerical scores in Table 23.3 depict relationship strength.

5. *Comparative evaluation of input customer requirements.* Section 5 of the house of quality matces two comparisons. First, the relative importance of each customer requirement is evaluated using a numerical scoring scheme. High values indicate that the customer requirement is important. Low values indicate a low priority. This evaluation can be used to guide the design of the proposed new product. Second, existing competitive products are evaluated relative to customer requirements. This helps to identify possible weaknesses or strengths in competing products that might be emphasized in the new design. A numerical scoring scheme might be used as before. (See Table 23.3.)

6. *Comparative evaluation of output technical requirements.* This is section 6 in Figure 23.9. In this part of the analysis, each competing product is scored relative to the output technical requirements. Finally, target values can be established in each technical requirement for the proposed new product.

TABLE 23.3 Numerical Scores Used For Correlations and Evaluations in Sections 3, 4, 5, and 6 of the QFD Matrix

Numerical Score	Strength of Relationship in Sections 3 and 4	Relative Importance in Section 5	Merits of Competing Product in Sections 5 and 6
0	No relationship	No importance	Not applicable
1	Weak relationship	Little importance	Low score
3	Medium-to-strong relationship	Medium importance	Medium score
5	Very strong relationship	Very important	High score

At this point in the analysis, the completed matrix contains much information about which customer requirements are most important, how they relate to proposed new product features, and how competitive products compare with respect to these input and output requirements. All of this information must be assimilated and assessed in order to advance to the next step in the QFD analysis. Those customer needs and product features that are most important must be stressed as the analysis proceeds through identification of technical requirements for components, manufacturing processes, and quality control in the succeeding QFD matrices.

EXAMPLE 23.1 Quality Function Deployment: House of Quality

We are engaged in a new product design project for a child's toy for children ages 3 to 9. It is a toy that could be used in a bathtub or on the floor. We want to construct the house of quality for such a toy (the initial matrix in QFD), first listing the customer requirements that might be obtained from one or more of

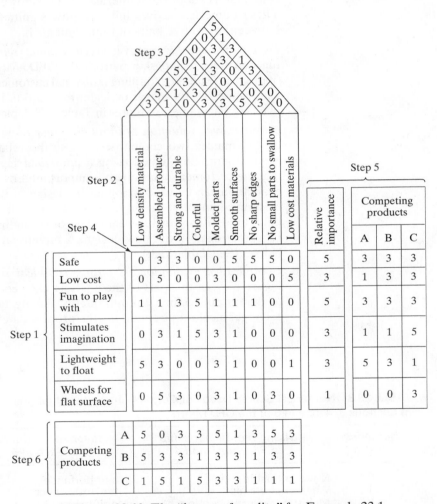

Figure 23.10 The "house of quality" for Example 23.1.

the methods listed in Table 23.2. We then want to identify the corresponding technical features of the product and develop the various correlations.

Solution: The first phase of the QFD analysis (the house of quality) is developed in Figure 23.10. Following the steps in our procedure generates the list of customer requirements in step 1 of the figure. Step 2 lists the corresponding technical features of the product that might be derived from these customer inputs. Step 3 presents the correlations among product features, and step 4 fills in the relationship matrix between customer requirements and product features. Step 5 indicates a possible comparative evaluation of customer requirements, and step 6 provides a hypothetical evaluation of competing products for the technical requirements.

REFERENCES

[1] AKAO, Y., Author and editor-in-chief, *Quality Function Deployment: Integrating Customer Requirements into Product Design*, English translation by G. H. Mazur, Productivity Press, Cambridge, MA, 1990.

[2] BAKERJIAN, R., and P. MITCHELL, *Tool and Manufacturing Engineers Handbook*, 4th ed., Volume VI, *Design for Manufacturability*, Society of Manufacturing Engineers, Dearborn, MI, 1992.

[3] BOSSERT, J. L., *Quality Function Deployment, A Practitioner's Approach*, ASQC Press, Milwaukee, WI, and Marcel Dekker Inc., NY, 1991.

[4] COHEN, L., *Quality Function Deployment*, Addison-Wesley Publishing Company, Reading, MA, 1995.

[5] EVANS, J. R., and W. M. LINDSAY, *The Management and Control of Quality*, 3d ed., West Publishing Company, St. Paul, MN, 1996.

[6] FINCH, B. J., "A New Way to Listen to the Customer," *Quality Progress*, May 1997, pp. 73–76.

[7] GOETSCH, D. L., and S. B. DAVIS, *Introduction to Total Quality*, 2d ed., Prentice Hall, Upper Saddle River, NJ, 1997.

[8] GROOVER, M. P., *Fundamentals of Modern Manufacturing: Materials, Processes, and Systems*, 3d ed., John Wiley & Sons, Inc., Hoboken, NJ, 2007.

[9] GROOVER, M. P., *Work Systems and the Methods, Measurement, and Management of Work*, Pearson/Prentice Hall, Upper Saddle River, NJ, 2007.

[10] GROOVER, M. P., and E. W. ZIMMERS, Jr., *CAD/CAM: Computer Aided Design and Manufacturing*, Prentice Hall, Inc., Englewood Cliffs, NJ, 1984.

[11] JURAN, J. M., and F. M. GRYNA, *Quality Planning and Analysis*, 3d ed., McGraw-Hill, Inc., NY, 1993.

[12] LEE, K., *Principles of CAD/CAM/CAE Systems*, Addison Wesley, Reading MA, 1999.

[13] MILLER, S., J. RICHMOND, and A. BOWMAN, "Streamlined from the Start," *Mechanical Engineering*, March 2006, pp. 30–32.

[14] SHIGLEY, J. E., and L. D. MITCHELL, *Mechanical Engineering Design*, 4th ed., McGraw-Hill Book Company, NY, 1983.

[15] THILMANY, J., "CAD meets CAE," *Mechanical Engineering*, October 1999, pp. 66–69.

[16] THILMANY, J., "Design with Depth," *Mechanical Engineering*, December 2005, pp. 32–34.

[17] THILMANY, J., "Pros and Cons of CAD," *Mechanical Engineering*, September 2006, pp. 38–40.

[18] USHER, J. M., U. ROY, and H. R. PARSAEI, Editors, *Integrated Product and Process Development*, John Wiley & Sons, Inc., NY, 1998.

[19] VAJPAYEE, S. K., *Principles of Computer-Integrated Manufacturing*, Prentice Hall, Englewood Cliffs, NJ, 1995.

REVIEW QUESTIONS

23.1 What are manufacturing support systems?

23.2 What are the six phases of the general design process?

23.3 What is computer-aided design?

23.4 Name four of the six reasons for using a CAD system to support the engineering design function.

23.5 Give some examples of engineering analysis software in common use on CAD systems.

23.6 What is rapid prototyping?

23.7 What is virtual prototyping?

23.8 What is computer-aided manufacturing?

23.9 Name four of the seven important applications of CAM in manufacturing planning.

23.10 What is the difference between CAD/CAM and CIM?

23.11 What is quality function deployment?

Chapter 24

Process Planning
and Concurrent Engineering

CHAPTER CONTENTS

The product design is the plan for the product and its components and subassemblies. A manufacturing plan is needed to convert the product design into a physical entity. The activity of developing such a plan is called process planning. It is the link between product design and manufacturing. *Process planning* involves determining the sequence of processing and assembly steps that must be accomplished to make the product. In the present chapter, we examine process planning and several related topics.

At the outset, we should distinguish between process planning and production planning, which is covered in the following chapter. Process planning is concerned with the technical details: the engineering and technological issues of how to make the product and its parts. What types of equipment and tooling are required to fabricate the parts and assemble the product? Production planning is concerned with the logistics issues of making the product: ordering the materials and obtaining the resources required to make the product in sufficient quantities to satisfy demand.

24.1 PROCESS PLANNING

Process planning consists of determining the most appropriate manufacturing and assembly processes and the sequence in which they should be accomplished to produce a given part or product according to specifications set forth in the product design documentation. The scope and variety of processes that can be planned are generally limited by the available processing equipment and technological capabilities of the company or plant. Parts that cannot be made internally must be purchased from outside vendors. The choice of processes is also limited by the details of the product design. This is a point we will return to later in Section 24.3.1.

Process planning is usually accomplished by manufacturing engineers (other titles include industrial engineers, production engineers, and process engineers). They must be familiar with the particular manufacturing processes available in the factory and be able to interpret engineering drawings. Based on the planner's knowledge, skill, and experience, the processing steps are developed in the most logical sequence to make each part. Following is a list of the many decisions and details usually included within the scope of process planning [9], [11]:

- *Interpretation of design drawings.* First, the planner must analyze the part or product design (materials, dimensions, tolerances, surface finishes, etc.).
- *Choice of processes and sequence.* The process planner must select which processes are required and their sequence, and prepare a brief description of all processing steps.
- *Choice of equipment.* In general, process planners must develop plans that utilize existing equipment in the plant. Otherwise, the company must purchase the component or invest in new equipment.
- *Choice of tools, dies, molds, fixtures,* and *gages.* The process planner must decide what tooling is required for each processing step. The actual design and fabrication of these tools is usually delegated to a tool design department and tool room, or an outside vendor specializing in that type of tooling.
- *Analysis of methods.* Workplace layout, small tools, hoists for lifting heavy parts, even in some cases hand and body motions must be specified for manual operations. The industrial engineering department is usually responsible for this area.
- *Setting of work standards.* Work measurement techniques are used to set time standards for each operation.
- *Choice of cutting tools* and *cutting conditions.* These must be specified for machining operations, often with reference to standard handbook recommendations. Similar decisions about process and equipment settings must be made for processes other than machining.

24.1.1 Process Planning for Parts

For individual parts, the processing sequence is documented on a form called a *route sheet* (sometimes known as an "operation sheet"). Just as engineering drawings are used to specify the product design, route sheets are used to specify the process plan. They are counterparts, one for product design, the other for manufacturing. A typical route sheet, illustrated in Figure 24.1, includes the following information: (1) all operations to be performed on the workpart, listed in the order in which they should be performed; (2) a brief description of each operation indicating the processing to be accomplished, with references to dimensions and tolerances on the part drawing; (3) the specific machines on which the work is to be done; and (4) any special tooling, such as dies, molds, cutting tools, jigs or fixtures, and gages. Some companies also include setup times, cycle time standards, and other data. It is called a route sheet because the processing sequence defines the route that the part must follow in the factory.

Decisions on processes to fabricate a given part are based largely on the starting material for the part. This starting material is selected by the product designer. Once the material has been specified, the range of possible processing operations is reduced considerably. The product designer's decisions on starting material are based primarily on functional requirements, although economics and ease of manufacture also play a role in the selection.

A typical processing sequence to fabricate an individual part consists of (1) a basic process, (2) secondary processes, (3) operations to enhance physical properties, and (4) finishing operations. The sequence is shown in Figure 24.2. A *basic process* determines the starting geometry of the workpart. Metal casting, plastic molding, and rolling of sheet metal are examples of basic processes. The starting geometry must often be refined by

Route Sheet		XYZ Machine Shop, Inc.				
Part no. **081099**	Part name **Shaft, generator**	Planner MPGroover	Checked by: N. Needed	Date 08/12/XX		Page 1/1
Material 1050 H18 Al	Stock size 60 mm diam., 206 mm length	Comments:				
No.	Operation description	Dept	Machine	Tooling	Setup	Std.
10	Face end (approx. 3 mm). Rough turn to 52.00 mm diam. Finish turn to 50.00 mm diam. Face and turn shoulder to 42.00 mm diam. and 15.00 mm length.	Lathe	L45	G0810	1.0 hr	5.2 min.
20	Reverse end. Face end to 200.00 mm length. Rough turn to 52.00 mm diam. Finish turn to 50.00 mm diam.	Lathe	L45	G0810	0.7 hr	3.0 min.
30	Drill 4 radial holes 7.50 mm diam.	Drill	D09	J555	0.5 hr	3.2 min.
40	Mill 6.5 mm deep x 5.00 mm wide slot.	Mill	M32	F662	0.7 hr	6.2 min.
50	Mill 10.00 mm wide flat, opposite side.	Mill	M13	F630	1.5 hr	4.8 min.

Figure 24.1 Typical route sheet for specifying the process plan.

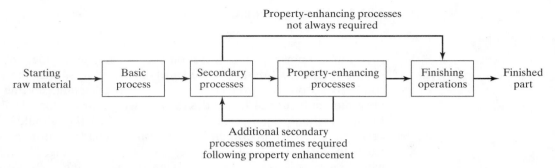

Figure 24.2 Typical sequence of processes required in part fabrication.

secondary processes, operations that transform the starting geometry into the final geometry (or close to the final geometry). The secondary processes that might be used are closely correlated to the basic process that provides the starting geometry. When sand casting is the basic process, machining operations are generally the secondary processes. When a rolling mill produces sheet-metal, stamping operations such as punching and bending are the secondary processes. When plastic injection molding is the basic process, secondary operations are often unnecessary, because most of the geometric features that would otherwise require machining can be created by the molding operation. Plastic molding and other operations that require no subsequent secondary processing are called *net shape processes*. Operations that require some minimal secondary processing, usually machining, are referred to as *near net shape processes*. Some impression die forgings are in this category. These parts can often be shaped in the forging operation (basic process) so that minimal machining (secondary processing) is required.

Once the geometry has been established, the next step for some parts is to improve their mechanical and physical properties. *Operations to enhance properties* do not alter the geometry of the part, only the physical properties. Heat-treating operations on metal parts are the most common example. Similar heating treatments are performed on glass to produce tempered glass. For most manufactured parts, these property–enhancing operations are not required in the processing sequence, as indicated by the alternative arrow path in Figure 24.2.

Finally, *finishing operations* usually provide a coating on the workpart (or assembly) surface. Examples include electroplating, thin film deposition techniques, and painting. The purpose of the coating is to enhance appearance, change color, or protect the surface from corrosion, abrasion, and other damage. Finishing operations are not required on many parts; for example, plastic moldings rarely require finishing. When finishing is required, it is usually the final step in the processing sequence.

Table 24.1 presents some typical processing sequences for common engineering materials used in manufacturing. In most cases, parts and materials arriving at the factory have completed their basic process. Thus, the first operation in the process plan follows the basic process that has provided the starting geometry of the part. For example, machined parts begin as bar stock or castings or forgings, which are purchased from outside vendors. The process plan begins with the machining operations in the company's own plant. Stampings begin as sheet-metal coils or strips bought from the rolling mill. These raw materials are supplied from outside sources so that the secondary processes,

TABLE 24.1 Some Typical Process Sequences

Basic Process	Starting Material	Secondary Processes	Final shape	Enhancing Processes	Finishing Processes
Sand casting	Sand casting	Machining	Machined part	(Optional)	Painting
Die casting	Die casting	(Net shape)	Die casting	(Optional)	Painting
Casting of glass	Glass ingot	Pressing, blow molding	Glassware	Heat treatment	(None)
Injection molding	Molded part	(Net shape)	Plastic molding	(None)	(None)
Rolling	Sheet-metal	Blanking, punching, bending, forming	Stamping	(None)	Plating, painting
Rolling	Sheet-metal	Deep drawing	Drawing	(None)	Plating, painting
Forging	Forging	(Near net shape) Machining	Machined part	(None)	Plating, painting
Rolling and bar drawing	Bar stock	Machining, grinding	Machined part	Heat treatment	Plating, painting
Extrusion of aluminum	Extrudate	Cutoff	Extruded part	(None)	Painting, anodizing
Atomize	Metal powders	Press	PM part	Sinter	Paint
Comminution	Ceramic powders	Press	Ceramic ware	Sinter	Glaze
Ingot pulling	Silicon boule	Sawing and grinding	Silicon wafer		Cleaning
Sawing and grinding	Silicon wafer	Oxidation, CVD, PVD, etching	IC chip		Coating

property-enhancing operations, and finishing operations can be performed in the company's own factory.

A detailed description of each operation is filed in the particular production department office where the operation is performed. It lists specific details of the operation, such as cutting conditions and tooling (if the operation is machining) and other instructions that may be useful to the machine operator. Sketches of the machine setup are often included with the description ("a picture is worth a thousand words"). Lean production, specifically the Toyota Production System, emphasizes the use of drawings and illustrations as communication aids (Section 26.4.2).

24.1.2 Process Planning for Assemblies

The type of assembly method used for a given product depends on factors such as (1) the anticipated production quantities; (2) complexity of the assembled product, for example, the number of distinct components; and (3) assembly processes used, for example, mechanical assembly versus welding. For a product that is to be made in relatively small quantities, assembly is generally accomplished at individual workstations where one worker or a team of workers perform all of the assembly tasks. For complex products made in medium and high quantities, assembly is usually performed on manual assembly lines (Chapter 15). For simple products of a dozen or so components, to be made in large quantities, automated assembly systems are appropriate. In any case, there is a precedence order in which the work must be accomplished, an example of which is shown in Table 15.4. The precedence requirements are sometimes portrayed graphically on a precedence diagram, as in Figure 15.5.

Process planning for assembly involves development of assembly instructions similar to the list of work elements in Table 15.4, but in more detail. For high production on an assembly line, process planning consists of allocating work elements to the individual stations of the line, a procedure called line balancing (Section 15.2.2). As in process planning for individual components, any tools and fixtures required to accomplish an assembly task must be determined, designed, and built, and the workstation arrangement must be laid out.

24.1.3 Make or Buy Decision

An important question that arises in process planning is whether a given part should be produced in the company's own factory or purchased from an outside vendor. If the company does not possess the technological equipment or expertise in the particular manufacturing processes required to make the part, then the answer is obvious: The part must be purchased because there is no internal alternative. However, in many cases, the part could either be made internally using existing equipment, or purchased externally from a vendor that possesses similar manufacturing capability.

In our discussion of the make or buy decision, it should be recognized at the outset that nearly all manufacturers buy their raw materials from suppliers. A machine shop purchases its starting bar stock from a metals distributor and its sand castings from a foundry. A plastic molding plant buys its molding compound from a chemical company. A stamping press factory purchases sheet metal either from a distributor or direct from a rolling mill. Very few companies are vertically integrated in their production operations all the way from raw materials to finished product. Given that a manufacturing company purchases some of its starting materials, it seems reasonable for the company to consider purchasing at least some of the parts that would otherwise be produced in its own plant. It is probably appropriate to ask the make or buy question for every component that is used by the company.

A number of factors enter into the make or buy decision. We have compiled a list of the factors and issues that affect the decision in Table 24.2. Cost is usually the most important factor in determining whether to produce the part or puchase it. If an outside vendor is more proficient than the company's own plant in the manufacturing processes used to make the part, then the internal production cost is likely to be greater than the purchase price even after the vendor has included a profit. However, if the decision to purchase results in idle equipment and labor in the company's own plant, then the apparent advantage of purchasing the part may be lost. Consider the following example.

EXAMPLE 24.1 Make or Buy Cost Decision

The quoted price for a certain part is $20.00 per unit for 100 units. The part can be produced in the company's own plant for $28.00. The cost components of making the part are as follows:

$$\text{Unit raw material cost} = \$8.00 \text{ per unit}$$
$$\text{Direct labor cost} = \$6.00 \text{ per unit}$$
$$\text{Labor overhead at } 150\% = \$9.00 \text{ per unit}$$
$$\text{Equipment fixed cost} = \underline{\$5.00 \text{ per unit}}$$
$$\text{Total} = \$28.00 \text{ per unit}$$

Should the component by bought or made in-house?

Solution: Although the vendor's quote seems to favor a buy decision, let us consider the possible impact on plant operations if the quote is accepted. Equipment fixed cost of $5.00 is an allocated cost based on an investment that was already made. If the equipment designated for this job is not utilized because of a decision to purchase the part, then the fixed cost continues even if the equipment stands idle. In the same way, the labor overhead cost of $9.00 consists of factory space, utility, and labor costs that remain even if the part is purchased. By this reasoning, a buy decision is not a good decision because it might cost the company as much as $20.00 + $5.00 + $9.00 = $34.00 per unit if it results in idle time on the machine that would have been used to produce the part. On the other hand, if the equipment in question can be used for the production of other parts for which the in-house costs are less than the corresponding outside quotes, then a buy decision is a good decision.

Make or buy decisions are not often as straightforward as in this example. The other factors listed in Table 24.2 also affect the decision. A trend in recent years, especially in the automobile industry, is for companies to stress the importance of building close relationships

TABLE 24.2 Factors in the Make or Buy Decision

Factor	Explanation and Effect on Make/Buy Decision
How do part costs compare?	This must be considered the most important factor in the make or buy decision. However, the cost comparison is not always clear, as Example 24.1 illustrates.
Is the process available in-house?	If the equipment and technical expertise for a given process are not available internally, then purchasing is the obvious decision. Vendors usually become very proficient in certain processes, which often makes them cost competitive in external-internal comparisons. However, there may be long-term cost implications for the company if it does not develop technological expertise in certain processes that are important for the types of products it makes.
What is the total production quantity?	The total number of units required over the life of the product is a key factor. As the total production quantity increases, this tends to favor the make decision. Lower quantities favor the buy decision.
What is the anticipated product life?	Longer product life tends to favor the make decision.
Is the component a standard item?	Standard catalog items (e.g., hardware items such as bolts, screws, nuts, and other commodity items) are produced economically by suppliers specializing in those products. Cost comparisons almost always favor a purchase decision on these standard parts.
Is the supplier reliable?	A vendor that misses a delivery on a critical component can cause a shutdown at the company's final assembly plant. Suppliers with proven delivery and quality records are favored over suppliers with lesser records.
Is the company's plant already operating at full capacity?	In peak demand periods, the company may be forced to augment its own plant capacity by purchasing a portion of the required production from external vendors.
Does the company need an alternative supply source?	Companies sometimes purchase parts from external vendors to maintain an alternative source to their own production plants. This is an attempt to ensure an uninterrupted supply of parts, e.g., as a safeguard against a wildcat strike at the company's parts production plant.

Source: Based on Groover [9] and other sources.

with parts suppliers. We will return to this issue in our later discussion of concurrent engineering (Section 24.3).

24.2 COMPUTER-AIDED PROCESS PLANNING

Manufacturing firms are very interested in automating the task of process planning using computer-aided process planning (CAPP) systems. The shop-trained people who are familiar with the details of machining and other processes are gradually retiring, and these people will be unavailable in the future to do process planning. An alternative way of accomplishing this function is needed, and CAPP systems are providing this alternative. CAPP is usually considered to be part of computer-aided manufacturing (CAM). However, this implies that CAM is a stand-alone system. In fact, a synergy results when CAM is combined with computer-aided design to create a CAD/CAM system. In such a system, CAPP becomes the direct connection between design and manufacturing. The benefits derived from computer-automated process planning include the following:

- *Process rationalization and standardization.* Automated process planning leads to more logical and consistent process plans than manual process planning. Standard plans tend to result in lower manufacturing costs and higher product quality.
- *Increased productivity of process planners.* The systematic approach and the availability of standard process plans in the data files permit more work to be accomplished by the process planners.
- *Reduced lead time for process planning.* Process planners working with a CAPP system can provide route sheets in a shorter lead time compared to manual preparation.
- *Improved legibility.* Computer-prepared route sheets are neater and easier to read than manually prepared route sheets.
- *Incorporation of other application programs.* The CAPP program can be interfaced with other application programs, such as cost estimation and work standards.

Computer-aided process planning systems are designed around two approaches: (1) retrieval CAPP systems and (2) generative CAPP systems. Some CAPP systems combine the two approaches in what is known as semi-generative CAPP [11].

24.2.1 Retrieval CAPP Systems

A retrieval CAPP system, also called a variant CAPP system, is based on the principles of group technology (GT) and parts classification and coding (Chapter 18). In this type of CAPP, a standard process plan (route sheet) is stored in computer files for each part code number. The standard route sheets are based on current part routings in use in the factory or on an ideal process plan that has been prepared for each family. Developing the data base of these process plans requires substantial effort.

A retrieval CAPP system operates as illustrated in Figure 24.3. Before the system can be used for process planning, a significant amount of information must be compiled and entered into the CAPP data files. This is what Chang et al. [4], [5] refer to as the "preparatory phase." It consists of (1) selecting an appropriate classification and coding scheme for the company, (2) forming part families for the parts produced by the company, and (3) preparing standard process plans for the part families. Steps (2) and (3) are ongoing as new parts are designed and added to the company's design data base.

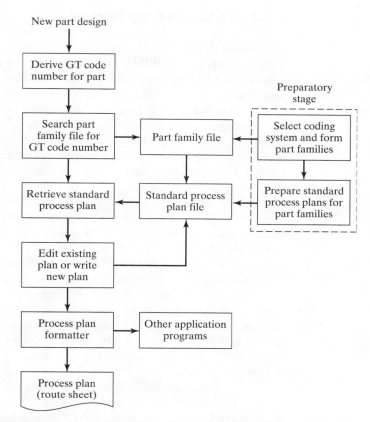

New part design

Figure 24.3 General procedure for using one of the retrieval CAPP systems.

After the preparatory phase has been completed, the system is ready for use. For a new component for which the process plan is to be determined, the first step is to derive the GT code number for the part. With this code number, the user searches the part family file to determine if a standard route sheet exists for the given part code. If the file contains a process plan for the part, it is retrieved (hence, the word "retrieval" for this CAPP system) and displayed for the user. The standard process plan is examined to determine whether any modifications are necessary. It might be that although the new part has the same code number, there are minor differences in the processes required to make it. The user edits the standard plan accordingly. This capacity to alter an existing process plan is what gives the retrieval system its alternative name: variant CAPP system.

If the file does not contain a standard process plan for the given code number, the user may search the computer file for a similar or related code number for which a standard route sheet does exist. Either by editing an existing process plan, or by starting from scratch, the user prepares the route sheet for the new part. This route sheet becomes the standard process plan for the new part code number.

The process planning session concludes with the process plan formatter, which prints out the route sheet in the proper format. The formatter may call other application programs into use, for example, to determine machining conditions for the various machine tool operations in the sequence, to calculate standard times for the operations (e.g., for direct labor incentives), or to compute cost estimates for the operations.

24.2.2 Generative CAPP Systems

Generative CAPP systems represent an alternative approach to automated process planning. Instead of retrieving and editing an existing plan contained in a computer data base, a generative system creates the process plan based on logical procedures similar to those used by a human planner. In a fully generative CAPP system, the process sequence is planned without human assistance and without a set of predefined standard plans.

Designing a generative CAPP system is usually considered part of the field of expert systems, a branch of artificial intelligence. An *expert system* is a computer program that is capable of solving complex problems that normally can only be solved by a human with years of education and experience. Process planning fits within the scope of this definition.

There are several necessary ingredients in a fully generative process planning system. First, the technical knowledge of manufacturing and the logic used by successful process planners must be captured and coded into a computer program. In an expert system applied to process planning, the knowledge and logic of the human process planners is incorporated into a so-called "knowledge base." The generative CAPP system then uses that knowledge base to solve process planning problems (i.e., create route sheets).

The second ingredient in generative process planning is a computer-compatible description of the part to be produced. This description contains all of the pertinent data and information needed to plan the process sequence. Two possible ways of providing this description are (1) the geometric model of the part that is developed on a CAD system during product design and (2) a GT code number of the part that defines the part features in significant detail.

The third ingredient in a generative CAPP system is the capability to apply the process knowledge and planning logic contained in the knowledge base to a given part description. In other words, the CAPP system uses its knowledge base to solve a specific problem—planning the process for a new part. This problem-solving procedure is referred to as the "inference engine" in the terminology of expert systems. By using its knowledge base and inference engine, the CAPP system synthesizes a new process plan from scratch for each new part it is presented.

24.3 CONCURRENT ENGINEERING AND DESIGN FOR MANUFACTURING

Concurrent engineering is an approach used in product development in which the functions of design engineering, manufacturing engineering, and other functions are integrated to reduce the elapsed time required to bring a new product to market. In the traditional approach to launching a new product, the two functions of design engineering and manufacturing engineering tend to be separated and sequential, as illustrated in Figure 24.4(a). The product design department develops the new design, sometimes without much consideration given to the manufacturing capabilities of the company. There is little opportunity for manufacturing engineers to offer advice on how the design might be altered to make it more manufacturable. It is as if a wall exists between design and manufacturing. When the design engineering department completes the design, it tosses the drawings and specifications over the wall, and only then does process planning begin.

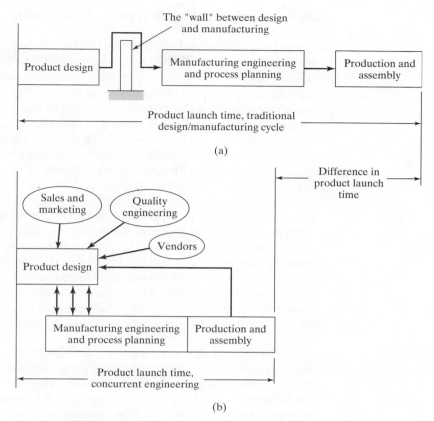

Figure 24.4 (a) Traditional product development cycle and (b) product development using concurrent engineering.

By contrast, in a company that practices concurrent engineering, the manufacturing engineering department becomes involved in the product development cycle early on, providing advice on how the product and its components can be designed to facilitate manufacture and assembly. It also proceeds with the early stages of manufacturing planning for the product. This concurrent engineering approach is pictured in Figure 24.4(b). The product development cycle also involves quality engineering, the manufacturing departments, field service, vendors supplying critical components, and in some cases the customers who will use the product. All of these groups can make contributions during product development to improve not only the new product's function and performance, but also its produceability, inspectability, testability, serviceability, and maintainability. Through early involvement, as opposed to reviewing the final product design after it is too late to conveniently make any changes, the duration of the product development cycle is substantially reduced.

Concurrent engineering includes several elements: (1) design for manufacturing and assembly, (2) design for quality, (3) design for cost, and (4) design for life cycle. In addition, certain enabling technologies such as rapid prototyping, virtual prototyping, and organizational changes are required to facilitate the concurrent engineering approach in a company.

24.3.1 Design for Manufacturing and Assembly

It has been estimated that about 70% of the life cycle cost of a product is determined by basic decisions made during product design [13]. These design decisions include the choice of material for each part, part geometry, tolerances, surface finish, how parts are organized into subassemblies, and the assembly methods to be used. Once these decisions are made, the ability to reduce the manufacturing cost of the product is limited. For example, if the product designer decides that a part is to be made of an aluminum sand casting but the part possesses features that can be achieved only by machining (such as threaded holes and close tolerances), the manufacturing engineer has no alternative except to plan a process sequence that starts with sand casting followed by the sequence of machining operations needed to achieve the specified features. In this example, a better decision might be to use a plastic molded part that can be made in a single step. It is important for the manufacturing engineer to have the opportunity to advise the design engineer as the product design is evolving, to favorably influence the manufacturability of the product.

Terms used to describe such attempts to favorably influence the manufacturability of a new product are design for manufacturing (DFM) and design for assembly (DFA). Of course, DFM and DFA are inextricably linked, so let us use the term *design for manufacturing and assembly* (DFM/A). Design for manufacturing and assembly involves the systematic consideration of manufacturability and assemblability in the development of a new product design. This includes (1) organizational changes and (2) design principles and guidelines.

Organizational Changes in DFM/A. Effective implementation of DFM/A involves making changes in a company's organizational structure, either formally or informally, so that closer interaction and better communication occurs between design and manufacturing personnel. This can be accomplished in several ways: (1) by creating project teams consisting of product designers, manufacturing engineers, and other specialties (e.g., quality engineers, material scientists) to develop the new product design; (2) by requiring design engineers to spend some career time in manufacturing to witness first-hand how manufacturability and assemblability are impacted by a product's design; and (3) by assigning manufacturing engineers to the product design department on either a temporary or full-time basis to serve as producibility consultants.

Design Principles and Guidelines. DFM/A also relies on the use of design principles and guidelines to maximize manufacturability and assemblability. Some of these are universal design guidelines that can be applied to nearly any product design situation, such as those presented in Table 24.3. In other cases, there are design principles that apply to specific processes, for example, the use of drafts or tapers in casted and molded parts to facilitate removal of the part from the mold. We leave these more process-specific guidelines to texts on manufacturing processes, such as Groover [9].

The guidelines sometimes conflict with one another. For example, one of the guidelines in Table 24.3 is to "simplify part geometry; avoid unnecessary features." But another guideline in the same table states that "special geometric features must sometimes be added to components" to design the product for foolproof assembly. And it may also be desirable to combine features of several assembled parts into one component to minimize the number of parts in the product. In these instances, a suitable compromise must be found between design for part manufacture and design for assembly.

TABLE 24.3 General Principles and Guidelines in DFM/A

Guideline	Interpretation and Advantages
Minimize number of components	Reduced assembly costs.
	Greater reliability in final product.
	Easier disassembly in maintenance and field service.
	Automation is often easier with reduced part count.
	Reduced work-in-process and inventory control problems.
	Fewer parts to purchase; reduced ordering costs.
Use standard commercially available components	Reduced design effort.
	Fewer part numbers.
	Better inventory control possible.
	Avoids design of custom-engineered components.
	Quantity discounts are possible.
Use common parts across product lines	Group technology (Chapter 18) can be applied.
	Quantity discounts are possible.
	Permits development of manufacturing cells.
Design for ease of part fabrication	Use net shape and near net shape processes where possible.
	Simplify part geometry; avoid unnecessary features.
	Avoid making surface smoother than necessary since additional processing may be needed.
Design parts with tolerances that are within process capability	Avoid tolerances less than process capability (Section 20.3.2).
	Specify bilateral tolerances.
	Otherwise, additional processing or sortation and scrap are required.
Design the product to be foolproof during assembly	Assembly should be unambiguous.
	Components designed so they can be assembled only one way.
	Special geometric features must sometimes be added to components.
Minimize flexible components	These include components made of rubber, belts, gaskets, electrical cables, etc.
	Flexible components are generally more difficult to handle.
Design for ease of assembly	Include part features such as chamfers and tapers on mating parts.
	Use base part to which other components are added.
	Use modular design (see following guideline).
	Design assembly for addition of components from one direction, usually vertically; in mass production this rule can be violated because fixed automation can be designed for multiple direction assembly.
	Avoid threaded fasteners (screws, bolts, nuts) where possible, especially when automated assembly is used; use fast assembly techniques such as snap fits and adhesive bonding.
	Minimize number of distinct fasteners.
Use modular design	Each subassembly should consists of 5-15 parts.
	Easier maintenance and field service.
	Facilitates automated (and manual) assembly.
	Reduces inventory requirements.
	Reduces final assembly time.
Shape parts and products for ease of packaging	Compatible with automated packaging equipment.
	Facilitates shipment to customer.
	Can use standard packaging cartons.
Eliminate or reduce adjustments	Many assembled products require adjustments and calibrations.
	During product design, the need for adjustments and calibrations should be minimized because they are often time consuming in assembly.

Source: Groover [9].

24.3.2 Other Product Design Objectives

To complete our coverage of concurrent engineering, let us briefly discuss the other design objectives: design for quality, cost, and life cycle.

Design for Quality. It might be argued that DFM/A is the most important component of concurrent engineering because it has the potential for the greatest impact on product cost and development time. However, the importance of quality in international competition cannot be minimized. High quality does not just happen. Procedures for achieving it must be devised during product design and process planning. Design for quality (DFQ) refers to the principles and procedures employed to ensure that the highest possible quality is designed into the product. The general objectives of DFQ are [1]: (1) to design the product to meet or exceed customer requirements; (2) to design the product to be "robust," in the sense of Taguchi (Section 20.7.1), that is, to design the product so that its function and performance are relatively insensitive to variations in manufacturing and subsequent application; and (3) to continuously improve the performance, functionality, reliability, safety, and other quality aspects of the product to provide superior value to the customer. Our discussion of quality in Part V is certainly consistent with the focus of design for quality, but our emphasis in those chapters was directed more at the operational aspects of quality during production.

Design for Product Cost. The cost of a product is a major factor in determining its commercial success. Cost affects the price charged for the product and the profit made by the company producing it. Design for product cost (DFC) refers to the efforts of a company to specifically identify how design decisions affect product costs and to develop ways to reduce cost through design. Although the objectives of DFC and DFM/A overlap to some degree, since improved manufacturability usually results in lower cost, the scope of design for product cost extends beyond manufacturing in its pursuit of cost savings. It includes costs of inspection, purchasing, distribution, inventory control, and overhead.

Design for Life Cycle. To the customer, the price paid for the product may be a small portion of its total cost when life cycle costs are considered. Design for life cycle refers to the product after it has been manufactured and includes factors ranging from product delivery to product disposal. Other life cycle factors include installability, reliability, maintainability, serviceability, and upgradeability. Some customers (e.g., the federal government) include consideration of these costs in their purchasing decisions. The producer of the product is often obliged to offer service contracts that limit customer liability for out-of-control maintenance and service costs. In these cases, accurate estimates of these life cycle costs must be included in the total product cost.

24.4 ADVANCED MANUFACTURING PLANNING

Advanced manufacturing planning emphasizes planning for the future. It is a corporate-level activity that is distinct from process planning because it is concerned with products being contemplated in the company's long-term plans (2–10-year future), rather than products currently being designed and released. Advanced manufacturing planning involves working with sales, marketing, and design engineering to forecast the new products that will be introduced and to determine what production resources will be needed to

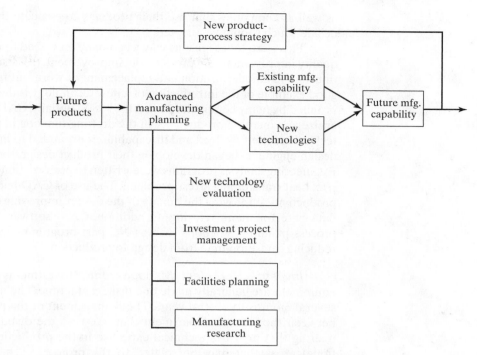

Figure 24.5 Advanced manufacturing planning cycle.

make those future products. Future products may require manufacturing technologies and facilities not currently available within the firm. In advanced manufacturing planning, the current equipment and facilities are compared with the processing needs of future planned products to determine what new facilities should be installed. The general planning cycle is portrayed in Figure 24.5. Activities in advanced manufacturing planning include (1) new technology evaluation, (2) investment project management, (3) facilities planning, and (4) manufacturing research.

New Technology Evaluation. Certainly one of the reasons a company may consider installing new technologies is because future product lines require processing methods not currently used by the company. To introduce the new products, the company must either implement new processing technologies in-house or purchase the components made by the new technologies from vendors. For strategic reasons, it may be in the company's interest to install a new technology internally and develop staff expertise in that technology as a distinctive competitive advantage for the company. These issues must be analyzed, and the processing technology itself must be evaluated to assess its merits and demerits.

A good example of the need for technology evaluation has occurred in the microelectronics industry, whose history spans only the past several decades. The technology of microelectronics has progressed very rapidly, driven by the need to include ever-greater numbers of devices in smaller and smaller packages. As each new generation has evolved, alternative technologies have been developed both in the products themselves and the required processes to fabricate them. It has been necessary for the companies in this industry, as well as companies that use their products, to evaluate the alternative technologies and decide which should be adopted.

There are other reasons why a company may need to introduce new technologies: (1) quality improvement, (2) productivity improvement, (3) cost reduction, (4) lead time reduction, and (5) modernization and replacement of worn-out facilities with new equipment. A good example of the introduction of a new technology is the CAD/CAM systems that were installed by many companies during the 1980s. Initially, CAD/CAM was introduced to modernize and increase productivity in the drafting function in product design. As CAD/CAM technology itself evolved and its capabilities expanded to include 3-D geometric modeling, design engineers began developing their product designs on these more powerful systems. Engineering analysis programs were written to perform finite-element calculations for complex heat transfer and stress problems. The use of CAD had the effect of increasing design productivity, improving the quality of the design, improving communications, and creating a data base for manufacturing. In addition, CAM software was introduced to implement process planning functions such as NC part programming (Section 7.6) and CAPP, thus reducing transition time from design to production.

Investment Project Management. Investments in new technologies or new equipment are generally made one project at a time. The duration of each project may be several months to several years. The management of the project requires a collaboration between the finance department that oversees the disbursements, manufacturing engineering that provides technical expertise in the production technology, and other functional areas that may be related to the project. Each project typically includes the following sequence of steps: (1) proposal to justify the investment is prepared, (2) management approvals are granted for the investment, (3) vendor quotations are solicited, (4) order is placed to the winning vendor, (5) vendor progress in building the equipment is monitored, (6) any special tooling and supplies are ordered, (7) the equipment is installed and debugged, (8) operators are trained, (9) responsibility for running the equipment is turned over to the operating department.

Facilities Planning. When new equipment is installed in an existing plant, the facility must be altered. Floor space must be allocated to the equipment, other equipment may need to be relocated or removed, utilities (power, heat, light, air, etc.) must be connected, safety systems must be installed if needed, and various other activities must be accomplished to complete the installation. In extreme cases, an entire new plant may need to be designed to produce a new product line or expand production of an existing line. The planning work required to renovate an existing facility or design a new one is carried out by the plant engineering department (or similar title) and is called facilities planning. In the design or redesign of a production facility, manufacturing engineering and plant engineering must work closely to achieve a successful installation.

Manufacturing Research and Development. To develop the required manufacturing technologies, the company may find it necessary to undertake a program of manufacturing research and development (R&D). Some of this research is done internally; in other cases projects are contracted to university and commercial research laboratories specializing in the associated technologies. Manufacturing research can take various forms, including the following:

- *Development of new processing technologies.* This R&D activity involves the development of new processes that have never been used before. Some of the processing technologies developed for integrated circuits fabrication fall into this category. Other recent examples include rapid prototyping techniques (Section 23.1.2).

- *Adaptation of existing processing technologies.* A manufacturing process may exist that has never been used on the type of products made by the company, yet it is perceived that there is a potential for application. In this case, the company must engage in applied research to customize the process to its needs.

- *Process fine-tuning.* This involves research on processes used by the company. The objectives of a given study can be any of the following: (1) improve operating efficiency, (2) improve product quality, (3) develop a process model, (4) achieve better control of the process, or (5) determine optimum operating conditions.

- *Software systems development.* These are projects involving development of customized manufacturing-related software for the company. Possible software development projects might include cost estimation software, parts classification and coding systems, CAPP, customized CAD/CAM application software, production planning and control systems, work-in-process tracking systems, and similar projects. Successful development of a good software package may give the company a competitive advantage.

- *Automation systems development.* These projects are similar to the preceding except they deal with hardware or hardware/software combinations. Studies related to applications of industrial robots (Chapter 8) in the company are examples of this kind of research.

- *Operations research and simulation.* Operations research involves the development of mathematical models to analyze operational problems. The techniques include linear programming, inventory models, queuing theory, and stochastic processes. In many problems, the mathematical models are too complex to be solved in closed form. In these cases, discrete event simulation can be used to study the operations. A number of commercial simulation packages are available for this purpose.

Manufacturing R&D is applied research. The objective is to develop or adapt a technology or technique that will result in higher profits and a distinctive competitive advantage for the company.

REFERENCES

[1] BAKERJIAN, R., and P. MITCHELL , *Tool and Manufacturing Engineers Handbook*, 4th ed., Volume VI, *Design for Manufacturability*, Society of Manufacturing Engineers, Dearborn, MI, 1992.

[2] BANCROFT, C. E., "Design for Manufacturability: Half Speed Ahead," *Manufacturing Engineering*, September 1988, pp 67–69.

[3] BOOTHROYD, G., P. DEWHURST, and W. KNIGHT, *Product Design for Manufacture and Assembly*, 2d ed., Marcel Dekker, NY, 2001.

[4] CHANG, T-C. and R. A. WYSK, *An Introduction to Automated Process Planning Systems*, Prentice Hall, Inc., Englewood Cliffs, NJ, 1985.

[5] CHANG, T-C, R. A. WYSK, and H. P. WANG, *Computer-Aided Manufacturing*, 3d ed., Pearson/Prentice Hall, Upper Saddle River, NJ, 2006.

[6] EARY, D. F., and G. E. JOHNSON, *Process Engineering for Manufacturing*, Prentice-Hall, Inc., Englewood Cliffs, NJ, 1962.

[7] FELCH, R. I., "Make-or-Buy Decisions," *Maynard's Industrial Engineering Handbook*, 4th ed., William K. Hodson (ed.), McGraw-Hill. Inc., NY, 1992, pp. 9.121–9.127.

[8] GROOVER, M. P., "Computer-Aided Process Planning—An Introduction," *Proceedings*, Conference on Computer-Aided Process Planning, Provo, UT, October 1984.

[9] GROOVER, M. P., *Fundamentals of Modern Manufacturing: Materials, Processes, and Systems*, 3d ed., John Wiley & Sons, Inc., Hoboken, NJ, 2007.

[10] GROOVER, M. P., and E. W. ZIMMERS, Jr., *CAD/CAM: Computer-Aided Design and Manufacturing*, Prentice Hall, Englewood Cliffs, NJ, 1984.

[11] KAMRANI, A. K., P. SFERRO, and J. HANDLEMAN, "Critical Issues in Design and Evaluation of Computer-Aided Process Planning," *Computers & Industrial Engineering*, Volume 29, No. 1–4, pp. 619–623, 1995.

[12] KUSIAK, A. (ed.), *Concurrent Engineering*, John Wiley & Sons, Inc., NY, 1993.

[13] NEVINS, J. L., and D. E. WHITNEY (eds.), *Concurrent Design of Products and Processes*, McGraw-Hill Publishing Company, NY, 1989.

[14] PARSAEI, H. R., and W. G. SULLIVAN (eds.), *Concurrent Engineering*, Chapman & Hall, London, UK, 1993.

[15] TANNER, J. P., *Manufacturing Engineering*, Marcel Dekker, Inc., NY, 1985.

[16] TOMPKINS, J. A., J. A. WHITE, Y. A. BOZER, and J. M. A. TANCHOCO, *Facilities Planning*, 3d ed., John Wiley & Sons, Inc., Hoboken, NJ, 2003.

[17] WANG, H. P., and J. K. LI, *Computer-Aided Process Planning*, Elsevier, Amsterdam, The Netherlands, 1991.

REVIEW QUESTIONS

24.1 What is process planning?

24.2 Name four of the seven decisions and details that are usually included within the scope of process planning.

24.3 What is the name of the document that lists the process sequence in process planning?

24.4 A typical process sequence for a manufactured part consists of four types of operations. Name and briefly describe them.

24.5 What is a net shape process?

24.6 Name five of the eight factors that influence the make-or-buy decision.

24.7 Name three of the five benefits derived from computer-aided process planning.

24.8 Briefly describe the two basic approaches in computer-aided process planning.

24.9 What is concurrent engineering?

24.10 Design for Manufacturing and Assembly (DFM/A) includes two aspects: (1) organizational changes and (2) design principles and guidelines. Identify two of the organizational changes that might be made in implementing DFM/A.

24.11 Name five of the eleven universal design guidelines in DFM/A (Table 24.3).

24.12 Name three of the four categories of activities often included within the scope of advanced manufacturing planning.

Production Planning and Control Systems

CHAPTER CONTENTS

Production planning and control (PPC) is concerned with the logistics problems that are encountered in manufacturing, that is, managing the details of what and how many products to produce and when, and obtaining the raw materials, parts, and resources to produce those products. PPC solves these logistics problems by managing information. Computers are essential for processing the tremendous amounts of data involved to

From Chapter 25 of *Automation, Production Systems, and Computer-Integrated Manufacturing*, Third Edition.
Mikell P. Groover. Copyright © 2008 by Pearson Education, Inc. Publishing as Prentice Hall. All rights reserved.

define the products and the manufacturing resources to produce them, and for reconciling these technical details with the desired production schedule. In a very real sense, PPC is the integrator in computer-integrated manufacturing.

Planning and control in PPC must themselves be integrated functions. It is insufficient to plan production if there is no control of the factory resources to achieve the plan. And it is ineffective to control production if there is no plan with which to compare factory progress. Both planning and control must be accomplished, and they must be coordinated with each other and with other functions in the manufacturing firm, such as process planning, concurrent engineering, and advanced manufacturing planning (Chapter 24). Now that we have emphasized the integrated nature of PPC, let us nevertheless try to explain what is involved in each of the two functions, production planning and production control.

Production planning consists of (1) deciding which products to make, in what quantities, and when they should be completed; (2) scheduling the delivery and/or production of the parts and products; and (3) planning the manpower and equipment resources needed to accomplish the production plan. Activities within the scope of production planning include

- *Aggregate production planning.* This involves planning the production output levels for major product lines produced by the firm. These plans must be coordinated among various functions in the firm, including product design, production, marketing, and sales.
- *Master production planning.* The aggregate production plan must be converted into a *master production schedule* (MPS) which is a specific plan of the quantities to be produced of individual models within each product line.
- *Material requirements planning* (MRP). MRP is a planning technique, usually implemented by computer, that translates the MPS of end products into a detailed schedule for the raw materials and parts used in those end products.
- *Capacity planning.* This is concerned with determining the labor and equipment resources needed to achieve the master schedule.

Production planning activities divide into two stages: (1) aggregate planning, which results in the MPS, and (2) detailed planning, which includes MRP and capacity planning. Aggregate planning involves planning six months or more into the future, whereas detailed planning is concerned with the shorter term (weeks to months).

Production control consists of determining whether the necessary resources to implement the production plan have been provided, and if not, attempting to take corrective action to address the deficiencies. As its name suggests, production control includes various systems and techniques for controlling production and inventory in the factory. The major topics covered in this chapter are

- *Shop floor control.* Shop floor control systems compare the actual progress and status of production orders in the factory with the production plans (MPS and MRP schedule).
- *Manufacturing execution systems.* A manufacturing execution system (MES) is a computerized system that accomplishes shop floor control using automated data collection techniques.

- *Inventory control.* Inventory control includes a variety of techniques for managing the inventory of a firm. One of the important tools in inventory control is the economic order quantity formula.
- *Manufacturing resource planning.* Also known as MRP II, manufacturing resource planning combines MRP and capacity planning as well as shop floor control and other functions related to PPC.
- *Enterprise resource planning.* Abbreviated ERP, this is an extension of MRP II that includes all of the functions of the organization, including those unrelated to manufacturing.

The activities in a modern PPC system and their interrelationships are depicted in Figure 25.1. As the figure indicates, PPC ultimately extends to the company's supplier base and customer base. This expanded scope of PPC control is known as supply chain management.

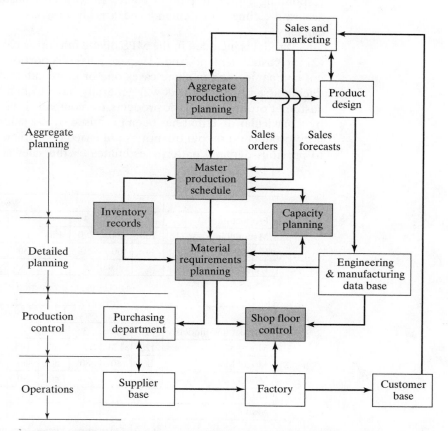

Figure 25.1 Activities in a PPC system (highlighted in the diagram) and their relationships with other functions in the firm and outside.

25.1 AGGREGATE PRODUCTION PLANNING AND THE MASTER PRODUCTION SCHEDULE

Aggregate planning is a high-level corporate planning activity. The *aggregate production plan* indicates production output levels for the major product lines of the company. The aggregate plan must be coordinated with the plans of the sales and marketing departments. Because the aggregate production plan includes products that are currently in production, it must also consider the present and future inventory levels of those products and their component parts. Because new products currently being developed will also be included in the aggregate plan, the marketing plans and promotions for current products and new products must be reconciled against the total capacity resources available to the company.

The production quantities of the major product lines listed in the aggregate plan must be converted into a very specific schedule of individual products, known as the *master production schedule* (MPS). It is a list of the products to be manufactured, when they should be completed and delivered, and in what quantities. A hypothetical MPS for a narrow product set is presented in Figure 25.2(b), showing how it is derived from the corresponding aggregate plan in Figure 25.2(a). The master schedule must be based on an accurate estimate of demand and a realistic assessment of the company's production capacity.

Products included in the MPS divide into three categories: (1) firm customer orders, (2) forecasted demand, and (3) spare parts. Proportions in each category vary for different companies, and in some cases one or more categories are omitted. Companies producing assembled products will generally have to handle all three types. In the case of customer orders for specific products, the company is usually obligated to deliver the item by a particular date that has been promised by the sales department. In the second category, production output quantities are based on statistical forecasting techniques applied to previous demand patterns, estimates by the sales staff, and other sources. For many

Product line	Week									
	1	2	3	4	5	6	7	8	9	10
M model line	200	200	200	150	150	120	120	100	100	100
N model line	80	60	50	40	30	20	10			
P model line							70	130	25	100

(a) Aggregate production plan

Product line models	Week									
	1	2	3	4	5	6	7	8	9	10
Model M3	120	120	120	100	100	80	80	70	70	70
Model M4	80	80	80	50	50	40	40	30	30	30
Model N8	80	60	50	40	30	20	10			
Model P1								50		100
Model P2							70	80	25	

(b) Master production schedule

Figure 25.2 (a) Aggregate production plan and (b) corresponding MPS for a hypothetical product line.

companies, forecasted demand constitutes the largest portion of the master schedule. The third category consists of repair parts that either will be stocked in the company's service department or sent directly to the customer. Some companies exclude this third category from the master schedule since it does not represent end products.

The MPS is generally considered to be a medium-range plan since it must take into account the lead times to order raw materials and components, produce parts in the factory, and then assemble the end products. Depending on the product, the lead times can range from several weeks to many months; in some cases, more than a year. The MPS is usually considered to be fixed in the near term. This means that changes are not allowed within about a six week horizon because of the difficulty in adjusting production schedules within such a short period. However, schedule adjustments are allowed beyond six weeks to cope with changing demand patterns or the introduction of new products. Accordingly, we should note that the aggregate production plan is not the only input to the master schedule. Other inputs that may cause the master schedule to depart from the aggregate plan include new customer orders and changes in sales forecast over the near term.

25.2 MATERIAL REQUIREMENTS PLANNING

Material requirements planning (MRP) is a computational technique that converts the master schedule for end products into a detailed schedule for the raw materials and components used in the end products. The detailed schedule identifies the quantities of each raw material and component item. It also indicates when each item must be ordered and delivered to meet the master schedule for final products. MRP is often thought of as a method of inventory control. It is both an effective tool for minimizing unnecessary inventory investment and a useful method in production scheduling and purchasing of materials.

The distinction between independent demand and dependent demand is important in MRP. *Independent demand* means that demand for a product is unrelated to demand for other items. Final products and spare parts are examples of items whose demand is independent. Independent demand patterns must usually be forecasted. *Dependent demand* means that demand for the item is directly related to the demand for some other item, usually a final product. The dependency usually derives from the fact that the item is a component of the other product. Component parts, raw materials, and subassemblies are examples of items subject to dependent demand.

Whereas demand for the firm's end products must often be forecasted, the raw materials and component parts used in the end products should not be forecasted. Once the delivery schedule for end products is established, the requirements for components and raw materials can be directly calculated. For example, even though demand for automobiles in a given month can only be forecasted, once the quantity is established and production is scheduled, we know that five tires will be needed to deliver the car (don't forget the spare). MRP is the appropriate technique for determining quantities of dependent demand items. These items constitute the inventory of manufacturing: raw materials, work-in-process (WIP), component parts, and subassemblies. That is why MRP is such a powerful technique in the planning and control of manufacturing inventories. For independent demand items, inventory control is often accomplished using order point systems, described in Section 25.5.1.

The concept of MRP is relatively straightforward. Its implementation is complicated by the sheer magnitude of the data to be processed. The master schedule provides the overall production plan for the final products in terms of month-by-month deliveries. Each product may contain hundreds of individual components. These components are produced from raw materials, some of which are common among the components. For example, several components may be made out of the same gauge sheet steel. The components are assembled into simple subassemblies, and these subassemblies are put together into more complex subassemblies, and so on, until the final products are assembled. Each step in the manufacturing and assembly sequence takes time. All of these factors must be incorporated into the MRP calculations. Although each calculation is uncomplicated, the magnitude of the data is so large that the application of MRP is practically impossible except by computer processing.

In our discussion of MRP that follows, we first examine the inputs to the MRP system. We then describe how MRP works, the output reports generated by the MRP computations, and finally the benefits and pitfalls that have been experienced with MRP systems in industry.

25.2.1 Inputs to the MRP System

To function, the MRP program needs data contained in several files. These files serve as inputs to the MRP processor. They are (1) the master production schedule, (2) the bill of materials file and other engineering and manufacturing data, and (3) the inventory record file. Figure 25.3 illustrates the flow of data into the MRP processor and its conversion into useful output reports. In a properly implemented MRP system, capacity planning also provides input to ensure that the MRP schedule does not exceed the production capacity of the firm. This concept is elaborated in Section 25.3.

The MPS lists what end products and how many of each are to be produced and when they are to be ready for shipment, as shown in Figure 25.2(b). Manufacturing firms generally work on monthly delivery schedules, but the master schedule in our figure uses weeks as the time periods. Whatever the duration, these time periods are called *time buckets*

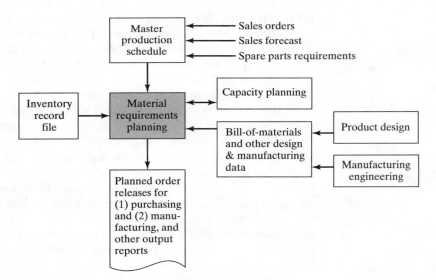

Figure 25.3 Structure of an MRP system.

in MRP. Instead of treating time as a continuous variable (which of course, it is), MRP makes its computations of materials and parts requirements in terms of time buckets.

The *bill of materials* (BOM) *file* provides information on the product structure by listing the component parts and subassemblies that make up each product. It is used to compute the raw material and component requirements for end products listed in the master schedule. The structure of an assembled product can be illustrated as in Figure 25.4. This is much simpler than most commercial products, but its simplicity will serve for illustration purposes. Product P1 is composed of two subassemblies, S1 and S2, each of which is made up of components C1, C2, and C3, and C4, C5, and C6, respectively. Finally, at the bottom level are the raw materials that go into each component. The items at each successively higher level are called the *parents* of the items feeding into it from below. For example, S1 is the parent of C1, C2, and C3. The product structure must also specify the number of each subassembly, component, and raw material that go into its respective parent. These numbers are shown in parentheses in our figure.

The inventory record file is referred to as the item master file in a computerized inventory system. The types of data contained in the inventory record are divided into three segments:

1. *Item master data.* This provides the item's identification (part number) and other data about the part such as order quantity and lead times.
2. *Inventory status.* This gives a time-phased record of inventory status. In MRP, it is important to know not only the current level of inventory, but also any future changes that will occur against the inventory. Therefore, the inventory status segment lists the gross requirements for the item, scheduled receipts, on-hand status, and planned order releases, as shown in Figure 25.5.
3. *Subsidiary data.* The third file segment provides subsidiary data such as purchase orders, scrap or rejects, and engineering changes.

25.2.2 How MRP Works

The MRP processor operates on data contained in the MPS, the BOM file, and the inventory record file. The master schedule specifies the period-by-period list of final products required, the BOM defines what materials and components are needed for each product,

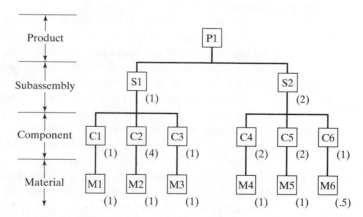

Figure 25.4 Product structure for product P1.

Period		1	2	3	4	5	6	7
Item: **Raw material M4**								
Gross requirements								
Scheduled receipts				40				
On hand	50	50	50	90				
Net requirements								
Planned order releases								

Figure 25.5 Initial inventory status of material M4 in Example 25.2.

and the inventory record file gives the current and future inventory status of each product, component, and material. The MRP processor computes how many of each component and raw material are needed each period by "exploding" the end product requirements into successively lower levels in the product structure.

EXAMPLE 25.1   MRP Gross Quantity Computations

In the master schedule of Figure 25.2, 50 units of product P1 are to be completed in week 8. Explode this product requirement into the corresponding number of subassemblies and components required.

Solution: Referring to the product structure in Figure 25.4, 50 units of P1 explode into 50 units of S1 and 100 units of S2. Similarly, the requirements for these subassemblies explode into 50 units of C1, 200 of C2, 50 of C3, 200 of C4, 200 of C5, and 100 of C6. Quantities of raw materials are determined in a similar manner.

Several complicating factors must be considered during the MRP computations. First, the quantities of components and subassemblies listed in the solution of Example 25.1 do not account for any of those items that may already be stocked in inventory or are expected to be received as future orders. Accordingly, the computed quantities must be adjusted for any inventories on hand or on order, a procedure called *netting*. For each time bucket, net requirements = gross requirements less on-hand inventories and less quantities on order.

Second, quantities of common use items must be combined during parts explosion to determine the total quantities required for each component and raw material in the schedule. *Common use items* are raw materials and components that are used on more than one product. MRP collects these common use items from different products to achieve economies in ordering the raw materials and producing the components.

Third, lead times for each item must be taken into account. The *lead time* for a job is the time that must be allowed to complete the job from start to finish. There are two kinds of lead times in MRP: ordering lead times and manufacturing lead times. *Ordering lead time* for an item is the time required from initiation of the purchase requisition to receipt of the item from the vendor. If the item is a raw material that is stocked by the vendor, the ordering lead time should be relatively short, perhaps a few days or a few weeks. If the item is fabricated, the lead time may be substantial, perhaps several months. *Manufacturing lead time* is the time required to produce the item in the company's own plant, from order release to completion, once the raw materials for the item are available. The scheduled delivery of end products must be translated into time-phased requirements for components and materials by factoring in the ordering and manufacturing lead times.

EXAMPLE 25.2 MRP Time-Phased Quantity Requirements

To illustrate these various complicating factors, let us consider the MRP procedure for component C4, which is used in product P1. This part also happens to be used on product P2 of the master schedule in Figure 25.2. The product structure for P2 is shown in Figure 25.6. Component C4 is made out of material M4, one unit of M4 for each unit of C4, and the inventory status of M4 is given in Figure 25.5. The lead times and inventory status for each of the other items needed in the MRP calculations are shown in the table below. Complete the MRP calculations to determine the time-phased requirements for items S2, S3, C4, and M4, based on the requirements for P1 and P2 given in the MPS of Figure 25.2. We assume that the inventory on hand or on order for P1, P2, S2, S3, and C4 is zero for all future periods except for the calculated values in this problem solution.

Item	Lead Time	Inventory
P1	Assembly lead time = 1 wk	No inventory on hand or on order
P2	Assembly lead time = 1 wk	No inventory on hand or on order
S2	Assembly lead time = 1 wk	No inventory on hand or on order
S3	Assembly lead time = 1 wk	No inventory on hand or on order
C4	Manufacturing lead time = 2 wk	No inventory on hand or on order
M4	Ordering lead time = 3 wk	See Figure 25.5.

Solution: The results of the MRP calculations are given in Figure 25.7. The delivery requirements for P1 and P2 must be offset by their 1 wk assembly lead time to obtain the planned order releases. These quantities are then exploded into requirements for subassemblies S2 (for P1) and S3 (for P2). These requirements

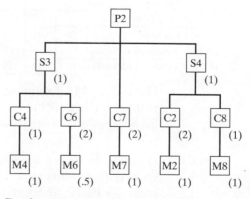

Figure 25.6 Product structure for product P2.

Period	1	2	3	4	5	6	7	8	9	10
Item: Product P1										
Gross requirements								50		100
Scheduled receipts										
On hand	0									
Net requirements								50		100
Planned order releases							50		100	
Item: Product P2										
Gross requirements							70	80	25	
Scheduled receipts										
On hand	0									
Net requirements							70	80	25	
Planned order releases						70	80	25		
Item: Subassembly S2										
Gross requirements							100		200	
Scheduled receipts										
On hand	0									
Net requirements							100		200	
Planned order releases						100		200		
Item: Subassembly S3										
Gross requirements							70	80	25	
Scheduled receipts										
On hand	0									
Net requirements							70	80	25	
Planned order releases						70	80	25		
Item: Component C4										
Gross requirements						70	280	25	400	
Scheduled receipts										
On hand	0									
Net requirements						70	280	25	400	
Planned order releases				70	280	25	400			
Item: Raw material M4										
Gross requirements				70	280	25	400			
Scheduled receipts				40						
On hand	50	50	50	90	20					
Net requirements				−20	260	25	400			
Planned order releases		260	25	400						

Figure 25.7 MRP solution to Example 25.2. Time-phased requirements for P1 and P2 are taken from Figure 25.2. Requirements for S2, S3, C4, and M4 are calculated.

are offset by their 1 wk assembly lead time and combined in week 6 to obtain gross requirements for component C4. Net requirements equal gross requirements for P1, P2, S2, S3, and C4 because of no inventory on hand and no planned orders. We see the effect of current inventory and planned orders in the time-phased inventory status of M4. The on-hand stock of 50 units plus scheduled receipts of 40 are used to meet gross requirements of 70 units of M4 in week 3, with 20 units remaining that can be applied to the gross requirements of 280 units in week 4. Net requirements in week 4 are therefore 260 units. With an ordering lead time of 3 wk, the order release for 260 units must be planned for week 1.

25.2.3 MRP Outputs and Benefits

The MRP program generates a variety of outputs that can be used in planning and managing plant operations. The outputs include (1) planned order releases, which provide the authority to place orders that have been planned by the MRP system; (2) reports of planned order releases in future periods; (3) rescheduling notices, indicating changes in due dates for open orders; (4) cancelation notices, indicating that certain open orders have been canceled because of changes in the MPS; (5) reports on inventory status; (6) performance reports of various types, indicating costs, item usage, actual versus planned lead times, and so on; (7) exception reports, showing deviations from the schedule, orders that are overdue, scrap, and so on; and (8) inventory forecasts, indicating projected inventory levels in future periods.

Of the MRP outputs listed above, the planned order releases are the most important because they drive the production system. Planned order releases are of two kinds, purchase orders and work orders. *Purchase orders* provide the authority to purchase raw materials or parts from outside vendors, with quantities and delivery dates specified. *Work orders* generate the authority to produce parts or assemble subassemblies or products in the company's own factory. Again, quantities to be completed and completion dates are specified.

Benefits reported by users of MRP systems include the following: (1) reduction in inventory, (2) quicker response to changes in demand than is possible with a manual requirements planning system, (3) reduced setup and product changeover costs, (4) better machine utilization, (5) improved capacity to respond to changes in the master schedule, and (6) aid in developing the master schedule.

Notwithstanding these claimed benefits, the success rate in implementing MRP systems throughout industry has been less than perfect. Some MRP systems have not been successful because (1) the application was not appropriate, usually because the product structure did not fit the data requirements of MRP; (2) the MRP computations were based on inaccurate data; and (3) the MPS was not coupled with a capacity planning system, so the MRP program generated an unrealistic schedule of work orders that overloaded the factory.

25.3 CAPACITY PLANNING

The original MRP systems that were developed in the 1970s created schedules that were not necessarily consistent with the production capabilities and limitations of the plants that were to produce the products. In many instances, the MRP system developed the detailed schedule based on a master production schedule that was unrealistic. A successful production schedule must consider production capacity. In cases where current capacity is inadequate, the firm must make plans for changes in capacity to meet the changing production requirements specified in the schedule. In Chapter 3, we defined production capacity and formulated equations to determine the capacity of a plant. *Capacity planning* consists of determining what labor and equipment resources are required to meet the current MPS as well as long-term future production needs of the firm (see Advanced Manufacturing Planning, Section 24.4). Capacity planning also identifies the limitations of the available production resources to prevent the MRP program from planning an unrealistic master schedule.

Capacity planning is typically accomplished in two stages, as indicated in Figure 25.8: first, when the MPS is established; and second, when the MRP computations are done. In the MPS stage, a *rough-cut capacity planning* (RCCP) calculation is made to assess the feasibility of the master schedule. Such a calculation indicates whether there is a significant violation of production capacity in the MPS. On the other hand, if the calculation shows no capacity violation, neither does it guarantee that the production schedule can be met. This depends on the allocation of work orders to specific work cells in the plant. Accordingly, a second capacity calculation is made at the time the MRP schedule is prepared. Called *capacity requirements planning* (CRP), this detailed calculation determines whether there is sufficient production capacity in the individual departments and in the work cells to complete the specific parts and assemblies that have been scheduled by MRP. If the schedule is not compatible with capacity, then either the plant capacity or the master schedule must be adjusted.

Capacity adjustments can be divided into short-term adjustments and long-term adjustments. Capacity adjustments for the short term include the following:

- *Employment levels.* Employment in the plant can be increased or decreased in response to changes in capacity requirements.
- *Number of temporary workers.* Increases in employment level can also be achieved by using workers from a temporary agency. When the busy period is passed, these workers move to positions at other companies where their services are needed.
- *Number of work shifts.* The number of shifts worked per production period can be increased or decreased.
- *Number of labor hours.* The number of labor hours per shift can be increased or decreased, through the use of overtime or reduced hours.
- *Inventory stockpiling.* This tactic might be used to maintain steady employment levels during slow demand periods.
- *Order backlogs.* Deliveries of the product to the customer could be delayed during busy periods when production resources are insufficient to keep up with demand.
- *Workload through subcontracting.* This involves the letting of jobs to other shops during busy periods, or the taking in of extra work during slack periods.

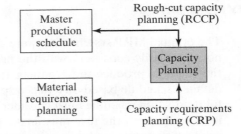

Figure 25.8 Two stages of capacity planning.

Capacity planning adjustments for the long term include changes in production capacity that generally require long lead times. These adjustments include the following actions:

- *Investing in new equipment.* This involves investing in more machines or more productive machines to meet increased future production requirements, or investing in new types of machines to match future changes in product design.
- *Constructing new plants.* Building a new factory represents a major investment for the company. However, it also represents a significant increase in production capacity for the firm.
- *Purchasing existing plants* from other companies.
- *Acquiring existing companies.* This may be done to increase productive capacity. However, there are usually more important reasons for taking over an existing company, such as to achieve economies of scale that result from increasing market share and reducing staff.
- *Closing plants.* This involves the closing of plants that will not be needed in the future.

Many of these capacity adjustments are suggested by the capacity equations and models presented in Chapter 3.

25.4 SHOP FLOOR CONTROL

Shop floor control (SFC) is the set of activities in production control that is concerned with releasing production orders to the factory, monitoring and controlling the progress of the orders through the various work centers, and acquiring current information on the status of the orders. A typical SFC system consists of three phases: (1) order release, (2) order scheduling, and (3) order progress. The three phases and their connections to other functions in the production management system are pictured in Figure 25.9. In modern implementations of shop floor control, these phases are executed by a combination of computer and human resources, with a growing proportion accomplished by computer-automated methods. The term *manufacturing execution system* (MES) is used to denote the computer software that supports SFC and that typically includes the capability to respond to on-line inquiries concerning the status of each of the three phases. Other functions often included in an MES are generation of process instructions, real-time inventory control, machine and tool status monitoring, and labor tracking. In addition, an MES usually provides links to other modules in the firm's information system, such as quality control, maintenance, and product design data.

25.4.1 Order Release

The order release phase of shop floor control provides the documentation needed to process a production order through the factory. The collection of documents is sometimes called the *shop packet*. It consists of (1) the route sheet, which documents the process plan for the item to be produced, (2) material requisitions to draw the necessary raw materials from inventory, (3) job cards or other means to report direct labor time devoted to the order and to indicate progress of the order through the factory, (4) move tickets to authorize the

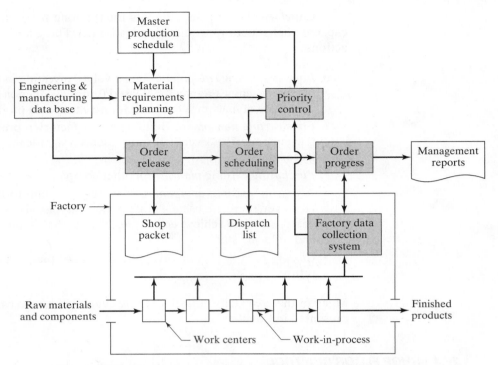

Figure 25.9 Three phases in a shop floor control system.

material handling personnel to transport parts between work centers in the factory if this kind of authorization is required, and (5) the parts list, if required for assembly jobs. In the operation of a conventional factory, which relies heavily on manual labor, these are paper documents that move with the production order and are used to track its progress through the shop. In a modern factory, automated identification and data capture technologies (Chapter 12) are used to monitor the status of production orders, rendering the paper documents (or at least some of them) unnecessary. We explore these factory data collection systems in Section 25.4.4.

The order release module is driven by two inputs, as indicated in Figure 25.9. The first is the authorization to produce that derives from the master schedule. This authorization proceeds through MRP, which generates work orders with scheduling information. The second input to the order release module is the engineering and manufacturing data base that provides the product structure and process plans needed to prepare the various documents that accompany the order through the shop.

25.4.2 Order Scheduling

The order scheduling module follows directly from the order release module and assigns the production orders to the various work centers in the plant. In effect, order scheduling executes the dispatching function in PPC. The order scheduling module prepares a dispatch list that indicates which production orders should be accomplished at the various work centers. It also provides information about relative priorities of the different jobs, for example, by showing due dates for each job. In current shop floor control practice, the dispatch list guides the shop foreman in making work assignments and allocating resources to different jobs to comply with the master schedule.

The order scheduling module in shop floor control is intended to solve two problems in production control: (1) machine loading and (2) job sequencing. To schedule a given set of production orders or jobs in the factory, the orders must first be assigned to work centers. Allocating orders to work centers is referred to as *machine loading*. The term *shop loading* is also used, which refers to the loading of all machines in the plant. Since the total number of production orders usually exceeds the number of work centers, each work center will have a queue of orders waiting to be processed. The remaining question is: In what sequence should these jobs be processed?

Answering this question is the problem in job sequencing. *Job sequencing* involves determining the sequence in which the jobs will be processed through a given work center. To determine this sequence, priorities are established among the jobs in the queue, and the jobs are processed in the order of their relative priorities. *Priority control* is a term used in production control to denote the function that maintains the appropriate priority levels for the various production orders in the shop. As indicated in Figure 25.9, priority control information is an important input in the order scheduling module. Some of the dispatching rules used to establish priorities for production orders in the plant include

- *First-come-first served.* Jobs are processed in the order in which they arrive at the machine. One might argue that this rule is the most fair.
- *Earliest due date.* Orders with earlier due dates are given higher priorities.
- *Shortest processing time.* Orders with shorter processing times are given higher priorities.
- *Least slack time.* Slack time is defined as the difference between the time remaining until due date and the process time remaining. Orders with the least slack in their schedule are given higher priorities.
- *Critical ratio.* The critical ratio is defined as the ratio of the time remaining until due date divided by the process time remaining. Orders with the lowest critical ratio are given higher priorities.

When an order is completed at one work center, it enters the queue at the next machine in its process routing. That is, the order becomes part of the machine loading for the next work center, and priority control is utilized to determine the sequence of processing among the jobs at that machine.

The relative priorities of the different orders may change over time. Reasons behind these changes include (1) lower or higher than expected demand for certain products, (2) equipment breakdowns that cause delays in production, (3) cancellation of an order by a customer, and (4) defective raw materials that delay an order. The priority control function reviews the relative priorities of the orders and adjusts the dispatch list accordingly.

25.4.3 Order Progress

The order progress module in shop floor control monitors the status of the various orders in the plant, WIP, and other characteristics that indicate the progress and performance of production. The function of the order progress module is to provide information that is useful in managing the factory. The information presented to production management is often summarized in the form of reports, such as

- *Work order status reports.* These reports indicate the status of production orders. Typical information in the report includes the current work center where each order is located, processing hours remaining before completion of each order, whether each job is on time or behind schedule, and the priority level of each order.
- *Progress reports.* A progress report is used to report performance of the shop during a certain time period (e.g., a week or month in the master schedule). It provides information on how many orders were completed during the period, how many orders should have been completed during the period but were not, and so forth.
- *Exception reports.* An exception report identifies deviations from the production schedule (e.g., overdue jobs) and similar exception information.

These reports are useful to production management in making decisions about allocation of resources, authorization of overtime hours, and other capacity issues, and in identifying problem areas in the plant that adversely affect achieving the MPS.

25.4.4 Factory Data Collection System

Various techniques are used to collect data from the factory floor. These techniques range from clerical methods that require workers to fill out paper forms that are later compiled, to fully automated methods that require no human participation. The *factory data collection system* (FDC system) consists of the various paper documents, terminals, and automated devices located throughout the plant for collecting data on shop floor operations, plus the means for compiling and processing the data. The factory data collection system serves as an input to the order progress module in shop floor control, as illustrated in Figure 25.9. It is also an input to priority control, which affects order scheduling. Examples of the types of data on factory operations collected by the FDC system include piece counts completed at each work center, direct labor time expended on each order, parts that are scrapped, parts requiring rework, and equipment downtime. The data collection system can also include the time clocks used by employees to punch in and out of work.

The ultimate purpose of the factory data collection system is twofold: (1) to supply status and performance data to the shop floor control system and (2) to provide current information to production foremen, plant management, and production control personnel. To accomplish this purpose, the factory data collection system must input data to the plant computer system. In current CIM technology, this is done using an on-line mode, in which the data are entered directly into the plant computer system and are immediately available to the order progress module. The advantage of on-line data collection is that the data file representing the status of the shop can be kept current at all times. As changes in order progress are reported, these changes are immediately incorporated into the shop status file. Personnel with a need to know can access this status in real-time and be confident that they have the most up-to-date information on which to base any decisions. Even though a modern FDC system is largely computerized, paper documents are still used in factory operations, and our coverage includes both manual (clerical) and automated systems.

Manual (Clerical) Data Input Techniques. Manually oriented techniques of factory data collection require production workers to read from and fill out paper forms indicating order progress data. The forms are subsequently turned in and compiled, using a combination of clerical and computerized methods. The paper forms include the following:

- *Job traveler.* This is a log sheet that travels with the shop packet through the factory. Workers who spend time on the order are required to record their times on the log sheet along with other data such as the date, piece counts, defects, and so forth. The job traveler becomes the chronological record of the processing of the order. The trouble with this method is its inherent incompatibility with the principles of real-time data collection. Since the job traveler moves with the order, it is not readily available for compiling current order progress.

- *Employee time sheets.* In the typical operation of this method, a daily time sheet is prepared for each worker, and the worker must fill out the form to indicate work that he/she accomplished during the day. Typical data entered on the form include order number, operation number on the route sheet, number of pieces completed during the day, and time spent. Some of these data are taken from information contained in the documents traveling with the order (e.g., engineering drawings and route sheets). The time sheet is turned in daily, and order progress information is compiled (usually by clerical staff).

- *Operation tear strips.* With this technique, the traveling documents include a set of preprinted tear strips that can be easily separated from the shop packet. The preprinted data on each tear strip includes order number and route sheet details. When a worker finishes an operation or at the end of the shift, the worker tears off one of the tear strips, records the piece count and time data, and turns in the form to report order progress.

- *Prepunched cards.* This is essentially the same technique as the tear strip method, except that prepunched computer cards are included with the shop packet instead of tear strips. The prepunched cards contain the same type of order data, and the workers must write the same kind of production data onto the card. The difference in the use of prepunched cards is that mechanized data processing procedures can be used to record some of the data to compile the daily progress report.

There are problems with all of these manually oriented data collection procedures. They all rely on the cooperation and clerical accuracy of factory workers to record data onto a paper document. There are invariably errors in this kind of procedure. Error rates associated with handwritten entry of data average about 3% (one error out of 30 data entries). Some of the errors can be detected by the clerical staff who compile the order progress records. Examples of detectable errors include wrong dates, incorrect order numbers (the clerical staff knows which orders are in the factory, and they can usually figure out when an erroneous order number has been entered by a worker), and incorrect operation numbers on the route sheet. (If the worker enters a certain operation number, but the preceding operation number has not been started, then an error has been made.) Other errors are more difficult to identify. If a worker enters a piece count of 150 pieces that represents the work completed in one shift when the batch size is 250 parts, this is difficult for the clerical staff to verify. If a different worker on the following day completes the batch and also enters a piece count of 150, then it is obvious that one of the workers overstated his/her production, but which one? Maybe both.

Another problem is the delay in submitting the order progress data for compilation. There is a time lapse between when events occur in the shop and when the paper data representing those events are submitted. The job traveler method is the worst offender in this regard. Here the data might not be compiled until the order has been completed, too late to take any corrective action. This method is of little value in a shop floor control system. The remaining manual methods suffer a one-day delay since the shop data are generally submitted at the end of the shift, and a summary compilation is not available until the following day. In addition to the delay in submitting the order data, there is also a delay associated with compiling the data into useful reports. Depending on how the order progress procedures are organized, the compilation may add several days to the reporting cycle.

Automated and Semi-Automated Data Collection Systems. To avoid the problems associated with the manual/clerical procedures, some factories use data collection terminals located around the factory. Workers input data relative to order progress using simple keypads or conventional alphanumeric keyboards. Data entered by keyboard are subject to error rates of around 0.3% (one error in 300 data entries), an order of magnitude improvement in data accuracy over handwritten entry. Also, error-checking routines can be incorporated into the entry procedures to detect syntax and certain other types of errors. Because of their widespread use in our society, PCs are becoming more and more common in factories, both for data collection and to present engineering and production data to shop personnel.

The data entry methods also include automatic identification and data collection (AIDC) technologies (Chapter 12) such as optical bar codes and radio frequency identification (RFID). Certain types of data such as order number, product identification, and operation sequence number can be entered with automated techniques using bar-coded or magnetized cards included with the shop documents (refer back to the bar-coded route sheet in Figure 12.7).

Some of the configurations of data collection terminals that can be installed in the factory include

- *One centralized terminal.* In this arrangement there is a single terminal located centrally in the plant. This requires all workers to walk from their workstations to the central location when they must enter the data. If the plant is large, this is inconvenient and inefficient. Also, use of the terminal tends to increase at time of shift change, resulting in significant lost time for the workers.
- *Satellite terminals.* In this configuration, there are multiple data collection terminals located throughout the plant. The number and locations are designed to strike a balance between minimizing the investment cost in terminals and maximizing the convenience of the plant workers.
- *Workstation terminals.* The most convenient arrangement for workers is to have a data collection terminal available at each workstation. This minimizes the time lost in walking to satellite terminals or a single central terminal. Although the investment cost of this configuration is the greatest, it may be justified when the number of data transactions is relatively large and when the terminals are also designed to collect certain data automatically.

The trend in industry is toward more automated factory data collection. Although they are called "automated," many of the techniques require the participation of human workers, as our coverage has indicated; hence, we have included "semi-automated" in the subtitle for this category of data collection system.

25.5 INVENTORY CONTROL

Inventory control attempts to compromise between two opposing objectives: (1) minimize the cost of holding inventory and (2) maximize customer service. Minimizing inventory cost suggests keeping inventory to a minimum, in the extreme, zero inventory. Maximizing customer service implies keeping large stocks on hand so that customer orders can immediately be filled.

The types of inventory of greatest interest in PPC are raw materials, purchased components, in-process inventory (WIP), and finished products. The major costs of holding inventory are (1) investment costs, (2) storage costs, and (3) cost of possible obsolescence or spoilage. The three costs are referred to collectively as *carrying costs* or *holding costs*. Investment cost is usually the largest component. When a company borrows money to invest in materials to be processed in the factory, it must pay interest on that money until the customer pays for the finished product. But many months may elapse between start of production and delivery to the customer. Even if the company uses its own money to purchase the starting materials, it is still making an investment that has a cost associated with it.

Companies can minimize holding costs by minimizing the amount of inventory on hand. However, when inventories are minimized, customer service may suffer, inducing customers to take their business elsewhere. This also has a cost, called the *stock-out cost*. Most companies want to minimize stock-out cost and provide good customer service. Thus, they are caught on the horns of an inventory control dilemma, balancing carrying costs against the cost of poor customer service.

In our introduction to MRP (Section 25.2), we distinguished between two types of demand, independent and dependent. Different inventory control procedures are used for independent and dependent demand items. For dependent demand items, MRP is the most widely implemented technique. For independent demand items, order point inventory systems are commonly used.

25.5.1 Order Point Inventory Systems

Order point systems are concerned with two related problems that must be solved when managing inventories of independent demand items: (1) how many units should be ordered? and (2) when should the order be placed? The first problem is often solved using economic order quantity formulas. The second problem can be solved using reorder point methods.

Economic Order Quantity Formula. The problem of deciding on the most appropriate quantity to order or produce arises when the demand rate for the item is fairly constant, and the rate at which the item is produced is significantly greater than its demand rate. This is the typical *make-to-stock* situation. The same basic problem occurs with dependent demand items when usage of the item is relatively constant over time due to a steady production rate of the final product with which the item is correlated. In this case,

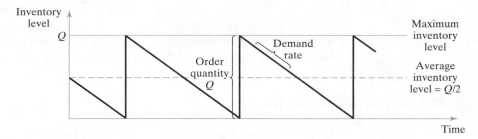

Figure 25.10 Model of inventory level over time in the typical make-to-stock situation.

it may make sense to endure some inventory holding costs so that the frequency of setups and their associated costs can be reduced. In these situations where demand remains steady, inventory is gradually depleted over time and then quickly replenished to some maximum level determined by the order quantity. The sudden increase and gradual reduction in inventory causes the inventory level over time to have a sawtooth appearance, as depicted in Figure 25.10.

A total cost equation can be derived for the sum of carrying cost and setup cost for the inventory model in Figure 25.10. Because of the sawtooth behavior of inventory level, the average inventory level is half the maximum level Q in our figure. The total annual inventory cost is therefore given by

$$TIC = \frac{C_h Q}{2} + \frac{C_{su} D_a}{Q} \tag{25.1}$$

where TIC = total annual inventory cost (holding cost plus ordering cost, \$/yr), Q = order quantity (pc/order), C_h = carrying or holding cost (\$/pc/yr), C_{su} = setup cost and/or ordering cost for an order (\$/setup or \$/order), and D_a = annual demand for the item (pc/yr). In the equation, the ratio D_a/Q is the number of orders or batches produced per year, which therefore gives the number of setups per year.

The holding cost C_h consists of two main components, investment cost and storage cost. Both are related to the time that the inventory spends in the warehouse or factory. As previously indicated, the investment cost results from the money the company must invest in the inventory before it is sold to customers. This inventory investment cost can be calculated as the interest rate paid by the company i (percent), multiplied by the value (cost) of the inventory.

Storage cost occurs because the inventory takes up space that must be paid for. The amount of the cost is generally related to the size of the part and how much space it occupies. As an approximation, it can be related to the value or cost of the item stored. For our purposes, this is the most convenient method of valuating the storage cost of an item. By this method, the storage cost equals the cost of the inventory multiplied by the storage rate, s. The term s is the storage cost as a fraction (percent) of the value of the item in inventory.

Combining interest rate and storage rate into one factor, we have $h = i + s$. The term h is called the holding cost rate. Like i and s, it is a fraction (percent) that is multiplied by the cost of the part to evaluate the holding cost of investing in and storing inventory. Accordingly, holding cost can be expressed as follows:

$$C_h = h C_{pc} \tag{25.2}$$

where C_h = holding (carrying) cost (\$/pc/yr), C_{pc} = unit cost of the item (\$/pc), and h = holding cost rate (rate/yr).

Setup cost includes the cost of idle production equipment during the changeover time between batches. The costs of labor performing the setup changes might also be added in. Thus,

$$C_{su} = T_{su}C_{dt} \tag{25.3}$$

where C_{su} = setup cost (\$/setup or \$/order), T_{su} = setup or changeover time between batches (hr/setup or hr/order), and C_{dt} = cost rate of machine downtime during the changeover (\$/hr). In cases where parts are ordered from an outside vendor, the price quoted by the vendor usually includes a setup cost, either directly or in the form of quantity discounts. C_{su} should also include the internal costs of placing the order to the vendor.

Eq. (25.1) excludes the actual annual cost of part production. If this cost is included then annual total cost is given by the equation

$$TC = D_aC_{pc} + \frac{C_hQ}{2} + \frac{C_{su}D_a}{Q} \tag{25.4}$$

where D_aC_{pc} = annual demand (pc/yr) multiplied by cost per item (\$/pc).

If the derivative is taken of either Eq. (25.1) or Eq. (25.4), the economic order quantity (EOQ) formula is obtained by setting the derivative equal to zero and solving for Q. This batch size minimizes the sum of carrying costs and setup costs:

$$Q = EOQ = \sqrt{\frac{2D_aC_{su}}{C_h}} \tag{25.5}$$

where EOQ = economic order quantity (number of parts to be produced per batch, pc/batch or pc/order), and the other terms have been defined previously.

EXAMPLE 25.3 Economic Order Quantity Formula

The annual demand for a certain item made-to-stock = 15,000 pc/yr. One unit of the item costs \$20.00, and the holding cost rate = 18%/yr. Setup time to produce a batch = 5 hr. The cost of equipment downtime plus labor = \$150/hr. Determine the economic order quantity and the total inventory cost for this case.

Solution: Setup cost C_{su} = 5 × \$150 = \$750. Holding cost per unit = 0.18 × \$20.00 = \$3.60. Using these values and the annual demand rate in the EOQ formula, we have

$$EOQ = \sqrt{\frac{2(15000)(750)}{3.60}} = 2500 \text{ units}$$

Total inventory cost is given by the TIC equation:

$$TIC = 0.5(3.60)(2,500) + 750(15,000/2,500) = \$9,000$$

Including the actual production costs in the annual total, by Eq (25.4) we have

$$TC = 15,000(20) + 9,000 = \$309,000$$

The economic order quantity formula has been widely used for determining so-called optimum batch sizes in production. More sophisticated forms of Eqs. (25.1) and (25.4) have appeared in the literature, for example, models that take production rate into account to yield alternative EOQ equations. Eq. 25.5 is the most general form and, in the author's opinion, quite adequate for most real-life situations. The difficulty in applying the EOQ formula is in obtaining accurate values of the parameters in the equation, namely (1) setup cost and (2) inventory carrying costs. These cost factors are usually difficult to evaluate; yet they have an important impact on the calculated economic batch size.

There is no disputing the mathematical accuracy of the EOQ equation. Given specific values of annual demand (D_a), setup cost (C_{su}), and carrying cost (C_h), Eq. (25.5) computes the lowest cost batch size to whatever level of precision the user desires. The trouble is that the user may be lulled into the false belief that no matter how much it costs to change the setup, the EOQ formula always calculates the optimum batch size. For many years in U.S. industry, this belief tended to encourage long production runs by manufacturing managers. The thought process went something like this: "If the setup cost increases, we just increase the batch size, because the EOQ formula always tells us the optimum production quantity."

The user of the EOQ equation must not lose sight of the total inventory cost (*TIC*) equation, Eq. (25.1), from which *EOQ* is derived. Examining the TIC equation, a cost-conscious production manager would quickly conclude that both costs and batch sizes can be reduced by decreasing the values of holding cost (C_h), and setup cost (C_{su}), The production manager may not be able to exert much influence on holding cost because it is determined largely by prevailing interest rates. However, methods can be developed to reduce setup cost by reducing the time required to accomplish the changeover of a production machine. Reducing setup time is an important focus in just-in-time production, and we consider it in Section 26.2.2.

Reorder Point Systems. Determining the economic order quantity is not the only problem that must be solved in controlling inventories in make-to-stock situations. The other problem is deciding when to reorder. One of the most widely used methods is the reorder point system. Although we have drawn the inventory level in Figure 25.10 as a very deterministic sawtooth diagram, the reality is that there are usually variations in demand rate during the inventory order cycle, as illustrated in Figure 25.11. Accordingly, when to reorder cannot be predicted with the precision that would exist if demand rate were a

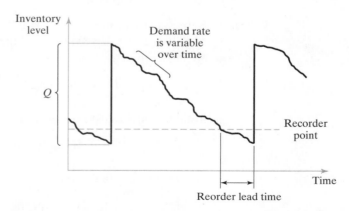

Figure 25.11 Operation of a reorder point inventory system.

known constant value. In a *reorder point system*, when the inventory level for a given stock item falls to some point specified as the reorder point, then an order is placed to restock the item. The reorder point is specified at a sufficient quantity level to minimize the probability of a stock-out between when the reorder point is reached and the new order is received. Reorder point triggers can be implemented using computerized inventory control systems that continuously monitor the inventory level as demand occurs and automatically generate an order for a new batch when the level declines below the reorder point.

25.5.2 Work-in-Process Inventory Costs

Work-in-process (WIP) represents a significant inventory cost for many manufacturing firms. In effect, the company is continually investing in raw materials, processing those materials, and then delivering them to customers when processing has been completed. The problem is that processing takes time, and the company pays a holding cost between start of production and receipt of payment from the customer after the goods are delivered. In Chapter 3, we showed that WIP and manufacturing lead time (MLT) are closely related. The longer the manufacturing lead time, the greater the WIP. This section presents a method for evaluating the cost of WIP and MLT. The method is based on concepts suggested by Meyer [5].

In Chapter 2, we indicated that production typically consists of a series of separate manufacturing steps or operations. Each operation consumes time, and that time has an associated cost. There is also time between each operation (at least for most manufacturing situations) that we have referred to as the nonoperation time. The nonoperation time includes material handling, inspection, and waiting in the queue before the next operation. There is also a cost associated with the nonoperation time. These times and costs for a given part can be graphically illustrated as shown in Figure 25.12. At time $t = 0$, the cost of the part is simply its material cost C_m. The cost of each processing step on the part is

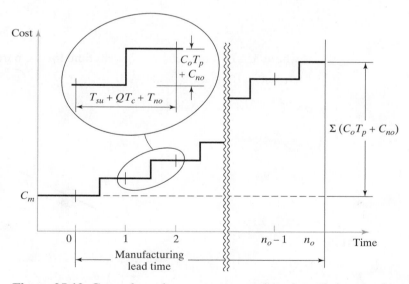

Figure 25.12 Cost of product or part as a function of time in the factory. As operations are completed, value and cost are added to the item.

the production time multiplied by the rate for the machine and labor. Production time T_p is determined from Eq. (3.3) and accounts for both setup time and operation time. Let us symbolize the rate as C_o. Nonoperation costs (e.g., inspection and material handling) related to the processing step are symbolized by the term C_{no}. Accordingly, the cost associated with each processing step in the manufacturing sequence is the sum

$$C_o T_p + C_{no}$$

The cost for each step is shown in Figure 25.12 as a vertical line, suggesting no time lapse. This is a simplification in the graph, justified by the fact that the time between operations spent waiting and in storage is generally much greater than the time for processing, handling, and inspection.

The total cost that has been invested in the part at the end of all operations is the sum of the material cost and the accumulated processing, inspection, and handling costs. Symbolizing this part cost as C_{pc}, we can evaluate it by using the equation

$$C_{pc} = C_m + \sum_{k=1}^{n_o} (C_o T_{pk} + C_{nok})$$

where k is used to indicate the sequence of operations, and there are a total of n_o operations. For convenience, if we assume that T_p and C_{no} are equal for all n_o operations, then

$$C_{pc} = C_m + n_o (C_o T_p + C_{no}) \qquad (25.6)$$

The part cost function shown in Figure 25.12 and represented by Eq. (25.6) can be approximated by a straight line as shown in Figure 25.13. The line starts at time $t = 0$ with a value $= C_m$ and slopes upward to the right so that its final value is the same as the final part cost in Figure 25.12. The approximation becomes more appropriate as the number of processing steps increases. The equation for this line is

$$C_m + \frac{n_o (C_o T_p + C_{no})}{MLT}$$

where MLT = manufacturing lead time for the part, and t = time in Figure 25.13.

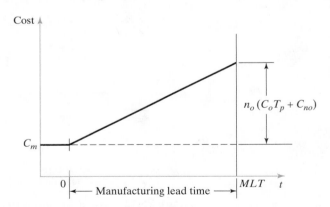

Figure 25.13 Approximation of product or part cost as a function of time in the factory.

As in our derivation of the economic order quantity formula, we apply the holding cost rate h to the accumulated part cost defined by Eq. (25.6), but substitute the straight-line approximation in place of the stepwise cost accumulation in Figure 25.12. In this way, we have an equation for total cost per part that includes the WIP carrying costs:

$$TC_{pc} = C_m + n_o(C_o T_p + C_{no}) + \int_0^{MLT} \left(C_m + \frac{n_o(C_o T_p + C_{no})}{MLT} t \right) h \, dt$$

Let us use a simpler form of this cost equation by making the following substitution for the $n_o(C_o T_p + C_{no})$ term just mentioned

$$C_p = n_o(C_o T_p + C_{no}) \tag{25.7}$$

Then,

$$TC_{pc} = C_m + C_p + \int_0^{MLT} \left(C_m + \frac{C_p t}{MLT} \right) h \, dt \tag{25.8}$$

Carrying out the integration, we have the following:

$$TC_{pc} = C_m + C_p + \left(C_m + \frac{C_p}{2} \right) h(MLT) \tag{25.9}$$

The holding cost is the last term on the right-hand side of the equation:

$$\text{Holding cost/pc} = \left(C_m + \frac{C_p}{2} \right) h(MLT) \tag{25.10}$$

Figure 25.14 shows the effect of adding the holding cost to the material, operation, and nonoperation costs of a part or product during production in the plant.

EXAMPLE 25.4 Inventory Holding Cost for WIP During Manufacturing

The cost of the raw material for a certain part is $100. The part is processed through 20 processing steps in the plant, and the manufacturing lead time is 15 wk. The production time per processing step is 0.8 hr, and the machine and

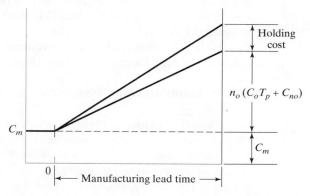

Figure 25.14 Approximation of product cost showing additional cost of holding WIP during the manufacturing lead time.

labor rate is \$25.00/hr. Inspection, material handling, and other related costs average to \$10 per processing step by the time the part is finished. The interest rate used by the company $i = 20\%$, and the storage rate $s = 13\%$. Determine the cost per part and the holding cost.

Solution: The material cost, operation costs, and nonoperation costs are by Eq. (25.6),

$$C_{pc} = \$100 + 20(\$25.00/\text{hr} \times .8 \text{ hr} + \$10) = \$700/\text{pc}$$

To compute the holding cost, first calculate C_p:

$$C_p = 20(\$25.00/\text{hr} \times .8 \text{ hr} + \$10) = \$600/\text{pc}$$

Next, determine the holding cost rate $h = 20\% + 13\% = 33\%$. Expressing this as a weekly rate $h = (33\%)/(52 \text{ wk}) = 0.6346\%/\text{wk} = 0.006346/\text{wk}$. According to Eq. (25.10),

$$\text{Holding cost/pc} = (100 + 600/2)(.006346)(15 \text{ wk}) = \$38.08/\text{pc}$$

$$TC_{pc} = 700.00 + 38.08 = \$738.08/\text{pc}$$

The \$38.08 in this example is more than 5% of the cost of the part; yet the holding cost is usually not included directly in the company's evaluation of part cost. Rather, it is considered as overhead. Suppose that this is a typical part for the company, and 5000 similar parts are processed through the plant each year; then the annual inventory cost for WIP of 5000 parts = 5000 × \$38.08 = \$190,400. If the manufacturing lead time could be reduced to half its current value, this would translate into a 50% savings in WIP holding cost.

25.6 EXTENSIONS OF MRP

The initial versions of MRP in the early 1970s were limited to the planning of purchase orders and factory work orders and did not take into account such issues as capacity planning or feedback data from the factory. MRP was strictly a materials and parts planning tool whose calculations were based on the master production schedule (MPS). Over time, it became evident that MRP should be tied to other software packages to create a more integrated PPC system. The PPC software packages that evolved from MRP have gone through several generations and variations, two of which are described in the following sections: (1) manufacturing resource planning and (2) enterprise resource planning.

25.6.1 Manufacturing Resource Planning (MRP II)

Manufacturing resource planning evolved from material requirements planning in the 1980s. It came to be abbreviated MRP II to distinguish it from the original abbreviation and to indicate that it was second generation, that is, more than just a material planning system. *Manufacturing resource planning* can be defined as a computer-based system for planning, scheduling, and controlling the materials, resources, and supporting activities needed to meet the master production schedule. MRP II is a closed-loop system that integrates and coordinates the major functions of the business involved in production. The term "closed-loop system" means that MRP II incorporates feedback of data on various aspects of operating performance so that corrective action can be taken in a timely manner; that is, MRP II includes a shop floor control system.

MRP II can be considered to consist of three major modules: (1) material requirements planning, or MRP, (2) capacity planning, and (3) shop floor control. These modules are discussed in Sections 25.2, 25.3, and 25.4, respectively. MRP accomplishes the planning function for materials, parts, and assemblies, based on the master production schedule. In so doing it also provides a schedule for factory operations. The capacity planning module interacts with the MRP module to ensure that the schedules created by MRP are feasible. Finally, the shop floor control module performs the feedback control function using its factory data collection system to implement the three phases of order release, order scheduling, and order progress.

Manufacturing resource planning represented an improvement over material requirements planning because it included production capacity and shop floor feedback in its computations. But MRP II was limited to the manufacturing operations of the firm. As further enhancements were made to MRP II systems, the trend was to consider all of the operations and functions of the enterprise rather than just manufacturing. The culmination of this trend in the 1990s was enterprise resource planning.

25.6.2 Enterprise Resource Planning (ERP)

Enterprise resource planning (ERP) is a computer software system that organizes and integrates all of the data and business functions of an organization through a single, central data base. The functions include sales, marketing, purchasing, operations, logistics, distribution, inventory control, accounting, finance, and human resources. The operations function can include service activities as well as production, so ERP can be used by service organizations – it's not just for manufacturing. Enterprise resource planning commonly runs as a client-server system, which means that users access and utilize the system through personal computers at their respective workplaces. ERP operates on a company-wide basis; it is not a plant-based system as MRP applications often are.

Because it uses a single data base, the organization avoids or minimizes problems such as data redundancy and conflicting data in different data bases, time delays in entering data, and communication difficulties between different data bases and the modules that operate on these data bases. Before ERP, departments within an organization would typically have their own data bases and computer systems. For example, the data base of the Human Resources department would contain the personal data about each employee and the reporting structure within each department. At the same time, the Payroll department would calculate the weekly wages of employees, using much of the same personal data but keeping it in a separate data base. When an employee left the company, or a new employee joined, both data bases would need to be updated.

In ERP, everyone in the organization has access to the same sets of data according to their individual job responsibilities (not all of the data can be accessed by all employees). When a customer orders a product and the order is entered into the ERP system, all of the business functions that are affected by the order, such as inventory records, purchasing, production schedules, shipping, and invoicing, are updated in the central database. Anyone requiring access to the data base, and this may include customers and external suppliers, has the latest information.

ERP systems are comprised of multiple software modules, each focused on a different business function or group of functions within the organization. The modules are integrated through the ERP framework to accomplish transactions that may affect several functional areas. For example, an engineering change in a part design would impact process planning and the production department in which the part is made. All modules access the same central data base, so data transactions accomplished in any given module are immediately accessible by all others. The engineering change notice would be immediately available to manufacturing engineering, industrial engineering, and other departments affected by the change. The modules in an ERP system can be classified into the following main groups [7]:

- *Production and materials management.* This module performs the transactions related to manufacturing resource planning (which includes material requirements planning, capacity planning, and shop floor control). It also supports other functions such as master production scheduling, process planning, inventory control, purchasing, and product costing. Finally, to carry out its computations related to production, it must interface with the company's design data base (e.g., bills of material, CAD part drawings).

- *Sales and marketing.* This module accomplishes transactions to provide support for customer relationship activities in the organization. These activities include customer contact and service, order input and processing, pricing, product availability, delivery, shipping, invoicing, product returns, and handling of customer complaints.

- *Finance and accounting.* This module includes functions such as capital budgeting, asset management, investment management, cost control, accounts payable, accounts receivable, and cash management. The finance and accounting module provides information on the financial effects of transactions in other modules in the ERP system.

- *Human resources.* This module addresses all of the personnel issues and activities. These include payroll, benefits, time and attendance, training, workforce planning, job applicant processing, job descriptions, employee performance appraisals, organization charts, and employee personal data.

Today's ERP systems feature open architecture and the software consists of individual modules that can be combined into one system. This means that a company can select certain modules from one software vendor and other modules from a different vendor, thus obtaining the best combination of modules for its own business. For example, if one vendor is recognized as having the best accounting module, and another has the best MRP module, these two software packages can be used within the company's ERP system, even though the ERP software itself was purchased from a completely different supplier. ERP systems that utilize the open architecture framework are called ERP II systems.

The success of enterprise resource planning depends on the accuracy and currency of its data base, which means that all transactions and events that affect the data base must be entered as they occur [2]. When a sales order comes in, that transaction must be entered immediately because it actuates other functions in the system such as order processing, purchasing, inventory records, factory work orders, production schedules, and so on. If a part is received by the plant and taken directly to the shop floor because it is urgently needed, but the receipt is never recorded due to negligence or oversight, then the ERP data base is missing that transaction, and this inaccuracy cascades throughout the other functions in the system. If similar discrepancies occur in other areas, then the accuracy of the ERP model representing plant status erodes over time. As this erosion occurs and is recognized within the organization, people gradually lose faith in the system and begin working around it instead of using it properly, exacerbating the problem. To summarize, as powerful a tool as ERP is, its successful implementation requires a discipline throughout the organization of recording events as they occur and making sure each record is accurate.

REFERENCES

[1] BAUER, A., R. BOWDEN, J. BROWNE, J. DUGGAN, and G. LYONS, *Shop Floor Control Systems*, Chapman & Hall, London, UK, 1994.

[2] BROWN, A. S., "Lies Your ERP System Tells You," *Mechanical Engineering*, March 2006, pp. 36–39.

[3] CHASE, R. B., and N. J. AQUILANO, *Production and Operations Management: A Life Cycle Approach*, 5th ed., Richard D. Irwin, Inc., Homewood, IL, 1989.

[4] KASARDA, J. B., "Shop Floor Control Must be Able to Provide Real-Time Status and Control," *Industrial Engineering*, November 1980, pp. 74–78, 96.

[5] MEYER, R. J., "A Cookbook Approach to Robotics and Automation Justification," *Technical Paper MS82-192*, Society of Manufacturing Engineers, Dearborn, MI, 1982.

[6] REID, R. D., and N. R. SANDERS, *Operations Management*, John Wiley & Sons, Inc., Hoboken, NJ, 2002.

[7] RUSSELL, R. S., and B. W. TAYLOR III, *Operations Management*, 4th ed., Prentice Hall, Upper Saddle River, NJ, 2003.

[8] SILVER, E. A., D. F. PYKE, and R. PETERSON, *Inventory Management and Production Planning and Control*, John Wiley & Sons, Inc., Hoboken, NJ, 1998.

[9] SIPPER, D., and R. L. BUFFIN, *Production: Planning, Control, and Integration*, McGraw-Hill Companies, Inc., NY, 1997.

[10] SULE, D. R., *Industrial Scheduling*, PWS Publishing Company, Boston, MA, 1997.

[11] SUZAKI, K., *The New Manufacturing Challenge: Techniques for Continuous Improvement*, Free Press, NY, 1987.

[12] VEILLEUX, R. F., and L. W. PETRO, *Tool and Manufacturing Engineers Handbook*, 4th ed., Volume V, *Manufacturing Management*, Society of Manufacturing Engineers, Dearborn, MI, 1988.

[13] VOLLMAN, T. E., W. L. BERRY, and D. C. WHYBARK, *Manufacturing Planning and Control Systems*, 2d ed., Richard D. Irwin, Inc., Homewood, IL, 1988.

REVIEW QUESTIONS

25.1 What is production planning?

25.2 Name three of the four activities within the scope of production planning identified in the text.

25.3 What is production control?

25.4 What is the difference between the aggregate production plan and the master production schedule?

25.5 What is material requirements planning (MRP)?

25.6 What is the difference between independent demand and dependent demand?

25.7 What are the three main inputs to the MRP processor?

25.8 What are common use items in MRP?

25.9 Name three benefits of a well-designed MRP system?

25.10 What is capacity planning?

25.11 Capacity adjustments can be divided into short-term adjustments and long-term adjustments. Name four of the capacity adjustments for the short term.

25.12 What is shop floor control?

25.13 What are the three phases of shop floor control? Provide a brief definition of each activity.

25.14 What does the term *machine loading* mean?

25.15 What are carrying costs in inventory control?

25.16 What is a reorder point system in inventory control?

25.17 What is the difference between material requirements planning (MRP) and manufacturing resource planning (MRP II)?

25.18 What is enterprise resource planning (ERP)?

PROBLEMS

Order-Point Inventory Systems

25.1 The annual demand for a certain part is 2,000 units per year. The part is produced in a batch model manufacturing system. Annual holding cost per piece is $3.00. It takes 2 hours to set up the machine to produce the part, and cost of system downtime is $150/hr. Determine (a) the most economical batch quantity for this part and (b) the associated total inventory cost.

25.2 Annual demand for a made-to-stock product is 60,000 units. Each unit costs $8.00 and the annual holding cost rate is 24%. Setup time to change over equipment for this product is 6 hr, and the downtime cost of the equipment is $120/hr. Determine (a) economic order quantity and (b) total inventory costs.

25.3 Demand for a certain product is 25,000 units/yr. Unit cost is $10.00. Holding cost rate is 30%/yr. Changeover (setup) time between products is 10.0 hr, and downtime cost during changeover is $150/hr. Determine (a) economic order quantity, (b) total inventory costs, and (c) total inventory cost per year as a proportion of total production costs.

25.4 A part is produced in batches of 3000 pieces. Annual demand is 60,000 pieces, and piece cost is $5.00. Setup time to run a batch is 3.0 hr, cost of downtime on the affected equipment is figured at $200/hr, and annual holding cost rate is 30%. What would the annual savings be if the product were produced in the economic order quantity?

25.5 In the previous problem, (a) how much would setup time have to be reduced in order to make the batch size of 3,000 pieces equal to the economic order quantity? (b) How much would total inventory costs be reduced if the EOQ = 3,000 units compared to the EOQ calculated in the previous problem? (c) How much would total inventory costs be reduced if the setup time were equal to the value obtained in part (a) compared to the 3.0 hr used in the previous problem?

25.6 A certain machine tool is used to produce several components for one assembled product. To keep in-process inventories low, a batch size of 100 units is produced for each component. Demand for the product is 3,000 units per year. Production downtime costs an estimated $150/hr. All parts produced on the machine tool are approximately equal in value: $9.00/unit. Holding cost rate is 30%/yr. In how many minutes must the changeover between batches be accomplished so that 100 units is the economic order quantity?

25.7 Annual demand for a certain part is 10,000 units. At present the setup time on the machine tool that makes this part is 5.0 hr. Cost of downtime on this machine is $200/hr. Annual holding cost per part is $1.50. Determine (a) EOQ and (b) total inventory costs for this data. Also, determine (c) EOQ and (d) total inventory costs if the changeover time could be reduced to six minutes.

25.8 A variety of assembled products are made in batches on a batch model assembly line. Every time a different product is produced, the line must be changed over which causes lost production time. The assembled product of interest here has an annual demand of 12,000 units. The changeover time to set up the line for this product is 6.0 hours. The company figures that the hourly rate for lost production time on the line due to changeovers is $200/hr. Annual holding cost for the product is $7.00 per product. The product is currently made in batches of 1000 units for shipment each month to the wholesale distributor. (a) Determine the total annual inventory cost for this product in batch sizes of 1000 units. (b) Determine the economic batch quantity for this product. (c) How often would shipments be made using this EOQ? (d) How much would the company save in annual inventory costs, if it produced batches equal to the EOQ rather than 1000 units?

25.9 A two-bin approach is used to control inventory for a certain low-cost hardware item. Each bin holds 500 units of the item. When one bin becomes empty, an order for 500 units is released to replace the stock in that bin. The order lead time is slightly less than the time it takes to deplete the stock in one bin. Accordingly, the chance of a stock-out is low and the average inventory level of the item is about 250 units, perhaps slightly more. Annual usage of the item is 6,000 units. Ordering cost is $40. (a) What is the imputed holding cost per unit for this item, based on the data given? (b) If the actual annual holding cost per unit is 5 cents, what lot size should be ordered? (c) How much does the current two-bin approach cost the company per year, compared with using the economic order quantity?

Work-In-Process Inventory Costs

25.10 A workpart costing $80 is processed through the factory. The manufacturing lead time for the part is 12 weeks, and the total time spent in processing during the lead time is 30 hours for all operations at a rate of $35 per hour. Nonoperation costs total $70 during the lead time. The holding cost rate used by the company for work-in-process is 26%. The plant operates 40 hours per week, 52 weeks per year. If this part is typical of the 200 parts per week processed through the factory, determine the following: (a) the holding cost per part during the manufacturing lead time, and (b) the total annual holding costs to the factory. (c) If the manufacturing lead time were to be reduced from 12 weeks to 8 weeks, how much would the total holding costs be reduced on an annual basis?

25.11 A batch of large castings is processed through a machine shop. The batch size is 20. Each raw casting costs $175. There are 22 machining operations performed on each casting at an average operation time of 0.5 hour per operation. Setup time per operation averages 5 hours. The cost rate for the machine and labor is $40 per hour. Nonoperation costs (inspection, handling between operations, etc.) average $5 per operation per part. The corresponding nonoperation time between each operation averages two working days. The shop works five 8-hour days per week, 52 weeks per year. Interest rate used by the company is 25% for investing in WIP inventory, and storage cost rate is 14% of the value of the item held. Both of these rates are annual rates. Determine the following: (a) manufacturing lead time for the batch of castings, (b) total cost to the shop of each casting when it is completed, including the holding cost, and (c) total holding cost of the batch for the time it spends in the machine shop as work-in-process.

Material Requirements Planning

25.12 Using the master schedule of Figure 25.2(b), and the product structures in Figures 25.4 and 25.5, determine the time-phased requirements for component C6 and raw material M6. The raw material used in component C6 is M6. Lead times are as follows: for P1, assembly lead time is 1 week; for P2, assembly lead time is 1 week; for S2, assembly lead time is 1 week; for S3, assembly lead time is 1 week; for C6, manufacturing lead time is 2 weeks; and for M6, ordering lead time is 2 weeks. Assume that the current inventory status for all of the above items is zero units on hand, and zero units on order. The format of the solution should be similar to that presented in Figure 25.7.

25.13 Solve previous Problem 25.12, but use the current inventory on hand and on order for S3, C6, and M6 as follows: for S3, inventory on hand is 2 units and quantity on order is zero; for C6, inventory on hand is 5 units and quantity on order is 10 for delivery in week 2; and for M6, inventory on hand is 10 units and quantity on order is 50 for delivery in week 2.

25.14 Material requirements are to be planned for component C2 given the master schedule for P1 and P2 in Figure 25.2(b), and the product structures in Figures 25.4 and 25.5. Assembly lead time for products and subassemblies (P and S levels) is 1 week, manufacturing lead time for components (C level) is 2 weeks, and ordering lead time for raw materials (M level) is 3 weeks. Determine the time-phased requirements for M2, C2, and S1. Assume there are no common use items other than those specified by the product structures for P1 and P2 (Figures 25.4 and 25.5), and that all on-hand inventories and scheduled receipts are zero. Use a format similar to Figure 25.7. Ignore demand beyond period 10.

25.15 Requirements are to be planned for component C5 in product P1. Required deliveries for P1 are given in Figure 25.2(b), and the product structure for P1 is shown in Figure 25.4. Assembly lead time for products and subassemblies (P and S levels) is 1 week, manufacturing lead time for components (C level) is 2 weeks, and ordering lead time for raw materials (M level) is 3 weeks. Determine the time-phased requirements for M5, C5, and S2 to meet the master schedule. Assume no common use items. On-hand inventories are: 100 units for M5 and 50 units for C5, zero for S2. Scheduled receipts are zero for these items. Use a format similar to Figure 25.7. Ignore demand for P1 beyond period 10.

25.16 Solve previous Problem 25.15 using the following additional information: scheduled receipts of M5 are 50 units in week 3 and 50 units in week 4.

Order Scheduling

25.17 It is currently day 10 in the production calendar of the XYZ Machine Shop. Three orders (A, B, and C) are to be processed at a particular machine tool. The orders arrived in the sequence A-B-C. The table that follows indicates the process time remaining and production calendar due date for each order. Determine the sequence of the orders that would be scheduled using the following priority control rules: (a) first-come-first-served, (b) earliest due date, (c) shortest processing time, (d) least slack time, and (e) critical ratio.

Order	Remaining Process Time	Due Date
A	4 days	Day 20
B	16 days	Day 30
C	6 days	Day 18

25.18 In previous Problem 25.17, for each solution, (a) through (e), determine which jobs will be delivered on time and which jobs will be tardy.

Index

788